Automated Technologies for the Development and Production of Radiopharmaceuticals

R. Michael van Dam • Yuji Kuge
Giancarlo Pascali
Editors

Automated Technologies for the Development and Production of Radiopharmaceuticals

Springer

Editors
R. Michael van Dam
Crump Institute for Molecular Imaging and Department of Molecular & Medical Pharmacology
University of California Los Angeles (UCLA)
Los Angeles, CA, USA

Yuji Kuge
Central Institute of Isotope Science
Hokkaido University
Sapporo, Japan

Giancarlo Pascali
Australian Nuclear Science and Technology Organisation
University of New South Wales, University of Sydney
NSW, Sydney, Australia

ISBN 978-3-031-84631-1 ISBN 978-3-031-84632-8 (eBook)
https://doi.org/10.1007/978-3-031-84632-8

This Springer imprint is published by the registered company Springer Nature Switzerland AG
The registered company address is: Gewerbestrasse 11, 6330 Cham, Switzerland

If disposing of this product, please recycle the paper.

Prologue

Radiopharmaceuticals are drugs labeled with radionuclides that are administered into the body to perform nuclear medicine imaging (diagnosis) and therapy procedures. In nuclear medicine imaging, radiopharmaceuticals are labeled with nuclides that emit γ-rays (for scintigraphy or single photon emission computed tomography) or positrons (for positron-emission tomography), and diagnostic images are obtained by detecting the radiation emitted from the radiopharmaceuticals accumulated in disease lesions. In nuclear medicine therapy (also called radioligand therapy, internal isotope radiation therapy, targeted radiotherapy, or radiopharmaceutical therapy), radiopharmaceuticals are labeled with nuclides that emit α-particles, β-particles, or Auger electrons, and the radiation emitted from the radionuclides kills tumor cells in the local vicinity of the disease lesions in which the radiopharmaceuticals accumulated, with minimal damage to healthy tissues.

Due to the short half-lives of radionuclides used for imaging and therapy, radiopharmaceuticals have a very short "shelf life" and must be manufactured repeatedly and often at multiple locations around the globe. Radionuclides are produced in particle accelerators, generators, or nuclear reactors, in which a "target" material undergoes a nuclear reaction to produce the desired radionuclide. The radionuclide must then be separated from the target material and processed into a chemical form suitable for labeling. In the radiolabeling step, the radionuclide is incorporated into the desired radiopharmaceutical molecule through a series of one or more chemical reactions. Subsequently, the crude reaction product is purified and formulated into an injectable solution. Finally, quality control (QC) testing is performed to assure the purity and safety of the formulation via a series of analytical measurements, and individual patient doses of the radiopharmaceutical can be dispensed, delivered, and administered to patients.

To ensure a consistent product across different production sites and to minimize radiation exposure to the radiochemists, these processes involve specialized instruments to provide automation and radiation protection. As the demand for, and variety of, radiopharmaceuticals in use continues to increase, there has been renewed research interest in improving the technologies for their routine production. For example, performing radiosyntheses using new microfluidic and flow chemistry techniques offers many advantages that can improve the yields and decrease the costs of producing radiopharmaceuticals. Recently, research in this area has been expanding to include upstream (e.g., radionuclide processing) and downstream processes (e.g., radiopharmaceutical purification, formulation, analysis). Other recent innovations include automated QC testing and modular approaches that increase equipment versatility. The successful application of radiopharmaceuticals in imaging and therapy has also fueled intense interest in the discovery of new radiopharmaceuticals with improved properties or targeting additional diseases, and there has also been significant innovation in developing improved tools for radiopharmaceutical discovery. For example, new techniques are becoming available for high-throughput experimentation in radiochemistry, and new microfluidic radioassay techniques are enabling evaluation of pharmacologic properties of radiopharmaceuticals in cell cultures, organoids, and single cells.

In light of the exciting developments in this field, we have been organizing, for a number of years, a focused symposium at the Pacifichem conference, entitled "Innovative technologies for production, purification, analysis, and in vitro evaluation of radiopharmaceuticals," most recently held in 2021. This symposium aimed to highlight the progress made to date, and foster discussions about future directions, in the use of innovative technologies in the radiopharmaceutical field, with a particular focus on miniaturization, microfluidics, flow chemistry, modular equipment, process integration, and disposable-based designs for cGMP production.

To share the remarkable experience of this symposium with interested researchers worldwide, and to provide an educational resource for new students and investigators in the field, we decided to publish this comprehensive book. Many of the chapters have been written by symposium participants, updated with progress since the symposium. Additional chapters were invited both to include a historical background on the key fields and to provide a more thorough coverage by adding new technical developments that have occurred during the last few years.

We aim that this book will serve as a comprehensive reference to better understand the unique challenges, concepts, and implementation of innovative technologies in radiopharmaceutical research and commercial practice. Each chapter provides extensive background information in addition to presenting cutting-edge developments and future perspectives, written at a level and style that we expect will be accessible to new and established investigators as well as technology innovators. We hope and anticipate that the presented techniques will vastly improve the cost, efficiency, and accessibility of radiopharmaceutical manufacturing for nuclear medicine diagnosis and therapy, as well as open up the field of radiopharmaceutical development and accelerate the pace of discovery of this incredible class of drug molecules.

Contents

Part I
Radionuclide Production and Purification Technologies

Cyclotron Radioisotope Production: Developments and Challenges in the New Era

Emiliano Cazzola, Chiara Favaretto, and Giancarlo Gorgoni

1 Introduction

Particle accelerators are machines that can generate a particle beam and steer it toward a target. The accelerated particles can penetrate the target's atoms to the nucleus, resulting in a nuclear reaction and the formation of radionuclides. These systems have been fundamental in achieving significant results and developing theories in physics. They enable scientists to study atomic structures and facilitate the production of new elements and isotopes. In particular, neutron-deficient radionuclides, which decay by electron capture (EC) or β^+-emission, are produced mainly by proton accelerators (cyclotrons). As a result, cyclotrons have been crucial for producing key isotopes that have become central in the field of nuclear medicine (Table 1).

A cyclotron consists of two semicircular electrodes, called dees, positioned back to back with a narrow gap between them and situated between the poles of a magnet (Fig. 1). The dees are contained within a chamber where a vacuum is maintained, and a radio-frequency oscillator generates an alternating electric field across the gap. Particles to be accelerated are introduced in the system from the source near the center of the cyclotron, within the gap. The electric field accelerates these particles into one of the dees, where the magnetic field curves their path into a semicircle. As the particles complete this semicircular path and re-enter the gap, the electric field, now reversed in polarity, accelerates them into the opposite dee. As a result, the particles are accelerated to energies in the order of MeV. The radio-frequency oscillator is tuned to a specific frequency to ensure this continuous acceleration.

E. Cazzola (✉) · C. Favaretto · G. Gorgoni
IRCCS Sacro Cuore Don Calabria Hospital, Negrar di Valpolicella, Verona, Italy
e-mail: emiliano.cazzola@sacrocuore.it

R. M. van Dam et al. (eds.), *Automated Technologies for the Development and Production of Radiopharmaceuticals*,
https://doi.org/10.1007/978-3-031-84632-8_1

Table 1 List of main radioisotopes that can be produced with a cyclotron, together with their decay properties and production routes (IAEA, n.d.-b)

Radionuclide	Half-life	Eα (MeV) (Iα (%))	$E\beta^{+}_{av}$ (keV) (Iβ (%))	Eγ (keV) (Iγ (%))	Nuclear reaction
F-18	110 min	–	249.8 (96.7)	–	$^{18}O(p,n)^{18}F$
C-11	20.4 min	–	385.7 (99.8)	–	$^{10}B(d,n)^{11}C$; $^{14}N(p,\alpha)^{11}C$
N-13	10.0 min	–	491.8 (99.8)	–	$^{12}C(d,n)^{13}N$; $^{13}C(p,n)^{13}N$
Tc-99m	6.0 h	–	–	140.5 (89)	$^{100}Mo(p,2n)^{99m}Tc$
In-111	2.8 days	–	–	171.3 (90.7), 245.4 (94.1)	$^{111}Cd(p,n)^{111}In$; $^{nat}Cd(p,xn)^{111}In$
I-123	13.2 h	–	–	159.0 (83.6)	$^{124}Xe(p,2n)^{123}Ce \rightarrow {}^{123}Xe \rightarrow {}^{123}I$; $^{123}Te(p,n)^{123}I$
Ga-67	3.3 days	–	–	93.3 (38.8), 184.6 (21.4), 300.2 (16.6)	$^{68}Zn(p,2n)^{67}Ga$; $^{nat}Zn(p,xn)^{67}Ga$; $^{nat}Zn(d,xn)^{67}Ga$
Ga-68	67.7 min	–	829.5 (87.7)	1077.3 (3.2)	$^{68}Zn(p,n)^{68}Ga$
Cu-64	12.7 h	–	278 (17.5)	–	$^{64}Ni(p,n)^{64}Cu$; $^{64}Zn(n,p)^{64}Cu$
Cu-61	3.33 h	–	500 (61)	282.9 (12.7), 656.0 (10.4)	$^{64}Zn(p,a)^{61}Cu$; $^{nat}Zn(p,a)^{61}Cu$
Zr-89	3.3 days	–	396 (22.7)	909.1 (99.0)	$^{89}Y(p,n)^{89}Zr$
Sc-44	4.04 h	–	630.2 (94.3)	1157 (99.9)	$^{44}Ca(p,n)^{44}Sc$; $^{43}Ca(d,n)^{44}Sc$; $^{45}Sc(p,2n)^{44}Ti \rightarrow {}^{44}Sc$; $^{47}Ti(p,a)^{44}Sc$
Sc-43	3.89 h	–	476 (88.1)	372.9 (22.5)	$^{44}Ca(p,2n)^{43}Sc$; $^{43}Ca(p,n)^{43}Sc$; $^{46}Ti(p,a)^{43}Sc$
Tb-155	5.32 days	–	–	87 (32), 105 (25)	$^{155}Gd(p,n)^{155}Tb$; $^{156}Gd(p,2n)^{155}Tb$; $^{159}Tb(p,5n)^{155}Dy \rightarrow {}^{155}Tb$
Ac-225	10.0 days	5637 (4.4), 5732 (8.0), 5790 (8.6), 5792 (18.1), 5830 (50.7)	–	–	$^{232}Th(p,x)^{225}Ac$; $^{226}Ra(p,2n)^{225}Ac$; $^{226}Ra(\gamma,n)^{225}Ra \rightarrow {}^{225}Ac$
At-211	7.3 h	5870 (41.8)	–	–	$^{209}Bi(\alpha,2n)^{211}At$

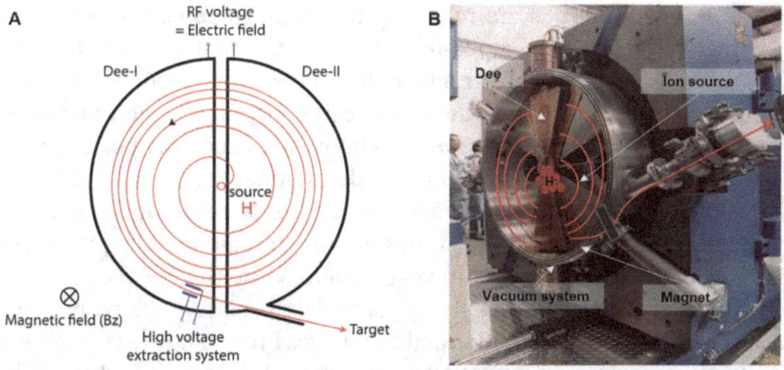

Fig. 1 Cyclotron concept scheme (**a**) and picture (**b**). The accelerated protons pathway, from the source outwards, is highlighted in red

Starting from E. Lawrence's early studies (Lawrence & Edlefson, 1930), numerous advancements have been made in cyclotrons, leading to the development of the current machines. The evolution of cyclotrons over the years is a fascinating subject, especially considering the shift in the 1990s from research-focused machines to commercially available cyclotrons (Synowiecki et al., 2018). These standard cyclotrons, offered at reasonable prices, proved essential for routine isotope production, meeting the growing demand for positron emission tomography (PET) radiopharmaceutical development.

Most of the radionuclides produced by cyclotrons involve the nuclear reaction: "Target Material (p,xn) Desired Radionuclide." The proton (p) irradiation of the target material converts it into the desired isotope by co-producing neutrons (n). Of course, different particles can be used to irradiate materials, and various particles can be emitted to maintain the energy balance of the nuclear reaction. Initially, carbon-11 was produced and used for labeling known molecules to study metabolic diseases and receptor interactions. However, the golden age of cyclotrons truly started with the realization of the potential uses of modified sugar labeled with fluorine-18. [^{18}F]Fluorodeoxyglucose ([^{18}F]FDG) was initially produced using electrophilic fluorine-18 generated by cyclotrons (Ido et al., 1977). However, the introduction of the nucleophilic production pathway significantly increased yields, enhanced reproducibility, and simplified production processes (Hamacher et al., 1986). Consequently, seeing the high efficacy of the process, the demand for cyclotrons surged, prompting various companies to develop commercial medical cyclotrons. Since then, cyclotrons have evolved to feature high-energy acceleration capabilities and the ability to accelerate different particles, making them versatile tools for numerous applications.

Recent advancements in accelerator physics are enhancing the capabilities of particle accelerators. Specifically, the use of high-energy deuterons as projectiles enables the production of several unique radiometals (Tárkányi et al., 2004). However, their availability is restricted in comparison to that of proton accelerators.

Alpha-particle beam cyclotrons are relatively rare, though some do exist. The yields of radionuclides relying on alpha-particle bombardment are significantly lower than for proton or deuteron bombardment due to low cross-section; however, alpha bombardment is critical for the production of certain medically relevant radionuclides (Qaim et al., 2016). However, several radionuclides can only be produced using alpha particles. For example, the radionuclide astatine-211 ($T_{1/2}$ = 7.3 h), an alpha-emitting heavy nuclide of significant interest in targeted alpha therapy (Table 1), has been produced in GBq quantities with high radionuclidic purity via the ^{209}Bi(α,2n)^{211}At reaction for the past 40 years (Qaim & Stcklin, 1983).

Linear electron accelerators (LINACs) are not widely used for isotope production yet because of their poor dissemination and challenges in operation (Starovoitova et al., 2014). In cyclotrons, transmutations are driven by protons, whereas in LINACs, electrons do not directly collide with the target nucleons. Instead, accelerated electrons impact a converter, producing Bremsstrahlung photons, which then induce photonuclear reactions in the target material. Typically, 20–40 MeV electron beams are used to generate a cone of Bremsstrahlung photons. Several feasibility studies have demonstrated the potential of LINACs for radionuclide production (Bindu et al., 2013; Danon et al., 2008).

2 Cyclotron Classification

Cyclotrons for radionuclide production can be schematically classified into four categories on the basis of the energy of the accelerated proton beam (Braccini, 2016). In the early days of cyclotron development, there were just two classifications:

- Research Cyclotrons (energy above 20 MeV)
  Research cyclotrons, with beam energies exceeding 20 MeV, were primarily installed in physics institutions for isotope production and matter studies. These high-energy cyclotrons were rarely used for routine medical isotope production due to their high operational costs and the complexities involved in setting up radiopharmaceutical production facilities within these institutions, especially given the increasing demand for pharmaceutical-grade production standards (Pupillo et al., 2023).
- Medical Cyclotrons (energy below 20 MeV)
  From the early 1990s, low-energy cyclotron installations have steadily increased over the decades, primarily driven by the rising demand for fluorine-18, spurred by the growing use of [^{18}F]FDG in PET imaging. This demand has been further propelled by the reduction in PET scanner costs and the opening of new markets, such as the Middle East and Eastern Europe. Multiple energy cyclotrons were present, but basically, two subclasses were identified. Low-energy cyclotrons for internal use (<15 MeV), which are machines ideal for the production of fluorine-18 and carbon-11 for internal use in hospitals. They enable single proton beam

irradiation, in some cases allowing dual beam targets, capable of producing approximately 50 GBq of fluorine-18, sufficient for preparing nucleophilic radiopharmaceuticals for a single nuclear medicine unit. Designed to minimize both initial and operational costs, these cyclotrons use low-volume enriched water for target loading and require low maintenance expenses. Predominantly installed in North America, Europe, and Asia, these systems are mainly suitable for hospitals with low patient volumes or those located in regions with low population density, thereby enhancing the territorial distribution of PET diagnostics. A second class includes high-energy medical cyclotrons (>16 MeV) designed for high production capabilities and versatility, responding to the high demand for [^{18}F]FDG in oncology diagnostics. These machines offer dual particle beam options (proton and deuteron) and can handle multiple applications, allowing institutions to produce large amounts of radiopharmaceuticals and distribute them to various hospitals. The first industrial Good Manufacturing Practice (GMP) facilities dedicated to fluorine-18 radiopharmaceutical production emerged alongside these high-productivity cyclotrons. These systems require more professional operators and are designed for flexible production capabilities. Additionally, they could be used for the development and production of other isotopes, such as iodine-124 and copper-64, meeting the evolving demands of both high-activity fluorine-18 production and new isotope requirements. Nowadays, there are more than 300 cyclotrons of this kind in operation, and their number is continuously increasing (Schmor, 2011).

- Starting in 2010, there was a shift in the energy and production capabilities required for cyclotrons due to shortages in isotope production and the need for new isotopes. According to an IAEA publication (IAEA, 2021), cyclotron manufacturers have anticipated market demands, resulting in the development of two additional cyclotron categories:
- Low-Energy Cyclotrons (energy below 10 MeV)
  These cyclotrons aim to meet customer demands for isotope production while reducing costs. They support the "dose on demand" philosophy, which emerged in the early 2010s. This approach integrates cyclotrons with microfluidic technology, enabling the production of single doses tailored for individual patients. Several companies and projects have developed prototypes, some of which are now installed in hospitals.
- High-Energy Cyclotrons (energy above 30 MeV)
  This category addresses the need for access to new, promising isotopes and seeks to manage emergency situations, such as those seen during the technetium-99m shortage. These cyclotrons can produce a variety of isotopes, including technetium-99m, actinium-225, astatine-211, copper-64, iodine-123, iodine-124, gallium-67, and indium-111, among others. The expansion of research cyclotrons to energies above 30 MeV facilitates the exploration and utilization of these isotopes for future medical applications. While small medical cyclotrons are typically commercial devices installed in nuclear medicine departments of hospitals, intermediate- and high-energy cyclotrons are located at large research facilities, such as at the Paul Scherrer Institute (Injector 2 separated-sector

Table 2 Cyclotron classification according to the beam energy and use

Category	Proton energy	Application	Main use[a]
Low-energy medical cyclotrons	7–11 MeV	F-18/C-11 production	Hospital internal
Medical cyclotrons	12–20 MeV	F-18/C-11 and PET radionuclide production	Hospital internal Commercial
Research cyclotrons	21–30 MeV	PET, SPECT, and α-nuclide production	Commercial Research (research institutes)
High-energy research cyclotrons	31–70 MeV	Research/exotic radionuclides production	Research (research institutes)

[a] Hospital internal = production for internal use in the nuclear medicine department; Commercial = production for distribution to multiple sites; Research = production for non-hospital use, specifically for research purposes

cyclotron, 72 MeV, Switzerland), Arronax (IBA Cyclone 70, 70 MeV, France), INFN (BEST 70, 70 MeV, Italy), and TRIUMF (PIF-BL1B, 500 MeV, Canada), among others (IAEA, n.d.-a; Goethals & Zimmermann, 2015). Depending on the type of cyclotron employed, the proton energy can be decreased to lower energies using specific materials in order to test alternative production routes (van der Meulen et al., 2020) in dedicated and separated beamlines.

The classification described in this section is summarized in the following tables. Table 2 outlines the current classification of cyclotrons based on energy levels and primary user requirements (Braccini, 2016).

The cyclotron business remains appealing for companies aiming to meet market demands, particularly by implementing new technical advancements to address emerging challenges and the growing need for isotopes. Table 3 provides an overview of the principal cyclotron models, listed by manufacturer, model, beam particles, energy, and target options. The table is not exhaustive, as companies are continuously developing their products to meet evolving market requirements.

3 Target Configurations

The cyclotron is not the only system that has undergone development and improvement over the years to meet evolving user demands. Depending on the type of target and the technology applied, different radioisotopes can be produced with varying yields, leading to a range of potential applications (Table 4). Target technology advancements have been crucial to address the increasing requirements, such as the high demand for fluorine-18 activity and the higher energy capacities of modern cyclotrons. These advancements have not only enhanced target performance but also aimed to reduce maintenance needs and improve operator safety.

Physically, the target could be directly connected to the cyclotron as shown in Fig. 1, or connected to a beamline (Fig. 2). The beamline is a vacuum tube where

Table 3 Cyclotrons currently available on the market, categorized by manufacturer, model, beam particles and energy, and target options. Cyclotron data sourced from IAEA (2008)

Manufacturer	Model	Beam particles	Max energy	Target options
Advanced Cyclotron Systems Inc. (ACSI)	TR-19	Protons/ deuterons	19 MeV/9 MeV	Liquid/gas/solid target/beamline (optional)
	TR-24	Protons	24 MeV	Beamline/liquid/ gas/solid target
	TR-30	Protons/ deuterons/ alphas	30 MeV/15 MeV	Beamline/liquid/ gas/solid target
Ion Beam Applications (IBA)	Cyclone KEY	Protons	9.5 MeV	Liquid
	Cyclone KUBE	Protons	18 MeV	Liquid/gas/solid target/beamline (optional)
	Cyclone IKON	Protons/ deuterons/ alphas	30 MeV	Beamline/liquid/ gas/solid target
	Cyclone 70	Protons/ deuterons/ alphas	70 MeV	Beamline/liquid/ gas/solid target
General Electric (GE)	MINItrace	Protons	9.6 MeV	Liquid/gas
	PETtrace 800	Protons/ deuterons	16.5 MeV/8.4 MeV	Liquid/gas/solid target/beamline (optional)
Team Best, Best Cyclotron System (BCS)	Best 15p Cyclotron	Protons	15 MeV	Liquid/gas/solid target
	Best 25p Cyclotron	Protons	25 MeV	Liquid/gas/solid target/beamline (optional)
	Best 35p Cyclotron	Protons	35 MeV	Beamline/liquid/ gas/solid target
	Best 70p Cyclotron	Protons	70 MeV	Beamline / liquid / gas / solid target
Alcen/PMB	iMiTrace Cyclotron	Protons	12 MeV	Liquid
Sumitomo	CYPRIS HM-12	Protons/ deuterons	12 MeV/6 MeV	Liquid/gas/solid target
	CYPRIS HM-20	Protons/ deuterons	20 MeV/10 MeV	Liquid/gas/solid target/beamline (optional)

the beam is magnetically confined, primarily used to move the beam to a different location away from the cyclotron chamber. The beamline is commonly used in cyclotrons operating at energies higher than 19 MeV, allowing for the precise management of the beam in different target bunkers or for defocusing and degrading the

Table 4 Target options available on medical cyclotrons and isotopes mainly produced using each of them

Target	Isotopes	Cyclotron category	Main use[a]
Gas	C-11	Low to medium energy	Hospital internal
	O-15	Low to medium energy	Hospital internal
	Electrophilic F-18	Medium energy	Hospital internal
Liquid	F-18	Low to medium energy	Hospital internal/commercial
	N-13	Low to medium energy	Hospital internal
	Ga-68	Medium to research energy	Hospital internal/commercial
	Cu-64/Cu-61	Medium to research energy	Hospital internal/commercial
Solid	Ga-68	Medium to research energy	Hospital internal/commercial
	Cu-64	Medium to research energy	Hospital internal/commercial
	I-124/I123	Medium to research energy	Hospital internal/commercial
	Zr-89	Medium to research energy	Hospital internal/commercial
	Ac-225	Research to high energy	Commercial

[a] Hospital internal = production for internal use in the nuclear medicine department; Commercial = production for distribution to multiple sites

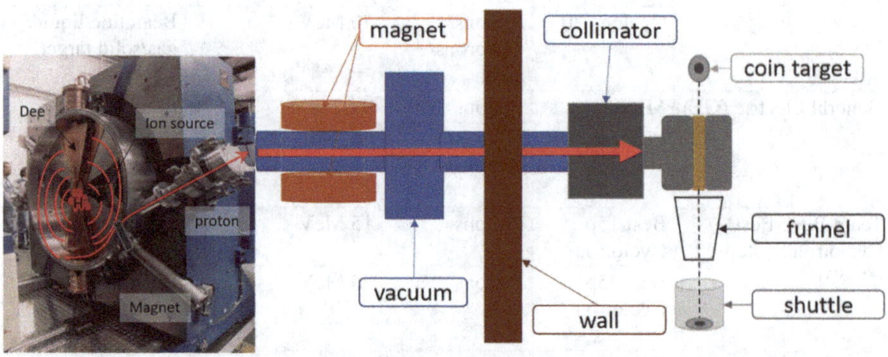

Fig. 2 Cyclotron beam line for solid target production application

beam at the desired shape and energy requested by the nuclear reaction. In low- and medium-energy cyclotrons, the beamline is specially used on solid target production, as it facilitates better beam shaping and easier recovery of the target materials out of the high-activity areas where the other targets are placed.

3.1 Gas Targets

Gas target systems on medical cyclotrons are frequently utilized to produce PET radionuclides such as fluorine-18, oxygen-15, and carbon-11. Figure 3 illustrates a gas target including filling and recovery of target gas as well as cooling flow. These systems are highly user-friendly, requiring no manual handling for routine

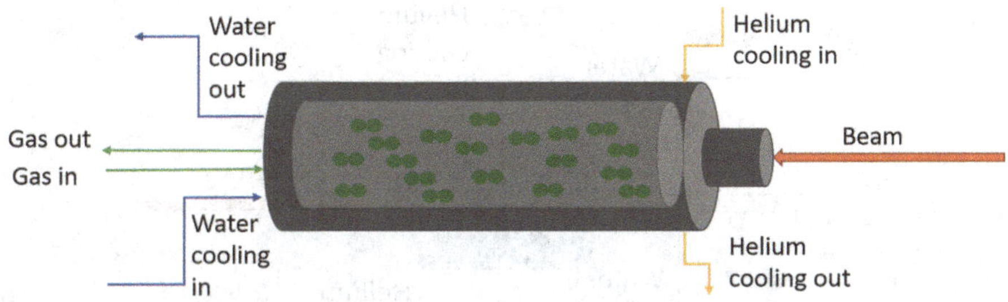

Fig. 3 Gas target representation

production processes. The gaseous state of the target material allows for efficient heat transfer and cooling. In addition, immediately after the end of bombardment (EOB), the produced radioactivity is transferred via small bore tubing to the automated synthesis module using pressurized inert gas, reducing the pre- and post-irradiation target preparation processes and facilitating target transfer. Consequently, gas target systems are characterized by their safety, robustness, and rapid operation (Steinbach, 1990; Synowiecki et al., 2018). For the mentioned reasons, gas targets were among the earliest types of targets developed during the initial phase of PET applications. In addition, throughout the 1980s and up until 2010, carbon-11 was one of the most significant radionuclides produced by cyclotrons. Carbon-11 is produced via the $^{14}N(p,\alpha)^{11}C$ nuclear reaction, which has a cross-section of 120 mb at 14 MeV energies, with the initial target material primarily being natural nitrogen gas (IAEA, 2009). These targets are typically housed in a cylindrical chamber with lengths measured in millimeters. Sometimes a conical design is used to maximize cooling and optimize the interaction between the gas and the beam, which becomes defocused as it passes through the window's foils and the target gas. As target technology advanced, the pressure within the target increased, enabling the production of up to 6 Ci of $[^{11}C]CO_2$ using the latest versions of carbon-11 gas targets, such as those developed by GE Healthcare (Norling et al., 2008). This activity level meets the demand for multiple doses of well-established radiopharmaceuticals based on carbon-11 or allows for the preparation of radiopharmaceuticals requiring multiple-step synthesis. Gas target technology has also been applied to fluorine-18 production, although it is a more time-consuming and maintenance-intensive process compared to liquid target systems (IAEA, 2012).

3.2 *Liquid Targets*

Isotope production using liquid targets represents another simple method in terms of manipulation, which is why medical cyclotron facilities were initially designed to handle and produce isotopes using this approach. The ease of installation and replacement of delivery lines, along with the convenience of delivering and

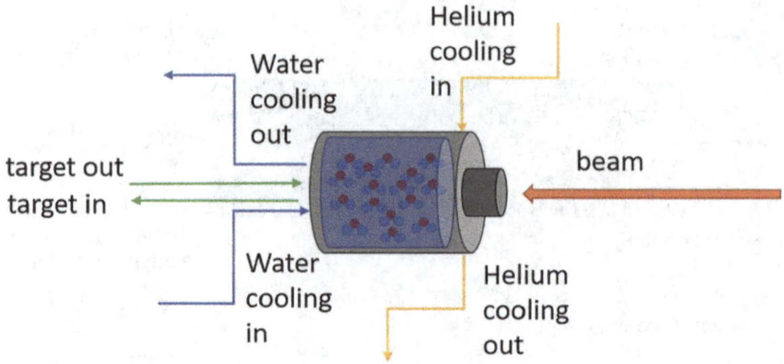

Fig. 4 Liquid target representation

managing isotopes in liquid form, has driven the development of robust, reliable, and easy-to-maintain targets. These targets typically feature a niobium chamber to contain the liquid target material, with a Havar foil to isolate it and allow beam collection (Fig. 4). Frontal helium cooling combined with rear water cooling ensures effective heat removal to prevent damage, as seen in systems known as single foil targets or water-cooled targets. To reduce possible causes of failures, frontal helium cooling could be removed, increasing the rear water-cooling flow; this solution simplifies the target management and prevents possible helium cooling leaks that are difficult to identify. The pressurization of liquid targets with inert gas prevents liquid boiling and, as well as for gas targets, can facilitate the transfer of the produced isotopes after irradiation (Alves et al., 2018). Filling volume, combined with modifications on cooling geometry are the main developed items over time to allow increased production capacity and align the cooling to the production needs. In addition to the development of targets for fluorine-18 production, new applications have emerged, focusing on the production of radiometals using liquid targets. Liquid targets, in which the target material is dissolved in an acid solution, offer the advantage of producing radiometals directly in liquid form using existing medical cyclotrons, thereby eliminating the need for investment in a solid target station (Pandey et al., 2014).

3.2.1 Fluorine-18 Production

Fluorine-18 is the most commonly produced and important isotope generated by cyclotrons worldwide. It is obtained via the $^{18}O(p,n)^{18}F$ nuclear reaction, starting from ^{18}O-enriched water (i.e., $[^{18}O]H_2O$), with a cross-section of 100–700 mb in the energy range of 4–16 MeV, peaking at 6 MeV (Ruth & Wolf, 1979). The demand for fluorine-18 continues to grow annually, driven by the increasing need for $[^{18}F]$FDG in oncological PET diagnostics and the emergence of ^{18}F-based radiopharmaceuticals for prostate cancer evaluation, which are entering the market and meeting the needs of prostate cancer patients. These new radiopharmaceuticals will further increase the demand for fluorine-18, necessitating maximum production efficiency.

Most installed cyclotrons are equipped with multiple fluorine-18 targets to maximize production capacity and provide redundancy. Due to advancements in target development and increases in cyclotron operation efficiency, fluorine-18 production can be scaled up to 740 GBq on the highest-performing cyclotrons, with [^{18}F]FDG production capabilities exceeding 500 GBq. While achieving this production level presents challenges in terms of radioprotection and maintenance strategies, it is attainable with current technology.

Starting from the initial general target layout, shown in Fig. 4, which includes frontal helium cooling, several configurations have been investigated in recent years. These studies aim to improve production yield by increasing the enriched water volume, optimizing beam-water interaction, and enabling higher beam currents. The filling volume has been expanded from 0.8 mL in the first targets to up to 5 mL, significantly boosting fluorine-18 production. A notable improvement in target performance was achieved by modifying the target shape. Advanced Cyclotron Systems Inc. (ACSI) developed a 20–30° oriented target (Fig. 5) to enhance beam interaction with the enriched water (Synowiecki et al., 2018). This design increases production yield, supports higher beam currents, and facilitates cooling due to the larger water surface area. In particular, in this setup there is an oblique angle between the beam and the target liquid reservoir, allowing the beam to interact with a greater volume of enriched water.

A major advancement in target maintenance and reliability was achieved by eliminating frontal helium cooling and related helium foil, which were a significant cause of leaks and foil damage during target aging. The new cooling approach focused on enhancing the backing water cooling, allowing for the removal of frontal helium cooling while providing multiple benefits. Advantages of these single-foil (water-cooled) targets include reduced gas usage during target operation (limited to pressurization and liquid transfer), prevention of gas leaks, and minimized beam interruptions.

Another advance in target shape was adding lead shielding into the target body. When used in routine high-activity production of fluorine-18, testing showed that shielded targets allow reduction of operator radiation exposure, and enable operators to perform maintenance operations closer to the target.

3.2.2 Radiometal Production on Liquid Targets

Liquid targets can be used for the production of radiometals like gallium-68, copper-64, copper-61, zirconium-89, scandium-44, yttrium-86, zinc-63, and technetium-99m (Alves et al., 2018; Do Carmo et al., 2020). Liquid target radiometal production represents a new application in liquid target technology, addressing the need for well-established radiometals in cyclotron facilities without requiring modifications to the bunker and laboratory layout, which is otherwise required for a solid target system implementation. The interest in cyclotron-based radiometal

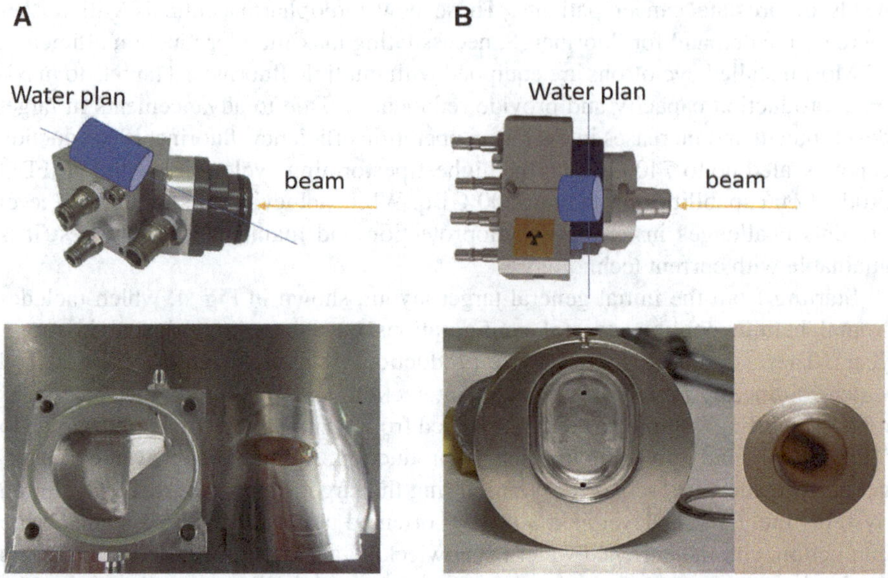

Fig. 5 Sloped (**a**) and vertical (**b**) liquid target representation. The upper sections represent the beam angle to the enriched water. The lower sections report pictures of the target body (enriched water reservoir) and the enriched water foils with the beam shape

production has increased due, for instance, to the rising costs of gallium-68 generators and the growing demand for gallium-68 radiopharmaceuticals. An alternative production route to the generators for gallium-68 is the $^{68}Zn(p,n)^{68}Ga$ nuclear reaction, traditionally conducted using solid targets with high production yields. However, this method is limited to facilities equipped with solid target systems, leaving many medical cyclotrons unable to participate in this production.

Alves et al. developed a routine procedure for producing gallium-68 from liquid targets by using a $Zn(NO_3)_2$ solution (Alves et al., 2018). This process generates a gallium-68 solution that can be easily transferred to a synthesis module for conversion into starting material or direct labeling with biomolecules. Generally, the preparation of the solution for liquid targets should allow that the saturation point of the target material in solution is not achieved, and thus its precipitation out of the solution does not occur. This generally results in a solution that is extremely acidic, such that it may be corrosive and cause damage to the liquid target station. Consequently, it is essential to select a target solution that strikes a balance between the acidity and saturation point of the target material. This requires extensive research on materials and target design, leading to the development of more efficient targets for liquid applications. As an example, the development of the production of gallium-68 using liquid targets involved stabilizing the solution concentration, acidity, and gas release, as well as eliminating potential metal contamination from the cyclotron. Various solutions were explored to minimize target pressure during irradiation and reduce secondary gas production. Plastic valves, connections, and all the parts used were implemented to prevent corrosion from the acidic solution used to dissolve the metal, and efforts were made to find the optimal solution concentration to avoid precipitation in the delivery line and ensure efficient radioisotope transfer to the hot cell. Solution density needed to be studied and evaluated as well to avoid issues in the delivery to and from the target. Finally, maintaining an acidic environment in all lines was crucial to prevent clogging from metal precipitation.

However, due to physical limitations on target volume and radioisotope concentration, liquid target production is generally less efficient than solid target methods. In addition, a major drawback of this technology is the need to dedicate targets exclusively for a single radiometal production, as switching between different isotopes in the same target is currently impractical due to residual metal and radiometal impurities. Nevertheless, it remains an attractive option for specific applications, such as gallium-68 production for R&D purposes, where high activity can be achieved in a reasonable time frame in comparison to a generator production in a clinical setting where priority is given to the patient dose. In addition, radiometal production with the liquid target approach generally involves less expensive target materials and no modification to the targetry layout of the cyclotron.

3.3 *Solid Targets*

Over the past decade, solid target isotope production has emerged as a crucial component in cyclotron setups. The shortage of isotopes such as technetium-99m, gallium-68, copper-64, scandium-44, and zirconium-89 has highlighted the necessity for advanced radiopharmaceutical preparation techniques (George et al., 2021). Nowadays, most neutron-deficient novel radiometals are produced in cyclotron facilities using solid target systems (Fig. 6).

A solid target station is depicted in Fig. 6a, hosting a solid target coin-shaped, which comprises a backing, target material, and occasionally a frontal isolation foil. However, various target shapes are available (Fig. 6b), depending on the underlying technology. The target material can be attached to the target body using different methods, some of the most common being (Stolarz, 2014):

- **Single metal foil** attached to the backing (Link et al., 2017).
- **Pellet** created by mechanically reshaping solid or powder materials by pressing or rolling (van der Meulen et al., 2020).
- **Electrodeposition**, depositing the target material in solution onto a target body through a potential difference (Kretowicz et al., 2023).
- **Sputtering**, where vaporized metal is deposited onto the backing (Skliarova et al., 2019).
- **Sintering**, which involves compacting and forming a solid mass of material by applying pressure or heat, without melting the material to the point of liquefaction. A particular type of this technique is spark plasma sintering (SPS), where

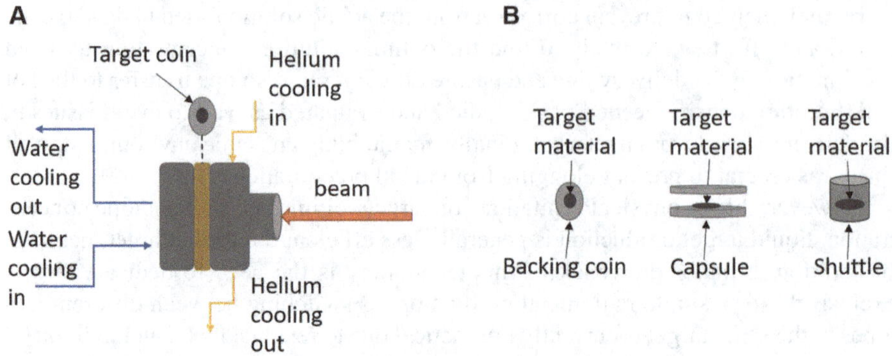

Fig. 6 Representation of solid target station (**a**) and types of target (**b**)

the heat generation is internal and generated from pulsed electrical current with low voltage (Cisternino et al., 2022).

- **Melting** of the target material and reshaping it by altering its state (Stolarz, 2014).
- **Sedimentation** of the target material suspended in a solution onto the backing (Stolarz, 2014).

In addition, the choice of target and backing material is crucial for effective heat transmission and to prevent contamination during the target dissolution. The target material must exhibit high thermal stability, a high melting point, and efficient heat conductivity at the interface between the target and the backing material. Additionally, targets must be mechanically stable, radiation-resistant, and chemically compatible with post-irradiation processing. The material must be as chemically and isotopically pure as possible, as impurities can lead to the creation of undesired byproducts, affecting the purity of the radionuclide product. Target uniformity, homogeneity, and the presence of impurities are highly dependent on the target production process. The thickness of the target is another critical factor, as irradiation energy degrades differently across the target based on its thickness and can lead to unwanted reactions. Optimal target thickness must be evaluated based on several factors. While increasing the amount of target material can enhance yield, it is important to consider the potential rise in production costs due to the often high expense of target materials. On the other hand, if the target material is too thin, the beam may penetrate through to the backing, potentially generating side products that could lead to contamination if the material is not inert. Additionally, a thicker target can reduce cooling efficiency, increasing the risk of overheating both the target station and the target itself (Talip et al., 2020).

3.3.1 Developments in Solid Target Stations

In contrast with liquid targets, which can be transported directly through dedicated lines, solid targets must be physically extracted from the target station and manually transferred to a hot cell for dissolution and further processing in liquid form. This not only complicates the workflow but also increases the potential for contamination during transfer. Furthermore, the increased handling required for solid targets elevates the operator's exposure to radiation, thereby raising safety concerns and the need for stringent protective measures. While solid targets present challenges in target management and implementation in existing facilities, they are essential for increasing the activity yield from target irradiation. Therefore, over the past 10 years, developments in solid target technology have encompassed not only the targets themselves but also the target stations, transfer options, and management systems, such as cooling systems that preserve the integrity of the irradiated target and target transfer mechanisms (Braccini et al., 2020; Dellepiane et al., 2021; Sciacca et al., 2021). These components have been standardized and improved to convert R&D activities into routine processes, similar to those for fluorine-18 production.

The target station is a crucial component of the isotope production process, as it needs to be easily accessible, have low activation levels, and allow for potential automation. Traditionally, especially in medical cyclotrons, target stations were designed as vertical or sloped target holders, where solid targets—often in coin shapes with attached target material—could be manually inserted and securely positioned during the irradiation (Fig. 7). In this system, an inclination of the target increases the amount of target material interacting with the beam, thereby improving the yield of the reaction. A significant advancement was introduced by Comecer with the Alceo automatic target station. This system integrates all steps by using a shuttle that encapsulates a pellet or a coin with the target material deposited on its surface, which can then be pneumatically transferred to and from the target station.

In addition, general constraints due to manual operations, the need for dedicated and expensive hot cells and equipment combined with the need for rapid recharging

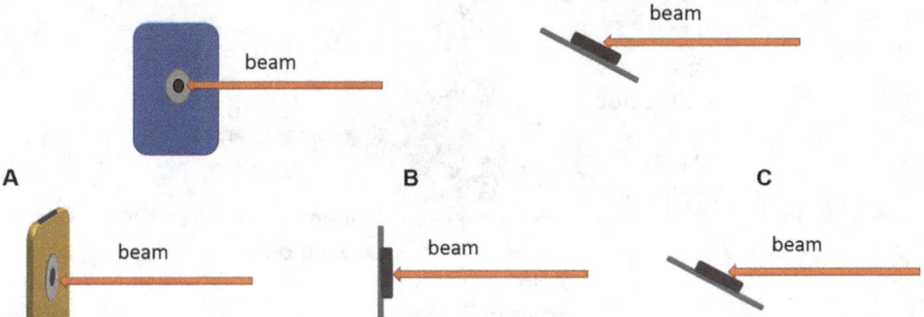

Fig. 7 (**a**) Representation of beam incidence on a solid target inserted on vertical target holder. (**b**) Vertical solid target beam incidence. (**c**) Sloped solid target beam incidence

of new targets have led to the development of projects like multiple coin loaders. These systems enable the automatic loading and delivery of several dedicated shuttles that incorporate the target material (Malinconico et al., 2018). ARTMS' QUANTM Irradiation System made an automatic procedure for multiple solid target capsules/coins in combination with a compact standalone dissolution system (Thisgaard et al., 2021). This approach is beneficial for routine procedures due to the integration of multiple steps and centralized control through a single software system. This represents a significant step forward in the standardization and efficiency of radiopharmaceutical production, even though it can limit its flexibility. Additionally, the automation of the process reduces the operator's dose exposure and minimizes the risk of contamination. However, revolving multiple target load systems are generally produced in-house and developed to allow multiple beam operation before interacting with the solid target station.

3.4 Hybrid Targets

Hybrid target technology combines the possibility to achieve the high production capability of solid targets with the easy management of liquid targets. Hybrid targets can be loaded with a pellet-shaped target in a multiple loading system, and after the irradiation time, an appropriate solution selectively dissolves the irradiated target material (Fig. 8). After dissolution, the liquid radiometal solution can be transferred via tubing to an appropriate hot cell for further operations. The limitation of hybrid target technology is related to the need to have multiple targets occupied for different isotopes produced, but of course, it is a very effective alternative pathway for radiometal production.

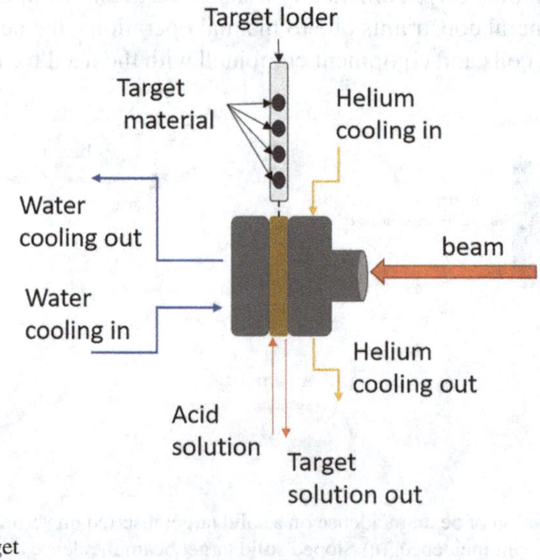

Fig. 8 Hybrid target

Currently, SYNIQ Hybrid target systems can be installed on existing cyclotrons with only a few limitations on production capability based on different cyclotron shielding options and space on multiple target holders. Gallium-68 was already produced in good quality using this type of target under varying irradiation conditions. All the batches produced achieved European Pharmacopoeia monograph requirements for cyclotron-produced gallium-68 (Nelson et al., 2022).

4 Outlook

Cyclotrons in the new century continue to be one of the most efficient ways to produce radionuclides for multiple applications at a reasonable cost. Over time, the demand for cyclotrons has evolved. Since the early 2000s, efforts have been focused on maximizing system performance for fluorine-18 production, but in the last decade, the demand for higher energy cyclotrons has increased due to the need for autonomy in isotope production, not only for fluorine-18 but also for technetium-99m and new alpha-emitting radionuclides. This has led to a growing demand for more powerful cyclotrons.

The cost of the high-energy cyclotrons combined with the knowledge and service required for their operation, along with a different country or regional organization for service, will likely result in an uneven geographical distribution of such machines. The lack of this kind of infrastructure could lead to poor isotope distribution and different isotope availability throughout the countries. Additional hubs will be necessary to meet demand if nuclear isotope properties permit their distribution, placing increased pressure on isotope transportation logistics. Therefore, sustainability studies for research and high-energy facilities should be encouraged to demonstrate the need for and support the implementation of such infrastructure.

Moreover, alternative technologies have become available and have been transferred to industrial applications for isotope production. Linear accelerator producers are entering this market segment to produce machines suitable for isotope production, offering significant advantages such as less system contamination and higher achievable beam energy.

One of the key points for future applications is the continuous and significant improvement of targets. This will facilitate easier management of isotopes after irradiation, reduce unavoidable backing activation, and enable the production of high-activity isotopes, all while enhancing operator safety.

References

Alves, V., et al. (2018). Automated purification of radiometals produced by liquid targets. *Instruments, 2*, 1–9. https://doi.org/10.3390/instruments2030017

Bindu, K. C., et al. (2013). Optimization of commercial scale photonuclear production of radioisotopes. *AIP Conference Proceedings, 1525*, 407–411. https://doi.org/10.1063/1.4802359

Braccini, S. (2016). Compact medical cyclotrons and their use for radioisotope production and multi-disciplinary research. In *CYC 2016—Proceedings of the 21st International Conference on Cyclotrons and their Applications* (pp. 229–234).

Braccini, S., et al. (2020). Novel irradiation methods for theranostic radioisotope production with solid targets at the Bern medical cyclotron. In *CYC 2019—Proceedings of the 22nd International Conference on Cyclotrons and Their Applications* (pp. 128–132). https://doi.org/10.18429/JACoW-Cyclotrons2019-TUA02

Cisternino, S., et al. (2022). Target manufacturing by spark plasma sintering for efficient 89Zr production. *Nuclear Medicine and Biology, 104–105*, 38–46. https://doi.org/10.1016/j.nucmedbio.2021.11.004

Danon, Y., et al. (2008). Medical isotope production using a 60 MeV linear electron accelerator. *Transactions of the American Nuclear Society, 98*, 894–895.

Dellepiane, G., et al. (2021). Instruments and methods for theranostic radioisotope production at the Bern medical cyclotron. In *TIPP2020—International Conference on Technology and Instrumentation in Particle Physics*.

Do Carmo, S. J. C., Scott, P. J. H., & Alves, F. (2020). Production of radiometals in liquid targets. *EJNMMI Radiopharmacy and Chemistry, 5*(1), 2. https://doi.org/10.1186/s41181-019-0088-x

George, K. J. H., Borjian, S., Cross, M. C., Hiks, J. W., Schaffer, P., & Kovacs, M. S. (2021). Expanding the PET radioisotope universe utilizing solid targets on small medical cyclotrons. *RSC Advances, 49*, 31098–31123. https://doi.org/10.1039/D1RA04480J

Goethals, P. E., & Zimmermann, R. (2015) *Cyclotrons used in nuclear medicine world market report & directory*. 2015 Ed.

Hamacher, K., Coenen, H. H., & Stocklin, G. (1986). Efficient stereospecific synthesis of no-carrier-added 2-[18F]-fluoro-2-deoxy-D-glucose using aminopolyether supported nucleophilic substitution. *Journal of Nuclear Medicine, 27*(2), 235–238.

IAEA. (2008). *Cyclotron produced radionuclides: Principles and practice*. IAEA technical reports series n. 465. Retrieved from https://www.iaea.org/publications/7849/cyclotron-produced-radionuclides-principles-and-practice

IAEA. (2009). *Cyclotron produced radionuclides: Physical characteristics and production methods*. IAEA technical reports series n. 468. Retrieved from https://www.iaea.org/publications/7892/cyclotron-produced-radionuclides-physical-characteristics-and-production-methods

IAEA. (2012). *Cyclotron produced radionuclides: Operation and maintenance of gas and liquid targets*. IAEA radioisotopes and radiopharmaceuticals reports n. 4. Retrieved from https://www.iaea.org/publications/8783/cyclotron-produced-radionuclides-operation-and-maintenance-of-gas-and-liquid-targets

IAEA. (2021). *Alternative radionuclide production with a cyclotron*. IAEA radioisotopes and radiopharmaceuticals reports n. 4. Retrieved from https://www.iaea.org/publications/13649/alternative-radionuclide-production-with-a-cyclotron

IAEA. (n.d.-a). *IAEA cyclotron database*. Retrieved from https://nucleus.iaea.org/sites/accelerators/Pages/Cyclotron.aspx

IAEA. (n.d.-b) *IAEA live chart of nuclides*. Retrieved from https://www-nds.iaea.org/relnsd/vcharthtml/VChartHTML.html

Ido, T., Wan, C. N., Fowler, J. S., & Wolf, A. P. (1977). Fluorination with molecular fluorine. A convenient synthesis of 2-deoxy-2-fluoro-D-glucose. *Journal Organic Chemistry, 42*, 2341–2342. https://doi.org/10.1021/jo00433a037

Kretowicz, M. N., Barrett, K. E., Barnhart, T. E., & Engle, J. W. (2023). Recycling of 52Cr electroplated targets for 52gMn production. *Applied Radiation and Isotopes, 200*(2023), 211–214. https://doi.org/10.1016/j.apradiso.2023.110924

Lawrence, E. O., & Edlefson, N. E. (1930). *Science, 72*, 376.

Link, J. M., Krohn, K. A., & O'Hara, M. J. (2017). A simple thick target for production of 89Zr using an 11 MeV cyclotron. *Applied Radiation and Isotopes, 122*, 211–214. https://doi.org/10.1016/j.apradiso.2017.01.037

Malinconico, M., Asp, A., Lang, C., Boschi, F., Tieu, W., Kuan, K., Guidi, G., & Takhar, P. (2018, May). 68Ga and 45Ti production on a GE PETtrace cyclotron using the ALCEO solid target. *Journal of Nuclear Medicine, 59*(Suppl 1), 664. https://doi.org/10.1016/j.apradiso.2017.01.037

Nelson, B. J. B., Andersson, J. D., Wuest, F., et al. (2022, October). Good practices for 68Ga radiopharmaceutical production'. *EJNMMI Radiopharmacy and Chemistry, 7*, 27. https://doi.org/10.1186/s41181-022-00180-1

Norling, J., et al. (2008). *High yield [11C]-CO_2 a commercial cyclotron enhances possibilities for clinical tracer usage*. WTTC12, April 2008, TRIUMF, Canada.

Pandey, M. K., et al. (2014). Production of 89Zr via the 89Y(p,n)89Zr reaction in aqueous solution: Effect of solution composition on in-target chemistry. *Nuclear Medicine and Biology, 41*(4), 309–316. https://doi.org/10.1016/j.nucmedbio.2014.01.006

Pupillo, G., et al. (2023). Cyclotron-based production of innovative medical radionuclides at the INFN-LNL: State of the art and perspective. *European Physical Journal Plus., 138*(12), 1095. https://doi.org/10.1140/epjp/s13360-023-04564-3

Qaim, S. M., & Stcklin, G. (1983). Production of some medically important short-lived neutron-deficient radioisotopes of halogens. *Radiochimica Acta, 34*(1–2), 25–40. https://doi.org/10.1524/ract.1983.34.12.25

Qaim, S. M., et al. (2016). Uses of alpha particles, especially in nuclear reaction studies and medical radionuclide production. *Radiochimica Acta, 104*(9), 601–624. https://doi.org/10.1515/ract-2015-2566

Ruth, T., & Wolf, A. (1979). Absolute cross sections for the production of 18F via the 18O(p, n)18F reaction. *Radiochimica Acta, 1*(26), 21. https://doi.org/10.1524/ract.1979.26.1.21

Schmor, P. (2011). *Review of cyclotrons for the production of radioactive isotopes for medical and industrial applications*. World Scientific Publishing Co. Pte. Ltd. https://doi.org/10.1142/9789814383998_0005

Sciacca, G., et al. (2021). A universal cassette-based system for the dissolution of solid targets. *Molecules, 26*(20), 6255. https://doi.org/10.3390/molecules26206255

Skliarova, H., et al. (2019). Medical cyclotron solid target preparation by ultrathick film magnetron sputtering deposition. *Instruments, 3*(1), 1–15. https://doi.org/10.3390/instruments3010021

Starovoitova, V. N., Tchelidze, L., & Wells, D. P. (2014). Production of medical radioisotopes with linear accelerators. *Applied Radiation and Isotopes, 85*, 39–44. https://doi.org/10.1016/j.apradiso.2013.11.122

Steinbach, J. (1990). The gas target technique for radionuclide production with small accelerators. *Isotopenpraxis Isotopes in Environmental and Health Studies, 26*(3), 102–106. https://doi.org/10.1080/10256019008624242

Stolarz, A. (2014). Target preparation for research with charged projectiles. *Journal of Radioanalytical and Nuclear Chemistry, 299*, 913–931. https://doi.org/10.1007/s10967-013-2652-2

Synowiecki, M. A., Perk, L. R., & Nijsen, J. F. W. (2018). Production of novel diagnostic radionuclides in small medical cyclotrons. *EJNMMI Radiopharmacy and Chemistry, 3*(1), 3. https://doi.org/10.1186/s41181-018-0038-z

Talip, Z., et al. (2020). A step-by-step guide for the novel radiometal production for medical applications: Case studies with ^{68}Ga, ^{44}Sc, ^{177}Lu and ^{161}Tb. *Molecules, 25*(966), 1–28. https://doi.org/10.3390/molecules25040966

Tárkányi, F., et al. (2004). Excitation functions for production of radioisotopes of niobium, zirconium and yttrium by irradiation of zirconium with deuterons. *Nuclear Instruments and*

Methods in Physics Research, Section B: Beam Interactions with Materials and Atoms, 217(3), 373–388. https://doi.org/10.1016/j.nimb.2003.11.084

Thisgaard, H., Kumlin, J., Langkjær, N., et al. (2021). Multi-curie production of gallium-68 on a biomedical cyclotron and automated radiolabelling of PSMA-11 and DOTATATE. *EJNMMI Radiopharmacy and Chemistry, 6*, 1. https://doi.org/10.1186/s41181-020-00114-9

van der Meulen, N. P., et al. (2020). Developments toward the implementation of ^{44}Sc production at a medical cyclotron. *Molecules, 25*(4706), 1–16. https://doi.org/10.3390/molecules25204706

Nuclear Reactor-Based Production of Medical Radionuclides

Izabela Cieszykowska, Wioletta Wojdowska, Dariusz Pawlak, and Renata Mikołajczak

1 Principle of Operation and Neutron Spectrum

The nuclear reactor is a facility in which a fissile atomic nucleus such as ^{235}U, ^{239}Pu, or ^{241}P absorbs a low-energy neutron and undergoes nuclear fission (Lundqvist, 2014). In the process, fast neutrons are produced with energies from about 10 MeV and below (the fission neutron spectrum). The neutrons are slowed down in a moderator (usually water), and the slowed-down neutrons start new fissions. By regulating this nuclear chain reaction, there will be a steady state of neutron production with a typical neutron flux of the order of 10^{14} neutrons $\cdot$ cm^{-2} $\cdot$ s^{-1}. Since neutrons have no charge and are, thus, unaffected by the Coulomb barrier, they can enter the target nucleus and cause a nuclear reaction. Thermal neutrons, with an energy of about 0.025 eV, usually have a much larger effective neutron absorption cross-section for a given nuclide than fast neutrons and can therefore be absorbed more easily by an atomic nucleus. However, some nuclear reactions, depending upon the cross-section, require fast neutrons with energy between 1 and 10 MeV (IAEA, 2003).

Most nuclear reactors in the world are used for energy production, but generally, only research reactors are flexible enough for radioisotope production. A research reactor's analytical and research capabilities are determined primarily by the available thermal neutron flux. The International Atomic Energy Agency (IAEA) Research Reactor Database (IAEA, 2022) categorizes research reactors as low flux, medium flux, or high flux reactors according to the following levels of thermal neutron flux:

I. Cieszykowska (✉) · W. Wojdowska · D. Pawlak · R. Mikołajczak
National Centre for Nuclear Research, Radioisotope Centre POLATOM, Otwock, Poland
e-mail: Izabela.Cieszykowska@polatom.pl

R. M. van Dam et al. (eds.), *Automated Technologies for the Development and Production of Radiopharmaceuticals*,
https://doi.org/10.1007/978-3-031-84632-8_2

- Low flux reactor ($\leq 1 \times 10^{12}$ n/cm²/s)
- Medium flux reactor ($>1 \times 10^{12}$ n/cm²/s and $<1 \times 10^{14}$ n/cm²/s)
- High flux reactor ($\geq 1 \times 10^{14}$ n/cm²/s)

While the usefulness of low neutron flux reactors for the production of medical isotopes is limited, medium and high neutron flux reactors are widely utilized for this purpose (IAEA, 2022).

A nuclear reaction in which a neutron is captured or absorbed by the nucleus, leading to a nuclear transformation, is called neutron activation (IAEA, 2003). The radioactivity of the produced radioisotope (A) can be calculated from the formula:

$$A = \phi * N * \sigma * \left(1 - e^{-\lambda t}\right)$$

where:

ϕ—the neutron flux
N—number of nuclei in the target
σ—cross-section
t—irradiation time
λ—decay constant of the produced radioisotope

Reactor production generally leads to neutron-rich radionuclides, which predominantly decay by β^- emission and are therefore especially suitable for radiotherapy. Various radionuclides can be produced depending on the target material used for irradiation and the cross-section of a given nuclear reaction. The examples of radionuclides of current interest in nuclear medicine are listed in Table 1.

1.1 Types of Nuclear Reactions

After neutron capture, the excited atomic nucleus can stabilize itself mainly by emitting gamma radiation in (n, γ) nuclear reaction, proton emission in (n, p) nuclear reaction, alpha particle emission (n, α), or fission (n, f) nuclear reaction. These nuclear reactions are briefly characterized below:

Nuclear Reaction (n,γ)

As a result of the *(n,γ)* nuclear reaction, a radioisotope of the same element as the target is produced, but with a mass number greater by 1 (Fig. 1):

$${}^{A}_{Z}X + n \rightarrow {}^{A+1}_{Z}X$$

where:

X—irradiated element
n—neutron
A—mass number
Z—atomic number

Table 1 Examples of reactor-produced radionuclides of current interest in nuclear medicine (cross-section values collected from Cohen et al., 2013; Depuydt & de Mevergnies, 1962; IAEA, 1987; Mughabghab, 2003; O'Brien, 1969; Al-Abyad et al., 2006)

Radionuclide	Nuclear reaction	Cross-section, b	Type of neutrons	$T_{1/2}$	E [keV] particle/photons per 100 decay
^{64}Cu	$^{63}Cu\ (n,\gamma)^{64}Cu$	4.50 ± 0.02	Thermal	12.7 h	β^+ 653 (17.5); β^- 579 (38.5); γ 511 (35.0)
	$^{64}Zn(n,p)^{64}Cu$	(34.5 ± 1.7) × 10^{-3}	Fast		
^{67}Cu	$^{67}Zn(n,p)^{67}Cu$	(0.82 ± 0.04) × 10^{-3}	Fast	2.7 days	β^- 562 (20), 468 (22), 377 (57); γ 93 (16.1), 185 (48.7)
^{169}Er	$^{168}Er(n,\gamma)^{169}Er$	2.74 ± 0.08	Thermal	9.4 days	β^- 343 (45), 351 (55)
^{166}Ho	$^{165}Ho(n,\gamma)^{166}Ho$	64.7 ± 1.2	Thermal	26.8 h	β^- 1773 (49.9), 1855 (48.8); γ 81 (6.6)
	$^{164}Dy(n,\gamma)^{165}Dy(n,\gamma)^{166}Dy \rightarrow {}^{166}Ho$	(2700 ± 75)/(3900 ± 300)			
^{125}I	$^{124}Xe(n,\gamma)^{125}Xe \rightarrow {}^{125}I$	165 ± 11	Thermal	59.41 days	γ 35.5(6.7)
^{131}I	$^{130}Te(n,\gamma)^{131g}Te \rightarrow {}^{131}I$	0.29 ± 0.06	Thermal	8.02 days	β^- 606 (89.2), 334 (7.2); γ 364 (81.5)
	$^{235}U(n,f)^{131}I$	582.2 ± 1.3	Fission		
^{177}Lu	$^{176}Lu(n,\gamma)^{177}Lu$	2090 ± 70	Thermal	6.65 days	β^- 177 (11.6), 498 (79.3); γ 113 (6.2), 208 (10.4)
	$^{176}Yb(n,\gamma)^{177}Yb \rightarrow {}^{177}Lu$	2.85 ± 0.05			
^{99}Mo	$^{98}Mo(n,\gamma)^{99}Mo$	0.130 ± 0.006	Thermal	2.75 days	β^- 436 (16.4), 1214 (82.2) γ 739.5 (12.3),
	$^{235}U(n,f)^{99}Mo$	582.2 ± 1.3	Fission		
^{32}P	$^{32}S(n,p)^{32}P$	(65 ± 3) × 10^{-3}	Fast		β^- 1710 (100%)
^{188}Re	$^{186}W(n,\gamma)^{187}W(n,\gamma)^{188}W \rightarrow {}^{188}Re$ (generator)	(37.0 ± 1.5)/(64 ± 10)	Thermal	17.0 h	β^- 965 (25.6), 2120 (71.1); γ 155 (15.2)
^{47}Sc	$^{46}Ca(n,\gamma)^{47}Ca \rightarrow {}^{47}Sc$	0.74 ± 0.07	Thermal	3.35 days	β^- 441 (68.5), 601 (31.5); γ 159 (68.1)
	$^{47}Ti(n,p)^{47}Sc$	(19.81 ± 0.90) × 10^{-3}	Fast		
^{153}Sm	$^{152}Sm(n,\gamma)^{153}Sm$	206 ± 6	Thermal	1.93 days	β^- 641 (31.3), 694 (49.4); γ 103 (29.2)
	$^{153}Eu(n,p)^{153}Sm$	(0.26 ± 0.04) × 10^{-3}	Fast		
^{89}Sr	$^{89}Y(n,p)^{89}Sr$	(0.91 ± 0.20) × 10^{-3}	Fast	50.56 days	β^- 1501 (100%)
^{161}Tb	$^{160}Gd(n,\gamma)^{161}Gd \rightarrow {}^{161}Tb$	0.77 ± 0.02	Thermal	6.89 days	β^- 460 (25.7), 522 (65.0); γ 26 (23.2), 49 (17.0), 75 (10.2)
^{90}Y	$^{90}Zr(n,p)^{90}Y$	(2.7 ± 0.5) × 10^{-3}	Fast	2.7 days	β^- 2280 (100%)
	$^{235}U(n,f)^{90}Sr \rightarrow {}^{90}Y$ (generator)	582.2 ± 1.3	Fission		

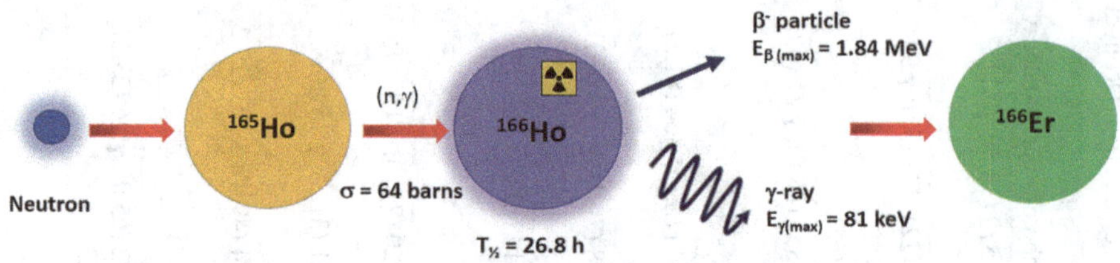

Fig. 1 Schematic representation of the ^{165}Ho (n,γ) ^{166}Ho nuclear reaction leading to carrier-added (c.a.) ^{166}Ho (Adapted from Hun et al., 2020)

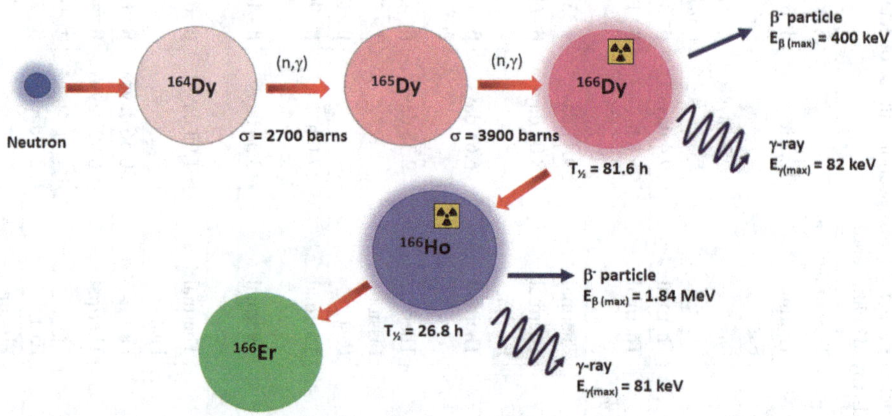

Fig. 2 Schematic representation of the ^{nat}Dy(n,γ)^{166}Dy → ^{166}Ho nuclear reaction

The reaction involves thermal or epithermal neutrons. The obtained radionuclides are characterized by relatively low specific activity carrier-added, (c.a.) since the radioactive product can't be separated from the target material, and stable atoms dilute the radioactive atoms. Due to rather high cross-sections, the yields are good. The higher the neutron flux, the cross-section of the reaction, and the isotopic enrichment of the target, the higher the specific activity that can be achieved. High chemical purity target materials are essential since impurities in the target are also irradiated and may contaminate the final product (Parus & Mikołajczak, 2012). Several radionuclides can be obtained in (n, γ) nuclear reactions, among them ^{64}Cu, ^{99}Mo, ^{153}Sm, ^{166}Ho, or ^{177}Lu (Table 1).

In certain cases, the (n, γ) nuclear reaction is followed by subsequent β⁻ decay, (n, γ) → β⁻ (Fig. 2). As a result, the nucleus of another element is formed with an atomic number greater by 1 and with a mass number also greater by 1:

$$ {}_{Z}^{A}X + n \rightarrow {}_{Z}^{A+1}X \xrightarrow{\beta^-} {}_{Z+1}^{A+1}Y $$

This new element can be separated from the target material; hence its specific activity is very high or no carrier added (n.c.a.). Examples of radionuclides that can be obtained in this way are ^{47}Sc, ^{125}I, ^{131}I, ^{161}Tb, ^{166}Ho, or ^{177}Lu (Table 1).

Nuclear Reaction (n,p)

As a result of this reaction, a radioisotope of the same mass number and with an atomic number lower by 1 is formed, which can then be obtained in the n.c.a. form:

$$^{A}_{Z}X + n \rightarrow ^{A}_{Z-1}Y + p$$

However, the (n,p) nuclear reaction requires a fast neutron flux. Reactions involving fast neutrons usually have cross-sections that are of the order of millibarns, which, coupled with the much lower neutron flux at higher energy relative to thermal neutron fluxes, leads to lower yields (Mikołajczak & Parus, 2005). This fact limits the use of this reaction to a few high-flux reactors in the world. Examples of radionuclides obtained by this process are ^{32}P, ^{47}Sc, ^{64}Cu, ^{67}Cu, ^{89}Sr, ^{90}Y, and ^{153}Sm (Table 1).

Fission Reaction (n, f)

Fission may take place in any of the heavy nuclei after the capture of a neutron. However, low-energy (slow or thermal) neutrons can cause fission only in those isotopes of uranium and plutonium whose nuclei contain odd numbers of neutrons (e.g., ^{233}U, ^{235}U, and ^{239}Pu). When a neutron passes near a heavy nucleus, for example, uranium-235, the neutron may be captured by the nucleus, and this may or may not be followed by fission. The creation of the fission fragments is followed almost instantaneously by the emission of several neutrons (typically 2 or 3), which enables the chain reaction to be sustained (Fig. 3). The sum of the fragments' mass will be close to the mass of the ^{235}U nucleus. In the case of ^{235}U, the average fission fragment mass is about 118, but in practice, they have unequal masses, and fragments' masses of 95 and 137 are most probable (Fig. 4). Fission almost always results in neutron-rich progeny nuclides that decay by beta emission. The most popular

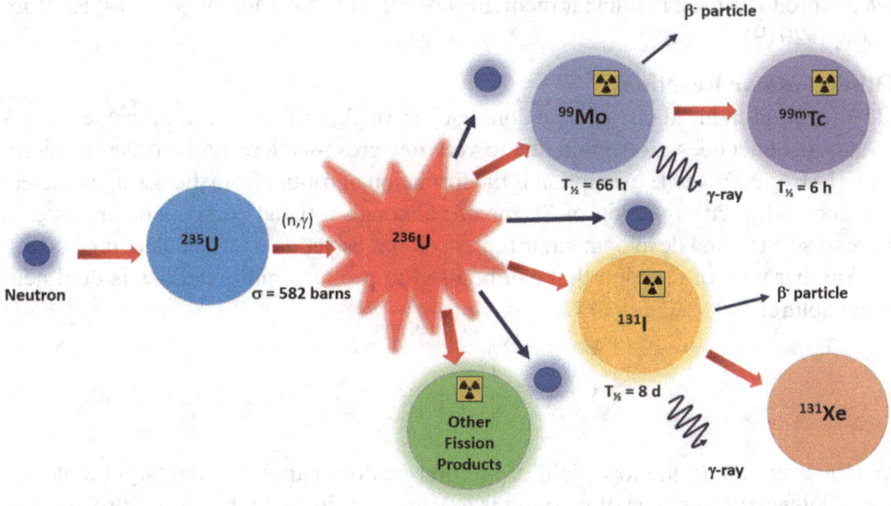

Fig. 3 Schematic representation of ^{235}U fission after thermal neutron capture

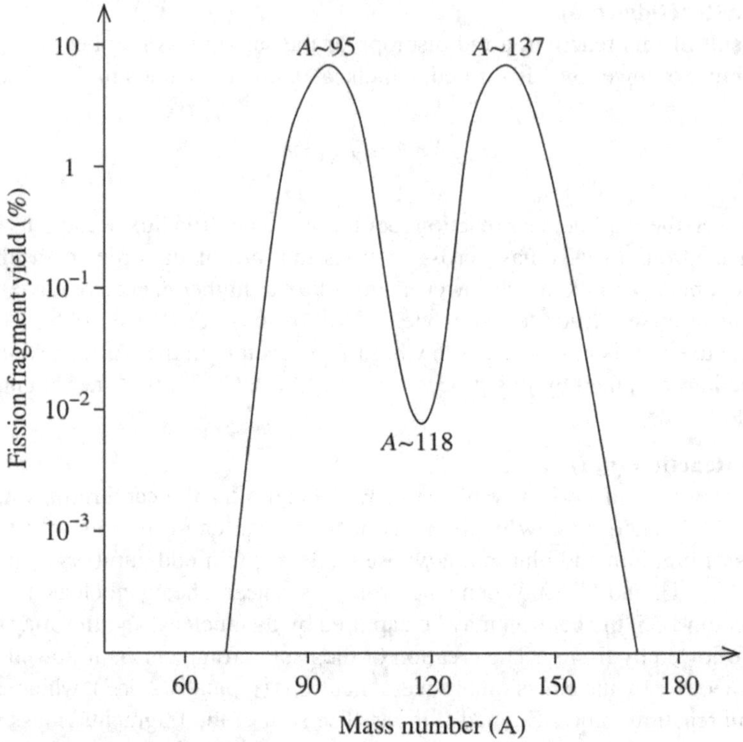

Fig. 4 Typical fission spectrum of uranium-235 (This figure was published in Ahmed (2014), p. 55, Copyright Elsevier)

example of this reaction is the production of ^{99}Mo from the fission of ^{235}U. Other fission products used in nuclear medicine are ^{131}I and ^{90}Sr (Lundqvist, 2014; Radford & Lapi, 2019).

Other Nuclear Reactions

Other neutron-induced nuclear reactions such as (n, 2n), (n, np), or (n, α) are less common in radionuclides production due to very low cross-sections of the order of microbarn (IAEA, 2003). One exemption is the formation of tritium from the ^{6}Li(n, α) nuclear reaction with a cross-section of 38 mbarn; although tritiated compounds are used in tracer discovery and development, tritium itself has as yet no use in nuclear medicine.

Another type of reaction that can be used to produce radionuclides is dual neutron capture:

$$^{A}_{Z}X + n \rightarrow {}^{A+1}_{Z}X + n \rightarrow {}^{A+2}_{Z}X$$

However, due to the low yield of the first neutron capture in the target material, the efficiency of this type of reaction is low and requires a high neutron flux reactor, such as in the case of irradiation of tungsten targets enriched in ^{186}W for production of tungsten-188 (^{188}W), the parent radionuclide of rhenium-188 (^{188}Re) (Knapp, 2017).

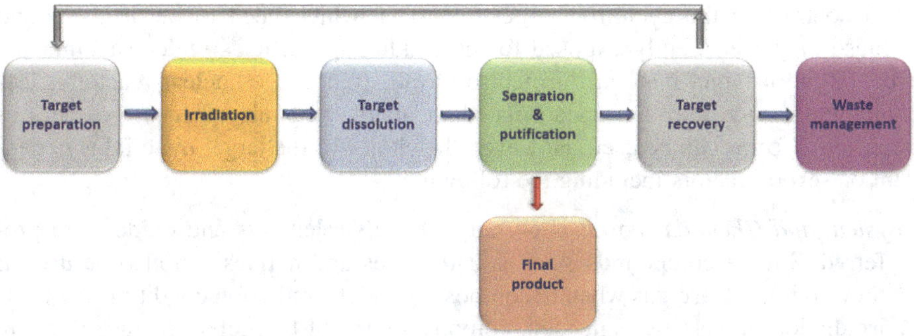

Fig. 5 Main steps of the production of medical radionuclides in a nuclear reactor

2 Practical Aspects of Radionuclide Production in Nuclear Reactors

The production of radionuclides for medical applications in nuclear reactors involves several steps, starting with the choice of nuclear reaction and the target material, target irradiation, and chemical separation of the radionuclide in question to obtain a product with high specific activity and high radionuclidic and chemical purity, suitable for labeling of biomolecules or other applications (Fig. 5). Moreover, target material recovery is often needed from the economic point of view. In the reactor production of radionuclides, automation is not as common as in cyclotron production. This is mainly because in the reactor, in most cases, radionuclides with a longer half-life are obtained, and a fast production process is not necessary, as for ^{18}F, for example. Frequently in the case of radionuclides obtained in a reactor, automation covers individual steps of the manufacturing process rather than the entire process (IAEA, 2004), such as target opener (devices for cutting aluminum containers or quartz ampoules containing target material); target crushing device; dry distillation equipment; automatic capsule filling system; or automatic dispensing system. In this section, some of these practical aspects are briefly discussed.

2.1 Targetry

Each nuclear reactor is characterized by the energy spectrum, the neutron flux, and the temperature at the irradiation position (Lundqvist, 2014). It produces a neutron cloud in which the target is placed so that it will be isotropically irradiated. Placing the target in different positions exposes it to neutrons of different energy. Usually, the reactor facility has a pneumatic system for placing targets at predefined positions. In planning how to use these positions, the heat that is generated in the reactor core has to be considered, since the temperature at some irradiation positions may easily reach 200 °C.

In contrast to the cyclotron target materials, which need to be firmly bound because they are directly exposed to the irradiating particles under vacuum, the preparation of target material for neutron irradiation in the nuclear reactor is less challenging since they are encapsulated in quartz ampoules; hence, powders or other loose forms can be used. However, the choice of the target material is dependent on several factors including the following:

Physical and Chemical Form: High-purity metals (elements and oxides) are preferred. Simple compounds such as carbonates and halides can also be used if they do not release gas when decomposing under irradiation conditions. If post-irradiation processing is needed, compounds might be preferred since they are easier to dissolve.

Chemical Stability: The target material must be able to withstand high temperatures (sometimes as much as 1000 °C). Targets are therefore expected to have good thermal conductivity and high melting points. At the thermal neutron flux, the irradiated materials may decompose. If decomposition is accompanied by the release of gases, their increased pressure could rupture the containment capsule. For these reasons, organic compounds are not used as target materials.

Purity: The higher the purity of the target material, the lower the risk of contamination with neutron-activated impurities; in addition, higher specific activities are obtained using enriched target materials. Therefore, enriched materials are preferred over natural ones.

Encapsulation: For the safety of reactor operation and to reduce the risk of cross-contamination, the target material is sealed in tight containers, which are leak-free and do not corrode or overheat during irradiation. Usually, there are two levels of sealing; their tightness is carefully checked before the irradiation: the direct ones, which are most often used, are high-purity quartz ampoules; these ampoules are then sealed in aluminum or magnesium secondary containers due to the small cross-sections for the formation of ^{28}Al and ^{27}Mg and hence the low activation of these short-lived impurities. Caution should be paid to avoid contamination of quartz ampoules with sodium, since the ^{23}Na(n,γ)^{24}Na reaction causes significant activation of the capsule and thus handling problems due to the high-energy gamma-ray (1369 keV) of ^{24}Na (Mirzadeh et al., 2003). Quartz ampoules and aluminum containers utilized in the Maria Research Reactor (Poland) are shown in Fig. 6.

Availability: Target materials are commercially available, although sometimes at a very high cost, in particular for certain enriched materials. This may impact the economy of the process; hence, the post-processing recovery of the target material is one of the ways to improve the sustainability of the production.

Target Transport Systems: Individual systems for placing the target in the reactor core are specifically developed. Before irradiation, fully encapsulated targets are placed into a special holder. Once the target has been prepared, it can be introduced to the neutron flux in a variety of ways. For most irradiations, the sample holder will be inserted into the reactor pool via a hydraulic system, allowing the sample to be retrieved during normal operation. Pneumatic tube systems are also commonplace at research reactors; however, these systems are generally reserved

Fig. 6 Quartz ampoules and aluminum containers used for target irradiation in the Maria Research Reactor in Poland

for short bombardments at a lower neutron flux and thus are not typically used for the production of radionuclides (Mirzadeh et al., 2003).

Target Burn-Up and Self-Shielding: The burn-up is the expression used to describe the reduction of the number of nuclei of the target material as a result of nuclear transformations induced by neutrons. As it was demonstrated for ^{177}Lu, if the number of nuclei before and after irradiation differs largely, its further use may be affected (Neves et al., 2002; Zhernosekov et al., 2008). The decrease in radionuclide production efficiency due to the burn-up effect is most often seen when the target material has a high cross-section, is irradiated for a long time, or is returned to re-irradiation. In certain cases, where the cross-section for the nucleus from which the contaminants arise is significantly greater than that for the main reaction, the burn-up effect can serve to reduce the level of radionuclide contamination. Usually, the irradiations in nuclear reactors are performed in cycles with a defined number of days/hours. Rarely is the continuous activation with a loop reported. The latter was used to produce ^{125}I in the following nuclear reaction:

$$^{124}_{54}\mathrm{Xe}(\mathrm{n},\gamma)^{125}_{54}\,\mathrm{Xe}\xrightarrow{\beta^-}{}^{125}_{54}\mathrm{I}$$

However, in addition to the main reaction, there is also a neutron capture by the nucleus of ^{125}I, resulting in the formation of ^{126}I, which is an undesirable contaminant of the main product. The cross-section for the formation of ^{126}I is 1200 barns, and for the main reaction, it is 200 barns. This results in the fact that during irradiation in the reactor, the desired product is burned off, and an undesirable impurity is created in its place. To prevent this, a continuous loop process was used. Gaseous ^{124}Xe, a target material, circulates in a closed loop, a fragment of which is located in the reactor core. Iodine-125 produced in the reactor is removed from the flux zone to prevent burn-up and captured in a special absorber outside the reactor core (Baker & Gerrard, 1972). The continuous loop process can be easily applied to gaseous and liquid target materials. For solid targets, it is much more difficult since appropriate solutions need to be used, which significantly affects the efficiency of the process. Radiolysis of these solutions is also a serious problem. For the above reasons, the use of this method is very limited.

The self-shielding of the target material is also associated with the burn-up. This phenomenon becomes dominant for target materials with high mass and large absorption cross-sections. Generally, target nuclei in the inner part of the target do not experience the desired neutron flux due to neutron absorption by the outer part of the target, resulting in reduced product yield.

2.2 *Radiochemical Processing*

It is worth emphasizing that indirect nuclear reactions usually have much lower neutron irradiation cross-sections than direct ones. For example, the cross-section of the $^{176}Lu(n,\gamma)^{177}Lu$ nuclear reaction is almost 3 orders of magnitude higher than that for the nuclear reaction $^{176}Yb(n,\gamma)^{177}Yb \rightarrow {}^{177}Lu$. In the latter, to obtain the same radioactivity of ^{177}Lu, much larger amounts of target material must be used. In consequence, separating nanogram or picogram amounts of the given radionuclide from macro-amounts (mg–g) of the target material is challenging (Radchenko et al., 2021). In addition, the final product containing the given radionuclide needs to be of high chemical purity and radioactivity concentration, allowing the use of a small volume of a suitable solvent (usually a moderately acidic solution) to facilitate efficient radiolabeling.

2.2.1 Target Dissolution

The irradiated target material needs to be dissolved for further processing. Usually, the container with the ampoule is transferred to a processing hot cell. For target dissolution, either acidic or alkaline conditions can be used, depending on the target material and further steps of processing. An elevated temperature can facilitate the process. The time and efficiency of dissolution are important for the process yield; however, the reactions should not be violent to avoid unnecessary risk of contamination and to possibly allow further automation. Because neutrons do not attenuate in a sample as rapidly as charged particles, targets for neutron irradiation tend to be larger than targets for accelerators. A large mass of the target may require a longer separation time. Therefore, the mass of irradiated targets should be as small as reasonably justified.

2.2.2 Separation Process

The purpose of radiochemical processing is to isolate a given radionuclide, free from other radionuclides and chemical contaminants, and to transform it into a solution suitable for further use. The adopted techniques are strongly related to the chemical form of the target material and the properties of the elements to be separated. The separation procedure should be relatively fast, especially in the case of short-lived radionuclides, and highly efficient, resulting in a product of high purity and high specific activity. Target material recovery also might be a constraint. The chemical purity of the final radionuclide solution is of great importance, since other metal ions may interact with the binding site, thus lowering the radiolabeling yield. These impurities may arise from the target material or from materials and reagents used for post-irradiation processing.

There are various techniques used: solvent extraction, extraction chromatography, ion exchange, electrolysis, and electrodeposition, as well as precipitation and sublimation, as presented graphically in Fig. 7. These processes are frequently combined in two or three steps, namely: primary separation to remove the bulk mass of target material after dissolution, fine purification to reduce the mass of dissolved target material to the range of a micro component and pre-conditioning to reduce the radionuclide solution volume and to prepare the radionuclide in the form suitable for radiolabeling (Radchenko et al., 2021). There is abundant literature presenting various combinations of methods for the separation of reactor-produced radionuclides. Selected examples of separation protocols developed for reactor-produced ^{47}Sc and ^{67}Cu are shown in Table 2.

Precipitation

Separation of two elements by precipitation is possible only when the solubility of their compounds differs significantly, and the solubility of one of them is exceeded. One needs to take into account the possible losses, e.g., caused by occlusion and entrapment of the adsorbed radionuclide inside the sediment during its growth. This technique is also difficult to operate remotely. For example, precipitation followed by anion exchange was shown to be useful for bulk target separation in the production of ^{47}Sc (Pawlak et al., 2019).

Solvent Extraction

Solvent extraction or liquid-liquid extraction is based on the selective extraction of radionuclide as a nonpolar compound or complex from an aqueous solution by an immiscible organic solvent. The method can be applied over a wide concentration range. The time-determining step is the separation of the two phases. The operations of extraction, back-extraction, and washing with proper solvents can be repeated to

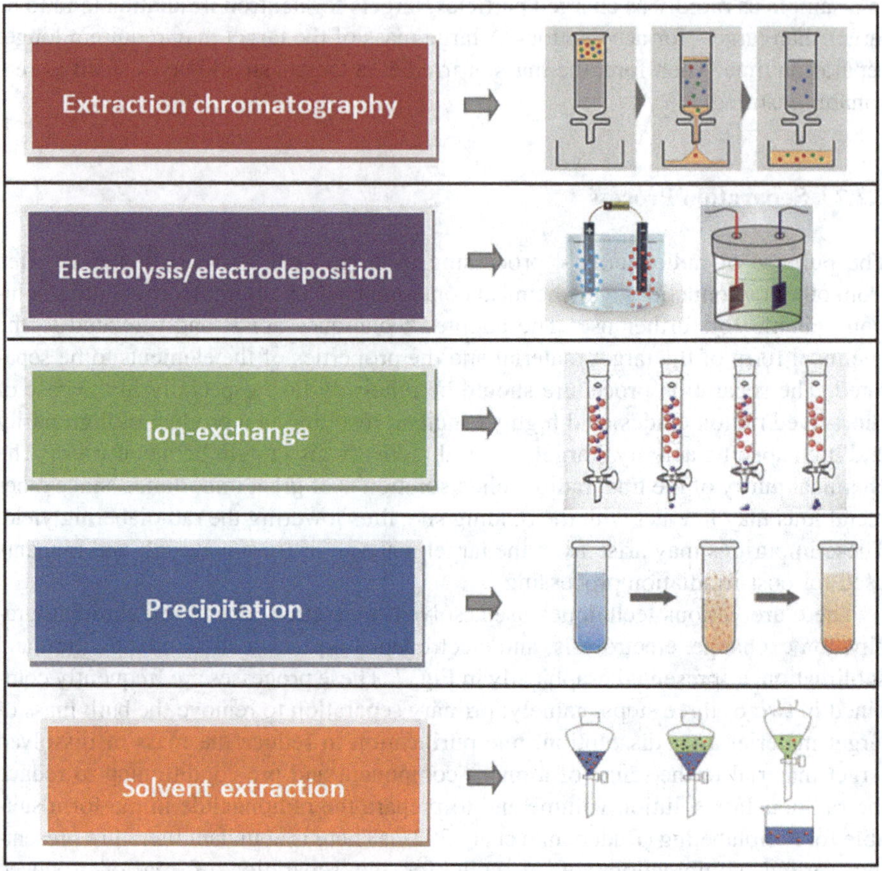

Fig. 7 Techniques used for separation of radionuclides from irradiated target material

improve the separation efficiency. Liquid-liquid extraction usually requires simple apparatus, such as a separator funnel, and may readily be adapted to remote or automated operation. The disadvantage of this method is the need for the utilization of radioactive waste containing toxic organic solvents. Solvent extraction was used for the separation of ^{67}Cu with dithizone (diphenylthiocarbazone) from bulk target material after irradiation through the $^{67}Zn(n,p)^{67}Cu$ nuclear reaction, followed by further purification using chromatographic methods (Gopalakrishna et al., 2018).

Extraction Chromatography

Extraction chromatography is a separation technique in which the hydrophobic extractant is fixed on a porous, inert solid support such as silica or various polymers. The mobile phase is usually a solution of mineral acid, although solutions of chelators are also used to improve the selectivity and efficiency of the separation. Thus, extraction chromatography is usually described as a technique that combines the selectivity of solvent extraction with the simplicity of handling that is characteristic

Table 2 Exemplary radiochemical procedures for the separation of the reactor-produced ^{47}Sc and ^{67}Cu

Separation method	Isolated element	Chemical form of a target material	Dissolution	Description	Yield	Reference
Precipitation	^{47}Sc	$CaCO_3$	HCl	Precipitation with 25% NH_4OH/AG50Wx4	80%	Pawlak et al. (2019)
Solvent extraction	^{67}Cu	Zn or ZnO or $ZnSO_4$	HCl or acidified water	Extractant: Dithizone; thenoyltrifluoroacetone, TTA; resacetophenone oxime in cyclohexanone	86–90%	Gopalakrishna et al. (2018), Uddin et al. (2014), IAEA (2003), Bhatki et al. (1969), Brown and Callahan (1972), Spahn et al. (2004)
	^{47}Sc	Ti	HCl/HF	Extractant: Tributyl phosphate, TBP	> 98%	Pietrelli et al. (1992)
Extraction chromatography	^{47}Sc	$Ca(NO_3)_2$ or $CaCO_3$	HCl	UTEVA® resin/AG50Wx4; DGA® resin; DGA® resin/SCX; DGA® resin/Dowex 50; TBP® resin/AG50Wx4	78–95%	Domnanich et al. (2017), Müller et al. (2014), Pawlak et al. (2019)
		Ti	HF/HCl	TBP/silica gel	97.7%	Pietrelli et al. (1992)
	^{67}Cu	Zn	HCl	6-Tungstocerate(IV) gel matrix	–	Hassanein et al. (2006)
Ion exchange	^{67}Cu	Zn or ZnO	HCl	Dowex 1 × 8 resin; Chelex 100 resin; AG1-X8 resin; AG1-X10 resin; Diaion SA-100	94–95%	Karimi et al. (2019), Mushtaq et al. (1990), O'Brien (1969), Shikata (1964)
	^{47}Sc	$CaCO_3$	HCl	Chelex 100 resin	95%	Gizawy et al. (2020)
		Ti TiO_2	HF or HF/HCl	AG MP-50; Dowex 1x8/Dowex 50Wx4 Dowex AG 50W-X4	>90%	Pietrelli et al. (1992), Bartoś et al. (2012), Połosak et al. (2013), IAEA (2003)
Electrolysis	^{67}Cu	Zn	HCl/HNO_3	Selective electrodeposition of Cu	80%	Mirzadeh et al. (1986)
		ZnO	H_2SO_4	Spontaneous electrochemical separation	95%	Mirzadeh and Knapp (1992)
	^{47}Sc	$CaCO_3$	HCl	Selective electro-amalgamation of Ca^{2+} ions	>80%	Chakravarty et al. (2017)

of chromatographic methods. An unquestionable advantage of extraction chromatography over solvent extraction is the high speed of the separation and the reduction of generated waste. An important limitation of this method is the relatively small sorption capacity of extraction chromatography columns compared to solvent extraction. Nonetheless, highly efficient separation of ^{47}Sc from bulk quantities of Ti was achieved using commercially available DGA (*N*,*N*,*N*′,*N*′-tetrakis-2-ethylhexyl-diglycolamide based) and ZR (hydroxamate-based) extraction chromatography resins in a scaled-up production process (McLain et al., 2023).

Ion Exchange

Ion exchange chromatography is based on differences in ion affinity between the solid phase (ion-exchange resin, cationite, or anionite) and the liquid phase, most often aqueous solutions of acids, bases, or chelators. The distribution of elements between these two phases depends on many factors, such as ionic form, electric charge of the ion, mass of the ion, dipole moment, and type of resin functional groups. In practice, the ions of various elements move on a column filled with ion exchange resin at different speeds, which allows their separation. In another scenario, several ionic species may be adsorbed together on the column and then selectively eluted with an appropriate eluent. This technique enables the separation within groups of very similar elements, such as lanthanides or actinides (Trautmann & Roesch, 2016), and is often used in the separation of radionuclides for medical applications due to its high efficiency and reproducibility. This technique relies on the charge interactions between the stationary phase (resin) and the analyte ions. The limited selectivity can lead to challenges in separating closely related ions with similar charges. Moreover, it is highly dependent on the pH of the mobile phase and has a finite capacity to bind ions. If the sample contains a high concentration of ions, the resin can become saturated, leading to breakthrough and reduced separation efficiency. In addition, ion-exchange resins require periodic regeneration and maintenance to ensure their optimal performance, which is time-consuming.

Electrolysis and Electrodeposition

In general, the electrochemical separation of two elements is based on the difference in their formal electrode potential in a given electrolyte solution. The ion to be separated can be selectively electrodeposited on either cathode or anode, while the other one remains in the electrolyte solution. In the next step, the deposit containing the element of interest is dissolved, usually in saline, through a reverse electrochemical reaction. Electrochemical separation was explored in the preparation of $^{90}Sr/^{90}Y$, $^{188}W/^{188}Re$, and $^{99}Mo/^{99m}Tc$ generators (Chakravarty et al., 2012). Electrochemical separation is also an effective concentration methodology that yields high radioactive concentrations of the daughter products. Recently, the electrochemical method was used for large-scale separation of n.c.a. ^{177}Lu and its further purification to obtain a clinically usable form (Patra et al., 2023). The simple and relatively inexpensive equipment as well as the use of aqueous medium and solid electrodes, which are easily amenable to automation, are advantages of this technique. However, this separation method applies only to elements differing in electrochemical potential. There is also a risk that electrode surfaces become fouled or coated with reaction products, impurities, or precipitates, which can reduce the efficiency of the separation process.

Thermal Separation
The differences in thermal volatility allow the separation of two elements by sublimation or distillation. This method is very popular for separating volatile radionuclides from a solid target. Reduced pressure may speed up the separation process. For example, sublimation is the most effective way to isolate ^{131}I from irradiated tellurium oxides. It was also used to distill the target material from the product, as was done to separate ^{32}P from the elemental sulfur target (Vimalnath et al., 2014). However, appropriate precautions are necessary to avoid the loss of volatile radioactive substances during the processing of irradiated targets.

2.3 *Target Recovery*

The cost of the target is high when the natural abundance of the isotope to be irradiated is low. Target retrieval may be necessary to complete the routine radionuclide production cycle. The recycling process is expected to provide high recovery efficiency, high purity of the recovered target material, and its appropriate chemical form, enabling re-irradiation. In certain cases, the time for the radioactive decay of contaminants may be required before re-irradiation of the recovered target material. ^{46}Ca-enriched target material used for production of ^{47}Sc is an example. This target material usually contains significant amounts of ^{44}Ca, which is activated to ^{45}Ca ($T_{1/2} = 162$ days) as a result of (n,γ) nuclear reaction. The latter decays to ^{45}Sc, which can lower the specific activity of ^{47}Sc and can be the source of long-lived ^{46}Sc ($T_{1/2} = 83.8$ days) (Jalilian et al., 2020).

2.4 *Requirements and Challenges in the Production of Medical Radionuclides*

Radionuclides for medical use should meet quality criteria relevant to their expected use. European Pharmacopoeia in the general monograph on Radiopharmaceutical Preparations (Ph.Eur. 0125) defines a radionuclide precursor as any radionuclide produced for radiolabeling of another substance before administration (EDQM, 2020). The specification includes such parameters as identity, radioactivity, specific activity, radionuclidic purity, radiochemical purity, and chemical purity. In certain cases, sterility and bacterial endotoxins may need to be assessed as well.

The specific activity of radionuclide is defined as radioactivity per unit mass of the product (Bq/g) (Coenen et al., 2017). The cross-section, target impurities, secondary nuclear reactions, target burn-up, or post-irradiation processing are factors affecting the specific activity of the produced radionuclide (Talip et al., 2020). High specific activity and n.c.a. radionuclides are desired in medical applications since they will allow the administration of low masses of radiolabeled preparation in a patient, thus minimizing potential adverse effects.

2.5 Selected Examples of Radionuclides Produced in Nuclear Reactors

2.5.1 Iodine-131

Iodine-131, with its energies $E_{\beta-}$ = 606, 334 keV, and E_γ = 364 keV, was the first radionuclide with true theranostic properties that can be used either for diagnosis or therapy depending on the administered radioactivity dose (Becker & Sawin, 1996). Iodine-131 was first produced by Livingood and Seaborg in 1938 in the cyclotron, and since 1941 it has been continuously used in nuclear medicine for the diagnosis and therapy of hyperthyroidism and thyroid cancer (McCready, 2017). Today, virtually all the ^{131}I produced in the world is obtained in a nuclear reactor using one of the two nuclear reactions (IAEA, 2003):

$$^{130}_{52}\mathrm{Te}(\mathrm{n},\gamma)^{131}_{52}\mathrm{Te}\xrightarrow{\beta^-}{}^{131}_{53}\mathrm{I}$$

$$^{235}_{92}\mathrm{U}(\mathrm{n},\mathrm{f})^{131}_{53}\mathrm{I}$$

In the first method, tellurium oxide is used as the target material. Iodine is separated from the irradiated target material by sublimation (dry distillation). The target material is heated to a temperature of about 700 °C, and the released iodine is transported by a stream of carrier gas to a scrubber filled with sodium hydroxide solution. This method is expected to produce significant amounts of stable ^{127}I and long-lived ^{129}I ($T_{1/2}$ = 1.57 · 10^7 years), which reduce the specific activity of ^{131}I. Another significant issue is the formation of large amounts of long-lived tellurium isotopes that hinder radioactive waste management. One solution to this problem is the use of highly enriched ^{130}Te as the target material. However, this is associated with a significant increase in the production cost of ^{131}I.

^{131}I is one of the fission products of ^{235}U. During fission, in addition to ^{131}I, several other radioisotopes will be formed, i.e., ^{99}Mo, ^{131}Cs, ^{90}Sr, or ^{99}Tc. The efficiency of the production of such a radioisotope with thermal neutrons in this process is 2.8%, and the produced ^{131}I is separated from other fission products by extraction. Due to the high radioactivity and complex chemical composition of fission products, the use of specialized separation methods is required. In the production process, first, the target material is dissolved, usually in an alkaline medium, and then the ^{131}I is separated by ion exchange chromatography. The specific activity of ^{131}I obtained in this method is higher than through the ^{130}Te(n,γ)^{131}I nuclear reaction. However, its significant limitation is the necessity to use highly enriched ^{235}U and the generation of a large amount of highly active radioactive waste.

2.5.2 Lutetium-177

Lutetium-177 belongs to the radio lanthanides and is ideally suited for therapeutic use due to its attractive physical characteristics and ready availability. It decays by β^- emission with energies of 177 and 498 keV. Furthermore, the emitted γ photons of 113 and 208 keV can be used for imaging the in vivo localization of the administered ^{177}Lu biomolecular targeting agent using SPECT scanners and facilitate dosimetry calculations (Table 1).

Lutetium-177 can be produced through (n,γ) reaction using two pathways:

$$^{176}_{71}\mathrm{Lu}\left(\mathrm{n},\gamma\right)^{177}_{71}\mathrm{Lu}$$

$$^{176}_{70}\mathrm{Yb}\left(\mathrm{n},\gamma\right)^{177}_{70}\mathrm{Yb}\xrightarrow{\beta^-}{}^{177}_{71}\mathrm{Lu}$$

The direct method involving the neutron irradiation of ^{176}Lu results in c.a. ^{177}Lu. The specific activity of ^{177}Lu depends on parameters such as neutron flux, ^{176}Lu enrichment, irradiation, and cooling time. The specific activity of ^{177}Lu that can be obtained is 740–1110 GBq/mg (20–30 Ci/mg) for medium-flux reactors and about 2800 GBq/mg (75 Ci/mg) for high-flux reactors (Cutler et al., 2013). The advantages of the direct production of ^{177}Lu are easy and fast processing of irradiated targets by dissolution in acidic media, high neutron capture cross-section of (2090 ± 70) barn (Table 1) leading to high yields, low volume of radioactive waste, and relatively low cost of production (Mikołajczak et al., 2019). The disadvantage is a long-lived isomer ^{177m}Lu ($T_{1/2}$ = 160 days), which is co-produced in the competing ^{176}Lu(n,γ)^{177m}Lu nuclear reaction.

The indirect method allows us to obtain n.c.a. ^{177}Lu with a very high specific activity of 4100 GBq/mg (111 Ci/mg) (Cutler et al., 2013). ^{177}Lu obtained in this route is free from contamination with ^{177m}Lu. However, due to the low cross-sections for the ^{176}Yb (n, γ) ^{177}Yb reactions of 2.1 and 8.1 barns for thermal and epithermal neutrons, respectively, only a small amount of radioactivity of ^{177}Lu is produced compared to the total mass of the ^{176}Yb target. To obtain 37 GBq (1 Ci) ^{177}Lu, corresponding to 9 μg Lu, it is necessary to irradiate about 100 mg of ^{176}Yb in a medium flux reactor (Parus & Mikołajczak, 2012). Consequently, the separation of the micro amount of ^{177}Lu from the macro amount of the ytterbium requires a method with a high separation factor (>10^5).

Several techniques, such as ion-exchange chromatography, solvent extraction, extraction chromatography, cementation, or electro-amalgamation, have been studied for the radiochemical separation of n.c.a. ^{177}Lu (Choppin & Silva, 1956; Hashimoto et al., 2003; Horwitz et al., 2005; Mirzadeh et al., 2004; Van So et al., 2008a, 2008b). In ion exchange chromatography, both Yb and Lu can be adsorbed onto a strong cation exchanger and then selectively eluted with a suitable complexing agent. α-HIBA (α-hydroxyisobutyric acid) is a commonly used eluent in the cation exchange system (Balasubramanian, 1994; Hashimoto et al., 2003; Marx et al., 2014), and HDEHP (di(2-ethylhexyl) orthophosphoric acid) was reported to be an excellent extractant for ^{177}Lu (Horwitz et al., 1976). However, liquid-liquid solvent extraction requires a multistage process, which constitutes the major impediment to using HDEHP that probably limits its wide-scale utility. This problem was solved by extraction chromatography. The LN resin with di(2-ethylhexyl) orthophosphoric acid (HDEHP) as extractant fixed on an inert polymeric support has been demonstrated to provide quantitative separation of n.c.a. ^{177}Lu (Knapp Jr et al., 2005; Mirzadeh et al., 2004). The multi-column separation with the use of LN and DGA resins with either *N*,*N*,*N*′,*N*′-tetra-*n*-octyldiglycolamide or *N*,*N*,*N*′,*N*′-tetrakis-2-ethylhexyl-diglycolamide as extractant allowed to obtain separation with an efficiency of 73% (Horwitz et al., 2005). Another technique described for the isolation of n.c.a. ^{177}Lu is based on selective electro-amalgamation of the ytterbium target material in a mercury-pool cathode with controlled potential (Bilewicz et al., 2009; Chakravarty et al., 2010; Cieszykowska et al., 2014; Lebedev et al., 2000). The process frequently involves multiple electrolysis cycles and allows for a high separation factor >10^6. However, special attention should be paid to the risk of contamination of the final product with mercury.

2.5.3 Scandium-47

Scandium-47 can be used as an alternative therapeutic radionuclide to ^{177}Lu. It has a half-life of 3.35 days and emits β^- particles (441 and 601 keV) and γ-ray (159 keV), allowing for visualization of the distribution of therapeutic dose with SPECT imaging (Müller et al., 2014). The advantage of ^{47}Sc over ^{177}Lu in terms of therapeutic application lies in the other scandium radioisotopes suitable for PET imaging, as ^{43,44}Sc, which together with ^{47}Sc create an ideal theranostic pair (Huclier-Markai et al., 2018; Mikołajczak et al., 2021).

Similarly to ^{177}Lu, in a nuclear reactor n.c.a. ^{47}Sc can be produced via two routes (Kolsky et al., 1998; Mausner et al., 1998; Müller et al., 2014; Srivastava, 2011):

$$^{47}_{22}\mathrm{Ti}(\mathrm{n,p})^{47}_{21}\mathrm{Sc}$$

$$^{46}_{20}\mathrm{Ca}(\mathrm{n},\gamma)^{47}_{20}\mathrm{Ca}\xrightarrow{\beta^-}{}^{47}_{21}\mathrm{Sc}$$

In both approaches, enriched targets are required to obtain quantities of ^{47}Sc sufficient for therapy. ^{47}Ti oxide is available with very high enrichment and at a reasonable cost, while ^{46}Ca is available with a maximum of 30% enrichment. The low natural abundance of ^{46}Ca (0.004%) makes ^{46}Ca target material rather expensive (Chakravarty et al., 2017). Hence, target material recycling should be considered. Production of approx. 2800 GBq of ^{47}Sc at EOB as a result of irradiation of a 10 g ^{47}Ti target was confirmed in an HFIR reactor (Oak Ridge National Laboratory) (Kolsky et al., 1998; Mausner et al., 1998). The disadvantage of ^{47}Sc production from the ^{47}Ti target is the limited number of facilities delivering fast neutrons and contamination due to ^{46}Sc co-production along with ^{47}Sc (Bartoś et al., 2012; Bokhari et al., 2009; Domnanich et al., 2017).

Various methods of ^{47}Sc separation from metallic and TiO_2 targets based on tributyl phosphate (TBP) extraction or extraction chromatography using TBP or DGA resin and ion exchange chromatography with AG MP-50, AG1-X8, or Dowex resins have been reported (Table 2).

Despite the high costs of the ^{46}Ca target, ^{47}Sc could be easily produced in nuclear reactors with thermal neutrons. In addition, the irradiated targets are easy to dissolve. The longer half-life of ^{47}Ca ($T_{1/2}$ = 4.5 days) compared with the half-life of ^{47}Sc ($T_{1/2}$ = 3.35 days) enables repeated separations of ^{47}Sc in a generator-like system (Domnanich et al., 2017; Pawlak et al., 2019). Numerous techniques for Sc separation from Ca have been investigated in this context. They include the precipitation of scandium hydroxide with ammonia, extraction chromatography with UTEVA resin (with DP[PP] (dipentyl pentylphosphonate) as extractant), or DGA resin and a combination of extraction resins with strong cation exchange resin (Table 2). Up to 2 GBq of ^{47}Sc has been produced by irradiation of enriched ^{46}Ca targets with high radionuclidic purity (>99.99%) (Domnanich et al., 2017).

2.5.4 Terbium-161

Terbium-161 has a half-life of $T_{1/2}$ = 6.89 days and decays by β^- emission to stable ^{161}Dy. It is a low-energy β^- (460 and 522 keV) and γ (26, 49, and 75 keV) emitter (Table 1). These physical properties make it similar to ^{177}Lu, but ^{161}Tb also emits a significant amount of conversion and Meitner-Auger electrons, which are expected to provide a more pronounced therapeutic effect compared to ^{177}Lu. In fact, Meitner-Auger electrons deposit their energy over nanometer-micrometer distances, resulting in high linear energy transfer (LET) that is potent for causing lethal damage in cancer cells. In the last decade, ^{161}Tb is gaining particular interest because, together with other terbium isotopes such as ^{149}Tb (alpha emitter), ^{152}Tb (positron emitter), and ^{155}Tb (gamma emitter), it can form a series of theranostic pairs (Lehenberger et al., 2011; Müller et al., 2012). Terbium-161 can be produced in a nuclear reactor by irradiation of ^{160}Gd with thermal neutrons; the resulting ^{161}Gd decays to ^{161}Tb with a $T_{1/2}$ = 3.66 min by β^- emission:

$$^{160}_{64}\text{Gd}(\text{n},\gamma)^{161}_{64}\text{Gd}\xrightarrow{\beta^-}{}^{161}_{65}\text{Tb}$$

To obtain ^{161}Tb with high specific activity (n.c.a.) and radionuclide purity, it is necessary to use highly enriched ^{160}Gd as a target material, since another gadolinium isotope, ^{158}Gd, if present in the target, is activated to ^{159}Gd, which decays ($T_{1/2}$ = 18.5 h) by β^- emission to stable ^{159}Tb. The latter gets activated to a long-lived ^{160}Tb ($T_{1/2}$ = 72.3 days) and creates a radiological hurdle for the safe use of ^{161}Tb in patients.

The separation of ^{161}Tb from macro amounts of gadolinium is challenging because of the high chemical similarity of these neighboring lanthanides. An ion exchange chromatography on a strong cation exchanger and selective elution of ^{161}Tb with a complexing agent α-HIBA (α-hydroxyisobutyric acid) is commonly used for this purpose (Gracheva et al., 2019; Müller et al., 2012, 2018). Other methods include the use of LN (or di(2-ethylhexyl)orthophosphoric acid (HDEHP)) extraction resin and various nitric acid concentrations (Aziz & Artha, 2016). UTEVA, Chelex 100, and DGA resins were employed to separate Sc from Ca; however, to date, DGA resin is the most commonly used system for the separation of Sc (Talip et al., 2020). Recently, the first attempts toward automation of the separation process were published (McNeil et al., 2022). These authors demonstrated that the separation of ^{161}Tb from Gd using small solid-phase extraction columns in an automated module is feasible; the method is cost-efficient and easy to operate, thus suitable for smaller centers and labs.

2.5.5 Copper-67

Copper-67 is a β^- emitter with energies of 562, 468, and 377 keV useful for treatments and strong γ emissions at 93 and 185 keV suitable for diagnostic imaging and dosimetry calculations (DeNardo et al., 1999; Junde et al., 2005). Its half-life of 2.7 days is long enough for use in vivo in slow pharmacokinetics, such as monoclonal antibodies (MoAbs) or large biomolecules. Copper-67 can also be paired with the β^+ emitters ^{64}Cu, ^{61}Cu, and ^{60}Cu to perform pre-therapy biodistribution determinations and dosimetry using positron emission tomography (PET) systems (Mou et al., 2022). ^{67}Cu can be produced in a nuclear reactor via the following reaction:

$$^{67}_{30}\mathrm{Zn}\left(\mathrm{n,p}\right)^{67}_{29}\mathrm{Cu}$$

After the separation process, it is possible to obtain ^{67}Cu in n.c.a. form (Mou et al., 2022). Enrichment of the ^{67}Zn target material is crucial to obtain significant amounts of ^{67}Cu. However, due to the low cross-section value, the ^{67}Zn(n,p)^{67}Cu nuclear reaction requires a high flux of fast neutrons exceeding 10^{14} n cm^{-2} s^{-1} (Kolsky et al., 1994; Mikołajczak & Parus, 2005; Mirzadeh & Knapp, 1992; Shikata, 1964).

Various methods have been employed for the separation of the radiocopper from the zinc target. They involve liquid-liquid extraction with dithizone, ion exchange methods using such resins as Chelex-100, Dowex 1x8, AG 1-X8, AG50W-X4, precipitation with $AgNO_3$, electrolysis, and sublimation (for detailed procedures see the summary in Table 2).

2.5.6 Medically Relevant α-Emitters

Alpha particle-emitting radionuclides are promising agents for targeted alpha therapy (TAT) of cancer. Unique properties of alpha particles, such as the limited range (50–80 μm) in tissue and the high linear energy transfer (LET), can provide an effective radiation dose to the tumor while sparing the healthy tissues (Guseva & Dogadkin,

2010). While the majority of medical isotopes today are sourced from nuclear reactors (IAEA, 2003), the potential for reactor-based production of alpha emitters is limited. They are mostly produced from uranium, actinium, or thorium legacy sources. Most of the alpha emitters of medical use can be produced from thorium sources: ^{223}Ra is produced from ^{227}Th, ^{224}Ra and ^{212}Pb are produced from ^{228}Th, and ^{225}Ac and ^{213}Bi are produced from ^{229}Th. Radium-223, used for the targeted therapy of bone metastases, is an interesting example of an alpha-emitter produced in a reactor. It is a progeny of ^{227}Ac decay and is produced via neutron irradiation of a ^{226}Ra target. The β^- decay of the ^{227}Ra ($T_{1/2}$ = 43 min) produced via the ^{226}Ra(n,γ)^{227}Ra nuclear reaction leads to ^{227}Ac, a parent nuclide of ^{227}Th, and subsequently ^{223}Ra (Fig. 8) (Kukleva et al., 2014). A high flux of thermal neutrons is preferable, as ^{226}Ra does have a fission cross-section for neutrons with energy above 1 MeV. Bearing in mind the long half-life of ^{226}Ra, which additionally decays by ejection of α-particles producing gaseous ^{222}Rn, the preparation and processing of the ^{226}Ra target is complex and challenging. To obtain ^{223}Ra, a ^{227}Ac/^{227}Th/^{223}Ra generator system is needed such as the one based on the Dipex-2/AG50Wx12 column system (Knapp & Dash, 2016).

The potential of ^{225}Ac and its progeny radionuclide ^{213}Bi for TAT in various cancers has been proven in recent years. The successive multiple neutron capture on ^{226}Ra in a nuclear reactor can be used for the production of ^{225}Ac. A mixture of ^{227}Ac, ^{228}Th, and ^{229}Th is produced, from which ^{229}Th can be directly obtained with a relatively high yield, and ^{225}Ac can be isolated by the ion exchange separation method using titanium phosphate adsorbent (Boll et al., 2005). The ^{225}Ac parent nuclide ^{225}Ra can also be produced in nuclear reactors via the ^{226}Ra(n, 2n)^{225}Ra reaction; however, this reaction requires an intense source of high-energy (>6.4 MeV) neutrons. Given that significantly lower energy neutrons would be present, these irradiations would be dominated by the co-production of ^{227}Ac, a long-lived ($T_{1/2}$ = 21.7 years) and highly toxic actinium nuclide, the presence of which in significant quantities may prevent the clinical approval of a pharmaceutical (Robertson et al., 2018).

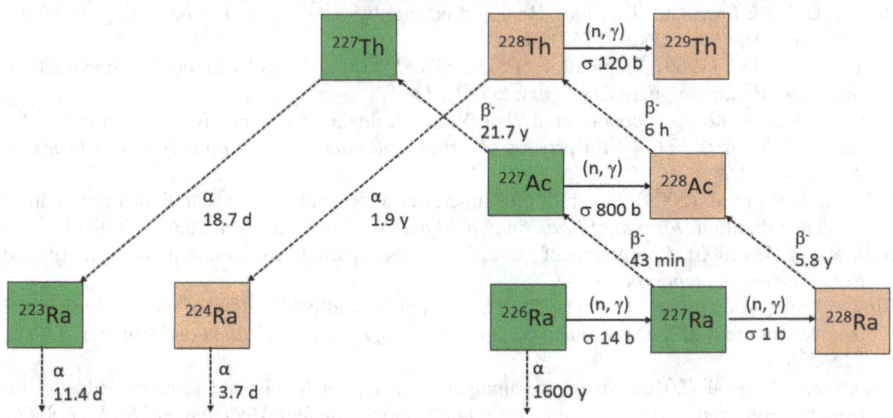

Fig. 8 Simplified reaction scheme of ^{226}Ra under neutron irradiation. (Adapted from Kukleva et al., 2014)

3 Summary

The role of nuclear reactors in the supply chain of radionuclides for medical applications is undeniable and the attractiveness of reactor-produced radioisotopes is increasing due to their theranostic applications. Novel carrier molecules are gaining interest and clinical indications for their administration to patients; this trend translates to the growing demand for reactor-produced radionuclides. However, research reactors are aging, and the number of irradiation sites is limited; hence, new initiatives are necessary to cover the existing and future needs. Automation of production processes apparently may lead to an increased supply of radionuclides, allowing process optimization, standardization of production protocols, and thus improved cost-effectiveness. One can expect that next to the developments already reported in this chapter, the coming years will bring significant progress in the radionuclide production processes and their automation.

References

Ahmed, S. N. (2014). *Physics and engineering of radiation detection* (2nd ed., pp. 1–64). Elsevier.

Al-Abyad, M., et al. (2006). Nuclear data for production of the therapeutic radionuclides ^{32}P, ^{64}Cu, ^{67}Cu, ^{89}Sr, ^{90}Y and ^{153}Sm via the (n,p) reaction: Evaluation of excitation function and its validation via integral cross-section measurement using a 14MeV d(Be) neutron source. *Applied Radiation and Isotopes, 64*(6), 717–724.

Aziz, A., & Artha, W. T. (2016). Radiochemical separation of ^{161}Tb from Gd/Tb matrix using Ln resin column. *Indonesian Journal of Chemistry, 16*(3), 283–288.

Baker, P. S., & Gerrard, M. (1972). *Iodine-125*. Isotopes Information Center, Oak Ridge National Laboratory.

Balasubramanian, P. S. (1994). Separation of carrier-free lutetium-177 from neutron irradiated natural ytterbium target. *Journal of Radioanalytical and Nuclear Chemistry, 185*, 305–310.

Bartoś, B., et al. (2012). New separation method of no-carrier-added ^{47}Sc from titanium targets. *Radiochimica Acta, 100*(7), 457–462.

Becker, D. V., & Sawin, C. T. (1996). Radioiodine and thyroid disease: The beginning. *Seminars in Nuclear Medicine, 26*(3), 155–164.

Bhatki, K. S., et al. (1969). Preparation of carrier-free ^{64,67}Cu nuclides by liquid–liquid extraction. *Journal of Radioanalytical Chemistry, 2*(1–2), 73–77.

Bilewicz, A., et al. (2009). Separation of Yb as $YbSO_4$ from the ^{176}Yb target for production of ^{177}Lu via the $^{176}Yb(n,\gamma)^{177}Yb \rightarrow ^{177}Lu$ process. *Journal of Radioanalytical and Nuclear Chemistry, 280*(1), 167–169.

Bokhari, T. H., et al. (2009). Separation of no-carrier-added radioactive scandium from neutron irradiated titanium. *Journal of Radioanalytical and Nuclear Chemistry, 283*(2), 389–393.

Boll, R., et al. (2005). Production of Ac-225 for alpha particle radioimmunotherapy. *Applied Radiation and Isotopes, 62*, 667–679.

Brown, L. C., & Callahan, A. P. (1972). The separation and purification of carrier-free copper isotopes for medical use. *The International Journal of Applied Radiation and Isotopes, 23*(11), 535–538.

Chakravarty, R., et al. (2010). An electro-amalgamation approach to isolate no-carrier-added ^{177}Lu from neutron irradiated Yb for biomedical applications. *Nuclear Medicine and Biology, 37*(7), 811–820.

Chakravarty, R., et al. (2012). Electrochemical separation is an attractive strategy for development of radionuclide generators for medical applications. *Current Radiopharmaceuticals, 5*(3), 271–287.

Chakravarty, R., et al. (2017). An electro-amalgamation approach to separate ^{47}Sc from neutron-activated ^{46}Ca target for use in cancer theranostics. *Separation Science and Technology, 52*(14), 2363–2371.

Choppin, G., & Silva, R. (1956). Separation of the lanthanides by ion exchange with alpha-hydroxy isobutyric acid. *Journal of Inorganic and Nuclear Chemistry, 3*(2), 153–154.

Cieszykowska, I., et al. (2014). Separation of ytterbium from $^{177}Lu/Yb$ mixture by electrolytic reduction and amalgamation. *SOP Transactions on Applied Chemistry, 1*(2), 6.

Coenen, H. H., et al. (2017). Consensus nomenclature rules for radiopharmaceutical chemistry—Setting the record straight. *Nuclear Medicine and Biology, 55*, 5–11.

Cohen, I. M., et al. (2013). The consistency of the data for neutron fission averaged cross-sections of threshold reactions: A study on the cross-section of $^{46}Ti(n,p)^{46}Sc$, $^{47}Ti(n,p)^{47}Sc$, $^{48}Ti(n,p)^{48}Sc$ and $^{64}Zn(n,p)^{64}Cu$ reactions. *Applied Radiation and Isotopes, 77*, 115–129.

Cutler, C. S., et al. (2013). Radiometals for combined imaging and therapy. *Chemical Reviews, 113*(2), 858–883.

DeNardo, S. J., et al. (1999). ^{67}Cu-21T-BAT-Lym-1 pharmacokinetics, radiation dosimetry, toxicity and tumor regression in patients with lymphoma. *Journal of Nuclear Medicine, 40*, 302–309.

Depuydt, H., & de Mevergnies, M. N. (1962). Average cross-section of the $^{32}S(n, p)^{32}P$ and $^{27}AI(n, \alpha)^{24}Na$ reactions for fission neutrons. *Journal of Nuclear Energy. Parts A/B. Reactor Science and Technology, 16*(9), 447–453.

Domnanich, K. A., et al. (2017). ^{47}Sc as useful beta$^{(-)}$-emitter for the radiotheragnostic paradigm: A comparative study of feasible production routes. *EJNMMI Radiopharmacy and Chemistry, 2*(1), 5.

EDQM. (2020). Radiopharmaceutical preparations, general monograph: 0125. In *European pharmacopeia*. Council of Europe.

Gizawy, M. A., et al. (2020). Feasibility study on production of Sc-47 from neutron irradiated Ca target for cancer theranostics applications. *Radiochimica Acta, 108*(3), 207–215.

Gopalakrishna, A., et al. (2018). Production, separation and supply prospects of ^{67}Cu with the development of fast neutron sources and photonuclear technology. *Radiochimica Acta, 106*, 549–557.

Gracheva, N., et al. (2019). Production and characterization of no-carrier-added ^{161}Tb as an alternative to the clinically-applied ^{177}Lu for radionuclide therapy. *EJNMMI Radiopharmacy and Chemistry, 4*(12), 1–16.

Guseva, L. I., & Dogadkin, N. N. (2010). A generator system for production of medical alpha radionuclides Ac-225 and Bi-213. *Journal of Radioanalytical and Nuclear Chemistry, 285*, 667–673.

Hashimoto, K., et al. (2003). Production of no-carrier-added ^{177}Lu via the $^{176}Yb(n,\gamma)^{177}Yb \rightarrow ^{177}Lu$ process. *Journal of Radioanalytical and Nuclear Chemistry, 255*(3), 575–579.

Hassanein, M., et al. (2006). Separation of carrier-free $^{64,67}Cu$ radionuclides from irradiated zinc targets using 6-tungstocerate(IV) gel matrix. *Journal of Radioanalytical and Nuclear Chemistry, 269*, 75–80.

Horwitz, E. P., et al. (1976). Radiochemical and isotope separations by high-efficiency liquid-liquid chromatography. *Journal of Chromatography, 125*, 203–218.

Horwitz, E., et al. (2005). A process for the separation of ^{177}Lu from neutron irradiated ^{176}Yb targets. *Applied Radiation and Isotopes, 63*(1), 23–36.

Huclier-Markai, S., et al. (2018). Promising scandium radionuclides for nuclear medicine: A review on the production and chemistry up to in vivo proofs of concept. *Cancer Biotherapy and Radiopharmaceuticals, 33*(8), 316–329.

Hun, Y. T., et al. (2020). Neutron-activated theranostic radionuclides for nuclear medicine. *Nuclear Medicine and Biology, 90–91*, 55–68.

IAEA. (1987). *Handbook on nuclear activation data*. Technical Reports Series No. 27. International Atomic Energy Agency.

IAEA. (2003). *Manual for reactor produced radioisotopes*. IAEA-TECDOC-1340. International Atomic Energy Agency.

IAEA. (2004). *Radioisotope handling facilities and automation of radioisotope production*. IAEA TECDOC-1430. International Atomic Energy Agency.

IAEA. (2022). *Research Reactor Database (RRDB)*. Retrieved from https://www.iaea.org/resources/databases/research-reactor-database-rrdb

Jalilian, A. J., et al. (2020). IAEA activities on ^{67}Cu, ^{186}Re and ^{47}Sc theranostic radionuclides and radiopharmaceuticals. *Current Radiopharmaceuticals, 14*(4), 306–314.

Junde, H., et al. (2005). *Nuclear Data Sheets, 106*, 159 (2005). Data extracted from the ENSDF database, National Nuclear Data Base. Brookhaven National Laboratory.

Karimi, Z., et al. (2019). Experimental production and theoretical assessment of ^{67}Cu via neutron induced reaction. *Annals of Nuclear Energy, 133*, 665–668.

Knapp, F. F. (2017). Continued availability of the tungsten-188/rhenium-188 generator to enhance therapeutic utility of ^{188}Re. *International Journal of Nuclear Medicine Research.*, Special Issue, 3–15.

Knapp, F. F., & Dash, A. (2016). Availability of alpha-emitting radioisotopes by reactor and accelerator production and via decay of naturally occurring parents. In F. F. Knapp & A. Dash (Eds.), *Radiopharmaceuticals for therapy* (pp. 159–166). Springer.

Knapp, F. F., Jr., et al. (2005). Production of therapeutic radioisotopes in the ORNL High Flux Isotope Reactor (HFIR) for applications in nuclear medicine, oncology and interventional cardiology. *Journal of Radioanalytical and Nuclear Chemistry, 263*, 503–509.

Kolsky, K. L., et al. (1994). Improved production, and evaluation of Cu-67 for tumor radioimmunotherapy. *Journal of Nuclear Medicine, 35*, 259.

Kolsky, K. L., et al. (1998). Radiochemical purification of no-carrier-added scandium-47 for radioimmunotherapy. *Applied Radiation and Isotopes, 49*, 1541–1549.

Kukleva, E., et al. (2014). Preparation of ^{227}Ac/^{223}Ra by neutron irradiation of ^{226}Ra. *Journal of Radioanalytical and Nuclear Chemistry, 304*(1), 263–266.

Lebedev, N. A., et al. (2000). Radiochemical separation of no-carrier-added ^{177}Lu as produced via the ^{176}Yb(n,γ)^{177}Yb→^{177}Lu process. *Applied Radiation and Isotopes, 53*(3), 421–425.

Lehenberger, S., et al. (2011). The low-energy β^- and electron emitter ^{161}Tb as an alternative to ^{177}Lu for targeted radionuclide therapy. *Nuclear Medicine and Biology, 38*(6), 917–924.

Lundqvist, H. O. (2014). Radionuclide production. In D. L. Bailey, J. L. Humm, A. Todd-Pokropek, & A. van Aswegenpp (Eds.), *Nuclear medicine physics: A handbook for teachers and students* (pp. 117–148). IAEA.

Marx, S., et al. (2014). *Method of manufacturing non-carrier added high-purity ^{177}Lu compounds as well as non-carrier added ^{177}Lu compounds*. United States Patent Application US20140294700A1.

Mausner, L. F., et al. (1998). Radionuclide development at BNL for nuclear medicine therapy. *Applied Radiation and Isotopes, 49*, 285–294.

McCready, V. R. (2017). Radioiodine—The success story of nuclear medicine. *European Journal of Nuclear Medicine and Molecular Imaging, 44*, 179–182.

McLain, D. R., et al. (2023). Evaluation of two extraction chromatography resins for scandium and titanium separation for medical isotope production. *Journal of Radioanalytical and Nuclear Chemistry, 332*(3), 553–562.

McNeil, S. W., et al. (2022). A simple and automated method for ^{161}Tb purification and ICP-MS analysis of ^{161}Tb. *EJNMMI Radiopharmacy and Chemistry, 7*, 31.

Mikołajczak, R., & Parus, J. L. (2005). Reactor produced beta-emitting nuclides for nuclear medicine. *World Journal of Nuclear Medicine, 4*(3), 184–190.

Mikołajczak, R., et al. (2019). Radiometals for imaging and theranostics, current production, and future perspectives. *Journal of Labelled Compounds and Radiopharmaceuticals, 62*(10), 615–634.

Mikołajczak, R., et al. (2021). Production of scandium radionuclides for theranostic applications: Towards standardization of quality requirements. *EJNMMI Radiopharmacy and Chemistry, 6*(1), 19.

Mirzadeh, S., & Knapp, F. F. (1992). Spontaneous electrochemical separation of carrier-free copper-64 and copper-67 from zinc targets. *Radiochimica Acta, 57*, 193–200.

Mirzadeh, S., et al. (1986). Production of no-carrier added ^{67}Cu. *International Journal of Radiation Applications and Instrumentation. Part A. Applied Radiation and Isotopes, 37*(1), 29–36.

Mirzadeh, S., et al. (2003). Reactor-produced medical radionuclides. In A. Vertes, S. Nagy, & Z. Klencsar (Eds.), *Handbook of nuclear chemistry* (Vol. 4, pp. 1–46). Kluwer Academic Publishers.

Mirzadeh, S., et al. (2004). *Method for preparing high specific activity ^{177}Lu.* US Patent 6,716,353.

Mou, L., et al. (2022). ^{67}Cu production capabilities: A mini review. *Molecules, 27*(5), 1501.

Mughabghab, S. F. (2003). *Thermal neutron capture cross-sections resonance integrals and G-factors.* INDC(NDS)-440. IAEA.

Müller, C., et al. (2012). A unique matched quadruplet of terbium radioisotopes for PET and SPECT and for α- and β-radionuclide therapy: An in vivo proof-of-concept study with a new receptor-targeted folate derivative. *Journal of Nuclear Medicine, 53*(12), 1951–1959.

Müller, C., et al. (2014). Promising prospects for ^{44}Sc/^{47}Sc-based theragnostics: Application of ^{47}Sc for radionuclide tumor therapy in mice. *Journal of Nuclear Medicine, 55*(10), 1658–1664.

Müller, C., et al. (2018). Scandium and terbium radionuclides for radiotheranostics: Current state of development towards clinical application. *The British Journal of Radiology, 91*(1091), 20180074.

Mushtaq, A., et al. (1990). Production of no-carrier-added ^{64}Cu and ^{67}Cu in a reactor. *Journal of Radioanalytical and Nuclear Chemistry, 141*, 261–269.

Neves, M., et al. (2002). Radionuclide production for therapeutic radiopharmaceuticals. *Applied Radiation and Isotopes, 57*(5), 657–664.

O'Brien, H. A. (1969). The preparation of ^{67}Cu from ^{67}Zn in a nuclear reactor. *The International Journal of Applied Radiation and Isotopes, 20*(2), 121–124.

Parus, J. L., & Mikołajczak, R. (2012). Beta-emitting radionuclides for peptide receptor radionuclide therapy. *Current Topics in Medicinal Chemistry, 12*(23), 2686–2693.

Patra, et al. (2023). Electrochemical separation and purification of no-carrier-added ^{177}Lu for radiopharmaceutical preparation: Translation from bench to bed. *Chemical Engineering Journal Advances, 14*, 100444.

Pawlak, D., et al. (2019). Comparison of separation methods for ^{47}Ca/^{47}Sc radionuclide generator. *Applied Radiation and Isotopes, 151*, 140–144.

Pietrelli, L., et al. (1992). Separation of carrier-free ^{47}Sc from titanium targets. *Journal of Radioanalytical and Nuclear Chemistry, 157*(2), 335–345.

Połosak, M., et al. (2013). Stability of ^{47}Sc-complexes with acyclic polyamino-polycarboxylate ligands. *Journal of Radioanalytical and Nuclear Chemistry, 295*(3), 867–872.

Radchenko, V., et al. (2021). Radiochemical aspects in modern radiopharmaceutical trends: A practical guide. *Solvent Extraction and Ion Exchange, 39*(7), 714–744.

Radford, L. L., & Lapi, S. E. (2019). Methods for the production of radionuclides for medicine. In J. S. Lewis, A. D. Windhorst, & B. M. Zeglis (Eds.), *Radiopharmaceutical chemistry* (pp. 63–84). Springer.

Robertson, A. K. H., et al. (2018). Development of ^{225}Ac radiopharmaceuticals: TRIUMF perspectives and experiences. *Current Radiopharmaceuticals, 11*(3), 156–172.

Shikata, E. (1964). Research of radioisotope production with fast neutrons, (VI). *Journal of Nuclear Science and Technology, 1*(5), 177–180.

Spahn, I., et al. (2004). Enhanced production possibility of the therapeutic radionuclides ^{64}Cu, ^{67}Cu and ^{89}Sr via (n,p) reactions induced by fast spectral neutrons. *Radiochimica Acta, 92*(3), 183–186.

Srivastava, S. C. (2011). Paving the way to personalized medicine: Production of some theragnostic radionuclides at Brookhaven National Laboratory. *Radiochimica Acta, 99*(10), 635–640.

Talip, Z., et al. (2020). A step-by-step guide for the novel radiometal production for medical applications: Case studies with ^{68}Ga, ^{44}Sc, ^{177}Lu and ^{161}Tb. *Molecules, 25*(4), 966.

Trautmann, N., & Roesch, F. (2016). Radiochemical separations. In F. Rösch (Ed.), *Modern applications* (Vol. 2, pp. 215–254). De Gruyter.

Uddin, M. S., et al. (2014). Radiochemical measurement of neutron-spectrum averaged cross sections for the formation of ^{64}Cu and ^{67}Cu via the (n,p) reaction at a TRIGAMark-II reactor: Feasibility of simultaneous production of the theragnostic pair ^{64}Cu/^{67}Cu. *Radiochimica Acta, 102*, 473–480.

Van So, L., et al. (2008a). Alternative chromatographic processes for no-carrier added ^{177}Lu radioisotope separation. Part I. Multi-column chromatographic process for clinically applicable. *Journal of Radioanalytical and Nuclear Chemistry, 277*(3), 663–673.

Van So, L., et al. (2008b). Alternative chromatographic processes for no-carrier added ^{177}Lu radioisotope separation. Part II. The conventional column chromatographic separation combined with HPLC for high purity. *Journal of Radioanalytical and Nuclear Chemistry, 277*(3), 675–683.

Vimalnath, K. V., et al. (2014). Reactor production of ^{32}P for medical applications: An assessment of ^{32}S(n,p)^{32}P and ^{31}P(n,γ)^{32}P methods. *Journal of Radioanalytical and Nuclear Chemistry, 301*, 555–565.

Zhernosekov, K. P., et al. (2008). Target burn-up corrected specific activity of ^{177}Lu produced via ^{176}Lu(n, γ)^{177}Lu nuclear reactions. *Applied Radiation and Isotopes, 66*(9), 1218–1220.

New Trends in the Automation of Solid Target Dissolution for Radionuclide Production: A Universal Target Dissolution System for Cassette-Based Radiosynthesis Modules

Gabriele Sciacca, Petra Martini, Sara Cisternino, Liliana Mou, Juan Esposito, and Emiliano Cazzola

1 Introduction

The increasing number of medical cyclotrons has significantly promoted the production of conventional and emerging radioisotopes (RIs) for medical applications (Bénard et al., 2014; Fonslet et al., 2017; Kasbollah et al., 2013; Martini et al., 2016, 2018; McCarthy et al., 1997; Obata et al., 2003; Pupillo et al., 2019, 2020; Pyles et al., 2021; Solbach et al., 2013; Synowiecki et al., 2018; Uccelli et al., 2019), due to the possibility they offer to produce RIs on-site and on-demand. In particular, the interest is growing for radiometals, which represent the main innovative radionuclides for enabling personalized medicine via a "theragnostic" approach. With this novel strategy, patients are selected by evaluating their positive response to the targeted treatment, an evaluation performed with a preliminary diagnostic imaging procedure using equivalent radiopharmaceutical product (Boschi et al., 2019; Qaim, 2019; Smilkov et al., 2014; Srivastava, 2013).

G. Sciacca
National Institute for Nuclear Physics - Legnaro National Laboratories (INFN-LNL), Legnaro, Italy

Department of Industrial Engineering, University of Padova, Padova, Italy

P. Martini (✉)
Department of Environmental and Prevention Sciences, University of Ferrara, Ferrara, Italy
e-mail: mrtptr1@unife.it

S. Cisternino · L. Mou · J. Esposito
Laboratori Nazionali di Legnaro—INFN, Legnaro, Italy

E. Cazzola
Radiopharmacy and Cyclotron Department, IRCCS Sacro Cuore Don Calabria Hospital, Negrar di Valpolicella, Italy

R. M. van Dam et al. (eds.), *Automated Technologies for the Development and Production of Radiopharmaceuticals*,
https://doi.org/10.1007/978-3-031-84632-8_3

While the most widespread RIs for positron emission tomography (PET) produced by cyclotrons, such as F-18 and C-11, are obtained starting from liquid or gas targets, radiometal production is instead primarily based on the use of solid precursor materials. Generally, in this kind of target, a pellet, usually made of a costly isotopically enriched element, is bonded to a supporting plate (backing) or encapsulated in a holding shell (Cisternino et al., 2022; Gagnon et al., 2012; Palmieri et al., 2019; Skliarova et al., 2018, 2019a, 2020). While a liquid target is preferred in the case of short-lived radionuclides, where fast irradiation and processing chemistry are needed (Pandey & DeGrado, 2021), for radiometals, it can lead to low production yield and to problems related to its corrosiveness or to gas release during irradiation (Oehlke et al., 2015; Pandey et al., 2014; Riga et al., 2018). On the other hand, the higher yields of a solid target-based production are counterbalanced by the higher technological and structural investments aimed at target manufacturing, target station, delivery systems, and post-irradiation target processing (Boschi et al., 2017, 2019), besides the necessity to recover the costly precursor material (Skliarova et al., 2019b).

For target processing, to the best of our knowledge, only four brands to date offer a device able to manage solid metallic target dissolutions: the IBA-Pinctada® Metal (IBA—Radiopharma Solutions, n.d.), the ARTMS QIS® (ARTMS, n.d.), the Comecer-ALCEO Metal (Comecer, n.d.), and the SynIQ hybrid target system for Ga-68 production (SYNIQ, n.d.). All of them are self-standing systems physically separated from the synthesis module, suited for the characteristics of the specific target station provided by the same company. If a radiopharmaceutical facility is already equipped with a dedicated solid target station and pneumatic transfer system, it would be difficult to adapt the existing solid target dissolution system to the available production line, eventually leading to the development of a custom-made system. Instead, having access to a compact dissolution system that is compatible with commercially available cassette-based synthesis modules would make it possible to perform the whole radiochemical processing, from dissolution to labeling steps, all at once by using a single, remotely controlled device in the same hot cell. This would in turn lower the contamination of the product as well as the radiation dose to the operator. Furthermore, the time saved in this way allows an increase in the recovery yield too.

In this chapter, an overview of solid target dissolution systems, both available on the market and developed for research purposes, is presented. Particular focus will be dedicated to the innovative universal solid target dissolution system developed in a collaboration between the LARAMED (LAboratory of RAdionuclides for MEDicine) group of the Legnaro National Laboratories of the National Institute for Nuclear Physics (INFN-LNL) in Italy, and the Sacro Cuore Don Calabria Hospital (SCDCH), located at Negrar di Valpolicella, VR, Italy (Esposito et al., 2018).

2 State of the Art

In this section, the main existing dissolution system equipment to manage irradiated solid targets, both available on the market and homemade for research purposes, are described.

2.1 Commercially Available Systems

The most tested modules spread in the market are the ones developed by some of the principal companies working in the field of radiopharmaceuticals production, namely IBA, ARTMS, Comecer, and SynIQ.

IBA has developed the Pinctada® Metal module (Poniger et al., 2015), devised to perform both dissolution and purification of metallic solid targets having the typical coin shape (i.e., target axis parallel to beam). Irradiated target coins can be interfaced with the module through a dedicated lodging present on the lateral side of the module. After arrival in the hot cell, the target is manipulated with the hot cell's tong and inserted from above on an appropriate slot through a small funnel. In this way, the target central axis is horizontally oriented during the dissolution. After that, a mechanism clamps the target to perform the chemical attack. Subsequently, the dissolved target solution is transferred to perform the following purification step. With such a configuration, target coins that match the size of the mentioned lodging (corresponding to the ones of the IBA's target stations) can be used. Albeit no details are provided about target pellet dissolution efficiency, a high-purity final product is reported, with no contamination coming from the target support. As an example, the system was used for Zr-89 and Cu-64 production from Y-89 and Ni-64 targets, respectively (IBA—Radiopharma Solutions, n.d.; Poniger et al., 2012). However, since the module alone occupies a whole standard-sized hot cell, the purified radioisotope has to be transferred to another hot cell, with an increased risk of contamination of both the solution and the operator.

The QIS® (QANTM® Irradiation System) by ARTMS Inc. is an entire assembly comprising the solid target irradiation and chemical processing. Indeed, it consists of a unique system where the dissolution module is part of the sending/receiving station, indivisible and completely integrated with the supplied irradiation station and transfer system. Therefore, the included module only allows one to perform the dissolution process of solid targets compatible with ARTMS specifications. The subsequent separation/purification steps have instead to be performed with additional devices, implying the need for an additional hot cell to accommodate them. In any case, in this module, the whole irradiated target shuttle has to be inserted in the dedicated dissolution chamber located on the upper part of the module (target axis vertically oriented) by using a pair of tongs. The shuttle serves as the reactor vessel, as it is automatically clamped in the module to confine the chemical attack. After the target is dissolved, the solution has then to be pumped to other modules for the following radiochemistry steps. This system can be used for Tc-99m, Zr-89, Ga-68, and Cu-64 production starting from Mo-100, Y-89, Zn-68, and Ni-64 irradiation, respectively (ARTMS, n.d.; Thisgaard et al., 2021), but it was also used for Er-165 production with a ^{nat}Ho target (Da Silva et al., 2021). No details are provided about its dissolution efficiency.

Similarly to the ARTMS QIS®, the ALCEO Metal (Comecer) system was developed as a solution for the complete automatic production of radiometals like Cu-64, Zr-89, and Ga-68. In the same way, the dissolution unit and the sending/receiving station for the target capsule are restricted to the use of the specific Comecer target shuttle, irradiation unit, and dedicated transfer system. On the other hand, with this system, the irradiated target capsule is automatically positioned for dissolution by a robotized system, without the need to use tongs. However, the dissolution and transfer unit occupies an entire hot cell compartment, therefore requiring additional equipment and hot cell space to perform the radioisotope purification step. Such a system is thus set up as a fully automated process, making the production easier and more reliable, although no flexibility is foreseen in the use of different targets other than the ones specified by the vendor. Two ALCEO Metal dissolution systems are provided: one dedicated to Zr-89 and Ti-45 production from Y-89 and Sc-45 targets, respectively, and one for Cu-64 and Ga-68 produced with Ni-64 and Zn-68. The system can also offer the possibility to automatically electroplate the target directly in the same unit, like in the case of Ni-64. However, sometimes the plating process was found to be unreliable in target manufacturing (Erdahl et al., 2015; Fiedler et al., 2018). In any case, this system has been reported to provide a highly pure radioisotope (Malinconico et al., 2018; Matarrese et al., 2010) with high yield (Leonte et al., 2020), even if the characterization of dissolution efficiency is not clearly presented in any work.

In the upgraded version (2.0) of the ALCEO Metal dissolution system, the dissolution unit can be located near the irradiation unit to allow for the transfer of only the dissolved target solution toward the dedicated hot cell. This version of the system is designed for existing facilities configured for irradiation of liquid targets, where the installation of a pneumatic transfer line for shuttling the solid target can be costly or technically hard to achieve. Indeed, this constraint can be overcome by exploiting the existing delivery line for the liquid target, which can also be used for the transfer of the dissolved target solution (Comecer, n.d.). Obviously, this approach hinders the possibility to properly verify the successful target dissolution process; while this capability is not required for well-established procedures, it would be useful in the case of optimizing production routines.

SynIQ Ltd., a new company founded in 2019, offers a solid/liquid hybrid system for Ga-68 production using a cyclotron (SYNIQ, n.d.). This system includes a solid target station compatible with various brands of cyclotrons, onto which a multi-target holder is manually loaded. Indeed, this system enables successive irradiations by switching on different targets through automated sliding or revolving of the target holder on the irradiation station. Different from the other commercial systems, this one involves in situ dissolution of the target following the irradiation, allowing for the transfer of the liquid product without transferring the solid target to the hot cell, hence the name “hybrid target system.” In this way, it allows one to leverage existing liquid target transfer lines in radiopharmacies with cyclotrons that were not originally designed for solid targets. This hybrid system enables the exploitation of the higher production yield achievable with solid targets while maintaining the simplicity of liquid target transfer without impacting the structural aspects of already established facilities. Once the target is dissolved and transferred, it can be processed by any synthesis module for purification and radiolabeling.

2.2 Research Systems

Parallel to the commercial products, some alternative “homemade” modules have been developed, aimed at meeting the needs related to specific configurations of existing radioisotopes production facilities. In particular, the system proposed by Gelbart and Johnson (Gelbart & Johnson, 2019) for high-power solid targets was devised to process the target directly in the irradiation site, in order to decrease the radiation dose to the operator. In their system, the target station’s manipulator places the irradiated target in contact with the dissolution reactor in such a way that the target’s backing seals the reactor vessel. The reactant for the chemical attack is delivered directly by the radiochemistry laboratory through the installed capillary lines. After dissolution, the obtained radiochemical solution is recovered via the capillary line, and the target is released in a shielded container for the decay of the induced activity in the backing. The dissolution reactor offers the possibility to heat the solution up to a temperature of 100 °C and to promote chemical attack by gas bubbling agitation.

For the same purpose, Beaudoin et al. (2019) developed a similar system dedicated to encapsulated coin targets irradiated by medical cyclotrons. The dissolution occurs inside the vault dedicated to target irradiation, where the target holder is opened by an automatic system based on vacuum clamps to allow the target coin to fall into the dissolution chamber. The latter is sealed with a Teflon valve, in order to leave the possibility to pressurize the chamber. When the chemical dissolution process is over, the dissolved solid target is transferred through the line dedicated to liquid targets.

Finally, Sciacca et al. (2021) presented a dissolution system for medical cyclotron solid targets, which may be integrated into different commercial radiosynthesis modules for liquid targets. Within the system, an open-bottomed vial acts as the reactor, which during the dissolution is pressed upon the backing to confine the chemical solution by means of an O-ring. The vial position is controlled by exploiting the module’s movement mechanism of syringe pumps, and the heater/cooler reactor already present in the modules is used for docking the target. In this way, with a simple adaptor for commercial modules, all the radiochemical steps necessary for the radiopharmaceutical product preparation can be performed with a single module, providing a compact system with a resulting enhanced product quality. In the following sections, the developed reactor design, functionalities, and system performance are described in detail.

3 A Universal Cassette-Based System

Dealing with the production of innovative radiometals usually involves the study of new procedures for target manufacturing and processing, strictly dependent on the precursor material to be irradiated. Therefore, target systems designed for specific and established production processes do not provide the flexibility required to manage novel production routines. Typically, commercial systems are "closed" and do not offer the possibility to reconfigure any parameters of the whole process chain including the chemical processing of the target. This has encouraged the development of custom systems to allow more control over the target processing. To minimize costs, it would be beneficial to perform studies by using at least part of the equipment already present in production facilities. Indeed, this was the case of the LARAMED research project, which many times had to perform proof-of-concept production of emerging radionuclides, making use of tools and equipment aimed at the radiochemical processing provided by the external radiopharmacy departments, like in the Radiopharmaceutical unit of SCDCH (Martini et al., 2022). In the latter, an ACSI TR19/300 cyclotron is installed, mainly dedicated to F-18 production on a routine basis by irradiation of liquid $[^{18}O]H_2O$ targets, but is also equipped with a solid target station to irradiate standard 24 mm diameter target coins and a pneumatic transfer system to the hot cells of the radiopharmacy laboratories. Although solid targets able to be irradiated by the ACSI TR19 were manufactured by INFN-LNL (Cisternino et al., 2022; Skliarova et al., 2019a, 2020), however, no equipment was available at SCDCH for safely processing them once irradiated. For this reason, a dedicated system for target dissolution was developed that could be adapted directly to the downstream synthesis module.

3.1 The Reactor Assembly

The reactor essentially consists of an open-bottom vial pushed onto the target to be processed (Palmieri et al., 2019). The target backing, made of a chemically inert material, acts as an enclosure of the reactor by confining the acidic solution used for the chemical attack of the target pellet. In this way the dissolution reaction takes place inside the vial as an O-ring seals the contact between target backing and vial, and thus only the target material will be dissolved. A mechanical adapter was designed to connect the vial to the syringe drive of commercial synthesis modules. Indeed, the drives are often present in these modules as they serve to pump solutions among the various radiochemical steps for the radiopharmaceutical synthesis. In the same way, the target can be held in position during the chemical dissolution with a proper adaptor for the module's thermal controller, allowing heating during the dissolution process. The adaptor is made from a material with high thermal conductivity, like aluminum or copper, fitting in place of the module's reaction vessel. A detailed description of the various components is hereafter reported.

3.1.1 Reactor Dissolution Vial

Since the dissolution vial has to be filled with strong acids/bases, it must be made of inert materials. Among them, PEEK can be machined with standard manufacturing processes. On the contrary, ceramic materials, like glass and quartz, require special care in manufacturing, especially with particular shapes like the one in Fig. 1a. In particular, the vial top is threaded to interface with a vial head for three tubing connectors (Fig. 1b). Indeed, such vial heads are already available as module accessories and can be conveniently exploited to make the reactor easily compatible with cassette-based synthesis circuits. At the same time, the vessel bottom part includes a cavity along its surface for O-ring insertion, and this further complicates the reactor's manufacturing also given the requirement of specific size tolerances. Therefore, if visual monitoring of the process is not required like in the case of well-established protocols, PEEK would be preferable. PEEK also provides higher resilience. Similarly, for the O-ring, it is necessary to pay attention to the choice of the material

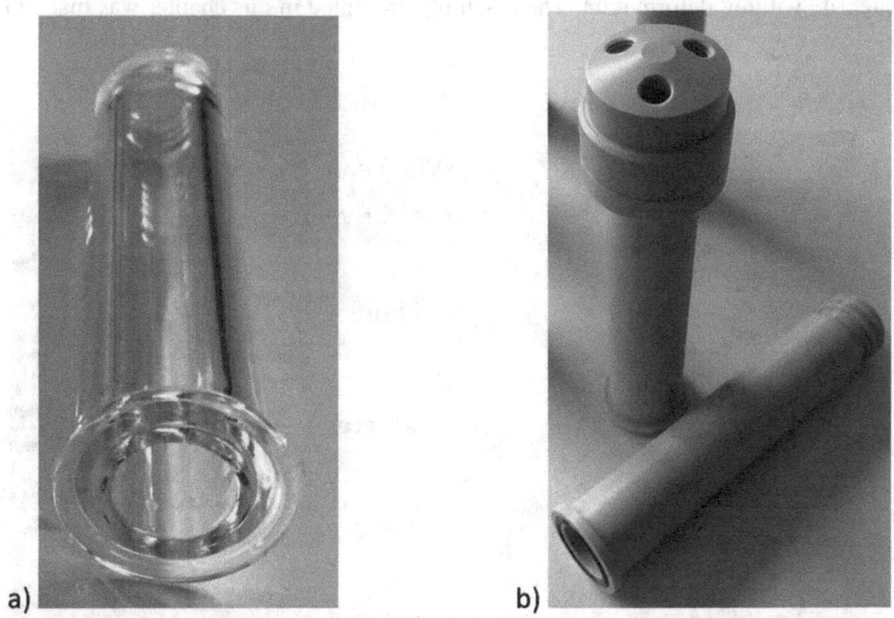

Fig. 1 (**a**) Reaction vial made of borosilicate glass manufactured by Intellion S.a.r.l (Paris); (**b**) PEEK vials, one with the PEEK vial head (upper side) and one with mounted O-ring on its bottom part (lower side)

too. In particular, the FKM fluoroelastomer Viton is recommended for its declared resistance to hydrochloric acid, but in certain cases, nitrile butadiene rubbers (NBR 70) can be acceptable.

3.1.2 Vial Connecting System

On synthesis modules, syringe drives usually present mechanical adapters for syringe plungers, which are connected to the module by bolted joints. This offers the opportunity to install a connecting system for the vial-holder directly to the stepper motor of the module to remotely close the reactor cylinder after positioning the target on the target holder. The downward force was applied via a vial holder (Fig. 2a) that was mechanically connected to the syringe drive (Fig. 2b).

The exact geometry of this holder can be varied to accommodate different synthesis modules. However, usually, the sizes of commercial modules are not known, and possible misalignments could be present between the target holder and vial. This can be prevented by including in the connection system appropriate slots and screws to enable fine-tuning of the *XYZ* positions. From this point of view, 3D computer-aided design tools, e.g., SolidWorks®, are valuable to determine the design of the adapter and ensure in advance that there are no collisions between the dissolution vial and parts of the synthesis module. Rigidity and mechanical resistance of the system are important factors too since the system must preserve its integrity without deformation. The assembly presented in this chapter was made in

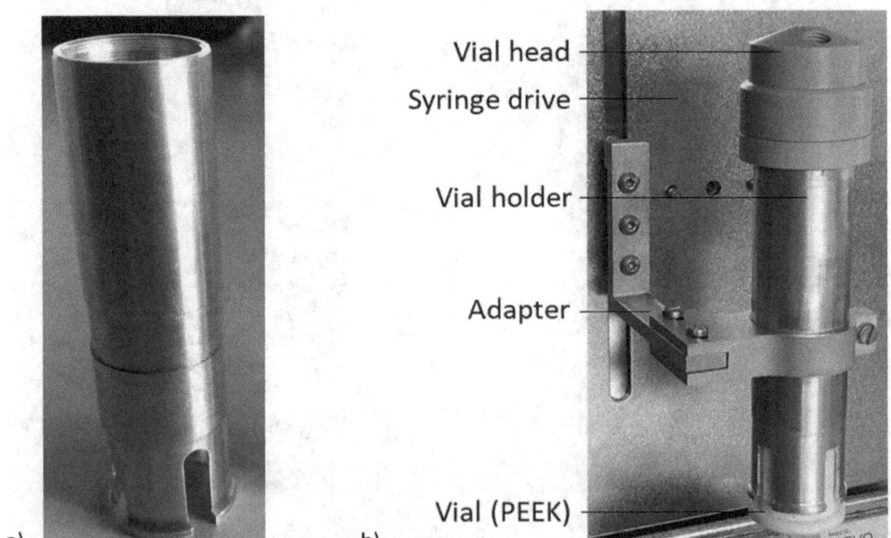

Fig. 2 (**a**) Vial holder; (**b**) assembled connection system mounted on an E&Z module

both machined aluminum alloy 6082 and stainless steel in relation to the required rigidity of the single connecting component.

3.1.3 Target Holder

To ensure the target remains constrained during the dissolution, a platform for target docking must be adopted. Conceiving it as able to be plugged into the heater/cooler of the module simplifies the design, providing for both target anchoring and controlling the process kinetics. The structure is illustrated in Fig. 3a, where the bottom is in the same shape as the reaction vessel normally used in the module. By using a material having a high thermal conductivity, the target holder acts as an extension of the reactor's heater while accommodating the target which is positioned in a cavity on its top. Finally, as shown in Fig. 3b, a groove surrounds the hollow on the top to confine possible losses of acid solution due to incorrect sealing of the reactor vessel.

3.2 Dissolution Operations

The dissolution operation starts with the positioning of the irradiated target, which arrives in the hot cell via pneumatic transfer, by using tongs for the safety of the operator. After the target is placed on the target holder with the target pellet pointing upward, the reactor vial is lowered to contact with the target. Once the vessel is sealed in this way, it is filled with the solvent for the chemical attack. Depending on the involved chemistry, the target holder is heated to the processing temperature for the required time needed for the solvent to react with the pellet. After that, the obtained solution can be cooled down and transferred through the exit at the vial's top to the following processing steps along the cassette manifold. Before each test,

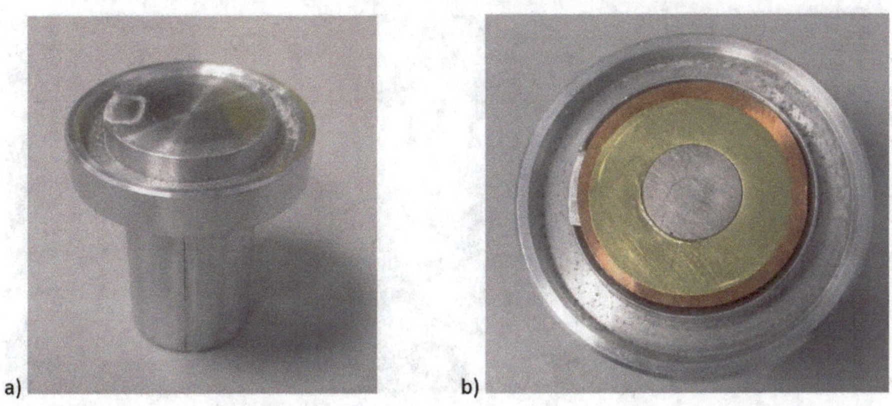

Fig. 3 (**a**) Target holder made for the Trasis AllinOne module shown in lateral view; (**b**) target holder housing a solid targets installed and ready for the chemical dissolution process

a leakage assessment is performed to evaluate the risk of losses during the dissolution, by filling the reactor with pure water.

The entire task is performed remotely with module's control unit, and all the operations to obtain the radiopharmaceutical can be done with a single synthesis module. Indeed, the dissolved solid target can be processed in the subsequent radiochemical operations as a liquid in the same cassette manifold to which the reactor is connected. Furthermore, given the system configuration, the target and/or backing thicknesses and its manufacturing procedures do not influence the process. In case of a different backing thickness it is only necessary to adjust the positioning distance from the syringe drive control unit interface. The O-ring available at the bottom of the vial allows for the confinement of the chemical reagents volume for the selective dissolution of the target material (i.e., the pellet) only. Finally, the transparency of the glass vial allows to monitor the process in the case of developing new procedures. Figure 4 shows the developed dissolution system implemented in two commercial modules based on cassette manifolds: Modular Lab Pharm Tracer by Eckert & Ziegler (E&Z) and the AllinOne by Trasis.

3.3 *Target Requirements*

As described above, the target's backing is used to close the bottom of the reactor vial during the dissolution process. In this configuration, inert backing materials must be selected to avoid contamination. Moreover, the target materials should also possess adequate thermal conductivity to dissipate the heating power deposited during the irradiation, preserve their properties during beam interaction, and have mechanical robustness for target delivery in the hot cell. Metallic materials have these properties, and gold also allows adequate inertness to the acidic attack during

a) b)

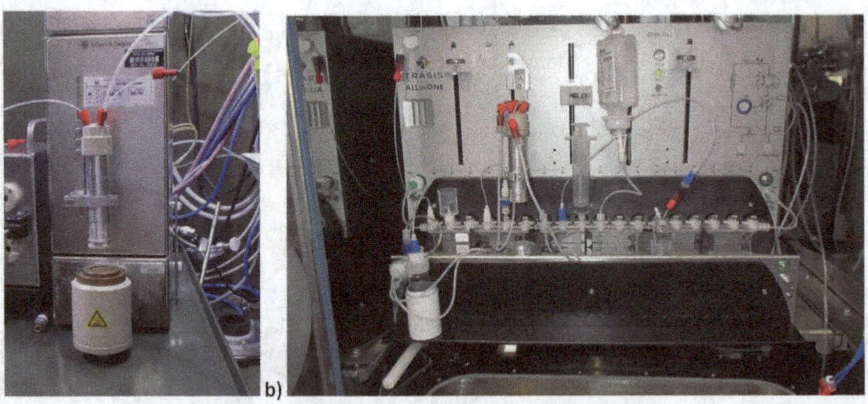

Fig. 4 Photographs of the solid target dissolution system mounted on: (**a**) an E&Z Modular Lab Pharm Tracer module; (**b**) a Trasis AllinOne module

dissolution. Alternatively, covering the backing with a thin Au layer permits the use of other materials that are less costly but have higher thermal conductivity, like copper, as the prototype shown in Fig. 3b. Alternatively, niobium can be used in place of Cu for reducing backing radioactivity and is also a relatively inert material depending on dissolution conditions. For example, in the case of Zr-89 production, the Y disc was directly bonded to an Nb disc, and high purity [^{89}Zr]Zr-oxalate could be achieved. While high-temperature resistant metal bonding may not be obtained by using electrodeposition, brazing, sputtering, or other methods such as spark plasma sintering (SPS) allows for fast and practical manufacturing of robust targets (Cisternino et al., 2022; Kotliarenko et al., 2023; Skliarova et al., 2020). This technology is based upon the simultaneous application of pressure and temperature, the latter achieved by the Joule effect due to the electric resistance of the high-intensity current passing through the sample (Anselmi-Tamburini & Groza, 2017). The results are sintered pellets of tailored size, starting from metal or oxide powders, that is, the form in which the isotopically enriched materials are usually supplied. The process takes place in a few minutes, and the high efficiency of this technique makes it quite advantageous in the case of expensive starting materials. Furthermore, a tight bond between different materials without the need for a filler can be achieved (Hu et al., 2020). Thus, the LARAMED team has started to exploit this technology for manufacturing a robust target that is (i) compatible with high current irradiations, (ii) easy to handle in a hot cell, and (iii) compatible with the dissolution system described.

3.4 Dissolved Target Analyses

The dissolved target can be analyzed in different ways to provide a demonstration of the dissolution quality, as done by Sciacca et al. (2021). To detect a possible incomplete dissolution, since the weight of the pellet is precisely known, the target/backing system is weighed before and after the chemical attack. The dissolution process could also lead to contaminants coming from the backing into the target solution; to check this possibility, gamma spectrometry is performed on an aliquot of the dissolution solution. Analysis of the obtained spectrum allows one to quantify the produced activity of the isotope of interest; this activity is rescaled considering the weight fraction of the aliquot with respect to the total solution. It is, therefore, possible to trace the activity produced at the end of the bombardment and compare this value with the theoretical one obtained with the online tool ISOTOPIA (International Atomic Energy Agency—Nuclear Data Section, n.d.). If the obtained values agree with the measured activity, this will give additional proof of the effective radioisotope recovery.

Finally, scanning electron microscopy coupled with energy-dispersive X-ray spectroscopy (SEM-EDS) analyses can be performed on the backing surface after a non-radioactive target dissolution. Indeed, it is possible to inspect if the target material is completely dissolved and qualitatively detect the residual target material that did not pass into the solution. This kind of analysis could be helpful for the evaluation of the dissolution efficiency during the setting up of experimental chemical procedures with new modules, as well as assessing the intactness of the inert layer material. In addition, these analyses could be useful when the target manufacturing process parameters are still in the optimization phase, and the adhesion between the target material and the backing is under investigation.

All these analyses were conducted to evaluate the performance of the dissolution reactor described above. In particular, this system was tested in the production of Mn-52, Tc-99m, and Zr-89 from chromium, molybdenum, and yttrium targets, respectively. The weight test revealed that the amount of dissolved weight corresponded to the amount of the original target material. Therefore, in all three cases, all the target material was efficiently dissolved and removed from the backing in a reproducible manner. Gamma-spectrometry analysis of the dissolved target revealed no radioactive contaminants coming from the backing material. Moreover, the extrapolation of the activities of the radioisotopes of interest from the spectra, rescaled at the end of beam (EOB), was in agreement with the predicted activity calculated at EOB by using the online tool ISOTOPIA. Furthermore, as additional confirmation of complete target dissolution, an additional chemical attack was performed with fresh solvents after the dissolved target removal from the reactor in order to detect any residual activity remaining undissolved on the backing. No significant activity was found in any of the three solutions. Finally, SEM/EDS analysis of the backing after dissolution clearly showed that the pellets were completely dissolved (Sciacca et al., 2021).

4 Conclusion

In this chapter, existing systems processing solid targets were briefly reviewed, and a novel adaptor for commercial radiosynthesis modules enabling the dissolution of solid targets was presented. The developed dissolution reactor was extensively validated under various operating conditions (vial material, irradiated targets, different modules), demonstrating high efficiency in the dissolution process. By coupling the target dissolution reactor onto the synthesis module, it is integrated directly with the isotope separation and purification process, as well as downstream synthesis process, and operated through the same PLC and software to guarantee the continuity of the process. Therefore, the presented system gives the opportunity to those who already own modules like E&Z Modular-lab or Trasis AllinOne to simply adapt their devices for solid target dissolution.

The realization of a flexible solid target dissolution system that is easily installable on commercial cassette-based synthesis modules is a fundamental improvement in the management of the radiochemistry for this kind of target. For the time being, the market is missing a product with these characteristics, while the presented system allows for the processing of solid targets with efficiency and simplicity. With the presented reactor, all the radiochemistry processes for obtaining a radiopharmaceutical starting from irradiated solid targets can be done in an automated fashion, minimizing contamination of both the product and the operator while increasing the yield. Indeed, using a solely automated system for dissolution, separation, purification, and then labeling sharply limits radiation exposure of operators in handling high radioactive materials and increases the reproducibility of the whole process, making it faster and more traceable, which remains a key factor for attaining a clinical-grade quality of the radiopharmaceutical. Furthermore, the developed system also permits one to process targets of different thicknesses, regardless of their manufacturing procedure. Due to its design consisting of an open-bottom vial and the inert protective layer covering the backing, the irradiated target material can be selectively dissolved without attacking its support (i.e., the backing). This simplifies the associated radiochemistry, thus obtaining a high-purity radiopharmaceutical.

The same design of target holder and vial connection system works with different modules, and these components may be modified to be compatible with commercial modules of other brands (e.g., GE, IBA radiopharma solutions, ORA-NEPTIS, SCINTOMICS, and IPHASE) with small adjustment in the design and/or structure of the system. In fact, their materials can be selected in relation to the needs of thermal conductivity, resistance, and chemical inertness to the solvents involved in the process. A transparent glass vial can be used for the visualization of the reaction process inside the vessel during the chemical attack, and the system is compatible also with PEEK vials. Since the vials are integrated with the cassette manifold through the vial cap, the manufacturing complexity of the vial is minimized, and high compatibility is provided for a wide range of synthesis modules.

Acknowledgments This research has been funded by the National Institute for Nuclear Physics (INFN) in the framework of the METRICS (Multimodal pET/mRI imaging with Cyclotron-produced $^{51/52}$Mn iSotopes) project. The work was carried out in collaboration with the Sacro Cuore Don Calabria Hospital (SCDCH) Radiopharmacy, within the LARAMED (LAboratory of RAdioisotopes for MEDicine) research program running at Legnaro National Laboratories (LNL) of INFN. We thank the LNL mechanical workshop for the realization of the reactor's components and Intellion S.a.r.l (Paris, France) for the manufacturing of glass vials. We are also grateful to all the staff of the SCDCH radiopharmacy for their precious support.

References

Anselmi-Tamburini, U., & Groza, J. R. (2017). Critical assessment 28: Electrical field/current application—A revolution in materials processing/sintering? *Materials Science and Technology, 33*(16), 1855–1862.

ARTMS. (n.d.). *QUANTM Irradiation System*™. Retrieved August 30, 2023, from https://www.artms.ca/qis-ecosystem/

Beaudoin, J.-F., et al. (2019). Target carrier and dissolution system for facilitating end-to-end in-vault cyclotron production of radiometals in large quantities. *Journal of Nuclear Medicine, 60*(Suppl 1), 1162–1162.

Bénard, F., et al. (2014). Implementation of multi-curie production of 99mTc by conventional medical cyclotrons. *Journal of Nuclear Medicine, 55*(6), 1017–1022.

Boschi, A., et al. (2017). Recent achievements in Tc-99m radiopharmaceutical direct production by medical cyclotrons. *Drug Development and Industrial Pharmacy, 43*(9), 1402–1412.

Boschi, A., et al. (2019). Interdisciplinary tasks in the cyclotron production of radiometals for medical applications. The case of ^{47}Sc as example. *Molecules, 24*(3), 444.

Cisternino, S., et al. (2022). Target manufacturing by spark plasma sintering for efficient ^{89}Zr production. *Nuclear Medicine and Biology, 104*, 38–46.

Comecer. (n.d.). *ALCEO solid target processing system.* Retrieved October 31, 2023, from https://www.comecer.com/alceo-solid-target-processing-system/

Da Silva, I., et al. (2021). A high separation factor for ^{165}Er from Ho for targeted radionuclide therapy. *Molecules, 26*(24), 7513.

Erdahl, C. E., Bender, B. R., & Dick, D. W. (2015). Practical experience implementing the Comecer ALCEO Metal solid targetry system. In *Conference Proceeding of the 15th International Workshop on Targetry and Target Chemistry*. Retrieved from https://nbn-resolving.org/urn:nbn:de:bsz:d120-qucosa-166465

Esposito, J., et al. (2018). LARAMED: A laboratory for radioisotopes of medical interest. *Molecules, 24*(1), 20.

Fiedler, L., et al. (2018). Fully automated production and characterization of ^{64}Cu and proof-of-principle small-animal PET imaging using ^{64}Cu-labelled CA XII targeting 6A10 Fab. *ChemMedChem, 13*(12), 1230–1237.

Fonslet, J., et al. (2017). Optimized procedures for manganese-52: Production, separation and radiolabeling. *Applied Radiation and Isotopes, 121*, 38–43.

Gagnon, K., et al. (2012). Cyclotron production of 99mTc: Recycling of enriched ^{100}Mo metal targets. *Applied Radiation and Isotopes, 70*(8), 1685–1690.

Gelbart, W. Z., & Johnson, R. R. (2019). Solid target system with in-situ target dissolution. *Instruments, 3*(1), 14.

Hu, Z.-Y., et al. (2020). A review of multi-physical fields induced phenomena and effects in spark plasma sintering: Fundamentals and applications. *Materials & Design, 191*, 108662.

IBA—Radiopharma Solutions. (n.d.) *More chemistry.* Retrieved August 30, 2023, from https://www.iba-radiopharmasolutions.com/more-chemistry

International Atomic Energy Agency—Nuclear Data Section. (n.d.) *Medical isotope browser.* Retrieved August 30, 2023, from https://www-nds.iaea.org/relnsd/isotopia/isotopia.html

Kasbollah, A., et al. (2013). Review on production of ^{89}Zr in a medical cyclotron for PET radiopharmaceuticals. *Journal of Nuclear Medicine Technology, 41*(1), 35–41.

Kotliarenko, A., et al. (2023). First results on zinc oxide thick film deposition by inverted magnetron sputtering for cyclotron solid targets production. *Materials, 16*(10), 3810.

Leonte, R. A., et al. (2020). Process validation for production of copper radioisotopes in a TR-19 variable energy cyclotron. In *AIP Conference Proceedings* (p. 020022). AIP Publishing LLC.

Malinconico, M., et al. (2018). ^{68}Ga and ^{45}Ti production on a GE PETtrace cyclotron using the ALCEO solid target. *Journal of Nuclear Medicine, 59*(Suppl 1), 664.

Martini, P., et al. (2016). A solvent-extraction module for cyclotron production of high-purity technetium-99m. *Applied Radiation and Isotopes, 118*, 302–307.

Martini, P., et al. (2018). In-house cyclotron production of high-purity Tc-99m and Tc-99m radiopharmaceuticals. *Applied Radiation and Isotopes, 139*, 325–331.

Martini, P., et al. (2022). Cyclotron-production of Mn-52: An update on the METRICS project. *Nuclear Medicine and Biology, 108*, S169–S170.

Matarrese, M., et al. (2010). Automated production of copper radioisotopes and preparation of high specific activity [^{64}Cu] Cu-ATSM for PET studies. *Applied Radiation and Isotopes, 68*(1), 5–13.

McCarthy, D. W., et al. (1997). Efficient production of high specific activity ^{64}Cu using a biomedical cyclotron. *Nuclear Medicine and Biology, 24*(1), 35–43.

Obata, A., et al. (2003). Production of therapeutic quantities of ^{64}Cu using a 12 MeV cyclotron. *Nuclear Medicine and Biology, 30*(5), 535–539.

Oehlke, E., et al. (2015). Production of Y-86 and other radiometals for research purposes using a solution target system. *Nuclear Medicine and Biology, 42*(11), 842–849.

Palmieri, V., et al. (2019). *Method for obtaining a solid target for radiopharmaceuticals production*. Patent WO/2019/053570.

Pandey, M. K., & DeGrado, T. R. (2021). Cyclotron production of PET radiometals in liquid targets: Aspects and prospects. *Current Radiopharmaceuticals, 14*(4), 325–339.

Pandey, M. K., et al. (2014). Production of ^{89}Zr via the ^{89}Y (p, n) ^{89}Zr reaction in aqueous solution: Effect of solution composition on in-target chemistry. *Nuclear Medicine and Biology, 41*(4), 309–316.

Poniger, S. S., et al. (2012). Automated production of 124 I and 64 Cu using IBA Terimo and Pinctada metal electroplating and processing modules. In *AIP Conference Proceedings* (pp. 114–119). American Institute of Physics.

Poniger, S., et al. (2015). Fully automated production of Zr-89 using IBA Nirta and Pinctada systems. *Internal Medicine Journal, 45*, 32–32.

Pupillo, G., et al. (2019). Production of ^{47}Sc with natural vanadium targets: Results of the PASTA project. *Journal of Radioanalytical and Nuclear Chemistry, 322*(3), 1711–1718.

Pupillo, G., et al. (2020). Production of ^{67}Cu by enriched ^{70}Zn targets: First measurements of formation cross sections of ^{67}Cu, ^{64}Cu, ^{67}Ga, ^{66}Ga, ^{69}mZn and ^{65}Zn in interactions of ^{70}Zn with protons above 45 MeV. *Radiochimica Acta, 108*(8), 593–602.

Pyles, J. M., et al. (2021). Production of ^{52}Mn using a semi-automated module. *Applied Radiation and Isotopes, 174*, 109741.

Qaim, S. M. (2019). Medical radionuclide production. In *Medical radionuclide production*. De Gruyter.

Riga, S., et al. (2018). Production of Ga-68 with a General Electric PETtrace cyclotron by liquid target. *Physica Medica, 55*, 116–126.

Sciacca, G., et al. (2021). A universal cassette-based system for the dissolution of solid targets. *Molecules, 26*(20), 6255.

Skliarova, H., et al. (2018). Innovative target for production of technetium-99m by biomedical cyclotron. *Molecules, 24*(1), 25.

Skliarova, H., Cisternino, S., et al. (2019a). Medical cyclotron solid target preparation by ultrathick film magnetron sputtering deposition. *Instruments, 3*(1), 21.

Skliarova, H., Buso, P., et al. (2019b). Recovery of molybdenum precursor material in the cyclotron-based technetium-99m production cycle. *Instruments, 3*(1), 17.

Skliarova, H., et al. (2020). Cyclotron solid targets preparation for medical radionuclides production in the framework of LARAMED project. In *Journal of Physics: Conference Series* (p. 012022). IOP Publishing.

Smilkov, K., et al. (2014). Preparation and first biological evaluation of novel Re-188/Tc-99m peptide conjugates with substance-P. *Applied Radiation and Isotopes, 92*, 25–31.

Solbach, C., et al. (2013). Production of Zr-89 on a PETtrace cyclotron. *Journal of Nuclear Medicine, 54*(Suppl 2), 1098.

Srivastava, S. C. (2013). A bridge not too far: Personalized medicine with the use of theragnostic radiopharmaceuticals. *Journal of Postgraduate Medicine, Education and Research, 47*(1), 31–46.

SYNIQ. (n.d.). *Hybrid target system for ^{68}Ga production*. Retrieved October 31, 2023, from https://syniq.hu/product/hybrid-target-system-for-68ga-production

Synowiecki, M. A., Perk, L. R., & Nijsen, J. F. W. (2018). Production of novel diagnostic radionuclides in small medical cyclotrons. *EJNMMI Radiopharmacy and Chemistry, 3*(1), 1–25.

Thisgaard, H., et al. (2021). Multi-curie production of gallium-68 on a biomedical cyclotron and automated radiolabelling of PSMA-11 and DOTATATE. *EJNMMI Radiopharmacy and Chemistry, 6*(1), 1–11.

Uccelli, L., et al. (2019). Therapeutic radiometals: Worldwide scientific literature trend analysis (2008–2018). *Molecules, 24*(3), 640.

New Trends in the Automation of Solvent Extraction for the Production of Radionuclides: From Macro- to Microscale Systems

Petra Martini, Teresa Ghirardi, Francesca Porto, and Alessandra Boschi

1 Introduction

In the cyclotron-based production cycle of radioisotopes, which requires a multidisciplinary approach, radiochemistry is responsible for handling the processing of the target after irradiation. Its tasks include isolating and purifying the desired radioisotope from the bulk of the target material and other contaminants. Additionally, radiochemistry plays a crucial role in the development and preparation of radiopharmaceuticals, and performing quality controls to verify that the product is safe and injectable. The progress achieved by nuclear medicine is intimately connected to the availability of new radionuclides and radiopharmaceuticals. As a result, there have been widespread efforts to improve cyclotron-based radionuclide product strategies, technologies, and automated systems. Full automation of radiochemical separation or radiopharmaceutical synthesis is required when passing from experimental activity levels (few tens of MBq) to clinical production levels of activity (several GBq). Automation undoubtedly offers many significant advantages, including shorter process time, higher efficiency, increased reliability, and repeatability, as well as improved safety and prevention of human error. Moreover, automation can

P. Martini (✉)
Department of Environmental and Prevention Sciences, University of Ferrara, Ferrara, Italy
e-mail: mrtptr1@unife.it

T. Ghirardi
Legnaro National Laboratories of the National Institute of Nuclear Physics, Legnaro, Italy

F. Porto
Department of Translational Medicine, University of Ferrara, Ferrara, Italy

A. Boschi
Department of Chemical and Pharmaceutical Sciences, University of Ferrara, Ferrara, Italy

R. M. van Dam et al. (eds.), *Automated Technologies for the Development and Production of Radiopharmaceuticals*,
https://doi.org/10.1007/978-3-031-84632-8_4

improve the quality of the product and also assist in establishing and complying with all regulatory aspects under GMP.

So far, the automation of radiochemical processes has mainly driven the availability of PET-RNs with short half-lives and high-energy emissions. The adoption of fully automated processes with remote control and graphic interfaces, as well as the utilization of sterile disposable cassettes, ensured improved control over chemical purity, as well as the sterility and endotoxin content of PET radiopharmaceuticals.

The evolution of automation technologies combined with the introduction of new chromatographic resins has facilitated the automation of RN production.

The current technological revolution aims to achieve even more compact, faster, cheaper, and more reliable assay systems, thus leading to the development of micro-scale reactors (the so-called lab-on-a-chip) in order to improve performance and minimize chemical and radiological risks.

Solvent extraction (SE) is an alternative simple technique for the separation of small quantities of solute from a mixture, is very cost-effective, and has a long history of validation in the chemistry field. Indeed, SE methods were widely used in the production of radionuclides in the past. SE involves the selective separation and extraction of specific radionuclides from a mixture using organic solvents. It was a commonly employed technique due to its ability to efficiently extract and purify radionuclides. However, SE methods gradually gave way to resin-based approaches for several reasons. One significant factor was the introduction of new chromatographic resins that offered improved selectivity and efficiency in radionuclide separation. These resins allowed for better control over the purification process and reduced the risk of contamination. Additionally, resin-based approaches provided greater convenience and ease of use, as they often involved prepacked columns or cartridges that could be readily utilized in automated systems. In contrast, SE methods often require manual handling and are more labor-intensive, making them less attractive for modern production facilities seeking high throughput and precision. Therefore, mainly due to its perceived difficulty of automation and poor reliability with continued use, SE has been mostly replaced with resin-based approaches that still dominate the field of radionuclide production due to their overall advantages in terms of automation, scalability, safety, and performance.

Recently though, the effort spent in optimizing SE-based automatic systems at both the macro and micro scales has solved practical issues of SE and has contributed to a rebirth of SE applied to the radioisotope/radiopharmaceutical production field. In particular, the use of new flow chemistry methodologies and microfluidic systems is contributing to an increase in the reliability and efficiency of SE processes in this field (Martini et al., 2019).

Tc-99m is a prime example of the revolutionary breakthrough that occurred in the radiometal production field in the last decade. Indeed, while generator production is currently the most common method of obtaining Tc-99m, it relies on the availability of Mo-99 (parent nuclide) produced in nuclear reactors. The generator supply chain has been demonstrated to be vulnerable to disruptions, such as reactor maintenance or unexpected shutdowns, as occurred in 2009 and 2010 when the

Chalk River reactor in Canada, one of the major global producers of Mo-99, experienced an unplanned shutdown. This prompted research focus on cyclotron production of Tc-99m and on related technological aspects, including target development, solid target station and dissolution modules, automation of radiochemical separation, and target recovery (Boschi et al., 2017). Cyclotron production, on the other hand, offers independence from external supply, high-specific activity, and reduced radionuclidic impurities but requires specialized infrastructure and may have limited distribution. A Tc-99m cyclotron-based production cycle was reported by a couple of research groups that focused on trying to offer the community a reliable alternative route to the traditional reactor-driven one (Gagnon et al., 2012; Martini et al., 2018; Schaffer et al., 2015; Skliarova et al., 2019).

The extraction and purification of Tc-99m from the irradiated metal target is a key step in the production cycle, as it is necessary to make the radioisotope suitable for further radiopharmaceutical processing and patient injection. This process must be simple, reproducible, and efficient in order to rapidly supply a radioisotope with high purity. Automation of this process is essential to ensure these requirements and to minimize operators' radiation exposure (Gumiela, 2018).

Several techniques are known for the separation of technetium from molybdenum; among them, the SE method with methyl ethyl ketone (MEK) proved to be one of the most efficient techniques, thanks to the selective affinity of the pertechnetate ($[^{99m}Tc]TcO_4^-$) for MEK and the affinity of the binegative molybdate ($[^{99}Mo]MoO_4^{2-}$) for the aqueous phase (Boschi et al., 2017; Chattopadhyay et al., 2010; Dallali et al., 2007; Noronha, 1986; Skuridin & Chibisov, 2010; Tachimori et al., 1971; Taskaev, 1995; Zykov et al., 2001). Due to its complex nature, variable sample properties, sensitivity to parameters, equipment compatibility, need for operator expertise, and safety considerations, the automation of solvent extraction is particularly challenging but has recently been achieved at both the macro and micro scales, as described in the following sections.

2 Automation of SE at the Macro Scale

The first researcher to tackle solvent extraction automation in the radiometals production field for medical applications was Chattopadhyay et al. (2010, 2012), who, in 2010, created a homemade automatic system principally designed for the decay-extraction process of Tc-99m from low to medium-specific activity Mo-99. SE is the only technique applicable in case of cyclotron-indirect Tc-99m production precisely because of the low-specific activity of the produced Mo-99. Indeed, the indirect production involves, for example, the irradiation of enriched Mo-98 target material to obtain Mo-99; however, Mo-99 is not chemically separable from the target material. This implies that if an alumina column is used to separate Tc-99m from Mo-99, its specific capacity must be higher than that of the columns typically used in the $^{99}Mo/^{99m}Tc$ generator system, and consequently, elution volumes would also increase with a reduction in activity concentration (MBq/mL). The automatic system proposed by Chattopadhyay has evolved through the years into a more sophisticated and computerized one, possibly applicable also to cyclotron-produced Tc-99m/Mo-100 extraction with some initial treatment modifications (Chattopadhyay et al., 2019).

The chemical process unit described by Chattopadhyay is composed mainly of two major components confined in a lead shield: (1) an extraction vessel, to extract Tc-99m into MEK from the ^{99}Mo-rich aqueous solution; (2) a purification system composed of two small alumina columns (neutral or basic and acidic), connected in tandem with the extraction vessel, to remove Mo-99 trace impurities and MEK.

Similarly, in the framework of the LARAMED project of the Legnaro National Laboratories of the INFN, an automatic module based on the solvent extraction technique was developed for the extraction, separation, and purification of cyclotron-produced Tc-99m from the molybdenum metal target enriched in Mo-100 (Esposito et al., 2018; Martini et al., 2016, 2018).

This system, shown in Fig. 1, exploits a handmade glass column vial (length = 10 cm, diameter = 1 cm, hereafter called the "separation column") specifically designed to host the biphasic system, composed of an alkaline aqueous solution (coming from the dissolution of the irradiated target in hydrogen peroxide and sodium hydroxide containing Tc, Mo, and contaminants) and an organic phase of MEK (Martini et al., 2016). Briefly, the organic solvent is injected slowly from the bottom into the separation column, where the alkaline solution is collected. Helium gas is bubbled through the mixture to enhance contact between the two phases. After a rest period, a clear phase separation occurs, with the organic solution (containing [^{99m}Tc]pertechnetate) positioned above the aqueous phase. The organic phase is collected from the separation column by applying pressure with a tube placed on the top of the column and using a second dip tube placed above the interphase. Next, the organic solution, containing [^{99m}Tc]pertechnetate, undergoes purification by passing through silica and alumina columns. This process helps remove residual molybdenum and organic solvent, respectively, ensuring the purification of Tc. To maximize the extraction yield, fresh MEK can be further added to the aqueous phase still present at the bottom of the separation column, and the extraction and purification procedure repeated. The whole procedure lasts approximately 70 min, including the target dissolution step (Martini et al., 2016). This first prototype of automatic module was optimized for preliminary research purposes to deal with a thin foil target configuration (about 30 mg).

Later, this system was modified and improved, enabling it to deal with molybdenum targets up to 500 mg, produced and bonded to Cu + Au backing materials using an SPS (Spark Plasma Sintering) manufacturing technique. The target was designed for the hospital cyclotron's solid target station and Ci-level (37–100 GBq) production scale. Compared to the first prototype, the new automatic module was improved by substituting some components, such as a longer handmade glass column vial (length = 15 cm, diameter = 1 cm) and a homemade reactor heater with an open-bottom vial fitting the target design, in order to also integrate the process of dissolving the target material (Martini et al., 2018).

Fig. 1 Automation of solvent extraction through helium bubbling in a column vial

The total time to complete the automated procedure was approximately 60 min, including the dissolution step, and the recovery yield was about 93 ± 3% decay corrected. The purity values of the final product in the saline solution were in agreement with the values imposed by European Pharmacopoeia for accelerator-produced pertechnetate. As shown in the graph (Fig. 2), impurities of Mo or Nb can be efficiently separated from the Tc isotopes through automated solvent extraction and the purification procedure developed. Indeed, Mo and Nb contaminants remain in the aqueous fraction (>99%), while about 93% of Tc isotopes are well isolated in the final physiological solution. A small percentage (<5%) of Tc is intentionally left on

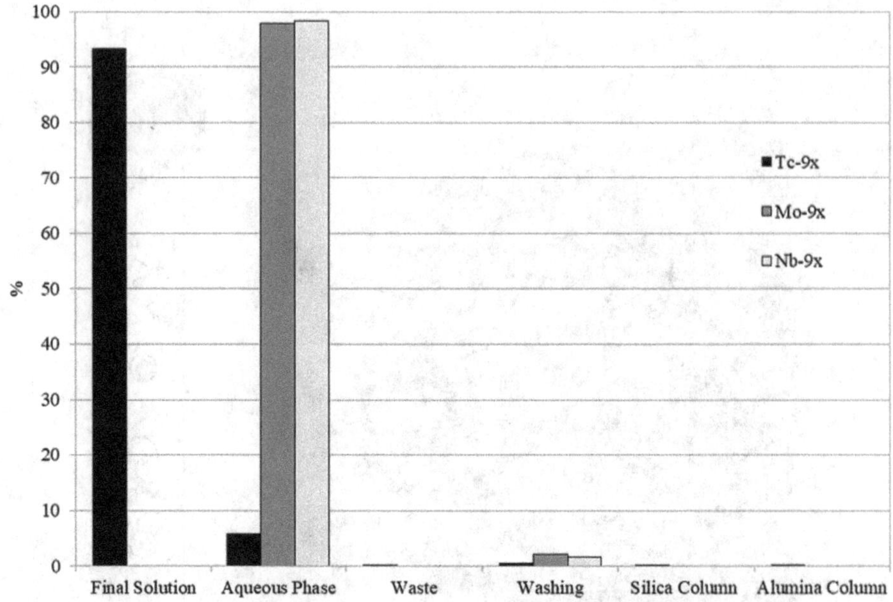

Fig. 2 Distribution of radionuclides in the different components of the automatic procedure

top of the aqueous phase in the separation column to avoid the withdrawal of Mn and Nb contaminants.

Compared with other chromatographic resin-based separation protocols and modules for cyclotron-produced Tc-99m, this solvent extraction-based system was the method with the highest recovery yield of Tc-99m. Automation of the solvent extraction process was a key development enabling usefulness of SE for radiometal separations (Boschi et al., 2017; Gumiela, 2018).

3 Automation of SE at the Micro Scale

Recently, a step forward in the automation of solvent extraction was achieved by performing the extraction and separation process of technetium from molybdenum in a micro-scale flow chemistry regime with the aid of a capillary loop and a membrane-based liquid–liquid separator (SEP-10, ZAIPUT Flow Technologies, Cambridge, MA, USA). This device was recently used for the radiochemical separation of radioisotopes of nuclear medical interest (Adamo et al., 2013; Ban et al., 2011; Hellé et al., 2012; Pedersen et al., 2018).

The extraction step, in which the two phases are put into intimate contact, allowing for solute transfer, takes place inside a capillary according to flow chemistry (Martini et al., 2019, 2021). The capillary allows the two coinjected phases to form an alternating series of phases technically called the slug-flow regime, a typical micro-arrangement of phases optimal for liquid–liquid extraction, as it allows the surface/volume ratio of the phases to be maximized. Capillary dimensions and flow rate were selected to maximize the contact between the two phases and therefore also the extraction yield. The downstream membrane-based separator device allows the two immiscible phases to be effectively separated by exploiting the interfacial tension and the affinity of one of the two phases for a microporous hydrophilic or hydrophobic membrane (PTFE).

The system was assembled, as shown in Fig. 3 (Martini et al., 2021). The liquid transfer is performed by means of two peristaltic pumps. In order to assess the efficiency of the extraction and separation system, radioactive tests were performed by adding [^{99m}Tc]TcO$_4$Na, eluted by a Mo-99/Tc-99m generator, to an alkaline Mo solution prepared by dissolving metallic molybdenum in hydrogen peroxide and sodium hydroxide aqueous solution. An equivalent volume of organic solvent (MEK) is coinjected in the separation system through a T-junction. The developed system is able to extract and separate quantitatively (91.0 ± 1.8%) and selectively the [^{99m}Tc]TcO$_4$Na from the molybdenum in 20 min (total process time). The residence time in the loop is approximately 2.5 min. After the phase separation, the organic phase containing the [^{99m}Tc]TcO$_4^-$ was passed through a silica column to remove trace molybdate and through an alumina column, which adsorbs the [^{99m}Tc]TcO$_4^-$, allowing the MEK solvent to flow through. Finally, pertechnetate is eluted from the alumina column with saline solution. This process is carried out in order to obtain a pharmaceutical product compliant with the Pharmacopoeia standards. This system can be efficiently involved in the separation of technetium from both low and high-specific activity molybdenum.

Similarly, an innovative platform adapting a continuous flow extraction and ZAIPUT membrane-based phase separator was used for the SE separation of Ti-45 from a cyclotron irradiated ^{nat}Sc (Pedersen et al., 2018). In this work, the Sc foil target material was dissolved in 12 M HCl and extracted with a 9:1 (v/v) guaiacol/anisole mixture. The optimal conditions resulted in >80% extraction of Ti-45 from

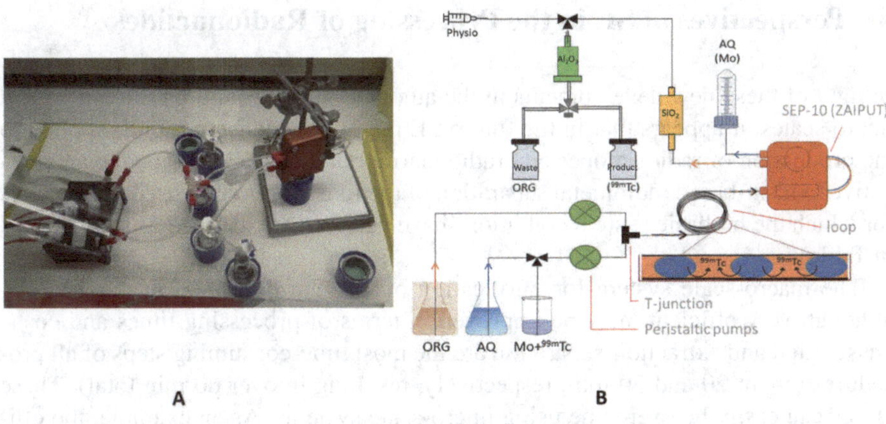

Fig. 3 (**a**) Picture and (**b**) schematic representation of the system

irradiated target material in a single flow SE process. Successful direct radiosynthesis of [^{45}Ti](salan)Ti(dipic) from the extracted Ti-45 was reported.

Recently, the same authors recognized that the cyclotron production of Ga-68 in a liquid target is compatible with flow chemistry-based extraction. The use of liquid targets, despite resulting in lower radionuclide yields compared to solid target production, offers greater automation potential. Pedersen et al. demonstrated the successful application of flow extraction to efficiently separate cyclotron-produced Ga-68 from enriched [^{68}Zn]$ZnCl_2$ salt solutions using a membrane-based liquid–liquid separator with integrated pressure control (Pedersen et al., 2019). For extractions from concentrated solutions of $ZnCl_2$ in 3 M HCl, a 1/2 (v/v) mixture of *i*-Pr_2O/trifluorotoluene (TFT) is preferred, extracting 76.3 ± 1.9% of Ga-68 and 1.9 ± 1.6% of Zn. When 1 M solutions of $ZnCl_2$ in 6 M HCl are used, the system of choice proved to be a 1/2 (v/v) mixture of *n*-BuOMe/TFT, extracting 95.7 ± 2.0% of Ga-68 and coextracting only 0.005 ± 0.003% of Zn. TFT plays a key role in controlling the interfacial tension between the aqueous and the organic phases and their immiscibility, which is critical for a successful operation of membrane-based separation of the liquid phases in flow.

Moreover, by coupling a liquid target-based radionuclide production step with a solvent extraction step, it is possible to construct a closed-loop system for continuous manufacturing of Ga-68 with continuously recycled isotopically enriched Zn-68. The continuous Ga-68 production process offers potential cost savings, reducing batch costs by $200–$700. It allows for flexible batch scaling and on-demand production. However, there is a limit to recirculation of Zn-68 due to yield loss and impurity accumulation. Currently, cyclotron-produced Ga-68 is produced in batch mode and purified from the Zn-68 salt solution by solid-phase extraction using commercial radiosynthesis modules, without directly recycling the Zn-68. The recycling of isotopically enriched Zn-68 target, if desired, needs to be performed in a separate step.

4 Perspectives of SE in the Processing of Radionuclides

In light of these new developments in the automation of SE at both the macro and micro scales, it appears that in the future SE might experience a rebirth applied to the production of radioisotopes and radiopharmaceuticals. SE can now be an alternative to resin-based radiometal separation of the most commonly used radiometals for which the biphasic system conditions have been already defined as summarized in Table 1.

The macro-scale system for purification of Tc-99m described in this chapter, although very efficient, may be improved in terms of processing times and costs. Dissolution and extraction-separation are the most time-consuming steps of all procedures (about 20 and 30 min, respectively, resulting in over 60 min total). These issues can easily be overcome using micro-scale systems. As an example, the efficiency of the MEK-based solvent extraction process of Tc-99m from a

Table 1 Examples of radioisotopes of medical interest purified via SE methods

Radiometal extracted	Originating mixture	Solvent extractor	No. of SE repetitions	Notes	Reference
Cu-67	Zn in HCl 0.5 M	0.01% dithizone in CCl_4	4	Back-extraction in 7.2 M HCl/H_2O_2 + purification from Ga and metal contaminants via isopropyl ether SE + anion exchange column	Dasgupta et al. (1991), Kastleiner et al. (1999)
		8-Mercaptoquinoline in toluene	–		Malinin et al. (1970)
		Thenoyl-trifluoroacetone in benzene		~90% extraction; back-extraction with 0.5 M HCl pre-equilibrated with benzene	Brown and Callahan (1972)
		1% solution of resacetophenone oxime (RAPOX) in cyclohexane	–		Bhatki et al. (1969)
Sc-47	Ti in 1.0 N HCl/0.3% H_2O_2	Cupferron in pure chloroform	5	Extraction by Ti(IV) complexation	Pietrelli et al. (1992)
		Tri-*n*-butyl phosphate (TBP)	Multiple repetition	Alternative: Impregnating TBP into a SiO_2 support and use column chromatography	Kolsky et al. (1998)
Zr-89	Y_2O_3 in HCl 9 M	3% (w/v) Triphenylphosphine oxide (TPPO) in chloroform	–	Zirconium oxalate back-extraction in 0.5% oxalic acid	Kandil et al. (2007)
Mn-52	Cr in HCl 8 M	0.8 M trioctylamine (TOA) in cyclohexane	3	Back-extraction in 5 mM NH_4OH	Lahiri et al. (2006), Lewis et al. (2015)
Th-227/Ac-225	Th-227/Ac-225 in HNO_3 0.001–0.01 M	Ionic liquid/di-2-ethylhexylphosphoric acid (HDEHP)	–	Separation factor $>10^7$ at 0.01 M HNO_3	Luo et al. (2012)

molybdenum-based liquid phased in a continuous flow micro-scale regime, using a compact ZAIPUT separator device, was demonstrated for the first time. The system allows to halve the time for the extraction and separation process. It can be used to purify technetium from molybdenum metal targets both in direct cyclotron Tc-99m production and in indirect Tc-99m production such as through the production of Mo-99 by irradiating natural molybdenum using a 14-MeV accelerator-driven neutron source (Capogni et al., 2018). Therefore, this system also helps to overcome the main issue of macro SE systems, namely their large size and space requirements.

Acknowledgments Work on Mo/Tc SE automation described in this chapter was performed in the framework of the LARAMED (LAboratory of RAdioisotopes for MEDicine) research program running at LNL-INFN.

References

Adamo, A., et al. (2013). Membrane-based, liquid–liquid separator with integrated pressure control. *Industrial & Engineering Chemistry Research, 52*(31), 10802–10808. https://doi.org/10.1021/ie401180t

Ban, Y., et al. (2011). Extraction of Am(III) at the interface of organic-aqueous two-layer flow in a microchannel. *Journal of Nuclear Science and Technology, 48*(10), 1313–1318. https://doi.org/10.1080/18811248.2011.9711821

Bhatki, K. S., Rane, A. T., & Kabadi, M. B. (1969). Preparation of carrier-free copper-64,67 nuclides by liquid-liquid extraction. *Journal of Radioanalytical Chemistry, 2*(1), 73–77. https://doi.org/10.1007/BF02518674

Boschi, A., et al. (2017). Recent achievements in Tc-99m radiopharmaceutical direct production by medical cyclotrons. *Drug Development and Industrial Pharmacy, 43*(9), 1402–1412. https://doi.org/10.1080/03639045.2017.1323911

Brown, L. C., & Callahan, A. P. (1972). The separation and purification of carrier-free copper isotopes for medical use. *The International Journal of Applied Radiation and Isotopes, 23*(11), 535–538. https://doi.org/10.1016/0020-708X(72)90136-6

Capogni, M., et al. (2018). 14 MeV neutrons for ^{99}Mo/^{99m}Tc production: experiments, simulations and perspectives. *Molecules, 23*(8), 1872. https://doi.org/10.3390/molecules23081872

Chattopadhyay, S., Das, S. S., & Barua, L. (2010). A simple and rapid technique for recovery of (99m)Tc from low specific activity (n,gamma)(99)Mo based on solvent extraction and column chromatography. *Applied Radiation and Isotopes: Including Data, Instrumentation and Methods for Use in Agriculture, Industry and Medicine, 68*(1), 1–4. https://doi.org/10.1016/j.apradiso.2009.07.024

Chattopadhyay, S., et al. (2012). A computerized compact module for separation of ^{99m}Tc-radionuclide from molybdenum. *Applied Radiation and Isotopes, 70*(11), 2631–2637. https://doi.org/10.1016/j.apradiso.2012.07.013

Chattopadhyay, S., et al. (2019). A compact solvent extraction based ^{99}Mo/^{99m}Tc generator for hospital radiopharmacy. *Applied Radiation and Isotopes, 143*, 41–46. https://doi.org/10.1016/j.apradiso.2018.10.017

Dallali, N., et al. (2007). Liquid-liquid extraction of ultra trace amounts of technetium produced by ^{100}Mo (p, 2n) ^{99m}Tc nuclear reaction in cyclotron. *Indian Journal of Chemistry—Section A Inorganic, Physical, Theoretical and Analytical Chemistry, 46*, 4.

Dasgupta, A. K., Mausner, L. F., & Srivastava, S. C. (1991). A new separation procedure for ^{67}Cu from proton irradiated Zn. *International Journal of Radiation Applications and Instrumentation. Part A. Applied Radiation and Isotopes, 42*(4), 371–376. https://doi.org/10.1016/0883-2889(91)90140-V

Esposito, J., et al. (2018). LARAMED: A laboratory for radioisotopes of medical interest. *Molecules, 24*(1), 20. https://doi.org/10.3390/molecules24010020

Gagnon, K., et al. (2012). Cyclotron production of ^{99m}Tc: Recycling of enriched ^{100}Mo metal targets. *Applied Radiation and Isotopes, 70*(8), 1685–1690. https://doi.org/10.1016/j.apradiso.2012.04.016

Gumiela, M. (2018). Cyclotron production of ^{99m}Tc: Comparison of known separation technologies for isolation of ^{99m}Tc from molybdenum targets. *Nuclear Medicine and Biology, 58*, 33–41. https://doi.org/10.1016/j.nucmedbio.2017.11.001

Hellé, G., Mariet, C., & Cote, G. (2012). Microfluidic tools for the liquid-liquid extraction of radionuclides in analytical procedures. *Procedia Chemistry, 7*, 679–684. https://doi.org/10.1016/j.proche.2012.10.103

Kandil, S. A., et al. (2007). A comparative study on the separation of radiozirconium via ion-exchange and solvent extraction techniques, with particular reference to the production of ^{88}Zr and ^{89}Zr in proton induced reactions on yttrium. *Journal of Radioanalytical and Nuclear Chemistry, 274*(1), 45–52. https://doi.org/10.1007/s10967-006-6892-2

Kastleiner, S., Coenen, H. H., & Qaim, S. M. (1999). Possibility of production of ^{67}Cu at a small-sized cyclotron via the (p,α)-reaction on enriched ^{70}Zn. *Radiochimica Acta, 84*(2), 107–110. https://doi.org/10.1524/ract.1999.84.2.107

Kolsky, K. L., et al. (1998). Radiochemical purification of no-carrier-added scandium-47 for radioimmunotherapy. *Applied Radiation and Isotopes: Including Data, Instrumentation and Methods for Use in Agriculture, Industry and Medicine, 49*(12), 1541–1549. https://doi.org/10.1016/s0969-8043(98)00016-5

Lahiri, S., Nayak, D., & Korschinek, G. (2006). Separation of no-carrier-added ^{52}Mn from bulk chromium: A simulation study for accelerator mass spectrometry measurement of ^{53}Mn. *Analytical Chemistry, 78*(21), 7517–7521. https://doi.org/10.1021/ac0607459

Lewis, C. M., et al. (2015). ^{52}Mn production for PET/MRI tracking of human stem cells expressing divalent metal transporter 1 (DMT1). *Theranostics, 5*(3), 227–239. https://doi.org/10.7150/thno.10185

Luo, H. M., et al. (2012). Facile solvent extraction separation of Th-227 and Ac-225 based on room-temperature ionic liquids. *Radiochimica Acta, 100*(10), 771–778. https://doi.org/10.1524/ract.2012.1950

Malinin, A. B., et al. (1970). Formation of ^{67}Cu according to the reaction ^{68}Zn(g,p) on a retarded beam of an electron accelerator with an energy of 25 MeV. *Soviet Radiochemistry, 12*, 748–751.

Martini, P., et al. (2016). A solvent-extraction module for cyclotron production of high-purity technetium-99m. *Applied Radiation and Isotopes, 118*, 302–307. https://doi.org/10.1016/j.apradiso.2016.10.002

Martini, P., et al. (2018). In-house cyclotron production of high-purity Tc-99m and Tc-99m radiopharmaceuticals. *Applied Radiation and Isotopes, 139*, 325–331. https://doi.org/10.1016/j.apradiso.2018.05.033

Martini, P., et al. (2019). Perspectives on the use of liquid extraction for radioisotope purification. *Molecules, 24*(2), 334. https://doi.org/10.3390/molecules24020334

Martini, P., et al. (2021). Highly efficient micro-scale liquid-liquid in-flow extraction of ^{99m}Tc from molybdenum. *Molecules, 26*(18), 5699. https://doi.org/10.3390/molecules26185699

Noronha, O. P. D. (1986). Solvent extraction technology of ^{99}Mo–^{99m}Tc generator system 1. An Indian experience: Process design considerations. *Isotopenpraxis Isotopes in Environmental and Health Studies, 22*(2), 53–57. https://doi.org/10.1080/10256018608623597

Pedersen, K. S., et al. (2018). Liquid–liquid extraction in flow of the radioisotope titanium-45 for positron emission tomography applications. *Reaction Chemistry & Engineering, 3*(6), 898–904. https://doi.org/10.1039/C8RE00175H

Pedersen, K. S., et al. (2019). Separation of radiogallium from zinc using membrane-based liquid-liquid extraction in flow: Experimental and COSMO-RS studies. *Solvent Extraction and Ion Exchange, 37*(5), 376–391. https://doi.org/10.1080/07366299.2019.1646982

Pietrelli, L., Mausner, L. F., & Kolsky, K. L. (1992). Separation of carrier-free ^{47}Sc from titanium targets. *Journal of Radioanalytical and Nuclear Chemistry, 157*(2), 335–345. https://doi.org/10.1007/BF02047448

Schaffer, P., Bénard, F., Bernstein, A., Buckley, K., Celler, A., Cockburn, N., Corsaut, J., Dodd, M., Economou, C., Eriksson, T., Frontera, M., Hanemaayer, V., Hook, B., Klug, J., Kovacs, M., Prato, F. S., et al. (2015). Direct production of ^{99m}Tc via ^{100}Mo(p,2n) on small medical cyclotrons. *Physics Procedia, 66*, 383–395. https://doi.org/10.1016/j.phpro.2015.05.048

Skliarova, H., et al. (2019). Recovery of molybdenum precursor material in the cyclotron-based technetium-99m production cycle. *Instruments, 3*(1), 17. https://doi.org/10.3390/instruments3010017

Skuridin, V. S., & Chibisov, E. V. (2010). Development of a small-size extractor for separation of the ^{99}Mo/^{99m}Tc couple. *Radiochemistry, 52*(1), 90–94. https://doi.org/10.1134/S1066362210010200

Tachimori, S., Amano, H., & Nakamura, H. (1971). Preparation of Tc-99m by direct adsorption from organic solution. *Journal of Nuclear Science and Technology, 8*(7), 357–362. https://doi.org/10.1080/18811248.1971.9735346

Taskaev, E. (1995). Extraction generator for [^{99m}Tc]sodium pertechnetate production. *Acta Chimica Slovenica* (February 1983), p. 4.

Zykov, M. P., et al. (2001). Use of extraction generator for preparing a ^{99m}Tc radiopharmaceutical. *Radiochemistry, 43*(3), 297–300.

Recent Advances in Liquid–Liquid Extraction in Flow of Radiometals for Positron Emission Tomography

Fedor Zhuravlev

1 Introduction

1.1 PET Radiometals: Decay Properties, Production Methods, and Implications for LLE-Based Purification

Over the past decades, positron emission tomography (PET) has become an essential research and clinical imaging modality (Matthews et al., 2012). Improved clinical outcomes for the patients translated into the growth of the global PET market, which is expected to reach ~2 billion USD by 2025 (Goethals & Zimmermann, 2020). Recent years have seen a dramatic increase in the use of PET radiopharmaceuticals based on radiometals (Brasse & Nonat, 2015). The key to the broad adoption of radiometals is their ability to probe a wide range of biological processes occurring on timescales from hours to days. Radiometals also lend themselves to a synergetic use of imaging-based diagnostics and radiotherapy (theranostics). Lastly, cold kit formulation allows for the simple preparation of radiopharmaceuticals by radiometal chelation, often performed as a last stage of radiosynthesis with little or no post-labeling purification (Lepareur, 2022). Table 1 lists clinically established as well as emerging PET radionuclides, with half-lives ranging from 1.3 min to 5.7 days. Among the listed PET radiometals, ^{64}Cu, ^{89}Zr, and, in particular, ^{68}Ga are experiencing growing preclinical and clinical use (Fig. 1). Therefore, new methods for the production, recovery, and processing of PET radiometals are attracting strong interest.

Most of the PET radiometals listed in Table 1 can be produced using medical cyclotrons. The radionuclides, however, pose a significant limitation regarding the

F. Zhuravlev (✉)
Ion Beam Application, Louvain-la-Neuve, Belgium
e-mail: Fedor.Zhuravlev@iba-group.com

R. M. van Dam et al. (eds.), *Automated Technologies for the Development and Production of Radiopharmaceuticals*,
https://doi.org/10.1007/978-3-031-84632-8_5

Table 1 Selected clinically established and emerging PET radionuclides

Entry	Radionuclide	$t_{1/2}$	β^+ (ratio, E_β)	Production	References
1	^{82}Rb	1.3 min	96%, 3.35 MeV	$^{82}Sr \rightarrow {}^{82}Rb$ (generator)	Brasse and Nonat (2015)
2	^{62}Cu	9.7 min	98%, 2.9 MeV	$^{62}Zn \rightarrow {}^{62}Cu$ (generator)	Brasse and Nonat (2015)
3	^{63}Zn	38.5 min	93%, 0.99 MeV	$^{63}Cu(p,n)^{63}Zn$ (cyclotron)	DeGrado et al. (2014)
4	^{68}Ga	67.7 min	89%, 1.9 MeV	$^{68}Ge \rightarrow {}^{68}Ga$ (generator) $^{68}Zn(p,n)^{68}Ga$ (cyclotron)	Brasse and Nonat (2015), Goethals and Zimmermann (2020)
5	^{45}Ti	3.09 h	85%, 1.04 Mev	$^{45}Sc(p,n)^{45}Ti$ (cyclotron)	Ramamoorthy (2020)
6	^{61}Cu	3.3 h	62%, 1.22 Mev	$^{61}Ni(p,n)^{61}Cu$ (cyclotron)	Fonseca et al. (2021)
7	^{44}Sc	3.93 h	94%, 632 keV	$^{44}Ti \rightarrow {}^{44}Sc$ (generator) $^{44}Ca(p,n)^{44}Sc$ (cyclotron)	Hernandez et al. (2014)
8	^{66}Ga	9.3 h	57%, 1.75 MeV	$^{nat}Zn(p,n)^{66}Ga$ (cyclotron) $^{66}Zn(p,n)^{66}Ga$ (cyclotron)	Aluicio-Sarduy et al. (2018)
9	^{64}Cu	12.7 h	19%, 653 keV	$^{64}Ni(p,n)^{64}Cu$ (cyclotron)	Aluicio-Sarduy et al. (2018)
10	^{90}Nb	14.6 h	53%, 0.66 MeV	$^{90}Zr(p,n)^{90}Nb$ (cyclotron)	Seyedi et al. (2021)
11	^{86}Y	14.7 h	33%, 600 keV	$^{86}Sr (p,n)^{86}Y$ (cyclotron)	Oehlke et al. (2015)
12	^{55}Co	17.5 h	76%, 570 keV	$^{58}Ni(p,\alpha)^{55}Co$ (cyclotron) $^{54}Fe(d,n)^{55}Co$ (cyclotron)	Aluicio-Sarduy et al. (2018)
13	^{152}Tb	17.5 h	20%, 1.14 MeV	p-induced spallation of Ta target	Aluicio-Sarduy et al. (2018)
14	^{69}Ge	39 h	21%, 1.2 MeV	$^{69}Ga(p,n)^{69}Ge$ (cyclotron)	Chakravarty et al. (2014)
15	^{89}Zr	78.4 h	23%, 900 keV	$^{89}Y(p,n)^{89}Zr$ (cyclotron)	Heskamp et al. (2017)
16	^{52}Mn	5.7 days	29%, 242 keV	$^{52}Cr(p,n)^{52}Mn$ (cyclotron)	Aluicio-Sarduy et al. (2018), Seyedi et al. (2021)

handling of radioactive material. As radiation protection requirements become more stringent, the field is gearing toward complete automation of radiopharmaceutical production, from targetry to formulation. Proton bombardment of solid targets is

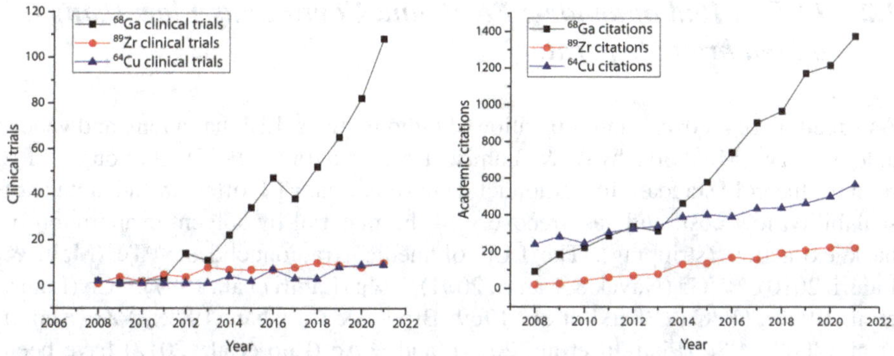

Fig. 1 The number of clinical trials (left, clinicaltrials.gov) and publications (right, Web of Science). The search string used was: "metal-ZZ OR metal ZZ OR ME-ZZ OR ZZME OR metalZZ OR MEZZ" (metal: gallium, copper, zirconium; ME: Ga, Cu, Zr; ZZ: 68, 89, and 64)

currently the most widely used method to produce PET radiometals. The solid target can be prepared as a foil, a pressed or sintered pellet, or as an electroplated metal layer of various thicknesses. The main advantage of using the solid targets is a high yield of the desired PET radiometal, which can, as in the case of ^{68}Ga, give multi-curie (37–194 GBq) amounts of the radionuclide (Thisgaard et al., 2021). In its automated implementation, solid targetry requires expensive equipment for target preparation, safe transportation, and post-processing. Importantly, the dissolution of the irradiated solid target material is always needed before any SPE- or LLE-based separation of the desired PET radiometal can be performed. The dissolution can be time-consuming, often requiring heating and generating toxic and corrosive acidic fumes. Recently, the production of PET radiometals in liquid targets has gained significant traction in clinics (Alves et al., 2018; do Carmo et al., 2020; Pandey et al., 2014; Pandey & DeGrado, 2021; Rodnick et al., 2020). The liquid targets normally produce less activity than the solid target. However, they do not require infrastructure investment and can use standard ^{18}F liquid targets, although specially designed liquid target models are also available from the cyclotron manufacturers. The cyclotron target solution is normally prepared and loaded as a solution of a nitrate salt in nitric acid. After irradiation, this solution can be sent for SPE or LLE purification without preprocessing. Therefore, the liquid target system is inherently compatible with LLE, including its LLE in flow (LLEF) implementation. Liquid targets are well suited for the production of PET radiometals with short and medium half-lives (Table 1, entries (3–7)) and are amenable to automation and process control (vide infra).

1.2 LLE of Radionuclides: Batch and Continuous Flow Using a Membrane Separator

As a method of recovery and separation of radionuclides, LLE has a long and venerable history (Mukhopadhyay & Lahiri, 1999; Navratil, 1986). Although SPE remains the gold standard for radionuclide purification, LLE offers the advantage of scalability, low cost, and easy recovery of the material by solvent evaporation or back extraction (stripping). The LLE of medical radionuclides ^{99m}Tc (Maiti & Lahiri, 2010), $^{66/67}$Ga (Nayak & Lahiri, 2001), ^{52}Mn (Lahiri et al., 1997), ^{55}Co (Lahiri et al., 1997), $^{64/67}$Cu (Bhatki et al., 1969; Brown & Callahan, 1972), ^{89}Zr (Kandil et al., 2008), ^{47}Sc (Pietrelli et al., 2005), and ^{225}Ac (Luo et al., 2012) have been reported and the field remains an active area of research (Martini et al., 2019). The early work used manual batch extraction, often with the help of a separatory funnel for phase separation (Brown & Callahan, 1972). While SPE methods benefit from straightforward integration into existing automated radiosynthesis modules, the difficulties of implementing the phase separation step of LLE into an automated workflow of radiopharmaceutical production stymied LLE adoption. The rise of continuous flow production of pharmaceuticals stimulated the development of new LLE-based purification techniques and their integration into a compact, modular, and automated end-to-end pharmaceutical production system (Bédard et al., 2018). In a general fluidic LLE process, aqueous and organic phases are mixed in a specially designed channel network, allowing substances to transfer from one phase to the other during the flow. At the end of this process, the heterogeneous mixture is separated back into distinct aqueous and organic phases (Martini et al., 2019).

In our lab, the membrane-based liquid–liquid separator developed for such a system by the researchers at MIT (Adamo et al., 2013) and later commercialized by Zaiput Flow Technologies has shown good performance in separating a highly acidic aqueous phase from an organic phase containing radioisotope titanium-45 (vide infra) (Pedersen et al., 2018; Søborg Pedersen et al., 2020). In the membrane separator, the organic phase permeated the PTFE membrane and passed through the permeate outlet. The aqueous phase was retained and passed through the retentate outlet (Fig. 2). The integrated PFA diaphragm provides backpressure over the retentate stream, ensuring clean phase separation.

We also anticipated that the membrane separator would serve as an effective LLE system. During operation, the two immiscible liquids within the separator form a particular flow pattern, where the dispersed phase forms droplets, also known as "slugs" (Fig. 2). The dimensions of slugs are subject to considerable variation, influenced by the specific conditions of the fluid environment. Slug flow leads to an efficient interfacial mass transfer to the extracting phase due to a constant exchange of the interface of each slug. In all experiments described below, the two phases were premixed by two 10-element PTFE static mixers and further mixed in a 108 cm long PFA tubing (1/16" OD, 0.03" ID), resulting in a steady slug flow at a total flow rate of 0.2–5 mL/min (Pedersen et al., 2019).

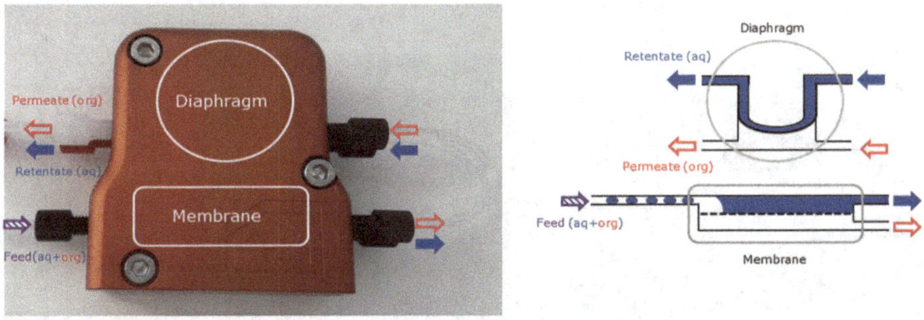

Fig. 2 A photograph (left) and a schematic (right) showing the commercially available SEP-10 (Zaiput) membrane separator with integrated pressure control. The aqueous phase (retentate) is shown in blue and the organic phase (permeate) is shown in white

2 LLEF of ^{45}Ti, ^{89}Zr, ^{64}Cu, and $^{66/67/68}$Ga

2.1 Titanium-45

Titanium-45, which can be produced in high yield by the proton bombardment of naturally monoisotopic scandium, possesses radioactive decay characteristics highly desirable for biomedical applications (Table 1, entry 5). The favorable imaging properties are offset by the daunting chemical challenges one faces trying to recover the ^{45}Ti from the irradiated Sc matrix. Ti^{4+} is highly oxophilic and will partially hydrolyze if the acid concentration falls below 8 M. (Kislik, 2002). We have previously isolated ^{45}Ti from its 12 M HCl solution by absorbing the radionuclide onto a PEG-functionalized polystyrene-based 1,3 diol and used it to prepare [^{45}Ti](salan)Ti(dipic) for subsequent PET imaging and radiotracing (Severin et al., 2015).

The aggressive nature of 12 M HCl and its partial solubility in many organic solvents made this system an ultimate testing ground for the membrane separator driven LLEF methodology. After irradiation and digestion in 12 M HCl, the Sc matrix solution, containing ^{45}Ti was combined with guaiacol (organic phase), through a tee and mixed with the help of two static mixers and mixing tubing (Fig. 3). Guaiacol, used as a Ti complexant, was able to extract 75% of ^{45}Ti activity and provide a clean phase separation in batch. However, a partial phase breakthrough was observed when running the experiment in flow. The COSMO-RS calculations (COSMOtherm, v.19.0.4, Cosmologic) showed that the addition of anisole could increase the interfacial tension of the system by reducing the contact area between the phases and, consequently, improve phase separation. Indeed, a 9/1 v/v guaiacol/anisole mixture, combined with a 0.2 μm (pore size) PTFE membrane and a 0.002″ (~0.0508 mm) diaphragm, led to complete phase separation (Pedersen et al., 2018).

For the recovery and processing of radionuclides, the total production time is a critical parameter. In this system, excellent extractions were achieved for the mixing tubing residence time as short as 13.7 s. We have also demonstrated excellent

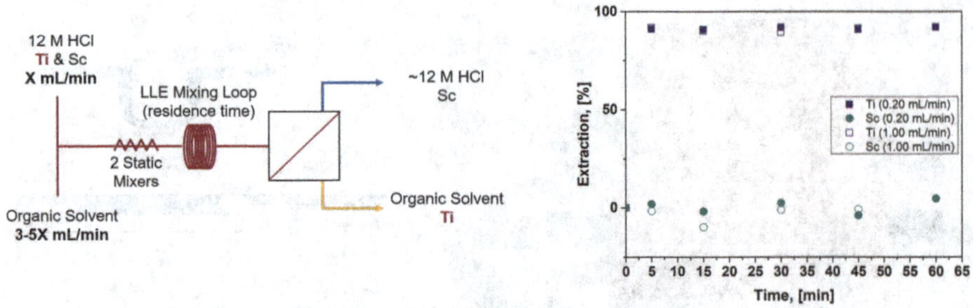

Fig. 3 Schematic of the setup used for continuous phase separation of ^{45}Ti from Sc. Aqueous phase: 12 M HCl, organic phase: guaiacol/anisole 9/1, v/v. Right: Extraction performance over time for a total flow rate of 0.20 mL/min (solid symbols) and for a fivefold scale-up at 1.00 mL/min (open symbols)

scalability and stability of extraction. When the total flow rates were scaled fivefold while keeping the same flow rate ratios and residence times, the maximum extraction of 90% (85% of ^{45}Ti) was still achieved at 9/1 guaiacol/anisole and a flow rate ratio of 1/3. The processing of ~6 mL of aqueous and ~30 mL of organic phase took around 30 min. The putative [^{45}Ti]*cis*-[TiCl$_2$(η^2-guaiacolato)$_2$] forming in the organic phase could be directly used for radiolabeling, obviating the need for solvent removal or stripping. A complete conversion (99%) into the [^{45}Ti](salan)Ti(dipic) within 15 min was observed by reacting the organic phase with salan and dipic (Scheme 1).

We have also used this methodology for the successful radiosynthesis of [^{45}Ti]salan-Ti-CA-PSMA (Søborg Pedersen et al., 2020). According to inductively coupled plasma optical emission spectroscopy (ICP-OES), the final formulation (3 mL) contained less than 8.3 μg of scandium, <5.0 μg of zinc, and <2.0 μg of iron.

Overall, the system provided a fast, stable, and scalable operation under very harsh experimental conditions (Fig. 4).

2.2 *Zirconium-89 and Copper-64*

Zirconium and titanium belong to the same group of the periodic table and share high oxophilicity and hydrolyzability. Zirconium-89 was produced by proton (<13 MeV) irradiation of naturally monoisotopic yttrium foil using a medical cyclotron (Table 1, entry 15). The aqueous phase was prepared by dissolving the Y foil in 12 M HCl. As expected, a guaiacol/anisole 9/1 v/v mixture turned out to be an excellent extractant of ^{89}Zr. Compared to titanium-45 LLE, lower flow rates and higher residence times in the mixing loop were needed to achieve good extraction (Zhuravlev et al., 2021) (Fig. 5, left).

To produce ^{64}Cu, we used the cyclotron irradiation of isotopically enriched ^{64}Ni (Table 1, entry 9). The LLEF of ^{64}Cu from 6 M HCl solution was performed with

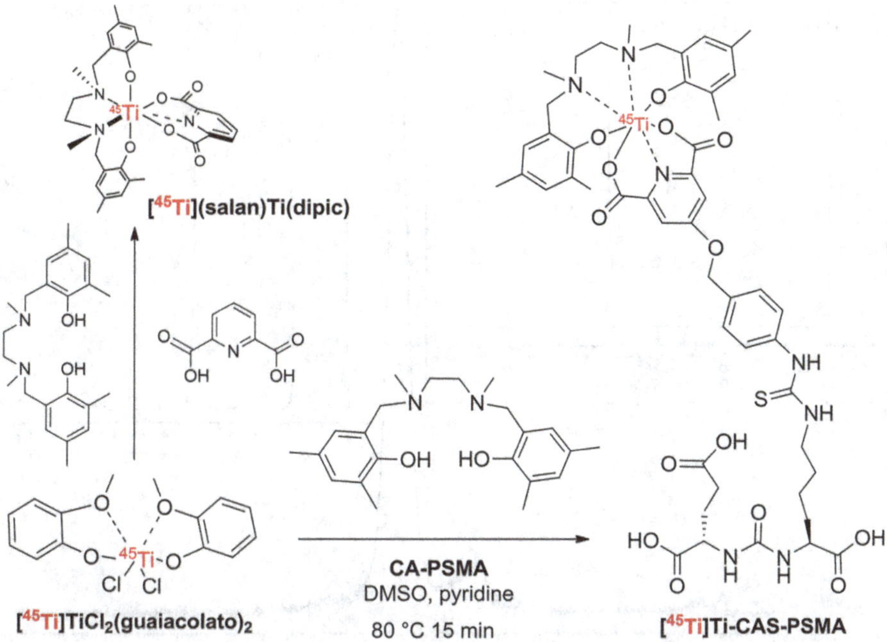

Scheme 1 Radiosynthesis of [^{45}Ti](salan)Ti(dipic) and [^{45}Ti]salan-Ti-CA-PSMA using the organic phase from LLEF of 12 M HCl solution of ^{45}Ti into a guaiacol/anisole mixture

0.1 M trioctylphosphine oxide (TOPO) in heptane. The highest extraction was obtained from 6 M HCl, where the neutral $CuCl_2$ species dominates the speciation diagram. This time, high flow rates gave higher extraction of the ^{64}Cu (Fig. 5, right). However, high selectivity of extraction was achieved with respect to nickel, TOPO co-extracted Co, Ag, Fe, and Zn. Therefore, a more selective extractant would be needed to make LLE suitable for radiopharmaceutical applications.

2.3 *Radiogallium (⁶⁷Ga and ⁶⁸Ga)*

Radiogallium occupies a special place in nuclear medicine. For many years, gallium-67 ($t_{1/2}$ = 78 h, used as ^{67}Ga citrate) scintigraphy and SPECT-CT have been used for imaging of Burkitt's, Hodgkin's, and non-Hodgkin's lymphomas (Bombardieri et al., 2003). Despite its nonspecificity, ^{67}Ga citrate showed excellent performance in detecting infection and inflammation of the lung and skeletal tissue (Vorster et al., 2018). The rise of PET and the subsequent FDA's approval of [^{68}Ga]Ga-DOTA-TATE (Netspot®) made gallium-68 (^{68}Ga, $t_{1/2}$ = 68 min) a radiometal of choice for neuroendocrine tumor diagnostics (Mojtahedi et al., 2014). The clinical success of ^{68}Ga continued with the development of prostate-specific membrane antigen (PSMA) ligands. The [^{68}Ga]Ga-HBED-PSMA-11 and various other

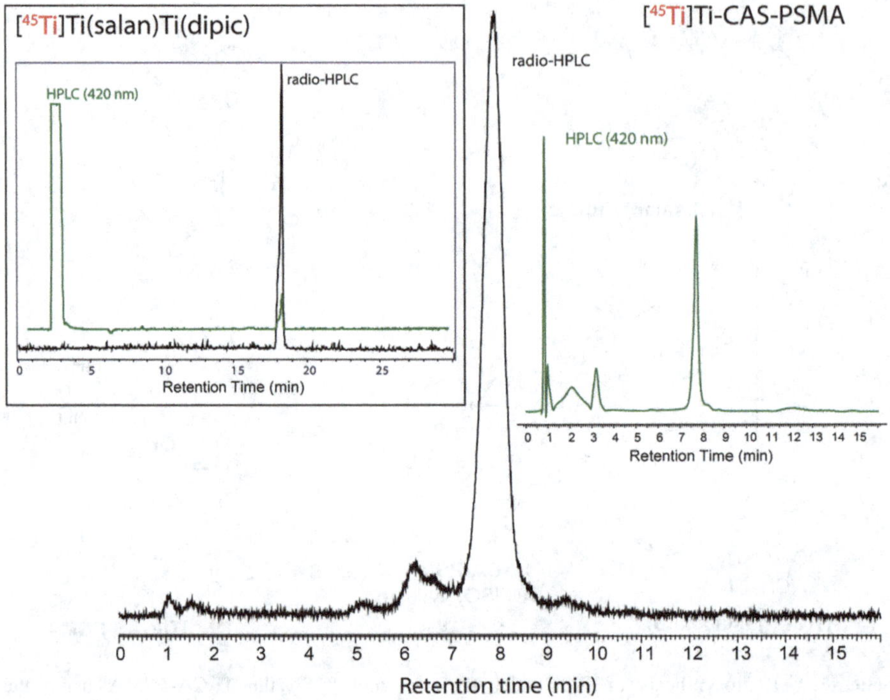

Fig. 4 HPLC (green trace) and radio-HPLC (black trace) of purified and formulated [^{45}Ti](salan) Ti(dipic) (inset) and [^{45}Ti]salan-Ti-CA-PSMA

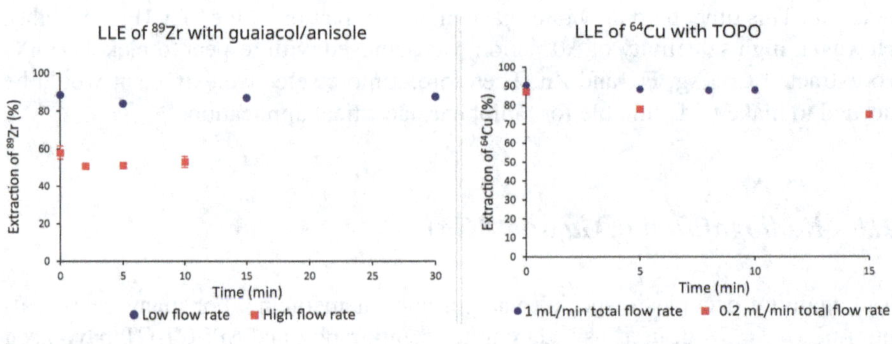

Fig. 5 Left: LLEF of ^{89}Zr in flow. Aqueous phase: 12 M HCl containing dissolved yttrium foil; organic phase: guaiacol/anisole. A flow rate ratio of 1:5 (aq:org) was applied. A total flow rate of 0.2 mL/min and a residence time of 2.5 min (blue), and a total flow rate of 1 mL/min and a residence time of 0.2 min (red) were tested. Right: The extraction efficiency of ^{64}Cu over time from the LLEF from 6 M HCl with 0.1 M TOPO in heptane. A total flow rate of 1 mL/min and a residence time in the mixing loop of 0.2 min (blue circles) and a total flow rate of 0.2 mL/min and residence time of 2.5 min in the mixing loop (red squares) were tested

^{68}Ga-labeled PSMA radiopharmaceuticals emerged as the gold standard for prostate cancer diagnostics (Witkowska-Patena et al., 2017). To meet the growing demand for ^{68}Ga, more clinics are adopting the cyclotron production route, and the corresponding European Pharmacopoeia monograph has been recently published (European Pharmacopoeia 10, 2020). In 2021, the first EU marketing authorization was granted for the distribution of the cyclotron produced gallium-68.

2.3.1 LLEF of $^{67/68}$Ga from Hydrochloric Acid Solutions

2.3.1.1 Cyclotron Production

The first production of ^{68}Ga in liquid targets was reported by Jensen and Clark (2011). Using an open target, they obtained [^{68}Ga]GaCl$_3$ by irradiation of a concentrated (1 g/mL) solution of zinc chloride in water. For our experiments, we adapted a flow target design, which could provide recirculation of the target solution and allowed us to collect and analyze fractions during irradiation (Fig. 6). A solution of 1 M [^{68}Zn]ZnCl$_2$ was prepared in 5.7 M HCl and irradiated at 10–30 μA for 30 min using a Nb liquid target equipped with a 25 μm HAVAR or 150 μm Nb foil to degrade the 16.3 MeV beam down to 13.5 MeV. Despite very aggressive conditions, both the Nb target and Nb front foil showed excellent corrosion resistance. ICP analysis of the liquid target at the end of bombardment (EOB) showed only 1–2 μg/mL of Nb. However, we noticed significant outgassing and chlorine evolution

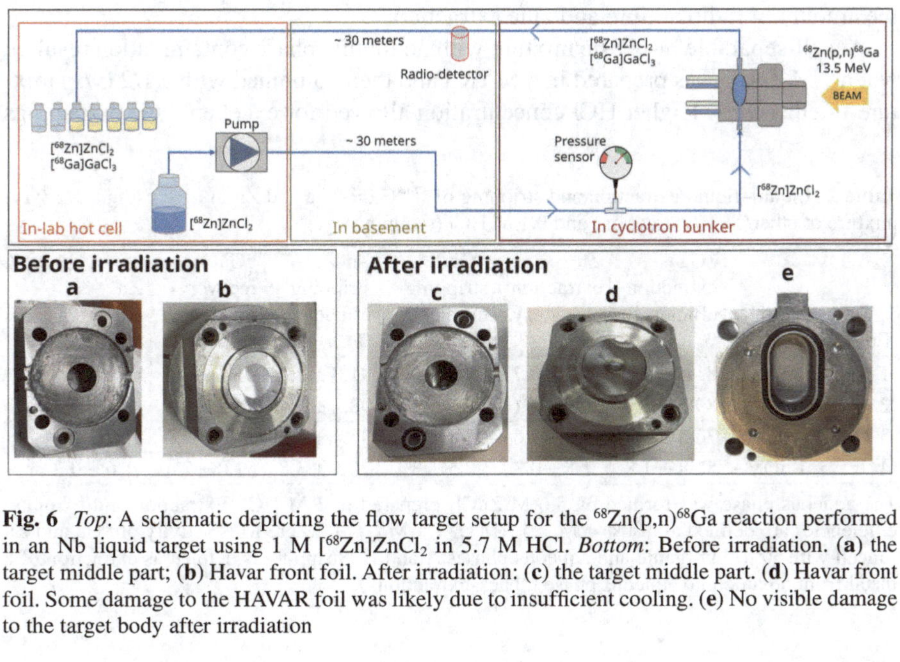

Fig. 6 *Top*: A schematic depicting the flow target setup for the ^{68}Zn(p,n)^{68}Ga reaction performed in an Nb liquid target using 1 M [^{68}Zn]ZnCl$_2$ in 5.7 M HCl. *Bottom*: Before irradiation. (**a**) the target middle part; (**b**) Havar front foil. After irradiation: (**c**) the target middle part. (**d**) Havar front foil. Some damage to the HAVAR foil was likely due to insufficient cooling. (**e**) No visible damage to the target body after irradiation

during irradiation. Nevertheless, up to 3.5 GBq of ^{68}Ga could be produced in a flow target in hydrochloric acid. The irradiation of zinc at the natural abundance level produces a mixture of ^{66}Ga, ^{67}Ga, and ^{68}Ga, which we used for most of the development work described below.

2.3.1.2 Extraction

From an extraction standpoint, hydrochloric acid is an ideal media: very high extraction efficiencies were reported for gallium extraction from 5 to 7 M hydrochloric acid solutions into dialkyl ethers without any need for a special extractant (Swift, 1924). For radiogallium extraction, we adopted the general setup used for LLEF of ^{45}Ti from Sc (Fig. 3). We found that the concentration of zinc chloride played an important role in phase equilibrium and gallium extraction efficiencies. High electrolyte concentrations led to the partial miscibility of the aqueous and organic phases and, consequently, to a phase breakthrough during the membrane separation step. To increase the interfacial tension and aid in phase separation, trifluorotoluene (TFT) was used as a diluent. At 5.6 M $ZnCl_2$, moderate-to-high ^{67}Ga extractions into diethyl ether (Et_2O), tetrahydropyran (THP), diisopropyl ether ($^iPr_2O$), and *n*-butyl methyl ether (*n*-BuOMe) were obtained (Table 2).

In all cases except *n*-BuOMe, more than 70% of radiogallium was extracted into the organic phase within the first 2 min. The *Ga can then be back-extracted (stripped) into the aqueous phase with 0.1 M HCl. The stripping efficiencies were uniformly high (>95%) across the series. However, little selectivity was observed for *Ga vs. Zn during stripping. The use of $^iPr_2O$ strikes the best balance between the amount of radiogallium and zinc extraction.

A well-separable bi-phasic mixture with no mutual phase contamination resulted when 1 M $ZnCl_2$ was prepared in 6 M HCl and then combined with a 1/2 (v/v) mixture of ether/TFT. Higher HCl concentration allowed for excellent *Ga extractions

Table 2 Liquid–liquid extraction and stripping of $^{67/68}Ga$ = *Ga and Zn in flow using a 1/2 (v/v) mixture of ether/TFT (extraction) and 0.1 M HCl (stripping)

Entry	Ether	*Ga extraction efficiency, %	Zn extraction efficiency, %	*Ga stripping efficiency, %	Zn stripping efficiency, %	*Ga recovery efficiency, %	Zn contamination, %
1	Et_2O	81.4 ± 2.6	4.2 ± 0.7	99.7 ± 0.7	91.0 ± 8.2	81.1 ± 2.6	3.8 ± 0.8
2	THP	89.6 ± 3.7	26.2 ± 8.6	99.3 ± 2.8	98.1 ± 4.4	88.9 ± 4.5	25.7 ± 8.5
3	$^iPr_2O$	76.3 ± 1.9	1.9 ± 1.6	97.8 ± 2.4	93.4 ± 6.3	74.6 ± 2.6	1.7 ± 1.5
4	*n*-BuOMe	51.8 ± 1.8	2.1 ± 0.3	98.5 ± 3.1	76.5 ± 3.8	51.0 ± 2.4	1.6 ± 0.2

The aqueous phase was formed by 5.6 M $ZnCl_2$ prepared in 3 M HCl. Extraction and stripping efficiencies are defined as 100% * $M_{phase\ A}/(M_{phase\ A} + M_{phase\ B})$, where M is activity of *Ga and Zn (traced with ^{65}Zn) in organic and aqueous phases A and B. Stripping is defined as extraction of a metal from organic into aqueous phase ("back-extraction")

Table 3 Liquid–liquid extraction of $^{67/68}Ga$ = *Ga and Zn in batch and in flow using a 1/2 (v/v) mixture of ether/TFT

		Batch		Flow	
Entry	Extractant	*Ga %	Zn %	*Ga %	Zn %
1	Et_2O	96.4 ± 1.7	0.3 ± 0.2	98.9 ± 0.2	0.2 ± 0.2
2	THP	98.0 ± 1.1	1.7 ± 0.5	97.4 ± 2.0	1.5 ± 0.5
3	$^iPr_2O$	93.1 ± 1.1	0.4 ± 0.5	97.5 ± 1.3	0.1 ± 0.1
4	*n*-BuOMe	98.0 ± 0.4	0.5 ± 0.7	95.7 ± 2.0	$5 \cdot 10^{-3} \pm 3 \cdot 10^{-3}$
5	*n*-HexOMe	93.9 ± 1.6	0.4 ± 0.2	90.6 ± 3.5	$2 \cdot 10^{-3} \pm 2 \cdot 10^{-4}$
6	Bu_2O	84.8 ± 4.9	0.3 ± 0.2	87.3 ± 0.4	0.01 ± 0.01
7	*n*-Am_2O	73.6 ± 1.4	0.5 ± 0.1	80.4 ± 1.5	$2 \cdot 10^{-3} \pm 1 \cdot 10^{-3}$

The aqueous phase was formed by 1 M $ZnCl_2$ prepared in 6 M HCl

both in batch and inflow (Table 3). Importantly, the co-extraction of zinc was one order of magnitude lower for the non-hydrophobic ethers (Table 3 entries 1–4).

The process did not extract Cu, Mn, and Co but did co-extract Fe.

To rationalize and build a predictive model of the extraction, we have used the conductor-like screening model for real solvents (COSMO-RS). Assuming chlorogallic acid, $HGaCl_4$, and $ZnCl_2$ as the main extractable species, the COSMO-RS theory predicts the solvation-driven mechanism for extraction of Ga and Zn. An insight into solvation of $HGaCl_4$ by $^iPr_2O$ could be obtained by examining their respective sigma profiles, which give a relative amount of surface with a given polarity σ. Figure 7 shows that the weakly negatively charged part of $HGaCl_4$ represented by peak *H* is well solvated by weakly positively charged C-H moieties of $^iPr_2O$ (peaks *F* and *G*). The positively charged H-atom of $HGaCl_4$ (peak *C*) is solvated by the negatively charged O-atom of $^iPr_2O$, (peak *J*). Therefore, the sigma profiles of $HGaCl_4$ and $^iPr_2O$ are complementary, translating into high free energy of solvation ($\Delta G_{solv} = -9.9$ kcal/mol) and high extractability into the organic phase (Pedersen et al., 2019).

To meet the stringent requirements for radiogallium purity set by the European Pharmacopeia ([Zn] < 10 μg/GBq), the separation of radiogallium from zinc was implemented as a 3-stage process (Fig. 8). During the extraction stage, radiogallium was extracted from the aqueous feed containing a solution of 1 M $ZnCl_2$ in 6 M HCl, into a 1/2 (v/v) mixture of *n*-BuOMe/TFT. In stage 2 (scrubbing), the organic phase, containing radiogallium, was treated with the 8 M HCl to selectively remove residual zinc from the organic phase. In the stripping stage, radiogallium was back-extracted from the organic into the aqueous phase with 0.1 M HCl. The 3-stage process allowed us to obtain radiogallium solution with as little zinc as 0.06 μg/mL.

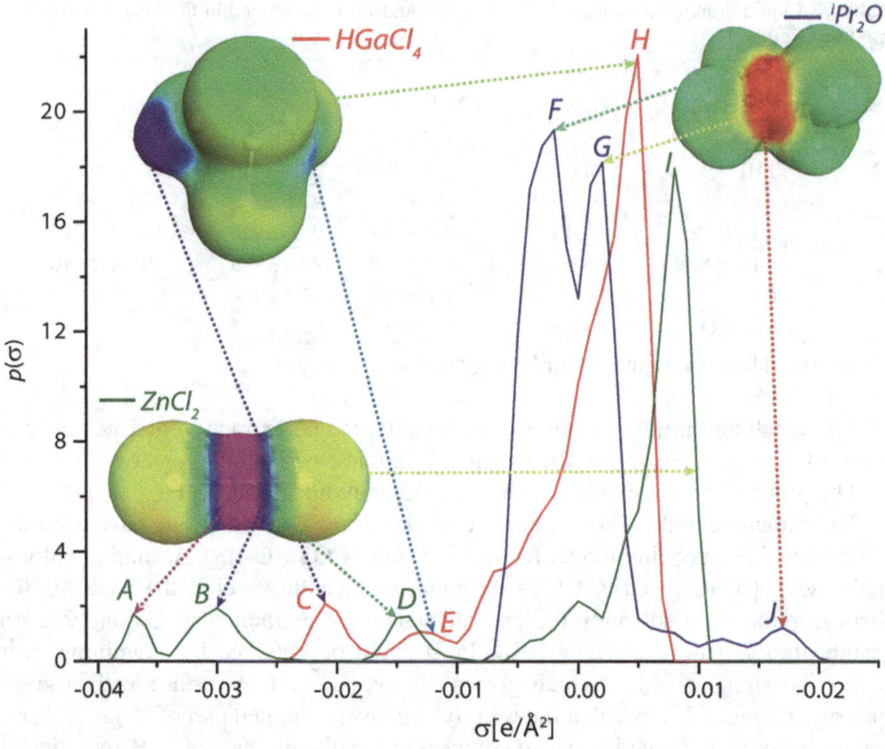

Fig. 7 The depiction of COSMO-RS sigma surfaces (polarization charge densities) and sigma profiles for $HGaCl_4$, $ZnCl_2$, and $^{i}Pr_2O$

2.3.2 LLEF of $^{67/68}Ga$ from Nitric Acid Solutions

Although we made important advances toward developing an efficient and automated process for the separation of radiogallium and recovery of zinc-68, the in-target chemistry of the $ZnCl_2$/HCl remained a challenge. While it was possible to combat rising in-target pressure using an open target, we noticed that outgassing was accompanied by chlorine production, resulting from the oxidation of chloride to chlorine during proton bombardment. The co-production of chlorine created potential problems with regard to the future implementation of the technology in a clinically relevant environment. Earlier, Pandey et al. (2019) demonstrated that the use of nitric acid could alleviate outgassing. As a result, the clinical production of gallium-68 using liquid targets has standardized on a zinc nitrate solution with concentrations ranging from 0.6 to 1.7 M, dissolved in nitric acid at concentrations between 0.01 and 1.5 M. Therefore, we turned our efforts toward the development of an efficient LLEF able to extract ^{68}Ga from $^{68}Zn(NO_3)_2/HNO_3$ solutions (Zhuravlev et al., 2022).

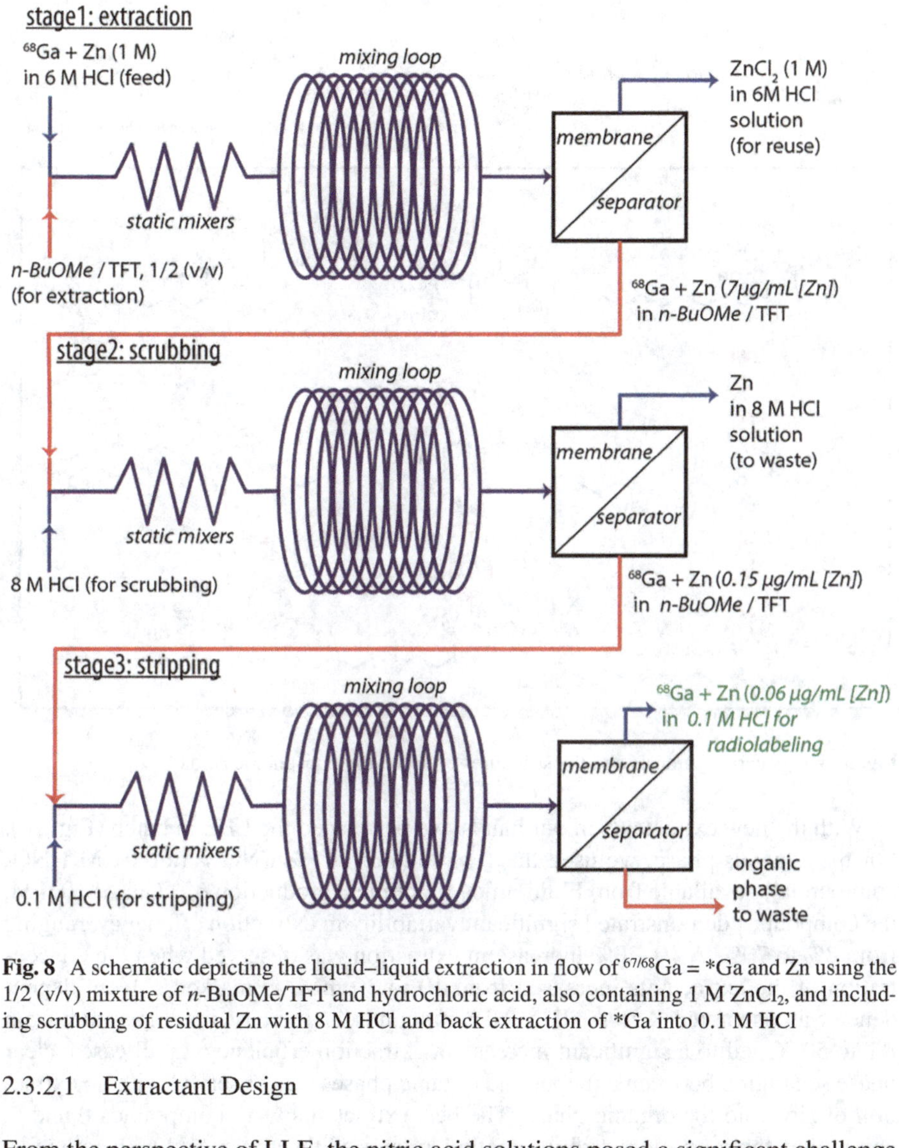

Fig. 8 A schematic depicting the liquid–liquid extraction in flow of $^{67/68}$Ga = *Ga and Zn using the 1/2 (v/v) mixture of *n*-BuOMe/TFT and hydrochloric acid, also containing 1 M $ZnCl_2$, and including scrubbing of residual Zn with 8 M HCl and back extraction of *Ga into 0.1 M HCl

2.3.2.1 Extractant Design

From the perspective of LLE, the nitric acid solutions posed a significant challenge because there was no known extractant capable of efficiently and selectively recovering gallium in the presence of a high concentration of zinc. For the design of a new extractant, we have chosen arylamino phosphonic acid scaffold, recognizing that its chelating strength and hydrophilicity can be carefully controlled by the judicious choice of the substituents on the aryl rings Ar_1 and Ar_2 and the phosphonic acid monoester moiety (Fig. 9). A one-pot, two-step synthesis, based on the Kabachnik–Fields reaction, yielded a series of new aminophosphonic mono-esters (**1**–**9**) in 40–50% yield.

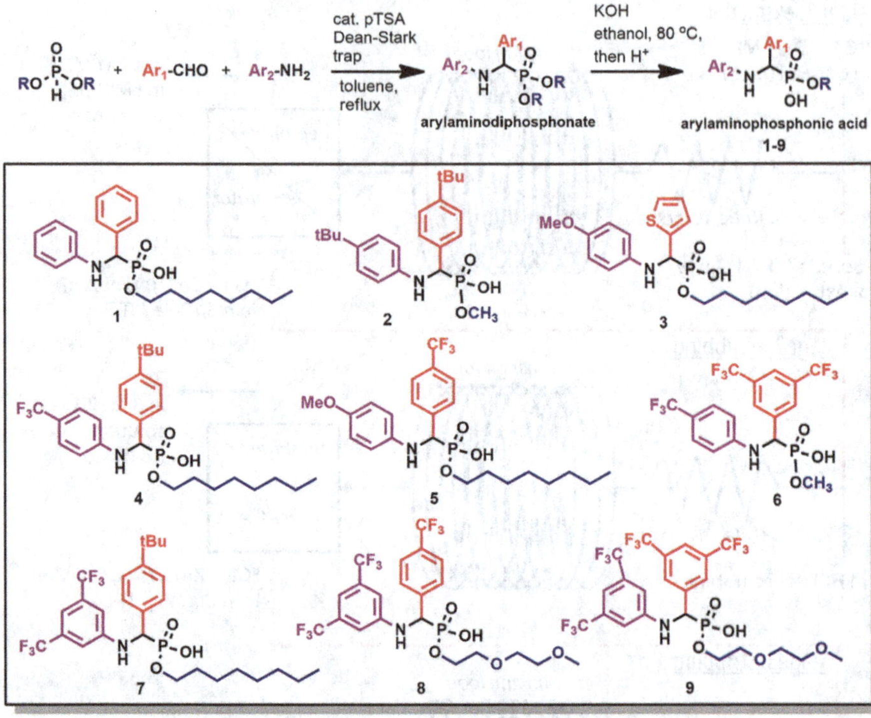

Fig. 9 The general synthesis and the structures of arylaminophosphonic acids **1–9**

With the new extractants in our hands, we performed the LLE in batch (Fig. 10). For the aqueous phase, we used the 1 M solution of $^{68}Zn(NO_3)_2$ in 0.01 M HNO_3 commercially available from Fluidomica for clinical production of ^{68}Ga. At 10 mM, the compounds demonstrated significant variability in extraction efficiency, ranging from 8% to 70%. A 10–20% increase in extraction was observed when the concentration of the extractants increased from 10 to 30 mM. A marked solvent dependence was also noted. Across all tested compounds, an increase in temperature from RT to 50 °C led to a significant increase in extraction efficiency. In all cases, clear phase separation between aqueous and organic phases was observed with no extraction of zinc into the organic phase. The best extractants were compounds **8** and **9**, combining three or more trifluoromethyl groups with ethylene glycol ponytails. These compounds demonstrate both high acidity (pKa calc = 0.7 and 0.4, correspondingly) and high aqueous solubility (1.60 and 1.73 mM), suggesting that the combination of low pKa and high hydrophilicity are key to high extraction efficiencies.

Having established the optimal conditions for the batch extractions, we translated the batch into a continuous flow. The experiment was performed in two stages: extraction and stripping. In the first stage, the two phases were mixed in a 1-meter mixing loop and sent into the glass microreactor maintained at 50 °C (Fig. 11). The

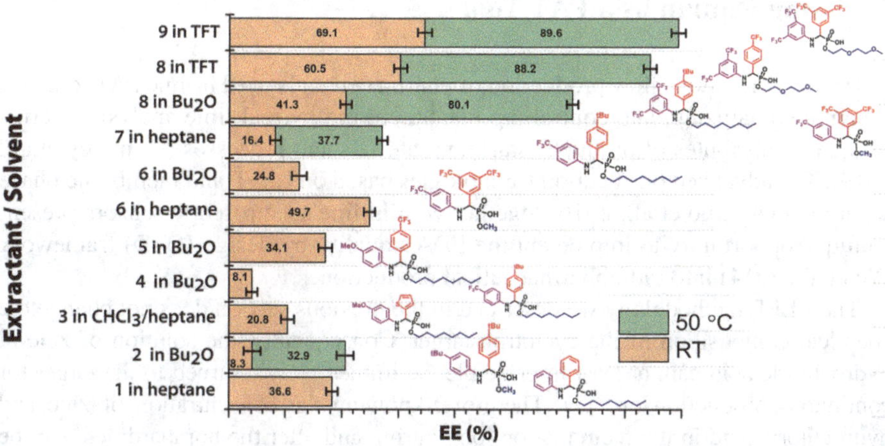

Fig. 10 Batch LLE of ^{68}Ga using extractants **1–9**. Extraction efficiency (EE) is defined in Table 2. The extractions were performed at room temperature (RT) and 50 °C, in batch. The organic solvent was chosen in such a way, as to provide the best solubility and phase separation for a given extractant

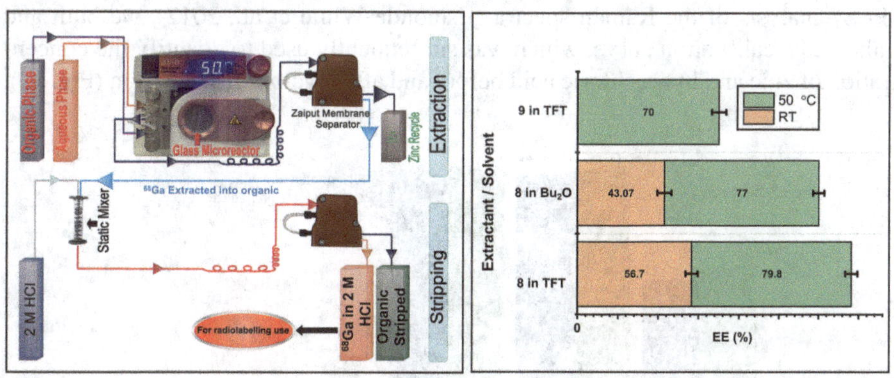

Fig. 11 Left: A diagram shows the two-stage LLEF. Right: ^{68}Ga EE(%) extracted from a 1 M $Zn(NO_3)_2$ solution in 0.01 M HNO_3 using the extractants **8** and **9** in TFT and Bu_2O at RT and 50 °C

reaction mixture was then passed through a membrane separator, which separated the aqueous phase containing zinc and the organic phase containing gallium. The zinc was kept for recycling. In the second stage, gallium-68 was back-extracted into 2 M hydrochloric acid at room temperature using the same setup. On average, the extraction efficiencies were 10% lower than in batch. Only 11 ppm of zinc was found in the stripped solution.

3 Using Raman as a PAT Tool

For years, continuous flow production of pharmaceuticals used in-line PAT tools for designing, analyzing, and controlling manufacturing by real-time analysis of critical quality attributes of products, starting materials and processes (Dumarey et al., 2019). The advanced purification technologies based on LLEF and membrane phase separation (Adamo et al., 2016) together with in-line spectroscopy sensors present a unique opportunity to implement the FDA's quality-by-design (QbD) framework (Yu et al., 2014) into radiopharmaceutical production.

The LLEF methodology we described in the previous section does not change the chemical composition of the cyclotron target. Consequently, the solution of zinc in hydrochloric acid can, at least in principle, be immediately returned to the target for continuous production of ^{68}Ga. Therefore, knowing the concentration of zinc and hydrochloric acid in the recirculation loop during and after the bombardment can be crucial for monitoring and controlling the yield of ^{68}Ga and its extraction efficiencies.

Raman spectroscopy is ideal for analyzing aqueous solutions. It uses laser light, which could be delivered into a flow cell via fiber optics. Therefore, in-line sampling can be seamlessly integrated into the automated production workflow. Using independently prepared samples and partial least squares projection to latent structures (PLS) analysis of the Raman spectra (Esmonde-White et al., 2017), we built and validated a calibration curve, which was subsequently used to quantify the concentration of zinc and hydrochloric acid before and after cyclotron irradiation (Fig. 12).

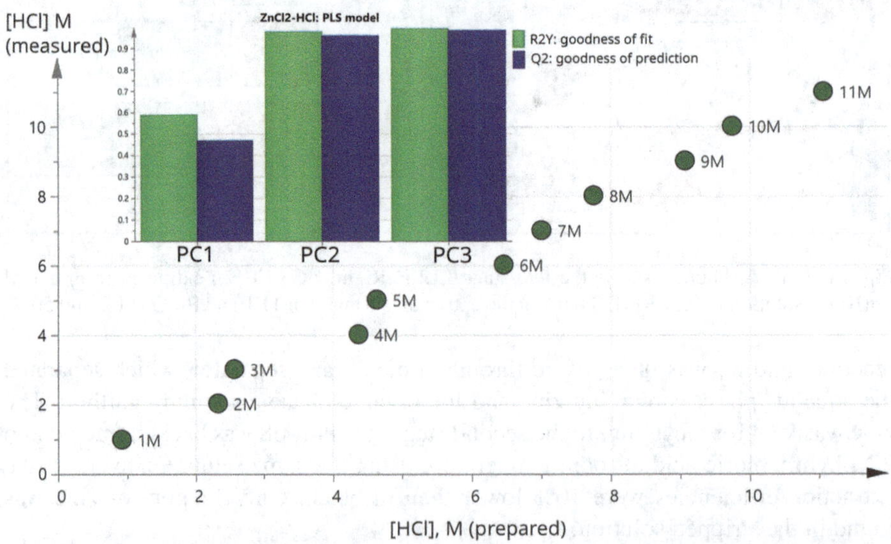

Fig. 12 A validation curve for determination of the concentrations of HCl in the presence of [^{68}Zn]$ZnCl_2$ in the cyclotron liquid target. The validation curve was obtained using independently prepared samples. Inset: a 3-component PLS model for the above curve (SIMCA 13.0). *PC* principal component

Table 4 The change in concentration of $^{68}ZnCl_2$ and HCl upon proton bombardment of the liquid target of two samples with different initial HCl concentrations (solutions A and B) as determined by Raman spectroscopy and analyzed using PLS calibration (Fig. 12)

Entry	Sample	[$^{68}ZnCl_2$], M	[HCl], M
1	Initial solution A	0.98	4.6
2	Solution A, 10 μA, EOB	0.98	4.4
3	Solution A, 30 μA, EOB	1.0	3.8
4	Initial solution B	1.0	5.3
5	Solution B, 10 μA, EOB	0.99	4.0
6	Solution B, 30 μA, EOB	1.0	3.1

Table 4 shows the results of four experiments performed by irradiating a 1 M solution of [^{68}Zn]$ZnCl_2$ prepared in 4.6 M HCl (solution A, entries 1–3) and 5.3 M HCl (solution B, entries 4–6) HCl. During the irradiation of solutions A the concentration of ^{68}Zn did not change, which indicated no acid or zinc loss from evaporation. However, the concentration of HCl in both solutions substantially decreased when the proton flux was increased from 10 to 30 μA. Extensive radiolysis of the chloride salts in closed, pressurized liquid targets was described elsewhere (Pandey et al., 2014). Our experiments with the open liquid target (vide supra) confirmed chloride oxidation accompanied by the generation of chlorine gas. Therefore, in-target radiolysis might necessitate external acidification to reach the optimal acidity required by the LLEF.

4 Perspectives

Cyclotron liquid targetry is compatible with fluidic chemistry and is well suited for the rapid production of short and medium-lived PET radionuclides ^{61}Cu, ^{62}Cu, ^{63}Zn, ^{68}Ga, ^{45}Ti, and ^{44}Sc (Table 1, entries 2–7). In the current batch production, the cyclotron target material is either discarded after workup or recovered in a separate process. This is time-consuming and leads to a loss of expensive isotopically enriched target material. By coupling the radionuclide production step with the liquid–liquid extraction and phase separation, it is possible to construct a closed-loop system for the continuous manufacturing of PET radionuclides where the target material is reused in-line. The continuous process for the production and separation of a generic PET radionuclide $^*M_2$ from the target metal M_1 can be envisioned as follows (Fig. 13). The liquid target **T**, connected to the separation module **S** is irradiated by the cyclotron beam producing PET radionuclide $^*M_2$ via the $M_1(p,n)^*M_2$ nuclear reaction. The irradiated solution is then delivered into **S**, where $^*M_2$ is extracted into the organic phase, then back-extracted into 0.1 M HCl, and finally, sent into the hot cell for further downstream processing and radiolabeling. The cyclotron target

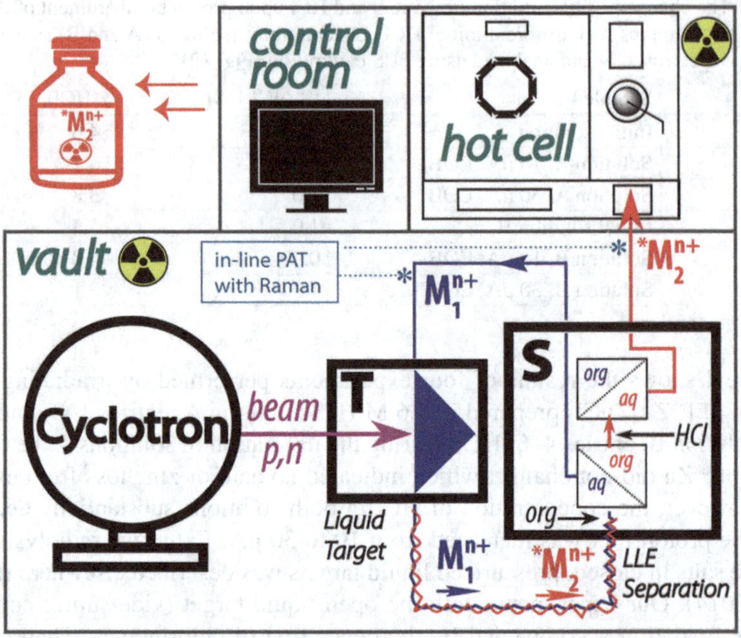

Fig. 13 The conceptual schematic of continuous production of a generic PET radionuclide $^*M_2$ using cyclotron solution target. The aqueous phase contains a solution of cyclotron target in the form of a salt M_1^{n+}. The organic phase contains a solution of an extractant in appropriate organic solvent. After irradiation (*T*) and LLE, the phase separation is performed using a membrane separator with integrated pressure control (*S*). The aqueous phase is returned to *T* for new irradiation, whereas $^*M_2^{n+}$ is back-extracted into 0.1 M HCl and is sent for downstream processing and radiolabeling

solution, containing M_1, is returned to **T** for new irradiation. In scenarios where the target volumes of liquid are several milliliters, and the LLE flow rates vary between 1:1 and 1:3, the entire extraction process can be completed in under 30 min.

In summary, we described a set of fluidic technologies that can enable continuous production of radiopharmaceuticals at the point of care. The cyclotron liquid targetry combined with liquid–liquid extraction in flow can provide scalability and the ability to run PET radionuclide production on demand. The target material can be recycled in-line, adding to sustainability of production. As a part of FDA's quality by design (QbD) framework Raman spectroscopy-based in-line sampling can be seamlessly integrated into the production workflow for real-time process control and quality assurance.

References

Adamo, A., Heider, P. L., Weeranoppanant, N., & Jensen, K. F. (2013). Membrane-based, liquid–liquid separator with integrated pressure control. *Industrial & Engineering Chemistry Research, 52*(31), 10802–10808. https://doi.org/10.1021/ie401180t

Adamo, A., Beingessner, R. L., Behnam, M., Chen, J., Jamison, T. F., Jensen, K. F., Monbaliu, J.-C. M., et al. (2016). On-demand continuous-flow production of pharmaceuticals in a compact, reconfigurable system. *Science (New York, N.Y.), 352*(6281), 61–67. https://doi.org/10.1126/science.aaf1337

Aluicio-Sarduy, E., Ellison, P. A., Barnhart, T. E., Cai, W., Nickles, R. J., & Engle, J. W. (2018). PET radiometals for antibody labeling. *Journal of Labelled Compounds & Radiopharmaceuticals, 61*(9), 636–651. https://doi.org/10.1002/jlcr.3607

Alves, V., do Carmo, S., Alves, F., & Abrunhosa, A. (2018). Automated purification of radiometals produced by liquid targets. *Instruments, 2*(3), 17.

Bédard, A.-C., Adamo, A., Aroh, K. C., Grace Russell, M., Bedermann, A. A., Torosian, J., Yue, B., Jensen, K. F., & Jamison, T. F. (2018). Reconfigurable system for automated optimization of diverse chemical reactions. *Science, 361*(6408), 1220–1225.

Bhatki, K. S., Rane, A. T., & Kabadi, M. B. (1969). Preparation of carrier-free copper-64,67 nuclides by liquid-liquid extraction. *Journal of Radioanalytical Chemistry, 2*(1), 73–77. https://doi.org/10.1007/BF02518674

Bombardieri, E., Aktolun, C., Baum, R. P., Bishof-Delaloye, A., Buscombe, J., Chatal, J. F., Maffioli, L., Moncayo, R., Mortelmans, L., & Reske, S. N. (2003). ^{67}Ga scintigraphy: Procedure guidelines for tumour imaging. *European Journal of Nuclear Medicine and Molecular Imaging, 30*(12), 1–1. https://doi.org/10.1007/s00259-003-1356-1

Brasse, D., & Nonat, A. (2015). Radiometals: Towards a new success story in nuclear imaging? *Dalton Transactions (Cambridge, England: 2003), 44*(11), 4845–4858. https://doi.org/10.1039/c4dt02911a

Brown, L. C., & Callahan, A. P. (1972). The separation and purification of carrier-free copper isotopes for medical use. *The International Journal of Applied Radiation and Isotopes, 23*(11), 535–538. https://doi.org/10.1016/0020-708X(72)90136-6

Chakravarty, R., Valdovinos, H. F., Chen, F., Lewis, C. M., Ellison, P. A., Haiming Luo, M., Meyerand, E., Nickles, R. J., & Cai, W. (2014). Intrinsically germanium-69-labeled iron oxide nanoparticles: Synthesis and in-vivo dual-modality PET/MR imaging. *Advanced Materials, 26*(30), 5119–5123. https://doi.org/10.1002/adma.201401372

DeGrado, T. R., Pandey, M. K., Byrne, J. F., Engelbrecht, H. P., Jiang, H., Packard, A. B., Thomas, K. A., Jacobson, M. S., Curran, G. L., & Lowe, V. J. (2014). Preparation and preliminary evaluation of ^{63}Zn-zinc citrate as a novel PET imaging biomarker for zinc. *Journal of Nuclear Medicine, 55*(8), 1348–1354. https://doi.org/10.2967/jnumed.114.141218

do Carmo, S. J. C., Scott, P. J. H., & Alves, F. (2020). Production of radiometals in liquid targets. *EJNMMI Radiopharmacy and Chemistry, 5*(1), 2. https://doi.org/10.1186/s41181-019-0088-x

Dumarey, M., Hermanto, M., Airiau, C., Shapland, P., Robinson, H., Hamilton, P., & Berry, M. (2019). Advances in continuous active pharmaceutical ingredient (API) manufacturing: Real-time monitoring using multivariate tools. *Journal of Pharmaceutical Innovation, 14*(4), 359–372. https://doi.org/10.1007/s12247-018-9348-7

Esmonde-White, K. A., Cuellar, M., Uerpmann, C., Lenain, B., & Lewis, I. R. (2017). Raman spectroscopy as a process analytical technology for pharmaceutical manufacturing and bioprocessing. *Analytical and Bioanalytical Chemistry, 409*(3), 637–649. https://doi.org/10.1007/s00216-016-9824-1

European Pharmacopoeia 10. (2020). *Council of Europe, European Pharmacopoeia Commission, European Directorate for the Quality of Medicines and Healthcare*. DeutscherApotheker Verlag.

Fonseca, A. I., Alves, V. H., Sérgio, J. C., do Carmo, A. F., Abrunhosa, A. J., & Alves, F. (2021). GMP-automated purification of copper-61 produced in cyclotron liquid targets: Methodological aspects. *Current Radiopharmaceuticals, 14*(4), 420–428. https://doi.org/10.2174/1874471013666201112141844

Goethals, P.-E., & Zimmermann, R. (2020). *MEDraysintell marketed radiopharmaceuticals report 2020 (part 1)*. MEDraysintell.

Hernandez, R., Valdovinos, H. F., Yang, Y., Chakravarty, R., Hong, H., Barnhart, T. E., & Cai, W. (2014). ^{44}Sc: An attractive isotope for peptide-based PET imaging. *Molecular Pharmaceutics, 11*(8), 2954–2961. https://doi.org/10.1021/mp500343j

Heskamp, S., Raavé, R., Boerman, O., Rijpkema, M., Goncalves, V., & Denat, F. (2017). ^{89}Zr-immuno-positron emission tomography in oncology: State-of-the-art ^{89}Zr radiochemistry. *Bioconjugate Chemistry, 28*(9), 2211–2223. https://doi.org/10.1021/acs.bioconjchem.7b00325

Jensen, M., & Clark, J. (2011). Direct production of Ga-68 from proton bombardment of concentrated aqueous solutions of [Zn-68] zinc chloride. In *Proceedings of 13th International Workshop on Targetry and Target Chemistry* (pp. 288–292). Danmarks Tekniske Universitet, Risø Nationallaboratoriet for Bæredygtig Energi.

Kandil, S. A., Scholten, B., Saleh, Z. A., Youssef, A. M., Qaim, S. M., & Coenen, H. H. (2008). A comparative study on the separation of radiozirconium via ion-exchange and solvent extraction techniques, with particular reference to the production of ^{88}Zr and ^{89}Zr in proton induced reactions on yttrium. *Journal of Radioanalytical and Nuclear Chemistry, 274*(1), 45. https://doi.org/10.1007/s10967-006-6892-2

Kislik, V. (2002). Competitive complexation/solvation theory of solvent extraction. II. Solvent extraction of metals by acidic extractants. *Separation Science and Technology, 37*(11), 2623–2657. https://doi.org/10.1081/SS-120004456

Lahiri, S., Banerjee, S., & Das, N. R. (1997). Liquid-liquid extraction of α-activation products of iron with HDEHP. *Journal of Radioanalytical and Nuclear Chemistry, 223*(1–2), 235–238. https://doi.org/10.1007/BF02223393

Lepareur, N. 2022. Cold kit labeling: The future of ^{68}Ga radiopharmaceuticals? *Frontiers in Medicine, 9*, 812050. Retrieved from https://www.frontiersin.org/articles/10.3389/fmed.2022.812050

Luo, H. M., Boll, R. A., Bell, J. R., & Dai, S. (2012). Facile solvent extraction separation of Th-227 and Ac-225 based on room-temperature ionic liquids. *Radiochimica Acta, 100*(10), 771–778. https://doi.org/10.1524/ract.2012.1950

Maiti, M., & Lahiri, S. (2010). Separation of ^{99}Mo and ^{99m}Tc by liquid–liquid extraction using trioctylamine as extractant. *Journal of Radioanalytical and Nuclear Chemistry, 283*(3), 661–663. https://doi.org/10.1007/s10967-009-0403-1

Martini, P., Adamo, A., Syna, N., Boschi, A., Uccelli, L., Weeranoppanant, N., Markham, J., & Pascali, G. (2019). Perspectives on the use of liquid extraction for radioisotope purification. *Molecules, 24*(2), 334. https://doi.org/10.3390/molecules24020334

Matthews, P. M., Rabiner, E. A., Passchier, J., & Gunn, R. N. (2012). Positron emission tomography molecular imaging for drug development. *British Journal of Clinical Pharmacology, 73*(2), 175–186. https://doi.org/10.1111/j.1365-2125.2011.04085.x

Mojtahedi, A., Thamake, S., Tworowska, I., Ranganathan, D., & Delpassand, E. S. (2014). The value of (68)Ga-DOTATATE PET/CT in diagnosis and management of neuroendocrine tumors compared to current FDA approved imaging modalities: A review of literature. *American Journal of Nuclear Medicine and Molecular Imaging, 4*(5), 426–434.

Mukhopadhyay, B., & Lahiri, S. (1999). Review article: Separation of the carrier free radioisotopes of second transition series elements. *Solvent Extraction and Ion Exchange, 17*(1), 1–21. https://doi.org/10.1080/07360299908934597

Navratil, J. D. (1986). Solvent extraction in nuclear technology. *Pure and Applied Chemistry, 58*(6), 885–888. https://doi.org/10.1351/pac198658060885

Nayak, D., & Lahiri, S. (2001). Alternative methods for the production of carrier-free ^{66,67}Ga. *Applied Radiation and Isotopes, 54*(2), 189–193. https://doi.org/10.1016/S0969-8043(00)00194-9

Oehlke, E., Hoehr, C., Hou, X., Hanemaayer, V., Zeisler, S., Adam, M. J., Ruth, T. J., et al. (2015). Production of Y-86 and other radiometals for research purposes using a solution target system. *Nuclear Medicine and Biology, 42*(11), 842–849. https://doi.org/10.1016/j.nucmedbio.2015.06.005

Pandey, M. K., & DeGrado, T. R. (2021). Cyclotron production of PET radiometals in liquid targets: Aspects and prospects. *Current Radiopharmaceuticals, 14*(4), 325–339.

Pandey, M. K., Engelbrecht, H. P., Byrne, J. F., Packard, A. B., & DeGrado, T. R. (2014). Production of ^{89}Zr via the ^{89}Y(p,n)^{89}Zr reaction in aqueous solution: Effect of solution composition on in-target chemistry. *Nuclear Medicine and Biology, 41*(4), 309–316. https://doi.org/10.1016/j.nucmedbio.2014.01.006

Pandey, M. K., Byrne, J. F., Schlasner, K. N., Schmit, N. R., & DeGrado, T. R. (2019, July). Cyclotron production of ^{68}Ga in a liquid target: Effects of solution composition and irradiation parameters. *Nuclear Medicine and Biology, 74–75*, 49–55. https://doi.org/10.1016/j.nucmedbio.2019.03.002

Pedersen, K. S., Imbrogno, J., Fonslet, J., Lusardi, M., Jensen, K. F., & Zhuravlev, F. (2018). Liquid–liquid extraction in flow of the radioisotope titanium-45 for positron emission tomography applications. *Reaction Chemistry & Engineering, 3*(6), 898–904.

Pedersen, K. S., Nielsen, K. M., Fonslet, J., Jensen, M., & Zhuravlev, F. (2019). Separation of radiogallium from zinc using membrane-based liquid-liquid extraction in flow: Experimental and COSMO-RS studies. *Solvent Extraction and Ion Exchange, 37*(5), 376–391. https://doi.org/10.1080/07366299.2019.1646982

Pietrelli, L., Mausner, L., & Kolsky, K. (2005). Separation of carrier-free ^{47}Sc from titanium targets. *Journal of Radioanalytical and Nuclear Chemistry, 157*(2), 335–345. https://doi.org/10.1007/bf02047448

Ramamoorthy, N. (2020). Compelling case for fostering titanium-45 for positron emission tomography tracer applications. *Indian Journal of Nuclear Medicine: IJNM: The Official Journal of the Society of Nuclear Medicine, India, 35*(3), 200–202. https://doi.org/10.4103/ijnm.IJNM_52_20

Rodnick, M. E., Sollert, C., Stark, D., Clark, M., Katsifis, A., Hockley, B. G., Christian Parr, D., et al. (2020, November). Cyclotron-based production of ^{68}Ga, [^{68}Ga]GaCl$_3$, and [^{68}Ga]Ga-PSMA-11 from a liquid target. *EJNMMI Radiopharmacy and Chemistry, 5*, 25. https://doi.org/10.1186/s41181-020-00106-9

Severin, G. W., Nielsen, C. H., Jensen, A. I., Fonslet, J., Kjær, A., & Zhuravlev, F. (2015). Bringing radiotracing to titanium-based antineoplastics: Solid phase radiosynthesis, PET and ex vivo evaluation of antitumor agent [^{45}Ti](Salan)Ti(Dipic). *Journal of Medicinal Chemistry, 58*(18), 7591–7595. https://doi.org/10.1021/acs.jmedchem.5b01167

Seyedi, H., Nikjou, A., Sharifian, M., & Sadeghi, M. (2021, November). Evaluation of nuclear reaction cross sections for optimization of production of the non-standard positron emitting radionuclide ^{90}Nb using proton induced reactions on Zr target. *Applied Radiation and Isotopes, 177*, 109917. https://doi.org/10.1016/j.apradiso.2021.109917

Søborg Pedersen, K., Baun, C., Nielsen, K. M., Thisgaard, H., Jensen, A. I., & Zhuravlev, F. (2020). Design, synthesis, computational, and preclinical evaluation of natTi/^{45}Ti-labeled urea-based glutamate PSMA ligand. *Molecules, 25*(5), 1104. https://doi.org/10.3390/molecules25051104

Swift, E. H. (1924). A new method for the separation of gallium from other elements. *Journal of the American Chemical Society, 46*(11), 2375–2381. https://doi.org/10.1021/ja01676a004

Thisgaard, H., Kumlin, J., Langkjær, N., Chua, J., Hook, B., Jensen, M., Kassaian, A., et al. (2021, January). Multi-curie production of gallium-68 on a biomedical cyclotron and automated radiolabelling of PSMA-11 and DOTATATE. *EJNMMI Radiopharmacy and Chemistry, 6*, 1. https://doi.org/10.1186/s41181-020-00114-9

Vorster, M., Buscombe, J., Saad, Z., & Sathekge, M. (2018). Past and future of Ga-citrate for infection and inflammation imaging. *Current Pharmaceutical Design, 24*(7), 787–794. https://doi.org/10.2174/1381612824666171129200611

Witkowska-Patena, E., Mazurek, A., & Dziuk, M. (2017). ^{68}Ga-PSMA PET/CT imaging in recurrent prostate cancer: Where are we now? *Central European Journal of Urology, 70*(1), 37–43. https://doi.org/10.5173/ceju.2017.947

Yu, L. X., Amidon, G., Khan, M. A., Hoag, S. W., Polli, J., Raju, G. K., & Woodcock, J. (2014). Understanding pharmaceutical quality by design. *The AAPS Journal, 16*(4), 771–783. https://doi.org/10.1208/s12248-014-9598-3

Zhuravlev, F., Pedersen, K. S., Imbrogno, J. M., Adamo, A., Jensen, K. F., & Fonslet, J. (2021). *Separation of radiometals*. European Union EP3834210A1, filed August 6, 2019, and issued June 16, 2021. Retrieved from https://patents.google.com/patent/EP3834210A1/en

Zhuravlev, F., Gulzar, A., & Falborg, L. (2022). Recovery of gallium-68 and zinc from HNO_3-based solution by liquid–liquid extraction with arylamino phosphonates. *Molecules, 27*(23), 8377. https://doi.org/10.3390/molecules27238377

Rapid and Effective Purification for Radioactive Cu and Ga by Means of Analytical Chemistry Approaches

Shin-Ichi Ohira, Yumi Sugo, and Masanobu Mori

1 Introduction

1.1 Introduction to Radiometal Isotopes

Metal radioisotope (RI) ions are widely used in medical diagnosis and as therapeutic agents. Some metal RIs, such as ^{67}Cu (β^-, $t_{1/2}$ = 2.58 days), ^{89}Sr (β^-, $t_{1/2}$ = 50.5 days), and ^{90}Y (β^-, $t_{1/2}$ = 2.67 days), have been used in therapy to target cancer, and the method is known as radioimmunotherapy. Typically, the antibodies and metal RIs are connected via chelators, such as 2,2′,2″,2‴-(1,4,7,10-tetraazacyclododecane-1,4,7,10-tetrayl) tetraacetic acid (DOTA). The β^- source metal RI is delivered to the target site, such as a cancer, and used for radiation. The methods are well established and have been approved by the United States Food and Drug Administration (US FDA) since 2002 (Grillo-López, 2002). Furthermore, metal RIs such as ^{68}Ga (β^+, $t_{1/2}$ = 67.7 m), ^{64}Cu (β^+, $t_{1/2}$ = 12.7 h), and ^{89}Zr (β^+, $t_{1/2}$ = 78.4 h) have been used as probes for positron emission tomography (PET) (Alirezapour et al., 2016; Pfeifer et al., 2012; Xiao et al., 2013). ^{68}Ga–DOTA–octreotide was approved by the US FDA in 2019 as a PET probe owing to its effectiveness for imaging and suitable half-life (Hennrich & Benešová, 2020). The first

S.-I. Ohira (✉)
Department of Chemistry, Kumamoto University, Kumamoto, Japan
e-mail: ohira@kumamoto-u.ac.jp

Y. Sugo
Department of Quantum-Applied Biosciences, Takasaki Institute for Advanced Quantum Science, National Institutes for Quantum Science and Technology (QST), Takasaki, Japan

M. Mori
Faculty of Science and Technology, Kochi University, Kochi, Japan

R. M. van Dam et al. (eds.), *Automated Technologies for the Development and Production of Radiopharmaceuticals*,
https://doi.org/10.1007/978-3-031-84632-8_6

European Union marketing authorization for the distribution of cyclotron-produced ^{68}Ga for human use was granted in 2021.

Short-half-life metal RIs require on-site cyclotron generation, followed by separation and purification. The pure metal RIs can then be used for probe synthesis. Rapid and highly effective methods are recommended for these purposes. Herein, newly developed methods for the separation and purification of metal RIs and synthesis of metal RI-based probes are discussed.

1.2 *Separation and Purification of Metal Radioisotopes (RIs) with Traditional Resin-Based Solid-Phase Extraction*

The current methods used to separate and purify radioactive Cu and Ga and synthesize probes are schematically shown in Fig. 1. Metal RIs are typically separated via solid-phase extraction using chemically functionalized resins as selective adsorbents. One of the most widely used separation methods for metal RIs, which involves the formation of chloro-complexes with anion-exchange resin adsorbents, was developed in 1953 (Kraus & Moore, 1953). Chloro-metal complex formation depends on the concentration of Cl^- ions. The irradiated metals are typically dissolved in a highly concentrated HCl solution (~6 M). Both RI and target metals form anionic chloro-complexes in the presence of HCl. The solution is then passed through an anion-exchange resin to collect the metal ions as chloro-complexes. Thereafter, a more dilute HCl solution is passed through the anion-exchange resin column to release the metals with lower chloro-complex formation constants and is then eluted from the column. Subsequently, further diluted HCl solution is passed through the anion-exchange resin column to elute all metals. The mixture is then mixed with a ligand solution to synthesize PET probes. Alternatively, specially

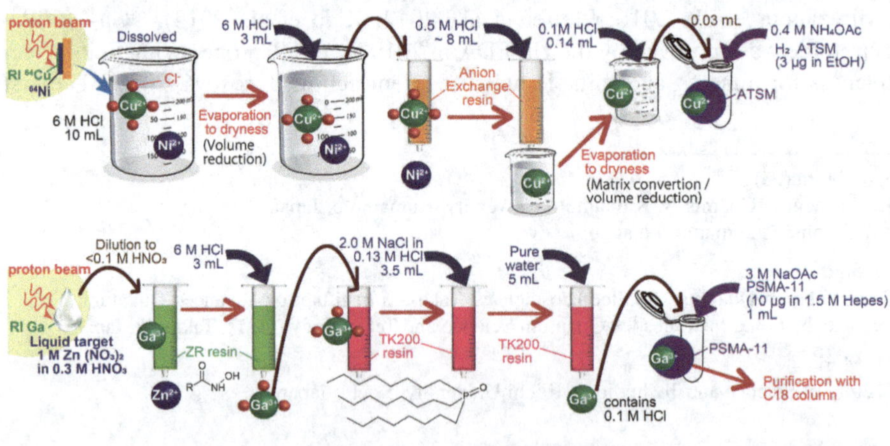

Fig. 1 Schematics of the current methods used to separate metal RIs and synthesize metal RI-based probes for Cu (top) (McCarthy et al., 1997) and Ga (bottom) (Rodnick et al., 2020)

designed solid phases have been used for separation and matrix conversion. These procedures are well established and can be performed using commercially available automated systems (Alves et al., 2018; Thisgaard et al., 2021). However, the use of highly concentrated acids and heating steps required for evaporation should be further improved. Concentrated acids may be contaminated with other metals, and the heating procedure is a time-consuming process. Rapid and compact separation/purification/synthesis systems will support the development of new types of PET probes. Therefore, we developed a new method for separating and purifying metal RIs and synthesizing metal RI-based probes. The developed system uses diluted acids (~30 mM) and also allows acid matrices to be exchanged without a heating procedure.

1.3 Separation and Determination of Metal Ions in Analytical Chemistry

"Analytical chemistry" studies instruments and methods used to separate, identify, and quantify matter. Separation is a critical step in chemical analysis. Therefore, some analytical chemistry approaches should be effective for separating and purifying metal RIs. However, it is challenging to determine whether the separation methods used in analytical chemistry can be used to separate and purify metal RIs because the concentration of metal RIs is low (less than nanograms per liter). In this study, we evaluated selective chelate formation, electrochemical separation, and electrodialytic ion transfer as methods for separation and purification of metal RIs, followed by synthesis of metal RI-based probes. The strategies of these separation methods are as follows.

Usually, the RI and target ions in the acid after dissolution are positively charged. Selective chelate formation can change the ionic charge polarity, and positively or negatively charged ions can be selectively removed, respectively, by cation and anion exchange resins. The formation of selective metal–ligand chelates can be controlled by tailoring the ligand concentration and solution pH. Theoretical calculations can also be performed to determine the optimal ligand concentration and solution pH. Furthermore, dilute organic ligands can be used to form complexes with trace metal RIs, and the formed complexes can then be decomposed using ultraviolet (UV) radiation.

Electrochemical metal detection is a highly sensitive method that has been used to analyze Hg or Bi thin films deposited on electrode surfaces. Typically, the metals are trapped on the film surface via reduction, followed by quantification using oxidation currents (Ohira et al., 2018). Furthermore, a highly effective felt-type electrode was used to recover Cu^{2+} ions from incineration ash (Mori et al., 2020). If one of the metal ions—the generated RI or target—is selectively deposited on the electrode surface, the separation of these metals can be achieved without any reagent except acids. The selectivity of electrochemical deposition methods can be controlled by adjusting the applied potential. Furthermore, the deposited metal can be eluted by electrochemical oxidation after the separation. Additionally, a flow-based electrochemical system was applied in the present method. The flow system can more effectively deposit the generated RI within a short time, especially using a felt-type electrode, and the relatively small amounts of the solution can be handled without any contamination.

Electrodialytic ion transfer has been developed by one of the authors (Ohira et al., 2012) and applied to the sample pretreatment for heavy metals in serum samples (Ohira et al., 2015a), soil extracts (Ohira et al., 2015b), and ultra-pure water samples (Okazaki et al., 2022). The ion transfer device (ITD) can quantitatively transfer ionic solutes from one to other solution phases. With this process, the metal RIs can be transferred from one acid to another, i.e., converting matrix acids. Furthermore, if the metal RIs are transferred into the ligand solution, the matrix conversion and probe synthesis can be simultaneously achieved.

In this study, we demonstrated that the analytical chemistry methods can be used for metal RI separation, matrix conversion, and probe synthesis in flow systems. The separations of Ga from Zn and Cu from Ni were performed as examples.

2 Materials and Methods

2.1 Radioisotope Preparation

The used radioactive Cu and Ga were prepared as follows.

^{64}Cu and ^{67}Ga were prepared via the ^{64}Ni(p,n)^{64}Cu and ^{68}Zn(p,2n)^{67}Ga nuclear reactions, respectively, at the Takasaki Ion Accelerators for Advanced Radiation Application (TIARA) at the National Institute for Quantum Science and Technology (Takasaki, Japan). The isotope-enriched [^{64}Ni]NiO (www.isoflex.com) and [^{nat}Zn]ZnO target materials were used as targets for ^{64}Cu and ^{67}Ga, respectively. The irradiated targets were dissolved into suitable acids and separated using a previously reported method (Watanabe et al., 2009a, 2009b), as needed.

2.2 Separation of Ga from Zn by Selective Chelate Formation with Cation-Exchange Resin

The separation of Ga from Zn has been performed with selective chelate formation of Ga, followed by cation exchange to adsorb Zn selectively. The sample solution was prepared using high-purity nitric acid (Tamapure-AA-100; www.tama-chem.co.jp) and ethylenediamine-*N*,*N*,*N*′,*N*′-tetra acetic acid (EDTA; www.dojindo.co.jp) as ligands. The solution pH and EDTA concentration were, respectively, adjusted to

1.5 and 1 mM. Under these optimized conditions, Ga selectively formed EDTA complex. Thus, Ga-EDTA and Zn are negatively and positively charged ions, respectively.

The flow diagrams of the separation process (Zn adsorption and desorption) are shown in Fig. 2a, b, respectively. In the adsorption process, the prepared sample solution was introduced into a sample loop (1.9 mL) connected to a six-port valve (6V-2). HNO_3 solutions (0.03 and 1 M) were continuously flowed at 0.70 mL/min using peristaltic pumps. A newly developed cation-trapping device (Fig. 2c) was connected to a six-port valve (6V-1). The prepared sample solution was introduced into the system via the sample loop by switching the 6V-2 valve. The 6V-1 valve was also switched to flow the sample solution into the cation-trapping device (Fig. 2a). The effluent of the cation-trapping device contained the Ga–EDTA complex. After 8 min of Zn adsorption, the valves were switched in the opposite direction for the desorption process, and a 1 M HNO_3 solution was introduced into the cation-trapping device from the backside to recover the adsorbed Zn (Fig. 2b). The cation-trapping device (Fig. 2c) was fabricated using cation-exchange membranes and a solution gasket. The solution gasket was prepared using a Parafilm® nylon screen (150 mesh, thickness 106 μm; Nippon Clever, Japan), with a solution channel (5 mm wide and 220 mm long) cut out using a CO_2 laser cutter (Beamo; flux-japan.jp). The Parafilm was then pressed onto the nylon screen using a heating press (500 kg, 42 °C, 15 min; three times). The prepared gasket was placed between the cation-exchange membranes (CMVN; www.agec.co.jp). The assembled system was placed between Al plates (250 mm × 15 mm × 5 mm) and tightened with screws. The solution inlet and outlet were prepared by inserting polyether ether ketone (PEEK) tubing (0.5 mm inner diameter). All tubing used was PEEK tubing.

The as-obtained Ga–EDTA complex after the separation was irradiated with UV light to decompose and recover Ga ion. The UV radiation reactor consisted of a quartz inner tube (8 mm i.d. × 10 mm o.d.) and a polytetrafluoroethylene (PTFE) outer tube (12 mm i.d. × 15 mm o.d.). A UV Hg lamp for ozone generation (80 μW/

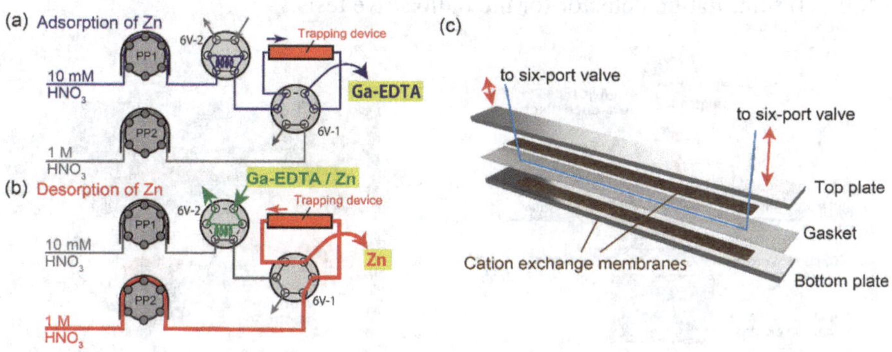

Fig. 2 Flow system for Zn adsorption (**a**) and Zn desorption (**b**), and cation-trapping device (**c**) used for separation of Ga and Zn with selective chelate formation

cm^2@30 cm for 254 nm, Analamp 81-1127-01; www.bhkinc.com) was inserted into the quartz tube. The bottom of the outer tube was inserted into a cut plastic syringe (10 mL) and sealed with PTFE seal tape. The male luer connector of the syringe was connected to a three-port stopcock. A 5 mL syringe was used to introduce the Ga-EDTA solution into the reactor via the stopcock. The solution was collected after 10 min of reaction time. Recovery was evaluated using the amounts of Ga^{3+} and Ga-EDTA separated using a cation-exchange resin, which selectively trapped Ga^{3+}.

2.3 *Electrochemical Separation of Cu from Ni*

Cu and Ni were separated via the selective electrochemical deposition of Cu on a carbon-felt electrode, followed by the oxidation to recover as Cu^{2+}. The flow electrolysis (FE) cell (EC FRONTIER) consisted of a carbon-felt working electrode (Φ18 mm, thickness 7.2 mm, effective surface area 1900 cm^2), a Pt wire counter electrode, and an Ag/AgCl reference electrode (Fig. 3a). The carbon-felt electrode was fabricated using high-density carbon fiber. The inset of Fig. 3a shows a scanning electron micrograph of the working electrode. All the solutions were introduced into the FE cell using a peristaltic pump at a flow rate of 0.5 mL/min. A 5 mL sample solution was first introduced into the FE by applying a reducing potential of −0.8 V (vs. the Ag/AgCl reference electrode), and the effluent is collected as fraction 1 (F1). A reducing potential was kept at −0.8 V, 20 mL of a 10 mM HNO_3 solution was used to flush the FE cell, and the effluent is collected as F2. Cu was desorbed as Cu^{2+} using 20 mL of a 10 mM HNO_3 solution and an applied voltage of +1.0 V, and the effluent is collected as F3. To condition the FE cell, 10 mL of 0.1 M HCl was introduced, and the effluent is collected as F4. The obtained effluents (F1–F4) were each analyzed using an inductively coupled plasma mass spectrometer or a flame atomic absorption spectrometer for the nonradioactive tests and a NaI (Tl) scintillation detector for the radioactive tests.

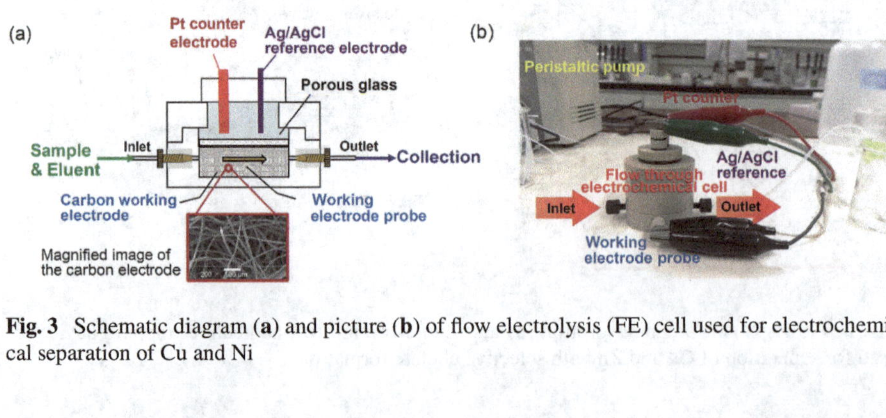

Fig. 3 Schematic diagram (**a**) and picture (**b**) of flow electrolysis (FE) cell used for electrochemical separation of Cu and Ni

2.4 Electrodialytic Matrix Conversion, Enrichment, and Probe Synthesis

Matrix conversion, enrichment, and probe synthesis were performed using an electrodialysis ITD. The structure of the ITD is described in detail in the literature (Ohira et al., 2012; Sugo et al., 2020). In brief, four solution layers were confined using a solution gasket and ion-permeable membranes (Fig. 4a). The solution gaskets were prepared as described in Sect. 2.2. Regenerated cellulose dialysis membranes (Spectra Por 1, molecular weight cutoff 6000–8000; www.spectrumchemical.com) and anion-exchange membranes (Selemion® AMV; www.agce.co.jp) were used as ion-permeable membranes. The gaskets and membranes were placed between polyvinyl chloride plates, which were tightened using screws (Fig. 4b). The solutions were continuously introduced into channels (A)-(D), as shown in Fig. 4c. The 30 mM HNO_3 introduced in channels (A) and (D), termed isolator channels, flowed at a rate of ~3.0 mL/min using miniature pumps (Unimor UPS-112E; Nitto Kohki, Japan). The donor (channel (B)) and acceptor (channel (C)) solutions (30 mM HNO_3) were introduced at 0.2 mL/min using a peristaltic pump (WPX1; www.welco-web.co.jp) and a syringe pump (Versa 3; www.norgren.co.jp), respectively. All of the solutions introduced into ITD except donor solution

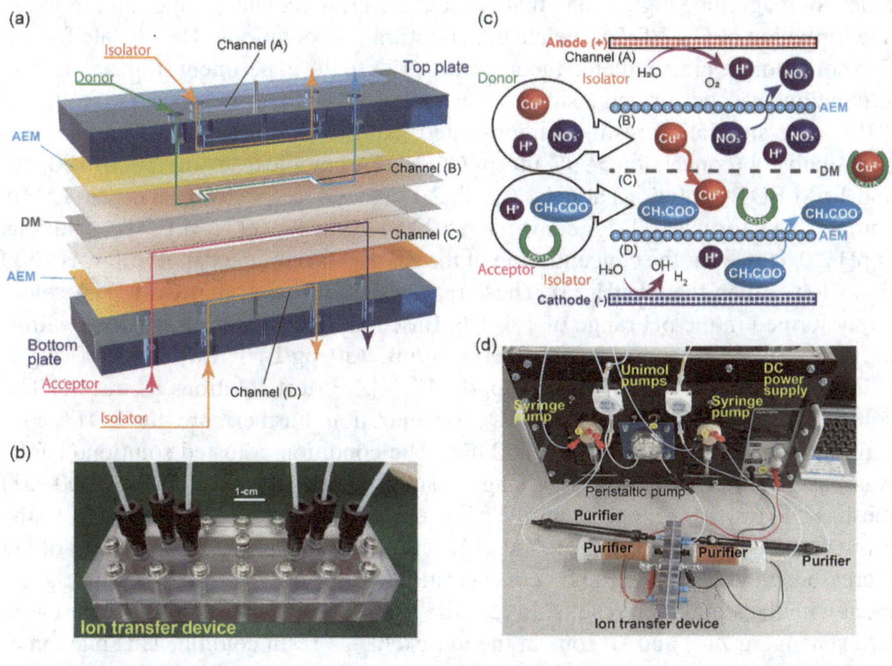

Fig. 4 Structure (**a**) and photo (**b**) of electrodialytic ion transfer device, schematic flow diagram (**c**), and photo of the whole system (**d**) used for matrix conversion, enrichment, and probe synthesis. *DM* dialysis membrane, *AEM* anion exchange membrane. Purifier columns are connected for in-line purification of the solutions

were purified with cation exchange resin filled in-line purifier column. The sample solution, which contained Cu^{2+} ions in an acid medium, was introduced via a six-port injection valve with a 1 or 2 mL sample loop. The acceptor solutions, which consisted of various acids or DOTA in acetic acid, were flowed at a rate of 0.10 mL/min. The photo of the whole system is shown in Fig. 4d. The effluents of the sample and donor channels were analyzed using a graphite furnace atomic absorption spectrometer or an inductively coupled plasma mass spectrometer for the nonradioactive tests and a NaI (Tl) scintillation detector for the radioactive tests.

3 Results

3.1 Separation of Ga from Zn with Selective Chelate Formation

Ga was separated from the target Zn via selective chelate formation, followed by cation trapping (Sugo et al., 2021). EDTA was used as the selected ligand, and the formation constants of the Ga–EDTA and Zn–EDTA chelates were $10^{20.27}$ and $10^{16.50}$, respectively. The formation constant of the Ga–EDTA chelate was ~3.8 orders of magnitude higher than that of the Zn–EDTA chelate, promoting the selective formation of Ga–EDTA chelate under optimized conditions. The chelate formation conditions were simulated theoretically prior to the experiment (Fig. 5a, b). The simulation was performed using the basic equilibrium theory (Christian et al., 2013). The simulation results demonstrated that the Ga^{3+} ions in the solution can quantitatively (conversion >99%) form Ga–EDTA complexes using 1000, 100, 10, and 1 μM EDTA solutions at pH 1.6, 1.8, 2.1, and 2.4, respectively. The Zn–EDTA complex was formed (EDTA conversion >99%) using 1–1000 μM EDTA solutions at pH 2.0. However, the concentration of the Zn–EDTA complex at pH 1.0 was 1000 times lower than that at pH 2.0. These results demonstrated that Ga-EDTA selectively formed in the pH range of 1.4–1.6. Based on the simulation results, preliminary tests were performed using 1.1 kBq/mL (50 pg/L) $[^{67}Ga]Ga^{3+}$, 100 mg/L $[^{nat}Zn]Zn^{2+}$, and 0.14 kBq/mL (450 pg/L) $[^{65}Zn]Zn^{2+}$ test solutions as tracers. The solution pH was adjusted by the HNO_3 concentration. Furthermore, the EDTA concentration was varied from 1 to 1000 μM. The condition-adjusted solution (1 mL) was passed through a cation-exchange resin column (Dowex 50W X8, 200–400 mesh, H^+-form, 1.0 g), and then, the ^{67}Ga and ^{65}Zn as the representative Zn in the effluent were analyzed using an NaI (Tl) scintillation counter. The recovery of Ga increased with increasing EDTA concentrations (Fig. 5c). This was in partial agreement with the simulation results. In real-life situations, the solution pH is decreased by exchanging Zn^{2+} and H^+ ions in the ion-exchange resin column. This phenomenon was not accounted for during the simulation. However, at low pH values, the equilibrium shifted toward a higher concentration of noncomplexed Ga^{3+}, which can be adsorbed using the cation-exchange resin column. On the other hand, with a

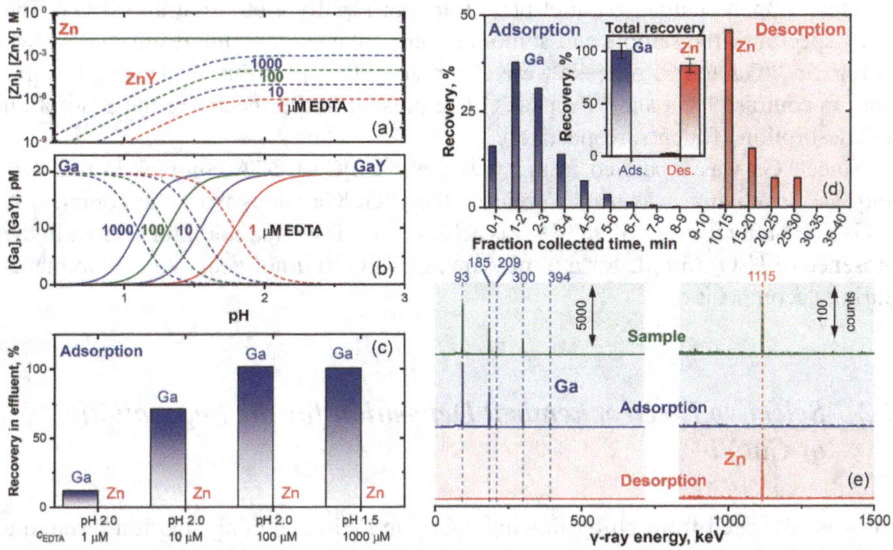

Fig. 5 Separation performances with a selective chelate formation method for Ga/Zn. Theoretical simulation of chelate formations for Zn (**a**) and Ga (**b**) with the different EDTA concentrations and solution pH. Preliminary RI experiment results as the recovery in cation exchange resin column effluent (**c**). Recovery of ^{67}Ga and ^{65}Zn in each fraction collected every 1 or 5 min for Ga and Zn, respectively (**d**). γ-Ray spectra for the solutions, sample, ion-trapping device effluent (adsorption), and Zn recovering solution (desorption) (**e**). All of the experiments were performed with the solution contained 1.1 kBq/mL (50 pg/L) [^{67}Ga]Ga^{3+}, 100 mg/L [^{nat}Zn]Zn^{2+}, and 0.14 kBq/mL (450 pg/L) [^{65}Zn]Zn^{2+} as a tracer

higher pH than 2.0, Zn effectively forms an EDTA complex that has a negative charge. Consequently, a pH of 1.5 and EDTA concentration of 1 mM were used to quantitatively separate Ga^{3+} ions from Zn^{2+} ions. Therefore, these experimental parameters were used in subsequent studies.

To achieve rapid separation in a closed system, a flow-injection system was prepared (Fig. 2a). The sample, which contained [^{65}Zn]Zn^{2+} and [^{67}Ga][Ga-EDTA]$^-$ ions, was passed through a cation-exchange trap to selectively remove Zn^{2+} ions from the solution. Subsequently, a 1 M HNO_3 solution was passed through the opposite side to recover the Zn^{2+} ions from the trap. The recoveries of ^{67}Ga and ^{65}Zn in the collected fractions were determined (Fig. 5d). During Zn^{2+}-ion removal (adsorption in Fig. 5d), only Ga was observed. The fraction collected at 1–2 min presented the highest recovery of Ga, and >95% of introduced Ga was collected within 5 min. Conversely, Zn was present in the fractions collected in the 10–30 min time interval of the recovery process (desorption in Fig. 5d). The peak dispersion was strongly affected by the type of cation-exchange trap used. Two types of traps, a resin-filled column and newly developed open-channel parallel-plate trap (Fig. 2b), were compared. For the resin-filled column, Ga and Zn were present in the fractions collected at 3–8 min and 15–30 min, respectively. The elution was delayed with the resin-filled column because of larger dead volumes. On the other hand, the newly developed open-channel parallel-plate trap can rapidly elute nontrapped Ga. The γ-ray spectra of the sample and solutions were obtained, and the results are shown in Fig. 5e. ^{67}Ga and ^{65}Zn peaks were observed in the spectrum of the sample solution. In contrast, ^{67}Ga and ^{65}Zn peaks were present in the spectra of the adsorption and desorption effluents, respectively.

Since ^{67}Ga was separated from Zn as the [^{67}Ga]Ga-EDTA complex in this system, UV irradiation was used to release the [^{67}Ga]Ga^{3+} ions from the complex. A [^{67}Ga]Ga^{3+}-ion recovery of 98.9% was achieved via UV irradiation for 10 min in the presence of H_2O_2 (50 μL for a 5 mL sample). H_2O_2 did not generate contaminants during decomposition.

3.2 *Selective Electrochemical Deposition for the Separation of Cu/Ni*

Cu was separated from Ni by flowing a Cu- and Ni-containing solution through a carbon-felt electrode cell. Unlike conventional electrodes, which use metal films or rods, large surface areas, and small solution volumes promote effective electrochemical reactions. The carbon-felt electrode facilitated the quantitative reduction of Cu^{2+} ions in the cell (Sugo et al., 2022).

In this study, Cu was selectively deposited on a carbon-felt electrode (Fig. 6a). A sample containing Cu^{2+} and Ni^{2+} ions in 0.1 M HCl was first introduced into the cell. By applying a negative potential, Cu^{2+} ions were reduced to Cu^0 on the electrode surface. In contrast, Ni^{2+} ions were not reduced and passed through the carbon-felt electrode. After the residual solution was flushed from the cell, the applied potential was switched to positive by flowing a 10 mM HNO_3 solution, and then, the deposited Cu^0 was eluted as Cu^{2+} ion in HNO_3. The effects of the applied potential on the recoveries of Cu^{2+} ions via deposition and elution are summarized in Fig. 6b, c, respectively. The yield of deposited Cu^{2+} ions reached >97% in the applied potential window of −0.8 to −0.5 V vs. the Ag/AgCl reference electrode. Conversely, the elution recovery was the highest (95 ± 5%) at the potential of +1.0 V vs. the Ag/AgCl reference electrode. These results suggest that the optimal potentials for Cu^{2+}-ion deposition and elution were −0.8 and +1.0 V vs. the Ag/AgCl reference electrode, respectively.

Radioactive Cu was separated from target Ni under optimized experimental conditions. Trace concentrations of [^{64}Cu]Cu^{2+} ions were selectively separated from Ni^{2+} ions with concentrations in the mM range (Fig. 6d). The recoveries of the Ni^{2+} and [^{64}Cu]Cu^{2+} ions in F1 and F3, which were obtained using applied potentials of −0.8 and +1.0 V vs. Ag/AgCl reference electrode, respectively, were >99%. In contrast, F2 and F4 did not contain both [^{64}Cu]Cu^{2+} and Ni^{2+} ions. The separation performance was further evaluated using the γ-ray spectra of the samples. The characteristic peaks of [^{64}Cu]Cu^{2+} and [^{57}Ni]Ni^{2+} ions were present in the γ-ray spectrum of the sample (^{57}Ni was added to the samples as a tracer). However, F1

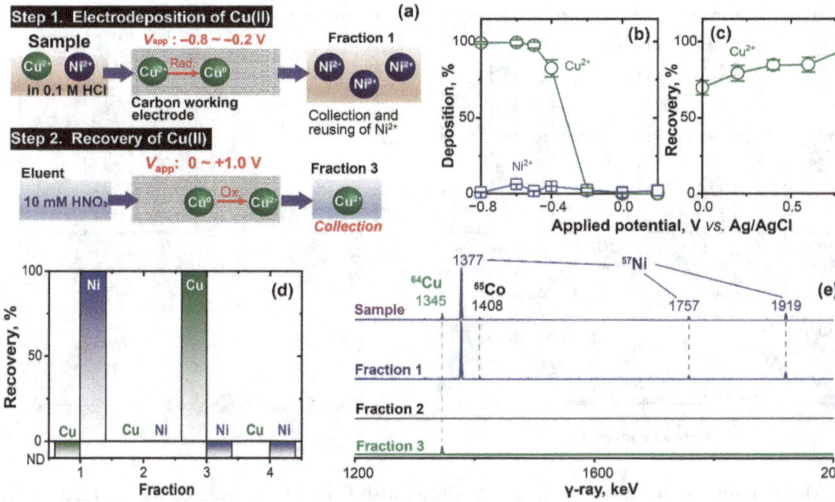

Fig. 6 Selective electrochemical deposition for the separation of Cu/Ni. Protocol of the separation (**a**). The applied potential effects on the Cu^{2+} deposition (**b**) and recovery efficiencies (**c**). These efficiencies were evaluated with the mixture of 200 μg/L Cu^{2+} and 200 μg/L Ni^{2+} in 0.1 M HCl. Recoveries of ^{64}Cu and Ni in each fraction (**d**). The γ-ray spectra of FE effluents were determined by a Ge semiconductor detector (**e**). These fraction studies were performed with the mixture, 20 kBq/mL (2.0 pM) [^{64}Cu]Cu^{2+}, 3.1 mM Ni^{2+}, 4 kBq/mL (1.2 pM) [^{57}Ni]Ni^{2+} as a tracer in 0.1 M HCl

and F3 contained only [^{57}Ni]Ni^{2+} and [^{64}Cu]Cu^{2+} ions, respectively. Separation was achieved within 30 min.

3.3 Electrodialytic Ion Transfer for Matrix Conversion, Enrichment, and Probe Synthesis

An electrodialysis ITD was used for electrodialytic matrix conversion (Sugo et al., 2020). A solution of Cu^{2+} ions in HCl was used as the sample, and then, Cu^{2+} ions were transferred into acids such as HNO_3, acetic acid, and citric acid (Fig. 7a). Cu^{2+} ions (5 μg/L) were quantitatively transferred into a 1 mM HNO_3 acceptor solution. Because the anions in the sample solution were also transferred into the acceptor solution, additional NO_3^- ions were supplied from a 10 mM HNO_3 isolator solution. Consequently, Cu^{2+} ions (5 μg/L), which were dissolved in 10 mM HCl, were added to 10 mM HNO_3. Similarly, solutions of Cu^{2+} ions in HCl were transferred into acetic and citric acids. Moreover, the anions in the acceptor solution were supplied by the isolator solution at the cathode. Although the transfer efficiencies of the acetic and citric acid acceptors were lower than that of HNO_3 (>75% and 50%, respectively), the acid matrices were converted from HCl to HNO_3, acetic acid, and citric

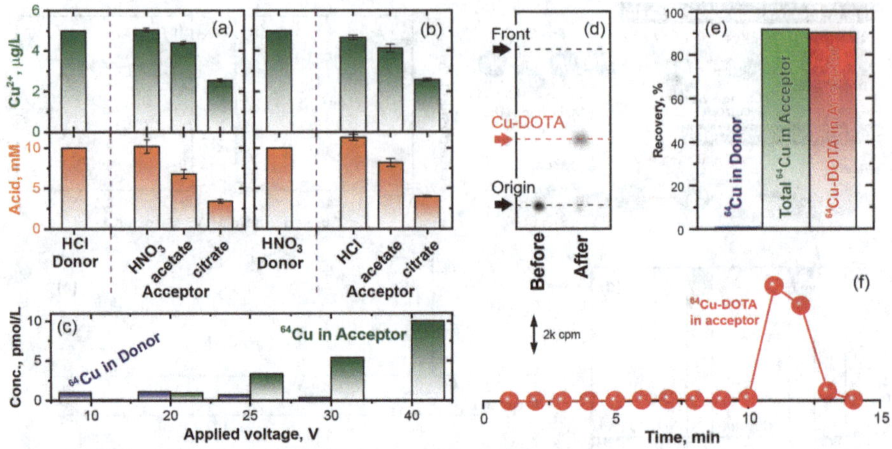

Fig. 7 Matrix conversion, enrichment, and synthesis with ITD. Matrix acid conversion from HCl to other acids (**a**) and HNO_3 to other acids (**b**). These evaluations were performed with 5 µg Cu/L solution. Initial concentration of the acceptor solution and cathode side isolator concentration were 1 and 10 mM, respectively. The effect of voltage on the [^{64}Cu]Cu^{2+} concentration in acceptor solution with the donor and acceptor flow rate ratio of 10 was performed (**c**). The synthesis of [^{64}Cu]Cu-DOTA complex with ITD was evaluated with radio thin layer chromatography (**d**), quantitative analysis (**e**), and radio high-performance liquid chromatography (**f**). All of these tests were performed with 10 kBq/L (1 pM) [^{64}Cu]Cu^{2+} in 10 mM HNO_3 as a sample solution. All of these experiments were performed with the donor, acceptor, and isolator solution flow rates of 0.1, 0.1, and 2.0 mL/min, respectively, except enrichment. The enrichment was performed with the donor flow rate of 1.0 mL/min

acid. Similar results were obtained for the transfer of Cu^{2+} ions from HNO_3 to HCl, acetic acid, and citric acid (Fig. 7b).

Electrodialytic enrichment was performed using [^{64}Cu]Cu^{2+}. Sample (10 kBq/L (1 pM) [^{64}Cu]Cu^{2+} in 10 mM HNO_3) and acceptor (1 mM HNO_3) solutions were supplied into the donor and acceptor channel of the ITD at flow rates of 1.0 and 0.1 mL/min, respectively. Upon achieving quantitative transfer, the same mole amount of [^{64}Cu]Cu^{2+} ion was dissolved in 1/10 of the volume of the acceptor solution after the enrichment. Therefore, the concentration of the solution was 10 times higher. In this study, the 10 kBq/L (1 pM) [^{64}Cu]Cu^{2+} ion solution used as the sample flowed under an increasing applied voltage (Fig. 7c). The transfer efficiency increased with increasing applied voltage. Consequently, quantitative transfer was achieved, and the concentration of [^{64}Cu]Cu^{2+} ions in the acceptor solution reached 10 pM, which was the same as the theoretical concentration calculated using the ratio of the flow rates of the sample and acceptor.

Furthermore, an ITD was used for probe synthesis. To synthesize a Cu^{2+}-based probe, the DOTA ligand, which is used to fabricate radiopharmaceuticals, was flowed into the acceptor channel as a diluted acetic acid solution. The sample solution comprising 10 kBq/mL (1 pM) [^{64}Cu]Cu^{2+} ions in 10 mM HNO_3 was added to the ITD, and [^{64}Cu]Cu^{2+} ions were transferred into the DOTA solution. The obtained acceptor solution contained [^{64}Cu]Cu^{2+} ions and DOTA, which formed a [^{64}Cu]Cu–DOTA complex. The obtained acceptor solution was analyzed using radio-thin-layer chromatography (Fig. 7d), and the results demonstrated that [^{64}Cu]Cu–DOTA was successfully synthesized using the ITD. Quantitative analysis was performed using a NaI (Tl) scintillation counter (Fig. 7e). The introduced ^{64}Cu was removed from the donor channel and introduced into the acceptor solution with an efficiency of >93%. Furthermore, 90% of the ^{64}Cu introduced into the donor channel was used to fabricate the [^{64}Cu]Cu–DOTA complex. These results were confirmed using radio-high-performance liquid chromatography. The effluent of the separation column in the HPLC system was collected every 1 min, followed by the analysis with an NaI (Tl) scintillation counter (Fig. 7f). The [^{64}Cu]Cu–DOTA complex was formed with high efficiency by passing the samples solution through the ITD once, which required ~10 s. The total process time for 1 mL solution was ~10 min with a flow rate of 0.1 mL/min.

4 Discussion

4.1 Separation of RI from Target Material

Separation of metal RIs from target materials is a critical process in metal RI-based pharmacotherapy. Currently, solid-phase extraction methods are the most widely used for metal RI purification. In most cases, these methods involve the use of highly concentrated acids, such as HCl. Even though HCl or HNO_3 can be removed via evaporation to dryness, complete removal is time consuming. The concentration of acids used in this study was low (~10 mM). Furthermore, universal methods that use EDTA as a chelating reagent or electrochemical deposition were successfully developed.

Selective chelation can be used to separate metal RIs from target materials based on differences in formation constants. In this study, Ga was separated from Zn via selective chelation. Moreover, selective chelation was used to separate ultra-trace amounts of metal RIs from large amounts of target materials (Zn/Ga concentration ratio of 10^{15}). Furthermore, selective chelation can be used to purify Ga solutions. The selectivity of Ga^{3+} over other metals was evaluated using a nonradioactive test under optimized conditions. A mixed solution of 100 μg/L Ga^{3+}, 100 mg/L Zn^{2+}, and 10 μg/L each of 23 other metals (ICP standard solution H; Kanto) was used in this study. After the solution pH and EDTA concentration were adjusted to 1.5 and 1 mM, respectively, the solution was passed through a cation-exchange resin column. The effluent contained a limited amount of metal ions. Oxoanions, As and Se, were not captured by the cation-exchange resin and were present in the effluent. In addition, Bi^{3+} and Fe^{3+} ions, which formed EDTA complexes with high formation constants (log $K_f = 27.9$ and 25.1, respectively), were present. However, the amount of Bi^{3+} ions in the sample was negligible. Moreover, Fe^{3+} ions can be removed from the sample by adding sulfite, which served as a reductant to the sample before separation. Furthermore, the concentration of Zn^{2+} ions in the effluent was only 2.3 μg/L. If all the introduced Ga^{3+} ions were assumed to be [^{68}Ga]Ga^{3+} ions, the concentration of Zn^{2+} ions was expected to be 15 pg/GBq, which was considerably lower than the concentration regulated by the European Pharmacopoeia (10 μg/GBq). Thus, a high-purity Ga^{3+}-ion solution was obtained via selective chelation.

Because the selective chelation method relies on the difference in formation constants, it can also be used to separate Ga from Zn and Zr from Y. Furthermore, the metal RI–ligand complexes can be decomposed via UV light irradiation to generate free metal RI ions (Sugo et al., 2021). These findings confirm that many types of radiopharmaceuticals can be synthesized via selective chelation.

Even though the selective chelate formation method can be used to separate many types of RIs, some RI–target combinations, such as Cu^{2+}–Ni^{2+}, are challenging to separate. The formation constants of the Cu^{2+}–EDTA and Ni^{2+}–EDTA complexes are comparable ($\log K_f = 18.80$ and 18.52, respectively). However, selective electrochemical deposition can be used to separate Cu^{2+} ions from Ni^{2+} ions. The selectivity of the electrochemical deposition method depends on the standard electrode potential ($E°$). The order of $E°$ is also known as the ionization tendency. The $E°$ values of Cu^{2+} and Ni^{2+} ions are +0.337 and −0.250 V, respectively. The polarities of the $E°$ values of Cu^{2+} and Ni^{2+} ions are opposite. Therefore, selective electrochemical deposition can be easily achieved by applying a negative potential to selectively reduce Cu^{2+} ions. Accordingly, Cu^{2+} ions were separated from Ni^{2+} ions using a highly efficient carbon-felt electrode. Trace ^{64}Cu was quantitatively recovered from a mixture of Cu^{2+} and Ni^{2+} ions using a diluted acid solution without evaporation. The total processing time was short (~30 min). Furthermore, expensive isotope-enriched ^{64}Ni was quantitatively recovered.

Both methods require small amounts of chemicals, nonheat, and short time and are effective for separating metal RIs, and in particular short-lifetime metal RIs. The universality of these methods is summarized in Table 1.

Table 1 The method application possibilities to radioactive metal separation from target element

metal RI[a]				Target			Possibility to apply	
Nuclide	$t_{1/2}$	K_f for EDTA[b]	E^{0c}	Reaction	K_f for EDTA	E^0	Selective chelate formation	Electrochemical deposition
[^{111}In]	67.392 h	24.95	−0.444	^{111}Cd(p,n)	16.46	−0.402	○	×
[^{67}Ga]	78.281 h	20.27	−0.549	^{67}Zn(p,n), ^{68}Zn(p,2n)	16.50	−0.762	○	×
[^{68}Ga]	1.13 h	20.27	−0.549	^{68}Zn(p,n)	16.50	−0.762	○	×
[^{64}Cu]	12.701 h	18.80	0.339	^{64}Ni(p,n)	18.62	−0.236	×	○
[^{86}Y]	14.74	18.09	−2.38	^{86}Sr(p,n)	8.6	−2.889	○	×
[^{89}Zr]	78.4 h	29.9	−1.45	^{89}Y(p,n)	18.09	−2.38	○	×

[a] Metal RIs for imaging and therapy are selected from the review (Bhattacharyya & Dixit, 2011)
[b] The formation constant values are cited from www.dojindo.co.jp/technical/protocol/log.pdf
[c] The standard reduction potential is cited from www.csun.edu/~hcchm003/321/Ered.pdf

4.2 Electrodialytic Ion Transfer for Matrix Conversion, Enrichment, and Probe Synthesis

Electrodialysis ITDs were used for sample pretreatment in ionic solute analyses (Ohira et al., 2012). Matrix conversion can be easily achieved using ITDs. The matrix of all analytical systems was the same. In this study, we successfully performed matrix conversion for metal RIs. The primary method used for matrix and acid conversion in metal RI treatment is evaporation. For ITD enrichment, the enrichment factor only depends on the flow rate and donor–acceptor ratio if the quantitative transfer is achieved. The enrichment of ionic solutes can be achieved using solid-phase extraction or evaporation. The proposed ITD method can simultaneously achieve matrix conversion and in-line enrichment. Furthermore, the method can be used to in-line synthesize probes. The introduced metal RI can be transferred to the ligand solution, and metal RI complexes can be synthesized. Unlike the traditional synthesis method, this method does not require mixing the solutions because ionic solutes can be directly introduced into the solution. The proposed method can simultaneously achieve matrix conversion, enrichment, and probe synthesis.

5 Conclusion

We have tried to establish theoretically new methods for separation, purification, and synthesis of RI metal ions. Some of these methods were initially established to analyze metal ions in environmental and biological samples. In the present, we have applied the methods to RI and proven that the system can work even at the trace levels. In the future, these methods will support a practical medical diagnosis with metal RIs.

Acknowledgments These works were supported by JSPS KAKENHI (Grant Numbers 21H02870, 20H03632, and 17K19137).

References

Alirezapour, B., Rasaee, M. J., Jalilian, A. R., Rajabifar, S., Mohammadnejad, J., Paknejad, M., Maadi, E., & Moradkhani, S. (2016). Development of [^{64}Cu]-DOTA-PR81 radioimmunoconjugate for MUC-1 positive PET imaging. *Nuclear Medicine and Biology, 43*, 73–80. https://doi.org/10.1016/j.nucmedbio.2015.07.012

Alves, V., do Carmo, S., Alves, F., & Abrunhosa, A. (2018). Automated purification of radiometals produced by liquid targets. *Instruments, 2*, 17. https://doi.org/10.3390/instruments2030017

Bhattacharyya, S., & Dixit, M. (2011). Metallic radionuclides in the development of diagnostic and therapeutic radiopharmaceuticals. *Dalton Transactions, 40*, 6112. https://doi.org/10.1039/c1dt10379b

Christian, G. D., Dasgupta, P. K., & Schug, K. A. (2013). *Analytical chemistry* (7th ed.). Wiley.

Grillo-López, A. J. (2002). Zevalin: The first radioimmunotherapy approved for the treatment of lymphoma. *Expert Review of Anticancer Therapy, 2*, 485–493. https://doi.org/10.1586/14737140.2.5.485

Hennrich, U., & Benešová, M. (2020). [^{68}Ga]Ga-DOTA-TOC: The first FDA-approved ^{68}Ga-radiopharmaceutical for PET imaging. *Pharmaceuticals, 13*, 38. https://doi.org/10.3390/ph13030038

Kraus, K. A., & Moore, G. E. (1953). Anion exchange studies. VI. [1,2] The divalent transition elements manganese to zinc in hydrochloric acid. *Journal of the American Chemical Society, 75*, 1460–1462. https://doi.org/10.1021/ja01102a054

McCarthy, D. W., Shefer, R. E., Klinkowstein, R. E., Bass, L. A., Margeneau, W. H., Cutler, C. S., Anderson, C. J., & Welch, M. J. (1997). Efficient production of high specific activity ^{64}Cu using a biomedical cyclotron. *Nuclear Medicine and Biology, 24*, 35–43. https://doi.org/10.1016/S0969-8051(96)00157-6

Mori, M., Honda, S., & Kozaki, D. (2020). Selective recovery of copper(II) from incineration ash produced from a municipal waste incineration facility using a flow-through type electrolysis method. *Analytical Sciences, 36*, 611–615. https://doi.org/10.2116/analsci.19SBN01

Ohira, S.-I., Kuhara, K., Kudo, M., Kodama, Y., Dasgupta, P. K., & Toda, K. (2012). Electrodialytic ion isolation for matrix removal. *Analytical Chemistry, 84*, 5421–5426. https://doi.org/10.1021/ac300982k

Ohira, S.-I., Hiroyama, Y., Nakamura, K., Koda, T., Dasgupta, P. K., & Toda, K. (2015a). Electrodialytic matrix isolation for metal cations. *Talanta, 132*, 228–233. https://doi.org/10.1016/j.talanta.2014.09.013

Ohira, S.-I., Nakamura, K., Shelor, C. P., Dasgupta, P. K., & Toda, K. (2015b). Simultaneous electrodialytic preconcentration and speciation of chromium(III) and chromium(VI). *Analytical Chemistry, 87*, 11575–11580. https://doi.org/10.1021/acs.analchem.5b03464

Ohira, S.-I., Okazaki, C., & Toda, K. (2018). Electrodialytic matrices isolation for determination of heavy metals in soil extracts by anodic stripping voltammetry. *Bunseki Kagaku, 67*, 761–766. https://doi.org/10.2116/bunsekikagaku.67.761

Okazaki, Y., Hoshi, S., Kato, T., Fukui, T., Toda, K., & Ohira, S.-I. (2022). Electrodialytic enrichment and matrix conversion for the determination of trace metals in ultra-pure water. *ACS Omega, 7*, 14082–14088. https://doi.org/10.1021/acsomega.2c00648

Pfeifer, A., Knigge, U., Mortensen, J., Oturai, P., Berthelsen, A. K., Loft, A., Binderup, T., Rasmussen, P., Elema, D., Klausen, T. L., Holm, S., von Benzon, E., Hojgaard, L., & Kjaer, A. (2012). Clinical PET of neuroendocrine tumors using ^{64}Cu-DOTATATE: First-in-humans study. *Journal of Nuclear Medicine, 53*, 1207–1215. https://doi.org/10.2967/jnumed.111.101469

Rodnick, M. E., Sollert, C., Stark, D., Clark, M., Katsifis, A., Hockley, B. G., Parr, D. C., Frigell, J., Henderson, B. D., Abghari-Gerst, M., Piert, M. R., Fulham, M. J., Eberl, S., Gagnon, K., & Scott, P. J. H. (2020). Cyclotron-based production of ^{68}Ga, [^{68}Ga]GaCl$_3$, and [^{68}Ga]Ga-PSMA-11 from a liquid target. *EJNMMI Radiopharmacy and Chemistry, 5*, 25. https://doi.org/10.1186/s41181-020-00106-9

Sugo, Y., Miyachi, R., Maruyama, Y., Ohira, S.-I., Mori, M., Ishioka, N. S., & Toda, K. (2020). Electrodialytic handling of radioactive metal ions for preparation of tracer reagents. *Analytical Chemistry, 92*, 14953–14958. https://doi.org/10.1021/acs.analchem.0c02456

Sugo, Y., Miyachi, R., Obata, S., Maruyama, Y., Manabe, H., Mori, M., Ishioka, N. S., Toda, K., & Ohira, S.-I. (2021). Rapid flow-based system for separation of radioactive metals by selective complex formation. *Analytical Chemistry, 93*, 17069–17075. https://doi.org/10.1021/acs.analchem.1c03866

Sugo, Y., Ohira, S.-I., Manabe, H., Maruyama, Y., Yamazaki, N., Miyachi, R., Toda, K., Ishioka, N. S., & Mori, M. (2022). Highly efficient separation of ultratrace radioactive copper using a flow electrolysis cell. *ACS Omega, 7*, 15779–15785. https://doi.org/10.1021/acsomega.2c00828

Thisgaard, H., Kumlin, J., Langkjær, N., Chua, J., Hook, B., Jensen, M., Kassaian, A., Zeisler, S., Borjian, S., Cross, M., Schaffer, P., & Dam, J. H. (2021). Multi-curie production of gallium-68 on a biomedical cyclotron and automated radiolabelling of PSMA-11 and DOTATATE. *EJNMMI Radiopharmacy and Chemistry, 6*, 1. https://doi.org/10.1186/s41181-020-00114-9

Watanabe, S., Iida, Y., Suzui, N., Katabuchi, T., Ishii, S., Kawachi, N., Hanaoka, H., Watanabe, S., Matsuhashi, S., Endo, K., & Ishioka, N. S. (2009a). Production of no-carrier-added ^{64}Cu and applications to molecular imaging by PET and PETIS as a biomedical tracer. *Journal of Radioanalytical and Nuclear Chemistry, 280*, 199–205. https://doi.org/10.1007/s10967-008-7443-9

Watanabe, S., Watanabe, S., Liang, J., Hanaoka, H., Endo, K., & Ishioka, N. S. (2009b). Chelating ion-exchange methods for the preparation of no-carrier-added ^{64}Cu. *Nuclear Medicine and Biology, 36*, 587–590. https://doi.org/10.1016/j.nucmedbio.2009.04.005

Xiao, L., Zhang, Y., Yue, W., Xie, X., Wang, J., Chordia, M. D., Chung, L. W. K., & Pan, D. (2013). Heptamethine cyanine based ^{64}Cu-PET probe PC-1001 for cancer imaging: Synthesis and in vivo evaluation. *Nuclear Medicine and Biology, 40*, 351–360. https://doi.org/10.1016/j.nucmedbio.2013.01.001

Part II
Radiolabeling Technologies

Basics and Progress of PET Radiopharmaceutical Synthesizers from a Company's Viewpoint

Jun Kato, Toshimitsu Watanabe, Satoshi Ueno, and Hiroki Higuchi

1 Introduction

1.1 Prevalence of PET Radiopharmaceutical Synthesizers

Positron emission tomography (PET) has been widely used as a tool for the noninvasive diagnosis of many major diseases. At present, pharmaceuticals labeled with radioactive isotopes are used in PET in many hospitals and laboratories around the world. The reason why the use of PET in diagnosis has spread so widely is that the radiopharmaceutical [^{18}F]FDG useful in cancer diagnosis is widely available (Boonyawan et al., 2021; Collins et al., 2017).

[^{18}F]FDG synthesis methods have been improved extensively since the method using [^{18}F]F_2 gas was published in 1976 (Adam, 1999). After Hamacher at the Julich Institute in Germany published the synthesis method by the nucleophilic substitution reaction using [^{18}F]F^- in 1986, the efficiency of synthesis was improved, and since around 2000, synthesis techniques under more moderate reaction conditions using alkaline hydrolysis were developed. At present, on-column hydrolysis techniques in which hydrolysis is performed on a column are mainly used (Hamacher et al., 1986; Fochtner et al., 1996; Lemaire et al., 2002). These improvements have been motivated by the growing usefulness of [^{18}F]FDG in cancer diagnosis (Reske, 2003; Higashi et al., 2003; Kitajima et al., 2009; Seltzer et al., 2002) and the consequent increasing demand for more efficient synthesis techniques (Culbert et al., 1995; Yuasa et al., 1997).

PET radiopharmaceutical synthesizers significantly evolved over this same period. In the early days, pharmaceuticals were synthesized using equipment such

J. Kato · T. Watanabe · S. Ueno · H. Higuchi (✉)
Industrial Equipment Division, Sumitomo Heavy Industries, Ltd., Niihama, Japan
e-mail: hiroki.higuchi@shi-g.com

R. M. van Dam et al. (eds.), *Automated Technologies for the Development and Production of Radiopharmaceuticals*,
https://doi.org/10.1007/978-3-031-84632-8_7

as experimental instruments consisting of scientific glassware (Mock et al., 1996; Chaly et al., 1990; Iwata et al., 1984; Ido et al., 1978). After this, there was an increasing demand for remotely controllable synthesizers from the viewpoints of improving working efficiency and reducing radiation exposure of workers, and a variety of synthesizers were developed (Kryza et al., 2008; Krasikova, 2007; Saxena et al., 2021; Boschi et al., 2013). The synthesizers were initially intended for specialized researchers (Iimori et al., 2011), mainly to validate the synthesis methods examined. Therefore, the synthesizers had simple structures (Toyohara et al., 2021; Naka et al., 2020) so that they could be modified by the researchers themselves without the need for special mechanical expertise. Eventually, the efficacy of the representative pharmaceutical [^{18}F]FDG was verified, and an increasing number of less specialized users required it. As a result, the focus shifted from the examination and validation of synthesis methods to easy and stable synthesis (Saxena et al., 2021), not requiring specialized knowledge. Recently, synthesizers that started as experimental equipment have been required to be not only easy to use but also fully usable in pharmaceutical manufacturing equipment (IAEA, 2021).

1.2 Features of PET Radiopharmaceutical Synthesizers

The PET radiopharmaceutical synthesizer is, as named, a piece of stand-alone equipment made up of different components that implement the various steps required to synthesize PET radiopharmaceuticals. Such systems have significant differences from other chemical synthesizers. Since the compounds produced by this device contain radioactive materials, it is imperative to prevent user exposure to radiation. Therefore, the synthesizer must be installed in an area enclosed by radiation shielding and operated remotely. Furthermore, the gas generated during the synthesis process may contain radioactive materials; therefore, a mechanism to prevent the release of this gas into the atmosphere is required. The synthesizer is a system in which a chemical reaction is induced. The following processes are performed in the synthesizer: (1) introduction of reagents; (2) heating under reduced pressure to evaporate reagents such as acetonitrile or water as a solvent to dryness; (3) heating to accelerate reactions; and (4) purification of the reaction mixture and transfer to the next step in the synthesis processes.

Other functions are added to these basic functions of the synthesizer in accordance with the method of synthesizing the required pharmaceutical. The additional functions include, for example, isolation by HPLC, solvent switching, and solid-phase extraction.

2 Synthesizer Components

Though there are many radiosynthesizers and manufacturers, including SHI, GE HealthCare, IBA RadioPharma Solutions, TRASIS, ORA, etc., most of them share common functional components (Petriev et al., 2016; Petroni et al., 2020). In this section, we describe the needed components and the most common implementations for each.

2.1 Reactor

There are three major factors to consider when designing the reactor: the heating mechanism, the reaction container material, and its capacity.

2.1.1 Heating Mechanism

The following three representative types of heating equipment are available: oil bath, air heater, and solid heater (Fig. 1). As for heating methods, aside from the three mentioned above, there are direct heating, optical heating, induction heating, and microwave heating (Hou et al., 2011). However, these are not frequently used in synthesizers, so they will not be discussed further in this chapter.

The features, advantages, disadvantages, and key points for the design of the heaters are discussed below.

2.1.1.1 Oil Bath

In the early-stage synthesizers, flasks were used as reaction containers and heated in oil baths, a setup where oil was poured into a container with an electric heating element, indirectly heating the reaction vessel. The controller was used to monitor and accurately maintain the oil temperature at the desired level. The advantage of using an oil bath was its ability to uniformly heat the reaction container due to the heating properties of the oil. However, because oil evaporates when heated, this system necessitated regular maintenance, including cleaning and replenishing the oil for proper operation. Additionally, both heating and cooling the oil were time-consuming processes.

The users (researchers) in the early days possessed specialized knowledge and understood the structure and principle of the equipment. Consequently, they handled it with care, performing periodic cleaning and monitoring oil levels to ensure consistent heating. However, in recent practice, synthesizers are often operated by less specialized users. Hence, heating systems relying on user maintenance should be phased out. At present, there are very few synthesizers that use oil baths.

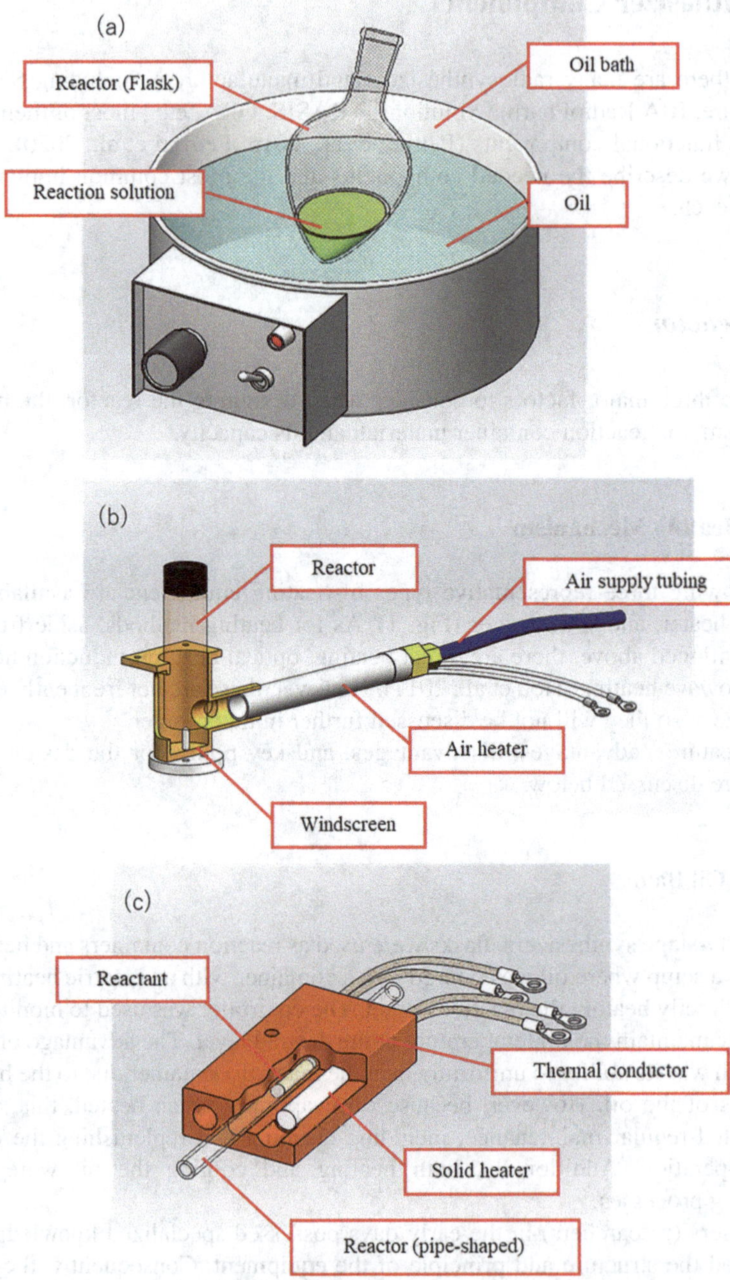

Fig. 1 Example of heaters. (**a**) Oil bath, (**b**) air heater, and (**c**) solid heater

2.1.1.2 Air Heater

The air heater blows heated compressed air through the reactor heater, thus providing a stream of hot air to heat the object indirectly. The heating medium, i.e., compressed air, does not contaminate the surroundings even when it is blown into the reactor, and it necessitates minimal maintenance by operators. The advantage of using the air heater is that the inside of the reactor can be observed from the outside by using a camera if the reactor is made of a transparent material such as glass since compressed air is transparent. After heating is stopped, the flow of compressed air can be used to efficiently cool the reactor. However, there are several points to be careful of when using air heaters. First, the controller controls the temperature of the air from the air heater but does not control the temperature of the object to be heated. Therefore, depending on the distance between the air heater and the object to be heated or the airflow path, there may be a difference between the controlled and the heated temperature, which may prevent heating as planned. Consequently, it is necessary to determine the relationship between the set temperature and the actual temperature achieved by heating. Moreover, if the supply of compressed air is suspended during heating, the air heater may overheat and fail. Therefore, a mechanism to prevent the suspension of the compressed air supply during heating or to stop heating when the supply is suspended must be provided. Additionally, the air heater heats the object with a blast of high-pressure compressed air and generates streams of air around the equipment. Therefore, it may not be desirable in an environment typical of recent pharmaceutical manufacturing facilities where the particulate concentration is limited.

2.1.1.3 Solid Heater

The solid heater consists of a heat-transmitting body (normally made of a material with high heat conductivity such as copper or aluminum) into which vials or tubes are inserted. A heating element, often a cartridge heater, is embedded within this body. Unlike the oil bath and air heater, the solid heater uses a solid as a heating medium; therefore, it causes neither dispersion of oil nor generation of particles due to flowing air and can be used in a clean environment.

The narrower the gap between the reactor and the heat-transmitting body is, the higher the heat conductivity is. However, if the gap is too narrow, depending on the tolerance of the reactor, the reactor cannot be set in the heat-transmitting body or the reactor may be damaged owing to the difference in the coefficients of expansion during heating. The gap must be determined considering the manufacturing tolerances of the reactor to be used.

The heat-transmitting body has a certain heat capacity. Therefore, it takes time to cool it after heating is stopped. The cooling time can be reduced by feeding air to the heat-transmitting body to cool it actively and more efficiently. However, if the air feed rate is high, streams of air are generated around the equipment as in the case of the air heater. The cooling method must be devised carefully.

To examine a cooling method without active air feeding, the heat-transmitting body must be designed to have a fin or a similar structure to facilitate the release of heat or to be as small in volume as possible to reduce its heat capacity. The heating mechanisms have their respective advantages and disadvantages. Nowadays, in many cases, solid heaters are used to meet the requirements depending on the environment where the synthesizers are used.

2.1.2 Material of Reaction Container

The reaction container for this reaction is used not only as a container of reagents but also as a heating container. Therefore, the container material must be taken into consideration when selecting a reaction container.

The major material properties to consider include heat conductivity, temperature compatibility, chemical resistance, and transparency.

2.1.2.1 Glassy Carbon

Glassy carbon has the characteristics of both glass and ceramic. It is very hard, has high heat conductivity, and is resistant to chemicals; therefore, its characteristics are ideal for chemical reaction containers. These reactors are reused after cleaning because they are expensive. However, if the reactors are cleaned insufficiently, contamination that interferes with synthesis may occur. Accordingly, for precision synthesis, a cleaning method must be established based on cleaning validation. Since it is easily damaged by impact, it must be handled carefully and is not very suitable for inexperienced users. Another drawback is that the material is opaque, preventing visualization of the inside of the reaction cavity.

2.1.2.2 Glass

Glass reactors are highly resistant to chemicals and transparent and have high visibility. In the case of heating with an air heater, the inside of the reactor can be easily observed from the outside by using a camera. In addition, glass reactors are considerably cheaper than glassy carbon reactors, making them affordable. Moreover, glass reactors can be designed and produced in various shapes. However, if the shape is very complicated, the production cost will increase, and, as with the glassy carbon reactors, the reactors must be cleaned and reused. In this case, cross-contamination, which interferes with synthesis, may be caused by insufficient cleaning. Recently, there has been an increasing number of synthesizers whose flow paths, including the reactors consist of disposable parts. In many cases where the parts are disposed of after a single use, mass-produced general-purpose reactors in simple shapes or existing standard glass vials are used as reactors.

2.1.2.3 Other Materials

In addition to glassy carbon and glass reactors, some manufacturers use plastic reactors. Plastic containers have the advantage of low cost because of the possibility of mass production. Many of the materials used are engineering plastics or super-engineering plastics. In particular, polycarbonate (PC), polyphenylene sulfide (PPS), polysulfone (PSU), polyether ether ketone (PEEK), and fluorinated polymers are often used. However, some of these materials are not so resistant to certain types of organic solvents used for synthesis, and all materials have the limitation on the temperature for heating. Thus, it is necessary to watch for contamination of the synthesized final pharmaceutical with leached materials. When using not only a plastic container but also any plastic parts, we must consider the possibility of the leakage of gas containing radioactive materials through these parts.

2.1.3 Container Capacity

Normally, the amount of the reagent required for synthesis is very small because the amount of the radioactive material itself is very small.

However, the reaction system of any synthesizer includes dead volume, and a loss of a certain amount of chemical solution cannot be avoided. Therefore, for stable synthesis, the amounts of the reagent used in the synthesis should be adjusted to reduce the effect of dead volume.

In many automated protocols in the literature, reagent volumes are most commonly 0.5–2 mL. The exception is when collecting the crude product from the reactor, in which up to 5–10 mL may be added for dilution.

2.1.4 Additional Suggestions

In Sumitomo's synthesizer, where the flow path consists of fixed tubes, the reactor is designed to be replaced for each synthesis. Once the synthesis is completed and the flow paths have been cleaned, the reactor is replaced with another that has been cleaned and dried separately. The tubing for introducing reagents and draining the reaction solution is placed in a joint called the flange, and the reactor is attached to the flange (Fig. 2).

2.2 Flow Path

The flow path is used mainly to transfer a reagent to the reactor, discharge the reaction solution to the purification system or product vial, and transfer carrier gas. For chemical reactions, it is important to prevent cross-contamination in the flow paths through which the reagents are fed. In many synthesizers, cross-contamination is

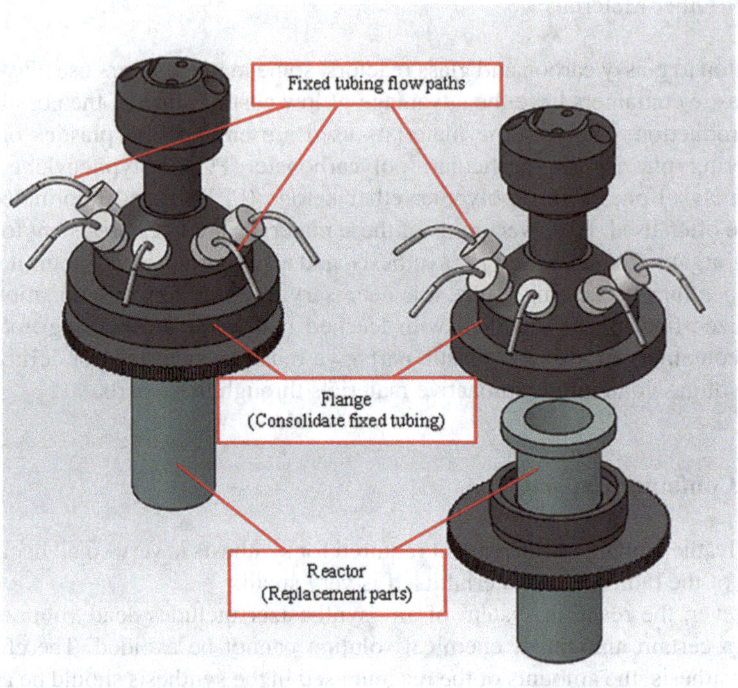

Fig. 2 Reactor and tubing construction

minimized by cleaning them before and after synthesis. However, recently, it has been required to ensure the cleanliness of the flow path through cleaning validation if the tubing flow path is reused after cleaning (FDA, 2014). As a result of this, it takes additional time to prepare for synthesis and subsequent operations, and the user must have a high level of skill and knowledge. To avoid this complexity, the chemical solution flow paths of most of the recent synthesizers consist of disposable parts. When examining the flow path parts for any method, it is necessary to select parts considering their characteristics, such as chemical resistance, pressure resistance, radiation resistance, and permeability, as well as the characteristics of the fluid to be carried. In this section, we explain the fixed tubing flow path and the disposable tubing flow path.

2.2.1 Fixed Tubing Flow Path

The fixed tubing flow path is conventionally used. Plastic tubing is generally used, and the flow path is controlled by solenoid valves. The fixed tubing flow path consists of tubing, solenoid valves, and fittings (Fig. 3).

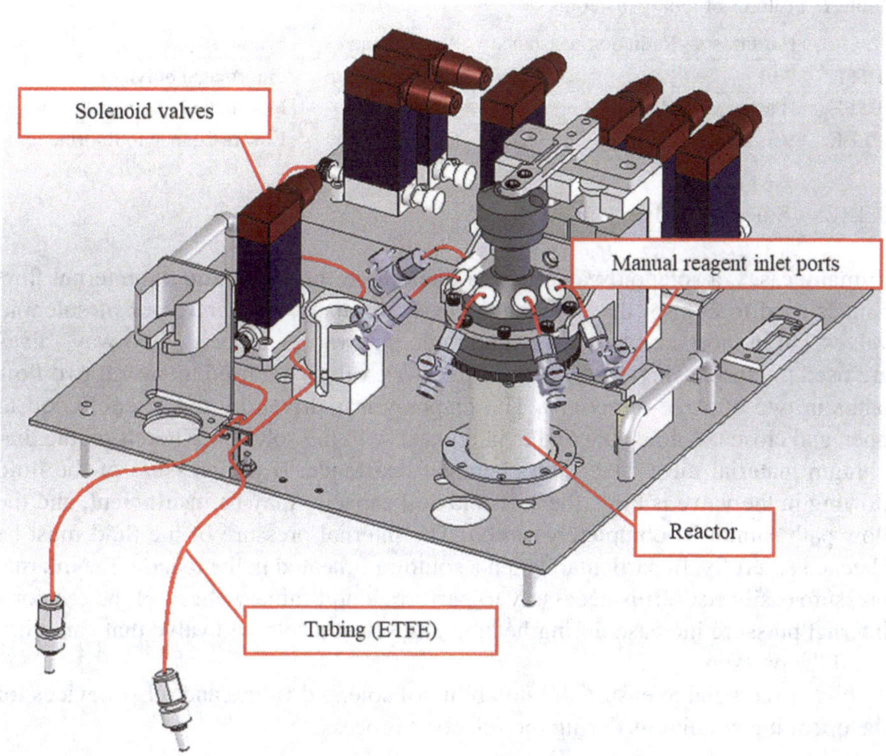

Fig. 3 Example of fixed tubing flow path (FCN-MPS100 F-Tray, Sumitomo Heavy Industries)

2.2.1.1 Tubing

In many cases, resins with high chemical resistance, represented by fluorinated polymers and PEEK, are selected as tubing materials. Polytetrafluoroethylene (PTFE) has higher flexibility than other fluorinated polymers and is easy to use for tubing, but it is less resistant to radiation and easily deteriorates upon prolonged exposure to high-intensity radiation. Ethylene tetrafluoroethylene (ETFE) has high radiation resistance but is less flexible than PTFE. Therefore, if ETFE is used for tubing of more than a certain thickness, care must be taken not to bend the tubing with a small radius. PEEK has good chemical and radiation resistances, but if fluorine gas (used for the synthesis of some radiopharmaceuticals) is fed into thin PEEK tubing, the flexibility of the tubing will significantly decrease, and fracturing will occur, even with a small external force. These materials have distinct features, and they must be selected in accordance with the conditions of use (Table 1).

Table 1 Features of tubing materials

	Hardness	Radiation resistance	Appearance	Features
PTFE	Soft	Low	Opaque (white)	Fluorinated polymer
ETFE	Hard	High	Translucent	Fluorinated polymer
PEEK	Very hard	High	Opaque	Low resistance to fluorine gas

2.2.1.2 Solenoid Valve

In many cases, a solenoid valve with a diaphragm for separating the internal flow path is used to control the flow path. There are many different types of solenoid valves, but in most cases, two-way and three-way valves are used. Two-way valves are used to close a single flow path. Three-way valves are used to switch two flow paths in two of three directions. The diaphragm is driven by the solenoid coil to open and close the flow path and is in contact with the solution. Therefore, the diaphragm material must have high chemical resistance. If the pressure of the fluid flowing in the valve is high, the solenoid coil capacity may be insufficient, and the flow path cannot be completely closed. The internal pressure of the fluid must be checked carefully. In particular, when a solution is heated in the reactor, the internal pressure easily rises. It is necessary to perform a preliminary check of the reactor's internal pressure increase during heating and select a solenoid valve that can withstand the pressure.

It is also crucial to ensure the suitability of solenoid valves and other devices for the operating conditions during the selection process.

2.2.1.3 Fittings

Since tubing and solenoid valves are standardized and easily available, they are connected using fittings employed for HPLC, for example. The fitting comprises a nut and a ferrule. For the ferrule, which is a wetted part, ETFE with high chemical resistance is normally used. For the nut, which does not come in contact with reagents, any material may be used. However, it is recommended to use a material with high chemical resistance because the chemical solution may leak due to unexpected problems.

2.2.2 Disposable Parts of Tubing Flow Path

Recently, there have been an increasing number of synthesizers whose flow paths, including the reactors, consist of disposable parts because the preparation procedure of such parts is simplified and cleaning validation is not required (Fig. 4). Generally, the disposable tubing comprises inexpensive flexible tubes that are combined to form the flow path, and the flow path is controlled by a mechanism on the outside of the tubing that will not come in contact with reagents.

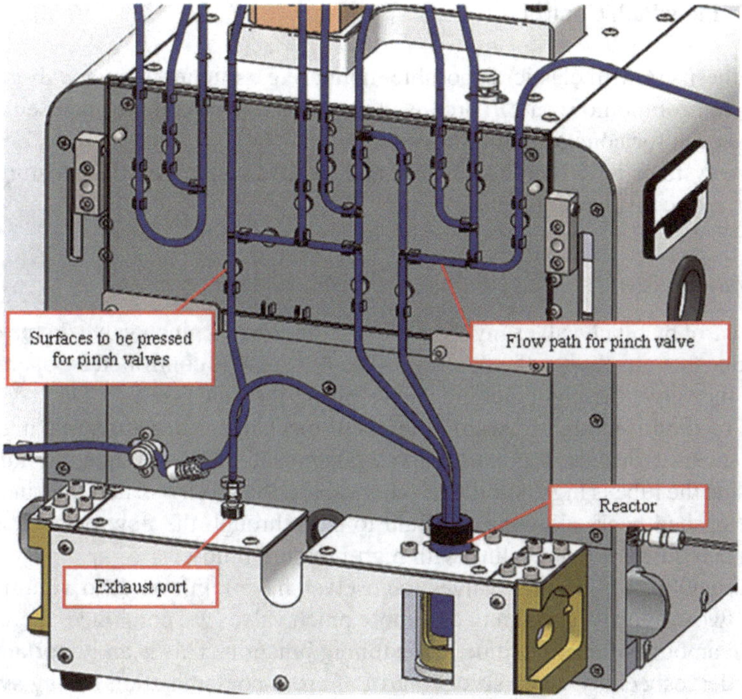

Fig. 4 Example of disposable tubing flow path used in the F300 synthesizer (Sumitomo Heavy Industries)

2.2.2.1 Tubing

Silicone tubing, a low-cost flexible tubing, is often used, but other materials such as PTFE, ETFE, perfluoroalkoxy alkane (PFA), and fluorinated ethylene propylene (FEP) are also used. Each material has its desirable characteristics, such as temperature compatibility, chemical resistance, and strength, so it is necessary to select the appropriate tubing material in accordance with the reagents used, temperature, and other synthesis conditions. There is a range of chemically resistant tubing options designed for peristaltic pumps, and it is advisable to choose disposable tubing for synthesizers from among these options. The gas produced during synthesis may contain radioactive materials, raising the possibility of radioactive leakage. To reduce radioactive leakage, measures such as covering the outer surface of the tube to prevent gas leakage can be taken.

2.2.2.2 Flow Path Control

For synthesizers with elastic disposable tubing (e.g., silicone), pinch valves can be used to control the flow path. Three-way stopcock valves can also be used and are effective even for nonelastic tubing systems.

These systems have their advantages and disadvantages, and their features must be taken into consideration.

Pinch Valve System

Flow control by pinch valves involves the installation of a pinch valve on the outside of the tube that constitutes the flow path. The flow path is controlled by opening and closing the valve, thereby pinching and releasing the tube (see Fig. 5).

This method is straightforward in terms of mechanical structure, making it easy to implement. However, it comes with certain considerations. When the fluid pressure inside the tube is high, similar to other valves, there is a risk that the pinch valve may be pushed back, allowing the fluid to pass through the flow path. Therefore, caution is required when dealing with high-pressure fluids.

Additionally, while pinch valves themselves have minimal dead volume when used as two-way valves, when two or more pinch valves are combined, there can be a certain amount of dead volume at the tubing junction. This is an important factor to consider, especially when strict control of cross-contamination during synthesis is necessary.

Furthermore, some tube materials have low elasticity. If the pinch valve system is used for a flow path consisting of low-elasticity tubing, the flow path tubing may

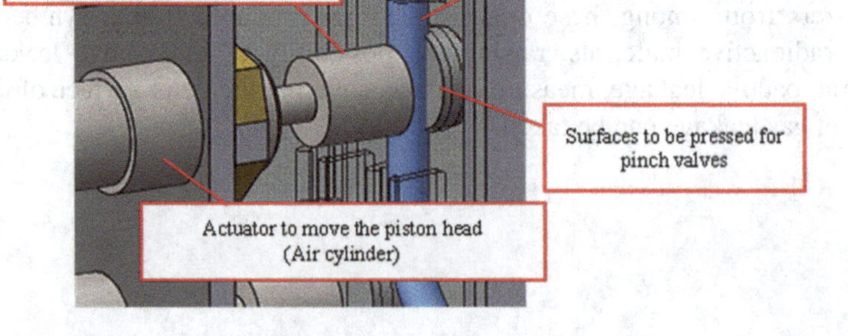

Fig. 5 Control of flow paths by pinch valves

remain closed even though the pinch valve is actuated to open the flow path. Hence, it is necessary to consider the elasticity of the tubes.

Three-Way Stopcock System

The three-way stopcock system is designed to control the flow path by turning the three-way stopcock using an external mechanical device. The rotary drive system to turn the three-way stopcock is more complicated and larger than the pinch valve system. When using a three-way stopcock system, it is important to be cautious, as excessive internal pressure can lead to leaks at connections or result in liquid or gas flowing in the wrong direction. Typically, a single three-way stopcock is used to switch between two directions, i.e., connecting the common port to either outlet #1 or outlet #2 (Fig. 6).

When using a commercially available three-way stopcock, attention must be paid to its material. Three-way stopcocks are available in materials such as polycarbonate, polypropylene, and polyethylene. Polycarbonate has limited chemical resistance to certain organic solvents, while polypropylene and polyethylene exhibit better solvent resistance. The selection of the three-way stopcock material should be based on compatibility with the reagents used in the synthesis.

2.3 Reagent Container

The reagent container is used to store the reagent to be used for synthesis. When fixed tubing is selected for the flow path, glass reagent containers are often provided in the flow path and filled with the chemical solutions to be used in the synthesis. In

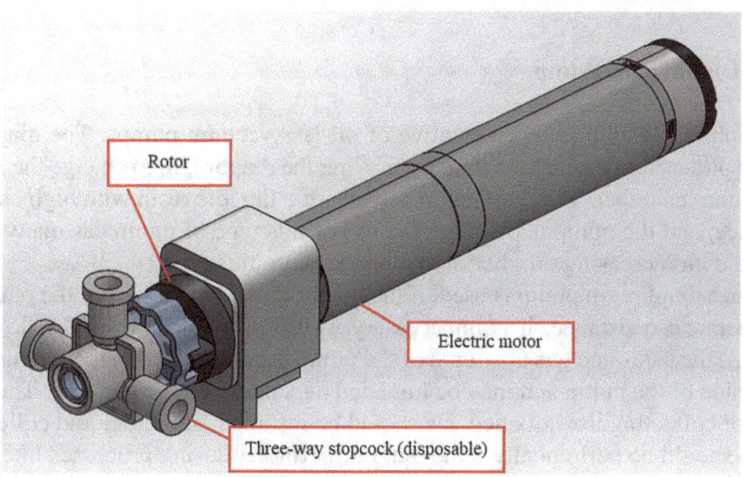

Fig. 6 Control of flow paths using three-way stopcocks

this case, the reagent containers must be cleaned after each synthesis in addition to the fixed tubing flow path. Commercially available vials may be used as reagent containers. In this case, each reagent is weighed in advance and poured into the vial, dissolved with the appropriate solvent if needed, and the vial is set in the synthesizer. In many cases, glass containers are used as reagent containers. However, glass containers are not suitable for storing alkaline solutions for a long period. If it is known that alkaline solutions will be stored for a long period, it is desirable to use plastic containers.

2.4 Vacuum Pump

In many cases, a vacuum pump is used to generate negative pressure for the transfer of reagents and evaporation of solvents during synthesis. Various types of vacuum pump are available, including rotary pumps and diaphragm pumps.

2.4.1 Rotary Pump

Rotary pumps are general vacuum pumps that are inexpensive. The rotor in the pump scoops out the gas in it to reduce the pressure. The gas generated during the drying process of synthesis is mixed with oil when it passes through the pump, resulting in damage to the pump and performance degradation. This issue can be effectively prevented by installing a cold trap in the middle of the tubing to remove solvents. However, the cold trap requires troublesome maintenance, including the disposal of liquid and the replacement of refrigerant, after each synthesis. As a result, the use of oil-less vacuum pumps has been increasing in recent times.

2.4.2 Diaphragm Pump

The diaphragm pump is representative of oil-less vacuum pumps. The diaphragm pump aspirates and discharges fluid by moving the diaphragm to change the volume in the pump chamber. The diaphragm is made of a flexible resin with high chemical resistance, and the pump is oil-free. The use of this type of pump has many advantages, and an increasing number of synthesizers use this type.

When a diaphragm pump is used, if the diaphragm, a major part of the pump, has high chemical resistance, it is unnecessary to install a cold trap in the flow path. However, the aspirated gas upstream of the pump is compressed in the tubing downstream side of the pump and may be liquefied depending on the gas type. Therefore, a trap for collecting the liquefied gas should be installed, and the liquid collected in the trap should be periodically removed. Furthermore, during processes like evaporation, the gas being expelled, particularly in cases involving radioactive substances, should be collected and directed to a waste tank or bag for proper disposal.

For general chemical synthesis, in many cases, low-power, small, and inexpensive pumps will fully meet the functional requirements. There is no need to pay special attention to pump performance.

However, when a small and low-performance diaphragm pump is used, the pump may become locked owing to the pressure balance between the upstream and downstream sides of the pump. The specifications for the pump to be used must be thoroughly checked before use.

2.5 Liquid Transfer

The methods of supplying reagents from reagent containers to a synthesizer can be broadly categorized into three types: pressure feeding using a carrier gas, suction via negative pressure, or a syringe drive system.

2.5.1 Pressure Feeding

This method involves supplying a carrier gas to the reagent container. It introduces the carrier gas into the reactor for gas replacement during synthesis and to restore pressure in the reactor after distillation under reduced pressure. Typically, chemically inert gases like helium or nitrogen are used. It is important to note that separate control is necessary for the carrier gas supply tubing and the solution transfer tubing, which can complicate the setup, especially when dealing with numerous reagents.

2.5.2 Negative Pressure Suction

In contrast, negative pressure suction requires only one tubing line to transfer solutions from reagent containers, resulting in a simpler flow path design. However, if the initial pressure in the reagent containers is low, the solution may not aspirate. To prevent this, it is advisable to prepare an air vent flow path for the reagent containers in advance. Caution should be exercised when using this method with volatile chemical solutions, as they may evaporate. When providing an air vent, it is advisable to include a filter or a similar component to prevent foreign matter from entering the reagent container from the outside.

2.5.3 Syringe Drive System (Fig. 7)

In this method, the reagent is transferred by pulling or pushing the syringe. However, it is crucial to consider the combination of the liquid volume and the syringe size when using this system. Handling a large volume with a small syringe may result in

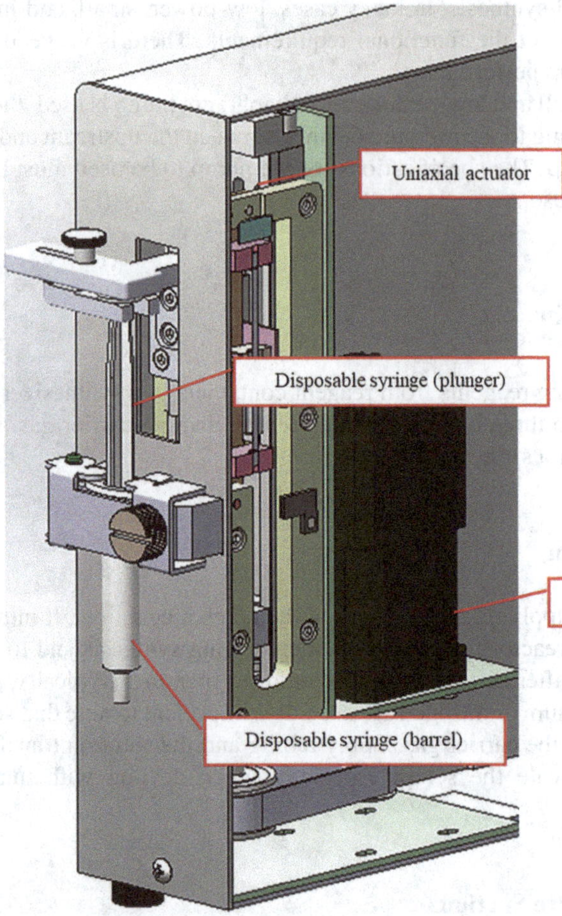

Fig. 7 Syringe drive system

numerous drive cycles and extended transfer times. Conversely, handling a small volume with a large syringe can lead to issues like dead volume losses. Nevertheless, the syringe drive system allows precise transfer of the desired amount of liquid by presetting the syringe stroke. When the drive stroke of the syringe drive system can be controlled, you have the flexibility to transfer any amount of liquid. However, the syringe drive system has a complex structure, making it challenging to install multiple mechanisms in a single synthesizer. Therefore, it is often preferable to use one syringe drive for multiple reagents. In such cases, a thorough examination of the synthesis method is necessary to prevent cross-contamination during the synthesis reactions.

2.6 Purification

In the synthesis steps of radioactive pharmaceuticals, the reaction mixture contains various components, including by-products, organic solvents, catalysts, and more. A process to remove them must be implemented at the final stage and also sometimes during the synthesis.

If the object to be removed is an organic solvent with a low boiling point, it can be distilled away by heating. However, other chemical components must be removed by other methods.

Solid-phase extraction is a well-known chemical technique for the removal of impurities. In particular, ion exchange and hydrophobic interaction (reversed phase) methods are commonly used. In ion exchange methods, ions in solution are removed by adsorption on an ion exchange resin. In the hydrophobic interaction method, hydrophobic substances (mainly organic substances) present in the solution are adsorbed onto the resin. Subsequently, the resin is typically washed, and the organic substances are eluted using an organic solvent, typically ethanol.

When using the ion exchange method, the following points should be considered. (1 The cartridge should have sufficient exchange capacity because if the amount of impurities to be adsorbed exceeds the ion exchange capacity (equivalent) of the cartridge, incomplete adsorption will occur, resulting in breakthrough. (2) The choice of the functional group should be considered because the impurities to be adsorbed in the ion exchange method will act as counter ions and the ions desorbed from the functional groups in the resin will flow into the liquid to be recovered. (3) Ion exchange resins have various functional groups suitable for strong or weak and positive or negative ions, so it is necessary to select a resin suitable for the impurity to be removed.

The following points should be considered in hydrophobic interaction methods. (1) A suitable stationary phase should be selected to adsorb hydrophobic substances present in the solution (C18 is commonly selected). (2) If the solution has a high organic solvent ratio during absorption, it will not be adsorbed by the resin and will pass through it, so if a solution with a high organic solvent ratio is used, it must be diluted to reduce the ratio beforehand.

This solid-phase extraction method is not limited to the removal of impurities and is sometimes used in a multistep process where, once the target substance is adsorbed, other components are discarded and only the adsorbed target substance is finally extracted.

To remove solid foreign particles, a fine particle filter or the like is installed immediately before the final product container. This fine particle removal filter has micropores with a diameter of about 0.2 μm and also functions as a sterilizing filter.

2.7 Other Functions

The above sections describe the functions and mechanisms necessary for the [^{18}F]FDG synthesizers that have the highest demand in the market. In addition to the [^{18}F]FDG synthesizers, various other types of synthesizers are available, and functions other than those described above are required depending on the method of synthesizing the target pharmaceutical. These additional functions are described in this section.

2.7.1 HPLC Unit

Isolation by HPLC may be required as an intermediate or final process depending on the pharmaceutical to be synthesized. Such processes require the synthesizer to include HPLC pumps, UV detectors, and radiation detectors. The key point of selection is the size so that the needed hot cell space can be as small as possible. Typically, systems use remotely controllable HPLC pumps whose flow rate can be set to up to 10 mL/min and UV detectors whose wavelength can be set from 200 to 600 nm.

To introduce a sample into a preparative HPLC unit, we use a method in which the sample is introduced into the sample loop fitted to a six-port valve by a syringe-driven system or pressure-driven system, and the six-port injection valve flow path is switched to introduce the sample into the HPLC flow path. This enables remote injection operation, thereby reducing the exposure of operators to radioactivity.

A UV detector and a radioactivity detector are installed on the HPLC column outlet flow path to monitor eluted materials, and a flow path switch function is provided to collect the sample when the target substance is eluted.

When an HPLC unit is used, the flow path must be resistant to high pressure; thus, disposable parts cannot be used for the flow path or its switching mechanism. Therefore, PEEK or stainless-steel or thick-walled PTFE tubing is used.

In this case, the flow path tubing is fixed and repeatedly used for synthesis. Recently, it has been required to establish a cleaning method to remove residues of reagents and avoid cross-contamination and to validate the cleaning method (Gillings et al., 2021).

2.7.2 Formulation Unit

The formulation unit is a device to remove solvent from the reaction solution and redissolve the reaction solution in physiological saline to obtain the finished pharmaceutical.

For solvent displacement, the evaporator method and the solid-phase extraction method are available.

2.7.2.1 Evaporator Method

Sometimes, a method of re-dissolution in physiological saline after solvent distillation in the evaporator is used for the formulation of labeled PET radiopharmaceuticals.

Formulation using an evaporator is a highly versatile method that can reduce the quantities of residues of organic solvents. In principle, the design is similar to that of a normal evaporator, and solenoid valves and other devices are incorporated to remotely control the flow paths for transferring the solution. However, when producing labeled PET radiopharmaceuticals, a rapid process is imperative, necessitating thorough consideration of solvent distillation conditions to prevent thermal decomposition and bumping, which poses a risk of product loss. Additionally, conditions for recovery from the flask with a solution must be examined. The evaporator method is less effective in certain cases: (1) if using high-boiling point solvents like DMSO or (2) if using after HPLC, and there is a high content of certain salts in the HPLC mobile phase.

Many evaporator formulation units use flask rotating mechanisms and special glassware, and therefore, it is difficult to use disposable parts. Cross-contamination in this unit directly affects the quality of the finished pharmaceutical. Great care must be paid to flow path cleaning.

2.7.2.2 Solid-Phase Extraction Method

The solid-phase extraction method is available as a formulation method without the need to use an evaporator. This method involves diluting the solution to be displaced with solvent, passing the solution through a solid-phase extraction resin column for hydrophobic interaction to capture the target substance, and then feeding a small quantity of eluate, such as ethanol, to the solid-phase extraction resin to elute the target substance and finally dilute it with saline to reduce the ratio of the ethanol contained. This formulation approach is suitable for hydrophobic compounds; however, it is difficult to use for hydrophilic compounds because it is difficult to retain such species in the solid-phase resin column.

The method has the advantages that the target compound can be collected in a disposable solid-phase column, so special glass containers and rotating mechanisms are not required. Moreover, disposable parts can be used for all flow path components, although it is necessary to consider whether the compound can be eluted with a small amount of eluate.

3 Progress of Synthesizer Designs

The mechanisms used in synthesizers are described above. Although the control standards for pharmaceuticals were ambiguous when the use of PET radiopharmaceuticals began, recently, management based on GMP has been required, as with general pharmaceuticals (Schwarz et al., 2019; Gillings et al., 2021). GMP is a set of proper manufacturing standards for pharmaceuticals managed by facilities, and the synthesizers have been improved to be easily controlled in accordance with these standards.

Using example synthesizers from Sumitomo Heavy Industries, we demonstrate, in this section, the evolution of synthesizer design to meet the changing requirements of regulators and end users.

3.1 [^{18}F]FDG Synthesizers

F100 (Fig. 8) and F300 (Fig. 9) are [^{18}F]FDG synthesizers developed around 2000 and 2010, respectively.

Developed in the early days of PET radiopharmaceutical synthesizers, F100 was mostly a laboratory instrument with the following features.

- Fixed tubing flow path.
- Fixed glass reservoir connected to a flow path.
- Oil bath heater that requires regular maintenance.
- Each synthesis operation requires washing and cleaning of the tubing and reactor.

The performance of F100 depended on the skill of the user, as it had several features that required maintenance. On the other hand, F300, which was developed with the expansion of PET radiopharmaceuticals, the acquirement of knowledge, and the development of techniques for synthesis, has the following features.

- Air heaters with easy maintenance.
- Disposable tubing for the flow path.
- Disposable parts of the flow path are replaced with new ones after each synthesis operation; no cleaning is required.
- Commercial vials are used as reagent containers.

The user's workload for cleaning and daily maintenance was reduced, and synthesis failures due to inadequate cleaning were eliminated by using disposable parts for the flow path. These features enable the synthesizer to produce radiopharmaceuticals of consistent quality, independent of the skill of the operator (Table 2).

In addition to systems specifically aimed at [^{18}F]FDG production, multipurpose synthesizers for ^{18}F- and ^{11}C-labeling have also been developed and manufactured for many years.

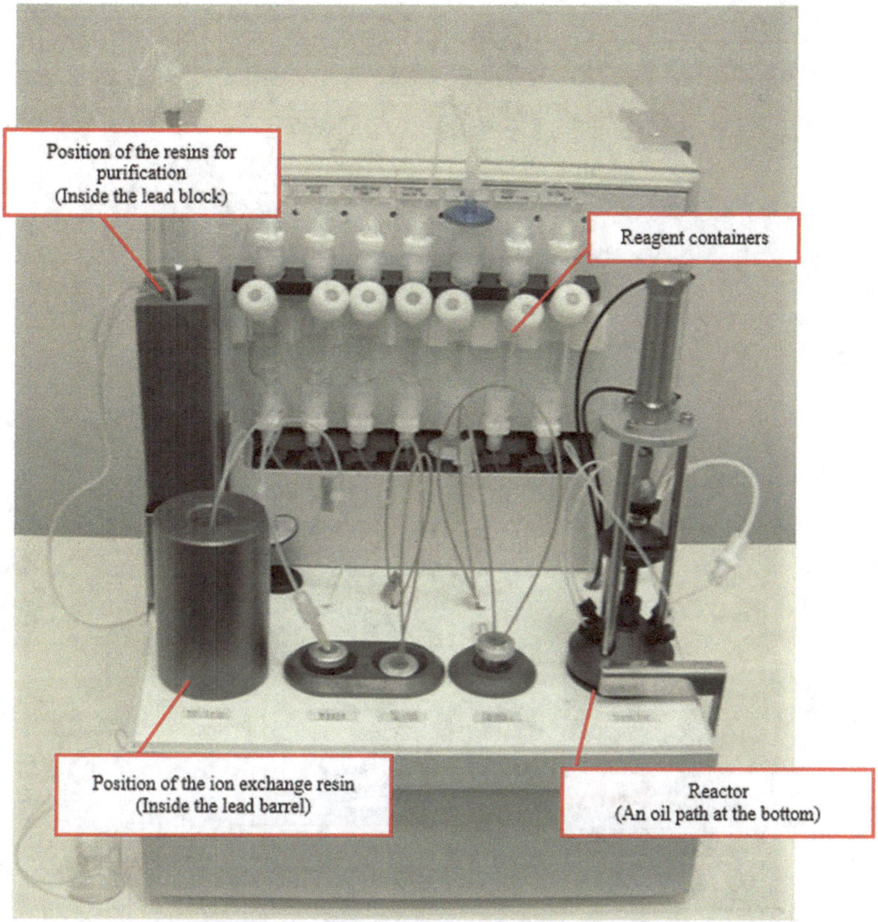

Fig. 8 F100 synthesizer for [^{18}F]FDG (Sumitomo Heavy Industries)

3.2 ^{18}F-Labeling Multipurpose Synthesizer

We started the manufacture of the ^{18}F-labeled compound synthesizer (F1 synthesizer) on the basis of the [^{18}F]FDG synthesizer with the technical assistance of Julich Institute. This synthesizer was characterized by the use of a glassy carbon reaction container to reduce the loss caused by the adsorption of [^{18}F]F$^-$ on the reaction container and an oil bath heater of the reaction container.

Subsequently, to meet the demands of users who want to perform the fluorination labeling reaction and the deprotection reaction using separate reactors and add a purification process by solid-phase extraction method before and after synthesis, we increased the number of reaction container heaters and the number of reagents that can be introduced, improved the synthesizer to add a tubing route for the

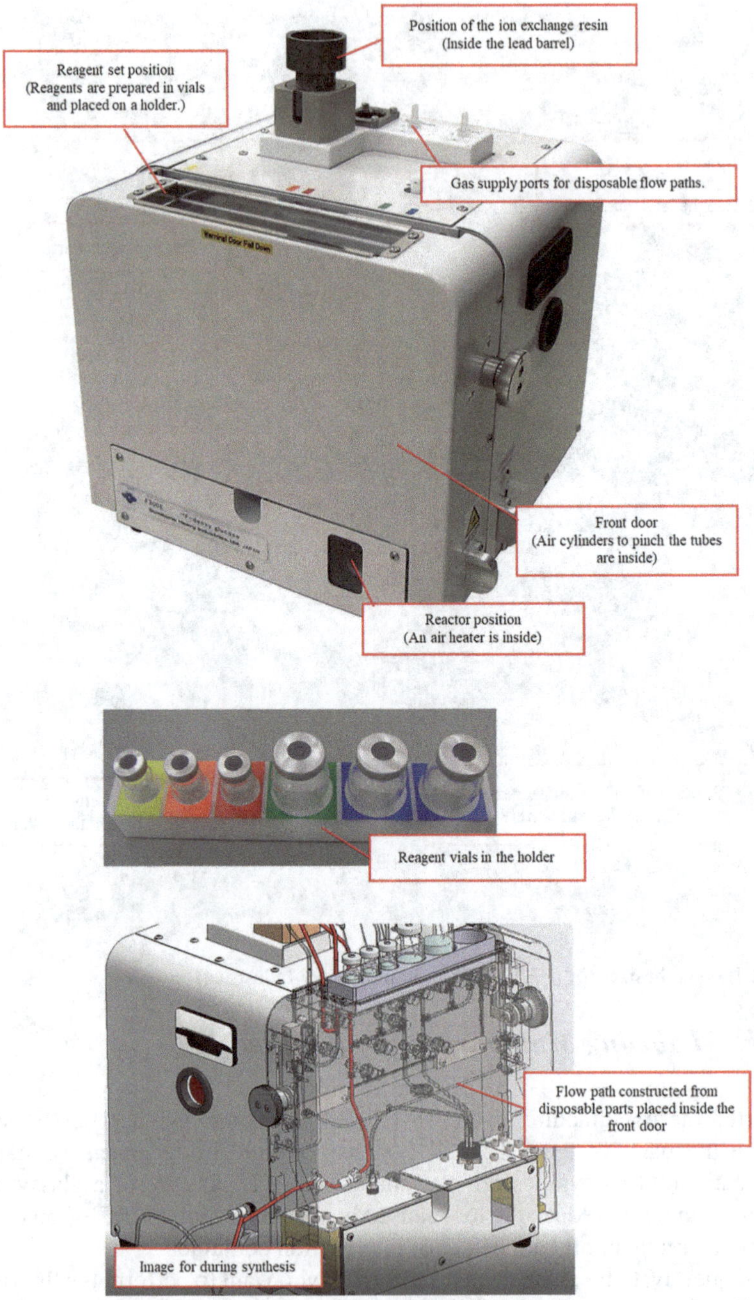

Fig. 9 F300 synthesizer for [^{18}F]FDG (Sumitomo Heavy Industries)

Table 2 Summary of features, advantages, and disadvantages of F100 and F300

	F100	F300
Type of flow path tubing	Fixed tubing	Disposable tubing
Flow path control	Solenoid valve	Pinch valve
Cleaning after synthesis	Required	Not required (cassette replacement method)
Heater	Oil bath	Air heater
Reactor	Glassy carbon	Glass vial
Supply of chemical solution	Fixed glass reservoir	Vial
Synthesis parameter adjustment	Possible	Impossible

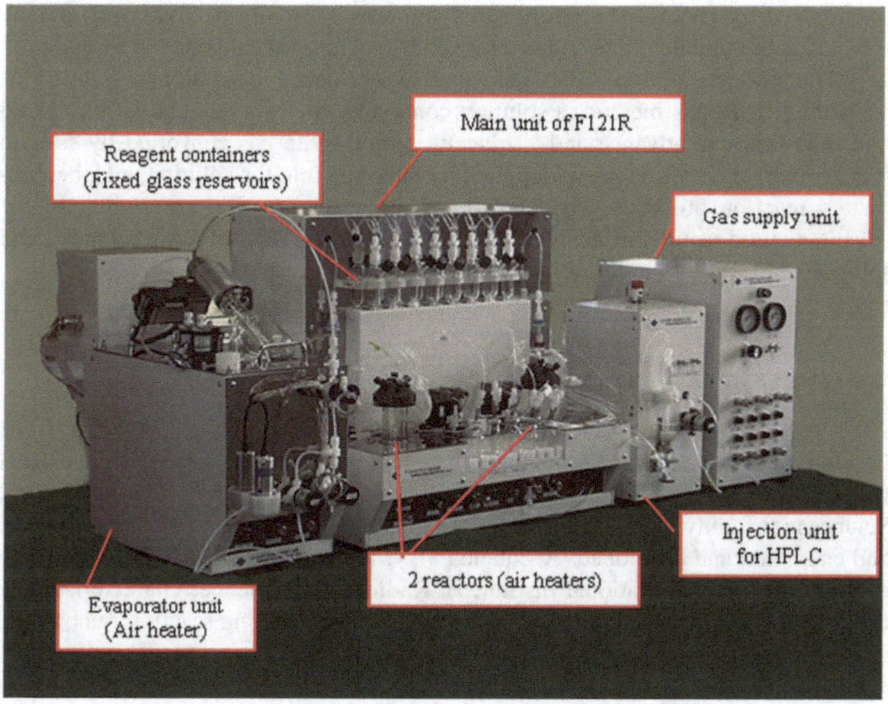

Fig. 10 F121R multipurpose synthesizer (Sumitomo Heavy Industries)

solid-phase extraction process. The multipurpose synthesizers for ^{18}F-labeling applicable to more than one synthesis process (Fig. 10) were then released.

In addition to the above-mentioned processes and change of the heating method for the reaction container and the evaporator flask from an oil bath heater to an air heater, these multipurpose synthesizers are different from the early synthesizers in terms of the cleanliness inside the hot cell.

The F121R synthesizer consists of two reactors heated by air heaters and reagent reservoirs for up to 8 reagents. Each reactor is also equipped with an air cooler capable of cooling the reactor to −10 °C.

In addition to the main unit, the system also combines an HPLC unit and a formulation unit (evaporator). The main unit can perform chemical reactions with F-18 in two ways. One is with $[^{18}F]F^-$ and the other is with $[^{18}F]F_2$ gas. In the $[^{18}F]F^-$ process, $[^{18}F]F^-$ is introduced into the reactor by ion exchange, mixed with K_2CO_3 and K2.2.2 from the reagent reservoir, dried, and then fluorinated by introducing a solution containing the raw material. This process remains the same even for different radiopharmaceuticals, except that the raw material added to the reactor is different. In subsequent processes, the reactions are carried out by hydrolysis, intermediate purification, vacuum concentration, or other methods, depending on the compounds to be synthesized. The reaction solution is then transferred from the main unit and fractionated by HPLC. The fractionated solution is usually a mixture of organic solvents in the mobile phase and is removed by solvent exchange in the formulation unit. The product solution is then recovered as an injectable solution.

In the $[^{18}F]F_2$ gas method, a solution containing the raw material is introduced into the reactor in advance and $[^{18}F]$acetyl hypofluorite gas converted by sodium acetate hydrate or $[^{18}F]F_2$ gas is supplied by bubbling into the solution, and the fluorinated reaction proceeds. The procedures after this reaction are similar to the $[^{18}F]F^-$ method using various reactions, preparative HPLC, solvent exchange, and solution recovery by the formulation unit.

3.3 *^{11}C-Labeling Multipurpose Synthesizer*

In addition to developing ^{18}F-synthesizers, ^{11}C-labeling multipurpose synthesizers were also developed with the assistance of the National Institute of Radiological Sciences (currently National Institutes for Quantum Science and Technology) that had examined and developed ^{11}C-labeled alkyl iodide synthesizers. The synthesizers developed by the National Institute of Radiological Sciences characteristically consisted of devices common to the automation of the labeling reaction, such as the gas supply unit, heating unit, and cooling unit, which are connected to the main unit, and parts that were contaminated with reagents and radioactive nuclides, such as the reagent tubing and reaction container, which are connected to in the replaceable synthesis unit. Therefore, the synthesizers could be repeatedly used for the synthesis of the same labeled compound in a short cycle and other processes of labeling synthesis simply by replacing the synthesis unit after the decay of radioactivity to a certain level (Fig. 11).

A brief description of a typical synthesis method with C-MPS200 is given below.

The synthesis of ^{11}C-labeled compounds is often carried out by synthesizing $[^{11}C]CH_3I$ from $[^{11}C]CO_2$ produced by an accelerator and using it to synthesize various radiolabeled compounds.

Fig. 11 C-MPS200 multipurpose ^{11}C-synthesizer (Sumitomo Heavy Industries)

C-MPS200 is equipped with a cold trap system for the concentration of [^{11}C]CO_2. The concentration method using a cold trap allows for rapid gas recovery from the target at high flow rates while reducing the gas volume. This reduction in gas volume enables the use of smaller amounts of reagents such as LiAlH/THF or Grignard reagents, leading to time savings and potentially improved specific radioactivity. The [^{11}C]CO_2 gas produced by the accelerator is passed through pipes immersed in a refrigerant to liquefy and concentrate it. Then, the pipe is removed from the refrigerant and the [^{11}C]CO_2 in the pipe is vaporized and transferred to the synthesizer. Liquid argon should be used as the refrigerant for this cold trap as it has a lower boiling point than CO_2 and a higher boiling point than liquid nitrogen.

Liquid argon and liquid nitrogen can be used as refrigerants; however, in the case of liquid nitrogen, because the target gas is nitrogen, part of it may become liquefied, resulting in an increase in volume upon vaporization. Considering the boiling points of each gas, liquid argon is more preferable, but it is also possible to choose liquid nitrogen based on availability and price considerations.

A tetrahydrofuran (THF) solution containing lithium aluminum hydride is added to the reactor and bubbled through the [^{11}C]CO_2 produced and collected in the lithium aluminum hydride. Since the volume of the THF solution used is usually small (about 0.5 mL) and its boiling point is low, the solution is cooled by an air cooler during collection to prevent the loss of the solvent during gas collection. After the collection, the THF solution is heated and removed by evaporation, and hydroiodic acid (HI) is introduced into the reactor. The water content in HI reacts with the lithium aluminum hydride, which has absorbed [^{11}C]CO_2, to produce methanol, which reacts with HI to produce [^{11}C]methyl iodide.

The [^{11}C]methyl iodide produced is used as the main material in many ^{11}C-radiolabeled compounds. Some compounds require high reactivity to be synthesized. In this case, [^{11}C]CH_3I is converted to the more reactive [^{11}C]methyl triflate by passing it through silver trifluoro methane sulfonate and then used for synthesis. The compounds that have undergone the various reactions are purified as described in section F121R using an HPLC system or a formulation unit as appropriate and recovered as the product solutions.

3.4 CFN Labeling Multipurpose Synthesizer

3.4.1 CFN-MPS100

The CFN (carbon, fluorine, and nitrogen) synthesizer is designed for labeling reactions with [^{11}C]CO_2 gas, [^{18}F]F_2 gas, and [^{18}F]F^- ions and for the production of [^{13}N]NH_3.

Originally, Sumitomo Heavy Industries supplied separate ^{18}F- and ^{11}C-labeling multipurpose synthesizers. Later, after receiving feedback from the market that customers wish to synthesize both ^{18}F- and ^{11}C-radiopharmaceuticals but cannot introduce separate synthesis units because of cost and space considerations, we developed synthesizers applicable to both ^{18}F- and ^{11}C-radiopharmaceuticals.

CFN-MPS100 is a labeled-compound synthesizer based on the concept of the ^{11}C-labeled compound synthesizer and can be applied to various kinds of nuclide and more than one reaction process because trays for flow paths of various syntheses are optionally available for one base unit having common functions, such as heating, cooling, and gas supply (Fig. 12).

We designed the base unit on the basis of the functions of the ^{11}C-labeling multipurpose synthesizer and added various functions to increase the number of reagents that can be introduced to the reaction container and a pressure gauge to monitor the pressure in the reaction container.

We then developed "trays" of components that have different synthesis pathways. We developed another tray for syntheses of ^{18}F-radiopharmaceuticals and manufactured a synthesizer applicable to both F-18 and C-11 by changing the trays (Yamamoto et al., 2022; Ohnishi et al., 2014; Akhilesh et al., 2022).

The description of a synthesis method using the unit developed for the ^{18}F-Standard, which is exchangeable, is given below. This unit can synthesize using both the [^{18}F]F^- and [^{18}F]F_2 routes.

For both synthesis procedures, the chemical reactions and basic methods are almost the same as those described for F121R in Sect. 3.2.

That is, in the [^{18}F]F^- method, [^{18}F]F^- is introduced into the reactor by ion exchange, mixed with K_2CO_3 and K2.2.2 from the reagent reservoir, dried, and then

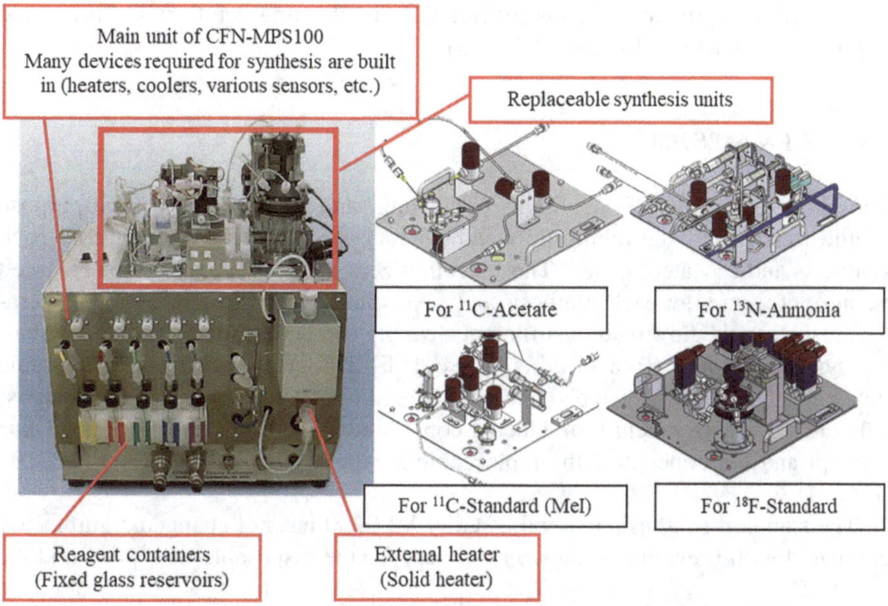

Fig. 12 CFN-multipurpose synthesizer CFN-MPS100 (Sumitomo Heavy Industries)

fluorinated by introducing a solution containing the raw material. After adding various reagents to the solution, fractionation by HPLC and recovery after solvent exchange by the formulation unit are carried out. The [^{18}F]F$_2$ gas method is also almost identical to the F121R method. [^{18}F]F$_2$ gas is collected by bubbling into the solution containing the raw material. Then various reagents are added to the solution, and fractionation by HPLC and recovery after solvent exchange by the formulation unit are carried out. In CFN-MPS100, the problems associated with F121R were alleviated. Typical ways to alleviate the problems include the following.

F121R was developed to allow various synthesis methods to be designed in accordance with the user's requirements. Therefore, reagent containers were connected in a manifold. However, depending on the synthesis method designed, this could lead to cross-contamination during synthesis. When designing the synthesis method for CFN-MPS100, manifolds in the flow path for introducing reagents into the reactor were eliminated and the number of branches was decreased to reduce the risk of cross-contamination.

In the reactor of F121R, tubes extend to the bottom of the reactor as a flow path for exporting solution from the reactor or for bubbling. However, when the reactor is sealed off and heated, as in fluorination, the reaction solution is taken up through these tubes as the internal pressure of the reactor increases.

As a result, the solution cannot be sufficiently heated and chemical reactions in some synthesis processes are inefficient. Therefore, the lifting function is incorporated into the reactor flange of CFN-MPS100. This allows the tubes to be lowered to the bottom of the reactor only when exporting solution from the reactor or for bubbling. This eliminates the loss of liquid during heating due to raised tubes and improves the reaction efficiency (Fig. 13).

3.4.2 CFN-MPS200

The above-described ^{18}F-labeling, ^{11}C-labeling, and CFN-MPS100 multipurpose synthesizers have fixed tubing flow paths, which are cleaned and dried after each synthesis and repeatedly used. These synthesizers have fewer parts to be replaced because of wear after each synthesis and are economically attractive, but the operator must clean the flow path; insufficient cleaning and drying may often cause labeling synthesis failure. To avoid this failure, [^{18}F]FDG synthesizers with disposable flow paths have been developed and commercialized. We have developed a disposable-cassette-replacement-type labeled compound synthesizer on the basis of this concept and the concept of the replaceable synthesis flow path of CFN-MPS100 (Fig. 14).

The standard synthesis method of CFN-MPS200 has not changed significantly compared with previous synthesizers, except for the disposable flow path, and the

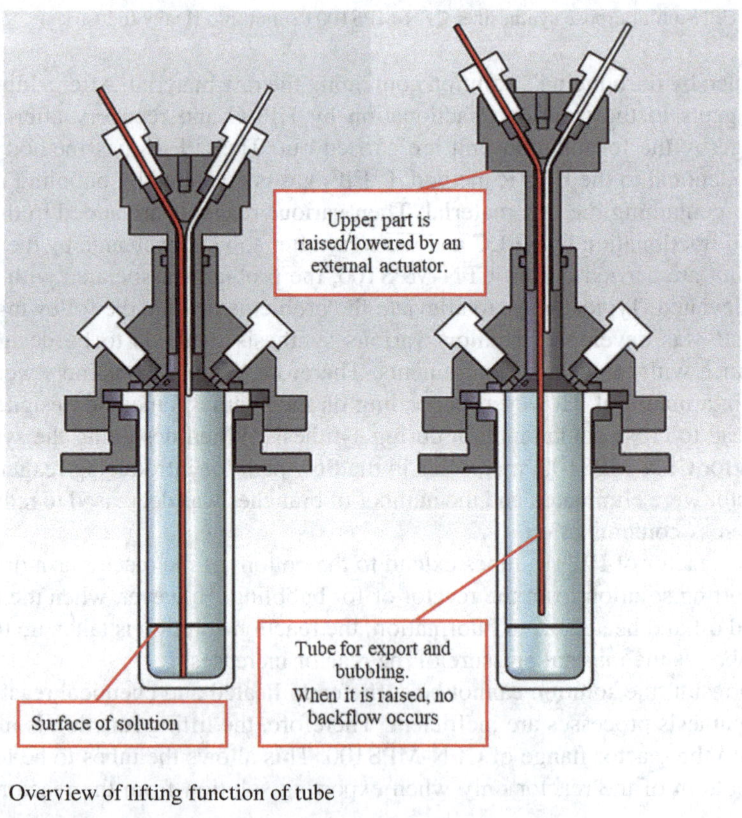

Fig. 13 Overview of lifting function of tube

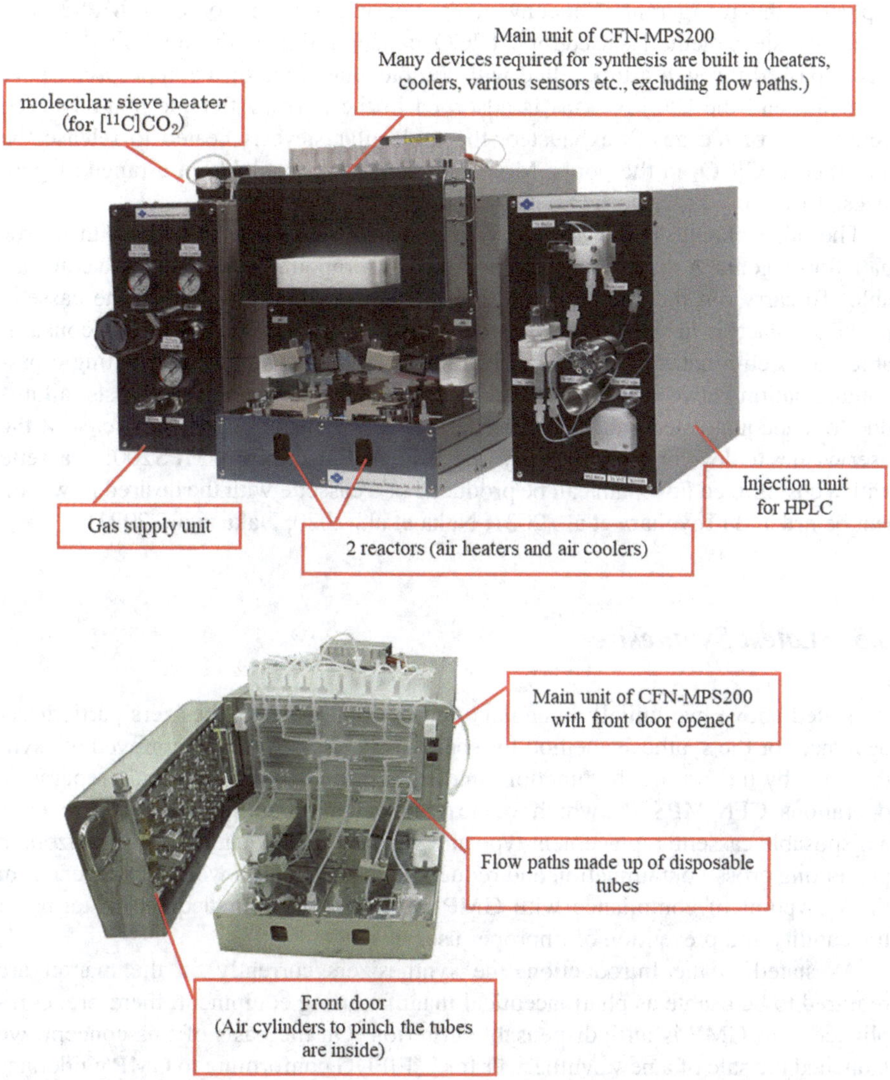

Fig. 14 CFN-multipurpose synthesizer CFN-MPS200 (Sumitomo Heavy Industries)

reaction and purification procedure are basically the same as those of previous synthesizers. Therefore, the specific synthesis method is omitted in this section. There are two major changes in this synthesizer compared with the previous synthesizers.

The first one is the $[^{11}C]CO_2$ concentration method. The method of $[^{11}C]CO_2$ concentration using refrigerants is presented in Sect. 3.3 for C-MPS 200. This method may fail owing to refrigerant loss if synthesis takes a long time after the refrigerant has been prepared. Moreover, some users argue that preparing and replacing the refrigerant is inconvenient. For this reason, for CNF-MPS200, a molecular sieve method is used for [^{11}C]CO_2 concentration. In this method, [^{11}C]CO_2 gas is passed through a tube filled with a molecular sieve of the appropriate pore diameter, and the [^{11}C]CO_2 gas is adsorbed in the pores. After the non-adsorbed component of the gas is exhausted, the molecular sieve is heated to release the adsorbed [^{11}C]CO_2 in the pores. Molecular sieves are available in a range of pore sizes; 13X is used to adsorb [^{11}C]CO_2.

The other change is the previously mentioned disposable cassette with a flow path for reagents. A disposable cassette for each compound is supplied as a consumable. To carry out the synthesis, all the operator needs to do is open the cassette package, place it in the synthesizer, and set the reagents and additional consumables. This eliminates the need for cleaning after each synthesis, preventing cross-contamination between syntheses and reducing the occurrence of synthesis failures due to inadequate cleaning. The cassettes can also be supplied piecewise, so if the user wishes to develop a new drug synthesis method using CFN-MPS200, a cassette with a customized flow path can be produced or a cassette with the desired flow path can be ordered (Toyohara et al., 2021; Naka et al., 2020; Naka et al., 2021).

3.5 *Latest Synthesizer*

As stated above, we initially manufactured multipurpose synthesizers particularly designed for the synthesis method for specific nuclides. We then improved the synthesizers by integrating the functions and simplifying the cleaning and preparation operations. CFN-MPS200, which we currently sell as a multipurpose synthesizer, is a disposable cassette replacement type and superior to conventional synthesizers in preventing cross-contamination and reducing the preparation work. However, from the viewpoint of compliance with GMP guidelines, it is inadequate in terms of traceability and prevention of improper use.

As stated in the Introduction, the synthesizers currently on the market are required to be usable as pharmaceutical manufacturing equipment; therefore, compliance with GMP is an indispensable criterion. On the basis of this concept, we launched the sale of a new synthesizer for [^{18}F]FDG conforming to GMP guidelines (Gillings et al., 2021; Schwarz et al., 2019) in 2022 (Fig. 15).

This synthesizer has the following features.

- **Prevention of cross-contamination by using disposable cassettes.** Cassettes are produced in a cleanliness-controlled factory and are individually checked for integrity. The chemical flow paths are concentrated in the cassettes, and cross-contamination between synthesis operations is prevented by using new cassettes after each operation.
- **Traceability of special disposable cassettes through ID management.** A radio-frequency identification (RFID) tag attached to a cassette can be read to check usage history, etc.

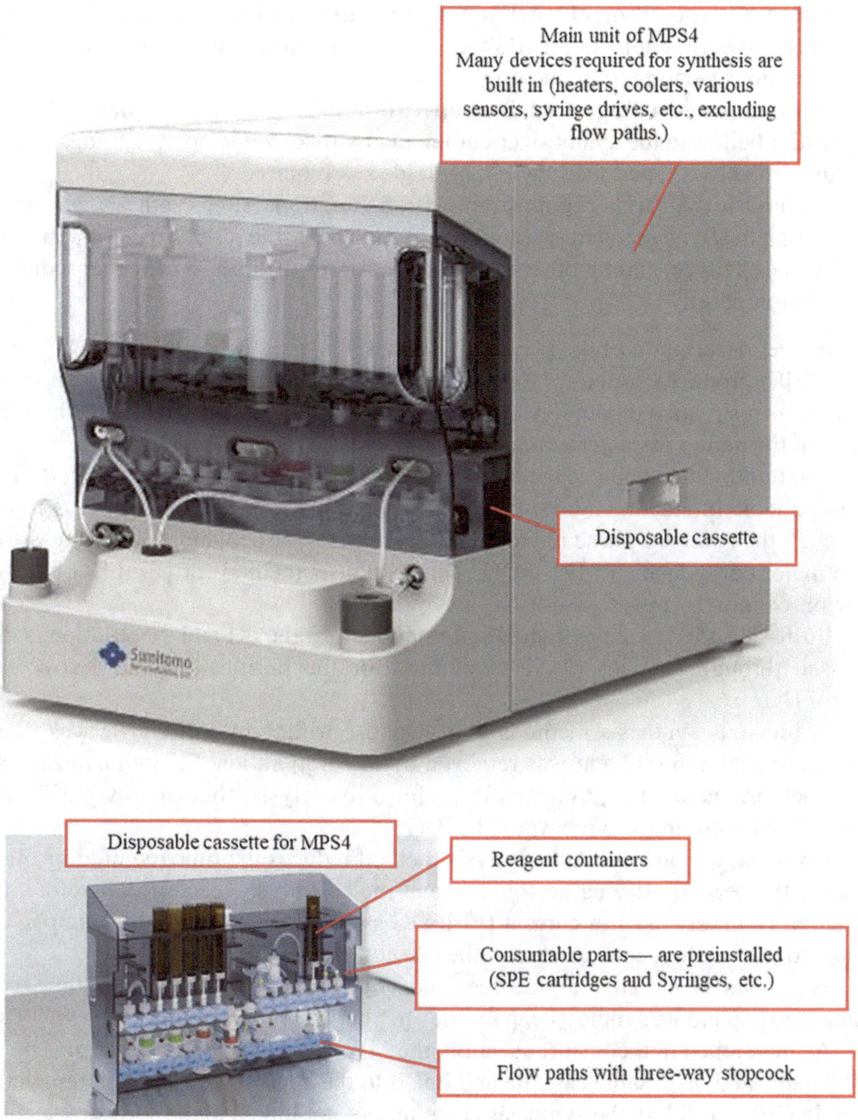

Fig. 15 Latest synthesizer MPS4 (Sumitomo Heavy Industries)

- **Prevention of operation failure by using special software.** Synthesizers for multiple radiopharmaceuticals are planned for future development. To synthesize multiple radiopharmaceuticals, cassettes and synthesis software must be prepared for each compound. To avoid the error of an unsuitable combination of the cassette and the software, the IDs of the two are linked to each other.

- **Automatic recording of work log through user ID management.** The synthesizer software requires user IDs for login management and keeps a record of the operations performed by the user.
- **Automatic recording of various operation logs of the synthesizer.** The controller built into the synthesizer automatically records the operation logs of each device included, and the logs can be loaded as required.
- **Automatic synthesis report preparation function.** After synthesis, a report is automatically generated with information about the user, trend graphs, and synthesis times, among others. This is helpful for the user to create a production control record.

A brief description of the [^{18}F]FDG synthesis method is given below.

[^{18}F]F^- produced in the cyclotron is introduced into the reactor using an anion-exchange resin, mixed with K_2CO_3 and K2.2.2., dried, and then fluorinated by introducing the main raw material mannose triflate dissolved in acetonitrile.

The fluorinated mannose triflate is adsorbed onto a reversed-phase resin through hydrophobic interactions to remove impurities. Then, sodium hydroxide is passed through the reversed-phase resin to hydrolyze the intermediate and thereby extract the major components of [^{18}F]FDG. This synthesis method for [^{18}F]FDG is called the on-column hydrolysis method.

Buffer is added to the extracted solution to adjust the pH and the solution is then passed through a series of SPE columns for the purification and recovery of [^{18}F]FDG.

In previous synthesis methods, the mannose triflate in acetonitrile was fluorinated, and then the solvent was removed by the evaporation. Sodium hydroxide or hydrochloric acid was introduced to perform hydrolysis, followed by purification through SPE columns to recover [^{18}F]FDG.

However, the on-column hydrolysis method reduces the time required by eliminating the need for drying acetonitrile and the hydrolysis reaction. Therefore, this method is widely used in current [^{18}F]FDG synthesis as it is a stable, simple, and easy-to-use method compared with the previous one.

We intend to add auxiliary devices and synthesis software to this synthesizer in stages to expand its synthesis capabilities to pharmaceuticals other than [^{18}F]FDG.

As mentioned in the features section, the MPS4 synthesizer can recognize the ID of a dedicated disposable cassette and link it to the appropriate software for synthesis. In CFN-MPS200, the synthesis cassettes are not linked to the software, and it is theoretically possible to run the incorrect synthesis software for different cassettes. This means that the prevention of synthesis errors due to the use of different software programs depends on the operator. However, the functions of MPS4 described above link the synthesis cassette to the appropriate software, so the cassette can be used without relying on operator confirmation.

This is in line with the requirement for recent synthesizers to minimize human error. Currently, the synthesizer only has a cassette and software for synthesizing [^{18}F]FDG, but in the future, this linking function will be used to prevent human error, even when synthesizing multiple compounds.

An extension unit (HPLC function, SPE function, etc.) is being developed for the same system, which will enable the synthesis of all the compounds in CFN-MPS200 in the future.

There are also plans to provide software with a developer mode and cassette parts to enable users to construct and test cassettes with customized flow paths if they wish to develop original synthesis methods.

References

Adam, M. J. (1999). Synthesis and preliminary evaluation of [18F]2-deoxy-2,2-difluoro-glucose as a potential PET imaging agent. *Journal of Labelled Compounds and Radiopharmaceuticals, 42*(8), 809–813.

Akhilesh, S. K., Shanker, N., Subhash, K. C., Sanjay, G., & Dixit, M. (2022). Fully automated synthesis of nitrogen-13-NH_3 by SHIs HM-18 cyclotron and dedicated module for routine clinical studies: Our institutional experiences. *Indian Journal of Nuclear Medicine, 37*(1), 50–53.

Boonyawan, T., Wimolwattanasarn, K., Phumruamjai, J., Kalyanamitra, K., Ekmahachai, M., & Phongsri, J. (2021). Radiopharmaceuticals production and background of the PET/CT & Cyclotron Center, Chiang Mai University. *Chiang Mai Medical Journal, 60*(1), 125–159.

Boschi, S., Lodi, F., Malizia, C., Cicoria, G., & Marengo, M. (2013). Automation synthesis modules review. *Applied Radiation and Isotopes, 76*, 38–45.

Chaly, T., Mattacchieri, R., Velez, J. W., Dahl, J. R., & Margouleff, D. (1990). A large scale manual production of [^{18}F]FDG using a synthetic unit made of sterile disposable components and operated by a master slave manipulator. *International Journal of Radiation Applications and Instrumentation. Part A. Applied Radiation and Isotopes, 41*, 29–34.

Collins, J., Waldmann, C. M., Drake, C., Slavik, R., Ha, N. S., Sergeev, M., Lazari, M., Shen, B., Chin, F. T., Moore, M., Sadeghi, S., Phelps, M. E., Murphy, J. M., & Michael Van Dam, R. (2017). Production of diverse PET probes with limited resources: 24 18F-labeled compounds prepared with a single radiosynthesizer. *Proceedings of the National Academy of Sciences of the United States of America, 114*(43), 11309–11314.

Culbert, P. A., Adam, M. J., Hurtado, E. T., Huser, J. M. A., Jivan, S., Lu, J., Ruth, T. J., & Zeisler, S. K. (1995). Automated synthesis of [18F]FDG using Tetrabutylammonium bicarbonate. *Applied Radiation and Isotopes, 46*(9), 887–891.

Fochtner, F., Steinbach, J., Mading, P., & Johannsen, B. (1996). Basic hydrolysis of 2-[18F]Fluoro-1,3,4, 6-tetra-O-acetyl-D-glucose in the preparation of 2-[8F]Fluoro-2-deoxy-D-glucose. *Applied Radiation and Isotopes, 47*, 61–66.

Food and Drug Administration. (2014). Validation of cleaning processes (7/93).

Gillings, N., Hjelstuen, O., Ballinger, J., Behe, M., Decristoforo, C., Elsinga, P., Ferrari, V., Peitl, P. K., Koziorowski, J., Laverman, P., Mindt, T. L., Neels, O., Ocak, M., Patt, M., & Todde, S. (2021). Guideline on current good radiopharmacy practice (cGRPP) for the small-scale preparation of radiopharmaceuticals. *EJNMMI Radiopharmacy and Chemistry, 6*(1), 8.

Hamacher, K., Coenen, H. H., & Stöcklin, G. (1986). Efficient stereospecific synthesis of no-carrier-added 2-[18F]-fluoro-2-deoxy-D-glucose using aminopolyether supported nucleophilic substitution. *Journal of Nuclear Medicine, 27*(2), 235–238.

Higashi, K., Matsunari, I., Ueda, Y., Ikeda, R., Guo, J., Oguchi, M., Tonami, H., & Yamamoto, I. (2003). Value of whole-body FDG PET in management of lung cancer. *Annals of Nuclear Medicine, 17*(1), 1–14.

Hou, S., Phung, D. L., Lin, W.-Y., Wang, M.-W., Liu, K., & Shen, C. K.-F. (2011). Microwave-assisted one-pot synthesis of N-succinimidyl-4-[18F]fluorobenzoate ([18F]SFB). *Journal of Visualized Experiments, 52*, 2755.

Ido, T., Wan, C. N., Casella, V., Fowler, J. S., Wolf, A. P., Reivich, M., & Kuhl, D. E. (1978). Labeled 2-deoxy-D-glucose analogs. 18F-labeled 2-deoxy-2-fluoro-D-glucose, 2-deoxy-2-fluoro-D-mannose and 14C-2-deoxy-2-fluoro-D-glucose. *Journal of Labelled Compounds and Radiopharmaceuticals, 14*(2), 175–183.

Iimori, H., Hashizume, Y., Sasaki, M., Kajiwara, Y., Sugimoto, Y., Sugiyama, Y., Watanabe, Y., & Senda, M. (2011). First automatic radiosynthesis of 11C labeled Telmisartan using a multipurpose synthesizer for clinical research use. *Annals of Nuclear Medicine, 25*(5), 333–337.

International Atomic Energy Agency. (2021). IAEA-TECDOC-1968; Production and Quality Control of Fluorine-18 Labelled Radiopharmaceuticals.

Iwata, R., Ido, T., Takahashi, T., & Monma, M. (1984). Automated synthesis system for production of 2-Deoxy-2-[18F]Fluoro-D-glucose with computer control. *The International Journal of Applied Radiation and Isotopes, 35*(6), 445–454.

Kitajima, K., Murakami, K., Kanegae, K., Tamaki, N., Kaneta, T., Fukuda, H., Nakajima, K., Fujii, H., Tateishi, U., Kubota, K., Suga, T., & Nakamoto, Y. (2009). Clinical impact of whole body FDG-PET for recurrent biliary cancer: A multicenter study. *Annals of Nuclear Medicine, 23*(8), 709–715.

Krasikova, R. (2007). Synthesis modules and automation in F-18 Labeling. *Ernst Schering Research Foundation Workshop, 62*, 289–316. https://doi.org/10.1007/978-3-540-49527-7_11

Kryza, D., Tadino, V., Filannino, M. A., Villeret, G., & Lemoucheux, L. (2008). Fully automated [18F]fluorocholine synthesis in the TracerLab MXFDG coincidence synthesizer. *Nuclear Medicine and Biology, 35*(2), 255–260.

Lemaire, C., Ph Damhaut, B., Lauricella, C., Mosdzianowski, J. L., Morelle, M., Van Monclus, J., Naemen, E., Mulleneers, J., Aerts, A., Plenevaux, C. B., & Luxen, A. (2002). Fast [18F]FDG synthesis by alkaline hydrolysis on a low polarity solid phase support. *Journal of Labelled Compounds and Radiopharmaceuticals, 45*(5), 435–447.

Mock, B. H., Vavrek, M. T., & Mulholland, G. K. (1996). Back-to-Back "one-pot" [18F]FDG syntheses in a single Siemens-CTI chemistry process control unit. *Nuclear Medicine and Biology., 23*, 497–501.

Naka, S., Watabe, T., Kurimoto, K., Uemura, M., Soeda, F., Neels, O. C., Kopka, K., Tatsumi, M., Kato, H., Nonomura, N., Shimosegawa, E., Cardinale, J., Giesel, F. L., & Hatazawa, J. (2020). Automated [18F]PSMA-1007 production by a single use cassette-type synthesizer for clinical examination. *EJNMMI Radiopharmacy and Chemistry, 5*(1), 8.

Naka, S., Watabe, T., Lindner, T., Cardinale, J., Kurimoto, K., Moore, M., Tatsumi, M., Mori, Y., Shimosegawa, E., Valla, F., Kato, H., & Giesel, F. L. (2021). One-pot and one-step automated radio-synthesis of [18F]AlF-FAPI-74 using a multi purpose synthesizer: A proof-of-concept experiment. *EJNMMI Radiopharmacy and Chemistry, 6*(1), 28.

Ohnishi, A., Senda, M., Yamane, T., Sasaki, M., Mikami, T., Nishio, T., Ikari, Y., Nishida, H., Shukuri, M., Takashima, T., Mawatari, A., Doi, H., Watanabe, Y., & Onoe, H. (2014). Human whole-body biodistribution and dosimetry of a new PET tracer, [11C]ketoprofen methyl ester, for imagings of neuroinflammation. *Nuclear Medicine and Biology, 41*(7), 594–599.

Petriev, V. M., Tishchenko, V. K., & Krasikova, R. N. (2016). 18F-FDG and other Labeled glucose derivatives for use in radionuclide diagnosis of oncological diseases (review). *Pharmaceutical Chemistry Journal, 50*(4), 209–220.

Petroni, D., Menichetti, L., & Poli, M. (2020). Historical and radiopharmaceutical relevance of [18F]FDG. *Journal of Radioanalytical and Nuclear Chemistry, 323*(3), 1017–1031.

Reske, S. N. (2003). PET and restaging of malignant lymphoma including residual masses and relapse. *European Journal of Nuclear Medicine and Molecular Imaging, 30*(SUPPL. 1), S89–S96.

Saxena, P., Singh, A. K., Dixit, M., Kheruka, S. C., Mahmood, T., & Gambhir, S. (2021). Establishing the [18F]-FDG production via two different automated synthesizers for routine clinical studies: Our institutional experiences of 4 years. *Indian Journal Of Nuclear Medicine, 36*(2), 120–124.

Schwarz, S. W., Decristoforo, C., Goodbody, A. E., Singhal, N., Saliba, S., Ruddock, P. S., Zukotynski, K., & Ross, A. A. (2019). Harmonization of U.S., European Union, and Canadian first-in-human regulatory requirements for radiopharmaceuticals: Is this possible? *Journal of Nuclear Medicine, 60*(2), 158–166.

Seltzer, M. A., Yap, C. S., Silverman, D. H., Meta, J., Schiepers, C., Phelps, M. E., Gambhir, S. S., Rao, J., Peter, E. V., & Czernin, J. (2002). The impact of PET on the management of lung cancer: The referring physician's perspective. *Journal of Nuclear Medicine, 43*, 752–756.

Toyohara, J., Nishino, K., Sakai, M., Tago, T., & Oda, T. (2021). Automated production of [18F]MK-6240 on CFN-MPS200', applied radiation and isotopes. *Applied Radiation and Isotopes, 168*, 109468. https://doi.org/10.1016/j.apradiso.2020.109468

Yamamoto, Y., Tago, T., Toyohara, J., Saito, Y., & Yamamoto, F. (2022). Radiosynthesis and in vivo and ex vivo evaluation of isomeric [^{11}C]-methoxy Analogs of Nimesulide as brain Cyclooxygenase-2-targeted imaging agents. *Biological & Pharmaceutical Bulletin, 45, Issue 1*, 94–103.

Yuasa, M., Yoshida, H., & Hara, T. (1997). Computer-controlled synthesis of [^{18}F]FDG by the Tetrabutylammonium method: Achievement of high yield, purity, reproducibility, reliability, and safety. *Applied Radiation and Isotopes, 48*(2), 201–205.

3D Printed Microreactors for Radiochemical Reactions

Giancarlo Pascali, Andrew Hall, and David Zhara

1 Introduction

The field of radiochemistry has seen tremendous growth over the past few decades, driven largely by advances in positron emission tomography (PET) for biomedical imaging and nuclear medicine. PET relies on the use of radiotracers—compounds labeled with short-lived positron-emitting radionuclides, such as ^{11}C, ^{13}N, ^{15}O, and $^{18}F^-$ to non-invasively image biochemical processes in vivo. The development and routine production of PET radiotracers, however, face several challenges due to the unique constraints imposed by working with radioactive materials. These challenges include the need to minimize radiation exposure, perform reactions rapidly before radioactive decay, achieve high molar activity, work with the small starting quantities of radionuclides produced at cyclotrons, and automate radiosynthesis for reliable results. For labeling with ^{11}C and ^{18}F, these challenges are considered even greater (e.g., increased chemical resistance requirements) and what is described in the following article could provide substantial benefits for these two nuclides in particular but could also be applied for other nuclides (e.g., radiometals), depending on the application of interest.

Microfluidic and flow chemistry technologies have shown great promise in addressing these challenges, offering improved mixing control, heat transfer, safety,

G. Pascali (✉)
Australian Nuclear Science and Technology Organisation, Sydney, NSW, Australia

School of Chemistry, University of New South Wales, Sydney, NSW, Australia

Brain and Mind Centre, University of Sydney, Sydney, NSW, Australia
e-mail: gianp@ansto.gov.au

A. Hall · D. Zhara
Australian Nuclear Science and Technology Organisation, Sydney, NSW, Australia

R. M. van Dam et al. (eds.), *Automated Technologies for the Development and Production of Radiopharmaceuticals*,
https://doi.org/10.1007/978-3-031-84632-8_8

automation, and reproducibility compared to conventional batch synthesis (Watts & Haswell, 2005). By precisely controlling reaction parameters in small fluid volumes, microfluidic reactors can enable faster reactions, higher yields, and improved product purity. Despite these clear advantages, the wider adoption of flow chemistry and microfluidics in radiochemistry has been hindered by the high costs of specialized commercial equipment and incompatibility between traditional microreactor component materials and radiochemical reaction mixtures. Polymers like polydimethylsiloxane (PDMS), while easy to prototype and useful for biological reactions, degrade and swell in many of the solvents used in radiolabeling chemistry (Wang et al., 2010). Glass and silicon are chemically resistant but expensive and challenging to machine.

In recent years, additive manufacturing or 3D printing has emerged as an enabling technology for rapidly fabricating custom microfluidic devices with complex geometries without the need for cleanroom facilities or photolithography. In particular, stereolithography (SLA)-based 3D printing using photocurable resins can achieve features on the order of 150 μm with a layer resolution down to 25 μm (depending on the material) on some systems, making it well suited for creating the micron-sized fluidic channels needed in microreactors. The materials palette for SLA has also expanded greatly to include engineered polymers and composites with enhanced thermal and chemical stability. Combined with the precise spatial control and design freedom of 3D printing, these advances provide new opportunities to address the specialized requirements of radiochemistry. While not exhaustive, this chapter will provide basic information on the most used printing techniques, select examples of chemistry applications, a worked example of microreactor manufacturing, and the current status in radiochemical processing. The reader is directed to more comprehensive reviews on 3D printed microfluidics to gain a wider perspective on the topic (Au et al., 2016).

1.1 3D Printing Techniques and Materials

Several 3D printing technologies have been applied to fabricate components for flow chemistry (Fig. 1), the most prominent being fused deposition modeling (FDM), PolyJet/inkjet, and SLA (Parra-Cabrera et al., 2018). Each technique offers distinct advantages along with certain limitations in resolution, materials selection, and processing conditions that must be considered when designing microreactors for radiochemistry.

Inkjet 3D printing uses inkjet deposition of photocurable resins followed by UV curing to build parts additively, and commercial printers (e.g., Stratasys (Eden Prairie, MN, USA), 3D Systems (Rock Hill, SC, USA)) can now achieve layer resolution as small as a few tens of μm using proprietary photopolymers (Waheed et al., 2016). A notable advantage is the ability to print rigid, multimaterial components with fine details, which can be used to obtain complex networks of small fluidic structures. However, the resins utilized in this technique degrade in many organic

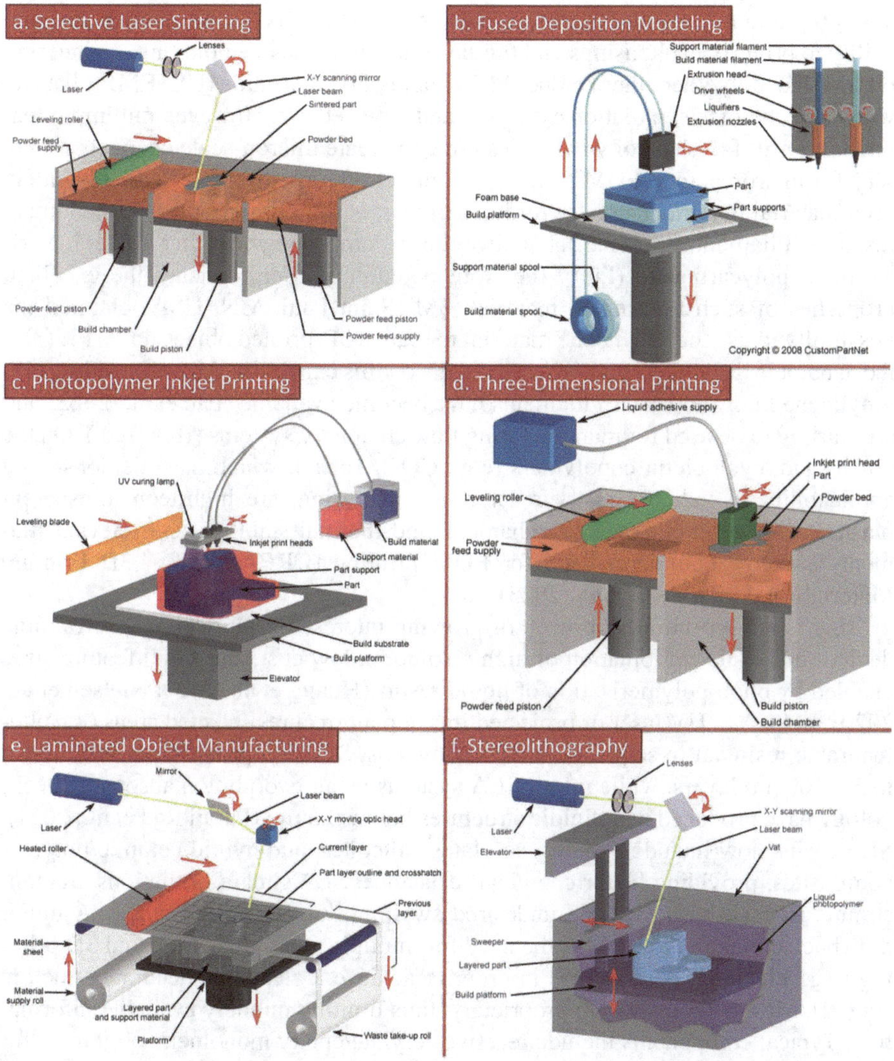

Fig. 1 Schematic of various techniques used for 3D printing microfluidic systems. (Adapted with permission from Au et al., 2016)

solvents, limiting chemical compatibility, and equipment costs are high; hence, this powerful approach has seen limited application in flow and microfluidics chemistry.

FDM works by selectively depositing molten thermoplastic filament layer-by-layer based on a 3D computer-aided design (CAD) model. Widely used desktop FDM printers can achieve down to ~200 μm resolution with polylactic acid (PLA) and acrylonitrile butadiene styrene (ABS) filaments (Kitson et al., 2012). Other systems, such as those produced by Ultimaker B.V. (Utrecht, Netherlands), are even capable of achieving down to 20 μm layer resolution; however, they are still limited to ~200 μm in *x–y* resolution. Key benefits of FDM are its low equipment cost and ability to print robust housings and fluidic connections, also employing engineering plastics like polyether ether ketone (PEEK) and polyetherimide (ULTEMP) that can withstand chemicals, radiation exposure, and repeated use. However, minimum feature sizes still fall short of what is desirable to create micron-scale channels necessary for microreactors. FDM printing of polytetrafluoroethylene (PTFE), a widely used material in radiochemistry for its inertness, is sometimes reported by filament resellers (Filament2Print, n.d.), but always as a composite with other harder materials (e.g., polycarbonate (PC)), therefore potentially compromising the excellent properties of such fluorinated material; 3M (Saint Paul, MN, USA) claimed the possibility to use their Dyneon® material to realize 3D printed objects in PTFE (3M, n.d.), but it is difficult to find updated details on this capability. More recently, polyvinylidene fluoride (PVDF) filaments have become available (Fan et al., 2022) and are starting to be used for manufacturing flow chemistry systems (IKA, n.d.). On the other hand, cycloolefin copolymers (e.g., COC, Topas), which have demonstrated applicability in radiochemistry reactor manufacturing, are high-tech transparent plastics with good chemical, mechanical, and thermal stability, and relevant filaments are now commercialized for FDM printing (CREAMELT – 3D Printing Materials, n.d.; Zwyssig et al., 2023).

SLA is a 3D printing approach of growing interest for fabricating microfluidic devices due to its combination of high resolution, low cost, and small feature sizes enabled by photopolymerization of liquid resins (Huang et al., 2020; Nielsen et al., 2020). In SLA, a UV laser or projected image pattern cures selected areas of a photocurable resin vat in sequential layers. Low-cost SLA 3D printers can print down to 50–100 μm layers, while micro-SLA systems using two-photon absorption technology have produced nanofluidic structures below 0.5 μm (Faraji Rad et al., 2021). SLA resins now include epoxies, acrylates, silicones, and hybrid ceramic/polymer composites, providing better chemical resistance. SLA's main limitations are constraints on resin viscosity and undesired swelling due to solvent vapors. A minor drawback compared to FDM is the need for multiple steps after the actual 3D printing (e.g., cleaning, post-curing) in order to achieve a clean and accurate structure (Fig. 2). Most SLA resins are proprietary, thus limiting publicly available information. Typical components include reactive acrylate/epoxy monomers, oligomers for mechanical properties, a photoinitiator, absorbers, and inhibitors to control polymerization kinetics and rheology (Andreu et al., 2021). Adjusting the type and ratio of these components enables tuning of properties like cure depth, resolution, toughness, swelling, and solvent compatibility. For example, Fluoropor resins developed by Mayoussi et al. (2021) use a fluorinated methacrylate monomer, a porogen mixture, cross-linker, and photoinitiator that form porous polymer networks with superhydrophobicity and chemical resistance. Dual cure systems with heat/latent thermal curing have also been reported, like vitrimer epoxy resins, allowing further functionalization after 3D printing to enhance stability (Gerdroodbar et al., 2023). However, in the context of flow chemistry, permeation or leaching of organic solvents and reactants into the bulk resin (i.e., out of the microchannels) during

Fig. 2 Formlabs (Somerville, MA, USA) modules allowing a complete SLA process, comprising 3D printing (left), model wash (center), and model curing (right). (Adapted from Formlabs (n.d.) under CCC license)

reactions remains a major concern for long-term durability, material integrity, and sample contamination. Advanced resins like fluoropolymers (3M, n.d.; Kotz et al., 2018) with increased chemical resistance still have limited commercial availability and high cost due to their small market size.

When selecting 3D printing approaches for radiochemical microreactors, biocompatibility (i.e., traditionally looked after in the numerous biological applications) is less important than chemical stability, radiation and mechanical hardness, and inertness of materials. While SLA may provide the best trade-offs between resolution, cost, and customizability to meet demanding microfluidic design needs, FDM may provide a slight advantage for what regards choice of materials with the appropriate properties for radiochemistry.

Microreactor design software platforms have become essential tools to streamline device prototyping. Free packages like FreeCAD© (FreeCAD, n.d.) allow constructing models for 3D printing by extruding and subtracting primitives; higher-end software, such as SolidWorks from Dassault Systèmes SE (Vélizy-Villacoublay, France) (SOLIDWORKS, n.d.), allow a larger freedom of design and are available as site subscription at most academic sites. Free online tools are also available, such as Flui'device from Eden Tech (Paris, France) (Eden Tech, n.d.) and 3DµF (Sanka et al., 2019); however, while useful to understand the basic features and requirements for designing a microfluidic chip, they offer limited design possibilities. The 3D CAD models are most commonly exported into the stereolithography file format (STL) containing mesh surface representations, although other file formats may be required depending on the printer requirements. This file is then loaded into a slicer program which slices the model into layers and generates the low-level instructions (typically in G-code) for the 3D printer. Before printing physical prototypes, advanced programs like Ansys Fluent (Ansys Inc. (Canonsburg, PA, USA)) and COMSOL Multiphysics ® (Comsol AB (Stockholm, Sweden)) can simulate fluid flow, mixing, and heat transfer behaviors to guide refinement of channel geometries virtually, and therefore help in identifying the best designs to print. A number of companies exist that offer fast prototyping of good quality components (Parallel Fluidics, n.d.), although using such services goes against the key motive for moving into 3D printing of reactors (i.e., flexible/fast/cost-effective in-house prototyping).

2 Examples of Applications

2.1 Chemistry Systems

3D printing has enabled fabrication of microreactors for a wide range of chemical syntheses, though applications in radiochemistry remain relatively scarce so far. Kitson and co-workers have thoroughly demonstrated the utility of 3D printed microreactors for common organic reactions like imine/enamine syntheses, Diels–Alder reactions, and online reductions using multichambered mixers and packed beds of immobilized catalysts (Kitson et al., 2016). In the same publication, the authors have also coupled 3D printed flow reactors with inline analytics for reaction monitoring and automated self-optimization using machine learning algorithms.

An interesting flow chemistry system developed at University College London (Penny et al., 2021) and currently commercialized by IKA Works Inc. (Wilmington, NC, USA) is based on 3D printed circular disk polypropylene (PP) reactors featuring an internal meandering microfluidic channel, and with liquid reagents delivered by means of precisely metered gas overpressure from two storage reservoirs (Fig. 3).

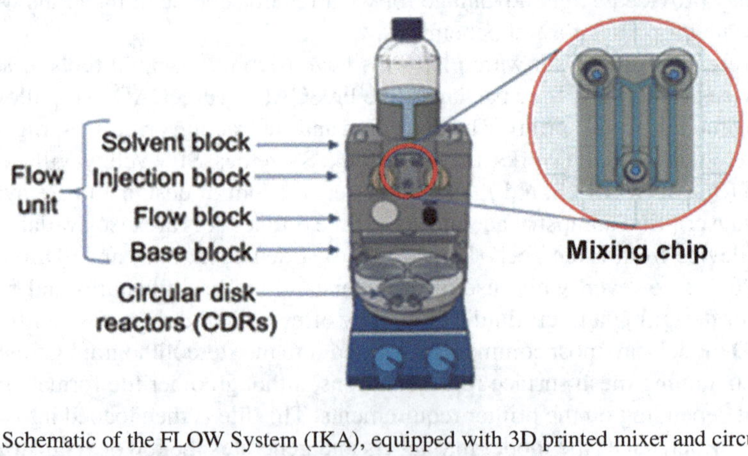

Fig. 3 Schematic of the FLOW System (IKA), equipped with 3D printed mixer and circular disk reactor components. (Adapted with permission from Penny et al., 2021)

This probably represents the cheapest commercial system that allows interfacing with custom-designed 3D printed microreactors, providing interconnectivity via standard ¼-28″ flat bottom inlet/outlet ports and respecting the circular shape of the heating component. In fact, although guaranteeing chemical resistance, this commercial option does not have a high temperature or pressure rating and is likely limited for some applications. Therefore, more resilient reactors can be 3D printed and adapted to the framework to improve these features.

Another useful approach has been reported by Christie and co-workers (Price et al., 2021), wherein they attempted to fulfill the different needs for chemical inertness (i.e., required in a chemical process) and mechanical strength (i.e., required to create leak-free interfaces). They resolved this dichotomy by creating multi-material connectors, in which the outside structure and the fitting heads were 3D printed using a strong plastic (e.g., PLA), while the internal "wetted" structure was printed with PP. This approach proved successful and achieved pressure ratings of up to 53 bar; however, the manufactured systems were only fluid tested and not used to perform chemical reactions. The authors also provide a quick software tool to draw custom flow microreactors, with a fixed shape philosophy (i.e., serpentine network) but with modifiable channel features (e.g., diameter, length).

More recently, Neumaier and co-workers have introduced a modular platform for 3D printed microreactors that could be manufactured both in PP and PEEK (Neumaier et al., 2019; Menzel et al., 2020). Of particular note, their approach provides solutions that allow separating tubing interface, core fluidic channel, and other elements, thus realizing systems in which different characteristics of the materials (e.g., chemical and mechanical resistance) are exploited to realize the best functionality. Neumaier and co-workers also report the 3D printing of other elements, such as back pressure regulators, liquid–liquid separator, and syringe pumps, and the integration with the required electronic components.

2.2 *Worked Example: In-House Manufacturing and Testing 3D Printed Microreactors for Radiolabeling*

In order to better understand the questions and challenges linked to producing in-house 3D printed microreactors, we provide a brief report on the preliminary results of our group in 3D printing a microreactor with the potential use in radiochemical reactions.

2.2.1 Print Protocols

We used an Ultimaker 3 FDM 3D printer, which has a variety of standard materials to choose from. Using a roll of filament from Ultimaker allows the system to automatically set default print parameters to match the loaded filament. For simple materials such as PLA, this is generally good enough for most applications and environments; however, for more challenging materials, print settings may need to be adjusted to produce the desired print quality. This is particularly the case when wishing to print with 100% infill, but it also highly depends on the environment (humidity, temperature, drafts) in which the printer is located.

For the test build shown, we used primarily the Ultimaker defaults for polypropylene (PP). A full enclosure was applied to the top of the printer to keep the chamber warm and minimize warping and lifting from the build plate, and a brim was used to further assist with build plate adhesion.

2.2.2 Design

The fluidic network chosen was a simple serpentine path with eight straight sections, featuring two inlets merged by a T-junction, and one outlet. Three of the eight sections were designed with internal passive mixing baffles, while the remaining five increased the residence time (Fig. 4, left). The channels had a square profile, with internal dimensions of 1x1 mm, and each straight section was 50 mm long. The softness of the PP meant we could not securely fasten the connector by cutting a thread and screwing the connector directly into the microreactor. Therefore, this design instead features locations to lock in position a hexagonal steel nut, inspired by previous works (Neumaier et al., 2019). The connector can then be screwed into the nut and securely seated onto the device inlet with sufficient force to create a seal against the PP surface. However, this approach is not infallible, as its reliability does heavily depend on the print quality and finish of the surface.

The internal structure of this initial design was drawn using SolidWorks® in such a way that a secondary user should be able to easily adjust the design parameters to test various features, such as the length of the channel as well as the length of the turbulent mixing baffles. Given the use of coned one-piece 10–32 fittings, optimization of the female cone angle on the chip was required to achieve a leak-free connection, and we found that 30° worked best. A crude sawtooth structure was added

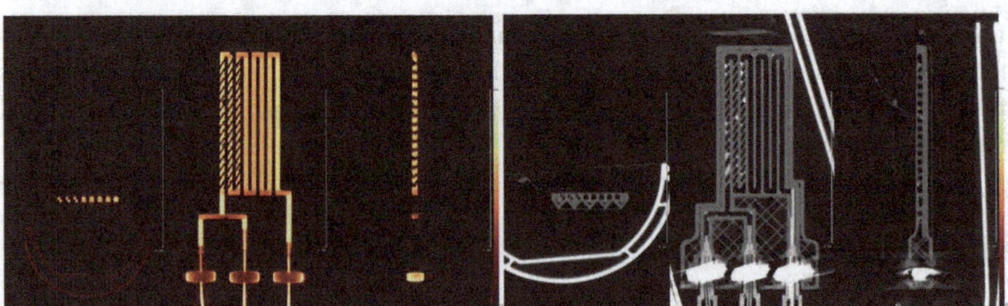

Fig. 4 CT images of: microfluidic reactor filled with contrast agent, to visualize connectedness of channels (left); the same reactor after being used in reactions, visualizing leftover salts (right). The supporting sawtooth structure can be seen in axial view (right)

below the main body, to provide rigidity to the microreactor structure and minimize the amount of support material required.

2.2.3 X-Ray CT Imaging

The microreactor was tested by filling it with a computed tomography (CT) contrast agent and performing X-ray CT scanning using a NanoScan PET/CT scanner (Mediso (Budapest, Hungary)) (Fig. 4, left). Images were acquired at 50 kVp, 980 μA, for 170 ms per projection over 720 projections for a full 360 degrees of rotation. The volume was reconstructed using the default filtered back-projection (FBP) reconstruction algorithm with a cosine filter applied during reconstruction and a voxel resolution of 0.117 mm^3. We were also able to verify that the channels (void area) were all connected using standard image processing software. We think that using CT imaging, typically available in nuclear imaging centers, would allow quick control of the quality of the device printed, especially for assessing the perviousness of the channels and the lack of major printing defects. In addition, it can be used to assess the status of the device after its usage and guide improved designs (Fig. 4, right).

2.2.4 Functional Testing

In order to assess the absence of leaks, we connected one device inlet to a nitrogen source and pressurized the reactor by blocking the other inlet/outlets. We submerged the device in water and gradually increased the gas pressure. We deemed the performance as "optimal" when no visible bubbles were formed with 5 bar pressure; performance was deemed "acceptable" when, under the same pressure, bubbles were formed only at the fitting interface. In any cases where bubbles were generated from the body of the reactor, the performance was deemed "unacceptable." To verify the mixing capability of the baffle network, we also performed a liquid flow test wherein yellow and blue dyes were delivered into the two inlets and complete color change was expected after the mixing zone (Fig. 5). The initial design with square profile channels and corners provided the most reliable, while attempts to realize a round channel profile mostly provided leaky devices; it is likely that such a feature could be realized, but extensive optimization of the printing parameters will be required.

2.2.5 Testing of Microfluidic Reactor Via [^{68}Ga]Ga-PSMA-11 Labeling

The best-performing device was tested using a model ^{68}Ga-labeling reaction; the reaction mixture was analyzed by radio-TLC, and the result compared to a similarly executed process in a 10 mL vessel.

For the vessel reaction, the standard MultiSyn cassette (MSH-120, iPhase (Rowville, VIC, Australia)) setup was used. 4 μg of PSMA-11 precursor dissolved

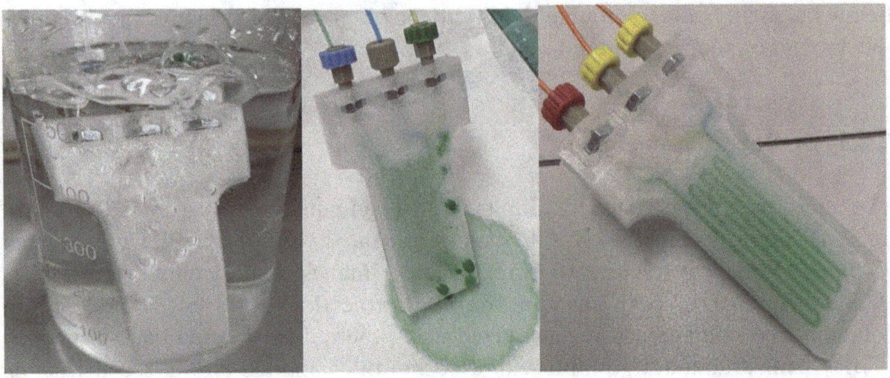

Fig. 5 Examples of gas pressure testing (unacceptable, left) and liquid flow testing (unacceptable, middle; optimal, right)

in 800 μL of 1.0 M sodium acetate was added to the reaction vessel. A syringe containing 4 mL of 0.05 M HCl was used to elute the gallium-68 generator directly into the reaction vessel. The radiolabeling was conducted at room temperature for over 5 min, after which the reaction mixture was transferred into a product vial, and a further 5 mL of saline was used to ensure complete transfer. The mixture was immediately spotted onto a TLC plate to evaluate the radiolabeling efficiency.

For the microfluidic process, a MultiSyn cassette, a dual syringe driver, and the microreactor were used (Fig. 6). A 1 mL syringe containing 4 μg of PSMA-11 precursor dissolved in 800 μL of 1.0 M sodium acetate was placed in the syringe driver and connected directly to one of the two inlet ports on the microreactor. A 5 mL syringe containing 4 mL of 0.05 M HCl was placed in the syringe driver and connected such that the gallium-68 generator eluent was directed into the remaining inlet port on the microreactor via the cassette. Due to equipment requirements, the syringe drivers were set at the same step speed for both syringes, resulting in flows of 1 mL min^{-1} for the 5 mL syringe and ~0.2 mL •min^{-1} for the 1 mL syringe; the two solutions were driven through the microreactor over 4 min. The product was transferred directly into a product vial connected to the outlet of the microreactor; at the conclusion of the 4-min process, 5 mL of additional saline was delivered from the same 5 mL syringe to rinse the microreactor into the product vial. The mixture was immediately spotted onto a TLC plate to evaluate the radiolabeling efficiency.

Radio-TLC was performed using iTLC-SG TLC plates (Agilent (Santa Clara, CA, USA)) and a 7.7% NH_4OAc in 50% v/v MeOH running buffer. Plates were scanned using a Scan-Ram system (Lablogic (Chantilly, VA, USA)) with a plastic scintillator detector, and the results were analyzed in the Laura TLC software (ver. 6.3.2.36, Lablogic).

The microreactor performed as expected, with no detectable leaks or internal breaks. The radiolabeling efficiency (Fig. 7) at room temperature was much higher in the microfluidic environment (14.7%) as opposed to the control reaction using the "vessel style" reaction (0.9%).

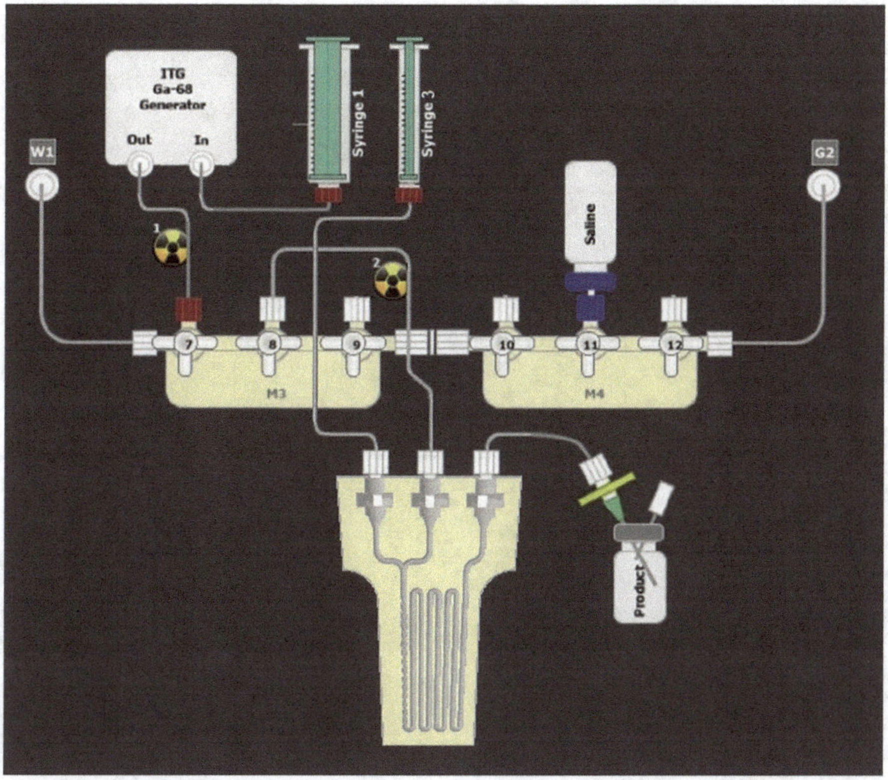

Fig. 6 Plumbing scheme for process implementation on a MultiSyn module (iPhase (Australia)) connected to the 3D printed microfluidic reactor

Room temperature labeling of PSMA-11 was anticipated to result in poor yields as such reaction is typically performed at elevated temperature, although the difference in labeling efficiency in the above experiment demonstrates the substantial positive impact the microfluidic environment can have, even in the absence of heating. Further testing is planned to determine if this difference persists upon mild heating and if the microreactor might enable radiolabeling reactions to be conducted at biocompatible temperatures (e.g., 37 °C), hence allowing the use of precursors that would degrade at the higher temperatures conventionally employed.

2.3 *Radiochemistry Systems*

Given the recent adoption of 3D printed reactors in chemistry, the impact within the radiopharmaceutical field is not yet substantial; herein we outline a few examples of its application within radiochemistry.

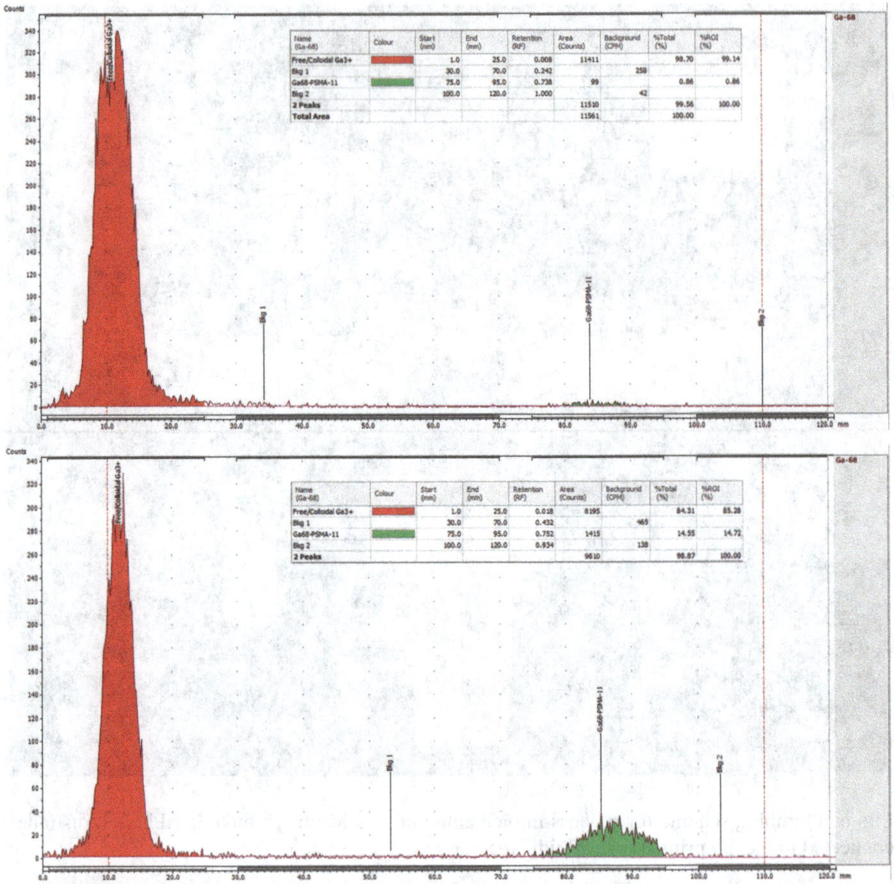

Fig. 7 Radiolabeling efficiencies for Gallium-68 reactions at room temperature performed in vessel (top) and microfluidic (bottom) environments

The first related example reported in the literature dealt with the manufacturing of entire automated synthesis units (ASU) out of 3D printed components (Amor-Coarasa et al., 2019). The primary target of the authors was to build up an ASU substantially cheaper than the commercially available ones; for this reason, they have printed the casing, holders, and adapters using hard plastics (e.g., nylon, ABS) and complemented them with off-the-shelf electronic components. Utilizing this framework and standard radiopharmaceutical cassette philosophy, they demonstrated its applicability in various types of processes, from ^{11}C-labeled fatty acid synthesis to ^{18}F-labelled tracers (also employing a dual reactor process) and to ^{68}Ga labeling. This design strategy (Fig. 8) allowed them to avoid dealing with the issue of chemical resistance of the 3D printed components, as all the fluidics were managed by commercially available consumables (e.g., manifolds, syringes, glass vials). In the cases where high temperatures were required, they resorted to 3D printed

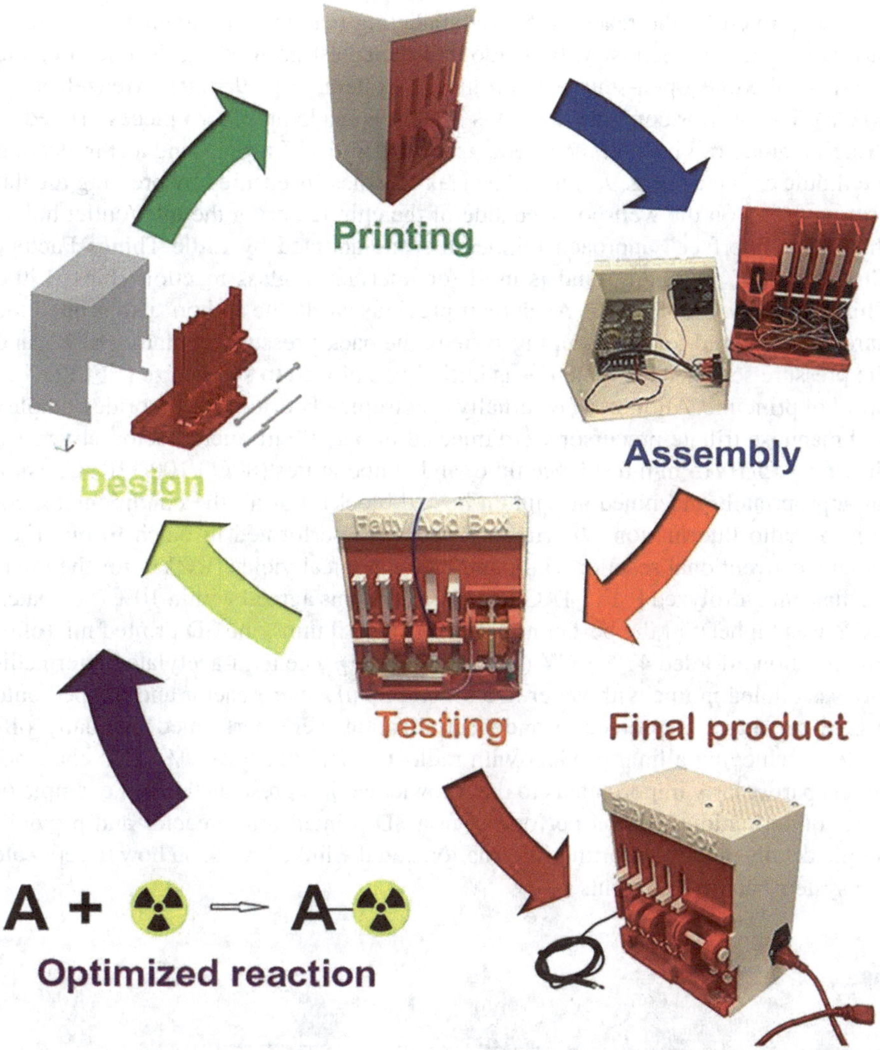

Fig. 8 3D printed ASU for radiopharmaceutical manufacturing. (Adapted from Amor-Coarasa et al. (2019)) under CCC license)

porcelain heating jackets. Of particular note, these systems could also include a Peltier cooling and a millifluidic flow reactor (i.e., coiled plastic tubing). The systems presented achieved the authors' aim, i.e., reducing the estimated cost of each ASU below the 10,000 US$ mark and maintaining the cost of spare components below the $50 USD mark. However, it is surprising how such systems have not been reported again in the literature.

Neumaier et al. are the only group reporting on the use of 3D printed wetted parts, i.e., microfluidic reactors, for radiolabeling reactions. Following their previous experience in chemistry, they adopted their best-performing design to create FOMSy (flexible open-source microfluidic system; Fig. 9, left) (Menzel et al. 2023a). The reactor component of this system is made up of two pieces printed in PEEK: a guide rail hosting the thread grooved for ¼-28″ fittings and a chip hosting the fluidic channels (Fig. 9, right). The leak-tightness is ensured by pressing the flat bottom fitting on the well-polished side of the chip featuring the inlet/outlet holes; this "edge-interface" approach mimics the one adopted by Little Things Factory GmbH (Elsoff, Germany) and is used for interfacing glass reaction chips (Little Things Factory GmbH, n.d.). As in their previous work, the authors also printed the hardware required for the pumping system, the back-pressure regulator (BPR), and the pressure sensor. This system was initially employed to synthesize [^{18}F] FDG as proof of principle. Aliquots of externally azeotropically dried [^{18}F] fluoride complex and mannose triflate precursor were injected into a 900 μl microreactor, at various flow rates (5/10/15 min residence time) and temperatures (80/90/100/110 °C, using an appropriately machined aluminum heated block). For all the conditions tested, control radio fluorination experiments were also performed in batch format (i.e., using a conventional reaction vial), and radiochemical yields (RCYs) for the intermediate unhydrolyzed [^{18}F] FTAG from both systems agreed within 10% (i.e., batch RCY was higher); in the best conditions (90 °C, 10 min), the 3D printed microfluidic reaction afforded 47% RCY (vs 52% in batch). The tetra-acetylated intermediate was diluted in line with water in a second 50 μL microreactor and trapped onto a C18 cartridge. Deprotection and neutralization were performed manually off-system, achieving a final product with radiochemical purity >99%. This contribution is particularly important, as to our knowledge, it represents the first example of a radiofluorination reaction performed in a 3D printed microreactor and provides ample details in the supporting information and the linked work on how to replicate the systems and experiments.

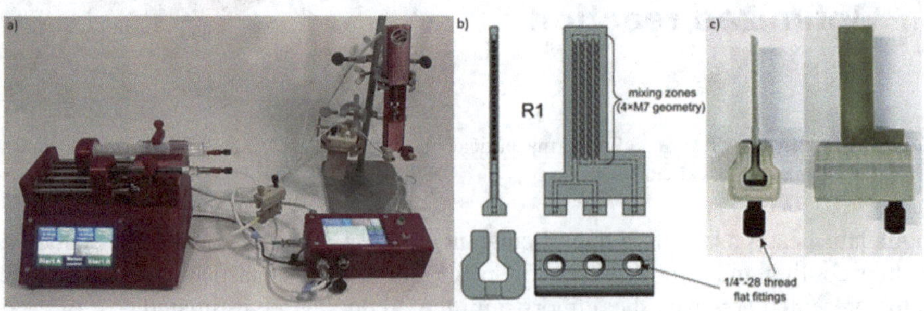

Fig. 9 (**a**) 3D printed microreactor system (FOMSy), featuring a dual syringe pump, a pressure sensor, a back pressure regulator (with its control unit) and PEEK microreactor. (Adapted from Menzel et al. (2023a) under CCC license). 3D printed PEEK microreactor design (**b**) and printed prototype (**c**) used for radiofluorination reactions. (Adapted from Menzel et al. (2020) under CCC license)

The same group recently reported a further example employing this 3D printed microreactor concept for a further set of radiofluorination reactions (Menzel et al. 2023a). The main aim of this contribution was to identify the best method to prepare a dry radiofluorination solution that could be of use in streamlined microfluidic systems. Their solution was to use [^{18}F] sulfonyl fluoride intermediates (Pascali et al., 2017; Zhang et al., 2019) and design a stand-alone automated system to afford a solution trapping the gaseous [^{18}F] triflyl fluoride reagent (Pees et al., 2018). The obtained solution was then successfully used to radiolabel three fluoroethyl synthons: 2-[^{18}F] fluoroethanol, 2-[^{18}F] fluoroethyl triflate, and 2-[^{18}F] fluoroethyl tosylate.

Flow and temperature optimization was performed, and the radiochemical conversion (RCC) and RCY compared to batch reactions executed in analogous procedures. Across all experiments, the 2-[^{18}F] fluoroethyl tosylate provided the lowest yield, with no difference between microfluidic and control labeling reactions (RCY: 53%, RCC: 86%). On the other hand, both 2-[^{18}F] fluoroethanol and 2-[^{18}F] Fluoroethyl triflate provided yields >15% higher in flow, with an RCY of 88% and 79%, respectively (Fig. 10).

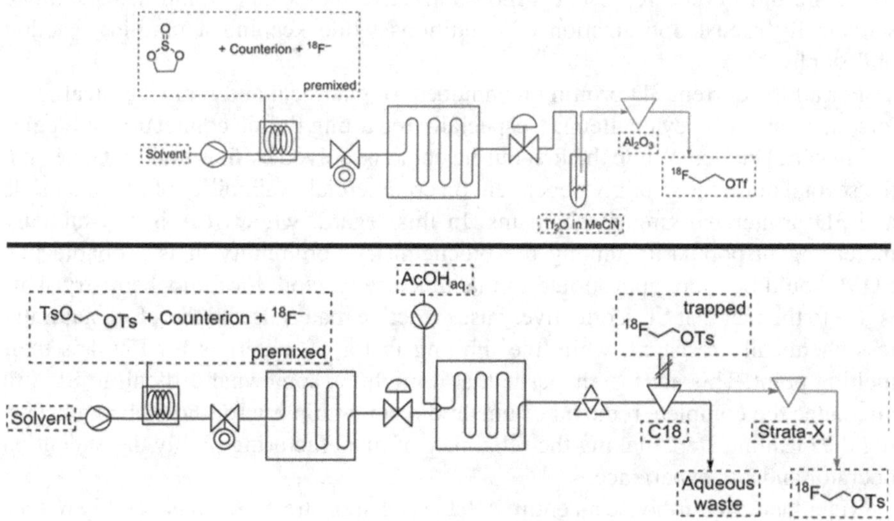

Fig. 10 Microfluidic setup and synthesis scheme for production of [^{18}F]FEtOTf (top). Setup and synthesis scheme for production of [^{18}F]FEtOTs (bottom). For synthesis of 2-[^{18}F] FEtOH, the reaction mixture was collected prior to addition of Tf_2O. The green lines highlight off-system operations. (Adapted from Menzel et al. (2023b) under CCC license)

Both the tosylate and triflate synthons were tested for their capacity of O-, S-, N-, and carboxylic alkylation in manual batch reactions at 90 °C for 10 or 20 min. Surprisingly, only the triflate afforded acceptable yields using N,N-diisopropylethylamine (DIPEA) as a base, but no alkylation was recorded on the phenol precursor tested. However, these coupling results should be regarded as preliminary, given how few precursors were tested and the limited reaction condition optimization.

3 Conclusion and Perspectives

3D printing has unlocked new avenues to fabricate customized microfluidic devices on demand, bringing enhanced flexibility and accessibility where traditional cleanroom microfabrication has posed high barriers to entry. The unique benefits of improved mixing, heat transfer, automation, and safety offered by miniaturized flow reactors are especially compelling for radiochemistry. However, it is also possible to forecast that 3D printing intelligently designed parallel batch reaction systems (Nevolab, n.d.) could represent a more attractive target that would also facilitate statistically robust optimization of conditions while keeping a familiar reaction philosophy.

For all the current 3D printing techniques for chemical and radiochemical reactors, there are still key challenges, especially regarding fluidic connection strategies and material choice. While the first hurdle could be solved by fine-tuning the design, the second hurdle is strongly dependent on commercial availability of raw materials and 3D printer working mechanisms. In this regard, while SLA is substantially increasing in popularity among the biochemistry community, it is probable that FDM would be more appropriate to manufacture reactors for radiochemistry. This is due to the fact that SLA effectively uses reactive materials and therefore naturally less chemically resistant, while the limiting factor of materials for FDM is their melting point. This adds to the issue that it might be somewhat difficult in SLA to guarantee the complete removal of uncured resin within microfluidic channels during the cleaning step, making the reliability of manufacturing highly dependent on operator and site experience.

Manufacturing in-house an entire ASU was demonstrated to considerably reduce the financial investment needed to equip a radiochemistry laboratory. A lighter version of this concept is already normal practice in many sites, whereas small components (e.g., holders, actuator heads) are already 3D printed instead of purchased from the original supplier, to expand the capabilities of existent radiosynthesizers. Added to the now-demonstrated possibility to manufacture microreactors compatible with radiochemical reactions, all these aspects should contribute to lowering the entry barrier of academic and budget-conscious labs in the fields of radiochemistry and nuclear imaging.

However, it is wise to remember that different skillsets are required to 3D print synthesizer components. One of the key knowledge areas is the capacity to create and modify 3D CAD designs in a streamlined, fast, and potentially modular manner. Another key is related to understanding which functional tests to perform to verify the quality and performance of the manufactured component. To these, additional electronics and programming skills are needed if the project also includes their in-house design. It is also worth noting that such an approach allows one to test in an efficient and fast manner different prototypes and reaction technologies, but the best-performing system needs to be translated to more robust manufacturing methods in order to become available at a commercial scale.

Radiochemists must leverage the design freedom of 3D printing in creative ways, matching functionality to desired reactions since simply shrinking batch chemistry equipment does not fully utilize potential benefits. Innovation of reactor geometries and features, for instance, to improve mixing, enable new reaction modalities (e.g., photocatalysis), integrate purification modules, and automate multistep sequences will be vital. Collaboration with engineers and computer simulation will no doubt play an important role in optimizing designs as well. Complementary to this, the bourgeoning field of AI could also play a role in creating the best designs, and already one group has reported on ChatGPT-aided CAD design of microfluidic helical and serpentine reactors, valves, and T-junctions (Nelson et al., 2023).

The growth of 3D printing technologies for chemical synthesis and the beginning of adoption in radiochemical research are encouraging signs toward realizing compact, automated systems that can take an elementary radionuclide source and convert it to purified, formulated product, all within microfluidic "lab-on-a-chip" devices. The tantalizing possibility of 3D printing entire radionuclide generators or self-contained radioisotope production modules also seems within reach. We are surely just witnessing the beginnings of a transformative fusion between fields poised to shape the future landscape of radiochemistry.

References

3M. (n.d.). A new dimension of opportunity. Retrieved January 11, 2024, from https://multimedia.3m.com/mws/media/1547162O/3m-dyneon-fluoropolymers-3d-printing-brochure.pdf

Amor-Coarasa, A., Kelly, J. M., & Babich, J. W. (2019). 3D-printed automation for optimized PET radiochemistry. *Science Advances, 5*(9), 4762–4775. https://doi.org/10.1126/SCIADV.AAX4762

Andreu, A., et al. (2021). 4D printing materials for vat photopolymerization. *Additive Manufacturing, 44*, 102024. https://doi.org/10.1016/J.ADDMA.2021.102024

Au, A. K., et al. (2016). 3D-printed microfluidics. *Angewandte Chemie International Edition, 55*(12), 3862–3881. https://doi.org/10.1002/ANIE.201504382

CREAMELT—3D Printing Materials. (n.d.). 'New high performance filament: CREAMELT® COC'. Retrieved January 11, 2024, from https://creamelt.com/wp/new-high-performance-filament-creamelt-coc/?v=18a5f226e926

Eden Tech. (n.d.). FLUI'DEVICE. Retrieved January 11, 2024, from https://eden-microfluidics.com/fluidevice/microfluidic-calculator/

Fan, J., et al. (2022). Effect of additive manufacturing on β-phase poly(vinylidene fluoride)-based capacitive temperature sensors. *Advanced Engineering Materials, 24*(11), 2200485. https://doi.org/10.1002/ADEM.202200485

Faraji Rad, Z., Prewett, P. D., & Davies, G. J. (2021). High-resolution two-photon polymerization: The most versatile technique for the fabrication of microneedle arrays. *Microsystems & Nanoengineering, 7*(1), 1–17. https://doi.org/10.1038/s41378-021-00298-3

Filament2Print. (n.d.). PC/PTFE. Retrieved January 11, 2024, from https://filament2print.com/gb/advanced/901-pc-ptfe.html

Formlabs. (n.d.). Form wash and form cure: Automated post-processing. Retrieved January 12, 2024, from https://formlabs.com/asia/post-processing/wash-cure/

FreeCAD. (n.d.). 'Your own 3D parametric modeler'. Retrieved January 11, 2024, from https://www.freecad.org/

Gerdroodbar, A. E., et al. (2023). Vitrimer chemistry for 4D printing formulation. *European Polymer Journal, 197*, 112343. https://doi.org/10.1016/J.EURPOLYMJ.2023.112343

Huang, J., Qin, Q., & Wang, J. (2020). A review of stereolithography: Processes and systems. *Processes, 8*(9), 1138. https://doi.org/10.3390/PR8091138

IKA. (n.d.). FLOW system flow chemistry. Retrieved April 16, 2024, from https://www.ika.com/en/Products-LabEq/FLOW-System-pg919/

Kitson, P. J., et al. (2012). Configurable 3D-printed millifluidic and microfluidic "lab on a chip" reactionware devices. *Lab on a Chip, 12*(18), 3267–3271. https://doi.org/10.1039/C2LC40761B/

Kitson, P. J., et al. (2016). 3D printing of versatile reactionware for chemical synthesis. *Nature Protocols, 11*(5), 920–936. https://doi.org/10.1038/nprot.2016.041

Kotz, F., et al. (2018). Highly fluorinated Methacrylates for optical 3D printing of microfluidic devices. *Micromachines, 9*(3), 115. https://doi.org/10.3390/MI9030115

Little Things Factory GmbH. (n.d.). 'Home'. Retrieved January 12, 2024, from https://ltf-gmbh.com/

Mayoussi, F., et al. (2021). Facile fabrication of micro-/nanostructured, superhydrophobic membranes with adjustable porosity by 3D printing. *Journal of Materials Chemistry A, 9*(37), 21379–21386. https://doi.org/10.1039/D1TA03352B

Menzel, F., et al. (2020). 3D-printed PEEK reactors and development of a complete continuous flow system for chemical synthesis. *Reaction Chemistry & Engineering, 5*(7), 1300–1310. https://doi.org/10.1039/D0RE00206B

Menzel, F., Cotton, J., Klein, T., et al. (2023a). FOMSy: 3D-printed flexible open-source microfluidic system and flow synthesis of PET-tracer. *Journal of Flow Chemistry, 13*(3), 247–256. https://doi.org/10.1007/S41981-023-00267-Z

Menzel, F., Cotton, J., Ziegler, T., et al. (2023b). Open-source flow setup for rapid and efficient [18F]fluoride drying for automation of PET tracer syntheses. *Journal of Labelled Compounds and Radiopharmaceuticals, 67*(2), 40–58. https://doi.org/10.1002/JLCR.4080

Nelson, M. D., Goenner, B. L., & Gale, B. K. (2023). Utilizing ChatGPT to assist CAD design for microfluidic devices. *Lab on a Chip, 23*(17), 3778–3784. https://doi.org/10.1039/D3LC00518F

Neumaier, J. M., et al. (2019). Low-budget 3D-printed equipment for continuous flow reactions. *Beilstein Journal of Organic Chemistry, 15*(1), 558–566. https://doi.org/10.3762/BJOC.15.50

Nevolab. (n.d.). 'Innovative heating and cooling for better chemistry. Retrieved January 12, 2024, from https://www.nevolab.de/en/xelsius-reactor/

Nielsen, A. V., et al. (2020). 3D printed microfluidics. *Annual Review of Analytical Chemistry, 13*, 45–65. https://doi.org/10.1146/ANNUREV-ANCHEM-091619-102649

Parallel Fluidics. (n.d.). 'Rapid Manufacturing for Microfluidics'. Retrieved January 11, 2024, from https://www.parallelfluidics.com/

Parra-Cabrera, C., et al. (2018). 3D printing in chemical engineering and catalytic technology: Structured catalysts, mixers and reactors. *Chemical Society Reviews, 47*(1), 209–230. https://doi.org/10.1039/C7CS00631D

Pascali, G., et al. (2017). Sulfur-fluorine bond in PET radiochemistry. *EJNMMI Radiopharmacy and Chemistry, 2*, 1–18. https://doi.org/10.1186/s41181-017-0028-6

Pees, A., et al. (2018). Fast and reliable generation of [^{18}F]triflyl fluoride, a gaseous [^{18}F]fluoride source. *Chemical Communications, 54*(72), 10179–10182. https://doi.org/10.1039/C8CC03206H

Penny, M. R., Tsui, N., & Hilton, S. T. (2021). Extending practical flow chemistry into the undergraduate curriculum via the use of a portable low-cost 3D printed continuous flow system. *Journal of Flow Chemistry, 11*(1), 19–29. https://doi.org/10.1007/S41981-020-00122-5/FIGURES/18

Price, A. J. N., et al. (2021). An open source toolkit for 3D printed fluidics. *Journal of Flow Chemistry, 11*(1), 37–51. https://doi.org/10.1007/S41981-020-00117-2/FIGURES/10

Sanka, R., et al. (2019). 3DμF—interactive design environment for continuous flow microfluidic devices. *Scientific Reports, 9*(1), 1–10. https://doi.org/10.1038/s41598-019-45623-z

SOLIDWORKS. (n.d.). 3D CAD design software & PDM systems. Retrieved January 11, 2024, from https://www.solidworks.com/

Waheed, S., et al. (2016). 3D printed microfluidic devices: Enablers and barriers. *Lab on a Chip, 16*(11), 1993–2013. https://doi.org/10.1039/C6LC00284F

Wang, M.-W. W., et al. (2010). Microfluidics for positron emission tomography probe development. *Molecular Imaging, 9*(4), 175–191. https://doi.org/10.2310/7290.2010.00027

Watts, P., & Haswell, S. J. (2005). The application of microreactors for small scale organic synthesis. *Chemical Engineering & Technology, 28*(3), 290–301. https://doi.org/10.1002/CEAT.200407124

Zhang, B., et al. (2019). [^{18}F]Ethenesulfonyl fluoride as a practical Radiofluoride relay reagent. *Chemistry—a European Journal, 25*(32), 7613–7617. https://doi.org/10.1002/chem.201900930

Zwyssig, D., et al. (2023). 3D printed microfluidic devices using TOPAS filament for THz spectroscopic measurements. *Optical Materials Express, 13*(4), 1031–1040. https://doi.org/10.1364/OME.477708

Democratizing Radiopharmaceutical Production: The RAPID Chip

H. Charles Manning, Federica Pisaneschi, Robert Ta, Mai Lin, and Leon M. Bellan

1 Introduction

Radiopharmaceuticals (RPs) are essential linchpins of cancer diagnosis, staging, and therapy. For diagnosis, the sensitivity and quantitative nature of positron emission tomography (PET) imaging, coupled with the ability to produce biologically active RPs bearing positron-emitting isotopes, render PET uniquely capable of detecting tumors and quantifying their actionable molecular features. To date, PET imaging has shown great promise for diagnosis and prognosis of cancer (Young et al., 1999; Fischer et al., 2001; Kubota, 2001; Gambhir, 2002; Lardinois et al.,

H. C. Manning (✉)
Division of Diagnostic Imaging, Department of Cancer Systems Imaging, University of Texas MD Anderson Cancer Center, Houston, TX, USA

Division of Diagnostic Imaging, Department of Cyclotron Radiochemistry Facility, University of Texas MD Anderson Cancer Center, Houston, TX, USA
e-mail: HCManning@mdanderson.org

F. Pisaneschi · R. Ta
Division of Diagnostic Imaging, Department of Cancer Systems Imaging, University of Texas MD Anderson Cancer Center, Houston, TX, USA

M. Lin
Division of Diagnostic Imaging, Department of Cyclotron Radiochemistry Facility, University of Texas MD Anderson Cancer Center, Houston, TX, USA

L. M. Bellan (✉)
Department of Mechanical Engineering, Vanderbilt University, Nashville, TN, USA

Department of Biomedical Engineering, Vanderbilt University, Nashville, TN, USA

Vanderbilt Institute of Nanoscale Science and Engineering (VINSE), Vanderbilt University, Nashville, TN, USA
e-mail: leon.bellan@vanderbilt.edu

R. M. van Dam et al. (eds.), *Automated Technologies for the Development and Production of Radiopharmaceuticals*,
https://doi.org/10.1007/978-3-031-84632-8_9

2003; Vansteenkiste et al., 2004; Wiering et al., 2005; Facey et al., 2007; Perera et al., 2016) and other diseases (Baxter Jr. et al., 1985; Duara et al., 1986; Reiman et al., 1986; Eitzman et al., 1992; Playford et al., 1992; Di Carli et al., 1994; Goyer et al., 1994; Rauch et al., 1996). However, the impact of PET imaging on healthcare overall, and especially next-generation precision medicine, is limited by the breadth of tracers available clinically. RP production is highly resource-intensive, requiring skilled personnel, major equipment, and dedicated space. Consequently, the proliferation of PET tracers beyond [^{18}F]FDG has been severely limited, inhibiting the expansion of PET-based precision medicine. Moreover, recent successes in theranostics (Gomes Marin et al., 2020), which utilize targeted RP pairs for diagnosis and therapy, have led to improved outcomes in challenging clinical settings, such as neuroendocrine and prostate cancers (PCs; Hawkey et al., 2022, Ichikawa et al., 2022). Encouraged by these results and recent clinical approvals, pharma and academia alike are investing heavily in RP Research and Development (R&D) in pursuit of theranostics targeting other solid tumors, such as pancreatic (Ferreira et al., 2020; Jaidev et al., 2021), breast (Jokar et al., 2021), and lung cancer (Filippov et al., 2020). From a drug development standpoint, radiotherapeutic agents are prepared using largely analogous chemistries as to their diagnostic counterparts. Therapeutic isotopes [e.g., lutetium-177 (^{177}Lu), yttrium-90 (^{90}Y)] are more expensive and, from a radiation safety perspective, more challenging to handle than diagnostic isotopes such as gallium-68 (^{68}Ga) and fluorine-18 (^{18}F). Therefore, it is common to first develop radiochemistry technologies using less expensive and more readily available diagnostic isotopes and later extend the same technology to therapeutic isotopes as proposed here.

Currently, RPs for patient use are routinely produced on a batch-to-batch basis using costly (>$250 k) and large synthesis modules in highly specialized cGMP laboratory environments. To offset the infrastructure and personnel costs required, batch production necessarily must yield multiple patient doses per batch to reduce per-dose costs. While this “economies-of-scale” strategy is well suited for commercial distribution of widely utilized tracers such as [^{18}F]FDG, research applications or “boutique” tracers frequently require only a single unit dose per batch and, accordingly, suffer from prohibitively high dose-per-batch costs (10×–100× higher cost). These economics effectively curtail dissemination of “personalized” tracers.

In this context, a cost-effective “single dose-batch/patient” concept could revolutionize how RPs are deployed and democratize PET-based molecular imaging and radiotherapies for precision medicine (Fig. 1).

The RAPID (Radiopharmaceuticals As Precision Imaging Diagnostics) chip has been conceived and developed as an ultra-low-cost, disposable, single-dose microfluidic radiosynthesis device that can produce clinical grade tracers without additional off-chip purification or concentration steps. The device has been tested for the production of two commonly used PET radiotracers, [^{18}F]Fallypride (Zhang et al., 2018) and [^{68}Ga]Ga-PSMA-11 (Zhang et al., 2020).

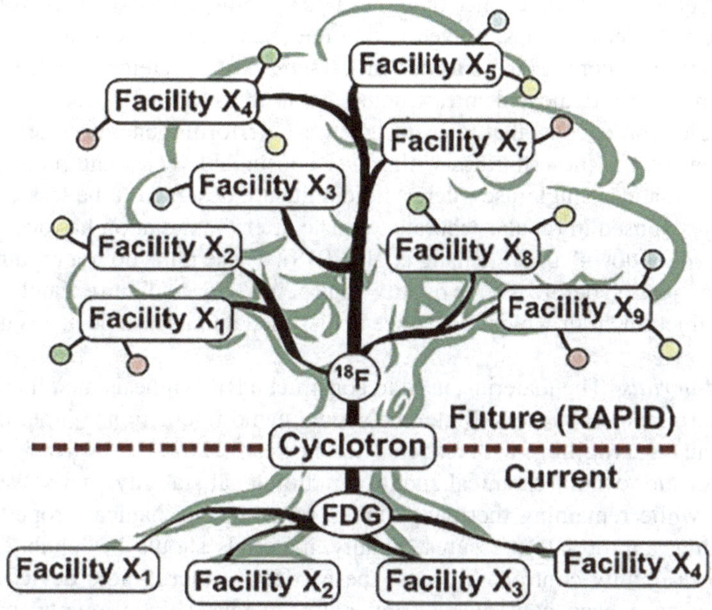

Fig. 1 A new model for RP distribution. Currently, centralized facilities distribute high-volume products (e.g., FDG) to be used at many sites. Enabled by the low-cost RAPID concept, starting material (e.g., ^{18}F) is broadly distributed and single dose-batch tracers (colored spheres) are produced on demand at distinct points-of-care

1.1 Benefits of Microfluidics for RPs

There is considerable interest in leveraging microfluidics to increase efficiency of radiosynthesis (Audrain, 2007; Elizarov, 2009; Miller, 2009; Wang et al., 2010; Pascali et al., 2011, 2013; Keng et al., 2012b; Watts et al., 2012; Fortt & Gee, 2013; Rensch et al., 2013); benefits include a decreased physical footprint (which in turn reduces laboratory/clinical space requirements), decreased reaction volumes and higher reaction concentrations (which decrease production times relative to macroscale production due to increased heat transfer, higher surface-to-volume ratio, and controllable mixing), and the ability to produce clinically meaningful doses from a single run. In turn, radiochemical yield, radiochemical purity, molar activity, and overall production reproducibility are all significantly enhanced.

Repeated Use Versus Single Use A key feature of RAPID that enables reduced cost of RP manufacturing is the fact that RAPID chips are single-use and disposable, similarly to conventional macroscale kit-based cassette systems. Disposable systems save valuable production time by eliminating the need to wait for radioactivity decay and cleaning cycles as required by reusable systems to eliminate cross-contamination (Pascali et al., 2011; Rensch et al., 2014). Additionally, single-use systems require less maintenance than reusable systems, where damage from radiation, repeated thermocycles, solvents, or other harsh reagents result in frequent requalification. In contrast to conventional cassette-based systems, which are expensive and require a dedicated infrastructure focused on a single tracer, single-use microfluidic approaches that are designed to perform generalizable synthetic schemes overcome these hurdles with lower consumable costs and reduced infrastructure demands. Single-use microfluidic units are designed to be less expensive and easily disposed in regular laboratory waste after the radiation has decayed (the so-called "cool down" time) (Zhang et al., 2018). These units do not require cleaning or qualification before use, as quality control is performed immediately after the tracer production, after which the device is disposed and a fresh device is used.

Device Materials The materials used to construct a RP synthesis module represent a key determinate of success. Materials should avoid interactions (either chemical or physical, e.g., swelling) with reagents (Zacheo et al., 2011). Construction materials must ensure overall chemical stability, radiological stability, and solvent compatibility while remaining thermally stable, retaining mechanical properties, and withstanding appropriate pressures. Finally, materials should be reliably sourced with known quality controls. To meet these criteria, microfluidic devices for RP production have been constructed from glass, metals, silicon, quartz, polymers, elastomers, plastics, ceramics, inorganic materials, and combinations of the above. Early systems used various glasses, but were limited by fluoride adsorption (Ungersboeck et al., 2011) which led to the use of plastics and polymers such as polydimethylsiloxane (PDMS), polytetrafluoroethylene (PTFE or Teflon™), fluorinated ethylene propylene (FEP), polyether ether ketone (PEEK), and SU-8 (Lu et al., 2004; Lee et al., 2005; Gillies et al., 2006; Wester et al., 2009; Zacheo et al., 2011; Zizzari et al., 2011; Zhang et al., 2018). PDMS is one of the most popular materials for producing microfluidic devices, having been used in many systems due to its low cost and simple fabrication while also allowing for solvent exchange via evaporation through thin gas permeable films. Its elastomeric properties permit it to conform and easily seal to other surfaces either reversibly or irreversibly, depending on surface treatment (Duffy et al., 1998; McDonald et al., 2000; McDonald & Whitesides, 2002; Lee et al., 2005; Maltezos et al., 2007). A possible limitation of PDMS for synthesis includes solvent-induced swelling with lengthy exposure (Lee et al., 2003), leading some to explore alternatives, especially when employing harsh reaction conditions. Examples include SU-8, an epoxy-based negative photoresist which is easily micropatterned and has promising chemical and mechanical properties (Zacheo et al., 2011; Zizzari et al., 2011) or cyclic olefin copolymers (COCs) (Rensch et al., 2014). These materials are more difficult to adapt to radiochemistry-appropriate architecture compared to PDMS, which increases the cost and complexity, especially at scale. Within the context of RP production, many groups (including ours) have reported impressive results when using PDMS devices, showing that swelling and other interactions with PDMS have negligible impact on tracer production in these platforms (Zhang et al., 2018, 2020). In fact, when devices are considered single-use, the effects of long-term solvent exposure on the device are negligible (Lee et al., 2005; Elizarov et al., 2010; Salvador et al. 2017a, 2017b; Zhang et al., 2018). Furthermore, a recent study found that interactions between [^{18}F]fluoride and PDMS as a function of radionuclide quantity, timeframe, temperature, and fluidic geometry demonstrated retention of <1% [^{18}F]fluoride introduced into a PDMS chip for production after a single elution (Fernandez-Maza et al., 2019). This agrees with our prior studies employing PDMS devices and highlights the suitability of PDMS for RAPID devices.

1.2 Prior Designs

Channel/Continuous Flow Microreactors The first microfluidic design conceptualized for radiosynthesis application was the continuous flow (CF) or "channel" microreactor, consisting of branching channel networks made with materials resistant to high pressure (Lu et al., 2004; Zizzari et al., 2011). Movement through the system is typically flow-controlled, as the reaction proceeds through the capillary reaction module (Wester et al., 2009; Elizarov et al., 2010; Bejot et al., 2011; Rensch et al., 2012, 2014; Lebedev et al., 2013). While there are some drawbacks associated with the CF design, the major advantages in the CF approach lie in its high-pressure containment microreactor, which allows the use of solvents at higher temperature, and its capabilities to rapidly vary reaction parameters between runs, thus promoting fast optimization of many different reaction conditions (Zheng et al., 2015; Lee et al., 2005).

Chamber Microreactors Batch mode or "chamber" microreactors (CMR) were the second system developed for radiosynthesis applications; these often incorporate on-chip valves that allow for isolation of incompatible processes and controlled direction of reagents, but render the chip design and fabrication far more complex, and require a large number of interconnects with external pneumatic apparatus. In comparison to CF systems, CMRs offer advantages including improved concentration and lower reagent volumes. Although the CMR has shown its ability in support clinical trial setting for production of [^{18}F]Fallypride (Lebedev et al., 2013), there remains a limitation of all-in-one chip capabilities.

Digital Microfluidics Instead of using mechanical valves, pumps, and channels, approaches employing the concept of "digital microfluidics" on electrowetting-on-dielectric (EWOD) devices utilize electric potentials applied either between two parallel surfaces (a "closed" system) or between electrodes on a single plate (an "open" system) to move, combine, and split individual droplets of liquid (Abdelgawad & Wheeler, 2009; Choi et al., 2012; Keng et al., 2012a; Chen et al., 2014). The plates are generally composed of an indium tin oxide (ITO) layer on both the top and bottom glass plates with various gold, silicon nitride, and amorphous fluoroplastic (Teflon) components and coatings (Ding et al., 2012; Chen et al., 2014). This allows for diverse chemical synthesis processes to be carried out with a single type of EWOD chip (Chen et al., 2014). A further embodiment of this approach includes passive droplet manipulation, relying on geometric designs to optimize the motion of the droplets while utilizing the properties of hydrophobicity and hydrophilicity to drive movement (Wang et al., 2017). To its credit, the passive droplet microfluidic device is elegant in its simplicity of design, requiring minimum actively controlled actuators, and appears to be less expensive than its predecessors.

2 RAPID-F for the Synthesis of [^{18}F]Fallypride

[^{18}F]Fallypride is a potent dopamine receptor antagonist with high affinity (33 pmol/L) to D2/D3 receptors in healthy humans and is a widely used radiotracer in PET imaging, especially in the monitoring of neurogenerative diseases (Mach & Luedtke, 2018). [^{18}F]Fallypride is clinically produced with classic synthetic modules such as the TracerLab™ FXFN series (GE Healthcare) or the TRASIS AllinOne (Brichard et al., 2012; Li et al., 2016), via a typical [^{18}F]fluorine tosyl-displacement synthetic scheme. In brief, after a suitable drying cycle with Kryptofix-222 in basic conditions, [^{18}F]-fluoride is reacted with the tosylated Fallypride precursor at high temperature. [^{18}F]Fallypride is then purified by high-performance liquid chromatography (HPLC) or solid-phase extraction (SPE) and reformulated in physiological-sterile conditions for animal or human use. Despite being usually synthesized by classic macroscale radiochemistry, [^{18}F]Fallypride has been previously produced by microfluidic chemistry, using a commercial coiled-tube microreactor (NanoTek; Advion) (Lu et al., 2009), and a microfluidic batch reactor (Lebedev et al., 2013). More recently, a multi-GBq production of the radiotracer has been reported by using a droplet microreactor (Wang et al., 2020) (Fig. 2).

We reported the synthesis of [^{18}F]Fallypride via RAPID-F, a RAPID device tailored to the needs of [^{18}F] radiochemistry. The goals of this highly integrated RAPID-F microfluidic platform include the following: (1) The entire production process should be fast and efficient, minimizing the decay of radionuclide and consuming minimal reagents. (2) The shield needed to protect users from radiation on the microfluidic chip should be simply constructed using lead bricks, avoiding the need for expensive commercially large, bulky hot cells. (3) The microfluidic chip should be low cost and disposable, and a fresh chip should be used for each

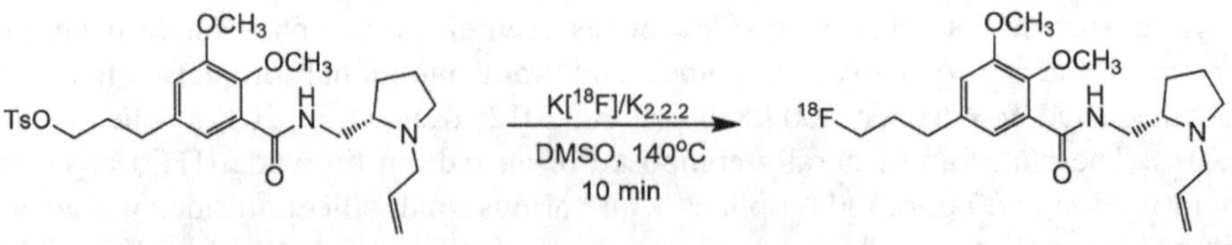

Fig. 2 Synthesis process for the production of [^{18}F]Fallypride

individual radiotracer production run. (4) The microfluidic chip should be able to produce purified products in sufficient doses, ready for animal or human imaging. (5) The technique to produce the on-chip columns should be flexible enough to enable the production of various PET radiotracers by simply switching the materials packed inside the microchannel and the reagents introduced into the chip.

RAPID-F integrates several modules on a single chip to accomplish all principal steps for radiotracer production. To demonstrate the utility of this microfluidic chip, [^{18}F]Fallypride was used as a model compound. Instead of relying on complicated on-chip flow control components that require multiple levels of lithography and feature alignment (e.g., elastomer valves formed via aligned multilayer PDMS) (Elizarov et al., 2010), we employed simple off-chip mechanical valves to provide convenient and reliable control of reagent transfer between various modules. We also used an on-chip ion-exchange column that can concentrate a quantity of fluoride sufficient for several human doses. Inspired by previous designs of microfluidic batch reactors for radiosynthesis, we used an on-chip cavity that facilitates rapid drying and enables subsequent fluorination reaction. However, unlike previous efforts where HPLC or commercial C_{18} cartridges were utilized for purification of the final product, we employed an on-chip C_{18} column that was integrated with the previous modules. Along with excellent throughput and radiochemical purity, the low-cost microfluidic chip is also disposable, allowing the usage of fresh chip for each PET tracer synthesis and continuous production by simply switching chips and reagents. Using this microfluidic device, a small shielded space composed of several lead bricks would satisfy the requirements for radiation shielding, eliminating the need for a standard hot cell. Moreover, further automation of the pumps and valves could accelerate the entire process and shorten the overall synthesis time period, minimizing the loss of imaging reagents due to the decay of radionuclides. It is also noteworthy that the simple microfluidic platform exhibits the potential to be generalized to produce various radiotracers on demand.

2.1 Device Description and Fabrication

The microchannel pattern was fabricated using facilities within the cleanroom affiliated with the Vanderbilt Institute of Nanoscale Science and Engineering (VINSE). A laser writer (Heidelberg μPG 101) was utilized to create patterns on a silicon wafer using a 60-μm-thick layer of mr-DWL 40 resist. Then, liquid polydimethylsiloxane (PDMS) (Sylgard 184, part A and part B mixed in 10:1 ratio and degassed) was poured onto the resist-patterned silicon wafer located in a petri dish to produce a 1-cm-thick layer. After curing in an oven (Thermo Scientific Lindberg/Blue M) at 65 °C for 12 h, the PDMS layer was peeled off the resist mold, and holes were punched at the inlet and outlet of the microchannels using a 1.5-mm internal diameter punch (Ted Pella). The patterned PDMS layer, along with a glass microscope slide (Fisher Scientific, Premium Plain Glass Microscope Slides), was exposed to an oxygen plasma (Harrick Plasma, PDC-001) for 60 s. Then, the two plasma-treated surfaces were bonded together and baked on a hot plate at 60 °C for 1 h before use. Next, the top surface of the microfluidic chip and a 0.2-mm-thick PDMS film with a 1.5-mm-size hole in the center were plasma treated again. The square PDMS film was manually aligned such that the hole on the PDMS film was located at the center of reaction cavity and subsequently bonded to the top of the reaction cavity. A coned NanoPort assembly (IDEX, N-333) was aligned and placed on top of the PDMS layer. Well-mixed PDMS was used as an adhesive and applied on the edge of the coned assembly and then baked in an oven at 65 °C for 2 h, such that the NanoPort was firmly attached to the top of the PDMS film.

Figure 3 shows both a schematic illustration and image of the microfluidic chip used for [^{18}F]Fallypride synthesis. Our device was composed of three main modules that served to concentrate the diluted [^{18}F]fluoride mixture, perform the heat-assisted fallypride fluorination reaction, and purify the synthesized [^{18}F]Fallypride (Zhang et al., 2018). The on-chip concentration column was 5 mm in width, 10 mm in

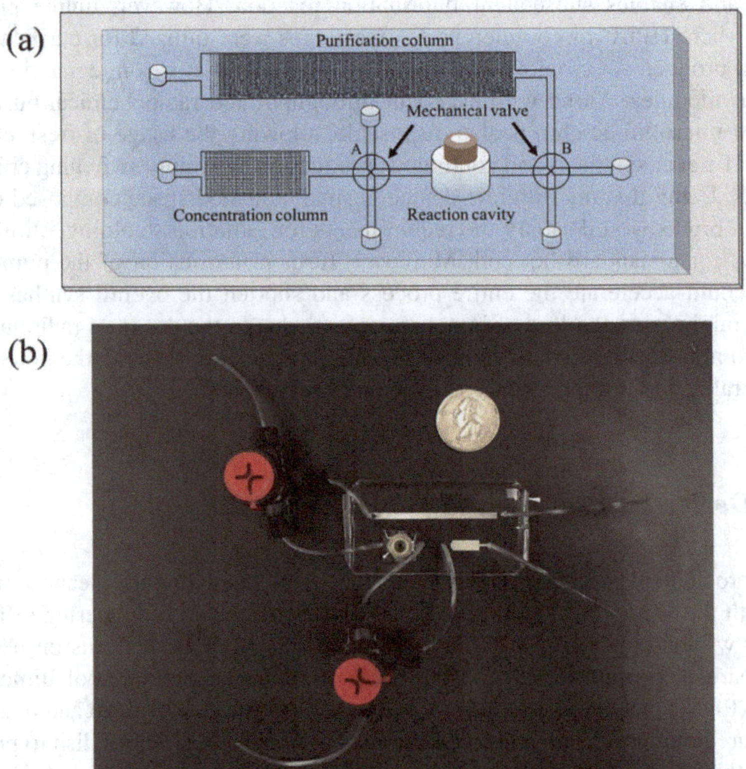

Fig. 3 **(a)** A schematic illustration of microfluidic chip employed for [^{18}F]Fallypride production composed of [^{18}F]fluoride concentration column, fluorination reaction cavity, and [^{18}F]Fallypride purification column. **(b)** A photograph of microfluidic chip with an American 25 cent coin placed above. (Reproduced with permission from the Royal Society of Chemistry, originally published by Zhang et al. (2018))

length, and 60 μm in height. Anion exchange beads (Source 15Q, GE healthcare) were packed into the concentration column by introducing a 20% ethanol solution containing suspended beads into the microfluidic channel. The desired quantity (3 μL) of anion exchange beads was trapped by a triple row of square PDMS pillars with 10-μm gap near the outlet of microchannel. Then, the anion exchange beads were activated using 1.0 M of $KHCO_3$ (0.2 mL) followed by flushing with 0.5 mL DI water (18 MΩ, Milli-Q Integral ultrapure water). The central reaction cavity was fabricated using an 8-mm inner diameter punch (Ted Pella). As our device was composed of a 1-cm thick PDMS top layer, the maximum volume allowed inside the cavity was 500 μL. The fallypride purification column was prepared using the same technique used to form the concentration column, inserting a triple row of square PDMS pillars with 40-μm gap near the outlet of microchannel. C_{18} microparticles (Fisher Scientific, Acros Organics, size 40–63 μm), suspended in pure ethanol, were introduced through the inlet of the purification microchannel and were trapped inside to form the desired column. The dimension of the fallypride purification column was 4.8 cm in length, 4 mm in width, and 60 μm in height, allowing 7.2 μL (~20 mg) of C_{18} microparticles to be packed inside. The purification column was activated before the experiment by flowing 0.5 mL DI water through it. Off-chip mechanical valves (Upchurch Valves, V-100D) were used to control the transit of reagents among these three modules. A syringe pump (New Era Pump Systems, NE-300) was used to control flow of the various reagents. A hot plate (IKA, Control-VISC) placed under the microfluidic chip was used to heat the entire device during solvent evaporation and fluorination reaction steps.

2.2 Radiosynthetic Steps

Figure 4 illustrates the steps required for [^{18}F]Fallypride radiosynthesis on our microfluidic chip. Radioactive [^{18}F]fluoride in ^{18}O-enriched water was obtained immediately after cyclotron irradiation (Zhang et al., 2018). As indicated in Fig. 4a, the [^{18}F]fluoride mixture was introduced into the inlet of the on-chip concentration column using the syringe pump with a 30 μL/min loading rate followed by 0.1 mL of air at a loading rate of 50 μL/min allowing all liquid to be pushed through the tubing connection and column. The liquid coming out from the concentration column was collected in a glass vial and typically contained a negligible amount of radioactivity (less than 0.074 MBq). We employed PDMS pillars to form physical barriers near the end of the microchannel, thus trapping desired anion exchange microparticles inside the concentration column, as shown in Fig. 5a.

The radioactivity of the initial [^{18}F]fluoride loading, captured inside the concentration column, and the [^{18}F]fluoride residue left in the syringe were measured separately via a dose calibrator. The efficiency of the concentration column was calculated after accounting for the decay of the radioisotope. In our experiment to test the [^{18}F]fluoride trapping and releasing efficiency of the on-chip column, we

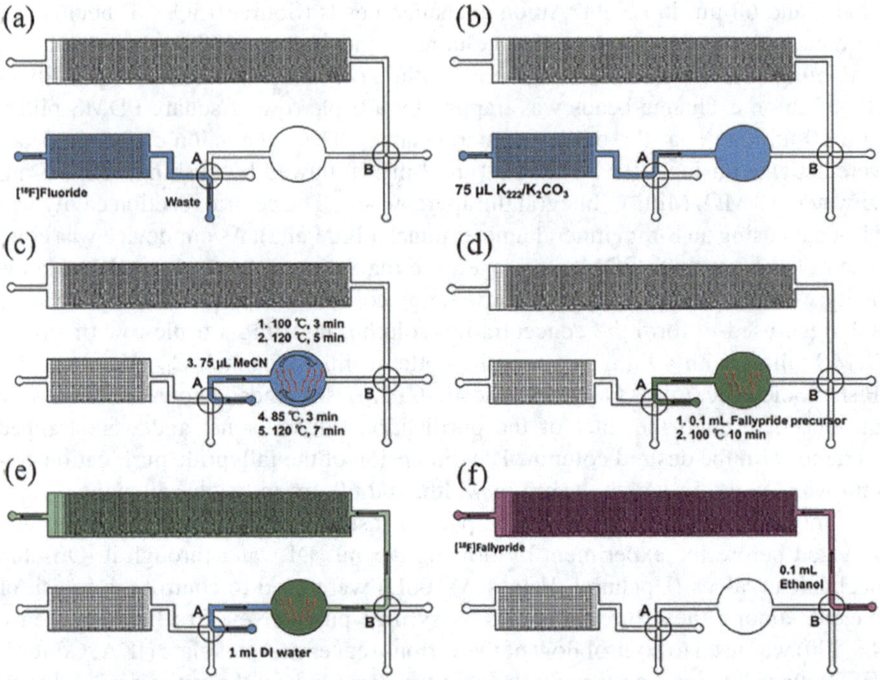

Fig. 4 Schematic representation of the steps used for the production of [^{18}F]Fallypride on our integrated microfluidic chip. The six steps shown are: (**a**) concentrate [^{18}F]fluoride using anion-exchange column, (**b**) release trapped [^{18}F]fluoride off the column and transfer into reaction cavity, (**c**) evaporate to achieve an anhydrous environment by heating and N_2 drying, (**d**) [^{18}F]Fallypride fluorination, (**e**) isolate undesired products using purification column, (**f**) collect purified [^{18}F]Fallypride off-chip. (Reproduced with permission from the Royal Society of Chemistry, originally published by Zhang et al. (2018))

have observed that the maximum quantity of [^{18}F]fluoride that could be captured by the concentration column was related to the volume of anion exchange beads packed inside the microchannel. Considering that the goal of our microfluidic chip is to synthesize multiple doses of radiotracer sufficient for human imaging, the on-chip concentration module must be capable of capturing radioactivity in the range of 3.7 GBq (100 mCi). To determine the relationship between the volume of anion exchange resins and the captured radioisotope, we fabricated several versions of ion-exchange columns with varying dimensions. Raw [^{18}F]fluoride with radioactivity values as high as 4.8 GBq (130 mCi), between 100 and 300 μL, was then loaded into each column to evaluate the trapping efficiency and determine the optimal dimension. Table 1 summarizes the results of these experiments. It turned out that 3 μL (~10 mg) of anion exchange resin captured over 3.7 GBq (100 mCi) [^{18}F]fluoride with the efficiency near 98% over a loading period of ~6 min (Zhang et al., 2018). However, a reduced timespan may be achieved with additional optimization of column geometry (i.e., wider microchannel) to allow a higher reagent loading rate.

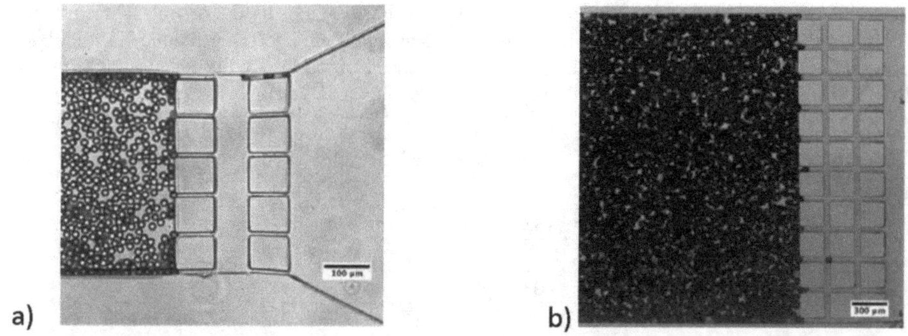

Fig. 5 Optical images of microfluidic chip concentration and purification column. (**a**) A photograph of anion exchange beads packed inside a microchannel by 10-μm gap PDMS pillars. (**b**) A photograph of reverse-phase C_{18} microparticles trapped inside a microchannel with 40-μm gap PDMS pillars. (Reproduced with permission from the Royal Society of Chemistry, originally published by Zhang et al. (2018))

Table 1 The relationship between the volume of anion exchange beads inside the concentration column and the radioactivity trapped on the microfluidic chip. All listed values are decay-corrected to the time point of initial radioactivity in the syringe. Reproduced with permission from the Royal Society of Chemistry, originally published by Zhang et al. (2018)

Volume of beads (μL)	Initial syringe radioactivity (MBq)	Final syringe radioactivity (MBq)	Waste vial radioactivity (MBq)	Radioactivity trapped on chip (MBq)	Efficiency
0.62	381	22.2	33.3	330	90.1%
0.62	322	22.2	25.9	278	92.2%
0.62	1325	66.5	466	796	65.0%
0.62	1446	55.6	529	773	59.7%
3	3541	12.6	0.37	3441	97.7%
3	4758	178	3.7	4551	95.9%

To release the captured [^{18}F]fluoride, 75 μL of K_{222}/K_2CO_3 eluting solution was introduced into the inlet of the concentration column at a 30-μL/min loading rate (Fig. 4b). At the same time, mechanical valve A, connected between the concentration column and batch reactor, was set to direct the released reagent into the reaction cavity. Another 0.1 mL of air was injected from the inlet of the concentration column to push all the remaining liquid into the reaction cavity. In our attempt to reduce the overall synthesis time the volume of K_{222}/K_2CO_3 eluting solution required to fully release all [^{18}F]fluoride off the concentration column was optimized. A large portion of the production time is devoted to achieving a completely anhydrous state prior to the synthesis reaction and thus optimizing the volume of eluting solution helps promote the overall production efficiency and minimizes the decay of radioisotope. We compared the release efficiency with various volumes, and the results indicated that 75 μL of K_{222}/K_2CO_3 was the optimal amount for releasing [^{18}F]fluoride with radioactivity loss less than 5% (Table 2) (Zhang et al., 2018).

Table 2 The relationship between the volume of K_{222}/K_2CO_3 used for releasing trapped [^{18}F]fluoride off the concentration column and the releasing efficiency. The efficiency is calculated accounting for the decay of the radionuclide over time. Reproduced with permission from the Royal Society of Chemistry, originally published by Zhang et al. (2018)

Volume of K_{222} (μL)	On-chip radioactivity (MBq)	Released radioactivity (MBq)	Radioactivity trapped on chip (MBq)	Efficiency
50	278	178	62.9	74.0%
50	170	133	34.8	77.0%
75	157	142	8.5	95.0%
100	229	203	3.7	98.0%

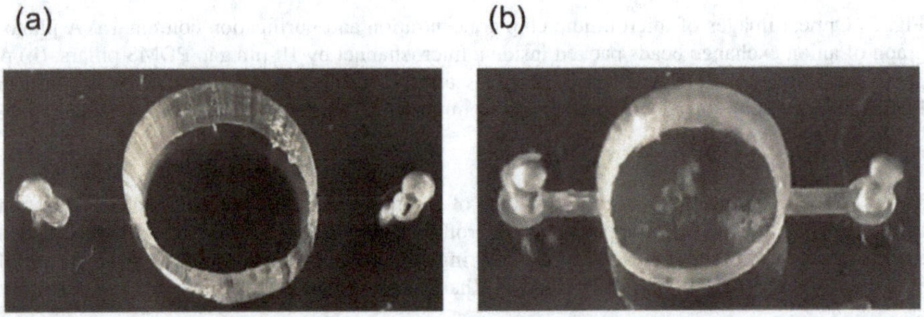

Fig. 6 Photograph of on-chip reaction cavity with no reagent inside (**a**) and after water evaporation (**b**). (Reproduced with permission from the Royal Society of Chemistry, originally published by Zhang et al. (2018))

Because the fluorination reaction requires anhydrous conditions, the microfluidic chip was first placed on a hot plate at 100 °C for 3 min and then at 120 °C for 5 min. This is the preferred method over direct heating at extremely high temperature to avoid violent boiling, which may result in the liquid splashing out from the upper hole. A slow nitrogen flow was introduced via the inlet of the reaction cavity to facilitate the evaporation process. The heated water inside the reaction cavity was evaporated through the 1.5-mm hole on the top. The microfluidic chip was then removed from the hot plate for 2 min, allowing the entire chip to cool down below the boiling point of acetonitrile. To completely remove residual moisture inside the reaction cavity, 75 μL of anhydrous acetonitrile was loaded from the inlet of the reaction cavity, and the entire chip was placed on another hot plate at 85 °C for 3 min and then at 120 °C for 7 min. Constant N_2 flow at 0.2 psi was applied through the inlet of the reaction cavity to facilitate the removal of residual moisture during the entire evaporation process. After these drying steps, solid residue was observed at the bottom of the cavity. In conventional macroscopic radiotracer production, two to three cycles of azeotropic distillation are typically performed to guarantee an anhydrous environment. In contrast, with our microfluidic platform, only one cycle of azeotropic distillation is sufficient to enable the subsequent nucleophilic substitution reactions, thus shortening the required production timeframe. Figure 6 shows an optical image of the solid salt deposited on the bottom of the reaction cavity after evaporation. The on-chip radioactivity after evaporation was measured, and the loss of radioactivity was less than 5% (Zhang et al., 2018).

After achieving the anhydrous conditions necessary for fallypride fluorination, the reaction itself is performed after introducing the fallypride precursor (dissolved in a polar aprotic solvent), into the reaction cavity containing the concentrated, dried fluoride-kryptofix complex. As shown in Fig. 4d, 0.1 mL of dimethyl sulfoxide (DMSO) containing fallypride precursor (3 mg), vortex-mixed for 30 min in advance, was injected from the inlet of the reaction cavity, and the cavity was immediately sealed by screwing a male nut (Fisher Scientific) into the NanoPort, minimizing the loss of radioactivity during the fluorination process. The chip was heated at 100 °C for 10 min to produce the radioactive fallypride. The successful synthesis of [^{18}F]Fallypride was validated via radio-HPLC analysis, the radioactive peak eluting at 11–12 min matching the reference standard.

In this approach, instead of a low-boiling-point solvent, such as acetonitrile, DMSO was used for dissolving the fallypride precursor. Though acetonitrile has been widely used in clinical production and has the benefit of being easily removed later, the required temperature for nucleophilic substitution reactions with [^{18}F]fluoride (usually around 100 °C), exceeds the boiling point of acetonitrile, which would increase the pressure inside the sealed cavity and possibly lead to leakage. Employing the high boiling point solvent DMSO avoids these issues. Based on a report by Javed et al., the concentration of fallypride precursor plays a significant role in final crude radiochemical yield in a batch microfluidic reactor (Javed et al., 2014). To optimize the yield of [^{18}F]Fallypride in our microfluidic chip, we further investigated the relationship between the concentration of fallypride precursor and the overall conversion efficiency. Similar to previous reports, we observe a rise of fluorination efficiency as the concentration of fallypride precursor increases (Fig. 7).

To obtain chemically pure [^{18}F]Fallypride suitable for PET imaging, the crude production material was allowed to cool to approximately room temperature and purified using the integrated on-chip column containing monodisperse C_{18} stationary phase. The purification column was fabricated in a manner identical to the concentration column. Figure 5b shows an optical image of C_{18} microparticles—packed inside a microchannel forming the purification column. Before transferring all liquid through the on-chip purification column, 0.1 mL of DI water was loaded from the inlet of the purification column to fully activate the C_{18} gel. Subsequently, valve A was closed, and the valve B was open, creating unidirectional crude product solution flow from the reaction cavity to the purification column (Fig. 4e). We employed a flow rate of 50 μL/min for this step. The tight seal of the reaction cavity provided by the male nut allowed the synthesized compound to be injected into the purification column without leakage. Undesired compounds, including unreacted fallypride precursor, any remaining [^{18}F]fluoride, and various by-products were thereby pushed off-chip, only leaving the desired [^{18}F]Fallypride trapped on-chip. Next, another 0.1 mL of DI water was passed over the purification column to completely remove unreacted [^{18}F]fluoride that would lower the overall radiochemical purity and confound imaging procedures. Finally, trapped [^{18}F]Fallypride was eluted off the chip using 0.1 mL of pure ethanol and collected in a glass vial (Fig. 4f). The

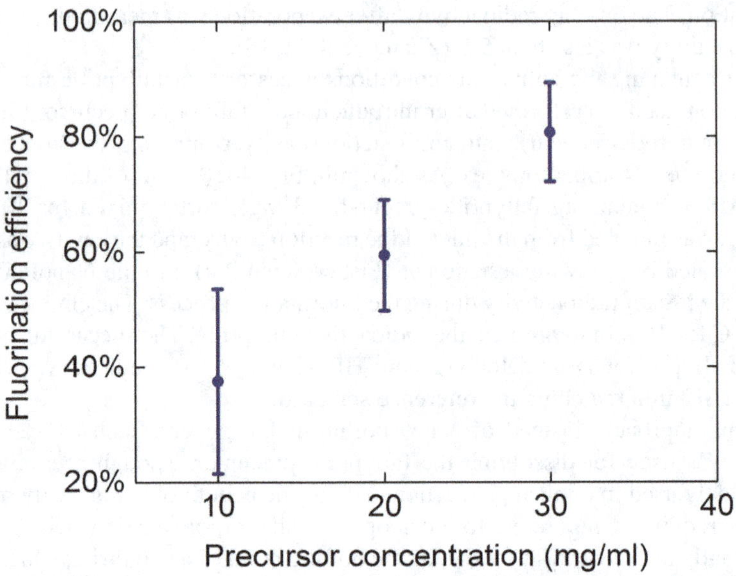

Fig. 7 The relationship between precursor concentration and fluorination efficiency. (Reproduced with permission from the Royal Society of Chemistry, originally published by Zhang et al. (2018))

(a)

(b)

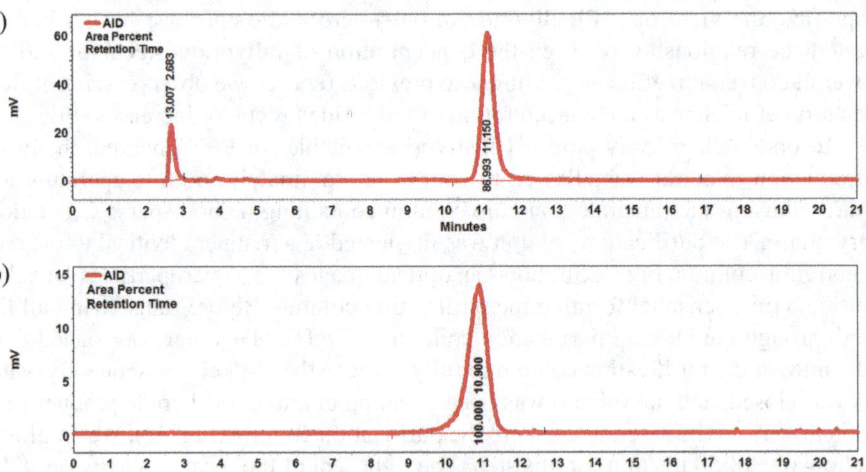

Fig. 8 (**a**) Crude radio-HPLC of [^{18}F]Fallypride synthesized inside the reaction cavity. (**b**) Radio-HPLC of [^{18}F]Fallypride eluted from the on-chip purification column. (Reproduced with permission from the Royal Society of Chemistry, originally published by Zhang et al. (2018))

successful synthesis of [^{18}F]Fallypride was confirmed via HPLC analysis, which shows radio-HPLC analysis of the eluted [^{18}F]Fallypride to have 100% radiochemical purity (Fig. 8b).

To minimize any potential loss of [^{18}F]Fallypride, the volume ratio of DMSO to water was tuned to 1:10 such that it was able to carry all of the crude radiochemical mixture in the reaction cavity but did not wash the captured [^{18}F]Fallypride off the on-chip purification column. HPLC analysis of the outlet waste indicated negligible [^{18}F]Fallypride loss during this step (Fig. 9).

The result indicated that the labeling efficiency during fluorination process ranging from 65 to 87% ($n = 5$), which is comparable to data typically obtained using conventional automated methodologies. The overall production efficiency (10 ± 3%, $n = 5$) is calculated as the ratio of the decay-corrected radioactivity of the resulting [^{18}F]Fallypride released off the chip, divided by the initial radioactivity of [^{18}F]fluoride loaded into the microfluidic chip. The whole process is completed in approximately 60 min starting from the loading of [^{18}F]fluoride to the collecting of purified [^{18}F]Fallypride; this could likely be further shortened using automated external valves and programmable syringe pumps.

To further verify the utility of the resulting [^{18}F]Fallypride, we produced (using our integrated microfluidic chip) a dose of 11.1 MBq with 723.35 GBq/μmol (19.55 Ci/μmol) of specific activity and 98.62% of radiochemical purity, and obtained PET data from a rat brain. The rat was brought to the imaging lab to habituate 3 h prior to imaging. Figure 10 shows the resulting PET imaging data. The accumulation and pharmacokinetics of [^{18}F]Fallypride were as expected when evaluated in the rat brain (Hawkey et al., 2022). Experiments in our group are currently underway to further optimize our production process and produce [^{18}F]Fallypride that satisfies the stringent criteria for human imaging.

We have developed a simple, economic, and efficient microfluidic platform to aid both radiochemistry research efforts and clinical efforts by rapidly producing ultrapure radioactive fallypride on demand. This highly integrated configuration enables all essential steps needed for [^{18}F]Fallypride production without any off-chip treatment, starting from reaction reagents and ending with a purified product. These disposable chips may allow high-throughput investigations of novel radiotracer chemistry, as well as the mass production of established PET tracers. By using this microfluidic device, we envision that clinicians may overcome a significant bottleneck imposed by limited access to an on-site cyclotron, as they could

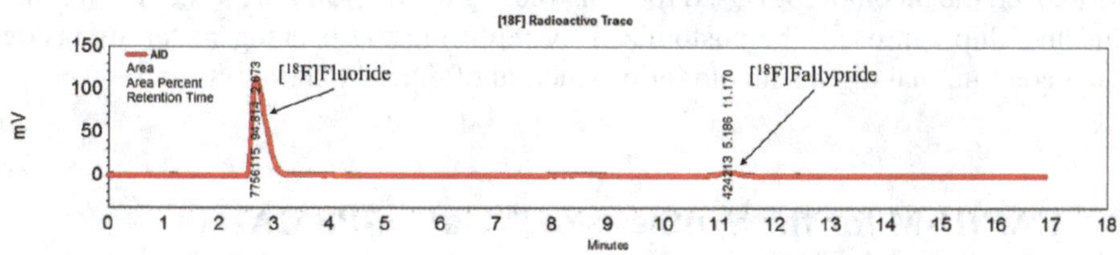

Fig. 9 Radio-HPLC analysis of the waste mixture passed through the C_{18} column. This injection was taken using the mixture solution flowing out from the purification column after labeling reaction. Ninety-five percent of [^{18}F]Fluoride was washed off the purification column, and only 5% [^{18}F]Fallypride was lost at this step. (Reproduced with permission from the Royal Society of Chemistry, originally published by Zhang et al. (2018))

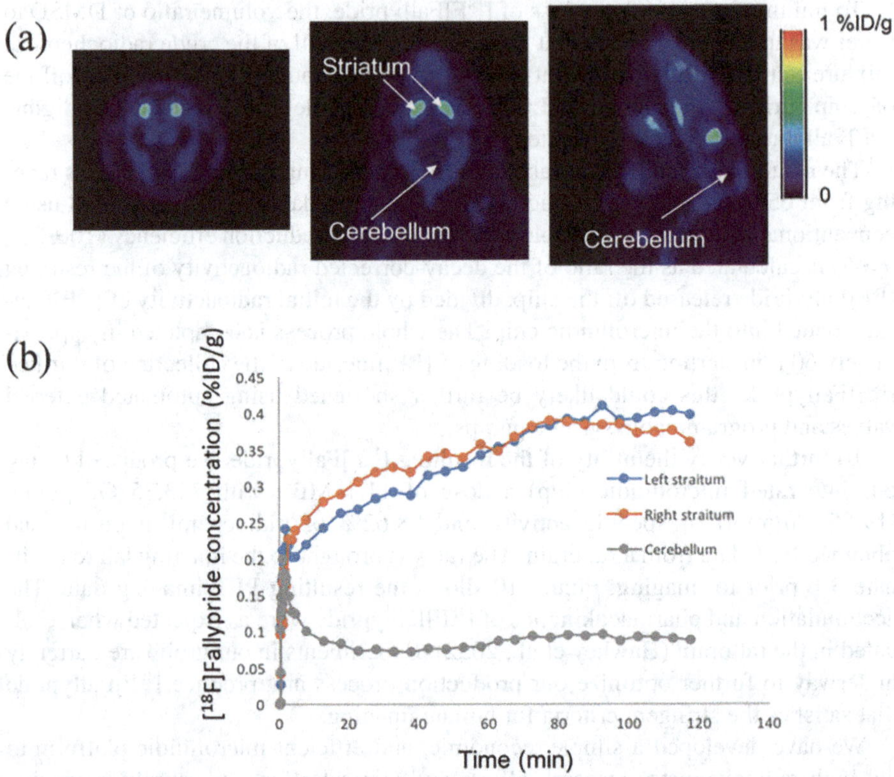

Fig. 10 (**a**) PET image of rat brain using [^{18}F]Fallypride produced in a microfluidic chip. (**b**) The variation over time of [^{18}F]Fallypride concentration imaged in the micro-PET for 2 h. (Reproduced with permission from the Royal Society of Chemistry, originally published by Zhang et al. (2018))

easily utilize delivered radionuclides and produce imaging reagents as needed without relying on centralized facilities. Another advantage offered by this simple microfluidic approach lies in the reduced need for highly trained radiochemical technicians, previously heavily involved in reagent transit and product analysis. Moreover, the methodology used for constructing the microfluidic system is simple, and the chip can easily be customized by replacing the packing materials inside each column, making it suitable for production of other PET tracers.

3 RAPID-M for the Synthesis of [^{68}Ga]Ga-PSMA-11

Prostate cancer (PC) is the second leading cause of cancer related deaths among adult males worldwide and the most common cancer in men (Siegel et al., 2020). Because of its high expression on the surface of prostate cancer cells, prostate-specific membrane antigen (PSMA) has been identified as a target for PET molecular imaging and for targeted radioligand therapy. The concept of PSMA ligands serving as theranostic agents has a long history (Miyahira & Soule, 2022), starting with the development of the monoclonal antibody 7E11-C5 (Horoszewicz et al., 1987; Lopes et al., 1990) that led to the FDA approval of [^{111}In]In-capromab pendetide (ProstaScint; EUSA Pharma), the first molecular imaging agent for PC. In 2001, Kozikowski's group (Kozikowski et al., 2001) developed a series of urea-based inhibitors targeting the neurotransmitter regulator GCP-II (NAALADase), a central nervous system version of PSMA, as potential neuroprotective agents. Pomper et al. recognized the potential of the urea-based PSMA-targeted ligands to be used as imaging agents for prostate cancer. The first compound, [^{11}C]MCG, was published in 2002 (Pomper et al., 2002). Most of the work on PSMA-PET derives from these early studies which led the way to FDA's 2020 and 2021 approval of two different PSMA PET agents; one, a kit-based [^{68}Ga]Ga-PSMA-11 PET agent known with trade name Illuccix®, and the other is [^{18}F]DCFPyL, commercially available as piflufolastat F 18 and trade name Pylarify®. More recently, the FDA has approved flotufolastat F 18 injection (Posluma), the first radiohybrid PSMA-targeted PET imaging agent for positron emission tomography of prostate-specific membrane antigen–positive lesions in men with prostate cancer.

PSMA-PET outperforms the previously widely used [^{11}C/^{18}F]Choline (Hara et al., 1998; Mapelli et al., 2016) and [^{18}F]Fluciclovine (Shoup et al., 1999) agents (Olivier et al., 2022), which, however, report on different biological pathways, namely choline and amino-acid metabolism, respectively, and can still be used to provide useful information for personalized medicine. Ga-68 is a widely used isotope for PET imaging. Importantly, Ga-68 can be conveniently obtained from a ^{68}Ge/^{68}Ga generator instead of relying on an on-site cyclotron, making it an attractive isotope and more affordable for many facilities.

PSMA ligands have significant potential as theranostic agents in addition to their use for purely diagnostic purposes. For targeted radiotherapy, PSMA ligands labeled with therapeutic isotopes, such as the beta-emitter Lu-177 and the alpha-emitter Ac-225 (Alam et al., 2022), represent a promising approach in the treatment of patients with recurrent and metastatic PC. As such, the FDA has recently granted approval to [^{177}Lu]Lu-PSMA-617 (Pluctivo) for the treatment of patients with PSMA–positive metastatic castration-resistant prostate cancer (mCRPC) who have shown resistance to other therapies (androgen-receptor inhibitors or taxane-based chemotherapy).

Inspired by our success with the RAPID-F concept, RAPID-M was developed as an inexpensive, high-yielding and multiuse microfluidic device that enables radiosynthesis of [^{68}Ga]Ga-PSMA-11 at substantially reduced cost. Conceptually, RAPID-M can be employed for the synthesis of any radiopharmaceutical produced by chelation of radiometals, including therapeutic isotopes. In other words, RAPID-M could aid the fast and efficient dissemination of several different theranostic agents. As in the RAPID-F concept, the microfluidic system consists of several miniaturized modules that together achieve all critical steps of ^{68}Ga-radiolabeling. These modules include a radioisotope concentration unit, passive on-chip mixing and heat-aided labeling, and final reformulation achieved by solid-phase extraction (SPE).

3.1 Microfluidic Device Description and Fabrication

Soft lithography was employed to pattern channels in polydimethylsiloxane (PDMS), which was then bonded to a glass slide after exposing to oxygen plasma. Off-chip mechanical valves (Upchurch Valves, V-100D) were used to control the transit of reagents between modules. Syringe pumps (New Era Pump System, NE-300) was used to load solution at controlled flow rate. A hot plate (IKA, Control-VISC) was used to heat the entire microfluidic device during on-chip mixing and labeling reaction.

As schematically illustrated in Fig. 11, the microfluidic chip was constructed with three key components: an on-chip strong cation exchange (SCX) column, an on-chip strong anion exchange (SAX) column, and a passive in-plane mixing/reaction module. Specifically, it was designed to sequentially concentrate the [^{68}Ga]GaCl$_3$ solution delivered from the ^{68}Ge/^{68}Ga generator, mix the eluted [^{68}Ga]GaCl$_3$ with the PSMA-11 precursor, and (together with a hot plate) enable a heated labeling reaction.

The on-chip SCX/SAX column was 5 mm in width, 20 mm in length, and 100 μm in height. Maxi-Clean SCX/SAX resins (GRACE, 600 mg, particle size 50 c) were suspended in ethanol (5 w/v%) and loaded into each column using syringe pumps

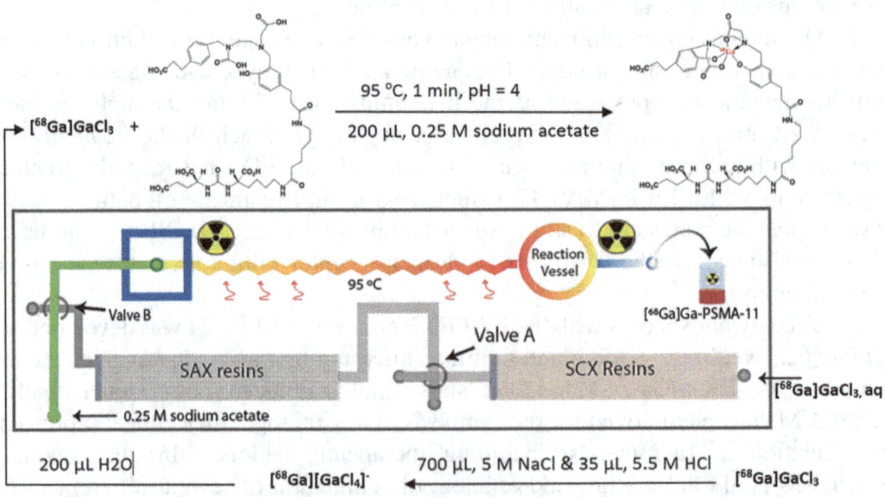

Fig. 11 Schematic representation of a chip designed for the synthesis of [^{68}Ga]Ga-PSMA-11. A schematic illustration of a synthesis circuit designed for the production of [^{68}Ga]Ga-PSMA-11, including an SCX column, SAX column, passive mixing module, and reactor for the labeling reaction

(loading rate: 100 μL/min). All resins can be trapped by rows of PDMS obstructions, thus allowing for the construction of columns with the desired functionalities. The in-plane passive mixing module adopted a flow-focusing geometry at the entrance. The dimension of zigzag mixing channel was 500 μm in width, 34 mm in length, and 100 μm in height. In addition, the width and height of diamond-shaped obstructions were 50 μm and 100 μm, respectively. The sizes of all the mixers were equal. The on-chip reaction cavity was fabricated using an 8-mm-diameter punch (Ted Pella), and a 1-mm-thick PDMS membrane was aligned and plasma bonded on top of the reactor, allowing for a maximum volume of 500 μL.

3.2 On-Chip Concentration and Release of [^{68}Ga]GaCl$_3$

The goal of the concentration step is to effectively capture the majority of [^{68}Ga]GaCl$_3$ from a full elution of the generator (4 mL) using miniaturized microfluidic columns. While numerous methods have already been developed and employed, most of these methods primarily rely on the use of commercially available cation/anion exchange cartridges (Mueller et al., 2016). More importantly, marginal effort has been devoted to determining advantages and disadvantages of these approaches when applied in a microfluidic system.

Previous studies have reported several different strategies that are routinely used for efficient [^{68}Ga]GaCl$_3$ concentration in practical production (Mueller et al., 2012, 2016). Since the goal here is to process a full elution of [^{68}Ga]GaCl$_3$ from the original generator, we employed a two-step approach that relies on combined effects of cationic concentration and anionic purification. This approach is extremely advantageous when applied in a microfluidic system due to the following aspects: first, it does not involve the addition of buffer solution or organic solvents, thus eliminating the necessity of purification; second, the combination of the SCX and SAX provides the capability to handle a full elution of [^{68}Ga]GaCl$_3$, demonstrating the capability to produce several imaging doses; third, the [^{68}Ga]GaCl$_3$ is released in small volume of pure water and exhibits high radionuclide and chemical quality, and low acidity, thereby allowing for direct use in a variety of radiolabeling reactions.

In our work, both on-chip SCX and SAX columns were constructed by trapping different functional resins inside microchannels based on previously reported strategy (Mueller et al., 2016). Both resins were obtained from Maxi-Clean SCX/SAX (GRACE, 600 mg, particle size 50 μm). On-chip SCX and SAX modules were preconditioned using 1 mL of 5.5 M HCl, followed by rinsing with 5 mL of DI water (18 MΩ, Milli-Q Integral ultrapure water). Notably, all experiments were performed using reagents with the lowest possible metal content to avoid unfavorable interference with [^{68}Ga]Ga-PSMA-11 labeling reaction.

Figure 12b and c show optical images of SCX and SAX resins densely packed by the end of microchannels, respectively. To examine the performance of on-chip columns, we carried out trapping and releasing trials separately on each column. Based

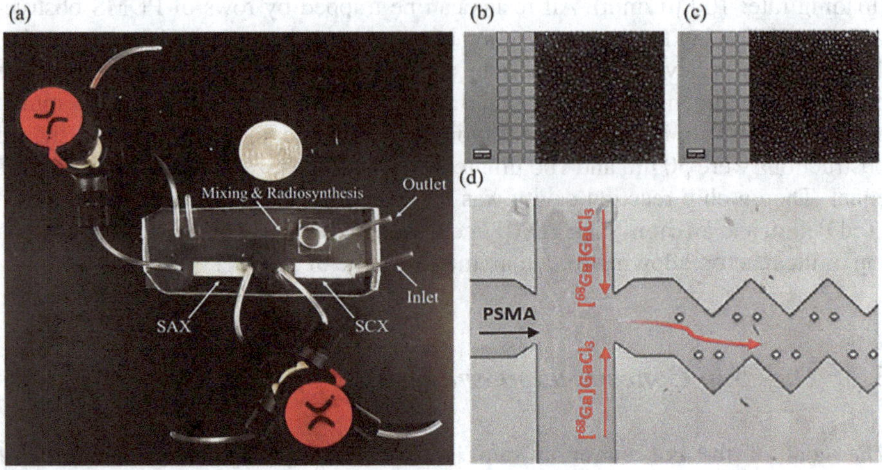

Fig. 12 Photographs of the fabricated microfluidic device for the radiosynthesis of [^{68}Ga]Ga-PSMA-11 (**a**), SCX (**b**), and SAX (**c**) resins trapped inside microchannels, and the entrance of mixing module (**d**) with diamond-shaped obstructions inserted throughout the zigzag microchannel. (Reprinted with permission from Springer Nature publication, originally published by Zhang et al. (2020))

on the measured radioactivity using a dose calibrator, the radioactivity concentration efficiency (RCE) and radioactivity release efficiency (RRE) were determined by:

$$\mathrm{RCE} = \frac{\text{On} - \text{chip trapped radioactivity}}{\text{Initially loaded radioactivity}}$$

$$\mathrm{RRE} = \frac{\text{Released radioactivity}}{\text{On} - \text{chip trapped radioactivity}}$$

Additionally, several operating parameters, including mass of loaded resins, solution loading rate, and volume of releasing buffer, were comprehensively studied. Since the underlying mechanism for [^{68}Ga]GaCl$_3$ concentration relies on the fact that [^{68}Ga]Ga^{3+} eluted off the generator should be absorbed on the surface of silica-based SCX resins, it is thus critical to determine the correlation between mass of loaded resins and concentration efficiency. As shown in Fig. 13, rigorous testing revealed that 8 mg of the SCX resins (RCE = 96%, n = 5) and 10 mg of the SAX resins (RCE = 91%, n = 5) were required to achieve efficient trapping performance over a loading duration of 5 min.

The amount of radioactivity of eluted [^{68}Ga]GaCl$_3$ used in experiments was 1102.6 ± 44.4 MBq (29.8 ± 1.2 mCi, n = 5) in 4 mL of 0.05 M HCl. The concentration step, achieved by syringe pump with an 800 μL/min loading rate, is schematically represented in Fig. 14a.

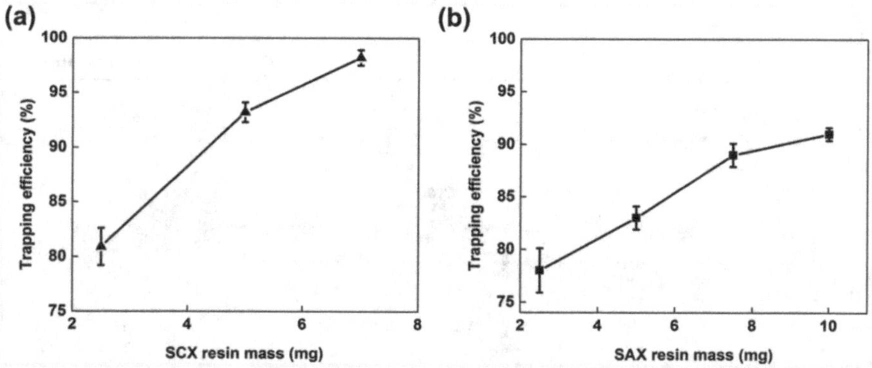

Fig. 13 Correlations between mass of loaded resins and trapping efficiencies where (**a**) shows that 8 mg of the SCX resins (RCE = 96%, n = 5) and (**b**) shows that 10 mg of the SAX resins (RCE = 91%, n = 5) required to achieve efficient trapping performance over a loading duration of 5 min. (Reprinted with permission from Springer Nature publication, originally published by Zhang et al. (2020))

To ensure that all liquid completely passed through the column, 0.1 mL of air was introduced at the same flow rate, and the isolated waste was collected in a glass vial. The distribution of activity was immediately measured to assess concentration efficiency. The following step, schematically represented in Fig. 14b, was the elution of [^{68}Ga]Ga^{3+} off the SCX column and onto the SAX column. This was achieved by employing a method described by Mueller et al. where a concentrated NaCl solution containing a very low amount of HCl was used to convert the cationic [^{68}Ga]Ga^{3+} to the anionic [^{68}Ga][$GaCl_4$]$^-$ (Mueller et al., 2012). As illustrated in Fig. 14b, the eluate released off the SCX column was immediately directed to pass through the SAX column, where [^{68}Ga][$GaCl_4$]$^-$ was absorbed. Optimization of elution conditions was necessary to enable both efficient desorption of [^{68}Ga]Ga^{3+} and efficient adsorption of [^{68}Ga][$GaCl_4$]$^-$. By adjusting the volume ratio of 5.5 M HCl relative to 5 M NaCl, we found that an eluate consisting of 700 μL of 5 M NaCl and 35 μL of 5.5 M HCl enabled the best performance (SCX RRE = 90.8%, n = 5 and SAX RCE = 89.5%, n = 5) (Table 3). As the solution was slowly loaded through the SAX column, the excess of HCl can be directly removed from the microfluidic chip.

As summarized in Table 3, levels of radioactivity were measured at critical steps to assess the efficiency of each process and identify potential areas for optimization. Starting from a full [^{68}Ga]$GaCl_3$ elution from the generator, the total trapping efficiency of the combination of SCX and SAX columns is 73.7%. The total time period required for completing gallium-68 concentration is ~12 min. It is worth noting that additional optimization of factors such as the selection of SCX and SAX materials (e.g., use of resins with higher porosity and larger surface area) and microchannel design could lead to better overall RCE performance.

[^{68}Ga]Ga^{3+}, at high chemical and radiochemical purity, was finally eluted from the SAX resin with 200 μL of pure water and mixed to the PSMA-11 precursor for the labeling step (Fig. 14c).

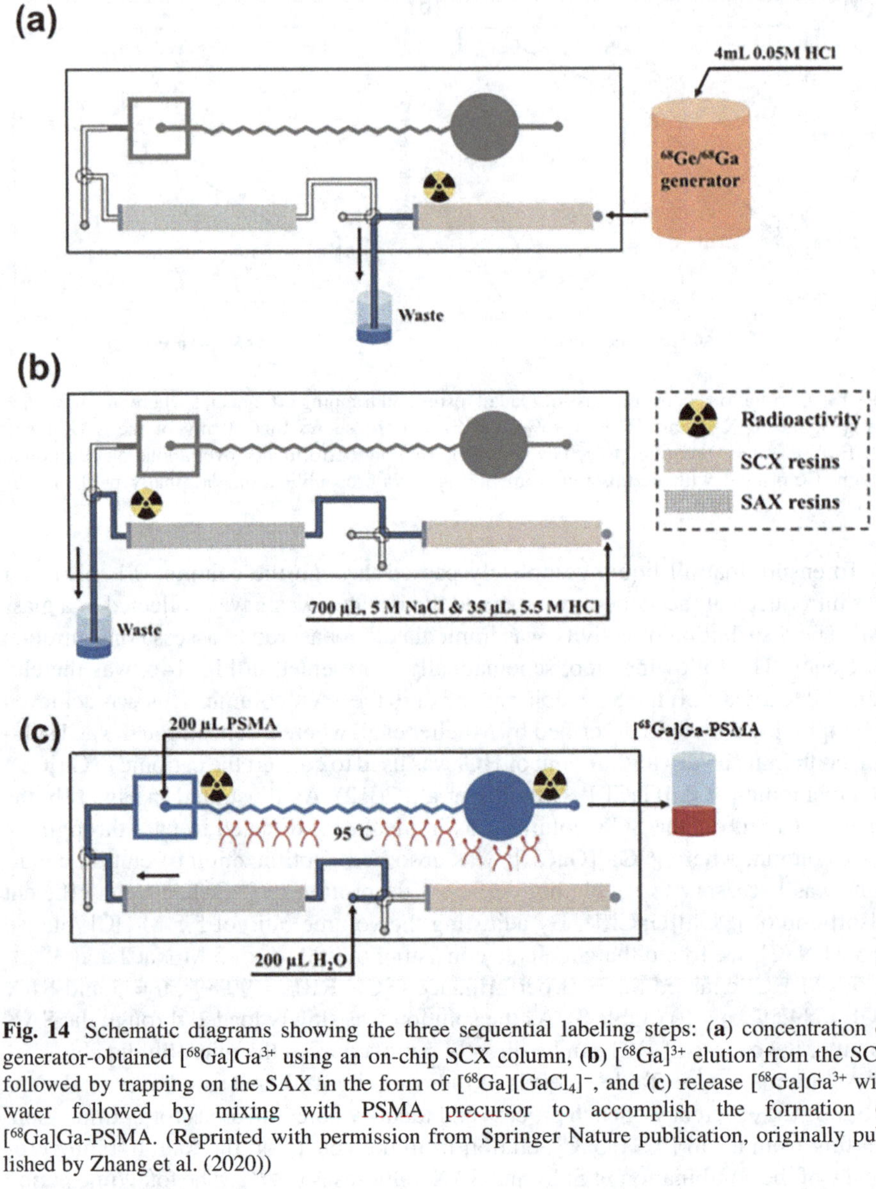

Fig. 14 Schematic diagrams showing the three sequential labeling steps: (**a**) concentration of generator-obtained [^{68}Ga]Ga^{3+} using an on-chip SCX column, (**b**) [^{68}Ga]$^{3+}$ elution from the SCX followed by trapping on the SAX in the form of [^{68}Ga][GaCl$_4$]$^-$, and (**c**) release [^{68}Ga]Ga^{3+} with water followed by mixing with PSMA precursor to accomplish the formation of [^{68}Ga]Ga-PSMA. (Reprinted with permission from Springer Nature publication, originally published by Zhang et al. (2020))

3.3 Labeling

The released gallium-68 was simultaneously mixed with the PSMA-11 precursor by flowing through an in-plane mixing channel. The precursor solution was prepared by dissolving 10 µg of PSMA-11 ligand in 1 mL of 0.25 M sodium acetate buffer, resulting in a 10 µg/mL concentration. Next, the microfluidic chip was heated to

Table 3 Experimental results of the efficiency of miniaturized SCX and SAX columns at critical steps. All efficiencies are calculated based on decay-corrected values. Reprinted with permission from Springer Nature publication, originally published by Zhang et al. (2020)

Parameter	Run 1	Run 2	Run 3	Run 4	Run 5	Average
Loaded activity (MBq)	1117	1051	1162	1077	1114	1103 ± 44
SCX RCE (%)	96.8	97.0	97.1	97.9	93.8	96.5 ± 1.6
SCX RRE (%)	87.3	90.4	91.0	92.5	92.7	90.8 ± 2.2
SAX RCE (%)	87.6	93.3	92.3	87.3	87.2	89.5 ± 3.0
SAX RRE (%)	93.0	94.2	95.0	93.5	93.7	93.9 ± 0.8
Total loss (%)	31.2	22.9	22.5	26.1	29.0	26.3 ± 3.8
Trapping efficiency (%)	68.8	77.1	77.5	73.9	71.0	73.7 ± 3.8

95 °C on a preheated hot plate. Because of the short time period required for gallium-68 incorporation reaction in a microfluidic system (Fig. 11), it was crucial to achieve thorough mixing prior to heating. It is known that mixing of several streams of solution inside a microfluidic channel depends primarily on diffusion, which occurs in an unfavorably slow manner due to the low Reynolds number ($Re \ll 1$). In this study, we employed a passive microfluidic mixing system that combined the effect of in-plane obstructions inside microchannels and repetitively wavy microchannels (Fig. 12). Since passive mixing neither complicates device fabrication nor requires additional power, this approach provides a simple yet efficient fashion to achieve enhanced mixing performance. As shown in Fig. 12d, a flow-focusing geometry was used at the entrance of mixing section by simultaneously loading released $[^{68}Ga]^{3+}$ from the top and bottom channels and PSMA-11 precursor from the middle channel. In order to characterize mixing performance, fluorescent beads (20 nm yellow-green FluoSpheres diluted 100×, Invitrogen) suspended in water were injected from the middle channel, while pure water was injected at the same time from the other two channels. The solution was loaded using two syringe pumps, each set at the flow rate of 100 μL/min. Fluorescence microscopy (Zeiss Axio Observer Z.1, Hamamatsu Flash 4.0) was used to record and visualize the mixing performance along the flow direction at different positions.

Figure 15a–c shows fluorescent images captured at the inlet, middle, and outlet of the mixing channel. After visually observing changes in width of the middle fluorescent stream, we concluded that sufficient mixing is achieved near the end of channel. Moreover, the mixing performance was quantified by measuring the fluorescent intensity perpendicular to the flow direction (illustrated by red dashed lines). Figure 15 depicts measured results of the normalized fluorescent intensity versus the width of fluorescent stream. Additionally, COMSOL Multiphysics numerical simulation was performed to predict the mixing tendencies of two solutions based on devised micro-mixers. Mixture model (transport of diluted species) was used to extract the concentration profile of PSMA-11 precursor throughout the mixing zone. The dispersion of two miscible streams in a laminar-flow channel was driven by the rate of the flow. The initial loading flow rate from all three inlets was set to be 100 μL/min. The initial concentration of PSMA-11 precursors and gallium-68

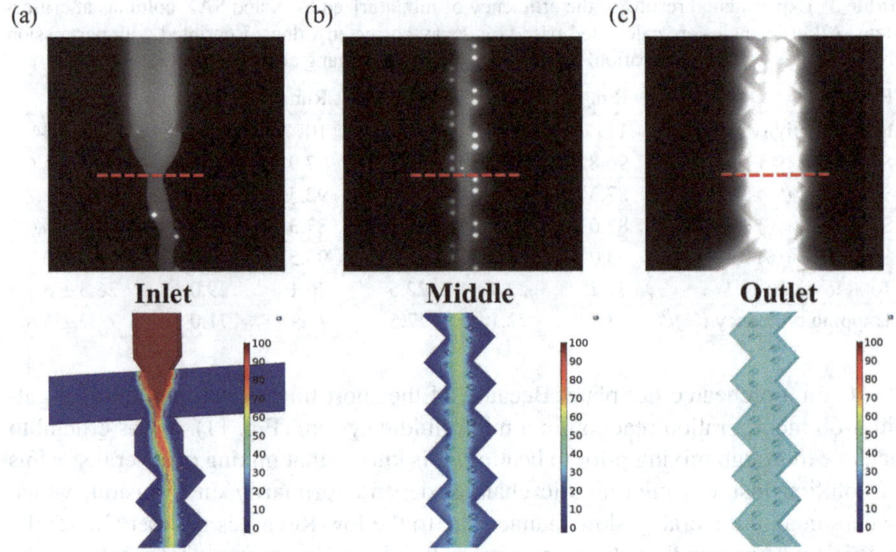

Fig. 15 Experimental evaluation of the mixing performance using the combination of inserted obstructions and zigzag microchannels. The captured fluorescent images indicate mixing results at inlet (**a**), middle (**b**), and outlet (**c**), respectively, which are in good agreement with corresponding numerical simulation. (Reprinted with permission from Springer Nature publication, originally published by Zhang et al. (2020))

were set to be 100 and 0 on the concentration profile scale, respectively. This means that both precursor and gallium-68 solutions are at 100% nominal concentration at the inlets. To simplify the model, the properties of two streams were defined as incompressible water, and the boundary condition for the inner walls of microchannels was set as a no slip condition. The red stream injected from the middle channel represents PSMA-11 precursor, and the blue streams injected from two side channels represent concentrated gallium-68. The numerical simulation variation of concentration profiles at various positions along the microchannel are shown in Fig. 15 as well. As expected, simulated mixing patterns are in good agreement with experimental results, shown by the even distribution of precursor and gallium-68 at the downstream position of the channel, thus indicating the realization of complete mixing after flowing through the micro-mixer.

Because of enhanced mass and heat transfer in a microfluidic system, traditional reaction conditions that are routinely adopted for macroscale synthesis are not necessarily optimal in a microfluidic context. In order to improve microfluidic-based RCY without affecting labeling efficiency, we conducted a systematic study by examining a variety of reaction parameters such as reaction time, reaction temperature, pH, and precursor concentration. In our preliminary investigation, we started from well-established macroscale protocols, specifically heating the mixture consisting of 100 μL of concentrated [^{68}Ga]GaCl$_3$ and 100 μL of 10 μg/mL PSMA-11 precursor in a sealed vial at 100 °C for 10 min. Subsequently, we only altered one reaction factor at a time, leaving the other factors unchanged. Therefore, the impact of targeted parameter can be determined using radio-TLC analysis.

In our efforts to reduce total synthesis time, we first investigated the correlation between reaction time and resulting reaction efficiency as shown in Fig. 16a. In this experiment, 200 μL of DI water was passed through the SAX column to release the [^{68}Ga]$GaCl_3$ and mixed with the 200 μL of PSMA-11 precursor solution (5 μg of PSMA-11 in 1 mL of 0.25 M sodium acetate buffer) injected at the mixing channel port of the preheated microfluidic device. The reaction time is designated as the total reaction time in which 200 μL of concentrated [^{68}Ga]$GaCl_3$ and 200 μL of PSMA-11 precursor mixture are completely loaded and transferred through the mixing channel to the reactor of the microfluidic device, heated at 95 °C, until the time the heating is turned off.

We found that our microfluidic system can obtain high conversion efficiency at substantially reduced duration. Specifically, labeling efficiency of ~99% was achieved by heating at 95 °C for 1 min, which consumed one-tenth of the time taken

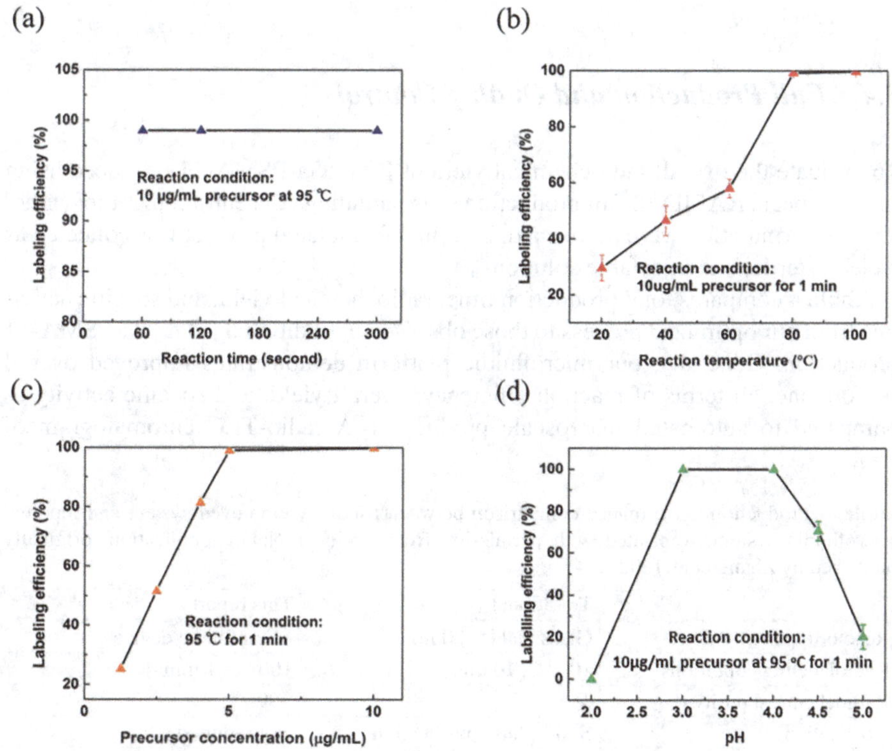

Fig. 16 Investigation on optimal reaction conditions for ^{68}Ga-labeling reaction in the microfluidic device. Several parameters including reaction time (**a**), reaction temperature (**b**), precursor concentration (**c**), and pH (**d**) are studied. (Reprinted with permission from Springer Nature publication, originally published by Zhang et al. (2020))

by current reactors employed to make PSMA-11. Afterward, we studied the impact of reaction temperature by performing radiosynthesis at varying temperatures ranging from 20 to 100 °C for 1 min. As shown in Fig. 16b, it is clear that sufficient heating is indispensable in order to obtain exceptionally high labeling efficiency. Interestingly, we observed that there might be a threshold temperature (~80 °C) above which labeling efficiency of 99% can be reliably achieved.

Apparent molar activity of PET tracers is one of the most critical factors that are typically used to evaluate the portion of nonradioactive form of the tracer, which can saturate rare biological receptors and worsen image quality. In our attempt to achieve high apparent molar activity, experiments focusing on titration of the concentration of PSMA-11 precursor were conducted, while maintaining the other parameters unaltered (heating at 100 °C for 1 min). As shown in Fig. 16c, it was found that 5 μg/mL of PSMA-11 precursor is enough to enable ~99% labeling efficiency. The apparent molar activity of the isolated product was observed to be greater than that of routinely produced clinical doses (Zhang et al., 2020). Additionally, we determined that the range of pH that enabled high-efficiency reaction was 3–4 (Fig. 16d), similar to macroscale production.

3.4 Full Production and Quality Control

To evaluate the overall radiochemical yield of [^{68}Ga]Ga-PSMA-11 produced using this approach, RAPID-M full production runs including concentration, radiosynthesis, and formulation were performed. The final formulated product was collected as a clear, sterile, and injectable solution.

Table 4 compares total production time, radiochemical yield, and specific activities from our optimized process to those observed in traditional [^{68}Ga]Ga-PSMA-11 production. Generally, our microfluidic platform demonstrates improved overall performance in terms of reaction efficiency, overall yield, and specific activity as compared to automated macroscale production. A radio-TLC chromatogram of

Table 4 Production performance comparison between routinely employed system and reported microfluidic system. Reprinted with permission from Springer Nature publication, originally published by Zhang et al. (2020)

	Traditional	This report
Reactor type	Glass vial (5–20 mL)	Microfluidic device
Radiolabeling conditions	100 °C, 10 min	100 °C, 1 min
Radiochemical purity (%)	>95	>99
Purification	Solid-phase purification	Not required
Total time (min)	25–20	15
Average RCY (%)	90	70
Specific activity (MBq/μg)	46.3	740.0

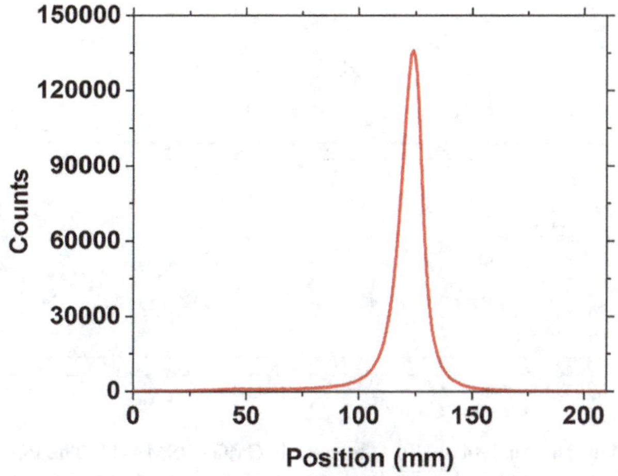

Fig. 17 Analytical TLC profile of [^{68}Ga]Ga-PSMA-11 synthesized using the microfluidic chip, showing quantitative conversion. Origin: 60 mm. Solvent front: 170. (Reprinted with permission from Springer Nature publication, originally published by Zhang et al. (2020))

synthesized product is shown in Fig. 17, and the ^{68}Ga-labeled peptide has an approximate retardation factor of 0.8–1.0.

Using our microfluidic device, incorporation efficiency of 100% can be consistently achieved as determined by radio-HPLC chromatogram (Fig. 18).

The only peak at 7.88 min corresponds to the retention time for synthesized [^{68}Ga]Ga-PSMA-11, indicating no remaining free [^{68}Ga]. It important to note that such high chemical and radiochemical purities at this stage allow for further simplification of synthesis procedures, this eliminates the need of HPLC purification. The stability of final product at room temperature was examined by radio-HPLC for a duration of up to 4 h. Based on our results (Fig. 18), no discrepancy on retention time or product purity was observed, suggesting good stability. After adding 2 mL of sodium acetate buffer into isolated reaction mixture, the final pH of formulated product was 6–7. After sterile filtration, the reaction mixture exhibited a radiochemical purity of >99% that satisfied standard release criterion; as such, the final product may simply be diluted for dose administration to any desired volume for animal or clinical use. The formulated solution was examined by standard quality control tests to assess its compliance with US Food and Drug Administration requirements for injectable PET tracers (Table 5).

In conclusion, RAPID-M is a high-yielding and cost-effective microfluidic synthetic platform that can be easily adaptable to clinical environment in synthesis of [^{68}Ga]Ga-PSMA-11 for molecular imaging and can likely be extended to the synthesis of [^{177}Lu]Lu-PSMA for targeted therapy. Without the need of any additional procedures, isolated final products met or exceeded QC requirements, thus allowing for potential direct administration to patients.

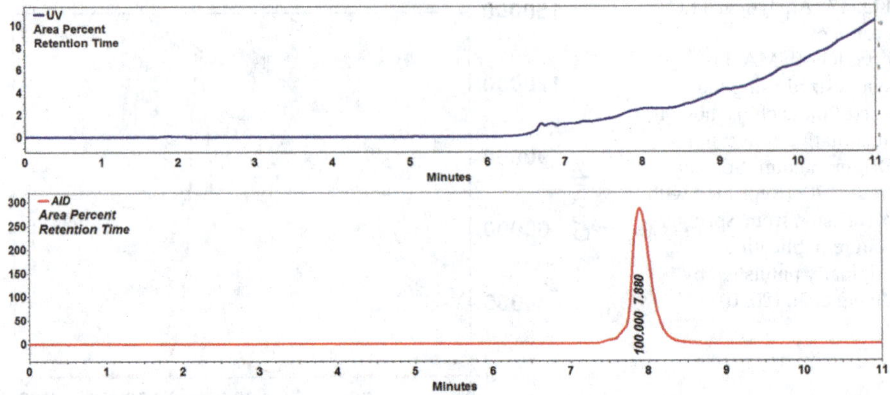

Fig. 18 Analytical HPLC trace of [^{68}Ga]Ga-PSMA-11. The top blue curve and bottom red curve indicate chemical and radiochemical purity, respectively. The retention time of [^{68}Ga]Ga-PSMA is 7.88 min. (Reprinted with permission from Springer Nature publication, originally published by Zhang et al. (2020))

Table 5 Quality control tests of synthesized [^{68}Ga]Ga-PSMA-11 using developed microfluidic system. Reprinted with permission from Springer Nature publication, originally published by Zhang et al. (2020)

Clinical QC test	Clinical acceptance criteria	Analysis of on-chip synthesized PET probes
Appearance	Clear and particle-free	Clear and particle-free
Filter integrity	Meet pressure specify by manufacture	Meet pressure specify by manufacture
pH	4–8	4
Radiochemical purity (%)	>90	>99
Identity	Matches retention time of standard (TLC)	Matches retention time of standard (TLC)
Half-life (min)	64–72	68
Sterility	No growth observed for 14 days	No growth observed for 14 days
Ge-68 breakthrough	<0.01% of the total radioactivity is Ge-68 at the time of product expiration	<0.01% of the total radioactivity is Ge-68 at the time of product expiration
Residual solvents	Dependent on methodology	Not required

4 Future Developments

If radiopharmaceuticals are to truly enhance and contribute to next-generation precision medicine, it is an essential requirement for manufacturers to deliver multiple unique RPs in a single day. This requires innovations in technology and business arrangements in the way in which RPs are produced and made available to end users. Current major barriers, however, include enormous infrastructure and labor costs, as well as key logistical challenges associated with producing more than a small number of unique tracers in a single day. Representing a major paradigm shift in RP production, the RAPID model will reduce the financial barrier-to-entry in radiotherapy and PET imaging research and allow significantly broader participation in multicenter radiotherapy and PET studies, because, regardless of whether a center has a cyclotron and/or a dedicated radiochemistry facility, participating centers will be able to produce (in a dose-on-demand fashion) a patient dose of a given RP by commercially sourcing the appropriate radionuclide for labeling, obtaining the necessary precursor reagents, and acquiring the RAPID device. Our RAPID concept starkly contrasts contemporary commercial models, where tracers (namely, [^{18}F]FDG) are mass-produced in large, dedicated facilities and widely distributed to multiple imaging centers. This precludes further development and commercialization of innovative tracers due to the economics of supply and demand, currently a major limitation within the field of molecular imaging and radionuclide therapy. However, when marketed as kit-based products, many radiotracers with limited distribution potential could easily become commercially viable due to the reduced start up and development costs and the fact that production only occurs on demand at the point of care (Fig. 1). To enable this new RP production strategy, we continue to develop the RAPID platform and are in the process of incorporating new concepts (e.g., optimized flow paths, on-chip heating, automation) that will further enhance throughput and ease-of-use of the system. Additionally, we are working toward adapting RAPID to more complex radiosynthetic schemes, thereby further broadening the applicability of the RAPID concept.

References

Abdelgawad, M., & Wheeler, A. R. (2009). The digital revolution: A new paradigm for microfluidics. *Advanced Materials, 21*, 920–925.

Alam, M. R., Singh, S. B., Thapaliya, S., Shrestha, S., Deo, S., & Khanal, K. (2022). A review of 177lutetium-PSMA and 225actinium-PSMA as emerging theranostic agents in prostate cancer. *Cureus, 14*, e29369.

Audrain, H. (2007). Positron emission tomography (PET) and microfluidic devices: A breakthrough on the microscale? *Angewandte Chemie, International Edition, 46*, 1772–1775.

Baxter, L. R., Jr., Phelps, M. E., Mazziotta, J. C., Schwartz, J. M., Gerner, R. H., Selin, C. E., & Sumida, R. M. (1985). Cerebral metabolic rates for glucose in mood disorders. Studies with positron emission tomography and fluorodeoxyglucose F18. *Archives of General Psychiatry, 42*, 441–447.

Bejot, R., Elizarov, A. M., Ball, E., Zhang, J., Miraghaie, R., Kolb, H. C., & Gouverneur, V. (2011). Batch-mode microfluidic radiosynthesis of N-succinimidyl-4-[^{18}F]fluorobenzoate for protein labelling. *Journal of Labelled Compounds and Radiopharmaceuticals, 54*, 117–122.

Brichard, L., Ferrari, V., Smith, R., & Aigbirhio, F. I. (2012). Synthesis of [18F]-Fallypride. In *Radiochemical syntheses* (pp. 95–102).

Chen, S., Javed, M. R., Kim, H.-K., Lei, J., Lazari, M., Shah, G. J., Michael van Dam, R., Keng, P.-Y., & Kim, C.-J. C. J. (2014). Radiolabelling diverse positron emission tomography (PET) tracers using a single digital microfluidic reactor chip. *Lab on a Chip, 14*, 902–910.

Choi, K., Ng, A. H. C., Fobel, R., & Wheeler, A. R. (2012). Digital microfluidics. *Annual Review of Analytical Chemistry, 5*, 413–440.

Di Carli, M. F., Davidson, M., Little, R., Khanna, S., Mody, F. V., Brunken, R. C., Czernin, J., Rokhsar, S., Stevenson, L. W., Laks, H., et al. (1994). Value of metabolic imaging with positron emission tomography for evaluating prognosis in patients with coronary artery disease and left ventricular dysfunction. *The American Journal of Cardiology, 73*, 527–533.

Ding, H., Sadeghi, S., Shah, G. J., Chen, S., Keng, P. Y., Kim, C.-J. C. J., & van Dam, R. M. (2012). Accurate dispensing of volatile reagents on demand for chemical reactions in Ewod chips. *Lab on a Chip, 12*, 3331–3340.

Duara, R., Grady, C., Haxby, J., Sundaram, M., Cutler, N. R., Heston, L., Moore, A., Schlageter, N., Larson, S., & Rapoport, S. I. (1986). Positron emission tomography in Alzheimer's disease. *Neurology, 36*, 879–887.

Duffy, D. C., Cooper McDonald, J., Schueller, O. J. A., & Whitesides, G. M. (1998). Rapid prototyping of microfluidic systems in poly(dimethylsiloxane). *Analytical Chemistry, 70*, 4974–4984.

Eitzman, D., Al-Aouar, Z., Kanter, H. L., vom Dahl, J., Kirsh, M., Deeb, G. M., & Schwaiger, M. (1992). Clinical outcome of patients with advanced coronary artery disease after viability studies with positron emission tomography. *Journal of the American College of Cardiology, 20*, 559–565.

Elizarov, A. M. (2009). Microreactors for radiopharmaceutical synthesis. *Lab on a Chip, 9*, 1326–1333.

Elizarov, A. M., van Dam, R. M., Shin, Y. S., Kolb, H. C., Padgett, H. C., Stout, D., Shu, J., Huang, J., Daridon, A., & Heath, J. R. (2010). Design and optimization of coin-shaped microreactor chips for PET radiopharmaceutical synthesis. *Journal of Nuclear Medicine, 51*, 282–287.

Facey, K., Bradbury, I., Laking, G., & Payne, E. (2007). Overview of the clinical effectiveness of positron emission tomography imaging in selected cancers. *Health Technology Assessment, 11*, iii-iv, xi-267.

Fernandez-Maza, L., Salvador, B., Orta, D., Corral, A., & Luque, A. (2019). Does PDMS really interact with [^{18}F]fluoride? applications in microfluidic reactors for 18F-radiopharmaceuticals. *Microfluidics and Nanofluidics, 23*, 109.

Ferreira, C. A., Ehlerding, E. B., Rosenkrans, Z. T., Jiang, D., Sun, T., Aluicio-Sarduy, E., Engle, J. W., Ni, D., & Cai, W. (2020). 86/90y-labeled monoclonal antibody targeting tissue factor for pancreatic cancer theranostics. *Molecular Pharmaceutics, 17*, 1697–1705.

Filippov, A., Bonjoc, K.-J. C., Chea, J., Bowles, N., Poku, E., & Chaudhry, A. (2020). Role of theranostics in thoracic oncology. *Journal of Thoracic Disease, 12*, 5140–5146.

Fischer, B. M., Mortensen, J., & Højgaard, L. (2001). Positron emission tomography in the diagnosis and staging of lung cancer: A systematic, quantitative review. *The Lancet Oncology, 2*, 659–666.

Fortt, R., & Gee, A. (2013). Microfluidics: A golden opportunity for positron emission tomography? *Future Medicinal Chemistry, 5*, 241–244.

Gambhir, S. S. (2002). Molecular imaging of cancer with positron emission tomography. *Nature Reviews. Cancer, 2*, 683–693.

Gillies, J. M., Prenant, C., Chimon, G. N., Smethurst, G. J., Dekker, B. A., & Zweit, J. (2006). Microfluidic technology for PET radiochemistry. *Applied Radiation and Isotopes, 64*, 333–336.

Gomes Marin, J. F., Nunes, R. F., Coutinho, A. M., Zaniboni, E. C., Costa, L. B., Barbosa, F. G., Queiroz, M. A., Cerri, G. G., & Buchpiguel, C. A. (2020). Theranostics in nuclear medicine: Emerging and re-emerging integrated imaging and therapies in the era of precision oncology. *Radiographics, 40*, 1715–1740.

Goyer, P. F., Andreason, P. J., Semple, W. E., Clayton, A. H., King, A. C., Compton-Toth, B. A., Schulz, S. C., & Cohen, R. M. (1994). Positron-emission tomography and personality disorders. *Neuropsychopharmacology, 10*, 21–28.

Hara, T., Kosaka, N., & Kishi, H. (1998). PET imaging of prostate cancer using carbon-11-choline. *Journal of Nuclear Medicine, 39*, 990–995.

Hawkey, N. M., Sartor, A. O., Morris, M. J., & Armstrong, A. J. (2022). Prostate-specific membrane antigen-targeted theranostics: Past, present, and future approaches. *Clinical Advances in Hematology & Oncology, 20*, 227–238.

Horoszewicz, J. S., Kawinski, E., & Murphy, G. P. (1987). Monoclonal antibodies to a new antigenic marker in epithelial prostatic cells and serum of prostatic cancer patients. *Anticancer Research, 7*, 927–935.

Ichikawa, Y., Kobayashi, N., Takano, S., Kato, I., Endo, K., & Inoue, T. (2022). Neuroendocrine tumor theranostics. *Cancer Science, 113*, 1930.

Jaidev, R. L., Chede, S. L., & Kandikattu, K. H. (2021). Theranostic nanoparticles for pancreatic cancer treatment. *Endocrine, Metabolic & Immune Disorders - Drug Targets, 21*, 203–214.

Javed, M. R., Chen, S., Lei, J., Collins, J., Sergeev, M., Kim, H.-K., Chang-Jin Kim, R., van Dam, M., & Keng, P. Y. (2014). High yield and high specific activity synthesis of [^{18}F]fallypride in a batch microfluidic reactor for micro-PET imaging. *Chemical Communications, 50*, 1192–1194.

Jokar, N., Velikyan, I., Ahmadzadehfar, H., Rekabpour, S. J., Jafari, E., Ting, H. H., Biersack, H.-J., & Assadi, M. (2021). Theranostic approach in breast cancer: A treasured tailor for future oncology. *Clinical Nuclear Medicine, 46*, e410.

Keng, P. Y., Chen, S., Ding, H., Sadeghi, S., Shah, G. J., Dooraghi, A., Phelps, M. E., Satyamurthy, N., Chatziioannou, A. F., Kim, C.-J., & Michael van Dam, R. (2012a). Micro-chemical synthesis of molecular probes on an electronic microfluidic device. *Proceedings of the National Academy of Sciences of the United States of America, 109*, 690–695.

Keng, P. Y., Esterby, M., & Michael van Dam, R. (2012b). Emerging technologies for decentralized production of PET tracers. In C.-H. Hsieh (Ed.), *Positron emission tomography—Current clinical and research aspects*. IntechOpen. https://doi.org/10.5772/31390

Kozikowski, A. P., Nan, F., Conti, P., Zhang, J., Ramadan, E., Bzdega, T., Wroblewska, B., Neale, J. H., Pshenichkin, S., & Wroblewski, J. T. (2001). Design of remarkably simple, yet potent urea-based inhibitors of glutamate carboxypeptidase II (NAALADase). *Journal of Medicinal Chemistry, 44*, 298–301.

Kubota, K. (2001). From tumor biology to clinical pet: A review of positron emission tomography (PET) in oncology. *Annals of Nuclear Medicine, 15*, 471–486.

Lardinois, D., Weder, W., Hany, T. F., Kamel, E. M., Korom, S., Seifert, B., von Schulthess, G. K., & Steinert, H. C. (2003). Staging of non-small-cell lung cancer with integrated positron-emission tomography and computed tomography. *The New England Journal of Medicine, 348*, 2500–2507.

Lebedev, A., Miraghaie, R., Kotta, K., Ball, C. E., Zhang, J., Buchsbaum, M. S., Kolb, H. C., & Elizarov, A. (2013). Batch-reactor microfluidic device: First human use of a microfluidically produced PET radiotracer. *Lab on a Chip, 13*, 136–145.

Lee, J. N., Park, C., & Whitesides, G. M. (2003). Solvent compatibility of poly(dimethylsiloxane)-based microfluidic devices. *Analytical Chemistry, 75*, 6544–6554.

Lee, C.-C., Sui, G., Elizarov, A., Shu, C. J., Shin, Y.-S., Dooley, A. N., Huang, J., Daridon, A., Wyatt, P., Stout, D., Kolb, H. C., Witte, O. N., Satyamurthy, N., Heath, J. R., Phelps, M. E., Quake, S. R., & Tseng, H.-R. (2005). Multistep synthesis of a radiolabeled imaging probe using integrated microfluidics. *Science, 310*, 1793–1796.

Li, S., Schmitz, A., Lee, H., & Mach, R. H. (2016). Automation of the radiosynthesis of six different 18F-labeled radiotracers on the allinone. *EJNMMI Radiopharmacy and Chemistry, 1*, 15.

Lopes, A. D., Davis, W. L., Rosenstraus, M. J., Uveges, A. J., & Gilman, S. C. (1990). Immunohistochemical and pharmacokinetic characterization of the site-specific immunoconjugate Cyt-356 derived from antiprostate monoclonal antibody 7e11-C5. *Cancer Research, 50*, 6423–6429.

Lu, S. Y., Watts, P., Chin, F. T., Hong, J., Musachio, J. L., Briard, E., & Pike, V. W. (2004). Syntheses of 11C- and 18F-labeled carboxylic esters within a hydrodynamically-driven micro-reactor. *Lab on a Chip, 4*, 523–525.

Lu, S., Anthony Giamis, M., & Victor Pike, W. (2009). Synthesis of [^{18}F]fallypride in a micro-reactor: Rapid optimization and multiple-production in small doses for micro-PET studies. *Current Radiopharmaceuticals, 2*, 49–55.

Mach, R. H., & Luedtke, R. R. (2018). Challenges in the development of dopamine D2- and D3-selective radiotracers for PET imaging studies. *Journal of Labelled Compounds and Radiopharmaceuticals, 61*, 291–298.

Maltezos, G., Garcia, E., Hanrahan, G., Gomez, F. A., Saurabh Vyawhare, R., van Dam, M., Chen, Y., & Scherer, A. (2007). Design and fabrication of chemically robust three-dimensional microfluidic valves. *Lab on a Chip, 7*, 1209–1211.

Mapelli, P., Incerti, E., Ceci, F., Castellucci, P., Fanti, S., & Picchio, M. (2016). 11C- or 18F-choline PET/CT for imaging evaluation of biochemical recurrence of prostate cancer. *Journal of Nuclear Medicine, 57*, 43s–48s.

McDonald, J. C., & Whitesides, G. M. (2002). Poly(dimethylsiloxane) as a material for fabricating microfluidic devices. *Accounts of Chemical Research, 35*, 491–499.

McDonald, J. C., Duffy, D. C., Anderson, J. R., Chiu, D. T., Wu, H., Schueller, O. J., & Whitesides, G. M. (2000). Fabrication of microfluidic systems in poly(dimethylsiloxane). *Electrophoresis, 21*, 27–40.

Miller, P. W. (2009). Radiolabelling with short-lived PET (positron emission tomography) isotopes using microfluidic reactors. *Journal of Chemical Technology & Biotechnology, 84*, 309–315.

Miyahira, A. K., & Soule, H. R. (2022). The history of prostate-specific membrane antigen as a theranostic target in prostate cancer: The cornerstone role of the Prostate Cancer Foundation. *Journal of Nuclear Medicine, 63*, 331–338.

Mueller, D., Klette, I., Baum, R. P., Gottschaldt, M., Schultz, M. K., & Breeman, W. A. (2012). Simplified NaCl based (68)Ga concentration and labeling procedure for rapid synthesis of (68)Ga radiopharmaceuticals in high radiochemical purity. *Bioconjugate Chemistry, 23*, 1712–1717.

Mueller, D., Breeman, W. A., Klette, I., Gottschaldt, M., Odparlik, A., Baehre, M., Tworowska, I., & Schultz, M. K. (2016). Radiolabeling of Dota-like conjugated peptides with generator-produced (68)Ga and using NaCl-based cationic elution method. *Nature Protocols, 11*, 1057–1066.

Olivier, P., Giraudet, A.-L., Skanjeti, A., Merlin, C., Weinmann, P., Rudolph, I., Hoepping, A., & Gauthé, M. (2022). Phase III study of ^{18}F-PSMA-1007 versus ^{18}F-fluorocholine PET/CT for localization of prostate cancer biochemical recurrence: A prospective, randomized, cross-over, multicenter study. *Journal of Nuclear Medicine, 64*(4), 579–585.

Pascali, G., Nannavecchia, G., Pitzianti, S., & Salvadori, P. A. (2011). Dose-on-demand of diverse 18F-fluorocholine derivatives through a two-step microfluidic approach. *Nuclear Medicine and Biology, 38*, 637–644.

Pascali, G., Watts, P., & Salvadori, P. A. (2013). Microfluidics in radiopharmaceutical chemistry. *Nuclear Medicine and Biology, 40*, 776–787.

Perera, M., Papa, N., Christidis, D., Wetherell, D., Hofman, M. S., Murphy, D. G., Bolton, D., & Lawrentschuk, N. (2016). Sensitivity, specificity, and predictors of positive (68)Ga-prostate-specific membrane antigen positron emission tomography in advanced prostate cancer: A systematic review and meta-analysis. *European Urology, 70*, 926–937.

Playford, E. D., Jenkins, I. H., Passingham, R. E., Nutt, J., Frackowiak, R. S., & Brooks, D. J. (1992). Impaired mesial frontal and putamen activation in Parkinson's disease: A positron emission tomography study. *Annals of Neurology, 32*, 151–161.

Pomper, M. G., Musachio, J. L., Zhang, J., Scheffel, U., Zhou, Y., Hilton, J., Maini, A., Dannals, R. F., Wong, D. F., & Kozikowski, A. P. (2002). 11C-MCG: Synthesis, uptake selectivity, and primate PET of a probe for glutamate carboxypeptidase II (NAALADase). *Molecular Imaging, 1*, 96–101.

Rauch, S. L., van der Kolk, B. A., Fisler, R. E., Alpert, N. M., Orr, S. P., Savage, C. R., Fischman, A. J., Jenike, M. A., & Pitman, R. K. (1996). A symptom provocation study of posttraumatic stress disorder using positron emission tomography and script-driven imagery. *Archives of General Psychiatry, 53*, 380–387.

Reiman, E. M., Raichle, M. E., Robins, E., Butler, F. K., Herscovitch, P., Fox, P., & Perlmutter, J. (1986). The application of positron emission tomography to the study of panic disorder. *The American Journal of Psychiatry, 143*, 469–477.

Rensch, C., Wängler, B., Yaroshenko, A., Samper, V., Baller, M., Heumesser, N., Ulin, J., Riese, S., & Reischl, G. (2012). Autoradiolysis suppression in microfluidic channels. *Journal of Nuclear Medicine, 53*, 574–574.

Rensch, C., Jackson, A., Lindner, S., Salvamoser, R., Samper, V., Riese, S., Bartenstein, P., Wängler, C., & Wängler, B. (2013). Microfluidics: A groundbreaking technology for PET tracer production? *Molecules, 18*, 7930–7956.

Rensch, C., Lindner, S., Salvamoser, R., Leidner, S., Böld, C., Samper, V., Taylor, D., Baller, M., Riese, S., Bartenstein, P., Wängler, C., & Wängler, B. (2014). A solvent resistant lab-on-chip platform for radiochemistry applications. *Lab on a Chip, 14*, 2556–2564.

Salvador, B., Luque, A., Quero, J. M., Fernández, L., Corral, A., Orta, D., & Fernández, I. (2017a). *Integrated anion-exchange cartridge for [^{18}F]F-preconcentration in a PDMS radiopharmacy chip*. Paper presented at the 2017 Spanish conference on electron devices (CDE), 8–10 February 2017.

Salvador, B., Luque, A., Fernandez-Maza, L., Corral, A., Orta, D., Fernández, I., & Quero, J. M. (2017b). Disposable PDMS chip with integrated [^{18}F]fluoride pre-concentration cartridge for radiopharmaceuticals. *Journal of Microelectromechanical Systems, 26*, 1442–1448.

Shoup, T. M., Olson, J., Hoffman, J. M., Votaw, J., Eshima, D., Eshima, L., Camp, V. M., Stabin, M., Votaw, D., & Goodman, M. M. (1999). Synthesis and evaluation of [^{18}F]1-amino-3-fluorocyclobutane-1-carboxylic acid to image brain tumors. *Journal of Nuclear Medicine, 40*, 331–338.

Siegel, R. L., Miller, K. D., & Jemal, A. (2020). Cancer statistics, 2020. *CA: A Cancer Journal for Clinicians, 70*, 7–30.

Ungersboeck, J., Philippe, C., Mien, L. K., Haeusler, D., Shanab, K., Lanzenberger, R., Spreitzer, H., Keppler, B. K., Dudczak, R., Kletter, K., Mitterhauser, M., & Wadsak, W. (2011). Microfluidic preparation of [^{18}F]Fe@Suppy and [^{18}F]Fe@Suppy:2—Comparison with conventional radiosyntheses. *Nuclear Medicine and Biology, 38*, 427–434.

Vansteenkiste, J., Fischer, B. M., Dooms, C., & Mortensen, J. (2004). Positron-emission tomography in prognostic and therapeutic assessment of lung cancer: Systematic review. *The Lancet Oncology, 5*, 531–540.

Wang, M. W., Lin, W. Y., Liu, K., Masterman-Smith, M., & Kwang-Fu Shen, C. (2010). Microfluidics for positron emission tomography probe development. *Molecular Imaging, 9*, 175–191.

Wang, J., Chao, P. H., Hanet, S., & Michael van Dam, R. (2017). Performing multi-step chemical reactions in microliter-sized droplets by leveraging a simple passive transport mechanism. *Lab on a Chip, 17*, 4342–4355.

Wang, J., Chao, P. H., Slavik, R., & Michael van Dam, R. (2020). Multi-Gbq production of the radiotracer [^{18}F]Fallypride in a droplet microreactor. *RSC Advances, 10*, 7828–7838.

Watts, P., Pascali, G., & Salvadori, P. A. (2012). Positron emission tomography radiosynthesis in microreactors. *Journal of Flow Chemistry, 2*, 37–42.

Wester, H.-j., Schoultz, B. W., Hultsch, C., & Henriksen, G. (2009). Fast and repetitive in-capillary production of [^{18}F]FDG. *European Journal of Nuclear Medicine and Molecular Imaging, 36*, 653–658.

Wiering, B., Krabbe, P. F., Jager, G. J., Oyen, W. J., & Ruers, T. J. (2005). The impact of fluor-18-deoxyglucose-positron emission tomography in the management of colorectal liver metastases. *Cancer, 104*, 2658–2670.

Young, H., Baum, R., Cremerius, U., Herholz, K., Hoekstra, O., Lammertsma, A. A., Pruim, J., & Price, P. (1999). Measurement of clinical and subclinical tumour response using [^{18}F]-fluorodeoxyglucose and positron emission tomography: Review and 1999 EORTC recommendations. European Organization for Research and Treatment of Cancer (EORTC) PET Study Group. *European Journal of Cancer, 35*, 1773–1782.

Zacheo, A., Arima, V., Pascali, G., Salvadori, P. A., Zizzari, A., Perrone, E., De Marco, L., Gigli, G., & Rinaldi, R. (2011). Radioactivity resistance evaluation of polymeric materials for application in radiopharmaceutical production at microscale. *Microfluidics and Nanofluidics, 11*, 35–44.

Zhang, X., Liu, F., Knapp, K. A., Nickels, M. L., Manning, H. C., & Bellan, L. M. (2018). A simple microfluidic platform for rapid and efficient production of the radiotracer [(18)F]fallypride. *Lab on a Chip, 18*, 1369–1377.

Zhang, X., Liu, F., Payne, A. C., Nickels, M. L., Bellan, L. M., & Charles Manning, H. (2020). High-yielding radiosynthesis of [68ga]Ga-PSMA-11 using a low-cost microfluidic device. *Molecular Imaging and Biology, 22*, 1370–1379.

Zheng, M.-Q., Collier, L., Bois, F., Kelada, O. J., Hammond, K., Ropchan, J., Akula, M. R., Carlson, D. J., Kabalka, G. W., & Huang, Y. (2015). Synthesis of [^{18}F]FMISO in a flow-through microfluidic reactor: Development and clinical application. *Nuclear Medicine and Biology, 42*, 578–584.

Zizzari, A., Arima, V., Zacheo, A., Pascali, G., Salvadori, P. A., Perrone, E., Mangiullo, D., & Rinaldi, R. (2011). Fabrication of Su-8 microreactors for radiopharmaceutical production. *Microelectronic Engineering, 88*, 1664–1667.

Progress in Droplet-Based Methods for Simple, Low-Cost Radiopharmaceutical Production

Yingqing Lu and R. Michael van Dam

1 Introduction

The current radiopharmaceutical supply chain relies on very expensive automated radiosynthesis equipment, operated inside radiation-shielded hot cells, further housed within a cGMP (current good manufacturing practice) facility with specialized equipment, procedures, and trained personnel. The high level of automation ensures consistent day-to-day production of these short-lived compounds, and the hot cells minimize radiation exposure to staff. Each batch must undergo quality-control (QC) testing to ensure safety and purity of individual doses before dispensing. Much of the equipment commercially available today is designed to be able to produce and handle the very largest of batches (i.e., for a large number of patient doses, and with sufficient excess to account for radioactive decay during production, QC testing, dose dispensing, and transport to the final clinical facility where the radiopharmaceutical is administered to the patient). Due to the short half-life (e.g., 110 min for ^{18}F-labeled positron-emission tomography (PET) imaging agents), transport distance is limited, and facilities must be replicated at multiple locations to reach the full patient population in a given country or region. For practical reasons, many production facilities will use a dedicated set of equipment/resources for each different radiopharmaceutical product that is produced.

Y. Lu · R. M. van Dam (✉)
Physics and Biology in Medicine Interdepartmental Graduate Program, University of California Los Angeles, Los Angeles, CA, USA

Crump Institute for Molecular Imaging, University of California Los Angeles, Los Angeles, CA, USA

Department of Molecular & Medical Pharmacology, David Geffen School of Medicine, University of California Los Angeles, Los Angeles, CA, USA
e-mail: mvandam@mednet.ucla.edu

R. M. van Dam et al. (eds.), *Automated Technologies for the Development and Production of Radiopharmaceuticals*,
https://doi.org/10.1007/978-3-031-84632-8_10

As a consequence of this resource-intensive manufacturing process, each radiopharmaceutical batch is very expensive (many thousands of $USD). Depending on the cost recovery mechanism, usually a minimum number of doses must be produced per batch for the radiopharmacy to break even. For widely used radiopharmaceuticals, for example, 2-[^{18}F]fluoro-2-deoxy-D-glucose ([^{18}F]FDG), with which millions of PET scans are performed per year, it is possible to pool together many doses (up to dozens or even hundreds) into a single batch production. With such economy of scale, the cost per dose can be significantly reduced (e.g., $100–200). With such ubiquity of production and low price, [^{18}F]FDG can readily be used in basic and clinical research studies. However, when the demand is lower, for example, for approved radiopharmaceuticals targeting more specialized clinical or research purposes, or compounds in earlier stages of development, such division of the production cost is not possible. Availability of such compounds tends to be extremely limited, and the costs are too high for most investigational purposes.

To address these challenges and change the economics, there has been extensive interest in developing new technologies for radiopharmaceutical manufacturing (i.e., microfluidics) that are less expensive, require less space and shielding, and need less training and expertise to operate. Numerous reviews have appeared during the last 16 years summarizing the various approaches pursued and the advantages and tradeoffs of each (Audrain, 2007; Elizarov, 2009; Elkawad et al., 2022; Knapp et al., 2020; Lebedev, 2013; Miller, 2009; Pascali & Salvadori, 2016; Rensch et al., 2013; Wang & van Dam, 2020). Broadly, these novel microscale technologies can be categorized as batch or flow microreactors. Batch microreactors use small glass vials, or microfluidic chips incorporating small cavities or chambers, as a reactor, and multistep radiosyntheses are carried out by sequentially adding radioisotope and reagents, evaporating solvents, and performing heated reactions to form the crude product. Droplet-based reactors are a specialized type of batch reactor, in which the reactions are confined to the boundaries of a droplet. Flow microreactors use syringe pumps to supply streams of radioisotope and precursor solutions through one or more reaction stages, each comprising an inline micromixer followed by heated capillary tubing or microchannels, for performing multistep radiosyntheses.

Microdevices can potentially transform radiopharmaceutical manufacturing in a number of important ways. (1) Small volume scales (Keng et al., 2016): The volume scale of current vial-based synthesizers (~1 mL) is mainly determined by the volumes of typical isotope sources and the volumes at which fluidic systems based on vials, valves, and tubing can be operated with minor residual losses, but is much larger than necessary to produce the very tiny quantities of radiopharmaceuticals produced in each batch (e.g., nanomole amount per patient for typical ^{18}F-labeled PET tracers). By operating at smaller volume scales, microfluidic devices can boost reactant concentrations and reaction rates, speed up processes such as heating or evaporation, reduce reagent costs, and simplify purification processes. (2) Disposability: Many types of microfluidic systems are inherently made from inexpensive materials using low-cost manufacturing processes. Many conventional vial-based systems are designed around the concept of disposable "cassettes" to simplify production in clinical settings, by eliminating the need to develop/validate/perform cleaning and providing the flexibility to synthesize multiple radiopharmaceutical products on a single radiosynthesizer. Disposability also reduces the need for maintenance by ensuring that radiation-sensitive components are frequently replaced. (3) Compact size: Microfluidic systems, with careful design, can be vastly smaller than conventional radiosynthesizers. It may be practical to operate them in a self-shielded manner, without the need for a conventional hot cell, greatly reducing the infrastructure needed for operation. (4) Simplification of cGMP facilities: Combined with technologies like automated QC testing (see chapter "Quality Control Automation Solutions for PET Drug Production"), microfluidics could revolutionize cGMP facilities, reducing size, cost, and operating complexity. (5) Dose-on-demand capability: Unlike in conventional synthesizers, achieving high molar activity in microfluidic reactions is feasible even with a relatively low starting activity, due to reduced isotopic contamination from reagents and materials (Sergeev et al., 2018), enabling efficient production with the amount of radioisotope scaled to the amount actually needed for imaging studies.

In this chapter, we focus on developments of droplet-based radiosynthesis technologies in particular, as they are capable of scaling down volumes and reagent amounts more significantly than in other platforms, maximizing the advantages associated with volume and reagent reduction. For example, some droplet-based techniques are able to perform reactions in volumes as small as ~1 μL, a 1000× reduction compared to conventional vial-based systems. Furthermore, it has been shown that extremely compact systems can be used to automate droplet reaction chips, creating the potential for new manufacturing paradigms for radiopharmaceuticals.

2 Droplet Reaction Systems and Applications

2.1 Overview

Numerous droplet-based technology platforms have been used in radiochemistry (Fig. 1, Table 1). In this section, we describe the structure, fabrication, and operation of each platform and summarize the applications of the technology to date. A comprehensive listing of radiotracers synthesized in droplet format, along with detailed performance metrics, is included in Table 2 at the end of this section.

2.2 Electrowetting-on-Dielectric (EWOD)

The first droplet-based approaches used in radiochemistry were based on "digital microfluidics" (DMF) devices using electrowetting-on-dielectric (EWOD) actuation methods (Fig. 1a). These chips used electric fields to move and mix

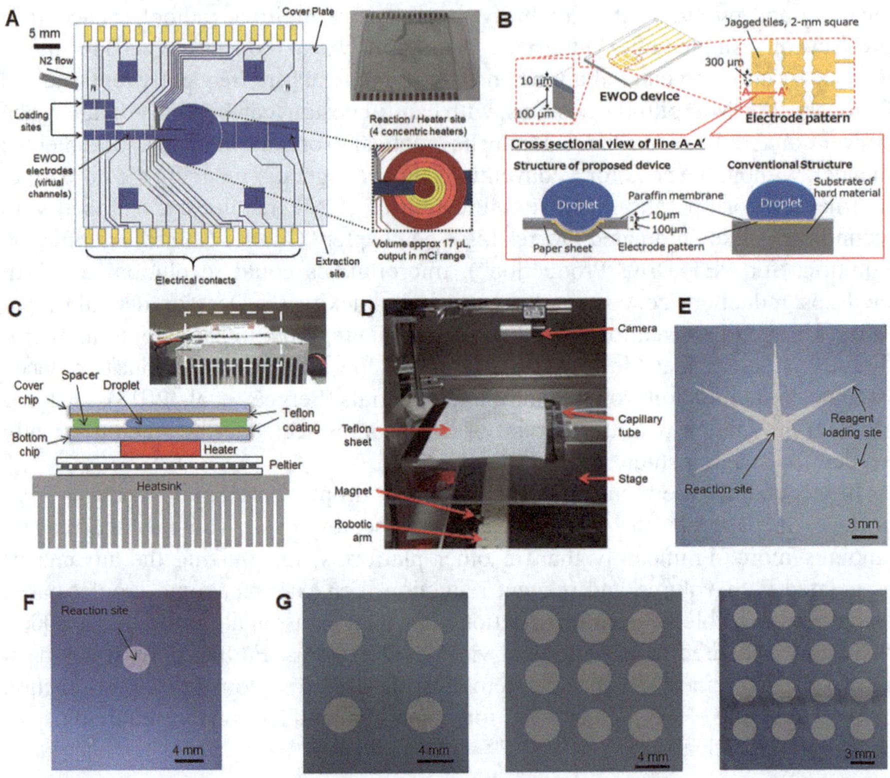

Fig. 1 Chips/substrates used in droplet-based radiochemistry platforms. (**a**) Schematic of electrodes on a glass EWOD microchip. Droplets deposited on the chip are toward the ~17 μL capacity reaction site by sequential electrode activation. The inset shows a magnified view of the concentric heater electrode design at the reaction site. Reproduced with permission from (Keng et al., 2012), copyright 2012 National Academy of Sciences. (**b**) Design of paper-based EWOD device, which uses a dimpled electrode structure to improve droplet-movement reliability. Reproduced from (Mogi et al., 2019) with permission from MDPI. (**c**) Schematic and photograph (inset) of the Teflon-coated glass chip atop a heating system. Adapted with permission from Lisova et al. (2018), copyright 2018 Elsevier. (**d**) Magnetic droplet microfluidics platform in which reactions take place in droplets deposited on top of a Teflon sheet. Droplets contain magnetic particles and are manipulated by a robotically moved magnet below the Teflon sheet. When needed, solvent is removed via suction by a capillary tube rather than by evaporation. Adapted with permission from Fiel et al. (2015), copyright 2015 American Chemical Society. (**e**) Photograph of passive transport (PT) chip comprising a patterned Teflon coating on a silicon substrate. Droplets deposited at the narrow end of a tapered pathway will spontaneously move toward the central reaction site. Reproduced from (Wang et al., 2017) with permission from the Royal Society of Chemistry. (**f**) Photograph of the surface-tension trap (STT) chip. Reagents are deposited directly to the reaction site. Adapted from Wang et al. (2019) with permission from the Royal Society of Chemistry. (**g**) Designs of multi-reaction STT chips for performing parallel reactions for synthesis optimization or reaction scale-up. (Adapted from Rios et al. (2019) with permission from the Royal Society of Chemistry)

Table 1 Comparison of platforms for droplet-based radiochemistry

Type	Materials	Chip cost	Open/ covered	Reagent addition	Product collection	Droplet actuation	Temp. control	Reagent compatibility/ temp. stability	Special features or limitations	Ref.
EWOD—Glass	Glass substrate; ITO or Au Electrodes; SiNx or parylene C dielectric; Cytop or Teflon AF hydrophobic coating	High	Covered	Pipette (manual) to edge of chip or prior to adding cover; or Syringe pump to edge of chip or prior to adding cover; or Pneumatic delivery through holes in the cover	Pipette (manual) after removing cover	EWOD	On-chip electrodes;or Heater under chip	High/high	Partial on-chip purification demonstrated using Al_2O_3 surface or beads (in solution or immobilized)	Keng et al. (2012); Ahmadi et al. (2022)
EWOD—Paper	Paper substrate; Conductive ink electrodes; Paraffin film (both dielectric and hydrophobic layers)	Low	Open	Not specified (manual)	Pipette (manual) or syringe (manual)	EWOD	None	Low/N/A	None	Mogi et al. (2021)
Magnetic	Teflon sheet substrate	Low	Open	Not specified (manual)	Syringe (manual)	Robotically-controlled magnet under substrate	None	Low/N/A	Requires magnetic particles/ carrier in droplets; A capillary tube with vacuum was used to remove waste liquid from beads	Fiel et al. (2015)

(continued)

Type	Materials	Chip cost	Open/ covered	Reagent addition	Product collection	Droplet actuation	Temp. control	Reagent compatibility/ temp. stability	Special features or limitations	Ref.
Teflon-coated glass	Glass substrate; Teflon AF coat	Low	Covered	Pipette (manual)	Pipette (manual) after removing cover	None	Heater under chip	High/high	None	Lisova et al. (2018)
Passive transport	Silicon substrate; Patterned Teflon AF coat	Low	Open	Piezoelectric dispenser (fixed in position)	Retractable dip tube	Spontaneous (passive) droplet movement to reaction site via patterned wettability	Heater under chip	High/high	None	Wang et al. (2017)
Surface-tension trap	Silicon substrate; Patterned Teflon AF coat	Low	Open	Pipette (manual) or piezoelectric dispenser (robotic)	Dip tube	None (External rotary motion used to add/ remove liquids)	Heater under chip	High/high	None	Wang et al. (2019)
Multi-reaction surface-tension trap	Silicon substrate; Patterned Teflon AF coat	Low	Open	Pipette (manual); or Piezoelectric dispenser and pipette (robotic)	Pipette (manual or robotic)	None (External robotic actuators used to add/ remove liquids)	Heater under chip	High/high	Arrays of 2 × 2, 3 × 3, and 4 × 4 reaction sites reported; Simultaneous operation of up to four chips reported using multiple heaters	Rios et al. (2019); Jones et al. (2023)

Abbreviations: *EWOD* electrowetting-on-dielectric; *ITO* indium tin oxide; *N/A* not applicable droplets of polar liquids to perform radiosynthesis processes (Keng et al., 2012). Typically, these devices contain a series of individually addressable tile-shaped electrodes patterned on a substrate to define droplet-movement paths, with dielectric and hydrophobic layers deposited on top. The droplets are sandwiched between this chip and a "cover plate" with a large ground electrode and hydrophobic layers deposited. The hydrophobic layer increases the native droplet contact angle on the surface, thereby increasing the magnitude of the contact angle change that occurs due to the electrowetting effect when an electrode is actuated. Droplets are loaded onto the chip on demand via pipette, syringe pump, or pneumatic delivery systems, or in some cases can be preloaded into on-chip "reservoirs." Droplet motion is carried out by actuating electrodes sequentially, each time pulling the droplet toward the activated electrode. Electrodes can also be used for localized heating and temperature sensing enabling closed-loop temperature control, or a heater can be placed under the chip. The final product is recovered by opening the chip and recovering the crude product via pipette. These devices are usually made from indium tin oxide (ITO)-coated glass wafers using cleanroom fabrication processes to pattern the ITO electrodes, followed by deposition of a dielectric layer and finally spin coating of hydrophobic material. These materials have high chemical and thermal stability, and EWOD chips have been used to synthesize [^{18}F]FDG (Ahmadi et al., 2022; Chen et al., 2014b; Dooraghi et al., 2013; Keng et al., 2012), [^{18}F]FLT (3'-deoxy-3'-[^{18}F]-fluorothymidine) (Chen et al., 2014b; Javed et al., 2014a), [^{18}F]Fallypride (Chen et al., 2014b; Javed et al., 2014b), and [^{18}F]SFB (N-succinimidyl 4-[^{18}F]fluorobenzoate) (Chen et al., 2014b; Kim et al., 2019) (Table 2).

Glass-based EWOD technology enables sophisticated droplet manipulations to be performed on chip with a compact all-electronic control system. The chip materials are inert, thermally stable, and compatible with a broad range of reactions and reaction conditions. In addition to droplet manipulations, additional features can be directly incorporated into the chips, including electrodes for heating, or chip structures that facilitate partial on-chip purification (Ahmadi et al., 2022; Chen et al., 2012, 2014a). Due to the requirement for a complex fabrication process in a cleanroom, EWOD chips can be costly and difficult to produce. More recently, Mogi et al. fabricated EWOD devices by inkjet printing of electrodes onto paper, laminating with a hydrophobic insulating membrane, and then embossing with dimple structures to improve actuation reliability (Fig. 1b). The devices were operated in an open format for aqueous, room temperature radiolabeling of [^{99m}Tc]Tc-DTPA (Mogi et al., 2021). While potentially reducing barriers to using EWOD technology, it is not clear if paper-based devices will have the same breadth of reagent/solvent and temperature compatibility as glass-based devices. For a more comprehensive discussion of EWOD-based approaches and applications, the reader is referred to chapter "Digital Microfluidics and Radiochemistry: Current State and Future Trends."

Table 2 Performance of droplet-based radiosyntheses with comparisons to representative results using macroscale radiosynthesis modules

Tracer	Format	Synthesis module	Automation	Reaction volume (μL)	Precursor amount (nmol)	Starting activity (SA, GBq)	Synthesis time (min)	Crude RCY (%)	Activity yield (%)	Molar activity (GBq/μmol)	Ref.
[^{18}F]FDG	Droplet	EWOD (glass)	Manual	16	416	0.074 ($n = 11$)	~75	32 ± 15	14 ± 5[a]	N.R.	Keng et al. (2012)
	Droplet	EWOD (glass)	Manual	N.R.	N.R.	N.R. ($n = 2$)	50 ± 3 (excluding purification)	45 ± 10	37 ± 13 (decay-corrected)	N.R.	Chen et al. (2014)
	Droplet	EWOD (glass)	Manual	4	416	0.019 ($n = 5$)	N.R.	72 ± 13	N.R.	N.R.	Dooraghi et al. (2013)
	Droplet	Teflon-coated glass	Manual	3	167	N.R.	N.R. ($n = 5$)	N.R.	71.2 (decay-corrected)	N.R.	Koag et al. (2014a)
	Droplet	EWOD (glass)	Full	30	N.R.	0.007 (n = N.R.)	19 ± 2	67 ± 3	N.R.	N.R.	Ahmadi et al. (2022)
	Droplet	PT chip	Full	2	104	0.007 ($n = 4$)	32	50 ± 8	29 ± 5[a]	N.R.	Wang et al. (2017)
	Vial	FASTlab (Phosphate cassette)	Full	N.R.	N.R.	59–263 ($n = 40$)	22	82.9 ± 17.4	58 ± 17	N.R.	Long et al. (2013)
[^{18}F]FLT	Droplet	EWOD (glass)	Manual	5–7	180	0.059 ($n = 5$)	63	N.R.	42 ± 3[a]	1800–2400	Javed et al. (2014a)
	Droplet	EWOD (glass)	Manual	N.R.	N.R.	N.R. ($n = 6$)	40 ± 4 (excluding purification)	74 ± 7	56 ± 8 (decay-corrected)	N.R.	Chen et al. (2014)
	Droplet	Teflon-coated glass	Manual	3	162	N.R. ($n = 5$)	N.R.	N.R.	71 (decay-corrected)	4600–6100	Koag et al. (2014b)
	Vial	TracerLab MX (GE)	Full	2200	12.1	0.185 ($n = 10$)	70.5 ± 10.5	N.R.	39 ± 3[a]	N.R.	Lee et al. (2007)
[^{18}F]FNB	Droplet	Teflon-coated glass	Manual	2–8	N.R.	0.74 ($n = 12$)	~90	54 ± 14 (conversion)	N.R.	840 ± 22	Sergeev et al. (2018)
	Vial	Vial	Manual	250	13.1	N.R.	N.R.	76–85	N.R.	N.R.	Vanbrocklin et al. (2001)

[18F]SFB	Droplet	EWOD (glass)	Manual	N.R.	N.R.	N.R. ($n = 3$)	58 ± 8 (excluding purification)	34 ± 10	19 ± 8 (decay-corrected)	N.R.	Chen et al. (2014)
	Droplet	EWOD (glass)	Manual	2	168	0.039 ($n = 4$)	~120	N.R.	18 ± 3	N.R.	Kim et al. (2019)
	Vial	Vial	Manual	1000	20,000	N.R. ($n = 10$)	<60	N.R.	43.8 ± 4.6	N.R.	Tang et al. (2008)
[18F]Fallypride	Droplet	EWOD (glass)	Manual	4	308	0.29 ($n = 7$)	~60	82 ± 6[a]	45 ± 4[a]	730	Javed et al. (2014b)
	Droplet	EWOD (glass)	Manual	N.R.	N.R.	N.R. ($n = 6$)	31 ± 1 (excluding purification)	84 ± 7	65 ± 11 (decay-corrected)	N.R.	Chen et al. (2014)
	Droplet	Teflon-coated glass	Manual	2–8	N.R.	0.74 ($n = 12$)	~90	71 ± 2 (via radio-TLC)	N.R.	673 ± 15	Sergeev et al. (2018)
	Droplet	PT chip	Full	2	154	0.01–0.04 ($n = 4$)	33	64 ± 6	33 ($n = 1$)[a]	185	Wang et al. (2017)
	Droplet	Multi-STT chip with [18F]fluoride concentrator	Full	8	616	Low SA: 0.01–0.17 ($n = 6$); Scale-up: 3.7–41 ($n = 11$)	50	Low SA: 72 ± 8 Scale-up: 65–25	18[a] ($n = 1$, for SA of 41 GBq)	81–270	Wang et al. (2020a)
	Droplet	STT chip	Full	6	231	0.02 ($n = 9$)	30	96 ± 3	65[a] ($n = 1$)	N.R.	Wang et al. (2019)
	Droplet	Multi-STT chip (3 mm diameter site)	Manual	6	234	0.01 ($n = 2$)	~35	93 ± 5	59[a] ($n = 1$)	N.R.	Rios et al. (2022)
	Droplet	Multi-STT chip (3 mm diameter site) (robotic automation)	Full (no purification)	6	230	0.004 ($n = 4$)	N.R.	92 ± 1	N.R.	N.R.	Jones et al. (2023)
	Vial	TracerLab FX_{FN} (GE)	Full	1000	3872	8.1–52 ($n = 42$)	51 ± 1	N.R.	49 ± 1[a]	140–192	Seok Moon et al. (2010)

(continued)

Table 2 (continued)

Tracer	Format	Synthesis module	Automation	Reaction volume (μL)	Precursor amount (nmol)	Starting activity (SA, GBq)	Synthesis time (min)	Crude RCY (%)	Activity yield (%)	Molar activity (GBq/μmol)	Ref.
3-Formyl-2,4,6-trimethylbenzene-sulfonyl [^{18}F]fluoride	Droplet	Magnetic	Full	60	3600	0.074–0.222 ($n = 3$)	10–15	N.R.	65–68[a]	N.R.	Fiel et al. (2015)
	Vial	Vial	Manual	100	6000	0.38–0.65 ($n = 4$)	20–25	N.R.	73 ± 7	105 ± 2	Inkster et al. (2012)
[^{18}F]AMBF$_3$-TATE	Droplet	Teflon-coated glass	Manual	5	5	Up to 3.7 ($n = 5$)	~40	17 ± 3 ($n = 15$)	12.4 ± 1.0[a] ($n = 5$)	~110 (for SA = 0.93–1.1 GBq)	Lisova et al. (2018)
	Vial	Trasis AllInOne module	Full	394	200	~150[a] ($n = 3$)	60	N.R.	10.3 ± 2.7[a]	435 ± 162	Lau et al. (2020)
[^{18}F]FET	Droplet	PT chip	Full (no purification)	10	60	0.04–0.74 ($n = 5$)	19	54 ± 6	N.R.	N.R.	Lisova et al. (2019)
	Droplet	STT chip	Manual	10	60	0.4 ± 0.1 ($n = 4$)	40	59 ± 7	43 ± 5[a]	48–119	Lisova et al. (2019)
	Droplet	STT chip	Semi	10	60	2.7 ± 0.4 ($n = 3$)	60	70 ± 9 (for SA < 0.02 GBq)	28 ± 14	420 ± 50	Lisova et al. (2021)
	Vial	TracerLab FX-FDG double (GE)	Full	1000	7400	1.5–21.7 ($n = 13$)	36	N.R.	43–52[a]	280–528 (for 4.6–10.8 GBq product)	Bogni et al. (2019)
[^{18}F]Florbetaben ([^{18}F]FBB)	Droplet	STT chip	Full	10	80	0.27 ± 0.1 ($n = 3$)	55	60 ± 5	35 ± 2[a]	338 ± 55	Lisova et al. (2020)
	Droplet	STT chip	Semi	10	80	3.2 ± 0.8 ($n = 6$)	60	~50	15 ± 4	480 ± 190	Lisova et al. (2021)
	Vial	ELIXYS FLEX/CHEM (Sofie)	Full	1800	13,000	15 ($n = 3$)	44	N.R.	45 ± 7[a]	N.R.	Collins et al. (2017)

[18F]FDOPA (iodonium salt)	Droplet	STT chip	Full	10	120	0.37 ($n = 3$)	31	5.2 ± 0.7	3.9 ± 1.8	>75	Wang et al. (2020a)
	Vial	ELIXYS FLEX/CHEM (Sofie)	Full	1000	16,800	15 ($n = 3$)	71	N.R.	2.9 ± 0.8[a]	30–33	Collins et al. (2017)
	Vial	Vial-based synthesizer	Manual	750	13,400	N.R.	117	N.R.	6.7 ± 1.9[a]	35 ± 4	Kuik et al. (2015)
[18F]FDOPA (Cu-mediated)	Droplet	Multi-STT chip (4 mm diameter site)	Manual	10	450	0.06–0.15 ($n = 7$)	25	60 ± 3	26 ± 3	N.R.	Lu and van Dam (2021a)
	Droplet	STT chip	Full	10	450	0.14–0.22 ($n = 2$)	25	N.R.	22 ± 2	>198	Lu and van Dam (2021b)
	Vial	TracerLab FXFN (GE)	Full	1000	4000	66.6 ($n = 3$)	110	N.R.	6 ± 1	140 ± 77	Mossine et al. (2019)
[18F]PBR06	Droplet	Multi-STT chip (3 mm diameter site)	Manual	8	160	0.014 ($n = 4$)	~35	94 ± 2	61[a] ($n = 1$)	N.R.	Rios et al. (2022)
	Vial	Custom module	Full	1000	1800	7.4–18.5 (n = N.R.)	50–60	N.R.	22–44[a]	37–222	Wang et al. (2011)
[18F]flumazenil ([18F]FMZ)	Droplet	Multi-STT chip (3 mm diameter site)	Manual	8	280	0.014 ($n = 4$)	~35	19.1 ± 0.6	9.3[a] ($n = 1$)	N.R.	Rios et al. (2022)
	Vial	Custom module	Semi	500–1000	11,900–25,000	N.R.	75–80	N.R.	18–19[a]	>185	Ryzhikov et al. (2005)
[18F]FEPPA	Droplet	Multi-STT chip (3 mm diameter site)	Manual	8	240	0.44 ($n = 1$)	30	N.R.	53	N.R.	Rios et al. (2022)
	Vial	Nuclear interface synthesizer (GE)	Full	500	12,000	9–13[a] ($n = 15$)	30	N.R.	38 ± 3	241 ± 13	Berroterán-Infante et al. (2018)

(continued)

Table 2 (continued)

Tracer	Format	Synthesis module	Automation	Reaction volume (μL)	Precursor amount (nmol)	Starting activity (SA, GBq)	Synthesis time (min)	Crude RCY (%)	Activity yield (%)	Molar activity (GBq/μmol)	Ref.
[^{18}F]FBnTP (Cu-mediated)	Droplet	Multi-STT chip (3 mm diameter site)	Manual	10	450	0.12 ($n = 3$)	42 ± 1	N.R.	49 ± 3	N.R.	Jones et al. (2023)
	Droplet	Multi-STT chip (4 mm diameter site)	Manual	25–35	1200–2400	0.9–3.7 ($n = 7$)	37–47	N.R.	43–49	339–877	Lu et al. (2024a)
	Vial	Vial-based synthesizer	Manual	850	4000	N.R. ($n = 2$)	N.R.	62 ± 1.4 (conversion)	N.R.	N.R.	Zhang et al. (2016)
	Vial	GE Tracerlab FX_N	Automated	N.R.	15,900	11.1–55.5 ($n = 7$)	60 ± 5	N.R.	6.2 ± 1.0	6.4 ± 1.0	Lin et al. (2023)
[^{18}F]YH149 (Cu-mediated)	Droplet	Multi-STT chip (4 mm diameter site)	Manual	10	280	0.2–1.45 ($n = 4$)	26	56 ± 3	43 ± 5[a]	77–854	Lu et al. (2023a)
	Vial	Vial-based synthesizer	Full	300	3700–5500	50–60 ($n = 5$)	75	N.R.	3.6–5.0[a]	100–308	He et al. (2022)
[^{18}F]AlF-NODA-Tz	Droplet	Multi-STT chip (4 mm diameter site)	Manual	15	5	0.02–0.07 ($n = 2$)	N.R.	86 ± 5	N.R.	N.R.	Lu et al. (2024b)
	Vial	Trasis AllInOne (Trasis)	Full	300–380	60	1 ($n = 3$)	26	N.R.	41 ± 1[a]	11.4 ± 2.0	Allott et al. (2017)
[^{18}F]AlF-FAPI-74	Droplet	Multi-STT chip (4 mm diameter site)	Manual	15	5	1.1–1.3 ($n = 4$)	16 ± 2	N.R.	72 ± 3	152–186	Lu et al. (2024b)
	Vial	Cassette-type CFN-MPS200	Full	630	80	32–40 ($n = 10$)	31	N.R.	37.0 ± 4.3	220 ± 45	Naka et al. (2021)

[18F]FPEB	Droplet	STT chip	Manual	10	200	0.037–0.055 ($n = 2$)	N.R.	~16	N.R.	N.R.	Laferriere-Holloway et al. (2023)
	Vial	Anatech robot system	Semi	500	8000	N.R. (n = N.R.)	120	N.R.	5	70.3 ± 7.4	Wang et al. (2007)
	Vial	GE TracerLab FX_{FN}	Full	1500	4010	37–111 ($n = 114$)	~90	N.R.	1.5	197 ± 91	Lim et al. (2014)
[99mTc]Tc-DTPA	Droplet	EWOD (paper)	Semi	20	1270	N.R. ($n = 3$)	5	99.7 ± 0.13	N.R.	N.R.	Mogi et al. (2021)
	Vial	Kit	Semi	2000–10,000	52,900	Up to 11.1 (n = N.R.)	15–30	≥95	N.R.	N.R.	"TechneScan® DTPA Kit," (n.d.)
[18F]fluorination of (2-((2-carboxyethyl) di-phenylphosphonio) phenyl) trifluoroborate	Droplet	Teflon-coated glass	Manual	10	5	0.2–1.0 ($n = 20$)	40	20 ± 8	N.R.	78 ± 30 ($n = 12$)	Lisova et al. (2017)
	Droplet	STT chip	Manual	10	5	0.07 ($n = 4$)	N.R.	29 ± 4	N.R.	N.R.	Lisova et al. (2021)

[a] These values were not directly reported but calculated based on information in the literature

2.3 *Teflon-Coated Glass*

During the development of EWOD-based microfluidics, a number of experiments were performed using simplified chips lacking electrodes and dielectric layer, comprising simply Teflon-coated glass (Fig. 1c). Such devices have the same thermal properties as glass-based EWOD chips, and same chemical properties (i.e., they have the same material—Teflon AF—in contact with the reagents), but are simpler and less expensive to fabricate.

Primarily, these chips have been used for initial optimization studies with protocols that were later implemented on EWOD chips for automated radiotracer production (Javed et al., 2014a, 2014b; Kim et al., 2019).

Under manual operation, these chips can also be used for production. Koag et al. used these chips to optimize the droplet-based synthesis of [^{18}F]FDG, in particular to explore the effect of using bulky alcohols in the fluorination step (Koag et al., 2014a). The optimal conditions—using 2,3-dimethyl-2-butanol (thexyl alcohol) with DMSO during radiofluorination—were then combined with hydrolysis and purification via a miniature cartridge to produce [^{18}F]FDG with radiochemical yield (RCY) of 71.2% ($n = 5$) in amounts up to 0.21 GBq. The overall synthesis time was not reported, but radiochemical purity (RCP) was >99%, and batches passed clinical QC testing. Using a similar approach, the same group developed an optimized synthesis of [^{18}F]FLT. The synthesis, including hydrolysis and micro-cartridge purification, exhibited an RCY of 71.3% ($n = 5$), RCP >99%, and quantities up to 0.20 GBq (Koag et al., 2014b). The batches exhibited very high molar activity (4600–6100 GBq/μmol) and passed clinical QC testing. Both results exhibited considerable improvement in terms of RCY and product quantity compared to studies on EWOD chips, and the performance compared favorably to conventional synthesis approaches.

Lisova et al. (2018) performed production of [^{18}F]AMBF$_3$-TATE (a radioligand for SSTR2 receptors) for preclinical imaging using Teflon-coated glass chips. The synthesis proceeds via a ^{19}F–^{18}F isotopic exchange mechanism, with chemically identical precursor and labeled product, and thus relies on minimization of the amount of precursor to ensure high molar activity. The droplet-based protocol required only 5 nmol of precursor in a 5 μL reaction volume, achieving an RCY of 16 ± 1% ($n = 5$) and an RCP of >99% in a ~40 min synthesis time. RCY was comparable to that (15 ± 4%, $n = 3$) of the automated conventional synthesis reported by Lau et al. (2020), but the droplet synthesis used 40× less precursor, and achieved high molar activity (110 GBq/μmol) with only ~1 GBq of starting activity compared to ~150 GBq needed in the vial-based reaction to achieve high molar activity (435 ± 162 GBq/μmol). Preparation of this compound enabled a preclinical imaging comparison to [^{68}Ga]Ga-DOTATATE and yielded dosimetry data.

Teflon-coated glass chips have also been used to study fundamental properties of droplet-based reactions. Sergeev et al. (2018) studied the impact of several reaction parameters on molar activity using the syntheses of [^{18}F]Fallypride and 1-[^{18}F]fluoro-4-nitrobenzene ([^{18}F]FNB) as model systems. Through investigations of starting activity (using different volumes of [^{18}F]F$^-$/[^{18}O]H_2O) and fluorination reaction volume, it was observed that droplet reactions had consistently high molar activity (estimated to be close to the molar activity of the [^{18}F]F$^-$ from the cyclotron), while vial-based reactions could only achieve high molar activities when using high starting activities, or exceptionally small reaction volumes. It is believed that contamination with stable fluoride ([^{19}F]F$^-$) is dramatically lower in droplet-based reactions. An implication of this work is that when producing small amount of product, droplet reactions can efficiently achieve high molar activity without needing to use high amounts of starting activity as needed in conventional vial-based reactions. With high molar activity [^{18}F]Fallypride produced using the setup, a study was conducted in mice to compare the influence on imaging of different molar activity levels.

The key limitation of these chips is the lack of droplet actuation mechanisms which necessitates manual handling of droplets, resulting in some radiation exposure to the operator and increasing the possibility of human error. Furthermore, the lack of droplet "containment" can sometimes cause problems, if the apparatus is not level or if there are any surface defects on the chip. For example, while evaporating the droplet of [^{18}F]F$^-$ and phase-transfer catalyst/base, the droplet can move around on the surface such that the dried residue is distributed across the surface in non-predictable ways, making it difficult to ensure the precursor droplet for the subsequent fluorination fully mixes with the radioisotope.

2.4 Magnetic Droplet Platform

An alternative method for conducting droplet-based radiosynthesis involves using magnetic actuation of droplets atop a small piece of hydrophobic Teflon sheet (Fig. 1d) (Fiel et al., 2015).

The one-step synthesis of an arylsulfonyl [^{18}F]fluoride prosthetic group (for peptide labeling) was reported using the magnetic droplet platform using the procedure illustrated in Fig. 2. Initially, a volume of [^{18}F]F$^-$/[^{18}O]H_2O solution (up to 1 mL) and reagent droplets (~50 μL) were manually deposited onto the Teflon sheet. Subsequently, the introduction of quaternary methyl ammonium (QMA)-functionalized magnetic beads provided a means to manipulate the droplets using a robotically actuated magnet positioned beneath the Teflon sheet. The magnetic particles captured a portion of liquid from the activity droplet and trapped the [^{18}F]F$^-$ on the beads, facilitating solvent exchange by movement of the beads and droplet to a small capillary located at one corner of the sheet where suction was applied to remove the solution. Repeating this process increases the amount of activity trapped on the beads and provides an elegant way to scale up the reaction if desired.

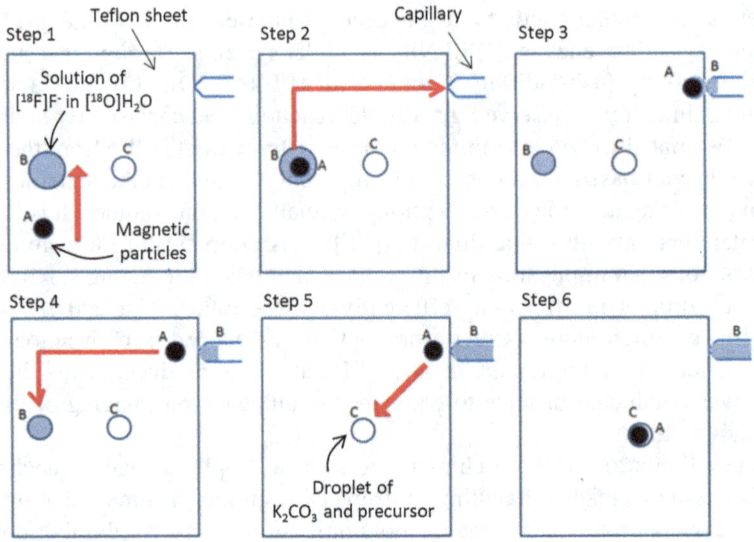

Fig. 2 Steps for radiosynthesis on the magnetic droplet platform. (Step 1) The magnetic particles (A) were moved toward the $[^{18}O]H_2O$ containing $[^{18}F]F^-$. The particles were looped in a circular motion within the solution to induce stirring. (Step 2) The particles, carrying a droplet of $[^{18}O]H_2O$ (~50 μL), were moved toward the capillary tube. (Step 3) The liquid within the $[^{18}O]H_2O$ droplet was collected by the capillary tube under suction using a syringe pump, leaving $[^{18}F]F^-$ bound to the QMA-functionalized magnetic beads on the Teflon sheet. (Step 4) The particles were moved back to the $[^{18}F]F^-/[^{18}O]H_2O$ solution to increase the activity trapped on the beads. Steps 2–4 were repeated continuously until $[^{18}F]F^-/[^{18}O]H_2O$ was completely consumed. (Step 5) The particles were then moved toward the mixed K_2CO_3 and precursor solution (C; ~100 μL) to release $[^{18}F]F^-$ ions from the magnetic particles and perform the fluorination reaction. The particles were again looped in a circular motion within the solution to induce stirring. (6) The crude product was collected via syringe for downstream purification and formulation. (Adapted with permission from Fiel et al. (2015), copyright 2015 American Chemical Society)

Following the full solvent exchange process, the beads with trapped $[^{18}F]F^-$ were mixed with eluent and precursor to initiate the reaction (aqueous fluorination), after which the product was manually collected using a syringe. The droplet-based synthesis performed similarly to a previous conventional synthesis (RCY of 72 ± 1%, $n = 3$ compared to 73 ± 7%, $n = 3$), but was completed in shorter time (10–15 min compared to 20–25 min) and used less precursor (Fiel et al., 2015; Inkster et al., 2012).

A limitation of this platform is that it lacks a method for applying heat, which greatly limits the range of reactions that can be successfully implemented. Furthermore, it is not clear if the magnetic beads can be used to reliably manipulate droplets composed of organic solvents rather than water. The substrate material (Teflon) may also adversely affect the molar activity for ^{18}F-labeled compounds due to radiolytic release of stable fluoride ions (Berridge et al., 2009).

2.5 *Passive Transport (PT) Chip*

To overcome the limitation of manual operation associated with Teflon-coated glass devices, Wang et al. (2017) developed a chip that enabled transport of droplets from loading sites to a reaction site without the need for electrodes (or dielectric layer) that complicate the fabrication of EWOD chips. The spontaneous droplet transport mechanism described by Ghosh et al. (2014), using tapered hydrophilic pathways on a hydrophobic surface as "channels," was adapted to create a device for radiochemical reactions. The chips were made of silicon, providing higher thermal conductivity than glass and thus faster and more accurate heating. A hydrophobic Teflon AF coating was applied and then patterned by deep reactive ion etching to create a hydrophilic circular reaction site (diameter 3 mm) surrounded by six inward-leading tapered hydrophilic pathways for reagent transport (Fig. 1e). When a small droplet is dispensed at one of the narrow ends of the delivery channels, it spontaneously moves toward the wider end (i.e., the reaction site). It was found that solvents commonly used in droplet reactions (including water, methanol, acetonitrile, and dimethyl sulfoxide) could move spontaneously from the reagent loading site to the reaction site when the taper angle was >4°. To operate the chip in an automated fashion, Wang et al. created a radiosynthesizer prototype, controlled by a LabView program, that includes subsystems for temperature control, reagent dispensing, and a product collection (Fig. 3a–c). The chip sits atop a ceramic heater and fan connected to a closed-loop temperature controller. Reagents (including [^{18}F]fluoride and precursor stock solutions) are loaded into V-vials connected to miniature, non-contact dispensers that can generate droplets ranging from 100s of nL to several μL by adjusting the valve opening time and pressure level. The system was designed to hold up to six dispensers, each positioned above one of the six loading sites of the passive transport chip. In the center of the same fixture, a retractable tube was installed that can be lowered close to the surface (~0.5 mm) to dip into the reaction droplet and transfer the crude product to a collection vial by applying a vacuum to the collection vial, or can be moved away from the surface (~50 mm) when not in use. Figure 3d–f illustrates how the setup is used for the synthesis of [^{18}F]Fallypride.

Leveraging the automated operation, this platform was used for preparing several radiopharmaceuticals. For example, Wang et al. (2017) demonstrated the production of [^{18}F]FDG (a radiotracer for imaging glucose metabolism) and [^{18}F]Fallypride (a tracer for imaging dopamine D2/D3 receptors) with isolated RCYs of 36 ± 6% ($n = 4$) and 46 ± 4% ($n = 4$), respectively. In addition to demonstrating feasibility, the results highlighted the minimal reagent use that is possible with droplet-based synthesis due to the tiny reagent volume. For [^{18}F]FDG, only 104 nmol of precursor was used in the 2 μL reaction, a ~400× reduction compared to a macroscale synthesis, which used 41,300 nmol in a 700 μL reaction volume (Lazari et al., 2014). Another important finding in this work was confirmation that droplet-based reactions can achieve high molar activity even starting with low initial starting activity. For starting activity of only 23.4 MBq, a molar activity of 185 GBq/μmol was reported for [^{18}F]Fallypride. The synthesis of the amino acid

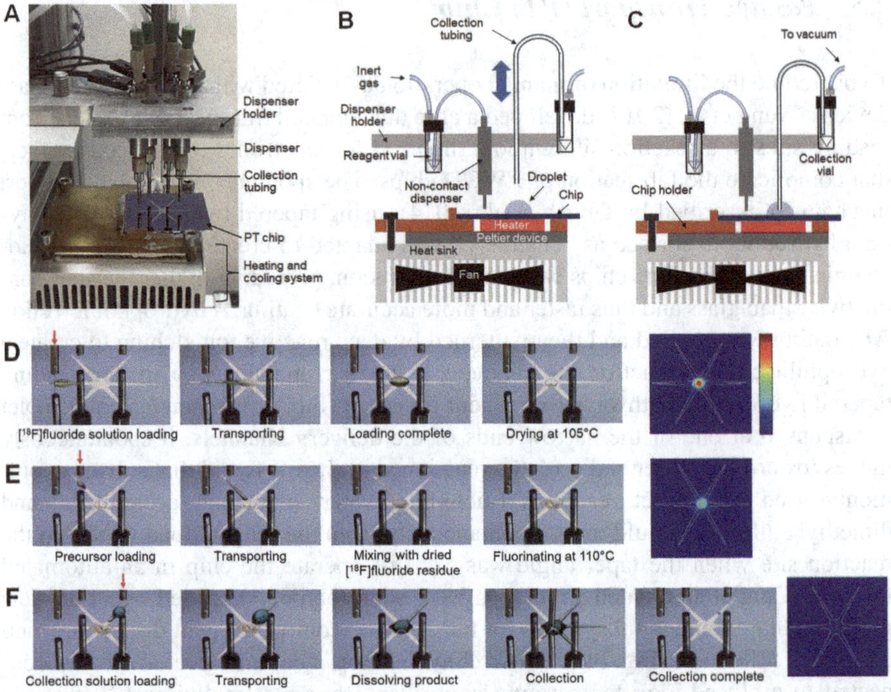

Fig. 3 Automated droplet radiochemistry platform based on passive transport. (**a**) Photograph of the PT chip-based radiosynthesis platform. (**b**) Schematic of apparatus comprising systems for reagent dispensing, chip temperature control, and product collection. (**c**) Schematic of setup during product collection, that is, with collection tubing lowered into the droplet. (**d–f**) Sequence of photographs of the microfluidic chip during the synthesis of [^{18}F]Fallypride. (**d**) An aqueous droplet of [^{18}F]TBAF (2 μL, dyed yellow) was loaded, spontaneously transported to the reaction site, and then the chip was heated to 105 °C to remove the solvent. (**e**) Next, two droplets of precursor dissolved in a 1:1 v/v mixture of MeCN and thexyl alcohol (1 μL, dyed red) were loaded from a separate inlet and transported to the reaction site in sequence, after which the droplet was heated to 110 °C to perform the fluorination reaction. (**f**) Next, two droplets of collection solution (9:1 v/v MeOH/water) (5 μL each, dyed blue) were loaded from a third inlet and transported to the center to dilute the reaction mixture. Finally, the collection tubing was lowered, and the droplet was collected into a vial with the aid of vacuum. Red arrows indicate which dispenser is active during liquid loading steps. Images at the right of each row are Cerenkov luminescence images showing the distribution of activity on the chip after performing the steps depicted in each row, with boundary of the hydrophilic region outlined in white. (Adapted from Wang et al. (2017) with permission from the Royal Society of Chemistry)

tracer [^{18}F]FET (via fluorination and deprotection) was also demonstrated on this platform (Lisova et al., 2019). High crude RCY of 54 ± 6% (n = 5) was achieved through on-chip synthesis in just 19 min, employing a minimal 60 nmol of precursor—significantly lower than the 7400 nmol used in vial-based approach (Bogni et al., 2019).

While the PT platform significantly reduced chip complexity and cost and better defined the location of the reaction site (compared to Teflon-coated glass chips), the detailed behavior of the droplet on the surface was very sensitive to the solvent type, temperature, and loaded volume. In some cases, there was unwanted spreading of the solution in the backward direction along the tapered reagent pathways on the chip, which could adversely affect synthesis performance and/or lead to inconsistent results (Wang et al., 2019). This sensitivity complicated the development of new synthesis protocols on this platform, necessitating significant expenditure of time and effort to optimize reagent volumes, solvents, loading protocols (timing), and other factors to achieve good performance. The low reaction volume (~2 μL) was also limiting in the amount of radiotracer product that could be produced. However, Wang and Chao et al. later included a mechanism for increasing the activity scale and loaded up to 41 GBq onto the chip (Wang et al., 2020a) by integrating a custom-built micro-cartridge-based concentrator, which was subsequently used to synthesize larger amounts of [^{18}F]Fallypride (see Sect. 3 for more details).

2.6 Surface-Tension Trap (STT) Chip

In a follow-up development, Wang et al. later simplified the chip design by removing the inward-leading tapered hydrophilic pathways, leaving only a circular hydrophilic reaction site, which acts as a surface-tension trap (STT) (Wang et al., 2019). The new design provided reliable and predictable confinement of droplets such that a wide range of solvent types and volumes can be loaded to the reaction site without overflowing the reaction site boundary. Furthermore, with the reaction site diameter increased from 3 to 4 mm, it was possible to load larger droplets (up to 30 μL of aqueous liquids), which assists with loading larger quantities of the isotope for scaling up reactions (see Sect. 3).

However, without any on-chip means to transport droplets to the reaction site, an external system was required to position different reagent dispensers directly over the reaction site. This was accomplished by incorporating a rotation stage beneath the fan and heatsink (Fig. 4a, b). With the chip aligned such that the reaction site is away from the center of rotation, the reaction site can be moved around a circular arc and positioned under any of 7 different reagent dispensers to add reagents on demand (Fig. 4c). The fixture to which all dispensers were connected was mounted on a vertical slide, so it could be raised above the surface to facilitate chip installation and removal, as well as dispenser cleaning. To transfer the crude product from the reaction site to the collection vial, a thin rigid tubing was installed in the dispenser fixture, with its end situated approximately 0.5 mm above the chip surface. When the crude product droplet is rotated underneath the collection tube, vacuum is applied to transfer it to the collection vial. All of these subsystems, including the rotary stage, temperature controller, reagent dispensers, movable fixture slide, and product collection system, were controlled through a LabView-based program. Figure 4d illustrates how the STT-chip system is used to synthesize [^{18}F]Fallypride.

Significant radiosynthesis improvement was observed when transitioning from the PT chip to the STT chip by avoiding the unwanted migration of liquid away

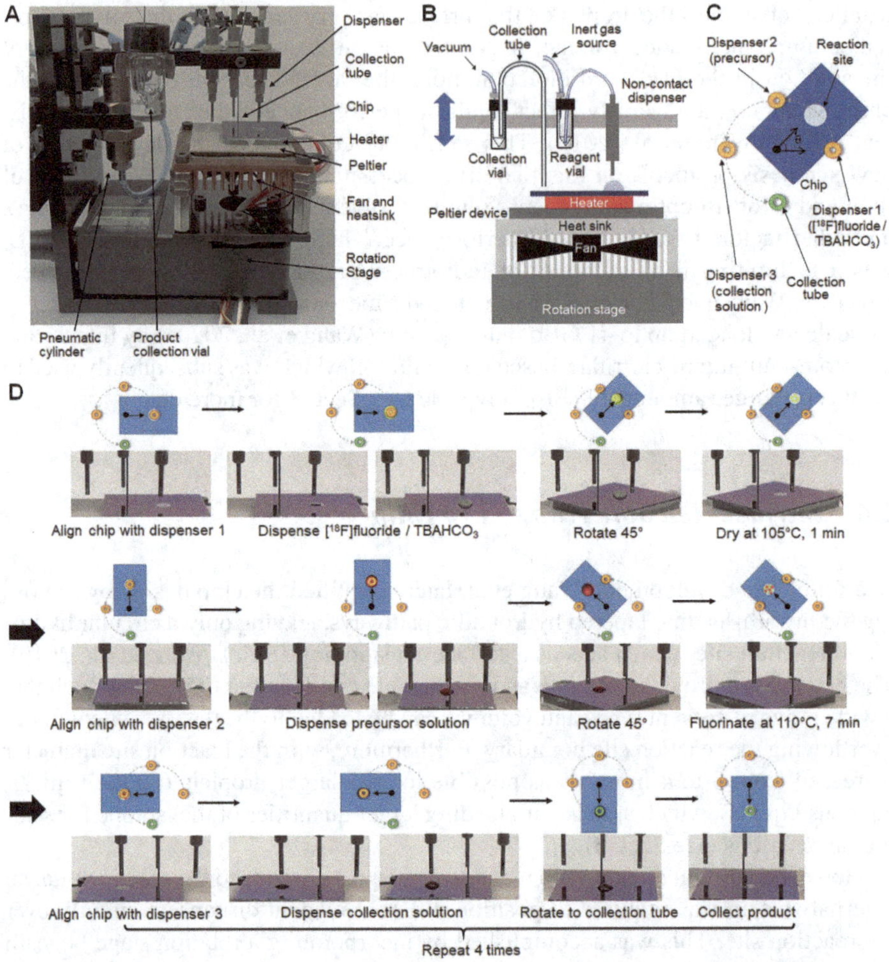

Fig. 4 Automation of surface-tension trap (STT) droplet radiochemistry. (**a**) Photograph of the STT-chip-based radiosynthesis platform. (**b**) Schematic of automated apparatus, comprising systems for reagent dispensing, crude product collection, chip temperature control, and rotation mechanism. (**c**) Top view schematic of the (movable) microfluidic chip and (fixed) locations of reagent dispensers and the collection tube. The angle marker shows the center of rotation. (**d**) Schematic (showing chip orientation) and photograph of the chip for each step of the [^{18}F]Fallypride synthesis process. First, the chip was rotated under the [^{18}F]fluoride/TBAHCO$_3$ dispenser, and a 10 μL droplet is loaded (dyed yellow). After rotation of 45°, the chip was heated to 105 °C to evaporate water and leave a dried residue of [^{18}F]TBAF. Next, the reaction site was rotated under the precursor dispenser and 10 μL of precursor dissolved in 1:1 v/v MeCN/thexyl alcohol (dyed red) was loaded. After a 45° rotation, the chip was heated to 110 °C for radiofluorination. Next, the reaction site was rotated under the collection solution dispenser, and 20 μL of 9:1 v/v MeOH/water (dyed blue) was loaded to dilute the reaction mixture. Finally, the reaction site was rotated under the collection tube, and the droplet was collected into the product vial using vacuum. The collection solution loading and collecting were repeated a total of four times to minimize the residue left behind at the reaction site. (Adapted from Wang et al. (2019) with permission from the Royal Society of Chemistry)

from the reaction site and simplifying the optimization process. For [^{18}F]Fallypride, crude RCY improved to 96 ± 3% (n = 9) on the STT platform in a slightly shorter preparation time of 30 min (Wang et al., 2019), compared to 64 ± 6% (n = 4) in 33 min using the PT chip (Wang et al., 2017). Similarly, performance of [^{18}F]FET synthesis on the STT chip had considerably higher crude RCY than on the PT reactor (70 ± 9%, n = 9 vs. 54 ± 6%, n = 5) (Lisova et al., 2019). Compared to EWOD chips, the open nature of the STT chip enables increased automation, because reagent and isotope addition and product collection can be fully automated. Furthermore, synthesis times are reduced due to the rapid evaporation of the uncovered droplet during the [^{18}F]fluoride drying step.

Full production processes (including purification) for multiple radiopharmaceuticals were further demonstrated on the STT platform, such as [^{18}F]Fallypride, [^{18}F]FET, [^{18}F]Florbetaben (for imaging β-amyloid deposits), and [^{18}F]FDOPA (for imaging dopaminergic function). Typically, compared to conventional vial-bases syntheses, activity yields achieved using the STT platform were comparable or higher than those obtained using conventional radiosynthesis modules, and synthesis processes were significantly shorter (Table 2). For example, with starting activity of 0.4 ± 0.1 GBq (n = 4), [^{18}F]FET was prepared using an automated STT setup in only 40 min with a high activity yield of 43 ± 5% (n = 4) and high molar activity (48–119 GBq/μmol), which were comparable to the conventional synthesis reported by Bogni et al. (2019), with activity yield of 43–52% (n = 13) and molar activity of 280–528 GBq/μmol (using a much higher starting activity of 4.6–10.8 GBq). Lisova and colleagues used this platform to prepare [^{18}F]FET with different molar activities to determine the impact on in vivo imaging (Lisova et al., 2019). With similar starting activity, [^{18}F]Florbetaben produced on the automated STT setup exhibited high activity yield (35 ± 2%, n = 3) and excellent molar activity (338 ± 55 GBq/μmol) (Lisova et al., 2020).

Similarly, preparation of [^{18}F]FDOPA from the iodonium salt precursor also showed improved performance compared to vial-based implementations (Wang et al., 2020b). Starting with 140× less precursor, the STT method provided a faster synthesis (31 vs. 71 min) and higher activity yield (3.9 ± 1.8%, n = 3 vs. 2.9 ± 0.8%, n = 3) compared to the automated vial-based approach reported by Collins et al. (2017). Although the activity yield was somewhat lower than the value (6.7 ± 1.9%) reported in an earlier manual macroscale synthesis by Kuik et al. (2015), the STT droplet method had the benefits of automation, reduced precursor consumption (110× less), shorter synthesis time (31 vs. 117 min), and significantly higher molar activity.

In a later study, Rios et al. (2022) successfully translated the radiosyntheses of multiple tracers from vial-based methods into droplet reactions using the STT platform, achieving improvements in synthesis performance (Rios et al., 2022). For example, [^{18}F]PBR06, a second-generation translocator protein (TSPO) radioligand, could be synthesized in a shorter time (~35 vs. 50–60 min) and with higher activity yield (61% (n = 1) vs. 22–44%) than the automated macroscale synthesis protocol (Wang et al., 2011). Good performance was also found in the droplet-based production of [^{18}F]FEPPA (a TSPO radioligand), which exhibited an activity yield of 53% (n = 1). Although there was no significant improvement in yield for the preparation of [^{18}F]Flumazenil (for imaging $GABA_A$ receptors), the total preparation time was significantly reduced (~35 vs. 75–80 min) compared to the conventional method (Ryzhikov et al., 2005). In another report, Laferriere-Holloway et al. (2023) prepared [^{18}F]FPEB (for quantifying metabotropic glutamate 5 receptors) from the nitro precursor using the STT platform, achieving a crude RCY of ~16% (n = 2). Although a full synthesis and RCY were not reported, the performance compares favorably to the low RCYs typically observed for this synthesis (Lim et al., 2014; Wang et al., 2007).

More recently, we have successfully extended the droplet-based technique to metal-mediated radiochemistry, such as copper-mediated radiofluorination (Lu et al., 2023a). As an illustrative example, the optimized droplet synthesis of [^{18}F]FDOPA, prepared via the Cu-mediated route from an organoboron precursor, exhibited a high crude RCY of 60 ± 3% (n = 7) when performed manually. In preliminary tests of automation on the STT platform, the activity yield was 22 ± 2% (n = 2), and synthesis time was just 25 min (Lu & van Dam, 2021a, 2021b), offering significant advantages compared to the reported macroscale radiosynthesis with activity yield of 6 ± 1% (n = 3) and synthesis time of 110 min (Mossine et al., 2019). To demonstrate the versatility of the droplet-based technique for other Cu-mediated radiosyntheses, we employed the STT method for the preparation of [^{18}F]FBnTP (a preclinical PET probe for myocardial perfusion imaging) and [^{18}F]YH149 (a novel tracer targeting monoacylglycerol lipase (MAGL)), both via Cu-mediated routes with corresponding bis(pinacolato)diboron (Bpin) precursors. For example, for the droplet-based synthesis of [^{18}F]YH149 (Lu et al., 2023a), a remarkable 9–12× improvement in activity yield was observed, with 13–20× less precursor required and 3× shorter preparation time than the vial-based synthesis (He et al., 2022).

We also demonstrated for the first time the feasibility of performing [^{18}F]AlF radiochemistry in a microdroplet reactor, using a tetrazine analog (NODA-Tz) as a model compound. The resulting crude RCY of [^{18}F]AlF-NODA-Tz was high (89 ± 2%, n = 2), in a synthesis time of only 5 min and using 30 nmol precursor. Even when employing a significantly reduced precursor quantity (6× less), the droplet-based approach still provided comparable synthesis results, yielding crude RCY of 86 ± 5% (n = 2) (Lu et al., 2023b). Our investigation extended to the clinically relevant PET probe [^{18}F]AlF-FAPI-74, targeting fibroblast activation protein (FAP), which is known to be overexpressed in the stroma of various cancer types. The droplet-based synthesis had an activity yield of 72 ± 3% (n = 4) with a total preparation time of 16 ± 2 min (n = 4) and used only 5 nmol of precursor (Lu et al., 2024b). The molar activity, starting with only 1.1–1.3 GBq of activity, was 152–186 GBq/μmol. In contrast, the macroscale method had much lower activity yield (37.0 ± 4.3%, n = 10), longer preparation time (31 min), required 16× more precursor, and required significantly higher starting activity (32–40 GBq) to obtain a comparable molar activity (Nakaoka et al., 2020). The low precursor amount is the critical advantage of droplet radiochemistry that enables high molar activity to be achieved even with only moderate starting activities, providing a more efficient way to produce smaller batches, such as those for preclinical imaging, where high molar activity is even more important than for clinical imaging (Hume et al., 1998).

2.7 *Multi-STT Reaction Chip*

To enable quicker translation of vial-based synthesis protocols into droplet reactions, our group developed an approach for performing multiple STT reactions in parallel by creating multiple reaction sites on the same chip. Multiple designs of these "multi-STT" reaction chip were developed (Fig. 1g), featuring different numbers of reaction sites. By enabling reactions to be performed simultaneously instead of sequentially, these array-based chips provide a means to greatly accelerate reaction optimization as described more thoroughly in chapter "High-Throughput Techniques in Radiochemistry." Such chips were used to optimize many of the syntheses reported in the previous section. For example, 2 × 2 arrays of 4-mm diameter reaction sites have been used to optimize the syntheses of [^{18}F]Fallypride (Rios et al., 2019) and [^{18}F]FDOPA (Cu-mediated) (Lu & van Dam, 2021a). Higher-throughput chips, with 3 x 3 arrays of 4-mm diameter reaction sites, have been used for optimization of [^{18}F]YH149 (Lu et al., 2023a) and [^{18}F]AlF-NODA-Tz (Lu et al., 2024b). And 4 x 4 arrays of 3-mm diameter reaction sites have been used for optimization of [^{18}F]Fallypride (Jones et al., 2023; Rios et al., 2022), [^{18}F]PBR06 (Rios et al., 2022), [^{18}F]Flumazenil (Rios et al., 2022), [^{18}F]FEPPA (Rios et al., 2022), and [^{18}F]FBnTP (Cu-mediated) (Jones et al., 2023). To further boost synthesis throughput, Rios et al. (2022) developed a high-throughput system with an array of four heaters, which allows up to 64 reactions to be run in parallel using four 4 × 4 multi-STT chips, and Jones et al. (2023) developed a robotic system to automate optimization studies. Both platforms are discussed in further detail in chapter "High-Throughput Techniques in Radiochemistry."

The ability to perform hundreds of reactions per week allows optimizations to consider many more reaction conditions than are normally practical. Taking [^{18}F]Flumazenil as an example, a broad range of synthesis parameters, including amount of base and precursor, reaction temperature and time, types of reaction solvent, base, and phase-transfer catalyst, could be manually explored within a few days using four 4 × 4 multi-STT chips (Rios et al., 2022) leveraging multilane radio-TLC via Cerenkov imaging as a high-throughput read-out method of reaction performance (Laferriere-Holloway et al., 2023; Wang et al., 2020c). Given that droplet-based optimization significantly reduces precursor and time consumption for each data point compared to vial-based reactions, it was possible to explore more imaginative conditions at a lower cost. For instance, Rios et al. experimented with extremely high radiolabeling temperatures, reaching up to 240 °C, in combination with various solvents with high boiling points (e.g., N-methylpyrrolidone, which is not commonly used in macroscale radiosynthesis). In an automated robotic system, significant optimization of the performance of the Cu-mediated synthesis of [^{18}F]FBnTP was accomplished in just one set of 64 experiments taking 4.5 h total. The optimized condition, using 450 nmol of Bpin precursor, offered high activity yield of 49 ± 3% ($n = 3$) in a synthesis time of just 42 ± 1 min, which was ~8 times higher than the activity yield obtained on a vial-based radiosynthesizer (6.2 ± 1.0%, $n = 7$) that required 15,900 nmol of precursor and 60 ± 5 min (Lin et al., 2023).

2.8 Summary

Recent advancements in droplet-based systems enable reliable, high-performance, and rapid production of many ^{18}F-labeled radiopharmaceuticals prepared via a wide range of labeling chemistries and can be extended to other isotopes, such as Tc-99m (Table 2). Droplet-based reactors demonstrate significant potential for cost reduction in on-demand PET radiopharmaceutical production. By operating at the microliter scale, synthesis processes are much shorter, and reagent consumption and costs are reduced by 1–3 orders of magnitude compared to conventional milliliter-scale radiosynthesizer technologies and microfluidic flow chemistry approaches. The small volume scale also minimizes the introduction of contaminants, particularly fluoride-19 derived from reagents (Sergeev et al., 2018), thereby ensuring that even small batches of ^{18}F-labeled tracers have high molar activities. This contrasts with conventional systems, in which a high starting activity is needed to achieve high molar activity. A further advantage of many droplet reaction technologies is the compact overall size of the automated platform, which points to the possibility of a self-shielded system that could avoid the need for specialized radiation-shielded facilities and equipment, such as hot cells or minicells.

3 Synthesis Scale-up in Droplet Systems

3.1 Overview

An important challenge in microvolume-based radiochemistry approaches is that the reaction volume (e.g., ~10 μL in droplet-based systems) is much smaller than the radioisotope source volume (e.g., 1–5 mL for [^{18}F]fluoride). In our lab, for EWOD, Teflon-coated glass, PT, STT, and multi-STT chip-based reactions, we typically directly use a portion of the 1.2 mL of [^{18}F]fluoride/[^{18}O]H_2O from the cyclotron. In the recent report by Ahmadi et al., [^{18}F]fluoride was first processed via a QMA cartridge with the eluted [^{18}F]KF/K_{222} provided in a 1.5 mL solution, from which portions were loaded into EWOD chips (Ahmadi et al., 2022). In such cases, clearly only a small fraction of the total activity can be loaded into each droplet reaction.

To produce batches that are sufficient for large preclinical or clinical studies, and/or more efficiently make use of the radioisotope, there is a need to scale up the amount of radioisotope loaded into the chip. Three successful scale-up strategies for droplet-based radiosynthesis (summarized in Fig. 5) have been reported to date: (1) pre-concentration of the radioisotope prior to introduction into the chip using

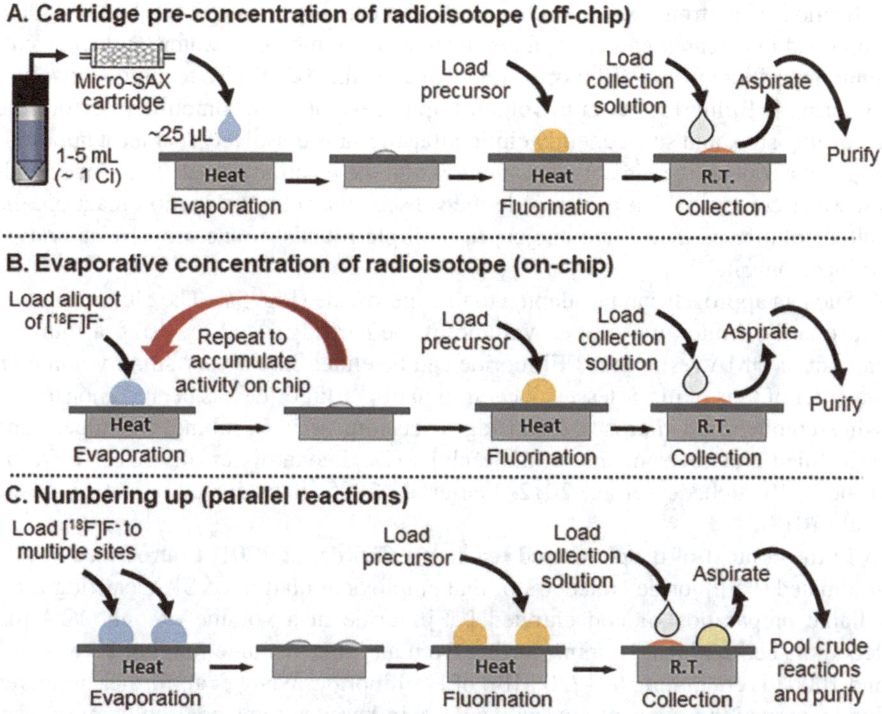

Fig. 5 Approaches for scale-up of radiopharmaceutical product amount in droplet-based radiosynthesis. (**a**) [^{18}F]fluoride is pre-concentrated using a miniature cartridge into a final volume that is compatible with a single reaction site. (**b**) Starting activity for a single droplet reaction is increased by repeated loading and evaporation of [^{18}F]fluoride aliquots on the droplet reaction chip prior to the fluorination reaction. (**c**) Multiple reaction sites are loaded with unconcentrated [^{18}F]fluoride and multiple droplet reactions are conducted in parallel. The crude reaction products are pooled prior to purification to increase the total product activity. (Figure reproduced from Lu et al. (2024a) with permission from the Royal Society of Chemistry)

miniature ion exchange cartridges (Chao et al., 2018; Wang et al., 2020a), (2) in situ droplet accumulation by repeated loading and evaporation of radioisotope aliquots on the chip (Chen et al., 2014b; Lisova et al., 2021), and (3) performing multiple reactions (numbering up). In this section, we introduce the operation of each method, provide synthesis examples, and summarize their advantages and limitations.

3.2 *Isotope Pre-concentration via Ion-Exchange Cartridge*

When performing ^{18}F-radiosyntheses in conventional synthesis modules, cartridge-based concentration is commonly used to reduce the volume of water in the isotope source prior to labeling. Taking [^{18}F]fluoride as an illustrative example, the utilization of a strong anion exchange (SAX) cartridge, such as QMA, is widely employed in research and clinical applications for removal of water (and some cationic impurities) from [^{18}F]fluoride (Haveman et al., 2023). This technique involves capturing [^{18}F]fluoride (even in volumes up to several mL), sometimes performing a washing step, and subsequently eluting it using salts dissolved in at least hundreds of μL of a mixed organic/aqueous solution into the reaction vessel. To reduce residual water content which can impede the subsequent radiofluorination reaction, the eluted solution is then often subject to multiple rounds of azeotropic evaporation with acetonitrile.

Such an approach can be adapted to the microscale (Fig. 5a). The elution volume is primarily limited by the bed volume of the cartridge, and by using a minimal amount of QMA resin, the [^{18}F]fluoride can be eluted into a very small volume on the order of tens of μL or less. Concentration of [^{18}F]fluoride has been demonstrated using commercial microscale cartridges, custom resin-in-tubing cartridges, and resin-filled chambers in microfluidic chips (De Leonardis et al., 2011; Elizarov et al., 2010; Lebedev et al., 2012; Lee et al., 2005; Salvador et al., 2017; Zhang et al., 2018).

In the context of droplet-based reactions, Chao et al. (2018) introduced a fully automated [^{18}F]fluoride concentrator that employs miniaturized SAX cartridges for reliable preparation of concentrated [^{18}F]fluoride in a volume of only 12.4 μL, accomplished within ~10 min, starting from an initial volume ranging between 60 and 1000 μL containing 9–1220 MBq of [^{18}F]fluoride. Wang et al. further improved and integrated this concentrator with a PT chip-based automated droplet microreactor setup for scaling up the synthesis of [^{18}F]Fallypride (Wang et al., 2020a). Upstream of the concentrator, the [^{18}F]fluoride was passed through a strong cation exchange (SCX) cartridge, which helped to remove contaminants and small particles and improve the reliability and trapping efficiency. Using 3 mg of QMA resin, the trapping efficiency was 96 ± 4% ($n = 2$). The subsequent release process involved initial elution using two portions (6.2 μL × 2) of 25 mM $TBAHCO_3$ and effectively recovered the majority of the activity (94 ± 3%, $n = 3$). The remaining activity was subsequently fully recovered through a dual rinse with two portions of water (6.2 μL × 2). To efficiently load the activity onto the PT chip, the concentrated activity was dispensed in 0.5 μL increments at 4 s intervals to a chip preheated to 100 °C, and an additional rinse of 25 μL DI water was transferred in the same fashion. The transfer process took a total of 6 min.

The optimized integrated setup was used to study the effects of activity scaling (from 3.7 to 41 GBq) on the subsequent droplet synthesis of [^{18}F]Fallypride. Up to 7.2 GBq of [^{18}F]Fallypride was produced in a single droplet reaction, a quantity sufficient for many human scans. Molar activity was high (81–270 GBq/μmol), and the overall production time (including the concentration process) was ~50 min. For starting activities up to ~10 GBq, crude RCY remained close to 65% but gradually decreased at higher activities. This reduction was attributed in part to a decrease in [^{18}F]fluoride trapping efficiency for activities higher than ~20 GBq, falling from ~94% to 63%, and a drop in fluorination conversion from roughly 90% to 40%. It was speculated the latter effect could be due to radiolysis or due to a marked decrease in precursor to fluoride ratio. It should be noted that while Wang et al. observed a reduction in cartridge efficacy above ~20 GBq (Wang et al., 2020a), the work by Lebedev et al. (though not combined with a droplet reactor) suggests that micro-cartridge trap and elute methods can be scaled up to at least 259 GBq of [^{18}F]fluoride (Lebedev et al., 2012). A notable advantage of this technique is that cartridge processing helps to remove impurities from the isotope, limiting their impact in the subsequent droplet reaction. However, the technique does add a substantial amount of time (~16 min) to the synthesis protocol. In addition, it is important to account for the elution procedure (specifically the amount of salts required for efficient elution of [^{18}F]fluoride from the resin) into any optimization studies for the subsequent droplet synthesis and can require extensive development efforts when performing reactions that are particularly base sensitive.

3.3 *In Situ Evaporative Isotope Concentration*

Another approach that has been used for increasing reaction scale is to concentrate the activity on-chip via evaporation.

For example, using a glass EWOD chip, Chen et al. showed the possibility to introduce a large volume of [^{18}F]fluoride (200 μL; activity not reported), premixed with the desired amount of phase-transfer catalyst and base, to an uncovered region of the chip, and evaporate most of the solvent by heating the chip (Chen et al., 2014b). By performing this step close to the edge of the covered part of the chip, when the volume had shrunk to ~5 μL, it could be pulled into the chip via electrode activation, whereupon it was transported to the reaction site and further dried to remove residual water. The efficiency of the process was not reported but the total time for evaporation was 12 min. This was only a feasibility study, and no subsequent labeling was performed. While apparently effective for moderate degrees of scaling, a potential drawback is the need for significant chip area for loading and evaporating the isotope.

Another approach is to sequentially load aliquots of [^{18}F]fluoride to the chip and dry them at the reaction site to incrementally build up an increasing amount of radioisotope (Fig. 5b). This approach was first used during initial attempts to scale up the synthesis in the PT reaction chip (Wang et al., 2017). Multiple 2 μL aliquots of [^{18}F]fluoride/[^{18}O]H_2O were sequentially dispensed to the loading site, followed by a single 2 μL aliquot of $TBAHCO_3$ (base), and the mixture was subsequently dried. Up to 8× increase in loaded activity could be achieved, but the demonstrated activity scale was still very small (2.3 MBq), and the crude RCY of [^{18}F]Fallypride synthesized from the concentrated activity was markedly reduced.

This technique was more effectively demonstrated in the context of STT chips by Lisova et al. (2021). On these larger capacity chips, up to 30 μL portions of [^{18}F]fluoride/[^{18}O]H_2O mixed with phase-transfer catalyst/base could be added per loading cycle. Evaporative drying was performed immediately after loading each droplet. The concentration of phase-transfer catalyst/base mixed with the [^{18}F]fluoride stock solution was adjusted to ensure the total amount after multiple loading and drying matched the desired quantity. Lisova et al. harnessed this technique to amplify the synthesis scale of [^{18}F]FET and [^{18}F]Florbetaben. Employing the same droplet reaction conditions derived from low-activity optimization studies (Lisova et al., 2019, 2020), but starting with 2–4 GBq of activity, this method enabled the production of batches of 0.6 ± 0.2 GBq (n = 3) of [^{18}F]FET and 0.5 ± 0.2 GBq (n = 6) of [^{18}F]Florbetaben, each sufficient for imaging in a human subject. Though the reaction performance of the scaled-up reactions were slightly lower than low-activity reactions, the overall RCY (36 ± 7%, n = 3 for [^{18}F]FET and 23 ± 3%, n = 6 for [^{18}F]Florbetaben) was still comparable to conventional methods (Bogni et al., 2019; Collins et al., 2017). The microscale syntheses consumed 100–150× less precursor than the macroscale approaches, and the reactions exhibited high molar activity (>400 GBq/μmol).

Rios et al. also used this approach to increase the synthesis scale for preparing [^{18}F]PBR06 on an STT chip, comparing the influence of starting activities from 0.014 to 3.17 GBq (Rios et al., 2022). A minor reduction in the crude RCY was observed with increasing activity scale. Later studies using different volumes of decayed [^{18}F]fluoride spiked with small amounts of activity suggested that the reduction was due in large part to a buildup of the concentration of impurities from the [^{18}F]fluoride source itself (Rios, 2022).

An important advantage of evaporative concentration is its simplicity: as it relies on using the existing radioisotope dispenser, it does not require additional components (i.e., cartridge, valves, control system) as is the case for the cartridge-based method. An additional noteworthy benefit is its compatibility with varying types and quantities of base, which can be added to the reaction site independently during or prior to the first [^{18}F]fluoride aliquot. This sets it apart from cartridge-based concentration method, where the nature and quantity of the base needed for high elution efficiency are intertwined with the downstream reaction conditions. Instead, the type and amount of base can be optimized in a low-activity study (i.e., to maximum reaction performance), without the concern whether the amount of base is sufficient to elute the anion exchange cartridge, and the scale can be increased simply by loading more activity aliquots while preserving the exact same amount of base. So far, scaling the starting activity up to ~4 GBq has been demonstrated, and potentially the method can be further scaled. However, there are likely practical limits to scale-up, including the additional time (1.5 min) needed for evaporation of each 30 μL aliquot. Furthermore, the reported reduction in reaction performance as the volume of [^{18}F]fluoride increases may limit the product scale that can be achieved.

3.4 Numbering up

An alternative strategy is to perform multiple droplet reactions in parallel (numbering up) and pool the crude products prior to purification (Fig. 5c).

Using multi-STT chips, Lu et al. (2024a) showcased the scalability of [^{18}F]FBnTP synthesis using this approach, achieving comparable performance for larger-scale pooled reactions as for lower activity optimization runs. Firstly, a short study on multi-STT chips allowed quick determination of the maximum practical volume of [^{18}F]fluoride per reaction where the performance matched the small-scale reaction performance. Volumes up to 20 μL (regardless of activity level) exhibited high performance (crude RCY 80–90%), while volumes of 25–30 μL exhibited moderate performance (crude RCY 60–80%), and higher volumes gave much lower and variable crude RCY. Subsequently, adhering to this constraint, the synthesis was scaled by performing multiple reactions in parallel, and the crude product from individual reaction was pooled together prior to purification to achieve the desired amount of product. In a 4-reaction demonstration, the observed RCY was 66 ± 1% ($n = 4$), with high RCP (>99%) and high molar activity (340–880 GBq/μmol). This method provided sufficient product for multiple clinical doses, ranging from 0.76 to 1.84 GBq, efficiently obtained from 1.6 to 3.7 GBq of [^{18}F]fluoride within a synthesis time of just 37–47 min.

This scale-up method offers notable advantages, particularly in terms of simplicity and consistent performance, in comparison to alternative techniques which have shown a decline in performance with increasing synthesis scale. In addition, the automation potential of this approach is promising and can be easily transferred to the existing STT-chip-based automated system using a multi-STT chip. While the numbering up method requires higher precursor consumption compared to radioisotope concentration methods that use only a single droplet reaction (Fig. 5a, b), the overall quantity remains lower than that required for the macroscale approach. For example, for [^{18}F]FBnTP, 1.2–2.4 μmol precursor was used for scale-up via 2–4 droplet reactions, while macroscale reactions consume 4.0–15.9 μmol (Lin et al., 2023; Zhang et al., 2016). Particularly beneficial for radiotracers in early-stage development or with low demand, this efficient droplet-based scale-up technique can be readily applied to prepare various radiotracers on demand, offering great flexibility in synthesis scale.

3.5 Summary

The three methods presented above have all successfully shown the ability to scale up droplet-based reactions with F-18 to multiple clinical dose levels but offer different tradeoffs in terms of hardware and process complexity, degree of scalability, overall synthesis time, and total reagent consumption (Table 3).

For production for preclinical research or for relatively small numbers of patients, scale-up via evaporative radioisotope concentration or the numbering up method are capable of producing the needed quantity and can be rapidly implemented without significant additional study after the low-activity synthesis optimization studies are complete. For larger amount of product, for example, for many doses of a widely used radiotracer, pre-concentration via a cartridge is likely necessary, and the research effort to co-optimize the cartridge concentration conditions and synthesis conditions can be justified.

Table 3 Comparison of scale-up methods reported for droplet-based radiochemistry

	Pre-concentration of isotope via cartridge	In situ concentration of isotope via evaporation	Numbering up (multiple reactions)
Compatible type(s) of droplet chips	Any	Any that can be heated	Any that can be implemented with multiple reaction sites (parallel reactions), or on which multiple sequential reactions can be rapidly performed
Added system complexity	Requires fluidic system to perform cartridge concentration (microvalves, micro-cartridge, control system)	None	Requires a mechanism to load reagents and collect crude product from multiple reactions
Processing time (volume of F-18)	~16 min (0.5–1 mL)	(1) 12 min (200 μL) (Chen et al., 2014b) (2) 1.5 min per 30 μL aliquot (Lisova et al., 2021)	1.5 min (20–30 μL/ reaction)
Reagent consumption	1×	1×	N×
Scalability	High	Moderate	Moderate
Limitations on scalability	(1) Higher amounts of radioisotope require larger cartridge mass and increased amount of eluent (2) Amount of eluent salts can impact downstream reaction (3) Portion of activity may be lost in the concentration system	(1) Impurities from radioisotope source accumulate and can interfere with some reactions (2) Each additional aliquot of source radioisotope requires more time to evaporate	(1) Each additional reaction increases reagent consumption (2) A little bit of extra time per reaction is needed for reagent loading and product collection (3) Each additional reaction increases total crude product volume and quantity of impurities, impacting purification
Advantages	(1) Removes impurities from radioisotope source, so they do not interfere with reaction (2) Isotope processing is similar to conventional macroscale processing	(1) Simple system setup (2) No additional optimization of elution protocol is required	(1) No additional optimization of elution protocol is required (2) Additional reactions can proceed largely in parallel on multi-STT chips (3) Synthesis performance remains consistent, as number of reactions is increased

4 Outlook

4.1 Range of Applications

The droplet-based radiosynthesis technique has proven to be quite versatile for labeling of clinically relevant and investigational radiotracers, demonstrating successful applications across a broad spectrum of chemistries. Notably, it has proven effective in facilitating aliphatic nucleophilic substitution of leaving groups, exemplified by the synthesis of a multitude of radiotracers including [^{18}F]FDG (Ahmadi et al., 2022; Chen et al., 2014b; Dooraghi et al., 2013; Keng et al., 2012; Koag et al., 2014a; Wang et al., 2017), [^{18}F]FET (Lisova et al., 2019, 2021), [^{18}F]FLT (Javed et al., 2014a; Koag et al., 2014b), [^{18}F]Florbetaben (Lisova et al., 2020, 2021), [^{18}F]Fallypride (Javed et al., 2014b; Jones et al., 2023; Rios et al., 2019, 2022; Wang et al., 2017, 2019), [^{18}F]PBR06 (Rios et al., 2022), and [^{18}F]FEPPA (Rios et al., 2022). Moreover, aromatic nucleophilic substitution of nitro groups has been demonstrated, with [^{18}F]flumazenil (Rios et al., 2022) and [^{18}F]FPEB (Laferriere-Holloway et al., 2023) serving as illustrative examples. Additionally, nucleophilic aromatic substitution of diaryliodonium salts has been shown, a prime example being the synthesis of [^{18}F]FDOPA (Wang et al., 2020b). Moreover, this technique excels in isotopic exchange reactions involving organotrifluoroborates, as evidenced by the synthesis of [^{18}F]AMBF$_3$-TATE (Lisova et al., 2018). The copper-mediated fluorination of organoboron compounds, including [^{18}F]FDOPA (Lu & van Dam, 2021a), [^{18}F]FBnTP (Jones et al., 2023; Lu et al., 2024a), and [^{18}F]YH149 (Lu et al., 2023a) have all been effectively performed in droplet format. In addition, compatibility with aluminum fluoride radiochemistry (Lu et al., 2024b) has recently been established. Droplet-based platforms have also been employed in the synthesis of prosthetic groups for peptide or protein labeling, including 1-[^{18}F]fluoro-4-nitrobenzene ([^{18}F]FNB) (Sergeev et al., 2018), 3-formyl-2,4,6-trimethylbenzene-sulfonyl [^{18}F]fluoride (Fiel et al., 2015), and N-succinimidyl-4-fluorobenzoate ([^{18}F]SFB) (Kim et al., 2019). In addition to the extensive applications to ^{18}F-labeled molecules, the radiometal chelation for preparation of the SPECT tracer [^{99m}Tc]Tc-DTPA has been demonstrated (Mogi et al., 2021).

This technique's potential extends far beyond these demonstrated applications and has generated considerable interest within the radiochemistry community for exploring additional chemical reactions as well as for other isotopes. Though preliminary work has been done with the radiometals Ga-68 (Chao, 2019) and Tc-99m (Mogi et al., 2021), more exploration is needed on the feasibility of using these and other isotopes on droplet-based platforms, and to better understand the advantages and disadvantages of droplet methods versus conventional techniques for applications beyond ^{18}F-labeling.

Though initial research using droplet-based radiochemistry focused on relatively small-scale reactions sufficient only for preclinical imaging, several methods have now been developed for increasing the scale to clinically relevant levels, including pre-concentration of the radioisotope via a miniature cartridge, in situ concentration by sequential addition and evaporation of aliquots of radioisotope solution, and performing multiple identical reactions to increase product scale (numbering up). Though scale-up demonstrations have all focused on F-18 to date, preliminary work by Chao et al. demonstrated that a promising cartridge-based method could concentrate large volumes of Ga-68 solution into 60 μL volume that would be compatible with droplet-based reactions (Chao, 2019). Mogi et al. speculated that scale-up for ^{99m}Tc-labeling on their paper EWOD chip could be achieved by numbering up via parallel or sequential reactions (Mogi et al., 2021), though this was not demonstrated.

4.2 Comparison of Droplet and Conventional Radiochemistry

Compared to conventional vial-based reactors, the improved efficiency of droplet-based reactors (Wang & van Dam, 2020) can largely be attributed to the use of microliter-scale volumes in combination with a high-performance chip heating and cooling mechanism. Volume reduction by ~100× not only enables a reduction in the consumption of reagents, particularly for those precursors that are costly or challenging to synthesize but also reduces the quantity of toxic reagents (e.g., phase-transfer catalysts, organic solvents) used and reduces the quantities of side-products formed. Due to smaller volumes, process steps such as evaporations and heating/cooling of the reaction mixture can be much quicker than a vial-based reactor. In the case of high-performance liquid chromatography (HPLC)-purified radiopharmaceuticals, the reduced crude product volume (<100 μL) and lower mass of impurities enable the use of analytical-scale HPLC (instead of semi-preparative) for purification, which reduces purification time and enables the purified fraction to be collected in a smaller volume (which shortens the formulation process). Small impurity masses may also be amenable to removal via on-chip techniques. Preliminary success has been shown for [^{18}F]fluoride scavenging techniques (Ahmadi et al., 2022; Chen et al., 2012, 2014a), removal of radio-impurities via on-chip cartridges (Tarn et al., 2013; Zhang et al., 2018, 2020), and rapid purification/formulation via preparative thin-layer chromatography (Laferriere-Holloway et al., 2022), which have the potential for further reductions in instrumentation complexity, apparatus size, and purification time.

Furthermore, the reduction of reagent amounts enables consistently high molar activity, even when commencing with low initial activity levels. This capability stands in stark contrast to conventional systems, which lack the capacity to generate small (low radioactivity) batches with high molar activities. In the case of ^{18}F-radiochemistry in particular, the reduction of reagent volume and the low surface area of droplet-chip contact minimizes contamination with non-radioactive fluoride, further boosting molar activities (Sergeev et al., 2018). The use of low reagent amounts is especially capable of ensuring high molar activity in isotopic exchange reactions (where excess precursor cannot be removed via purification) and chelation reactions (where excess precursor is often not removed via purification).

An additional remarkable advantage offered by the droplet-based approach is its exceptional high-throughput capability compared to conventional radiosynthesis technologies. Leveraging multi-STT chips with arrays of droplet reactions performed in parallel, data collection at the rate of hundreds of reactions per week is possible (Jones et al., 2023; Rios et al., 2022). After rapid optimization of labeling conditions, droplet-based platforms combined with scale-up techniques can immediately be used to produce moderate multi-GBq quantities to sustain small animal and human research studies.

Leveraging the small size of droplet reaction chips, another critical advancement has been the realization of compact designs for the overall automated microdroplet radiosynthesizer. For example, the STT-chip platform has been implemented in a device roughly the size of a small cup of coffee (~350 cm^3) (Wang et al., 2019), which is at least two orders of magnitude smaller than a typical conventional synthesis module (e.g., W × H × D = 63.2 × 37.3 × 56.6 cm^3 = 133,000 cm^3 for the ELIXYS FLEX/CHEM, Sofie, Inc.) (Lazari et al., 2013). Smaller size greatly reduces the cost and weight of shielding needed. In fact, this transformative development could enable standalone, self-shielded operation, rendering the need for a hot cell unnecessary, and the associated space, cost, and time needed for installation of such infrastructure. This gives the potential for more distributed and economical manufacturing of a wide variety of radiopharmaceuticals, especially for those where demand is lower or sporadic, such as in novel radiopharmaceutical development, preclinical research, small-scale human studies, and clinical use for diagnosis and treatment of rare diseases.

While some scalability of individual droplet reactions has been demonstrated, it is not clear if they will be suitable for the routine large-scale (>40 GBq) batch production that can be achieved with conventional vial-based radiosynthesizers in hot cells. As droplet-based reactions have been scaled up, some concerns have been noted about diminishing performance that may be attributed to increased impurity concentrations or radiolysis in the highly concentrated reactions, or other factors (Lisova et al., 2021; Wang et al., 2020a). Further studies of various scale-up approaches will ultimately reveal if further increases can be achieved. Interestingly, in a recent proof-of-concept study, Lu et al. showed that the optimized droplet-based conditions for the synthesis of [^{18}F]YH149 could be adapted to a vial reaction in a straightforward manner (Lu et al., 2023a), showing the potential for a seamless transition from micro- to macroscales that could be a future model of bridging the gap between moderate-scale and large-scale production.

4.3 *Further Development of Droplet Radiochemistry Technology*

In comparison to other microvolume platforms such as microvials, microreaction chambers, and some types of flow-based reactors, a major distinction of droplet-based methods is the open nature of the reaction (or semi-open nature, as in the case of covered geometries like glass-based EWOD chips or Teflon-coated glass chips).

Open droplets enable rapid solvent evaporation which accelerates steps, such as [^{18}F]fluoride drying, and enables easy access to droplets for adding reagents, collecting crude product, and taking analytical samples via manual or automated pipetting systems. But the open droplet surface can lead to unwanted evaporation during reaction steps, posing challenges in maintaining a consistent concentration and solvent level. This has largely been overcome by using higher boiling point solvents in reactions (Keng et al., 2012; Rios et al., 2022; Wang et al., 2017) and/or replenishing solvents periodically during reactions (Lu & van Dam, 2021a, 2021b). Due to the small reaction volumes, the total quantity of solvents used is low and thus using high-boiling solvents is not as problematic as it can be for macroscale reactions, where such solvents (e.g., DMSO) may need to be removed prior to purification. The open format of droplets can also make it difficult to handle reactions involving volatile reagents, intermediates, or products, potentially limiting the use of droplets for some specific classes of reactions. Sterility, safety, and regulatory concerns of open droplets can likely be overcome by enclosing the whole droplet systems within a sealed gas-tight container during operation, but there remains interest in exploring the feasibility of developing means to reversibly cover/enclose droplets during reaction steps in a way that is fully automated.

Among the different droplet-radiochemistry platforms, there are different capabilities and limitations (summarized in Table 1). Glass-based EWOD devices offer sophisticated droplet handling with multiple functions integrated into a single chip, including droplet actuation, heating, and even storage of preloaded reagents (except for radioactive or volatile reagents) (Ahmadi et al., 2022), promising a simple and compact overall radiosynthesis instrument. However, EWOD chips can be costly, and the procedure for collecting the crude product relies on manual handling. Developing approaches to interface reagent loading and product collection could provide a path to increased automation of these chips. Paper-based EWOD and magnetic bead approaches offer similar versatility of droplet manipulations as glass-based EWOD devices, but may have a limited range of applications due to the lack of heating capability and potential incompatibility of the platforms with organic solvents or harsh reagents. The need for magnetic beads within the reaction droplet in the magnetic platform could create incompatibilities for some types of reactions. Chips based on patterned wettability regions using Teflon-coated silicon (i.e., PT, STT, and multi-STT chips) may provide an ideal compromise of relatively low-cost chip fabrication, yet using inert and temperature stable materials to ensure versatility for a wide range of reactions. There is a tradeoff that platforms with simpler chips require more subsystems in the external fixed (non-disposable) part of the instrument. Overall, this difference is relatively small, however, because all droplet platforms currently require several external systems for processing and adding the radioisotope, collecting the crude product, and performing thorough purification and formulation.

To promote further adoption of these technologies, it will be necessary to demonstrate fully integrated prototypes and to make the instruments and chips available to investigators. Such development is underway for the STT-chip technology, including full integration of radioisotope processing (e.g., [^{18}F]fluoride concentration), cartridge- or HPLC-based purification, and cartridge formulation with the droplet-based synthesis process. Developing features akin to macroscale cassette-based radiosynthesizers, including reagent cassettes, disposable fluidics, and software that supports cGMP, would further be desired to improve usability and promote adoption of this technology for production of radiopharmaceuticals for human use.

Disclosures Dr. van Dam is a co-founder and shareholder of SOFIE Inc. and of DropletPharma Corp.

References

Ahmadi, F., Simchi, M., Perry, J. M., Frenette, S., Benali, H., Soucy, J.-P., Massarweh, G., & Shih, S. C. C. (2022). Integrating machine learning and digital microfluidics for screening experimental conditions. *Lab on a Chip, 23*, 81. https://doi.org/10.1039/D2LC00764A

Allott, L., Da Pieve, C., Turton, D. R., & Smith, G. (2017). A general [18 F]AlF radiochemistry procedure on two automated synthesis platforms. *Reaction Chemistry & Engineering, 2*, 68–74. https://doi.org/10.1039/C6RE00204H

Audrain, H. (2007). Positron emission tomography (PET) and microfluidic devices: A breakthrough on the microscale? *Angewandte Chemie International Edition, 46*, 1772–1775. https://doi.org/10.1002/anie.200603509

Berridge, M. S., Apana, S. M., & Hersh, J. M. (2009). Teflon radiolysis as the major source of carrier in fluorine-18. *Journal of Labelled Compounds and Radiopharmaceuticals, 52*, 543–548. https://doi.org/10.1002/jlcr.1672

Berroterán-Infante, N., Balber, T., Fürlinger, P., Bergmann, M., Lanzenberger, R., Hacker, M., Mitterhauser, M., & Wadsak, W. (2018). [18F]FEPPA: Improved Automated Radiosynthesis, Binding Affinity, and Preliminary in Vitro Evaluation in Colorectal Cancer. *ACS Medicinal Chemistry Letters, 9*, 177–181. https://doi.org/10.1021/acsmedchemlett.7b00367

Bogni, A., Laera, L., Cucchi, C., Iwata, R., Seregni, E., & Pascali, C. (2019). An improved automated one-pot synthesis of O-(2-[18F]fluoroethyl)-L-tyrosine ([18F]FET) based on a purification by cartridges. *Nuclear Medicine and Biology, 72–73*, 11–19. https://doi.org/10.1016/j.nucmedbio.2019.05.006

Chao, P. H.-S. (2019). *Novel microfluidic technologies for the concentration of radionuclides and radiotracers for positron emission tomography*. UCLA.

Chao, P. H., Lazari, M., Hanet, S., Narayanam, M. K., Murphy, J. M., & van Dam, R. M. (2018). Automated concentration of [18F]fluoride into microliter volumes. *Applied Radiation and Isotopes, 141*, 138–148. https://doi.org/10.1016/j.apradiso.2018.06.017

Chen, S., Lei, J., van Dam, R. M., Keng, P.-Y., & Kim, C.-J. C. J. (2012). Planar alumina purification of 18F-labeled radiotracer synthesis on EWOD chip for positron emission tomography (PET). In *Proceedings of the 16th international conference on miniaturized systems for chemistry and life sciences*. Presented at the MicroTAS, Okinawa, Japan, pp. 1771–1773.

Chen, S., Dooraghi, A., Lazari, M., van Dam, R. M., Chatziioannou, A. F., & Kim, C.-J. (2014a). On-chip product purification for complete microfluidic radiotracer synthesis. In *Proceedings of the 27th IEEE International Conference on Micro Electro Mechanical Systems (MEMS)*. Presented at the MEMS 2014, San Francisco, CA, pp. 284–287. https://doi.org/10.1109/MEMSYS.2014.6765631.

Chen, S., Rashed Javed, M., Kim, H.-K., Lei, J., Lazari, M., Shah, G. J., van Dam, R. M., Keng, P.-Y., & Kim, C.-J. C. J. (2014b). Radiolabelling diverse positron emission tomography (PET) tracers using a single digital microfluidic reactor chip. *Lab on a Chip, 14*, 902–910. https://doi.org/10.1039/C3LC51195B

Collins, J., Waldmann, C. M., Drake, C., Slavik, R., Ha, N. S., Sergeev, M., Lazari, M., Shen, B., Chin, F. T., Moore, M., Sadeghi, S., Phelps, M. E., Murphy, J. M., & van Dam, R. M. (2017). Production of diverse PET probes with limited resources: 24 ^{18}F-labeled compounds prepared with a single radiosynthesizer. *PNAS, 114*, 11309–11314. https://doi.org/10.1073/pnas.1710466114

De Leonardis, F., Pascali, G., Salvadori, P. A., Watts, P., & Pamme, N. (2011). On-chip preconcentration and complexation of [18F]fluoride ions via regenerable anion exchange particles for radiochemical synthesis of positron emission tomography tracers. *Journal of Chromatography A, 1218*, 4714–4719. https://doi.org/10.1016/j.chroma.2011.05.062

Dooraghi, A. A., Keng, P. Y., Chen, S., Javed, M. R., Kim, C.-J. C. J., Chatziioannou, A. F., & van Dam, R. M. (2013). Optimization of microfluidic PET tracer synthesis with Cerenkov imaging. *Analyst, 138*, 5654–5664. https://doi.org/10.1039/C3AN01113E

Elizarov, A. M. (2009). Microreactors for radiopharmaceutical synthesis. *Lab on a Chip, 9*, 1326–1333.

Elizarov, A. M., van Dam, R. M., Shin, Y. S., Kolb, H. C., Padgett, H. C., Stout, D., Shu, J., Huang, J., Daridon, A., & Heath, J. R. (2010). Design and optimization of coin-shaped microreactor chips for PET radiopharmaceutical synthesis. *Journal of Nuclear Medicine, 51*, 282–287. https://doi.org/10.2967/jnumed.109.065946

Elkawad, H., Xu, Y., Tian, M., Jin, C., Zhang, H., Yu, K., & He, Q. (2022). Recent advances in microfluidic devices for the radiosynthesis of PET-imaging probes. *Chemistry—An Asian Journal, 17*, e202200579. https://doi.org/10.1002/asia.202200579

Fiel, S. A., Yang, H., Schaffer, P., Weng, S., Inkster, J. A. H., Wong, M. C. K., & Li, P. C. H. (2015). Magnetic droplet microfluidics as a platform for the concentration of [18F]fluoride and radiosynthesis of sulfonyl [18F]fluoride. *ACS Applied Materials & Interfaces, 7*, 12923–12929. https://doi.org/10.1021/acsami.5b02631

Ghosh, A., Ganguly, R., Schutzius, T. M., & Megaridis, C. M. (2014). Wettability patterning for high-rate, pumpless fluid transport on open, non-planar microfluidic platforms. *Lab on a Chip, 14*, 1538–1550. https://doi.org/10.1039/C3LC51406D

Haveman, L. Y. F., Vugts, D. J., & Windhorst, A. D. (2023). State of the art procedures towards reactive [18F]fluoride in PET tracer synthesis. *EJNMMI Radiopharmacy and Chemistry, 8*, 28. https://doi.org/10.1186/s41181-023-00203-5

He, Y., Schild, M., Grether, U., Benz, J., Leibrock, L., Heer, D., Topp, A., Collin, L., Kuhn, B., Wittwer, M., Keller, C., Gobbi, L. C., Schibli, R., & Mu, L. (2022). Development of high brain-penetrant and reversible monoacylglycerol lipase PET tracers for neuroimaging. *Journal of Medicinal Chemistry, 65*, 2191–2207. https://doi.org/10.1021/acs.jmedchem.1c01706

Hume, S. P., Gunn, R. N., & Jones, T. (1998). Pharmacological constraints associated with positron emission tomographic scanning of small laboratory animals. *European Journal of Nuclear Medicine and Molecular Imaging, 25*, 173–176. https://doi.org/10.1007/s002590050211

Inkster, J. A. H., Liu, K., Ait-Mohand, S., Schaffer, P., Guérin, B., Ruth, T. J., & Storr, T. (2012). Sulfonyl fluoride-based prosthetic compounds as potential 18F labelling agents. *Chemistry – A European Journal, 18*, 11079–11087. https://doi.org/10.1002/chem.201103450

Javed, M. R., Chen, S., Kim, H.-K., Wei, L., Czernin, J., Kim, C.-J. C. J., van Dam, R. M., & Keng, P. Y. (2014a). Efficient radiosynthesis of 3′-Deoxy-3′-18F-fluorothymidine using electrowetting-on-dielectric digital microfluidic chip. *Journal of Nuclear Medicine, 55*, 321–328. https://doi.org/10.2967/jnumed.113.121053

Javed, M. R., Chen, S., Lei, J., Collins, J., Sergeev, M., Kim, H.-K., Kim, C.-J., van Dam, R. M., & Keng, P. Y. (2014b). High yield and high specific activity synthesis of [18F]fallypride in a batch microfluidic reactor for micro-PET imaging. *Chemical Communications, 50*, 1192–1194. https://doi.org/10.1039/C3CC47616B

Jones, J., Do, V., Lu, Y., & van Dam, R. M. (2023). Accelerating radiochemistry development: Automated robotic platform for performing up to 64 droplet radiochemical reactions in a morning. *Chemical Engineering Journal, 468*, 143524. https://doi.org/10.1016/j.cej.2023.143524

Keng, P. Y., Chen, S., Ding, H., Sadeghi, S., Shah, G. J., Dooraghi, A., Phelps, M. E., Satyamurthy, N., Chatziioannou, A. F., Kim, C.-J. C. J., & van Dam, R. M. (2012). Micro-chemical synthesis of molecular probes on an electronic microfluidic device. *PNAS, 109*, 690–695. https://doi.org/10.1073/pnas.1117566109

Keng, P. Y., Sergeev, M., & van Dam, R. M. (2016). Advantages of radiochemistry in microliter volumes. In Y. Kuge, T. Shiga, & N. Tamaki (Eds.), *Perspectives on nuclear medicine for molecular diagnosis and integrated therapy* (pp. 93–111). Springer Japan. https://doi.org/10.1007/978-4-431-55894-1_7

Kim, H.-K., Rashed Javed, M., Chen, S., Zettlitz, K. A., Collins, J., Wu, A. M., Kim, C.-J. C. J., van Dam, R. M., & Yuin Keng, P. (2019). On-demand radiosynthesis of N-succinimidyl-4-[18 F]fluorobenzoate ([18 F]SFB) on an electrowetting-on-dielectric microfluidic chip for 18 F-labeling of protein. *RSC Advances, 9*, 32175–32183. https://doi.org/10.1039/C9RA06158D

Knapp, K.-A., Nickels, M. L., & Manning, H. C. (2020). The current role of microfluidics in Radiofluorination chemistry. *Molecular Imaging and Biology, 22*, 463–475. https://doi.org/10.1007/s11307-019-01414-6

Koag, M. C., Kim, H.-K., & Kim, A. S. (2014a). Efficient microscale synthesis of [18F]-2-fluoro-2-deoxy-d-glucose. *Chemical Engineering Journal, 258*, 62–68. https://doi.org/10.1016/j.cej.2014.07.077

Koag, M. C., Kim, H.-K., & Kim, A. S. (2014b). Fast and efficient microscale radiosynthesis of 3′-deoxy-3′-[18F]fluorothymidine. *Journal of Fluorine Chemistry, 166*, 104–109. https://doi.org/10.1016/j.jfluchem.2014.07.033

Kuik, W.-J., Kema, I. P., Brouwers, A. H., Zijlma, R., Neumann, K. D., Dierckx, R. A. J. O., DiMagno, S. G., & Elsinga, P. H. (2015). In vivo biodistribution of no-carrier-added 6-18F-fluoro-3,4-dihydroxy-l-phenylalanine (18F-DOPA), produced by a new nucleophilic substitution approach, compared with carrier-added 18F-DOPA, prepared by conventional electrophilic substitution. *Journal of Nuclear Medicine, 56*, 106–112. https://doi.org/10.2967/jnumed.114.145730

Laferriere-Holloway, T. S., Rios, A., Carlucci, G., & van Dam, R. M. (2022). Rapid purification and formulation of radiopharmaceuticals via thin-layer chromatography. *Molecules, 27*, 8178. https://doi.org/10.3390/molecules27238178

Laferriere-Holloway, T. S., Rios, A., Lu, Y., Okoro, C. C., & van Dam, R. M. (2023). A rapid and systematic approach for the optimization of radio thin-layer chromatography resolution. *Journal of Chromatography. A, 1687*, 463656. https://doi.org/10.1016/j.chroma.2022.463656

Lau, J., Pan, J., Rousseau, E., Uribe, C. F., Seelam, S. R., Sutherland, B. W., Perrin, D. M., Lin, K.-S., & Bénard, F. (2020). Pharmacokinetics, radiation dosimetry, acute toxicity and automated synthesis of [18F]AmBF3-TATE. *EJNMMI Research, 10*, 25. https://doi.org/10.1186/s13550-020-0611-9

Lazari, M., Quinn, K. M., Claggett, S. B., Collins, J., Shah, G. J., Herman, H. E., Maraglia, B., Phelps, M. E., Moore, M. D., & van Dam, R. M. (2013). ELIXYS—A fully automated, three-reactor high-pressure radiosynthesizer for development and routine production of diverse PET tracers. *EJNMMI Research, 3*, 52. https://doi.org/10.1186/2191-219X-3-52

Lazari, M., Collins, J., Shen, B., Farhoud, M., Yeh, D., Maraglia, B., Chin, F. T., Nathanson, D. A., Moore, M., & van Dam, R. M. (2014). Fully automated production of diverse 18F-labeled PET tracers on the ELIXYS multireactor radiosynthesizer without hardware modification. *Journal of Nuclear Medicine Technology, 42*, 203–210. https://doi.org/10.2967/jnmt.114.140392

Lebedev, A. Y. (2013). Microfluidic devices for radio chemical synthesis. In X. (James) Li & Y. Zhou (Eds.), *Microfluidic devices for biomedical applications*, Woodhead Publishing Series in Biomaterials (pp. 594–633). Woodhead Publishing. https://doi.org/10.1533/9780857097040.4.594

Lebedev, A., Miraghaie, R., Kotta, K., Ball, C. E., Zhang, J., Buchsbaum, M. S., Kolb, H. C., & Elizarov, A. (2012). Batch-reactor microfluidic device: First human use of a microfluidically produced PET radiotracer. *Lab on a Chip, 13*, 136–145. https://doi.org/10.1039/C2LC40853H

Lee, S. J., Oh, S. J., Chi, D. Y., Kil, H. S., Kim, E. N., Ryu, J. S., & Moon, D. H. (2007). Simple and highly efficient synthesis of 3′-deoxy-3′-[18F]fluorothymidine using nucleophilic fluorination catalyzed by protic solvent. *European Journal of Nuclear Medicine and Molecular Imaging, 34*, 1406–1409. https://doi.org/10.1007/s00259-007-0391-8

Lee, C.-C., Sui, G., Elizarov, A., Shu, C. J., Shin, Y.-S., Dooley, A. N., Huang, J., Daridon, A., Wyatt, P., Stout, D., Kolb, H. C., Witte, O. N., Satyamurthy, N., Heath, J. R., Phelps, M. E., Quake, S. R., & Tseng, H.-R. (2005). Multistep synthesis of a radiolabeled imaging probe using integrated microfluidics. *Science, 310*, 1793–1796. https://doi.org/10.1126/science.1118919

Lim, K., Labaree, D., Li, S., & Huang, Y. (2014). Preparation of the metabotropic glutamate receptor 5 (mGluR5) PET tracer [18F]FPEB for human use: An automated radiosynthesis and a novel one-pot synthesis of its radiolabeling precursor. *Applied Radiation and Isotopes, 94*, 349–354. https://doi.org/10.1016/j.apradiso.2014.09.006

Lin, M., Pham, C., Mukhopadhyay, U., & Manning, H. C. (2023). CGMP compliant one-step, one-pot automated [18F]FBnTP production for clinical imaging of mitochondrial activity. *Journal of Nuclear Medicine, 64*, P22–P22.

Lisova, K., Pla, D., Sergeev, G., van Dam, G., van Dam, R. M., & Michael, R. (2017). Leveraging microvolume radiochemistry and trifluoroborate chemistry to synthesize 18F-labeled prosthetic groups via isotopic exchange with high specific activity. In *The 22nd International Symposium on Radiopharmaceutical Sciences. Presented at the ISRS 2017, Journal of Labelled Compounds and Radiopharmaceuticals* (p. S37). https://doi.org/10.1002/jlcr.3507

Lisova, K., Sergeev, M., Evans-Axelsson, S., Stuparu, A. D., Beykan, S., Collins, J., Jones, J., Lassmann, M., Herrmann, K., Perrin, D., Lee, J. T., Slavik, R., & van Dam, R. M. (2018). Microscale radiosynthesis, preclinical imaging and dosimetry study of [18F]AMBF3-TATE: A potential PET tracer for clinical imaging of somatostatin receptors. *Nuclear Medicine and Biology, 61*, 36–44. https://doi.org/10.1016/j.nucmedbio.2018.04.001

Lisova, K., Chen, B. Y., Wang, J., Fong, K. M.-M., Clark, P. M., & van Dam, R. M. (2019). Rapid, efficient, and economical synthesis of PET tracers in a droplet microreactor: Application to O-(2-[18F]fluoroethyl)-L-tyrosine ([18F]FET). *EJNMMI Radiopharmacy and Chemistry, 5*, 1. https://doi.org/10.1186/s41181-019-0082-3

Lisova, K., Wang, J., Chao, P. H., & van Dam, R. M. (2020). A simple and efficient automated microvolume radiosynthesis of [18F]Florbetaben. *EJNMMI Radiopharmacy and Chemistry, 5*, 30. https://doi.org/10.1186/s41181-020-00113-w

Lisova, K., Wang, J., Hajagos, T. J., Lu, Y., Hsiao, A., Elizarov, A., & van Dam, R. M. (2021). Economical droplet-based microfluidic production of [18F]FET and [18F]Florbetaben suitable for human use. *Scientific Reports, 11*, 20636. https://doi.org/10.1038/s41598-021-99111-4

Long, J. Z., Jacobson, M. S., & Hung, J. C. (2013). Comparison of FASTlab 18F-FDG production using phosphate and citrate buffer cassettes. *Journal of Nuclear Medicine Technology, 41*, 32–34. https://doi.org/10.2967/jnmt.112.112649

Lu, Y., & van Dam, R. M. (2021a). First microvolume metal-mediated synthesis of [18F]FDOPA in a microreactor [ABSTRACT]. *Nuclear Medicine and Biology, Abstracts of the Society of Radiopharmaceutical Sciences, 96–97*, S15–S16. https://doi.org/10.1016/S0969-8051(21)00291-2

Lu, Y., & van Dam, R. M. (2021b). *First metal-catalyzed synthesis of [18F]FDOPA in a microreactor*. Presented at the Pacifichem 2021, Virtual.

Lu, Y., He, Y., Schibli, R., Mu, L., & van Dam, R. M. (2023a). Proof-of-concept optimization of a copper-mediated 18F-radiosynthesis of a novel MAGL PET tracer on a high-throughput microdroplet platform and its macroscale translation. *Lab on a Chip, 23*, 4652–4663. https://doi.org/10.1039/D3LC00735A

Lu, Y., Sharma, G., Da Pieve, C., Kramer-Marek, G., Miller, P., & Van Dam, R. M. (2023b). *Proof of principle: [18F]AlF radiolabelling using a droplet microreactor*. Presented at the 25th international symposium on radiopharmaceutical sciences.

Lu, Y., Collins, J., Lin, K.-S., & van Dam, R. M. (2024a). Scalable droplet-based radiosynthesis of [18F]fluorobenzyltri-phenylphosphonium cation ([18F]FBnTP) via a numbering up approach. *Lab on a Chip, 24*, 728.

Lu, Y., Sharma, G., Da Pieve, C., Kramer-Marek, G., Miller, P., & Van Dam, R. M. (2024b). [18F]AlF radiolabelling using a droplet microreactor. *Nuclear Medicine and Biology, 126–127*, 108758.

Miller, P. W. (2009). Radiolabelling with short-lived PET (positron emission tomography) isotopes using microfluidic reactors. *Journal of Chemical Technology & Biotechnology, 84*, 309–315. https://doi.org/10.1002/jctb.2061

Mogi, K., Adachi, S., Takada, N., Inoue, T., & Natsume, T. (2019). Electrowetting on dielectric (EWOD) device with dimple structures for highly accurate droplet manipulation. *Applied Sciences, 9*, 2406. https://doi.org/10.3390/app9122406

Mogi, K., Kimura, H., Kondo, Y., Inoue, T., Adachi, S., & Natsume, T. (2021). Automatic radioisotope manipulation for small amount of nuclear medicine using an EWOD device with a dimple structure. *Royal Society Open Science, 8*, 201809. https://doi.org/10.1098/rsos.201809

Mossine, A. V., Tanzey, S. S., Brooks, A. F., Makaravage, K. J., Ichiishi, N., Miller, J. M., Henderson, B. D., Skaddan, M. B., Sanford, M. S., & Scott, P. J. H. (2019). One-pot synthesis of high molar activity 6-[18F]fluoro-L-DOPA by Cu-mediated fluorination of a BPin precursor. *Organic & Biomolecular Chemistry, 17*, 8701–8705. https://doi.org/10.1039/C9OB01758E

Naka, S., Watabe, T., Lindner, T., Cardinale, J., Kurimoto, K., Moore, M., Tatsumi, M., Mori, Y., Shimosegawa, E., Valla, F., Kato, H., & Giesel, F. L. (2021). One-pot and one-step automated radio-synthesis of [18F]AlF-FAPI-74 using a multi purpose synthesizer: a proof-of-concept experiment. *EJNMMI Radiopharmacy and Chemistry, 6*, 28. https://doi.org/10.1186/s41181-021-00142-z

Nakaoka, T., Uetake, Y., Kaneko, K., Niwa, T., Ochiai, H., Irie, S., Suezaki, Y., Otsuka, N., Hayashinaka, E., Wada, Y., Cui, Y., Maeda, K., Kusuhara, H., Sugiyama, Y., Hosoya, T., & Watanabe, Y. (2020). Practical synthesis of [18F]Pitavastatin and evaluation of hepatobiliary transport activity in rats by positron emission tomography. *Molecular Pharmaceutics, 17*, 1884–1898. https://doi.org/10.1021/acs.molpharmaceut.9b01284

Pascali, G., & Salvadori, P. A. (2016). Opportunities and challenges in the utilization of microfluidic technologies to the production of radiopharmaceuticals. *Chimica Oggi - Chemistry Today, 34*, 28–32.

Rensch, C., Jackson, A., Lindner, S., Salvamoser, R., Samper, V., Riese, S., Bartenstein, P., Wängler, C., & Wängler, B. (2013). Microfluidics: A groundbreaking technology for PET tracer production? *Molecules, 18*, 7930–7956. https://doi.org/10.3390/molecules18077930

Rios, A. (2022). *High-throughput technologies for the optimization of radiopharmaceuticals using microfluidics*. UCLA.

Rios, A., Wang, J., Chao, P. H., & van Dam, R. M. (2019). A novel multi-reaction microdroplet platform for rapid radiochemistry optimization. *RSC Advances, 9*, 20370–20374. https://doi.org/10.1039/C9RA03639C

Rios, A., Holloway, T. S., Chao, P. H., De Caro, C., Okoro, C. C., & van Dam, R. M. (2022). Microliter-scale reaction arrays for economical high-throughput experimentation in radiochemistry. *Scientific Reports, 12*, 10263. https://doi.org/10.1038/s41598-022-14022-2

Ryzhikov, N. N., Seneca, N., Krasikova, R. N., Gomzina, N. A., Shchukin, E., Fedorova, O. S., Vassiliev, D. A., Gulyás, B., Hall, H., Savic, I., & Halldin, C. (2005). Preparation of highly specific radioactivity [18F]flumazenil and its evaluation in cynomolgus monkey by positron emission tomography. *Nuclear Medicine and Biology, 32*, 109–116. https://doi.org/10.1016/j.nucmedbio.2004.11.001

Salvador, B., Luque, A., Fernandez-Maza, L., Corral, A., Orta, D., Fernández, I., & Quero, J. M. (2017). Disposable PDMS chip with integrated [18F]fluoride pre-concentration cartridge for radiopharmaceuticals. *Journal of Microelectromechanical Systems, 26*, 1442–1448. https://doi.org/10.1109/JMEMS.2017.2764121

Seok Moon, B., Hyung Park, J., Jin Lee, H., Sun Kim, J., Sup Kil, H., Se Lee, B., Yoon Chi, D., Chul Lee, B., Kyeong Kim, Y., & Eun Kim, S. (2010). Highly efficient production of [18F]fallypride using small amounts of base concentration. *Applied Radiation and Isotopes, 68*, 2279–2284. https://doi.org/10.1016/j.apradiso.2010.06.016

Sergeev, M., Lazari, M., Morgia, F., Collins, J., Javed, M. R., Sergeeva, O., Jones, J., Phelps, M. E., Lee, J. T., Keng, P. Y., & van Dam, R. M. (2018). Performing radiosynthesis in microvolumes to maximize molar activity of tracers for positron emission tomography. *Communications Chemistry, 1*, 10. https://doi.org/10.1038/s42004-018-0009-z

Tang, G., Zeng, W., Yu, M., & Kabalka, G. (2008). Facile synthesis of N-succinimidyl 4-[18F]fluorobenzoate ([18F]SFB) for protein labeling. *Journal of Labelled Compounds and Radiopharmaceuticals, 51*, 68–71. https://doi.org/10.1002/jlcr.1481

Tarn, M. D., Pascali, G., De Leonardis, F., Watts, P., Salvadori, P. A., & Pamme, N. (2013). Purification of 2-[18F]fluoro-2-deoxy-d-glucose by on-chip solid-phase extraction. *Journal of Chromatography. A, 1280*, 117–121. https://doi.org/10.1016/j.chroma.2013.01.032

TechneScan® DTPA Kit," (n.d.). https://perlamar.ie/wp-content/uploads/IE-4362-Technescan-DTPA-SPCFeb16.pdf

Vanbrocklin, H. F., O'Neil, J. P., Hom, D. L., & Gibbs, A. R. (2001). Synthesis of [18F]fluoroanilines: Precursors to [18F]fluoroanilinoquinazolines. *Journal of Labelled Compounds and Radiopharmaceuticals, 44*, S880–S882. https://doi.org/10.1002/jlcr.25804401309

Wang, J., & van Dam, R. M. (2020). High-efficiency production of radiopharmaceuticals via droplet radiochemistry: A review of recent progress. *Molecular Imaging, 19*, 1–21. https://doi.org/10.1177/1536012120973099

Wang, J.-Q., Tueckmantel, W., Zhu, A., Pellegrino, D., & Brownell, A.-L. (2007). Synthesis and preliminary biological evaluation of 3-[18F]fluoro-5-(2-pyridinylethynyl)benzonitrile as a PET radiotracer for imaging metabotropic glutamate receptor subtype 5. *Synapse, 61*, 951–961. https://doi.org/10.1002/syn.20445

Wang, M., Gao, M., Miller, K. D., & Zheng, Q.-H. (2011). Synthesis of [11C]PBR06 and [18F]PBR06 as agents for positron emission tomographic (PET) imaging of the translocator protein (TSPO). *Steroids, 76*, 1331–1340. https://doi.org/10.1016/j.steroids.2011.06.012

Wang, J., Chao, P. H., Hanet, S., & van Dam, R. M. (2017). Performing multi-step chemical reactions in microliter-sized droplets by leveraging a simple passive transport mechanism. *Lab on a Chip, 17*, 4342–4355. https://doi.org/10.1039/C7LC01009E

Wang, J., Chao, P. H., & van Dam, R. M. (2019). Ultra-compact, automated microdroplet radiosynthesizer. *Lab on a Chip, 2415–2424*, 2415. https://doi.org/10.1039/C9LC00438F

Wang, J., Chao, P. H., Slavik, R., & van Dam, R. M. (2020a). Multi-GBq production of the radiotracer [18F]fallypride in a droplet microreactor. *RSC Advances, 10*, 7828–7838. https://doi.org/10.1039/D0RA01212B

Wang, J., Holloway, T., Lisova, K., & van Dam, R. M. (2020b). Green and efficient synthesis of the radiopharmaceutical [18F]FDOPA using a microdroplet reactor. *Reaction Chemistry & Engineering, 5*, 320–329. https://doi.org/10.1039/C9RE00354A

Wang, J., Rios, A., Lisova, K., Slavik, R., Chatziioannou, A. F., & van Dam, R. M. (2020c). High-throughput radio-TLC analysis. *Nuclear Medicine and Biology, 82–83*, 41–48. https://doi.org/10.1016/j.nucmedbio.2019.12.003

Zhang, X., Basuli, F., Shi, Z.-D., Xu, B., Blackman, B., Choyke, P. L., & Swenson, R. E. (2016). Automated synthesis of [18F](2S,4R)-4-fluoroglutamine on a GE TRACERlab™ FX-N pro module. *Applied Radiation and Isotopes, 112*, 110–114. https://doi.org/10.1016/j.apradiso.2016.02.016

Zhang, X., Liu, F., Knapp, K.-A., Nickels, M. L., Manning, H. C., & Bellan, L. M. (2018). A simple microfluidic platform for rapid and efficient production of the radiotracer [18F]fallypride. *Lab on a Chip, 18*, 1369–1377. https://doi.org/10.1039/C8LC00167G

Zhang, X., Liu, F., Payne, A. C., Nickels, M. L., Bellan, L. M., & Manning, H. C. (2020). High-yielding radiosynthesis of [68Ga]Ga-PSMA-11 using a low-cost microfluidic device. *Molecular Imaging and Biology, 22*, 1370–1379. https://doi.org/10.1007/s11307-020-01515-7

Digital Microfluidics and Radiochemistry: Current State and Future Trends

Jay M. Pimprikar, Fatemeh Ahmadi, and Steve C. C. Shih

1 Need for Microscale Synthesis for Radiochemistry

Due to the high equipment and infrastructure costs associated with radioisotope synthesis and radiation shielding, the production of radiotracers, such as 2-[^{18}F]fluoro-2-deoxy-D-glucose ([^{18}F]FDG), used in Positron Emission Tomography (PET) scans, is typically centralized. A centralized approach provides a significant financial advantage for this tracer, since the expenses for producing this tracer can be spread among many patients with minimal additional costs incurred when scaling up the production. However, the limitation of centralized facilities is exacerbated, when the half-life is too short for transport (e.g., [^{11}C]Met) and when there is a need to obtain a diverse library of probes. Since most non-[^{18}F]FDG radiotracers are relatively low in demand, these tracers could not be provided by a centralized radiopharmacy at a reasonable price that is affordable by any laboratory. Therefore, this has prompted the development of decentralized solutions, such as microfluidic "lab-on-chip" platforms, where imaging centers (located in local hospitals) can synthesize individual doses of various types of tracers in an automated and time-sensitive manner themselves (Chen et al., 2014a, 2014b; Vaquero & Kinahan, 2015).

Several microfluidic, decentralized, solutions have been developed to robustly synthesize PET tracers. Each of these solutions can be classified as a subset of the microfluidic paradigm they leverage: continuous-flow (Liang et al., 2014a, 2014b; Calderwood et al., 2015; Selivanova et al., 2012; Kimura et al., 2016),

---

J. M. Pimprikar · F. Ahmadi · S. C. C. Shih (✉)
Centre for Applied Synthetic Biology, Concordia University, Montréal, QC, Canada

Department of Electrical and Computer Engineering, Concordia University, Montréal, QC, Canada
e-mail: steve.shih@concordia.ca

R. M. van Dam et al. (eds.), *Automated Technologies for the Development and Production of Radiopharmaceuticals*,
https://doi.org/10.1007/978-3-031-84632-8_11

batch-reactor-based (Zhang et al., 2018; Bejot et al., 2011; Lebedev et al., 2013; Elizarov et al., 2010; Lee et al., 2005), and digital microfluidics (DMF) (Keng et al., 2012; Chen et al., 2014a, 2014b; Javed et al., 2014; Wang et al., 2019; Mogi et al., 2021; Ahmadi et al., 2023). Continuous-flow solutions struggle with cross-contamination (small amounts of previous fluids remain on the channel and are introduced into a new sample), leakage, clogging, and the need for conventional purification (Gillies et al., 2006; Audrain, 2007). While batch-reactor-based solutions have made progress in overcoming some of the drawbacks of continuous-flow solutions, certain challenges continue to persist. The high vapor pressure of solvents, resulting from the required increase in temperature, along with reagent evaporation, complex chip design and fabrication, and the complexity of external valve and pump control systems, act as barriers for the widespread adoption of this method at localized facilities (Gillies et al., 2006; Lebedev et al., 2013; Elizarov et al., 2010).

2 Digital Microfluidic Automation and Operations

Digital microfluidics (DMF) is an electrode array-based platform that allows the manipulation of individual droplets through the application of an electric potential, eliminating the need for pumps or valves (Choi et al., 2012; Li, 2020; Liu et al., 2023). The benefits of DMF include "plug-and-play" automation, where the device can be directly connected to automation equipment, enabling programmable fluidic operations such as dispense, merge, mix, and split (Fig. 1a) (Moazami et al., 2019; Fobel et al., 2013) and reducing concerns of cross-contamination via routing of droplet movements along different electrode paths (Huang et al., 2010). The most common format for a DMF system is a two-plate system, where the droplet is sandwiched between the actuation (bottom) plate and the hydrophobically coated ground top plate (Fig. 1b). The bottom plate consists of metal-based electrodes where potentials are applied, upon which a dielectric insulator is coated on top to allow for charge insulation and polarization. This is followed by a hydrophobic layer to permit droplet mobility and reducing the friction between the droplet and the surface of the device. The architecture of the open DMF system is similar to that of the closed except the ground electrode is coplanar with the activated electrodes (Fig. 1b).

The mechanisms of DMF via applying an electric field to enable droplet actuation allows for pL to μL droplet mixing, merging, dispensing from on-chip reservoirs (i.e., electrodes that hold more volume than the typical actuation electrode), and splitting in an automated fashion (see review papers Choi et al., 2012; Li, 2020). Aside from these standard droplet operations, the radiochemistry applications performed on DMF require other non-droplet operations. First, temperature control is critical for many chemical processes, especially for radiochemical synthesis (Keng & van Dam, 2015; Chen et al., 2012a, 2012b, 2014a, 2014b; Ding et al., 2012; Javed et al., 2014; Keng et al., 2012; Mogi et al., 2021; Ahmadi et al., 2023), as the reactions need heat to accelerate the reaction kinetics and to limit the time for reactions involving isotopes with short half-lives. Generally, microheaters or thermoelectric

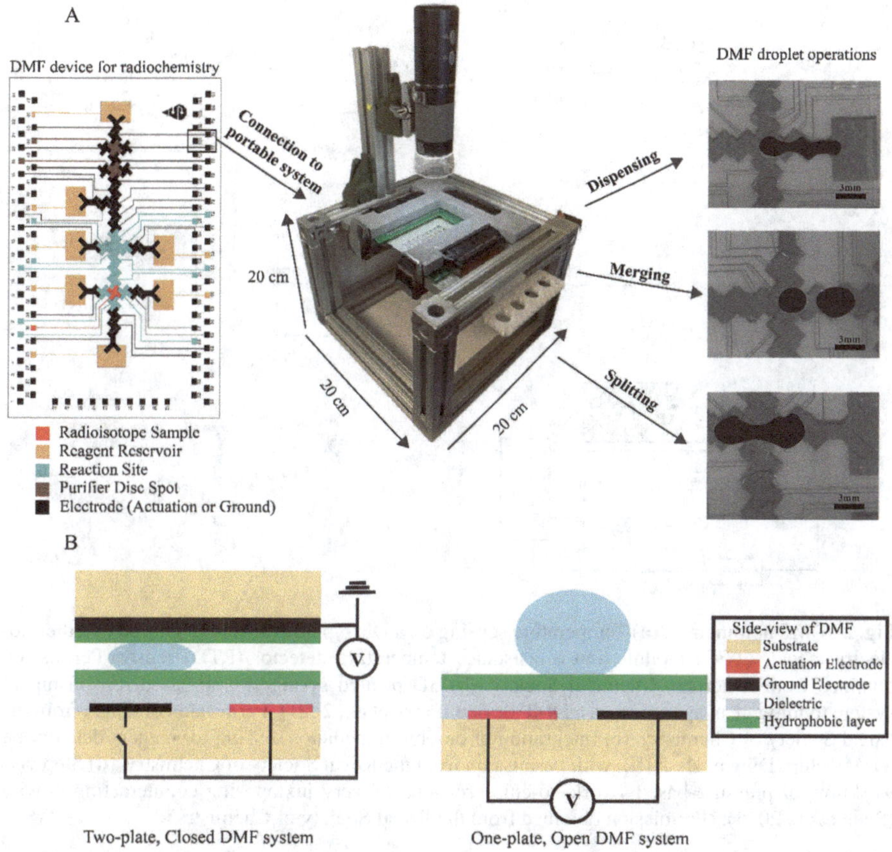

Fig. 1 Digital microfluidics for radiochemistry. (**a**) An automated digital microfluidic system for radiochemistry. To automate radiochemistry using DMF, a DMF device (left) requires a section for the radioisotope sample (red), the reaction site (cyan), and purification of the product (brown). Reagent reservoirs (gold) are used to dispense droplets of nL–μL volumes, which are then moved to different sections of the device by applying electric potentials to the electrodes (black). Application of the potentials are applied to devices via contact pads (rectangle) and are placed in a portable system (Perry et al., 2021) that can be used for decentralized synthesis. Using such a system, we can perform a number of droplet operations: mix, merge, and dispense as shown by the images from a movie (right) taken from Ahmadi et al., 2023 with permission from the Royal Society of Chemistry. (**b**) Side-view illustrations of two different DMF configurations. The two-plate, closed system includes a top, ground plate which is used to complete the actuation circuit. The one-plate system is open to atmosphere, and both ground and active electrodes are on the same plane

elements can be integrated on the bottom substrate along with a temperature sensor to continually monitor the temperature on the device. Reading the sensor temperature allows the execution of an algorithm, enabling precise temperature control through a closed-loop control system (Fig. 2a) (Moazami et al., 2019). Second, heating may cause sample evaporation and volume losses, and therefore a solution

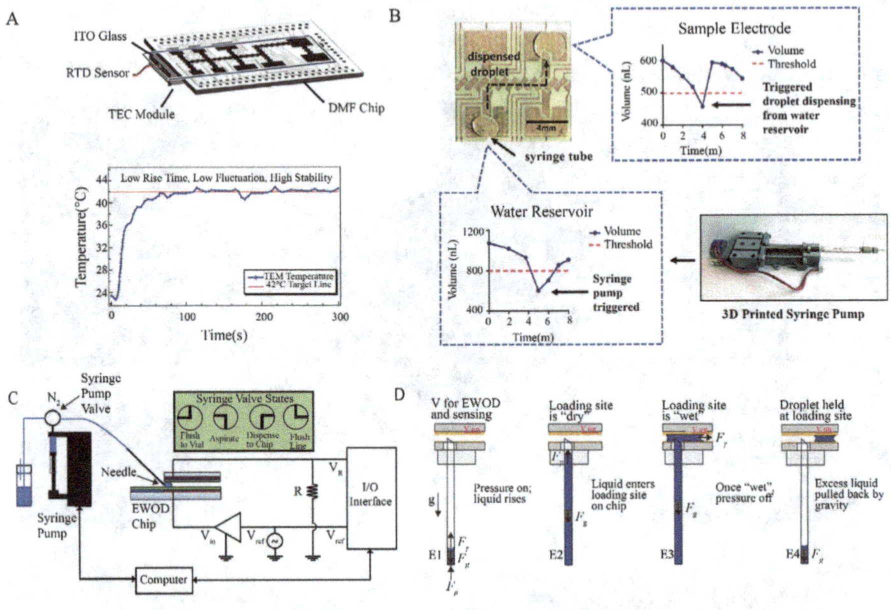

Fig. 2 DMF automation. (**a**) Temperature sensing on a DMF platform via integration of a thermoelectric cooler (TEC) module and a resistance temperature detector (RTD) sensor. Permission obtained from American Chemical Society. (**b**) 3D printed syringe pump-based replenishment system to mitigate evaporation on a DMF device (Perry et al., 2021). Permission obtained from the Royal Society of Chemistry. (**c**) Integration of on-chip impedance sensing for reagent delivery on a DMF chip (Ding et al., 2012) with permission from the Royal Society of Chemistry. (**d**) Stepwise workflow of pressure-based "world-to-chip" reagent delivery juxtaposing counteracting gravity (Shah et al., 2013). (Permission obtained from the Royal Society of Chemistry)

is to continually replenish and resupply samples with water to maintain the initial reaction concentrations. Several methods have been developed (Jebrail et al., 2015; Perry et al., 2021; Shah et al., 2013), but the most common method is to connect a syringe pump with the replenishment solution directly to an on-chip reservoir, which will then follow with typical droplet operations to replenish the sample volume (Fig. 2b). Third, evaporating solvent is a required operation for radiochemical synthesis which can be performed on the DMF device. The droplets are directly exposed to ambient air enables droplets being delivered to target areas where the droplets can evaporate directly on the device surface (Moon et al., 2006; Jebrail et al., 2010).

With its advanced automation capabilities, there exists significant potential in using DMF as a decentralized platform for radiotracer synthesis. This chapter is dedicated to a comprehensive exploration of the recent achievements and obstacles encountered in the integration of DMF into radiotracer synthesis processes. It particularly emphasizes the diverse array of operations achievable on DMF, encompassing both droplet and non-droplet techniques. Furthermore, we delve into the exciting prospects that lie ahead by synergizing DMF with the evolution of radiotracer development.

3 Current Works in DMF and Radiochemistry

Prior to performing synthesis on-chip, specific radiochemical reagent dispensing techniques have been implemented to move toward full automation capabilities. Ding et al. used a syringe pump system to automate the delivery of acetonitrile (MeCN) during [^{18}F]FDG synthesis. Since MeCN is a "wetting" liquid (contact angle less than 90°), storing it on a chip reservoir is not possible (Ding et al., 2012; Shah et al., 2013) due to its rapid evaporation rate. To account for evaporation that may take place between dispensing the MeCN and it reaching the chip via a needle, an on-chip liquid detector was generated based on impedance sensing between the loading electrode and the ground electrode. The sensing took place via the integration of a 1 kΩ resistor between the ground plate and the power supply ground, whereupon the activation of the loading electrode generates an alternating current and voltage drop across the resistor (Ding et al., 2012). Figure 2c shows the integration of the 1 kΩ resistor serving as an impedance sensor and how it can be implemented for delivering accurate volumes of volatile liquids commonly used in radiotracer synthesis.

Alternatively, to having a bulky syringe pump for loading each liquid, Shah et al. used a combination of compressed inert gas and gravity to exert driving and retracting forces for precise loading of wetting and non-wetting liquids on-chip via a feedback signal (Shah et al., 2013). Wetting liquids such as MeCN necessitate a "world-to-chip" solution for dispensing, and the authors curated a pneumatic liquid loading system against gravity. The liquids are drawn from pierced septum-capped vials and delivered from underneath the actuation plate, so that gravity serves as a counteracting force to the pressurized vial (Shah et al., 2013). Figure 2d provides a workflow schematic of pneumatic liquid delivery "world-to-chip" system.

The most widely automated PET radiotracer on DMF has been [^{18}F]FDG due to its broad use in cancer imaging (Ashraf & Goyal, 2023). The chemical reaction fluorinates the precursor (mannose triflate) to generate intermediate [^{18}F]FTAG, and, upon hydrolysis, [^{18}F]FDG is produced (Fig. 3a). However, many other ^{18}F-based tracers, such as [^{18}F]FLT, [^{18}F]fallypride, [^{18}F]SFB, and the Single Photon Emission Computed Tomography (SPECT) tracer [^{99m}Tc]Tc-DTPA have also been synthesized using a similar DMF platform (Keng & van Dam, 2015; Mogi et al., 2021). For example, Fig. 3b shows the synthesis steps for [^{18}F]fallypride, which also requires the necessary steps (liquid handling, heating, drying) for successful production. As illustrated in Fig. 3c, [^{18}F]fallypride and other tracers ([^{18}F]FDG, [^{18}F]SFB, [^{18}F]FLT) can be synthesized using a single DMF chip design, as demonstrated by Chen et al. (2014a, 2014b). In this design, patterned square electrodes are fabricated for automating the liquid handling steps, and a concentric electrode design is used for heating, while Ahmadi et al. (2023) use the patterned electrodes

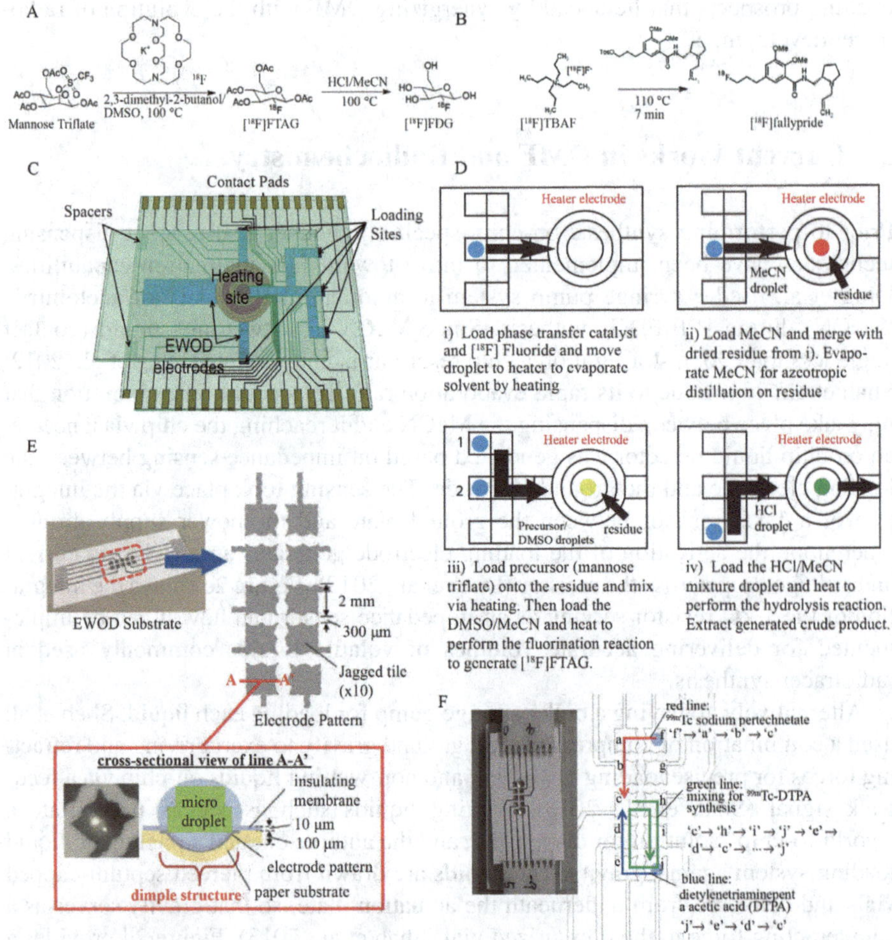

Fig. 3 Current DMF platforms for radiochemistry. (**a**) Schematic for multistep FDG synthesis via mannose triflate fluorination (Koag et al., 2014a, 2014b). Permission obtained from Elsevier. (**b**) Synthesis schematic of [^{18}F]fallypride (Wang et al., 2020). Permission obtained from the Royal Society of Chemistry. (**c**) Schematic of EWOD chip used for the synthesis of [^{18}F]FDG, [^{18}F]FLT, [^{18}F]fallypride, and [^{18}F]SFB (Chen et al., 2014a, 2014b). Permission obtained from the Royal Society of Chemistry. (**d**) Sample DMF FDG synthesis schematic adapted from Chen et al. (2014a, 2014b) and Keng and van Dam (2015) with permission from the Royal Society of Chemistry and the Proceedings of the National Academy of Sciences. (**e**) Schematic of a paper-based EWOD device showing electrode pattern and a cross-sectional view of the dimple structure (Mogi et al., 2021). Permission obtained from the Royal Society of Chemistry. (**f**) Schematic for the synthesis of [^{99m}Tc]Tc-DTPA on paper-based coplanar EWOD device (Mogi et al., 2021). (Permission obtained from the Royal Society of Chemistry)

exclusively for liquid handling and incorporate the heater externally to the substrate. In either case, droplets containing the precursors are dispensed, merged, mixed using the electrodes, and these droplets are further dried heated to expedite the fluorination labeling process at a target electrode location, and then extracted off the chip (Fig. 3d).

The above devices are generally fabricated by patterning metal on a glass substrate coated with a dielectric and a hydrophobic layer (Keng et al., 2012; Chen et al. 2014b; Ahmadi et al., 2023). Glass (and silicon) substrate devices pose a challenge, as they require fabrication in cleanroom environments, hindering scalability and rendering them economically impractical for low-cost decentralized synthesis. However, an alternative substrate that could address this issue is paper. Recently, paper microfluidics and DMF have been combined (Abadian et al., 2017; Ruecha et al., 2017), leveraging the automation capabilities of DMF along with the numerous advantages offered by paper microfluidics. An example of this is the work by Mogi et al., who created an alternative open-style DMF device comprising a primarily paper substrate incorporated with dimple-based electrode structures to minimize any erroneous movements during the synthesis of SPECT tracer [^{99m}Tc]Tc-DTPA (Mogi et al., 2021). Droplets were manipulated by voltage switching using a palm-sized controller and the device consisted of 2 mm × 2 mm electrode patterns inkjet printed with conductive ink (Mogi et al., 2021). Figure 3e demonstrates the electrode patterns and cross-sectional layout of the device. As shown, a 10-μm-thick insulating membrane was layered and embossed on top of the electrodes to create the dimple patterns (Mogi et al., 2021). The adhesion energy was determined to be equal to 1.3 times that of a flat surface, creating dimples with depth of 52.9 ± 2.5 μm. This adhesion energy provided the best balance between droplet mobility and stability in dimple structures. Droplet manipulation was subsequently possible for volumes between 5 and 30 μL (Mogi et al., 2021). Using such a device, the SPECT tracer [^{99m}Tc]Tc-DTPA was synthesized on chip by mixing 10 μL of DTPA saline solution with 10 μL of sodium pertechnetate and incubating the mixture for 5 min at room temperature, as shown in Fig. 3f. After incubation, the chelating efficiency of [^{99m}Tc]Tc-DTPA was evaluated using TLC and determined to be 99.7 ± 0.13% (Mogi et al., 2021).

3.1 Improving Fluorination Efficiency and Radiochemical Yield

The primary goal of miniaturizing radiosynthesis, which encompasses reducing synthesis costs, time, and enabling decentralization, also hinges on achieving both a high crude radiochemical yield (RCY) and a high fluorination efficiency percentage. A high fluorination efficiency indicates a significant proportion of [^{18}F]fluoride has been successfully labeled on the precursor molecule, which is assessed by performing radio-HPLC analysis (Walters et al., 2012). Generally, if the fluorination step is highly efficient, it will lead to a higher amount of the radiotracer product in the final synthesis, positively impacting the crude RCY. However, several parameters such as reagent concentration, droplet volume, phase transfer catalyst

(PTC)-to-precursor ratio, fluorination reaction temperature, and reaction time require optimization such that maximum fluorination efficiency can be achieved on microscale devices.

Each of these parameters was shown to be significant in altering fluorination percentage for the on-chip optimization of [^{18}F]FDG, [^{18}F]FLT, [^{18}F]fallypride, and [^{18}F]SFB (Keng & van Dam, 2015). In contrast to the macroscale, reagent concentrations need to be modestly increased to offset the ~100–1000× scale down in reaction size (Koag et al., 2014a, 2014b). For example, the maximum fluorination efficiency of [^{18}F]FDG synthesis was reached, when reagent concentrations were increased by at least five times (Keng & van Dam, 2015).

Similarly, a reduction in droplet volume increased fluorination efficiency, provided that there was sufficient volume for the reaction mixture to be completely solvated for the entire length of the reaction (Keng & van Dam, 2015). For instance, reducing the droplet volume from 4 to 2 μL (while maintaining a twofold increase in the reagent concentrations), Javed et al. demonstrated an increase in fluorination efficiency during the synthesis of [^{18}F]FLT from 79 to 94% (Javed et al., 2014). The main reason behind this increase is shortening the duration of the fluorination and hydrolysis steps, thereby minimizing the potential loss of radioactivity due to droplet volatility or on-chip stiction events. Another example is Kim et al., where reducing droplet size from 6 to 2 μL removed the need to have a drying step after hydrolysis, since there was minimal residual DMSO remaining after fluorination. As a result, Kim et al. were able to reduce the combined heating time of a three-step radiosynthesis of [^{18}F]SFB by avoiding the long DMSO evaporation step, which led to an increase in the crude RCY from 38 to 50% (Kim et al., 2019).

Additionally, another factor that significantly alters fluorination efficiency is the phase transfer catalyst (PTC) to precursor ratio. As shown in Fig. 4a, during the synthesis of [^{18}F]SFB, Kim et al. had to deviate from standard macroscale PTC to precursor ratios of 4–6 and instead found optimal fluorination efficiency with a ratio of 2.3 (Kim et al., 2019). Furthermore, varying precursor amounts alone can affect the fluorination step—that is, the production of the [^{18}F]FTAG intermediate in [^{18}F]FDG synthesis was altered by varying only the amount of precursor available during the fluorination reaction (Keng et al., 2012), as illustrated in Fig. 4c.

At the microscale, the fluorination reaction time and reaction temperature stand as the main significant factors influencing fluorination efficiency. Both Kim et al. (2019) and Keng et al. (2012) showed the effects of reaction time (Fig. 4b and 4e) and temperature (Fig. 4d) on fluorination efficiency. From Kim et al. (2019), reaction times less than 4.5 min were generally shown to have demonstrably lower fluorination efficiencies, however, increasing reagent concentration can improve the fluorination efficiency. Accordingly, Keng et al. (2012) illustrated that increasing the reaction time can increase [^{18}F]FTAG conversion. As shown, adding more time (7 min) can further increase the efficiency from 56 % (1 min) to > 80 % (7 min) at a temperature of 100 °C (Fig. 4e). Although increasing reaction temperature from 80 to 100 °C can have the largest step increase in the [^{18}F]FTAG conversion efficiency from ~45 to 85% for a static reaction time of 7 min (Fig. 4d). This follows

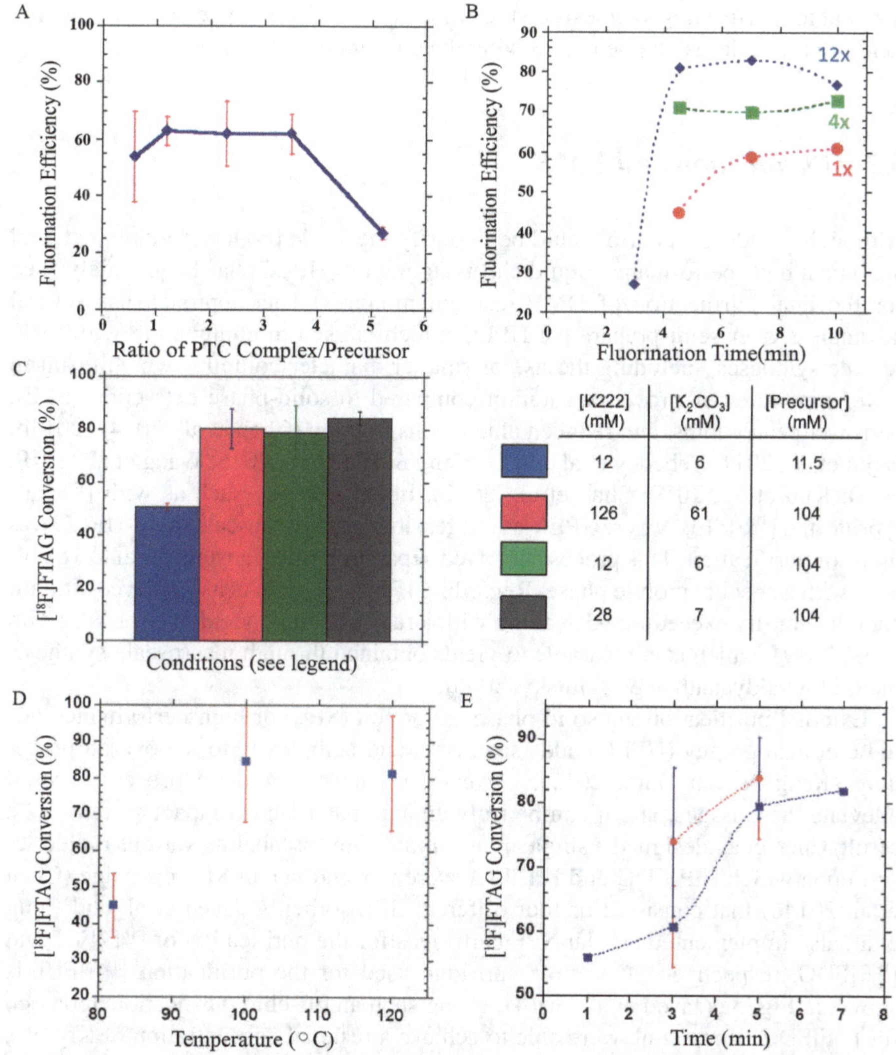

Fig. 4 Improving fluorination efficiency and radiochemical conversion. (**a**) Fluorination efficiency of [^{18}F]SFB as a function of PTC complex to precursor ratio (Kim et al., 2019). Permission obtained from the Royal Society of Chemistry. (**b**) Fluorination efficiency as a function of fluorination time for varied reagent concentrations (Kim et al., 2019). Permission obtained from the Royal Society of Chemistry. (**c**) Effects of reagent concentrations on fluorination efficiency during [^{18}F]FDG synthesis while temperature and reaction time remain constant at 90 °C and 5 min (Keng et al., 2012). Permission obtained from the Proceedings of the National Academy of Sciences. (**d**) Effect of temperature on [^{18}F]FTAG conversion efficiency for conserved reaction time of 5 min (Keng et al., 2012). Permission obtained from the Proceedings of the National Academy of Sciences. (**e**) Fluorination efficiency during [^{18}F]FDG intermediate synthesis for various reaction times (red circles indicate reactions performed at 100 °C, and blue squares indicate reactions performed at 120 °C) (Keng et al., 2012). (Permission obtained from the Proceedings of the National Academy of Sciences)

the goal for performing synthesis on the chip, which is to minimize the reaction time (and minimize decay) while maximizing the conversion efficiency.

3.2 *Purification and DMF*

Although the ideal scenario would be to purify the crude product on-chip, external analytical high-performance liquid chromatography (HPLC) has been widely used for the final purification of DMF reaction mixtures. This approach has several advantages over semi-preparative HPLC which is used in conjunction with macroscale syntheses, including the use of smaller diameter columns, which result in lower flow rates, improved separation compared to solid-phase extraction (SPE), narrower peak widths, and reduced eluent consumption (Chen et al., 2014a, 2014b; Bejot et al., 2011; Lebedev et al., 2013; Keng & van Dam, 2015; Wang et al., 2019, 2020; Kim et al., 2019; Shah et al., 2012). In some cases, such as with [^{18}F]fallypride and [^{18}F]SFB, where SPE was not feasible, analytical scale radio-HPLC was used for purification. This process involved separation using an analytical C-18 column with a specific mobile phase. Regarding [^{18}F]SFB, Kim et al. achieved a radiochemical purity exceeding 95% along with a radiochemical yield of 39 ± 7% (Kim et al., 2019), which is comparable to yields obtained through macroscale synthesis methods (Vaidyanathan & Zalutsky, 2006).

External purification via solid-phase extraction (SPE) or high-performance liquid chromatography (HPLC) adds size, cost, and complexity to the overall procedure (Keng & van Dam, 2015); however, miniaturizing these processes could alleviate these issues, and it can be truly an automated and compact system. As a result, Chen et al. designed a single microfluidic chip for labeling various radiotracers, upon which [^{18}F]FDG and [^{18}F]FLT were purified via an SPE cartridge (Chen et al. 2014b) that consisted of four different SPE sorbents. Javed et al. and Keng et al. also implemented similar SPE cartridges for the purification of [^{18}F]FLT and [^{18}F]FDG, respectively. A sample cartridge used for the purification [^{18}F]FLT is shown in Fig. 5a (Javed et al., 2014). Using such an off-chip purification approach for [^{18}F]FDG, Keng et al. were able to achieve a residual concentration of Krytofix less than 4 μg/mL as well as residual levels of DMSO, ethanol, and acetonitrile being 870 ppm, 115 ppm, and less than 20 ppm, respectively. All of these concentrations were well below the limit set by the United States Pharmacopeia. The overall radiochemical yield of [^{18}F]FDG was 22 ± 8% (Keng et al., 2012). Similarly, Javed et al. were able to obtain a RCY of 63 ± 5% for [^{18}F]FLT with over 99% radiochemical purity. The cartridge purification efficiency was 89 ± 2% (Javed et al., 2014). Nonetheless, the significant progress in miniaturization opens the possibility of integrating downstream processes on-chip, such as purification, which would play a crucial role in making decentralized PET production feasible in the laboratory setting.

Microscale purification solutions have been shown to completely automate the tracer synthesis workflow on chip with purification (Tarn et al., 2013; Chen et al.,

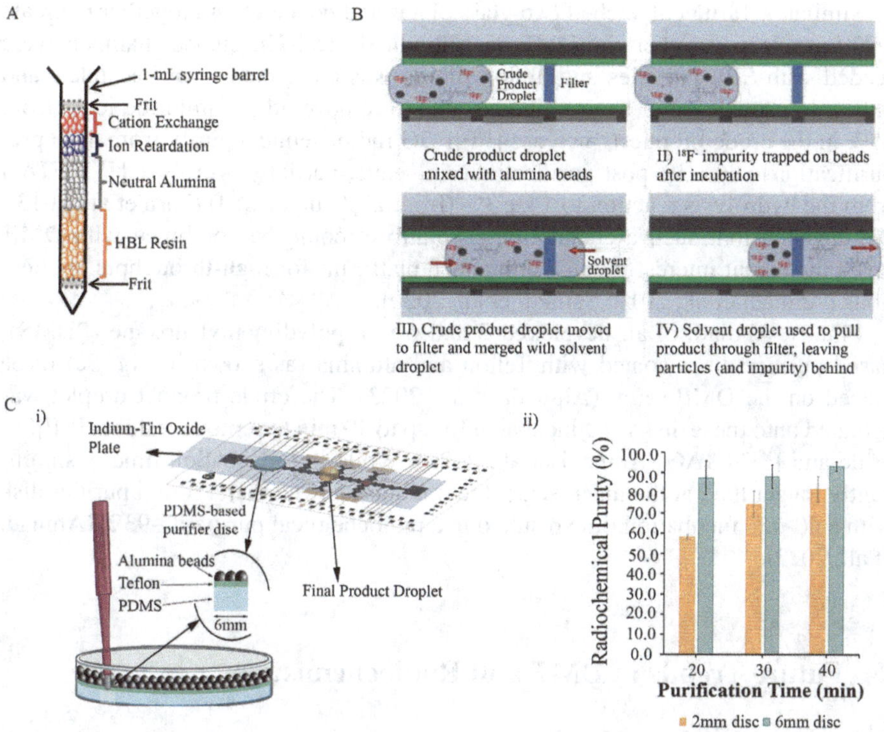

Fig. 5 Automating purification for on-chip synthesis. (**a**) Multilayered SPE purification cartridge for [^{18}F]FLT (Javed et al., 2014). Permission obtained from the Society of Nuclear Medicine and Molecular Imaging. (**b**) Workflow for on-chip SPE (Keng & van Dam, 2015). Permission obtained from Hindawi. (**c**) Using 2 and 6 mm PDMS disks for on-chip purification and their implications on RCY (Ahmadi et al., 2023). (Permission obtained from the Royal Society of Chemistry)

2014a, 2014b; Ahmadi et al., 2023). Primarily, Chen et al. demonstrated SPE on-chip via the integration (via photolithography) using KMPR® photoresist diamond shaped 140 μm tall pillars into top plate for the purification of [^{18}F]fallypride from [^{18}F]fluoride. These pillars served to mechanically separate the alumina particles, which were added after the reaction to scavenge unreacted [^{18}F]fluoride from the purified product (Chen et al., 2014a, 2014b). Although the particles and pillars served to eliminate [^{18}F]fluoride from the reaction mixture, they did not remove other impurities that may have been present. Figure 5b shows a methodological workflow for alumina bead separation using the fabricated pillars, and as shown, Chen et al. were able to achieve a removal of [^{18}F]fluoride to <1% (from the 16% in the crude product). In addition, the fluorination yield of 82% measured by Cerenkov imaging of activity in the purified product droplet versus activity trapped on the filtered alumina beads post-purification was consistent with the 83.9% measured radioactivity in the crude mixture via radio-TLC (Chen et al., 2014a, 2014b).

Similarly, Tarn et al. etched two glass plates and bonded them together to create a 300-μm-deep chamber non-DMF microfluidic device. The etched chambers were seeded with SPE particles and Teflon tubing was connected to drilled inlets and outlets to provide flow. Their chip was able to remove all [^{18}F]fluoride ions (from 25% in the crude product), increase [^{18}F]FDG radiochemical purity from 47% pre-purification to 86 ± 3% post-purification, and reduce partially hydrolyzed [^{18}F]FTAG from the hydrolysis reaction to 14 ± 3% (from 27% unpurified) (Tarn et al., 2013). We envision that such a channel-based platform could be combined with DMF, given the recent interest in integrating such platforms for high-throughput applications (Ahmadi et al., 2019; Samlali et al., 2020).

Finally, Ahmadi et al. developed 2 and 6 mm polydimethylsiloxane (PDMS)-based purifier disks coated with Teflon and alumina (as shown in Fig. 5c) to be placed on the DMF setup (Ahmadi et al., 2023). The crude product droplet was actuated onto these disks and incubated for up to 40 min to remove residual [^{18}F]fluoride and [^{18}F]FTAG (Ahmadi et al., 2023). While the incubation time is significantly longer than at the macroscale, the findings showed that a 6-mm purifier disk with a 40-min incubation time resulted in a radiochemical purity of ~93% (Ahmadi et al., 2023).

4 Future Trends in DMF and Radiochemistry

4.1 Machine Learning

Machine learning has emerged as a powerful tool for predicting the behavior of biochemical systems, enabling output predictions without the need for conducting an exhaustive library of experiments (Carbonell et al., 2019; Radivojević et al., 2020). Considering the necessity to finely adjust multiple parameters for radiotracer synthesis, the integration of machine learning offers a promising opportunity to improve radiolabeling efficiencies robustly. By inputting various previously measured conditions into a predictive model, the synthesis process can be optimized effectively.

In the past decade, considerable research efforts have focused on the use of machine learning and artificial intelligence to identify optimal sites for radiolabeling and reaction optimization (Bowden et al., 2019; Webb & Scott, 2021; Saboury et al., 2023). The advent of fully automated radiosynthesis systems has significantly expanded the capacity to conduct experiments and screen various conditions for optimization purposes. Recently, Ahmadi et al. presented a novel approach that combines radiochemistry, DMF, and machine learning to streamline the identification of a minimal number of experiments required for radiosynthesis optimization. The authors successfully demonstrated the integration of a one-factor at a time (OFAT) approach within the framework of radiochemistry, DMF, and machine learning. This innovative technique was employed to optimize the synthesis of

[^{18}F]FDG on a closed DMF platform, yielding promising results (Ahmadi et al., 2023). By obtaining a base level of reaction performance from literature for each factor, the value of one factor is changed, followed by measuring the output response (i.e., fluorination efficiency), and the process is repeated with another factor. After the values for each factor have been changed once, the output (fluorination efficiency in this case) of each case is compared to the base case via statistical analysis (t-test) to determine whether each of the parameters is significant or not. Figure 6a showcases eight experiments showing a value changed for each factor and the effect on the output fluorination efficiency for [^{18}F]FDG synthesis. The cumulative total of these trials, in addition to previously reported values, forms a training dataset. Finally, a multiple linear regression model was subsequently generated to predict optimal output values based on the training dataset generated (Ahmadi et al., 2023). The schematic of the model finding statistically significant parameters and optimizing synthesis conditions is shown in Fig. 6b.

Only five of the seven parameters showed a statistically significant impact on [^{18}F]FDG fluorination efficiency: precursor concentration, radiolabeling temperature, base concentration, deprotection time, and precursor-to-[^{18}F]fluoride volume ratio. Subsequently, the model determined the optimal values for each significant

A

	Mannose triflate volume / ^{18}F volume	Mannose triflate concentration (mM)	Radiolabeling temperature (°C)	Radiolabeling time (min)	NaOH concentration (M)	Deprotection temperature (°C)	Deprotection time (min)	Fluorination efficiency (%)
Base	1	30.00	85	5	1.00	25	5	61.27 ± 3.88 (n= 6)
	0.3	30.00	85	5	1.00	25	5	51.48 (0.0297)
	1	20.00	85	5	1.00	25	5	46.34 (0.0086)
	1	30.00	100	5	1.00	25	5	74.92 (0.0282)
	1	30.00	85	7	1.00	25	5	54.56 (0.0671)
	1	30.00	85	5	0.10	25	5	15.62 (0.0001)
	1	30.00	85	5	1.00	50	5	66.12 (0.2778)
	1	30.00	85	5	1.00	25	3	42.79 (0.0041)
Optimal	1	30.00	100	5	1.00	50	5	70.82 ± 1.54 (n=6)

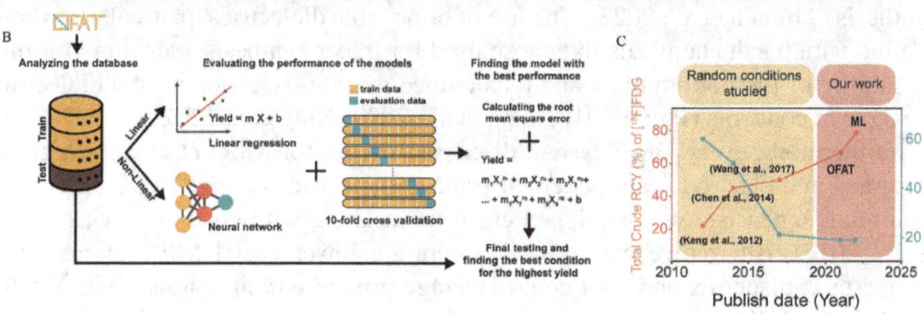

Fig. 6 Machine learning and DMF for radiochemistry. (**a**) Varying a single factor at a time (OFAT) and correlating each trial to fluorination efficiency (Ahmadi et al., 2023). Permission obtained from the Royal Society of Chemistry. (**b**) Schematic representation of using OFAT output to generate a predictive model to find significant parameters and optimal parameter values (Ahmadi et al., 2023). Permission obtained from the Royal Society of Chemistry. (**c**) Graphical illustration comparing the total crude radiochemical yield achieved by OFAT assay optimization (Ahmadi et al., 2023). (Permission obtained from the Royal Society of Chemistry)

parameter using a linear model equation predicted to obtain 85% fluorination efficiency, and the conditions were validated experimentally on-chip with a very similar (~84%) observed fluorination efficiency. Higher yields more than 85% were modeled, but the predicted experimental conditions (e.g., very high concentrations of NaOH) were not possible to be implemented on the chip (Ahmadi et al., 2023). However, using this optimization method, Fig. 6c demonstrates a decrease in synthesis time as well as an increase in synthesis yield compared to previous works. As this work was the first to integrate machine learning with radiotracer synthesis on DMF, the door opens to further leveraging machine learning either for optimization of [^{18}F]FDG using larger train-test datasets or the efficient synthesis of other tracers.

4.2 Moving DMF from Bench to Practice

Moving toward a decentralized platform for synthesis is to construct a DMF platform to be fully automated and miniaturized for easy adoption across various hospitals. Narahari et al. were able to create a fully encapsulated DMF system including all essential and peripheral electronics for patient sample processing in screening for Zika (Narahari et al., 2022). Their entire DMF platform was a box 26 cm in length by 22 cm in width allowing for portability and versatility (Narahari et al., 2022). The implementation of such an encompassing system for DMF radiotracer synthesis could aid in truly burgeoning adoption across testing facilities.

Similarly, although solutions have been demonstrated in creating a world-to-chip interface (Shah et al., 2013; Moazami et al., 2019; Perry et al., 2021), further development and integration of world-to-chip systems in encapsulated platforms like that of Narahari et al. are needed for an accurate and automated system to prevent manual handling of radioactive chemicals. In addition, the use of harsh chemicals deteriorating chip surfaces represents a barrier toward adoption of decentralized synthesis (Ahmadi et al., 2023). The use of removable dielectrics, particularly when dealing with harsh chemicals like those used for tracer synthesis, can allow for the reusability of the bottom plate and thus reduce reusability concerns and dielectric breakdown concerns (Freire 2016; Yang et al., 2009; Xing et al., 2021).

Furthermore, every DMF system discussed except for Mogi et al. uses a glass substrate, and electrodes are patterned prior to dielectric coating. These processes require an expensive, well-equipped clean room along with costly materials (Jain et al., 2017). Alternative methods of generating lower cost DMF systems with merge, mix, dispense, and split could leverage printed circuit boards (PCBs) with Parylene-C or PDMS as a dielectric (Abdelgawad & Wheeler, 2008; Gong & Kim, 2008; Jain et al., 2017; Perry et al., 2021) or thin-film transistor-based technologies (Anderson et al., 2021; Luo et al., 2022). As such, more cost-effective solutions would need to be leveraged to make DMF chips less expensive and more viable for mass production and readily available for automated radiotracer synthesis.

References

Abadian, A., Sepehri Manesh, S., & Jafarabadi Ashtiani, S. (2017). Hybrid paper-based microfluidics: Combination of paper-based analytical device (μPAD) and digital microfluidics (DMF) on a single substrate. *Microfluidics and Nanofluidics, 21*, 1–11.

Abdelgawad, M., & Wheeler, A. R. (2008). Low-cost, rapid-prototyping of digital microfluidics devices. *Microfluidics and Nanofluidics, 4*, 349–355.

Ahmadi, F., Samlali, K., Vo, P. Q., & Shih, S. C. (2019). An integrated droplet-digital microfluidic system for on-demand droplet creation, mixing, incubation, and sorting. *Lab on a Chip, 19*(3), 524–535.

Ahmadi, F., Simchi, M., Perry, J. M., Frenette, S., Benali, H., Soucy, J. P., Massarweh, G., & Shih, S. C. (2023). Integrating machine learning and digital microfluidics for screening experimental conditions. *Lab on a Chip, 23*(1), 81–91.

Anderson, S., Hadwen, B., & Brown, C. (2021). Thin-film-transistor digital microfluidics for high value in vitro diagnostics at the point of need. *Lab on a Chip, 21*(5), 962–975.

Ashraf, M. A., & Goyal, A. (2023). Fludeoxyglucose (18F). In *StatPearls*. StatPearls Publishing. Retrieved from https://www.ncbi.nlm.nih.gov/books/NBK557653/.

Audrain, H. (2007). Positron emission tomography (PET) and microfluidic devices: A breakthrough on the microscale? *Angewandte Chemie International Edition, 46*(11), 1772–1775.

Bejot, R., Elizarov, A. M., Ball, E., Zhang, J., Miraghaie, R., Kolb, H. C., & Gouverneur, V. (2011). Batch-mode microfluidic radiosynthesis of N-succinimidyl-4-[18F]fluorobenzoate for protein labelling. *Journal of Labelled Compounds and Radiopharmaceuticals, 54*(3), 117–122.

Bowden, G. D., Pichler, B. J., & Maurer, A. (2019). A design of experiments (DoE) approach accelerates the optimization of copper-mediated 18F-fluorination reactions of arylstannanes. *Scientific Reports, 9*(1), 1–10.

Calderwood, S., Collier, T. L., Gouverneur, V., Liang, S. H., & Vasdev, N. (2015). Synthesis of 18F-arenes from spirocyclic iodonium (III) ylides via continuous-flow microfluidics. *Journal of Fluorine Chemistry, 178*, 249–253.

Carbonell, P., Radivojevic, T., & Garcia Martin, H. (2019). Opportunities at the intersection of synthetic biology, machine learning, and automation. *ACS Synthetic Biology, 8*(7), 1474–1477.

Chen, S., Javed, R., Lei, J., Kim, H. K., Flores, G., van Dam, R. M., Keng, P. Y., & Kim, C. J. (2012a, June). Synthesis of diverse tracers on EWOD microdevice for positron emission tomography (PET). In *Technical digest of the solid-state sensor and actuator workshop, Hilton Head Island, SC, USA*, pp. 3–7.

Chen, S., Lei, J., van Dam, R. M., Keng, P. Y., & Kim, C. J. (2012b, October). Planar alumina purification of 18F-labeled radiotracer synthesis on EWOD chip for positron emission tomography (PET). In *16th international conference on miniaturized systems for chemistry and life sciences (mTAS)*, pp. 1771–1773.

Chen, S., Dooraghi, A. A., Lazari, M., van Dam, R. M., Chatziioannou, A. F., & Kim, C. J. C. (2014a). On-chip product purification for complete microfluidic radiotracer synthesis. In *2014 IEEE 27th international conference on micro electro mechanical systems (MEMS)* (pp. 284–287). IEEE.

Chen, S., Javed, M. R., Kim, H. K., Lei, J., Lazari, M., Shah, G. J., van Dam, R. M., & Keng, P. Y. (2014b). Radiolabelling diverse positron emission tomography (PET) tracers using a single digital microfluidic reactor chip. *Lab on a Chip, 14*(5), 902–910.

Choi, K., Ng, A. H., Fobel, R., & Wheeler, A. R. (2012). Digital microfluidics. *Annual Review of Analytical Chemistry, 5*, 413–440.

Ding, H., Sadeghi, S., Shah, G. J., Chen, S., Keng, P. Y., & van Dam, R. M. (2012). Accurate dispensing of volatile reagents on demand for chemical reactions in EWOD chips. *Lab on a Chip, 12*(18), 3331–3340.

Elizarov, A. M., van Dam, R. M., Shin, Y. S., Kolb, H. C., Padgett, H. C., Stout, D., Shu, J., Huang, J., Daridon, A., & Heath, J. R. (2010). Design and optimization of coin-shaped microreactor chips for PET radiopharmaceutical synthesis. *Journal of Nuclear Medicine, 51*(2), 282–287.

Fobel, R., Fobel, C., & Wheeler, A. R. (2013). DropBot: An open-source digital microfluidic control system with precise control of electrostatic driving force and instantaneous drop velocity measurement. *Applied Physics Letters, 102*(19).

Freire, S. L. (2016). Perspectives on digital microfluidics. *Sensors and Actuators A: Physical, 250*, 15–28.

Gillies, J. M., Prenant, C., Chimon, G. N., Smethurst, G. J., Dekker, B. A., & Zweit, J. (2006). Microfluidic technology for PET radiochemistry. *Applied Radiation and Isotopes, 64*(3), 333–336.

Gong, J., & Kim, C. J. (2008). Direct-referencing two-dimensional-array digital microfluidics using multilayer printed circuit board. *Journal of Microelectromechanical Systems, 17*(2), 257–264.

Huang, T. W., Lin, C. H., & Ho, T. Y. (2010). A contamination aware droplet routing algorithm for the synthesis of digital microfluidic biochips. *IEEE Transactions on Computer-Aided Design of Integrated Circuits and Systems, 29*(11), 1682–1695.

Jain, V., Raj, T. P., Deshmukh, R., & Patrikar, R. (2017). Design, fabrication and characterization of low cost printed circuit board based EWOD device for digital microfluidics applications. *Microsystem Technologies, 23*, 389–397.

Javed, M. R., Chen, S., Kim, H. K., Wei, L., Czernin, J., van Dam, R. M., & Keng, P. Y. (2014). Efficient radiosynthesis of 3′-deoxy-3′-18F-fluorothymidine using electrowetting-on-dielectric digital microfluidic chip. *Journal of Nuclear Medicine, 55*(2), 321–328.

Jebrail, M. J., Ng, A. H., Rai, V., Hili, R., Yudin, A. K., & Wheeler, A. R. (2010). Synchronized synthesis of peptide-based macrocycles by digital microfluidics. *Angewandte Chemie, 122*(46), 8807–8811.

Jebrail, M. J., Renzi, R. F., Sinha, A., Van De Vreugde, J., Gondhalekar, C., Ambriz, C., Meagher, R. J., & Branda, S. S. (2015). A solvent replenishment solution for managing evaporation of biochemical reactions in air-matrix digital microfluidics devices. *Lab on a Chip, 15*(1), 151–158.

Keng, P. Y., & van Dam, R. M. (2015). Digital microfluidics: A new paradigm for radiochemistry. *Molecular Imaging, 14*(12), 7290–2015.

Keng, P. Y., Chen, S., Ding, H., Sadeghi, S., Shah, G. J., Dooraghi, A., Phelps, M. E., Satyamurthy, N., Chatziioannou, A. F., Kim, C. J. C., & van Dam, R. M. (2012). Micro-chemical synthesis of molecular probes on an electronic microfluidic device. *Proceedings of the National Academy of Sciences, 109*(3), 690–695.

Kim, H. K., Javed, M. R., Chen, S., Zettlitz, K. A., Collins, J., Wu, A. M., van Dam, R. M., & Keng, P. Y. (2019). On-demand radiosynthesis of N-succinimidyl-4-[18F]fluorobenzoate ([18F]SFB) on an electrowetting-on-dielectric microfluidic chip for 18 F-labeling of protein. *RSC Advances, 9*(55), 32175–32183.

Kimura, H., Tomatsu, K., Saiki, H., Arimitsu, K., Ono, M., Kawashima, H., Iwata, R., Nakanishi, H., Ozeki, E., Kuge, Y., & Saji, H. (2016). Continuous-flow synthesis of N-succinimidyl 4-[18F]fluorobenzoate using a single microfluidic chip. *PLoS One, 11*(7), e0159303.

Koag, M. C., Kim, H. K., & Kim, A. S. (2014a). Efficient microscale synthesis of [18F]-2-fluoro-2-deoxy-d-glucose. *Chemical Engineering Journal, 258*, 62–68.

Koag, M. C., Kim, H. K., & Kim, A. S. (2014b). Fast and efficient microscale radiosynthesis of 3′-deoxy-3′-[18F]fluorothymidine. *Journal of Fluorine Chemistry, 166*, 104–109.

Lebedev, A., Miraghaie, R., Kotta, K., Ball, C. E., Zhang, J., Buchsbaum, M. S., Kolb, H. C., & Elizarov, A. (2013). Batch-reactor microfluidic device: First human use of a microfluidically produced PET radiotracer. *Lab on a Chip, 13*(1), 136–145.

Lee, C. C., Sui, G., Elizarov, A., Shu, C. J., Shin, Y. S., Dooley, A. N., Huang, J., Daridon, A., Wyatt, P., Stout, D., & Kolb, H. C. (2005). Multistep synthesis of a radiolabeled imaging probe using integrated microfluidics. *Science, 310*(5755), 1793–1796.

Li, J. (2020). Current commercialization status of electrowetting-on-dielectric (EWOD) digital microfluidics. *Lab on a Chip, 20*(10), 1705–1712.

Liang, S. H., Yokell, D. L., Jackson, R. N., Rice, P. A., Callahan, R., Johnson, K. A., Alagille, D., Tamagnan, G., Collier, T. L., & Vasdev, N. (2014a). Microfluidic continuous-flow radiosynthesis of [18F]FPEB suitable for human PET imaging. *Medicinal Chemistry Communication, 5*(4), 432–435.

Liang, S. H., Yokell, D. L., Normandin, M. D., Rice, P. A., Jackson, R. N., Shoup, T. M., Brady, T. J., Fakhri, G. E., Collier, T. L., & Vasdev, N. (2014b). First human use of a radiopharmaceutical prepared by continuous-flow microfluidic radiofluorination: Proof of concept with the tau imaging agent [18F]T807. *Molecular Imaging, 13*(8), 7290–2014.

Liu, X., Ma, D., Ye, H., Hou, Y., Bai, X., Xing, Y., Cheng, X., Lin, B., & Lu, Y. (2023). Electrowetting-based digital microfluidics: Toward a full-functional miniaturized platform for biochemical and biological applications. *TrAC Trends in Analytical Chemistry*, 117153.

Luo, Z., Xu, J., Pan, Z., Yin, H., Cao, L., Zhou, G., & Liu, S. (2022). Programmable high integration and resolution digital microfluidic device driven by thin film transistor arrays. *IEEE Access, 10*, 30573–30582.

Moazami, E., Perry, J. M., Soffer, G., Husser, M. C., & Shih, S. C. (2019). Integration of world-to-chip interfaces with digital microfluidics for bacterial transformation and enzymatic assays. *Analytical Chemistry, 91*(8), 5159–5168.

Mogi, K., Kimura, H., Kondo, Y., Inoue, T., Adachi, S., & Natsume, T. (2021). Automatic radioisotope manipulation for small amount of nuclear medicine using an EWOD device with a dimple structure. *Royal Society Open Science, 8*(5), 201809.

Moon, H., Wheeler, A. R., Garrell, R. L., & Loo, J. A. (2006). An integrated digital microfluidic chip for multiplexed proteomic sample preparation and analysis by MALDI-MS. *Lab on a Chip, 6*(9), 1213–1219.

Narahari, T., Dahmer, J., Sklavounos, A., Kim, T., Satkauskas, M., Clotea, I., Ho, M., Lamanna, J., Dixon, C., Rackus, D. G., & da Silva, S. J. R. (2022). Portable sample processing for molecular assays: Application to Zika virus diagnostics. *Lab on a Chip, 22*(9), 1748–1763.

Perry, J. M., Soffer, G., Jain, R., & Shih, S. C. (2021). Expanding the limits towards 'one-pot' DNA assembly and transformation on a rapid-prototype microfluidic device. *Lab on a Chip, 21*(19), 3730–3741.

Radivojević, T., Costello, Z., Workman, K., & Garcia Martin, H. (2020). A machine learning automated recommendation tool for synthetic biology. *Nature Communications, 11*(1), 4879.

Ruecha, N., Lee, J., Chae, H., Cheong, H., Soum, V., Preechakasedkit, P., Chailapakul, O., Tanev, G., Madsen, J., Rodthongkum, N., & Kwon, O. S. (2017). Paper-based digital microfluidic chip for multiple electrochemical assay operated by a wireless portable control system. *Advanced Materials Technologies, 2*(3), 1600267.

Saboury, B., Bradshaw, T., Boellaard, R., Buvat, I., Dutta, J., Hatt, M., Jha, A. K., Li, Q., Liu, C., McMeekin, H., & Morris, M. A. (2023). Artificial intelligence in nuclear medicine: Opportunities, challenges, and responsibilities toward a trustworthy ecosystem. *Journal of Nuclear Medicine, 64*(2), 188–196.

Samlali, K., Ahmadi, F., Quach, A. B., Soffer, G., & Shih, S. C. (2020). One cell, one drop, one click: Hybrid microfluidics for mammalian single cell isolation. *Small, 16*(34), 2002400.

Selivanova, S. V., Mu, L., Ungersboeck, J., Stellfeld, T., Ametamey, S. M., Schibli, R., & Wadsak, W. (2012). Single-step radiofluorination of peptides using continuous flow microreactor. *Organic & Biomolecular Chemistry, 10*(19), 3871–3874.

Shah, G. J., Lei, J., Chen, S., Kim, C. J., Keng, P. Y., & van Dam, R. M. (2012). Automated injection from EWOD digital microfluidic chip into HPLC purification system. In *Proceedings of international conference on miniaturized systems for chemistry and life sciences (mTAS)*, pp. 356–358.

Shah, G. J., Ding, H., Sadeghi, S., Chen, S., & van Dam, R. M. (2013). On-demand droplet loading for automated organic chemistry on digital microfluidics. *Lab on a Chip, 13*(14), 2785–2795.

Tarn, M. D., Pascali, G., De Leonardis, F., Watts, P., Salvadori, P. A., & Pamme, N. (2013). Purification of 2-[18F]fluoro-2-deoxy-d-glucose by on-chip solid-phase extraction. *Journal of Chromatography A, 1280*, 117–121.

Vaidyanathan, G., & Zalutsky, M. R. (2006). Synthesis of N-succinimidyl 4-[18F]fluorobenzoate, an agent for labeling proteins and peptides with 18F. *Nature Protocols, 1*(4), 1655–1661.

Vaquero, J. J., & Kinahan, P. (2015). Positron emission tomography: Current challenges and opportunities for technological advances in clinical and preclinical imaging systems. *Annual Review of Biomedical Engineering, 17*, 385–414.

Walters, L. R., Martin, K. J., Jacobson, M. S., Hung, J. C., & Mosman, E. A. (2012). goya. *Journal of Nuclear Medicine Technology, 40*(1), 52–56.

Wang, J., Chao, P. H., & van Dam, R. M. (2019). Ultra-compact, automated microdroplet radiosynthesizer. *Lab on a Chip, 19*(14), 2415–2424.

Wang, J., Chao, P. H., Slavik, R., & van Dam, R. M. (2020). Multi-GBq production of the radiotracer [18F]fallypride in a droplet microreactor. *RSC Advances, 10*(13), 7828–7838.

Webb, E. W., & Scott, P. J. (2021). Potential applications of artificial intelligence and machine learning in radiochemistry and radiochemical engineering. *PET Clinics, 16*(4), 525–532.

Xing, Y., Liu, Y., Chen, R., Li, Y., Zhang, C., Jiang, Y., Lu, Y., Lin, B., Chen, P., Tian, R., & Liu, X. (2021). A robust and scalable active-matrix driven digital microfluidic platform based on printed-circuit board technology. *Lab on a Chip, 21*(10), 1886–1896.

Yang, H., Luk, V. N., Abelgawad, M., Barbulovic-Nad, I., & Wheeler, A. R. (2009). A world-to-chip interface for digital microfluidics. *Analytical Chemistry, 81*(3), 1061–1067.

Zhang, X., Liu, F., Knapp, K. A., Nickels, M. L., Manning, H. C., & Bellan, L. M. (2018). A simple microfluidic platform for rapid and efficient production of the radiotracer [18F]fallypride. *Lab on a Chip, 18*(9), 1369–1377.

Recent Developments in Electrochemical Radiofluorination: Methods and Applications

Daniel Hernández-Valdés, Xueyi Wang, and Saman Sadeghi

1 Introduction

Positron emission tomography (PET) utilizes positron-emitting radioisotopes to non-invasively diagnose disease in clinical settings (Jacobson et al., 2015; Rong et al., 2021). Fluorine-18 is the most widely used PET radioisotope due to its relatively long half-life, low positron energy (0.635 MeV), and high positron decay ratio (97% β^+ decay) (Jacobson & Chen, 2010; Rong et al., 2021). Both electrophilic and nucleophilic fluorination reactions have been described in the literature for the synthesis of ^{18}F-based radiopharmaceuticals. However, nucleophilic radiofluorination is preferred to electrophilic radiofluorination because of the more reliable radioisotope production pathway and higher molar activity (A_m) of [^{18}F]F$^-$, compared to [^{18}F]F$_2$ and [^{18}F]F$_2$-derived reagents (Jacobson et al., 2015; Rong et al., 2021). The [^{18}F]F$^-$ ion is a strong nucleophile that can attack sites of low electron density and replace certain leaving groups via substitution (Jacobson & Chen, 2010; Jacobson et al., 2015).

Despite recent advancements in nucleophilic fluorination, these have not always been translated into development of new molecular imaging probes (Tang et al., 2011). The biggest roadblock for new probes innovation is that some precursors are not readily amenable to nucleophilic substitution reactions with [^{18}F]F$^-$ without some degree of positive charge at the site of fluorine labeling (Tredwell & Gouverneur, 2012). The unfavorable interaction between the negatively charged [^{18}F]F$^-$ and a high electron density reaction center requires the generation of an electrophilic center. Such reactions usually require suitably activated fluorination

D. Hernández-Valdés · X. Wang · S. Sadeghi (✉)
Department of Chemistry and Chemical Biology, McMaster University,
Hamilton, ON, Canada
e-mail: sadegs10@mcmaster.ca

R. M. van Dam et al. (eds.), *Automated Technologies for the Development and Production of Radiopharmaceuticals*,
https://doi.org/10.1007/978-3-031-84632-8_12

sites and proceed under harsh conditions (e.g., high pH and temperatures), which can hinder the scope and application of late-stage fluorination for small molecules. Late-stage radiofluorination is desired in PET radiochemistry, since fewer steps often translate into more robust synthetic protocols and higher radiochemical yield (RCY) due to reduced radioactive decay with time.

Anodic oxidation under controlled potential conditions presents an elegant strategy to overcome this reactivity challenge through abstraction of electrons and promoting electronegative fluorine atoms to more favorably approach the electron-rich labeling site (Sawamura et al., 2012; Yan et al., 2017). Electrochemical fluorination (ECF) has become widely used as an alternative for the fluorination of organic compounds. Previous reports have demonstrated the utility of electrochemistry as a versatile tool to drive organic chemical reactions, including arene couplings, C–N functionalization, C–H oxidation, decarboxylation, cycloaddition reactions, and heterocycle synthesis, all of which can be carried out effectively on an electrode surface (Yan et al., 2017, p. 200; Möhle et al., 2018; Wiebe et al., 2018). Electrochemistry has found extensive use in the preparation of non-radioactive fluorinated compounds, but its application in radiopharmaceuticals remains limited. The first efforts to use electrochemistry within ^{18}F-radiochemistry were to improve the separation of [^{18}F]Fluoride from the target [^{18}O]H_2O. Following these early developments, electrochemical radiofluorination emerged as a promising approach for late-stage radiofluorination, reducing synthesis steps, shortening production time, and achieving high RCY with regioselectivity (Hernández-Valdés & Sadeghi, 2021).

This chapter provides an overview of the application of electrochemistry in the radiofluorination of small molecules. Ongoing efforts to use electrochemistry as an alternative method for separation [^{18}F]F^- from the target [^{18}O]H_2O are also highlighted within this review. This chapter explores various approaches to electrochemical radiofluorination, encompassing both carrier-added and no-carrier-added radiofluorination methodologies. The discussion extends to the design of electrochemical cells and progress toward automating the process of electrochemical radiofluorination. Lastly, this chapter provides an outlook, examining potential future advancements and possibilities within the field of electrochemical radiofluorination.

2 Electrochemical Separation of [^{18}F]F^- from ^{18}O-Enriched Water

Despite the strong nucleophilicity of [^{18}F]F^-, its reactivity for nucleophilic substitution is diminished in aqueous solution due to the formation of [^{18}F]HF. Anhydrous conditions are critical to successfully achieve nucleophilic radiofluorination reactions. Since [^{18}F]F^- is obtained from the bombardment of [^{18}O]H_2O in a cyclotron, solvent exchange is required prior to most fluorination reactions. In traditional nucleophilic substitution, the obtained [^{18}F]F^- is captured in a strong anion-exchanged column/cartridge such as quaternary methyl ammonium (QMA), specifically designed to trap anionic analyte, while the unreacted [^{18}O]H_2O is recovered.

The trapped radioisotope is subsequently eluted by a solution containing water, organic solvents, and phase transfer agent (such as the complex formed by K_2CO_3 and Kryptofix 222). Cations, usually potassium and ammonium salts, coordinate with cryptands to form a lipophilic counterion which allows the recovery of $[^{18}F]F^-$ in a mixture consisting of a small portion of aqueous solution and organic solvent (Fig. 1a) (Yu, 2006). Residual water in the solvent can be eliminated by azeotropic distillation with anhydrous acetonitrile (MeCN) under argon or N_2 stream, and evaporation of residual water and MeCN would occur simultaneously (Yu, 2006). Once the $[^{18}F]$fluoride complex is dry, it can be reacted with the target molecules via aliphatic or aromatic nucleophilic substitution (S_N2 and S_NAr, respectively). This drying process is often performed automatically in commercial synthesizers, at the expense of additional time and reaction complexity, and therefore contributes to the decrease in activity yield (AY).

Electrochemistry offers an alternative method for separation of $[^{18}F]F^-$ from water in a three-step process (Fig. 1b). First, $[^{18}F]F^-$ ions are sequestered from the cyclotron target water solution using an electrodeposition technique. The electrochemical cell (E-Cell) is then dried using a stream of inert gas. Finally, the trapped $[^{18}F]F^-$ is then released into an anhydrous solvent (MeCN, dimethyl sulfoxide (DMSO), etc.) using a reverse potential. This approach avoids the necessity of multiple drying steps, while allowing the recovery of the unused $[^{18}O]H_2O$ target, which can be recycled for subsequent $[^{18}F]F^-$ productions.

Over the years, different electrochemical conditions have been tested to separate $[^{18}F]F^-$ from $[^{18}O]H_2O$ (Alexoff et al., 1989; Saito et al., 2001; Hamacher et al., 2002; Reischl et al. 2002a, b; Saito et al., 2007; Saiki et al., 2010; Wong et al., 2012; Sadeghi et al., 2013; He et al., 2014a, b; Kügler et al., 2014). Graphite and glassy carbon have been the most common choices for anode electrodes, but they often suffer from degradation after extended use, which results in inconsistent percentage of trapping and release of $[^{18}F]F^-$ activity (Saiki et al., 2010; He et al., 2014a, b). Various E-Cell designs, applied voltages, and temperature have also been investigated in previous reports to improve the activity recovery and the lifetime of the electrochemical systems.

Saiki et al. reported the use of a microfluidic electrochemical cell (16 μL volume) to concentrate $[^{18}F]F^-$ from $[^{18}O]H_2O$ target (Saiki et al., 2010). By applying 10 V at a flow rate of 700 μL/min, the authors were able to trap the activity and release it into 60 μL of MeCN containing Kryptofix 222–$KHCO_3$ complex using a reversed potential (−2.5 V) with an overall efficiency of 60%. In a follow-up paper, Wong et al. demonstrated the reactivity of $[^{18}F]F^-$ concentrated by electrochemical trap and release methods (Wong et al., 2012). The RCYs for the radiofluorination of $[^{18}F]$FDG, $[^{18}F]$FMISO, $[^{18}F]$flumazenil, and $[^{18}F]$fluoromethyl bromide using electrochemically concentrated $[^{18}F]F^-$ were comparable or superior to the ones obtained by conventional methods.

Kügler et al. have studied the effect of modifying the structure of the anode electrode in the no-carried-added electrochemical separation of $[^{18}F]F^-$ from $[^{18}O]H_2O$ (Kügler et al., 2014). A range of carbon-based anodes including various types of graphite, glassy carbon, amorphous diamond-like carbon films, and boron-doped diamonds were compared. They found that glassy carbon electrodes were superior in trapping and releasing the $[^{18}F]F^-$ in organic solvent. On the other hand,

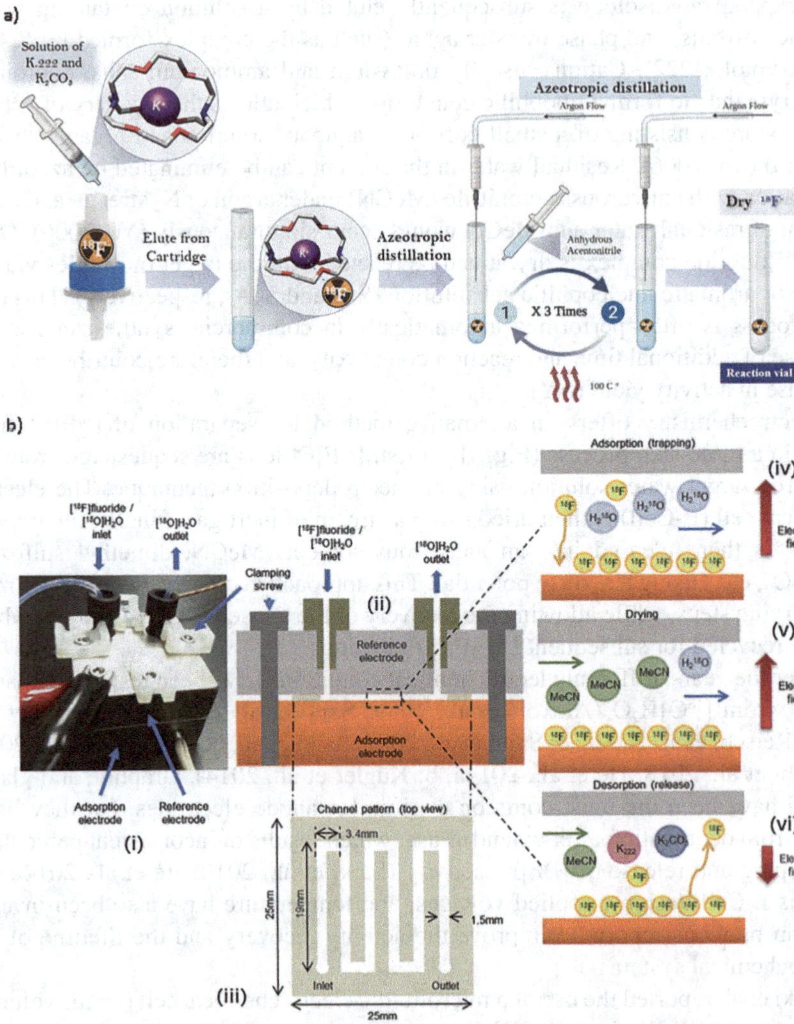

Fig. 1 (**a**) Traditional preparation of $[^{18}F]F^-$ required for nucleophilic substitution reactions including azeotropic distillation. Radioactivity captured in an ion-exchange column is dried and eluted using a solution containing Kryptofix 222 (K222) or bulky tetra-(butyl/ethyl) ammonium cation and counterions (potassium salt). Residual water in the reaction solvent is removed via azeotropic distillation with anhydrous MeCN under argon or N_2 stream, simultaneously evaporating residual water with MeCN. (**b**) Example of an electrochemical separation system. (i) Platinum-transition metal electrode cell assembled with fluidic and electrical connections. (ii) Cross-section of the layer defining the fluid path (iii) is sandwiched between an adsorption electrode and a reference electrode. Fluid follows a serpentine path to maximize the time within the electric field between the two electrodes. During trapping (iv), the negatively charged $[^{18}F]F^-$ ions are attracted to the positively charged adsorption electrode and adsorbed on the surface, while the $[^{18}O]H_2O$ continues to the outlet. (v) $[^{18}F]F^-$ is dried using an anhydrous solvent. (vi) $[^{18}F]F^-$ is released into a solution suitable for downstream radiochemical synthesis while applying a reverse potential. Part (b) reproduced with permission from reference (Sadeghi et al., 2013). Copyright Elsevier 2013

boron-doped diamond and amorphous diamond-like carbon had a low rate of [^{18}F]F^- fixation.

Other materials including 10–12 transition metals and alloys have been studied as anode electrodes in combination with platinum cathodes (Reischl et al. 2002a, b; Sadeghi et al., 2013). Van Dam et al. reported that a combination of a brass anode with a platinum-plated cathode could produce an overall trap and release efficiency of 84 ± 5% ($n = 7$) [95 ± 4% ($n = 7$) and 88 ± 3% ($n = 7$) trap and release efficiencies, respectively (Sadeghi et al., 2013)]. However, the main drawback was the decrease in reactivity and releasing efficiency after several runs. While electrochemical isotope separation and solvent exchange provide a feasible alternative to cartridge separation and azeotropic evaporation, further research and development are required to integrate the electrochemical platform into automated systems that are used in radiofluorination.

3 Electrochemical Fluorination

Introduction of a fluorine atom can confer biological advantages to an organic molecule, including increased stability and biological activity. However, selective C–F bond formation can be challenging, especially for functionalized aromatic carbons. Despite recent advances, both electrophilic and nucleophilic fluorination approaches have substrate limitations due to requirements such as the need for highly reactive substances (e.g., F_2) or poor nucleophilicity in protic solvents.

Electrochemical methods in organic synthesis offer an environmental-friendly alternative to conventional fluorination by replacing chemical reagents with electricity, resulting in reduced ecological footprints (Möhle et al., 2018). Recent advancements include integrating electrochemistry with organocatalysis, flow electrochemistry, and innovative strategies to control selectivity. The first application of electrochemistry in the preparation of organofluorine compounds was shown by Joshep H. Simons in 1949 (Simons, 1949; Ignat'ev, 2017). Various organic compounds including carboxylic acids, alcohol, ethers, amines, etc. were used as starting material to produce organofluorine products. When carboxylic acids were used, in addition to saturated fluorocarbons, other products of formula $C_nF_{2n+1}H$ were obtained. The Simons process is widely used in industry to produce highly fluorinated organic compounds. The process is carried out in the presence of anhydrous HF and using Ni-plated anodes. The main limitation of this method is the lack of selectivity for site-specific addition of the fluorine atom. Nagase et al. (1965) conducted comprehensive research on ECF of gaseous hydrocarbons, including methane, ethylene, ethane, and halogenated methane. These results have significantly expanded the scope of application of ECF for these type compounds. In the 1970s, T. Abe et al. explored the use of ECF to produce a diverse group of F-cyclic ethers and F-oxanes (Abe et al., 1980). Since the 1980s, nitrogen-containing F-carboxylic acids and heterocyclic atom have become attractive and versatile molecules for ECF, showcasing their high potential for exploration (Abe, 2000).

Selective electrochemical fluorination presents greater challenges, and only specific groups of precursors, such as aromatics, olefins, benzyl ketones, and α-phenylacetates, have demonstrated suitability for these types of reactions. The development of poly(HF) salt ionic liquids as electrolytes, particularly the anodically stable variants such as $Et_3N{\cdot}nHF$ ($n = 3–5$), and $Et_4NF{\cdot}nHF$ ($n = 3–5$), has led to intensive research on the selective electrochemical fluorination of various compounds (Konno & Fuchigami, 1997). Electrolytes such as $Et_3N{\cdot}nHF$ ($n = 3–5$) and $Et_4NF{\cdot}nHF$ ($n = 3–5$) support the charge transport between the two electrodes (cathode and anode) in solution. Additionally, the unique properties of poly(HF) salts enable the selective electrochemical fluorination of substrates with high oxidation potentials (Fuchigami & Inagi, 2011, 2020). Fuchigami and Inagi have recently summarized the effect of supporting electrolytes, solvents, and anode materials on the selectivity and efficiency of electrochemical fluorination (Fuchigami & Inagi, 2011, 2020).

Recent progress has expanded the scope of electrochemical fluorination methods for various substrates (Fuchigami & Inagi, 2020). Baran et al. used a nitrate additive and fluorine source (Select-fluor) with carbon-based electrode to achieve inactivated anodic C(sp3)–H fluorination (Takahira et al., 2019). Noel et al. optimized the anodic oxidation of thiols or disulfides with KF as fluoride source under aq. HCl solution to synthesize sulfonyl fluorides (Laudadio et al., 2019). Xu et al. demonstrated the simultaneous regioselective and chemoselective addition of a fluoride and an alkyltrifluoroborate group via electrochemical fluoroalkynylation in the presence of $Et_4NF{\cdot}3HF$ salt (Xiong et al., 2019). More recently, Waldvogel et al. have demonstrated that anodic oxidation in the presence of $Et_3N{\cdot}5HF$ produces a hypervalent iodine I(III) species, which mediates the cyclization of *N*-allylcarboxamide to form 5-fluoromethyl-2-oxazoline (Haupt et al., 2019).

ECF allows the selective oxidation of electron-rich organic precursors with a controlled potential, creating reactive intermediates that can overcome activation energy barriers. The general mechanism of electrochemical (radio)fluorination consists of a four-step process usually simplified as ECEC where the "E" represents each oxidation and the "C" represents a chemical reaction (Scheme 1) (Fuchigami et al., 1991, 1994). The initial step of the ECF consists of an oxidation of the precursor reducing the electron density at the reaction site. This is later followed by a nucleophilic attack by fluoride and a second oxidation step. A deprotonation step or a loss of a leaving group leads to the formation of the final product.

A variation to the general mechanism is proposed for the fluorination of thioethers, where after the initial oxidation step, an auxiliary base would act as the nucleophile that stabilizes the sulfonium cation-radical before a second anodic oxidation takes place. Typically, the high excess of fluoride works as an auxiliary base in the case of non-radioactive electrochemical fluorination or carrier-added (CA) radiofluorination. In the case of no-carrier-added (NCA) radiofluorination, the solvent or a weak base could play the role of auxiliary base (Allison et al., 2022). The fluoro-pummerer rearrangement (Scheme 1) is proposed as a key step of the mechanism. The sulfonium cation would be formed via the abstraction of hydrogen from the alpha carbon and the leaving of auxiliary base (Hernández-Valdés & Sadeghi, 2021). Finally, the addition of fluoride to the alkene cation results in the final product.

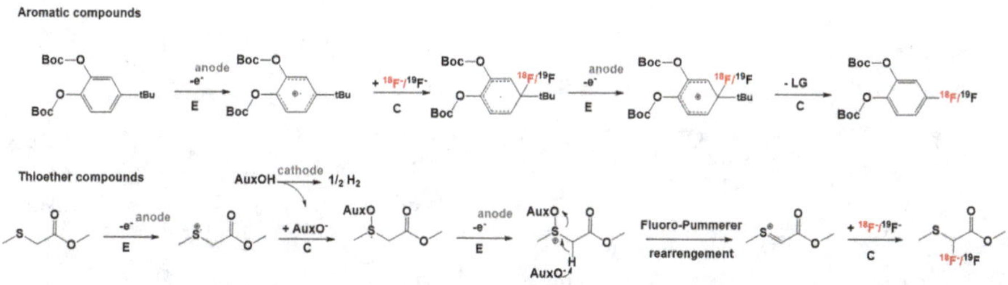

Scheme 1 General reaction mechanism for electrochemical (radio)fluorination of aromatic and thioether compounds exemplified with di-*tert*-butyl-(4-*tert*-butyl-1,2-phenylene)-dicarbonate and methyl(methylthio) acetate, respectively

4 Electrochemical Radiofluorination

Due to the success of electrochemical fluorination to synthesize non-radioactive organofluorinated compounds, there has been an interest in the use of ECF for late-stage radiofluorination. Electrochemical radiofluorination offers an alternative method for PET tracer production that minimizes the synthesis steps, shortens the production time, and promotes high RCY with regioselectivity. In PET imaging applications, maintaining adequate specific activity (A_m) is beneficial, as it helps prevent target saturation, ensuring optimal imaging contrast and accurate determination of the target concentration. A_m is influenced by the synthesis time (i.e., the decay of the radioactive product reduces A_m) and the presence of stable isotope (carrier) in the reaction (which lowers A_m). Electrolytes such as $Et_3N{\cdot}nHF$ ($n = 3–5$), $Et_4NF{\cdot}nHF$ ($n = 3–5$) and Bu_4NF are considered carrier-added. The presence of a high excess of carrier $[^{19}F]F^-$ from the electrolyte in electrochemical fluorination of organic molecules could potentially decrease A_m in ECF (Laudadio et al., 2019). Other alternative no-carrier-added electrolytes such as Bu_4NPF_6 and Bu_4NClO_4 can be used instead. Figure 2 shows a representation of the most common electrochemical cell designs, electrodes, and electrolytes used by various authors to perform electrochemical radiofluorination.

The first attempts to perform electrochemical radiofluorination on organic molecules were reported by Reischl et al. (2002a, b, 2003; Kienzle et al., 2005). The authors tested different electrolyte compositions and temperature to radiofluorinate simple aromatic compounds. RCYs were modest (0–16%) in all cases, and the use of high concentration of carrier $[^{19}F]F^-$ resulted in A_m on the order of 0.2–27 GBq/mmol which was lower than most clinical application requirements (Table 1). While the work done by Reischl et al. set a precedent for the field of electrochemical radiofluorination of small molecules, the results in RCY and A_m were lower than the ones obtained by other radiofluorination approaches (Furuya et al., 2011; Campbell & Ritter, 2014; Neumann & Ritter, 2015; Lee et al., 2019).

Sadeghi et al. performed the electrochemical radiofluorination of di-*tert*-butyl-(4-*tert*-butyl-1,2-phenylene)-dicarbonate using both CA and NCA conditions (He et al., 2014a, b, He et al., 2015). The di-*tert*-butyl-(4-*tert*-butyl-1,2-phenylene)-dicarbonate was chosen as a model for the PET tracer $[^{18}F]$FDOPA. For CA conditions, the use of 0.033 M $Et_3N{\cdot}HF$ and 0.05 M of Bu_4NPF_6 as electrolyte in combination

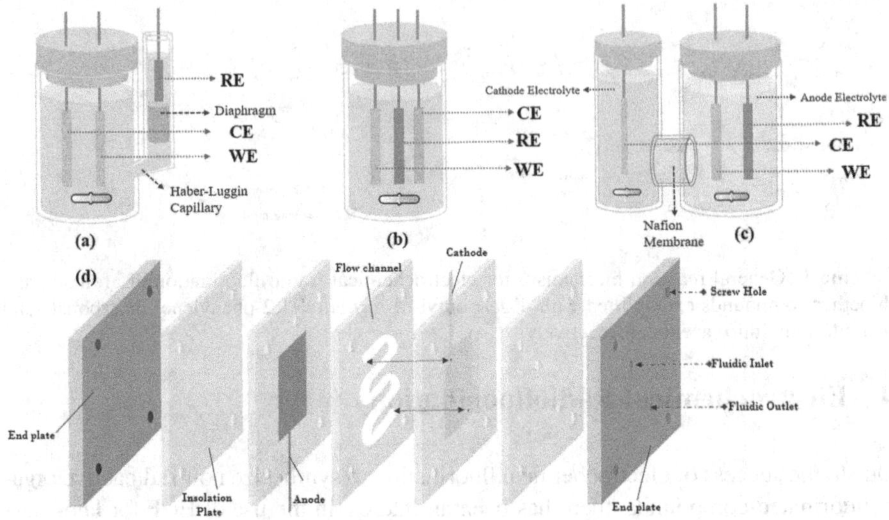

Fig. 2 Different electrochemical cell designs used in ECF. (**a**) Undivided cell with Haber-Luggin capillary; (**b**) undivided cell with three electrodes; (**c**) two-compartment cell with three electrodes; (**d**) undivided microfluidic cell

with 0.1 M of precursor at 0 °C yielded RCY of 10.4 ± 0.6%. The A_m was 43 GBq/mmol which was in the same order of magnitude of previous carrier-added radiofluorination experiments (He et al., 2014a, b). In an attempt to perform a NCA radiofluorination, the electrolyte was replaced by 0.01 M $Et_3N{\cdot}3HAc$ + 0.1 M Bu_4NClO_4. Under these conditions, the A_m increased by a factor of 300 up to 13 TBq/mmol (He et al., 2015). Replacing $[^{19}F]Et_3N{\cdot}HF$ with carrier-free electrolytes can enhance A_m, although it often comes at the cost of reducing RCY. The loss in stabilization effect of $[^{19}F]F^-$ led to nearly 50% RCY reduction in the abovementioned study.

Sadeghi et al. have also demonstrated the electrochemical preparation of a [18F]Celecoxib derivative as a COX-2 inhibitor (Lebedev et al., 2017). Previous attempts to radiolabel COX-2 inhibitors were limited to fluorine-18 added via a carbon chain or an aromatic position which resulted in low target specificity or in vivo instability (Laube et al., 2013; Dagallier et al., 2021). To enhance the in vitro and in vivo properties of a [18F]Celecoxib derivative, the authors attempted to radiofluorinate at the five-membered heteroatomic ring (Lebedev et al., 2017). Using ECF under CA conditions, Celecoxib was radiofluorinated with RCY of 0.8–2.0% and A_m of 111 GBq/mmol. Although CA conditions were required to ensure the formation of the product, preclinical studies showed high in vivo stability and brain uptake in healthy mice, demonstrating the possible application of ECF in radiopharmaceuticals. Furthermore, the investigation of ECF of Celecoxib under NCA conditions offers valuable opportunities to enhance the A_m of [18F]Celecoxib and therefore its future application as PET radiotracer.

Several CA and NCA attempts to radiofluorinate thioether derivatives have also been carried out (Table 1) (Balandeh et al., 2017, 2018; Waldmann et al., 2017;

Table 1 Comparison of different electrochemical radiofluorination experiments

Precursor	Conditions	Radiochemical yield (RCY%)	Molar activity (A_m)	Reference
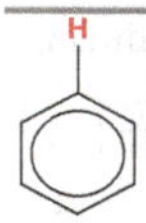	**E-Cell Type: A** **CA**, 0.1 M $Et_3N{\cdot}3HF$ **Solvent:** MeCN 2.0 V vs Ag/Ag^+, 50C 0 °C	0.4–0.9	0.2 GBq/mmol	Reischl et al. (2002a, b)
	CA, 0.033 M $Et_3N{\cdot}3HF$ + 0.066 M $Et_3N{\cdot}3HCl$ **Solvent:** MeCN 2.0 V vs Ag/Ag^+, 50C 0 °C	16 ± 9	27 GBq/mmol	
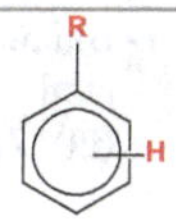 R = F, Cl, Br, tBu, -$COCH_3$	**E-Cell Type: A** **CA**, 0.033 M $Et_3N{\cdot}3HF$ + 0.066 M $Et_3N{\cdot}3HCl$ **Solvent:** MeCN 2.0 V vs Ag/Ag^+, 50C 0 °C		Not reported	Reischl et al. (2003)
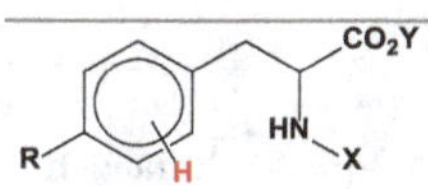 R = H, tBu X = TFA, Boc, Ac, DNP Y = Me, Et	**E-Cell Type: B** **CA**, 0.033 M $Et_3N{\cdot}3HF$ + Et_3NX + Bu_4NPF_6, X = Cl, Br, I **Solvent:** MeCN 1.5–2.0 V vs Ag/Ag^+ −10–0 °C	0–10.5	1.2 GBq/mmol	Kienzle et al. (2005)
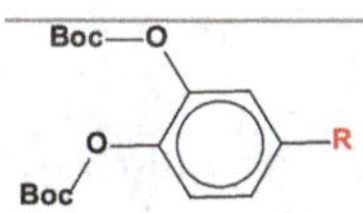 R = H, tBu	**E-Cell Type: B** **CA**, 0.033 M $Et_3N{\cdot}3HF$ + 0.050 M Bu_4NPF_6 **Solvent:** MeCN 2.7 V vs Ag/Ag^+ 0 °C **E-Cell Type: B**	10.4 ± 0.6	43 GBq/mmol	He et al. (2014a, b)
	NCA, 0.01 M $Et_3N{\cdot}3Hac$ + 0.1 M Bu_4NClO_4 **Solvent:** MeCN 2.58 V vs Ag/Ag^+ 0 °C	6.8 ± 1.3	13 TBq/mmol	He et al. (2015)
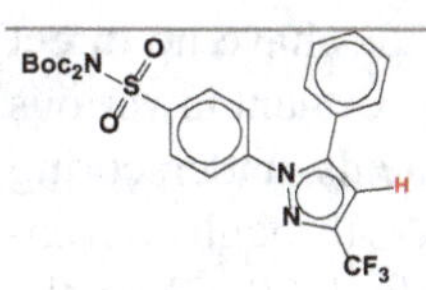	**E-Cell Type: B** **CA**, 0.02 M $Et_4NF{\cdot}4HF$ + 0.05 M Bu_4NClO_4 **Solvent:** MeCN 2.7 V vs Ag wire, 70 min r.t.	0.8–2.0	111 GBq/mmol	Lebedev et al. (2017)

(continued)

Table 1 (continued)

Precursor	Conditions	Radiochemical yield (RCY%)	Molar activity (A_m)	Reference
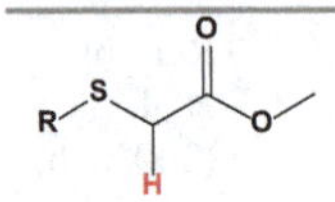 R = Phen, CH_3, CH_3CH_2	**E-Cell Type: B** **R = Phen**, automated synthesizer **CA**, 0.002 M $Et_4NF{\cdot}4HF$ + 0.05 M Bu_4NClO_4 **Solvent:** MeCN 1.95 V vs Ag wire, 2 h and 4 h r.t.	5.2 and 12.9	11 GBq/mmol and 26 GBq/mmol	Waldmann et al. (2017)
	E-Cell Type: B			
	R = Phenyl **CA**, 0.154 M Bu_4NF + 0.105 M TfOH **Solvent:** MeCN 1.4 V vs Ag/Ag^+, 30 min 60 °C	7 ± 1	Not reported	Balandeh et al. (2017)
	E-Cell Type: C			
	NCA, Cation Pool Technique **Solvent:** TFE + 2,6-di-tert-butyl-4-methyl-pyridine 1.6 V vs Ag wire −20 °C	5.7–20.6	42 TBq/mmol	Balandeh et al. (2018)
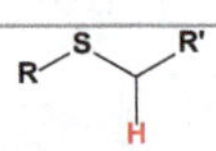 R = Phen, CH_3, CH_3CH_2 R' = CO_2Me, CN, $PO(OEt)_2$, $CONH_2$	**E-Cell Type: B** **NCA**, 0.05 M Bu_4NClO_4 **Solvent:** TFE 1.9 V vs Ag wire, 30 min 70 °C	8.3–93.5	4.7–5.3 TBq/mmol for Compound with R = CH_3, R' = CO_2Me	Allison et al. (2022)

Highlighted in red is the site where $[^{18}F]F^-$ substitution take places in the molecule. *CA* carrier-added, *NCA* no-carrier-added, *TFE* trifluoroethanol, *r.t.* room temperature. General schematic of cell types is illustrated in Fig. 2

Allison et al., 2022). While these model radiofluorinated molecules have no direct clinical applications, it is noteworthy that thioether moieties are prevalent in various bioactive compounds, including fatty acids and essential amino acids, underscoring their potential value as PET tracers. Radiofluorination of thioethers through conventional approaches is often challenging (Feng et al., 2016; Walsh, 2020). With the help of an automated synthesizer (Fig. 3), methyl 2-[^{18}F]fluoro-2-(phenylthio)acetate was prepared under CA conditions using 0.002 M $Et_4NF{\cdot}4HF$ + 0.05 M Bu_4NClO_4 as electrolyte. The electrochemical synthesis was performed at room temperature applying 1.95 V versus Ag wire as reference electrode for 2 h or 4 h. Both RCY and A_m improved with increasing electrolysis time. However, RCY and

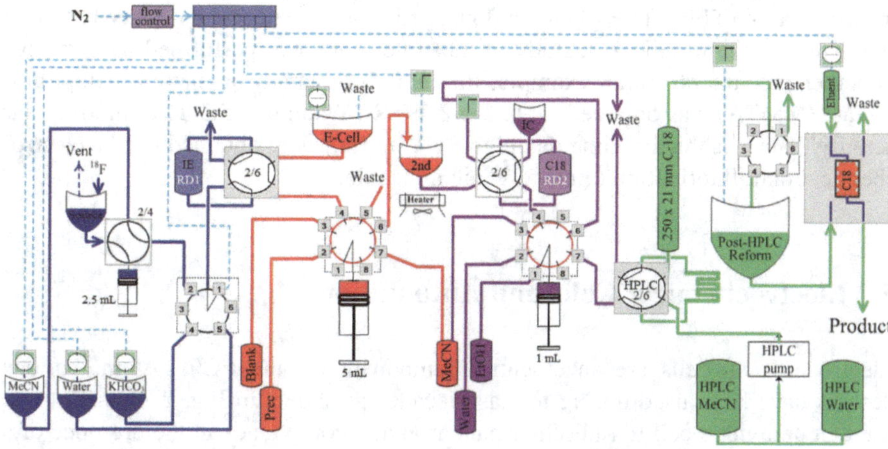

Fig. 3 Graphical user interface of the controlling software of an automated radiosynthesizer incorporating an electrochemical reactor. Color-coded subsystems: Processing of $[^{18}F]F^-$ (blue), chemical transformation (red), pre-HPLC purification (purple), and HPLC purification and reformulation (green). Reproduced with permission from Waldmann et al. (2017). Copyright Elsevier 2017

A_m obtained were modest (maximum RCY = 12.9%, A_m = 26 GBq/mmol after 4 h). An attempt to improve the reaction was performed by substituting the $Et_4NF{\cdot}4HF$/Bu_4NClO_4 electrolyte by a combination of Bu_4NF (0.154 M) and triflic acid (TfOH, 0.105 M), which resulted in a 7 ± 1% RCY. The use of Bu_4NF as electrolyte was only possible in the presence of TfOH, where the ratio TfOH/Bu_4NF = 0.68 was found to afford the highest RCY. The addition of TfOH is believed to form HF in situ, which plays a key role in stabilizing the cation-radical (Balandeh et al., 2017).

Since the use of CA conditions produces low A_m, investigators have looked for NCA alternatives that allow for obtaining A_m values in line with other reported NCA syntheses (Kim et al., 2010; Way & Wuest, 2014; Jacobson et al., 2015; Pike, 2018; Krasikova, 2020). A recent publication demonstrated the NCA electrochemical radiofluorination of thioethers using the cation pool technique. The application of the cation pool technique was first pioneered by Yoshida et al. to stabilize the carbocations formed after oxidation of organic molecules (Suga et al., 2002; Yoshida & Suga, 2002). Application of the cation pool technique for radiofluorination can decrease the need for high excess of $[^{19}F]HF$ salts required to stabilize the cation-radical formed after the first oxidation step (Scheme 1). Using the cation pool technique Sadeghi et al. were able to replace the previously employed CA electrolyte ($Et_4NF{\cdot}4HF$) by NCA electrolytes (i.e., Bu_4NClO_4, triflic acid, *p*-toluenesulfonic acid) and therefore increased the A_m of radiofluorinated thioether derivatives by three orders of magnitude (Table 1) (Balandeh et al., 2018). In a more recent paper, Sadeghi et al. have shown that addition of an auxiliary stabilizing agent could eliminate the need for excess of $[^{19}F]F^-$ salt to achieve electrochemical radiofluorination of thioethers without a cation pool technique (Allison et al., 2022). Reaction conditions included the use of 0.05 M Bu_4NClO_4 as electrolyte dissolved in trifluroethanol (TFE). TFE played a double role of solvent and auxiliary base that stabilize the cation-radical. A study in substrate scope was performed with various thioether precursors. As an example, the radiofluorination of methyl(methylthio) acetate (MMTA) was obtained in 92.7 ± 2.2% RCY and 4.7–5.3 TBq/mmol molar activity. This was the first time that high RCY and A_m were obtained for the electrochemical radiofluorination of any organic molecule.

5 Electrochemical Cells and Automation

Electrochemical cells are an essential component for the ECF system. Various designs have been used during the last decades as shown in Fig. 2. Reischl et al. used an undivided cell to radiofluorinate aromatic compounds in the presence of a high concentration of $[^{19}F]F^-$ carrier (Fig. 2a) (Reischl et al. 2002a, b, 2003; Kienzle et al., 2005). The working electrode (WE) was used as anode, and counter electrode (CE) as cathode, both made of platinum (Pt), while the pseudo-reference electrode (RE) was $Ag/AgNO_3$. $Et_3N{\cdot}3HF$ dissolved in MeCN was used as electrolyte and fluoride carrier.

Sadeghi et al. have used an undivided three-electrode cell and a divided three-electrode cell designs (Fig. 2b, c, respectively) (He et al., 2015; Balandeh et al., 2017, 2018; Lebedev et al., 2017; Allison et al., 2022). Several compounds including aromatic and thioether derivatives have been radiofluorinated using those E-Cells. Typically, Pt wire or mesh have been used as WE and CE, while Ag wire or Ag/Ag^+ have been employed as RE. Various electrolyte mixtures have been studied, involving both carrier-added or no-carrier-added conditions. Combinations of different concentrations of $Et_3N{\cdot}3HF$ or $Et_4NF{\cdot}4HF$ with other electrolytes were typically employed for carrier-added conditions, whereas $Et_3N{\cdot}3HAc$ and Bu_4NClO_4 have been applied as electrolytes in no-carrier-added radiofluorination.

The undivided cell with Haber-Luggin capillary (Fig. 2a) consists of two electrodes (anode and cathode) immersed in the same electrolyte solution. The Haber-Luggin capillary is used to measure the potential at a close proximity to the electrodes. Similarly, the undivided cell with three electrodes (Fig. 2b) includes an additional reference electrode for precise potential control. In both cases, the undivided cell design allows for direct contact between the electrodes which promote fast mass and charge transfer and therefore fast reaction kinetics. However, they may suffer from cross-contamination and interference between reactions. The two-compartment cell with three electrodes (Fig. 2c) addresses these issues by providing physical separation, allowing better control over reactions and measurements. In addition, the nafion membrane separating the anodic and cathodic chambers allows the use of strong acid media, not compatible with nucleophilic fluorination, in the cathodic chamber to promote rapid reduction. The ion-exchange membrane served to remove any redox reaction bottleneck from the counter electrode half-cell.

Due to the need for large-scale and high levels of activity in a clinical setting, automation of the synthesis, purification, and formulation of radiopharmaceuticals is required. Sadeghi et al. have shown that the electrochemical radiofluorination process can be integrated in an automated synthesizer with other required subsystems (Waldmann et al., 2017). The automatic platform enables the full preparation of radiotracers, starting from the processing of $[^{18}F]F^-$ obtained from the cyclotron to the synthesis, purification, and formulation of the final product. The reported automated electrochemical synthesis platform was able to perform multiples runs in a day which is helpful when optimizing radiolabeling conditions. Figure 3 shows the graphical interface of the automated synthesizer.

The integration of the electrochemical radiofluorination module in an automated synthesizer was tested using methyl 2-(phenylthio)acetate as a model precursor and taking advantage of the multirun capability of the system. Various electrochemical conditions such as electrolyte composition, applied voltage, and reaction time were investigated. In two different sets of experiments, the obtained product was purified and formulated with RCY of 5.2% and 12.9% in 223 min and 334 min, respectively. The A_m were calculated to be 11 GBq/mmol and 26 GBq/mmol, respectively. The low values for the obtained A_m correspond to the use of high excess of carrier in the electrochemical synthesis. Radiochemical purity was higher than 99% in both cases.

6 Outlook

The use of electricity has become an alternative path to perform organic chemical reactions that are otherwise challenging. In the radiopharmaceutical field, ECF represents an attractive alternative for the late-stage radiofluorination of small molecules, in particular for aromatic and thioether derivatives. The production of ^{18}F-based radiopharmaceuticals often requires the preparation of complex organic precursors that can be radiolabeled at the last stage. The synthesis and purification of these precursors involve, in most cases, lengthy and complicated chemical processes with the use of highly toxic chemical reagents and solvents. Using ECF to prepare ^{18}F-based radiopharmaceuticals could avoid the use and preparation of these complex organic precursors. Over the past decades, alternative electrochemical methods for late-stage nucleophilic radiofluorination of small molecules have emerged to minimize the use of non-environmentally friendly chemicals, shorten the synthesis steps and production time, and promote high RCY and A_m.

ECF offers advantages in the late-stage radiofluorination of electron-rich molecules compared to traditional electrophilic methods. ECF uses $[^{18}F]F^-$ produced directly from the cyclotron as a fluorine source which is easier to handle than $[^{18}F]F_2$ or other electrophilic reagents. Compared to conventional nucleophilic substitution methods, electrochemical radiofluorination does not require azeotropic distillation of $[^{18}F]F^-$ prior to a reaction.

Despite the advantages of ECF, there are several limitations and challenges that need to be addressed for the technique to find broader applicability in the radiopharmaceutical field. The implementation of a standardized electrochemical cell design may eliminate the differences arising from variations in E-Cell types and experimental methodologies across different research groups (Yan et al., 2018; Baran et al., 2021). One example of the commercially available E-Cell is IKA ElectraSyn, with a E-Cell design similar to the one depicted in Fig. 2b, which offers a range of volume and electrode materials. ECF of complex and large substrates with multiple functional groups can be difficult due to the likely formation of side products. This can be more challenging when functional groups with low oxidation potential are present in the structure. For such cases, protection of functional groups might be required. This could mean the need for deprotection steps post-radiofluorination, which would potentially increase the synthesis time and decrease RCY and A_m. Another important limitation is the frequent requirement for excess of fluoride salts which substantially impacts the A_m. Finding no-carrier-added ECF alternatives and expanding the scope of their applicability are ongoing topics of research in the field. As discussed here, high yield NCA radiofluorination reactions can be achieved with the use of the cation pool technique and/or by choosing solvents/auxiliary bases that can stabilize the cation-radical intermediates and support the proton abstraction. In addition, the use of bipolar electrode systems can effectively reduce the required concentration of supporting electrolyte (Miyamoto et al., 2019; Shida et al., 2019, 2021).

The application of emerging microflow reactor platforms may offer another technological solution to improve reaction yields and avoid side product formation. Microfluidic reactors have been reported to offer several advantages for radiopharmaceutical synthesis and characterization, including the ability to perform microscale (microliter or nanoliter volumes) reactions, shorten reaction times, and lead more precise control over reaction conditions (Rios et al., 2022). The development and applications of novel microfluidic systems is gaining traction for both the synthesis and quality control of radiopharmaceuticals (Ha et al., 2017). Microflow reactors can enhance electrochemical reactions in several ways: mixing is enhanced due to the short dimensions of a microfluidic cells; mass transfer is improved due to the greater surface-to-volume ratio, making electrode processes more efficient. Figure 2d shows a design of a microfluidic electrochemical cell.

Application of flow-chemistry to ECF could offer a method to scale up reactions by using continuous flow and parallel reactors. An important advantage of microfluidic cells stems from the fact that electrochemical fluorination depends on reactive oxidized intermediates, which are short-lived. Microflow reactors combine high surface area with low volume to rapidly oxidize the precursors and reduce the length of the time that molecules need to spend in the electrochemical cell. By quickly transferring reactive cations formed on the anode out of the E-Cell and into a downstream reaction channel, the short-lived reactive intermediates can be used for the subsequent reaction with the fluoride anion before decomposition. Electrochemical flow-cells with multistage reaction regions can benefit from a level of automation, as radiochemical experiments will need to be performed behind a shielded fume hood.

Development of automation and consideration of good manufacturing practice are also critical to clinical translation. Preliminary results discussed in this report have demonstrated the integration of simple electrochemical cells into an automated system with downstream purification and formulation, highlighting the feasibility of automation and a promising direction for ECF as a platform in radiopharmaceutical preparation.

The successful electrochemical radiofluorination of various radiotracers has set a precedent for future development of innovative designs of electrochemical platforms, methods, and applications. With further development and standardization, the ECF approach can be a useful methodology in the radiochemical toolbox to promote a more efficient synthesis of radiotracers for a variety of applications.

References

Abe, T. (2000). The electrochemical fluorination as a locomotive for the development of fluorine chemistry. *Journal of Fluorine Chemistry, 105*(2), 181–183. https://doi.org/10.1016/S0022-1139(00)00269-4

Abe, T., et al. (1980). Fluorination of 2-alkyl-substituted oxanes. The synthesis and purification of perfluoro(2-alkyl-substituted oxane)s. *Journal of Fluorine Chemistry, 15*(5), 353–380. https://doi.org/10.1016/S0022-1139(00)81770-4

Alexoff, D., Schlyer, D. J., & Wolf, A. P. (1989). Recovery of [18F] fluoride from [18O] water in an electrochemical cell. *International Journal of Radiation Applications and Instrumentation. Part A, Applied Radiation and Isotopes, 40*(1), 1–6.

Allison, N., et al. (2022). Electrochemical no-carrier-added radiofluorination of thioethers. *Journal of Fluorine Chemistry, 257–258*, 109988. https://doi.org/10.1016/j.jfluchem.2022.109988

Balandeh, M., et al. (2017). Electrochemical fluorination and radiofluorination of methyl(phenylthio)acetate using tetrabutylammonium fluoride (TBAF). *Journal of the Electrochemical Society, 164*(9), G99–G103. https://doi.org/10.1149/2.0941709jes

Balandeh, M., et al. (2018). Electrochemical flash fluorination and radiofluorination. *ChemElectroChem, 5*(22), 3353–3356. https://doi.org/10.1002/celc.201800973

Baran, P.S., Horn, E., & Waldmann, D. (2021) ElectraSyn 2.0. https://www.ika.com/en/Products-Lab-Eq/Electrochemistry-Kit-csp-516/.

Campbell, M. G., & Ritter, T. (2014). Late-stage fluorination: From fundamentals to application. *Organic Process Research & Development, 18*(4), 474–480.

Dagallier, C., et al. (2021). Development of PET radioligands targeting COX-2 for colorectal cancer staging, a review of in vitro and preclinical imaging studies. *Frontiers in Medicine, 8*, 675209.

Feng, M., et al. (2016). Sulfur containing scaffolds in drugs: Synthesis and application in medicinal chemistry. *Current Topics in Medicinal Chemistry, 16*(11), 1200–1216.

Fuchigami, T., & Inagi, S. (2011). Selective electrochemical fluorination of organic molecules and macromolecules in ionic liquids. *Chemical Communications, 47*(37), 10211. https://doi.org/10.1039/c1cc12414e

Fuchigami, T., & Inagi, S. (2020). Recent advances in electrochemical systems for selective fluorination of organic compounds. *Accounts of Chemical Research, 53*(2), 322–334. https://doi.org/10.1021/acs.accounts.9b00520

Fuchigami, T., Kayoko, Y., & Akinori, K. (1991). Electrolytic reactions of fluoro organic compounds. 8. Further study on anodic methoxylation and acetoxylation of aryl fluoroalkyl sulfides. *Tetrahedron, 47*(4–5), 625–634.

Fuchigami, T., et al. (1994). Electrolytic partial fluorination of organic compounds. 12. Selective anodic monofluorination of fluoroalkyl and alkyl sulfides. *The Journal of Organic Chemistry, 59*(20), 5937–5941.

Furuya, T., Kamlet, A. S., & Ritter, T. (2011). Catalysis for fluorination and trifluoromethylation. *Nature, 473*(7348), 470–477.

Ha, N. S., Sadeghi, S., & Van Dam, R. M. (2017). Recent progress toward microfluidic quality control testing of radiopharmaceuticals. *Micromachines, 8*(11), 337.

Hamacher, K., Hirschfelder, T., & Coenen, H. (2002). Electrochemical cell for separation of [18F] fluoride from irradiated 18O-water and subsequent no carrier added nucleophilic fluorination. *Applied Radiation and Isotopes, 56*(3), 519–523.

Haupt, J. D., Berger, M., & Waldvogel, S. R. (2019). Electrochemical fluorocyclization of N-allylcarboxamides to 2-oxazolines by hypervalent iodine mediator. *Organic Letters, 21*(1), 242–245. https://doi.org/10.1021/acs.orglett.8b03682

He, P., et al. (2014a). Advances in processes for PET radiotracer synthesis: Separation of [18F] fluoride from enriched [18O] water. *Applied Radiation and Isotopes, 91*, 64–70.

He, Q., et al. (2014b). Electrochemical nucleophilic synthesis of di-tert-butyl-(4-[18F]fluoro-1,2-phenylene)-dicarbonate. *Applied Radiation and Isotopes, 92*, 52–57. https://doi.org/10.1016/j.apradiso.2014.06.013

He, Q., Alfeazi, I., & Sadeghi, S. (2015). No-carrier-added electrochemical nucleophilic radiofluorination of aromatics. *Journal of Radioanalytical and Nuclear Chemistry, 303*(1), 1037–1040. https://doi.org/10.1007/s10967-014-3604-1

Hernández-Valdés, D., & Sadeghi, S. (2021). Electrochemical radiofluorination of small molecules: New advances. *The Chemical Record, 21*(9), 2397–2410. https://doi.org/10.1002/tcr.202100086

Ignat'ev, N. (2017). Electrochemical fluorination: A powerful tool for the preparation of organofluorine compounds. In *Modern synthesis processes and reactivity of fluorinated compounds* (pp. 71–123). Elsevier.

Jacobson, O., & Chen, X. (2010). PET designated flouride-18 production and chemistry. *Current Topics in Medicinal Chemistry, 10*(11), 1048–1059.

Jacobson, O., Kiesewetter, D. O., & Chen, X. (2015). Fluorine-18 radiochemistry, labeling strategies and synthetic routes. *Bioconjugate Chemistry, 26*(1), 1–18. https://doi.org/10.1021/bc500475e

Kienzle, G. J., Reischl, G., & Machulla, H. (2005). Electrochemical radiofluorination. 3. Direct labeling of phenylalanine derivatives with [18F] fluoride after anodic oxidation. *Journal of Labelled Compounds and Radiopharmaceuticals, 48*(4), 259–273.

Kim, D. W., et al. (2010). Recent trends in the nucleophilic [18 F]-radiolabeling method with no-carrier-added [18 F] fluoride. *Nuclear Medicine and Molecular Imaging, 44*, 25–32.

Konno, A., & Fuchigami, T. (1997). Electrolytic partial fluorination of organic compounds. 23.1 Regioselective anodic difluorination of sulfides using novel fluorine source Et4NF⊙ 4HF. *The Journal of Organic Chemistry, 62*(24), 8579–8581.

Krasikova, R. N. (2020). Nucleophilic synthesis of 6-l-[18F] FDOPA. Is copper-mediated radiofluorination the answer? *Molecules, 25*(19), 4365.

Kügler, F., et al. (2014). Optimizing the transfer of [18F] fluoride from aqueous to organic solvents by electrodeposition using carbon electrodes. *Applied Radiation and Isotopes, 91*, 1–7.

Laube, M., Kniess, T., & Pietzsch, J. (2013). Radiolabeled COX-2 inhibitors for non-invasive visualization of COX-2 expression and activity—A critical update. *Molecules, 18*(6), 6311–6355.

Laudadio, G., et al. (2019). Sulfonyl fluoride synthesis through electrochemical oxidative coupling of thiols and potassium fluoride. *Journal of the American Chemical Society, 141*(30), 11832–11836. https://doi.org/10.1021/jacs.9b06126

Lebedev, A., et al. (2017). Radiochemistry on electrodes: Synthesis of an 18F-labelled and in vivo stable COX-2 inhibitor. *PLoS One, 12*(5), e0176606. https://doi.org/10.1371/journal.pone.0176606. Edited by P. Garg.

Lee, S. J., et al. (2019). Copper-mediated aminoquinoline-directed radiofluorination of aromatic C–H bonds with K18F. *Angewandte Chemie International Edition, 58*(10), 3119–3122.

Miyamoto, K., et al. (2019). Development of a split bipolar electrode system for electrochemical fluorination of triphenylmethane. *ChemElectroChem, 6*(1), 97–100.

Möhle, S., et al. (2018). Modern electrochemical aspects for the synthesis of value-added organic products. *Angewandte Chemie International Edition, 57*(21), 6018–6041. https://doi.org/10.1002/anie.201712732

Nagase, S., Tanaka, K., & Baba, H. (1965). Direct preparation of partially fluorinated methane. *Bulletin of the Chemical Society of Japan, 38*(5), 834–838. https://doi.org/10.1246/bcsj.38.834

Neumann, C. N., & Ritter, T. (2015). Late-stage fluorination: Fancy novelty or useful tool? *Angewandte Chemie International Edition, 54*(11), 3216–3221.

Pike, V. W. (2018). Hypervalent aryliodine compounds as precursors for radiofluorination. *Journal of Labelled Compounds and Radiopharmaceuticals, 61*(3), 196–227.

Reischl, G., Ehrlichmann, W., & Machulla, H.-J. (2002a). Electrochemical transfer of [18F] fluoride from [18O] water into organic solvents ready for labeling reactions. *Journal of Radioanalytical and Nuclear Chemistry, 254*(1), 29–31.

Reischl, G., Kienzle, G., & Machulla, H.-J. (2002b). Electrochemical radiofluorination: Labeling of benzene with [18 F] fluoride by nucleophilic substitution. *Journal of Radioanalytical and Nuclear Chemistry, 254*(2), 409–411.

Reischl, G., Kienzle, G., & Machulla, H.-J. (2003). Electrochemical radiofluorination. Part 2. Anodic monofluorination of substituted benzenes using [18F] fluoride. *Applied Radiation and Isotopes, 58*(6), 679–683.

Rios, A., et al. (2022). Microliter-scale reaction arrays for economical high-throughput experimentation in radiochemistry. *Scientific Reports, 12*(1), 10263. https://doi.org/10.1038/s41598-022-14022-2

Rong, J., Haider, A., & Liang, S. (2021). Precision radiochemistry for Fluorine-18 labeling of PET tracers. In K. Szabó & N. Selander (Eds.), *Organofluorine chemistry* (1st ed., pp. 397–425). Wiley. https://doi.org/10.1002/9783527825158.ch12

Sadeghi, S., et al. (2013). Reusable electrochemical cell for rapid separation of [18F]fluoride from [18O]water for flow-through synthesis of 18F-labeled tracers. *Applied Radiation and Isotopes, 75*, 85–94. https://doi.org/10.1016/j.apradiso.2012.12.021

Saiki, H., et al. (2010). Electrochemical concentration of no-carrier-added [18F] fluoride from [18O] water in a disposable microfluidic cell for radiosynthesis of 18F-labeled radiopharmaceuticals. *Applied Radiation and Isotopes, 68*(9), 1703–1708.

Saito, F., et al. (2001). Recovery of 18F from [18O] water by electrochemical method. *Applied Radiation and Isotopes, 55*(6), 755–758.

Saito, F., et al. (2007). Electrochemical transfer of 18F from 18O water to aprotic polar solvent. *Applied Radiation and Isotopes, 65*(5), 524–527.

Sawamura, T., et al. (2012). Electrochemical fluorination using alkali-metal fluorides. *Angewandte Chemie International Edition, 51*(18), 4413–4416.

Shida, N., Zhou, Y., & Inagi, S. (2019). Bipolar electrochemistry: A powerful tool for electrifying functional material synthesis. *Accounts of Chemical Research, 52*(9), 2598–2608.

Shida, N., et al. (2021). Bipolar Electrochemical fluorination of triphenylmethane and bis (phenylthio) diphenylmethane derivatives in a U-shaped cell. *Electrochemistry, 89*(5), 476–479.

Simons, J. (1949). The electrochemical process for the production of fluorocarbons. *Electrochemical Society Journal, 95*.

Suga, S., Suzuki, S., & Yoshida, J. (2002). Reduction of a "cation pool": A new approach to radical mediated C–C bond formation. *Journal of the American Chemical Society, 124*(1), 30–31.

Takahira, Y., et al. (2019). Electrochemical C(sp3)–H fluorination. Synlett: Accounts and rapid Communications in Synthetic Organic. *Chemistry, 30*(10), 1178. https://doi.org/10.1055/s-0037-1611737

Tang, P., Wang, W., & Ritter, T. (2011). Deoxyfluorination of phenols. *Journal of the American Chemical Society, 133*(30), 11482–11484.

Tredwell, M., & Gouverneur, V. (2012). 18F labeling of arenes. *Angewandte Chemie International Edition, 51*(46), 11426–11437.

Waldmann, C. M., et al. (2017). An automated synthesizer for electrochemical 18F-fluorination of organic compounds. *Applied Radiation and Isotopes, 127*, 245–252. https://doi.org/10.1016/j.apradiso.2017.06.028

Walsh, C. T. (2020). *The chemical biology of sulfur*. Royal Society of Chemistry.

Way, J. D., & Wuest, F. (2014). Automated radiosynthesis of no-carrier-added 4-[18F] fluoroiodobenzene: A versatile building block in 18F radiochemistry. *Journal of Labelled Compounds and Radiopharmaceuticals, 57*(2), 104–109.

Wiebe, A., et al. (2018). Electrifying organic synthesis. *Angewandte Chemie International Edition, 57*(20), 5594–5619. https://doi.org/10.1002/anie.201711060

Wong, R., et al. (2012). Reactivity of electrochemically concentrated anhydrous [18F] fluoride for microfluidic radiosynthesis of 18F-labeled compounds. *Applied Radiation and Isotopes, 70*(1), 193–199.

Xiong, P., Long, H., & Xu, H.-C. (2019). electrochemical fluoroalkynylation of aryl alkenes with fluoride ions and alkynyltrifluoroborate salts. *Asian Journal of Organic Chemistry, 8*(5), 658–660. https://doi.org/10.1002/ajoc.201900020

Yan, M., Kawamata, Y., & Baran, P. S. (2017). Synthetic organic electrochemical methods since 2000: On the verge of a renaissance. *Chemical Reviews, 117*(21), 13230–13319.

Yan, M., Kawamata, Y., & Baran, P. S. (2018). Synthetic organic electrochemistry: Calling all engineers. *Angewandte Chemie (International ed. in English), 57*(16), 4149.

Yoshida, J., & Suga, S. (2002). Basic concepts of "cation pool" and "cation flow" methods and their applications in conventional and combinatorial organic synthesis. *Chemistry–A European Journal, 8*(12), 2650–2658.

Yu, S. (2006). Review of 18F-FDG synthesis and quality control. *Biomedical Imaging and Intervention Journal, 2*(4). https://doi.org/10.2349/biij.2.4.e57

Photoactivated Radiochemistry for the Construction of ^{18}F- and ^{11}C-Labeled Compounds

Manshu Li, Xuedan Wu, and Zibo Li

1 Introduction

Molecular imaging is a technology that provides in vivo distribution and functional information based on interactions at the molecular and cellular levels. Compared with conventional anatomical imaging technologies, molecular imaging can provide diagnostic information at an earlier stage of the disease, allows for real-time treatment monitoring, and reflects accurate biodistribution data in pharmacokinetic studies. Positron emission tomography (PET) is one especially important molecular imaging technology due to its high sensitivity (Ametamey et al., 2008; Cassidy & Radda, 2005; Weissleder & Mahmood, 2001; Pysz et al., 2010; Willmann et al., 2008). The accuracy and sensitivity of PET imaging depends on the properties of the PET imaging agents, which have been constructed through radiochemical synthesis with either radiometals (such as copper-64, gallium-68, zirconium-89) or lighter radionuclides (such as fluorine-18, carbon-11, nitrogen-13, oxygen-15).

The ^{18}F- and ^{11}C-labeled PET agents are particularly appealing because of their low positron energies, short half-lives for repetitive imaging, limited radiation exposure to the patient, and being bioisosteric/identical to existing drugs. To accommodate their short half-lives, rapid synthetic approaches have been developed to construct ^{18}F- and ^{11}C-labeled PET agents. However, many of these approaches used harsh conditions to increase the reactivity, which cannot be adopted for labile targeting vectors. The ideal radiosynthesis should be both fast and mild. The endeavors

Xuedan Wu and Zibo Li contributed equally to this work.

M. Li · X. Wu · Z. Li (✉)
Department of Radiology, Biomedical Research Imaging Center, Lineberger Comprehensive Cancer Center, University of North Carolina at Chapel Hill, Chapel Hill, NC, USA
e-mail: zibo_li@med.unc.edu

R. M. van Dam et al. (eds.), *Automated Technologies for the Development and Production of Radiopharmaceuticals*,
https://doi.org/10.1007/978-3-031-84632-8_13

and achievements of such method developments have been summarized in several comprehensive reviews (Deng et al., 2019; Jacobson et al., 2015; Krishnan et al., 2017; Halder & Ritter, 2021; Zhou et al., 2021; Eriksson et al., 2021). Among the newly developed methods, photoactivated radiosyntheses are especially interesting because of their mild conditions and high reactivities. This chapter focuses on the recent progress of photoactivated radiosyntheses using fluorine-18 and carbon-11 nuclides and provides an overview of their applications in PET probe constructions. The strategies include photoclick reactions, photoredox-catalyzed radical reactions, photoredox-catalyzed radical-polar crossover reactions with carbanion and carbocation intermediates, as well as organophotoredox-catalyzed reactions with reactive arene intermediates. These strategies involve both direct radiolabeling and utilization of radiolabeled prosthetic groups. In addition, the radiolabeling apparatus and irradiation light sources are also highlighted, providing a clear technical view of the feasibility of performing such photoactivated radiosynthesis.

2 Photoactivated Tetrazole Alkene 1,3-Dipolar Cycloaddition

Nitrile imine dipoles are highly reactive compounds that tend to form cyclized products with dipolarophiles through 1,3-dipolar cycloadditions (Fig. 1) (Bertrand & Wentrup, 1994). When an alkene is used as the dipolarophile, such reaction produces a pyrazoline product. A catalyst is not required for the typical nitrile imine dipolar cycloaddition due to the high reactivity of dipoles. This appealing feature provides an alternative click reaction approach, when residual metal toxicity is a concern (Jewett & Bertozzi, 2010).

Even though many stable heterosubstituted nitrile imines have been isolated, most non-heterosubstituted nitrile imines can only exist as reactive intermediates at room temperature and are trapped in situ with dipolarophiles (Bertrand & Wentrup,

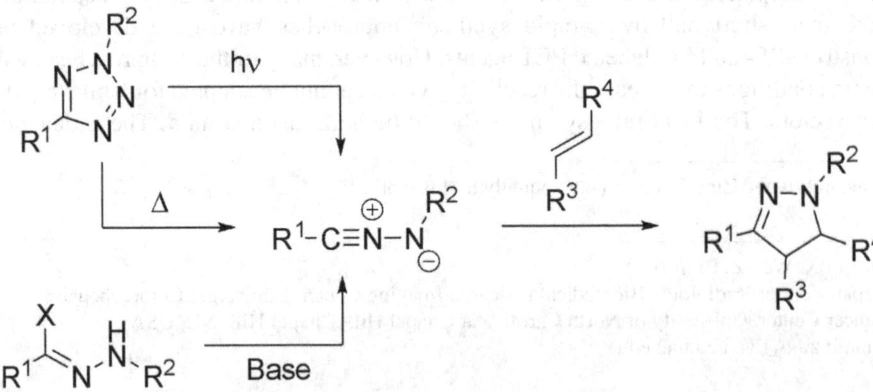

Fig. 1 In situ generation of nitrile imine dipoles and 1,3-dipolar cycloaddition with alkenes

1994). Typical in situ conditions generating nitrile imine intermediates include thermal decomposition, dehydrohalogenation under basic conditions, as well as photolysis of tetrazoles (Bertrand & Wentrup, 1994; Deepthi et al., 2022). Compared with the other two approaches, photolysis of tetrazoles provides a mild condition and therefore is usually the preferred approach when labile substrates are involved, especially in bioorthogonal reactions. After initially reported by Huisgen (Huisgen et al., 1962; Clovis et al., 1967), this photoactivated cyclization remained as an alternative approach in pyrazoline/pyrazole synthesis for decades (George & Angadiyavar, 1971; Padwa et al., 1978), with limited applications, despite the high reaction rates and high quantum yields reported from the mechanistic studies (Weinberg et al., 1989; Toubro & Holm, 1980). In 2007, Lin et al. revisited this strategy by optimizing the reaction conditions and envisioned its application with "protein target-guided in situ synthesis" (Wang et al., 2007). It was then applied as a bioorthogonal photoclick strategy (Lim & Lin, 2011; An et al., 2018; Song et al., 2008; Li et al., 2016; Holstein et al., 2015; Fan et al., 2013; He et al., 2013).

In 2011, Luxen et al. utilized this strategy in ^{18}F-radiochemistry (Fig. 2) (Thonon et al., 2011). Radiofluorinated diaryltetrazole **2** was prepared from its tosylate precursor **1** and [^{18}F]KF/K-222. This ^{18}F-labeled tetrazole **2** was subjected to photoclick conditions with diethylfumarate **3** under the irradiation of a "portable ultra-violet curing system" (wavelength not specified) (Thonon et al., 2011). The irradiation of a 400 W Spotlite curing system from UVP was filtered to remove photons in the IR range, mainly to reduce the heating. The reaction mixture was loaded into a quartz tube positioned only 5 mm away from the powerful light source. Robust reactivity of the ^{18}F-labeled tetrazole was observed: an irradiation time of 1 min was enough to activate all ^{18}F-labeled tetrazole in the solution of PBS buffer and acetonitrile, meanwhile a total reaction time of 5 min furnished 96% radiochemical conversion, when high concentration of diethylfumarate **3** was employed (10 mM and 1 mM concentrations). The radiochemical conversion decreased

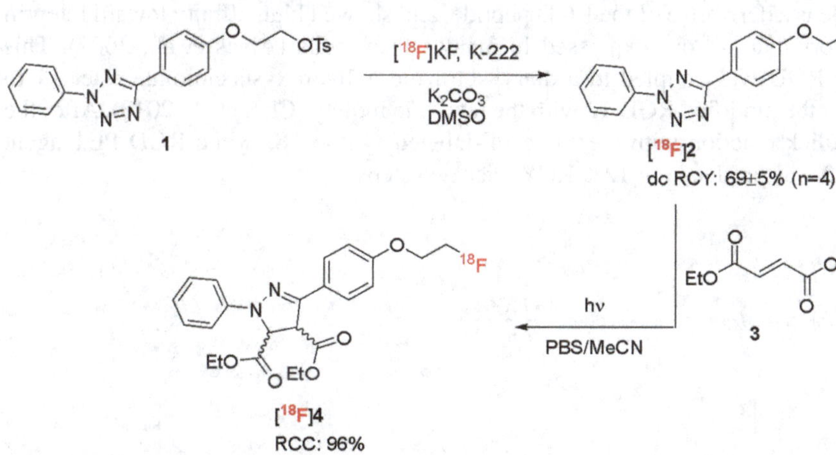

Fig. 2 Photoactivated tetrazole alkene cyclization with ^{18}F-labeled tetrazole synthon

significantly when lower concentration of diethylfumarate **3** was employed in the reaction mixture, due to the non-radioactive ^{19}F-substituted **2** competing with [^{18}F]**2** in the cyclization. Even though [^{18}F]**2** was prepared from no-carrier-added [^{18}F]fluoride, the accompanying non-radioactive [^{19}F]**2** was still observed in high concentration, and the measured molar activity of [^{18}F]**2** was low (7 to 25 GBq/μmol at EOB). The high concentration of [^{19}F]**2** consumed diethylfumarate **3** and affected the cyclization efficiency between [^{18}F]**2** and **3,** when limited concentration of **3** was used. By using only a portion of the purified [^{18}F]**2**, the concentration of the accompanied [^{19}F]**2** in the reaction mixture was reduced, and high radiochemical conversions can be achieved even at low dipolarophile concentrations. This, however, limited the scale of the radiosynthesis.

The method was then applied to the click coupling of an alkene-modified peptide **5** (Fig. 3) (Thonon et al., 2011). Prior studies identified electron-deficient alkenes as more reactive substrates in 1,3-dipolar cyclizations (Deepthi et al., 2022). Therefore, the peptide precursor was modified with an acrylic acid to tune the alkene moiety with appropriate electron density (Thonon et al., 2011). Decent radiochemical conversion (85%) was achieved, when 5.5 mM of the modified peptide **5** was used. The conversion decreased to 40% at lower alkene **5** loading (1 mM concentration). This observation was consistent with the synthesis of small molecule [^{18}F]**4**. Even though the efficiency of photoclick was limited by the low molar activity, this method is still appealing considering the rapid photoclick process.

In 2018, Jeong et al. further explored the method by employing the ^{18}F-labeled alkene [^{18}F]**8** as the radio synthon and utilized the method in the construction of a PET agent (Fig. 4) (Choi et al., 2018). After the synthesis of [^{18}F]**8** in 79% radiochemical yield (RCY), the photoclick was carried out with a diaryl-substituted tetrazole **9**. The irradiation from a 302 nm UV lamp lasted for 30 min and furnished [^{18}F]**10** in 36% RCY after HPLC purification.

This photoclick condition was then applied in the PET agent construction of a cyclic RGD tracer (Fig. 5) (Choi et al., 2018). The cyclic Arg-Gly-Asp-D-Tyr-Lys peptide is a derivative of the RGD peptide, and showed high affinity toward integrin receptors that are overexpressed in many cancer cells (Alves et al., 2007). This cyclic RGD was coupled to a diaryl tetrazole *N*-hydroxysuccinimide ester **11** to furnish the modified RGD **12** with the tetrazole moiety (Choi et al., 2018). After the photoclick reaction with the same ^{18}F-labeled synthon **8**, cyclic RGD PET agent [^{18}F]**13** was produced in 12% RCY over two steps.

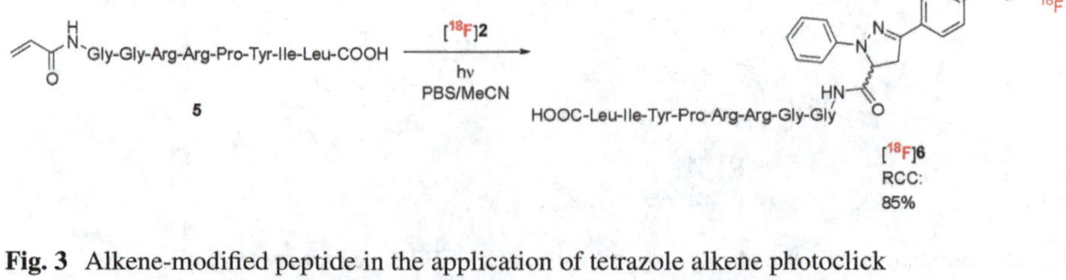

Fig. 3 Alkene-modified peptide in the application of tetrazole alkene photoclick

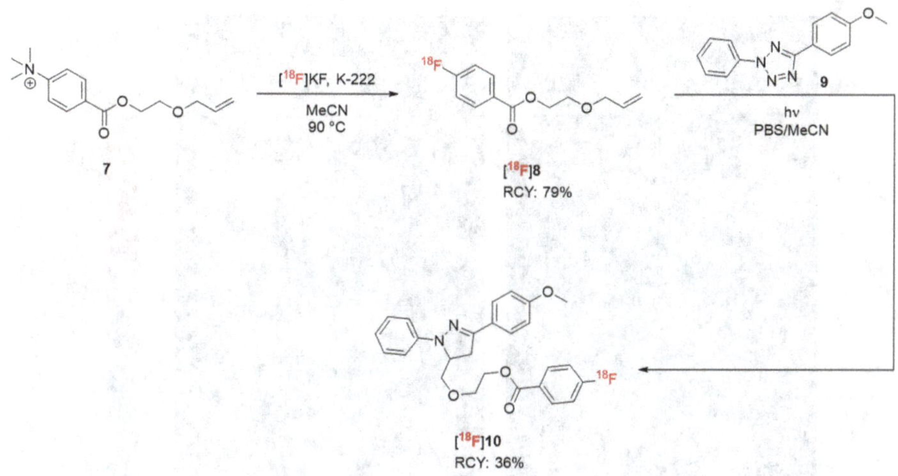

Fig. 4 Tetrazole alkene photoclick with radiolabeled alkene synthon

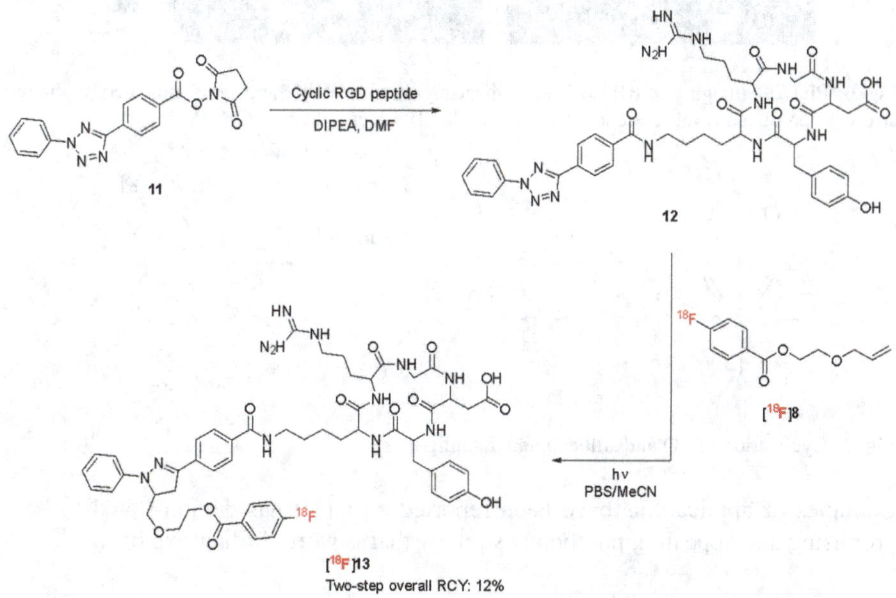

Fig. 5 Cyclic RGD PET agent construction through tetrazole alkene photoclick reaction

The total synthesis time of **[^{18}F]13** from [^{18}F]KF/K-222 was 95 min. An imaging study was then performed on RR1022 tumor-bearing rat. Cancer uptake with high tumor–background ratio was observed for **[^{18}F]13**, demonstrating the feasibility of this method in PET agent constructions (Fig. 6).

Overall, the photoactivated tetrazole alkene 1,3-dipolar cycloaddition is an efficient radioclick strategy in PET probe constructions. Even though only a few

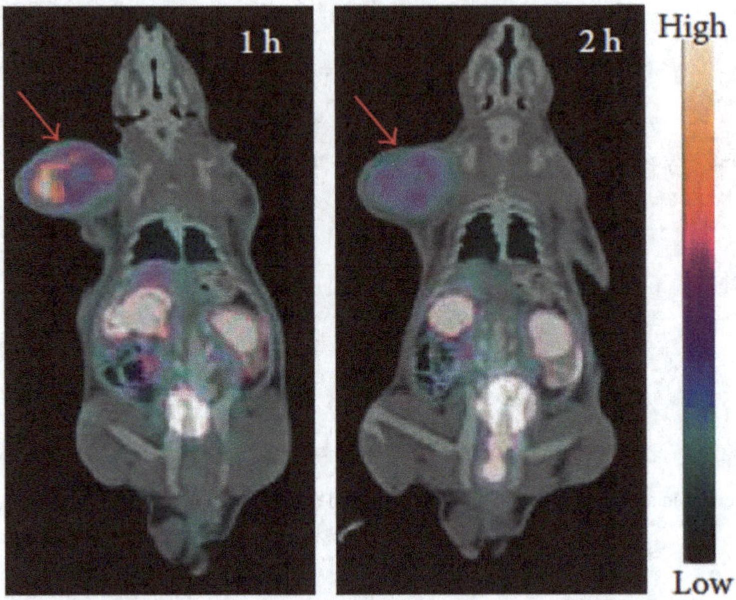

Fig. 6 PET/CT imaging of RR1022 tumor-bearing rat with [^{18}F]**13**, tumor is indicated by the red arrow. Reproduced from reference (Choi et al., 2018). Licensed under CC BY 4.0

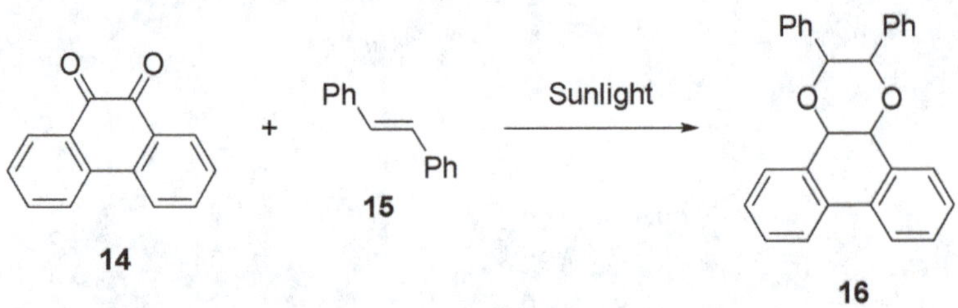

Fig. 7 Cyclization of PQ and stilbene under sunlight

examples of applications have been reported so far, it has demonstrated to be a promising and appealing photoclick strategy that is worth further exploring.

3 Phenanthrenequinone-Vinyl Ether [4 + 2] Cycloaddition

The cycloaddition between 9,10-phenanthrenequinone (PQ) **14** and stilbene **15** under sunlight was first discovered in 1944 by Schönberg and Mustafa (Fig. 7) (Schönberg & Mustafa, 1944). Besides stilbene, other types of alkene can also participate in such photoactivated cyclization (Rubin & Neuwirth-Weiss, 1972). The high reactivity of photoactivated 1,2-diketones was revealed in kinetic study

(Bohning & Weiss, 1966), and excited PQ has been applied in other types of transformations (Rubin & Zwitkowits, 1964; Kazuhiro et al., 1976; Rubin, 1963). However, further application of cycloaddition between PQ and alkenes remained scarce for decades.

The idea of utilizing this reaction as a bioorthogonal photoclick methodology was brought up in 2018 by Zhang et al. (Li et al., 2018). The photoexcited PQ undergoes cyclization with electron-rich vinyl ether substrates while remaining inert toward nucleophiles. This was considered as a major advantage over the photoactivated nitrile imines, which are facile to nucleophilic side reactions under in vivo conditions. In addition to several small molecule examples, a PQ derivative was also applied in the photoactivated cyclization of modified proteins. After treating cells with vinyl ether-modified antibody, such photoclick reaction can even occur on the surface of living cells (Li et al., 2018).

This exciting coupling method was then adopted by radiochemists. The photoclick flow reaction procedure between ^{18}F-labeled vinyl ethers and functionalized PQ was reported in 2021 (Fu et al., 2021). The reactivity of different vinyl ethers was studied initially using a conventional non-flow reaction setup with non-radioactive materials (Fig. 8). Good yields were recorded for most of the vinyl ether substrates. Under the irradiation with a 395 nm LED light, linear vinyl ethers reached full conversion within 180 s, while cyclic vinyl ethers showed an even faster reaction rate with full conversion within 90 s. This observation matched with their computational study that suggested cyclic vinyl ethers being more nucleophilic and therefore more reactive in these [4 + 2] reactions. The [^{19}F]alkylfluoride moiety in vinyl ether **17** and **19** did not interfere with the cycloaddition, demonstrating the possibility of its utilization in radiofluorination.

^{18}F-labeled synthons **[^{18}F]17** and **[^{18}F]19** were then prepared for the radiosynthesis from corresponding tosylate precursors (Fig. 9) (Fu et al., 2021). Moderate yields were obtained through rapid reaction process where the ^{18}F-substitution

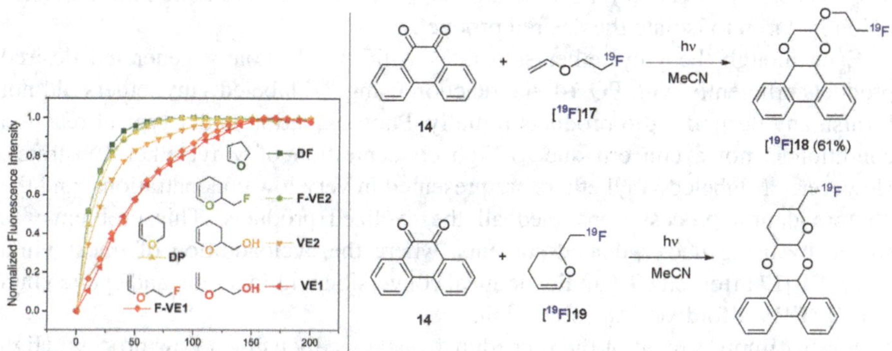

Fig. 8 Linear and cyclic vinyl ethers in cycloaddition. Reproduced from reference (Fu et al., 2021). Licensed under CC BY-NC-ND 4.0

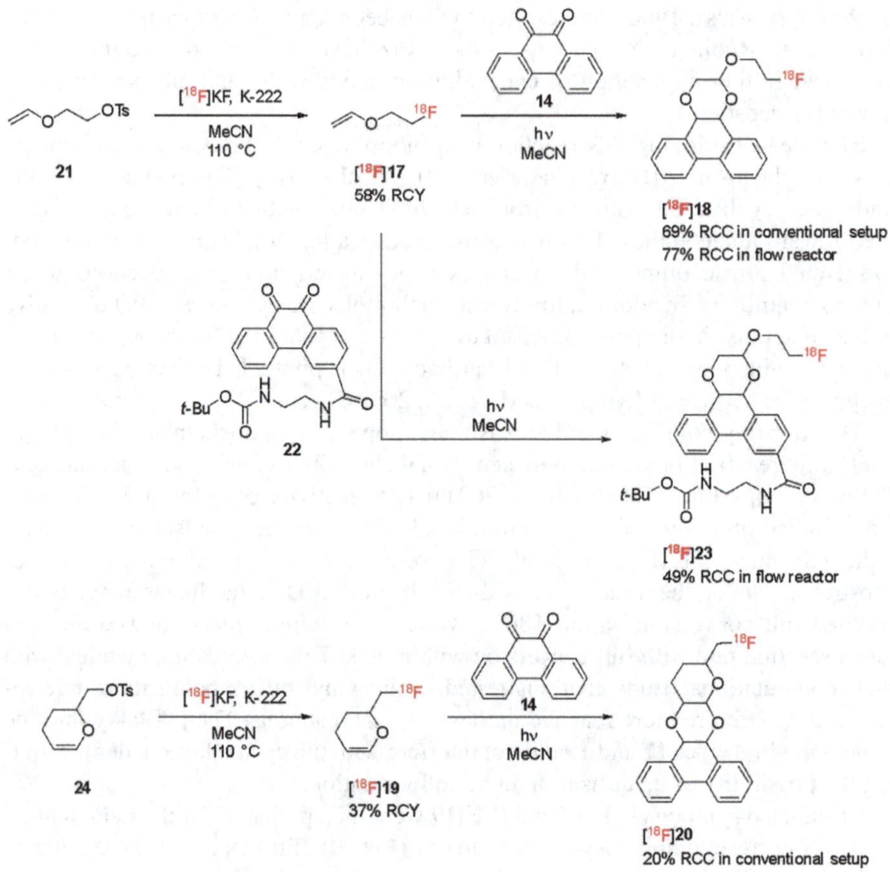

Fig. 9 Radiosynthesis of cyclized products with different ^{18}F-labeled synthons and PQ derivative

reactions only proceeded for 3 min, and the purifications were done through distillation in 2 min to isolate the desired product.

Even though the vinyl ether substrates with an ^{19}F-moiety generated desired products efficiently with PQ **14**, the reaction using ^{18}F-labeled vinyl ethers did not furnish any desired radio products initially. Photooxidation under typical reaction condition is not a concern due to high concentration of vinyl ether substrates. However, ^{18}F-labeled vinyl ethers are presented in very low concentrations, and the photooxidation process consumed all the cyclized products. This problem was solved by using deoxygenated solvents, where the cycloaddition of linear vinyl ether **[^{18}F]17** afforded 69% radiochemical conversion within 5 min, and cyclic vinyl ether **[^{18}F]19** afforded 20% conversion.

Further improvement of the procedures was done by using a flow process. High conversion (77%) can be achieved for linear **[^{18}F]17** within 60 s under the irradiated flow cyclization process. PQ derivative **22** with the amide moiety was also tested

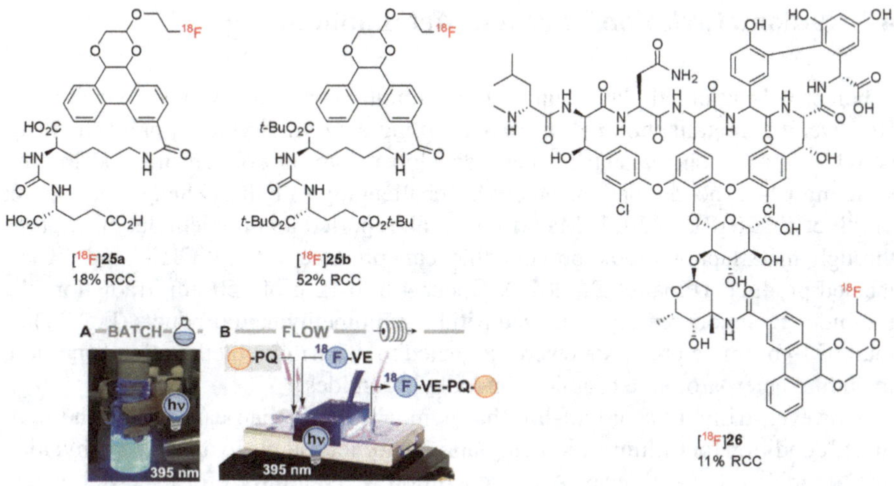

Fig. 10 Constructions of PET agents with conventional batch and flow processes. Reproduced from reference (Fu et al., 2021). Licensed under CC BY-NC-ND 4.0

under the flow reaction condition with **[¹⁸F]17** and produced **[¹⁸F]23** in 49% RCC in 1 min. This success demonstrated the possibility of functionalizing PQ with biologically active probes through PQ carboxylic acid/*N*-hydroxysuccinimide precursors and, thereafter, constructing the PET agents through 4 + 2 cycloaddition with functionalized PQ and ¹⁸F-labeled vinyl ether synthons.

Two prostate-specific membrane antigen (PSMA) binding agents were then modified into a PQ carboxylic acid amide. Vancomycin, an antibiotic, was also modified with the PQ motif. These substrates were subjected to the same flow process and generated the corresponding ¹⁸F-labeled PET agents (**[¹⁸F]25a, [¹⁸F]25b,** and **[¹⁸F]26**) (Fig. 10).

The flow process for this method was done in a Future Chemistry module equipped with a flow photomicroreactor (Fu et al., 2021). In both the conventional and flow reaction apparatus, a 395 nm hand lamp was used as the light source. This light source has 88 mW/cm² irradiation power and was placed right next to the reaction mixture.

The radiochemical synthesis efficiency of PQ-vinyl ether cycloaddition has been well demonstrated. However, as a PET agent construction strategy, the corresponding PET agents need to be further evaluated to determine how this prosthetic group affects the stability, targeting efficacy, and biocompatibility.

4 Radiomethylation Through Metallaphotoredox

Efficient ^{11}C-radiomethylation under mild conditions has always been a challenge. However, the metallaphotoredox cross-coupling between aryl/alkyl halide and carbon-12 methylation reagent has been developed into an efficient method in constructing C—C bonds that are otherwise challenging to build (Zhang et al., 2016; Smith et al., 2018). In 2021, MacMillan et al. reported a radiomethylation strategy through metallaphotoredox process that can produce CT_3-, $^{11}CH_3$-, and $^{14}CH_3$-labeled products (Pipal et al., 2021). Under 4 h to 12 h of 450 nm irradiation, the bromide precursors readily produced tritium radiomethylated products (Fig. 11). A total of 19 bromide precursors were subjected to this tritium methylation condition, including heteroaromatic bromides and alkyl bromides.

However, tritium has a half-life that is much longer than carbon-11. The optimized condition for tritium radiomethylation is not acceptable for the ^{11}C-methylation due to rapid carbon-11 decay. A new condition was optimized for [^{11}C]MeI, where addition of tetrabutylammonium iodide reduced the less reactive [^{11}C]CH_3Br intermediate and significantly promoted the reaction process (Fig. 12). Under this optimized condition, only 5 min of irradiation was required. ^{11}C-labeled Celebrex [^{11}C]**29** was produced in 48% RCY with a total synthesis time of 22.7 min. This metallaphotoredox method demonstrated high compatibility with various functional groups and compounds bearing high structural complexity ([^{11}C]**30a** to [^{11}C]**30e**). The value of high functionality compatibility was also demonstrated in the preparation of ^{11}C-methylated [^{11}C]**30d**, which was obtained from unprotected precursor in one step. In a prior protocol, [^{11}C]**30d** had to be synthesized from protected precursors followed by deprotection, featuring lower step economy (Koyama et al., 2017).

Besides the ^{11}C-radiomethylation of aryl bromides, alkyl bromides can also act as the methylation target. Under a slightly modified condition with different solvent and absence of the additive, three ^{11}C-labeled products were furnished from alkyl bromides where the functionality tolerance was demonstrated again. Unprotected phenol moiety was tolerated under this modified metallaphotoredox condition, enabling the ^{11}C-methylation of [^{11}C]**30 g** and [^{11}C]**30 h**. This is another example of improved efficiency over a prior protocol (Shoup et al., 2014).

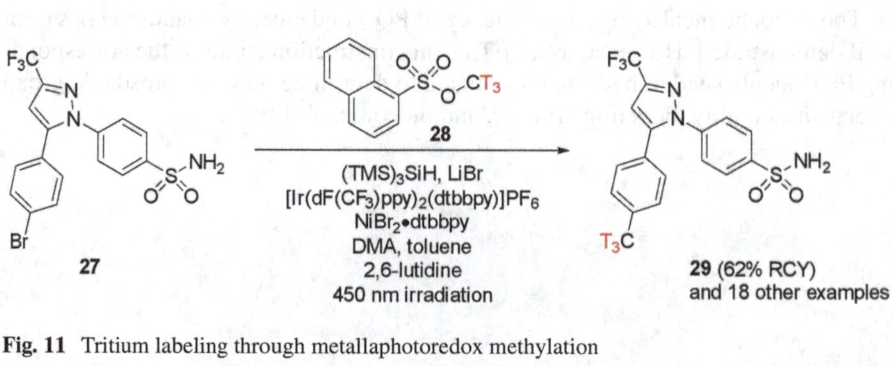

Fig. 11 Tritium labeling through metallaphotoredox methylation

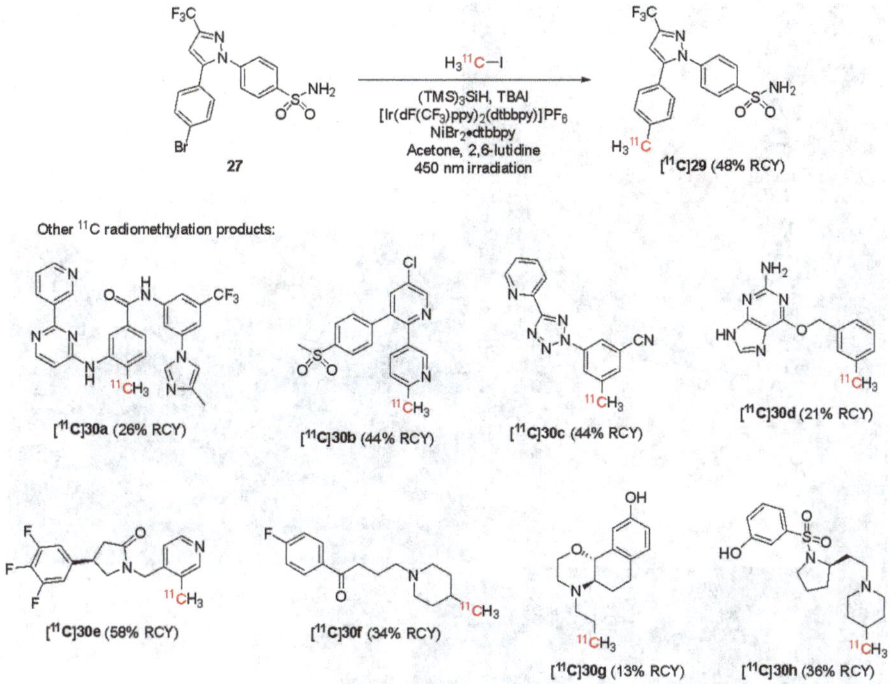

Fig. 12 Metallaphotoredox radiomethylation in ^{11}C-radiolabeling

These ^{11}C-labeling experiments were conducted in a Penn PhD Photoreactor with 450 nm light source (Fig. 13). It is noteworthy that the synthesis of **[^{11}C]30e** was also conducted by remote-controlled Siemens robotic manipulators to demonstrate the reproducibility where 72% RCY was obtained. Additionally, to accommodate the cGMP requirement in the clinical application scenario, a Synthra Melplus automated synthesis module was connected to the photoreactor and produced **[^{11}C]29** in 35% RCY with a 29 min total synthesis time.

5 Difluoromethylation with Sulfone Reagents

Difluoromethyl functionality has demonstrated significant impact on the properties of biologically active compounds (Purser et al., 2008; Meanwell, 2018). This moiety can act as a hydrogen-bond donor similar to hydroxyl group and thiol group, while maintaining a high lipophilicity (Sessler et al., 2017; Erickson & Mcloughlin, 1995; Zafrani et al., 2017). Several methods have been reported for the preparation of [^{18}F]CHF_2 labeled compounds. Decarboxylation radiofluorination with [^{18}F]selectfluor can produce both [^{18}F]CHF_2 and [^{18}F]CF_3 moieties but suffers from low molar activity (Mizuta et al., 2013; Sap et al., 2019). The same low molar

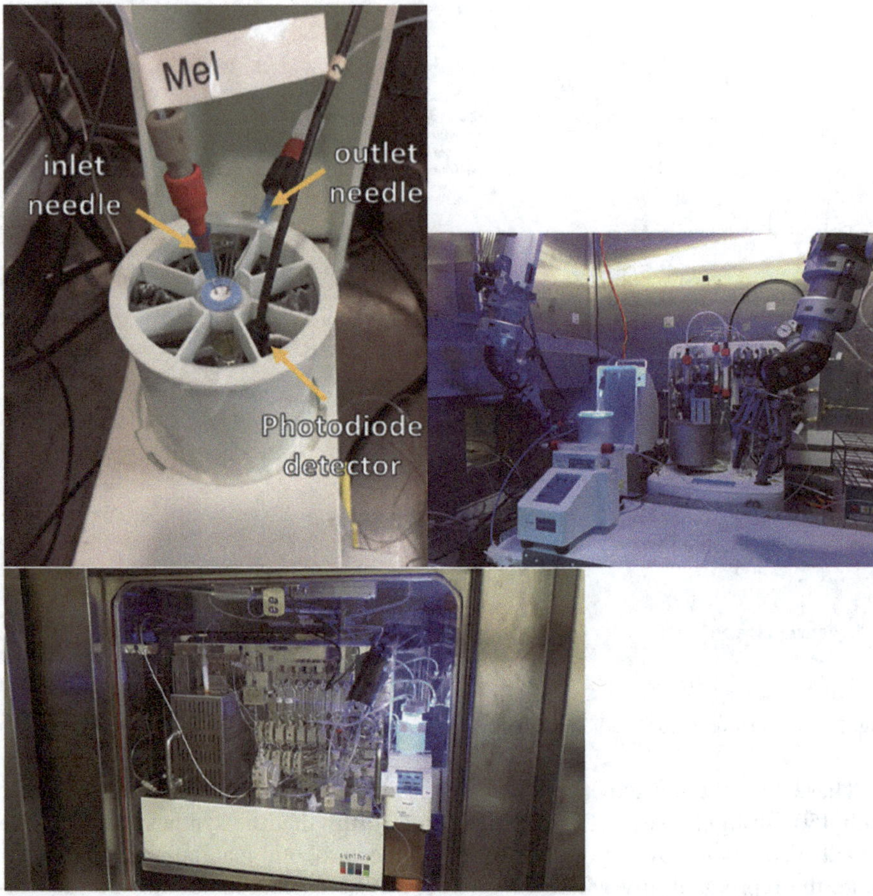

Fig. 13 Photoreactor and automated synthesis modules used for metallaphotoredox radiomethylation. Top left: The photoreactor. Top right: Remote-controlled robotic manipulators and the photoreactor. Bottom: Synthra MeIplus automated synthesis module and the photoreactor. Reproduced from reference (Pipal et al., 2021) with permission. Copyright 2020 Springer Nature

activity problem was also reported in the (pseudo) halogen atom exchange methods (Shi et al., 2016; Verhoog et al., 2016), as well as desulfurative fluorination (Newton et al., 2022). Moderate to good molar activities can be obtained through oxidative C—H activation fluorination, though this two-step radiosynthesis requires excessive heating for the second step (Yuan et al., 2017). Heating is also required for the reaction between [^{18}F]difluorocarbene and aryl boronic acid (Sap et al., 2022).

In 2019, an ^{18}F-difluoromethylation strategy through a photoredox process was reported by employing a sulfone reagent (Fig. 14) (Trump et al., 2019). This strategy features mild and rapid photoredox condition for the key step, with good specific activity. Although a two-step synthesis is required to prepare the ^{18}F-labeled sulfone [^{18}**F**]**33**, each step only takes 5 min, and only one final purification is needed.

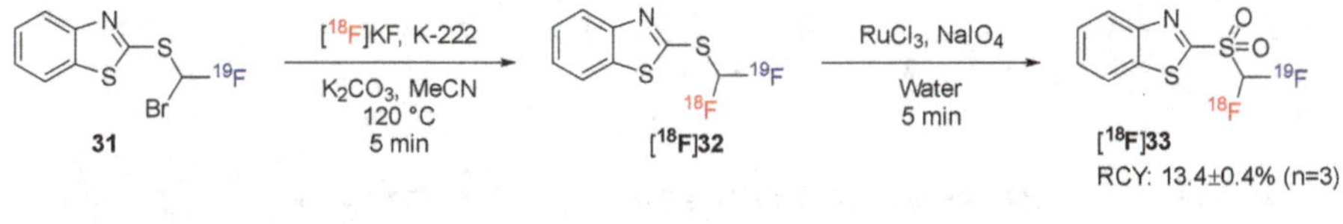

Fig. 14 Preparation of [18F]difluoromethyl sulfone **[18F]33**

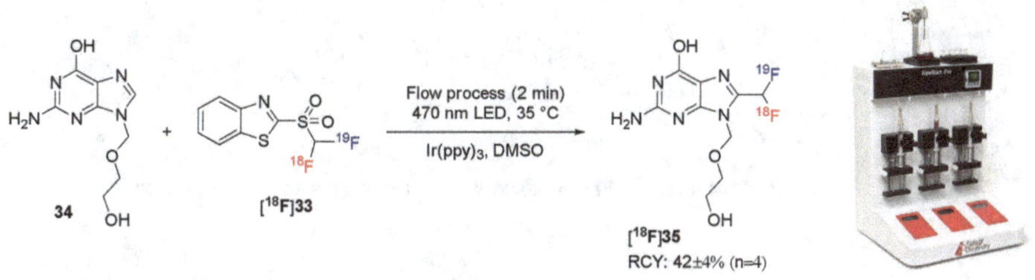

Fig. 15 Synthesis of radiodifluoromethylated acyclovir **[18F]35** using flow process with Futurechem FlowStart EVO system. Reproduced from reference (Trump et al., 2019). Licensed under CC BY 4.0

Therefore, the total synthesis time of sulfone **[18F]33** is only 45 min, with excellent molar activity. The limitation is the poor two-step RCY (13.4%), resulting mainly from the low efficiency of the first step (Halex reaction).

This 18F-labeled sulfone was applied in the radiodifluoromethylation of acyclovir under iridium-catalyzed photoredox condition (Fig. 15). To facilitate the reaction, a flow process was incorporated with 470 nm irradiation, which only took 2 min to complete at 35 °C. The **[18F]35** was furnished in 42% RCY through photoredox with good molar activity (44.4 ± 11.1 GBq/μmol). The total synthesis time of **[18F]35** was 90 min from [18F]KF/K-222.

More substrate scope scanning reviewed the broad tolerance of various heteroaromatic functionalities (14 small molecule examples). Examples of caffeine and its derivatives, various nucleic acid bases, as well as hypertension drug Moxonidine can all be labeled through this strategy (Fig. 16). To demonstrate the potential implication of the method, compound **[18F]36m** was synthesized in 25% RCY. **[18F]36m** is an analog of UCB, which is a family of PET agents targeting SV2A protein that is associated with central nervous system (CNS) disorders (Nabulsi et al., 2016; Warnier et al., 2016). Compared with other UCB PET agents, the difluoromethylated **[18F]36m** demonstrated the same level of affinity toward SV2A protein.

In 2020, Luxen et al. expanded the scope of [18F]difluoromethyl sulfones (Lemos et al., 2020). Following the same condition in Fig. 14, six sulfide precursors were directly labeled with [18F]KF/K-222 using a GE Healthcare FASTlab™ module (Fig. 17).

The subsequent oxidation reactions using sodium periodate and ruthenium chloride hydrate were much more efficient, generating the corresponding sulfones from sulfides. Three of these newly obtained 18F-labeled difluoromethyl sulfones were

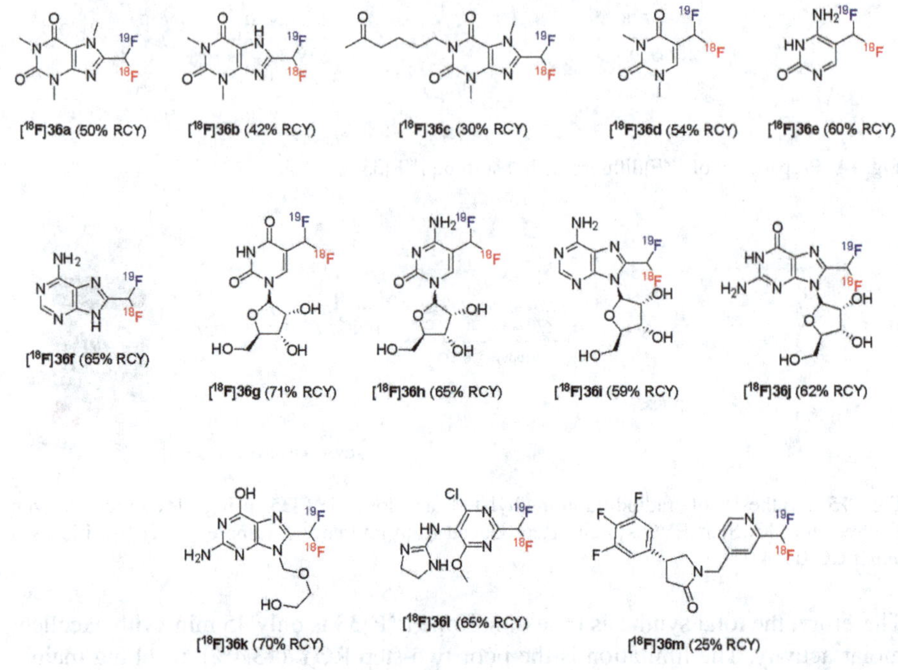

Fig. 16 Syntheses of radiolabeled difluoromethylated small molecules. Reaction condition is the same as Fig. 15

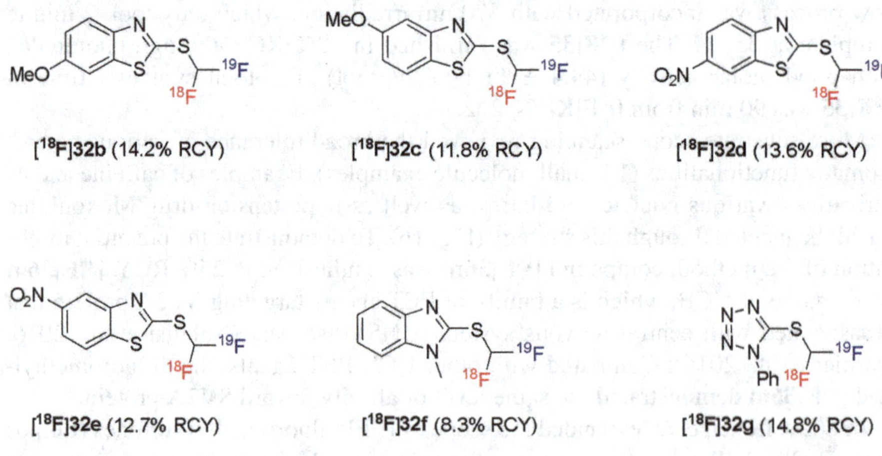

Fig. 17 Syntheses of [^{18}F]difluoromethyl sulfides

subjected to a photoredox flow process where reaction mixture was pumped into a microreactor (100 μL capacity) using a FlowStart Evo flow reactor from FutureChemistry. The microreactor was irradiated with 470 nm LED light. All three new ^{18}F-labeled sulfones promoted the difluoromethylation of acyclovir (Fig. 18),

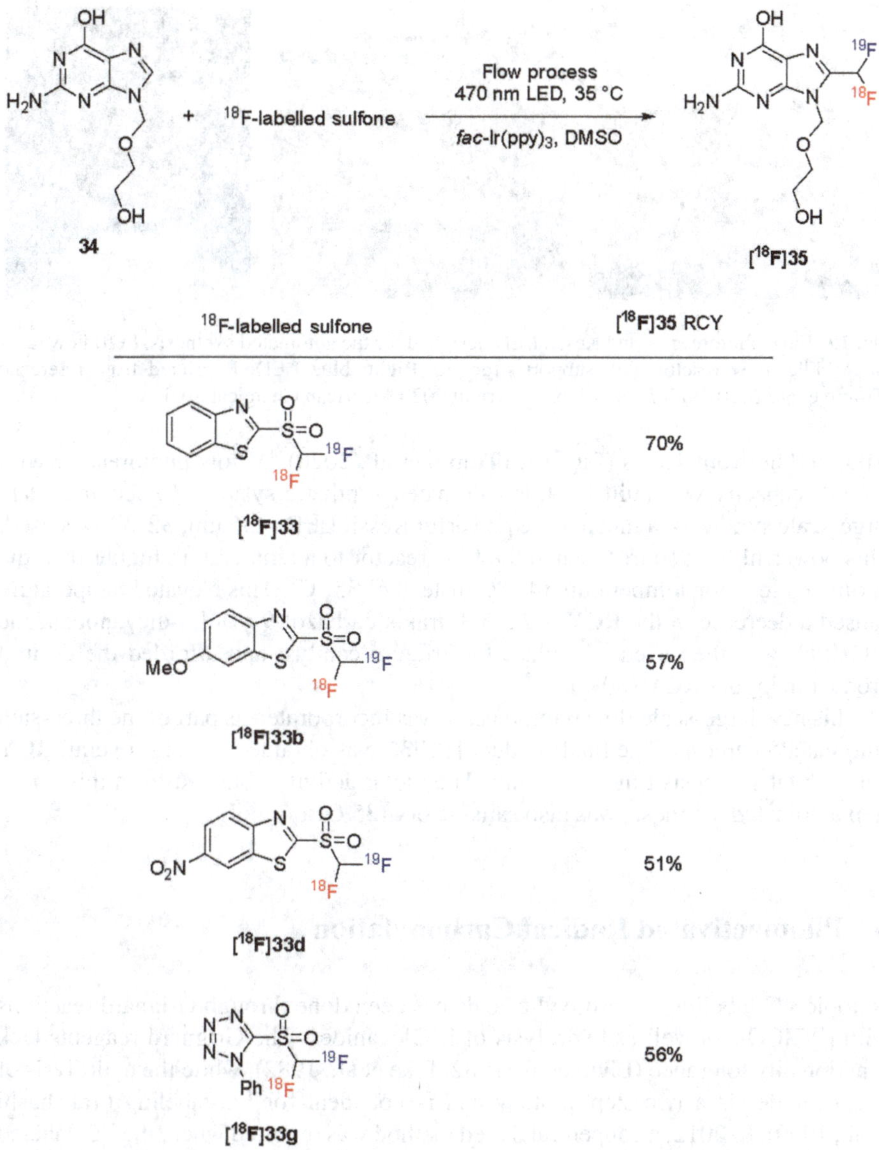

Fig. 18 Radiosyntheses of [^{18}F]difluoromethylated acyclovir [^{18}**F**]**35**

despite the lower efficiency compared to the sulfone [^{18}**F**]**33** in their 2019 work. More iridium photocatalyst and longer irradiation time (slower flow rate) both promoted the reaction efficiency.

The fully automated synthesis of ^{18}F-difluoromethylated acyclovir [^{18}**F**]**35** was made possible after the establishment of the automated photoredox process on the

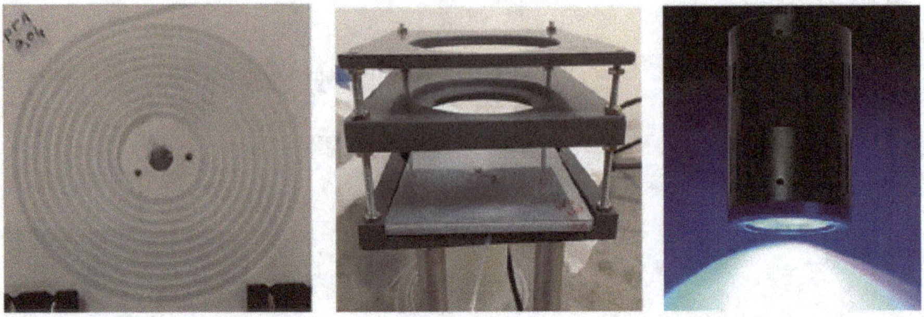

Fig. 19 Flow photoreactor and Kessil LED designed for the automated synthesis. Left: flow reactor. Middle: flow reactor with support structure. Right: blue LED. Reprinted from reference (Trump et al., 2020) with permission. Copyright 2020 American Chemical Society

AIO module from Trasis (Fig. 19) (Trump et al., 2020). A flow photoreactor with 1.2 mL capacity was built for this automated synthetic system. To accommodate large-scale synthesis, a much more powerful Kessil LED (455 nm, 32 W) was used. This powerful light source heated the flow reactor to a temperature higher than the optimized reaction temperature (45 °C instead of 35 °C). This elevated temperature caused a decrease in the RCY. 1,2,3,5-Tetrakis(carbazol-9-yl)-4,6-dicyanobenzene (4CzIPN) was then used to replace the iridium catalyst and afforded the desired product in higher RCY (80%).

This new large-scale flow photoreactor was incorporated as part of the three-step automated synthesis. The final product [^{18}F]**35** was obtained in 1.4% overall RCY with a total synthesis time of 95 min. The molar activity obtained from this three-step automated synthesis was also satisfactory (35 GBq/μmol).

6 Photoactivated Radical Carbonylation

Isotopic ^{11}C-labeling of carboxylic acids has been done through Grignard reactions with [^{11}C]CO_2, as well as hydrolysis of [^{11}C]cyanides. The Grignard reagents lack functionality tolerance (Livni et al., 1982; Pike et al., 1982), while the hydrolysis of [^{11}C]cyanides is a two-step strategy and is not ideal for ^{11}C-labeling (Takahashi et al., 1990). In 2012, a copper-catalyzed method was reported where the ^{11}C-labeled carboxylic acids can be generated from boronic acid ester precursors and [^{11}C]CO_2 (Riss et al., 2012). However, the reaction requires heating condition (90 °C to 100 °C).

A reported selenium- and palladium-catalyzed ^{11}C-carbonylation with [^{11}C]CO in the synthesis of [^{11}C]carbamates and [^{11}C]amides also requires harsh conditions (130 °C to 150 °C, 35 MPa pressure) (Kihlberg & Långström, 1999; Kihlberg et al., 2002). Even though efforts have been made over the decades to reduce the reaction temperature and pressure for the [^{11}C]CO carbonylation, the conditions required still cannot be considered as mild (Eriksson et al., 2021).

In 2004, Långström et al. reported a photoactivated ^{11}C-carbonylation in the synthesis of [^{11}C]amides (Fig. 20) (Itsenko et al., 2004). This method still requires high reaction pressure (35 MPa). However, the ambient reaction temperature was an improvement from the previous metal-catalyzed procedures. This procedure used a stainless-steel vessel to contain the high pressure and used a 400 W medium-pressure mercury lamp (model Philips HOK 4/120SE) for the irradiation process. The reaction mechanism was assumed to be the same as the non-radioactive version reported previously (Ryu et al., 1998).

Under the optimized reaction condition, only 400 s of the powerful mercury lamp irradiation was required for ^{11}C-labeling. The [^{11}C]amides can be generated rapidly from alkyl halides, amines, and the [^{11}C]CO. The conversion and RCY varied from the combination of alkyl halides and amines, where the highest 69% RCY was achieved in the synthesis of **[^{11}C]37a** (Fig. 21). Changing the solvents promoted the RCY in some examples.

In 2006, this method was expanded into the synthesis of [^{11}C]carboxylic acids (Fig. 22) (Itsenko et al., 2006). An automated Synthia system was applied in this protocol. Water was used as the nucleophile to generate carboxylic acids after the radical process. Primary, secondary, and tertiary alkyl halides all furnished desired products in good RCY. The 6 min UV irradiation promoted rapid reaction, which is ideal for ^{11}C-labeling. A slightly modified automated synthesis procedure can also be applied in the synthesis of [^{13}C]carboxylic acids.

The UV irradiation (280–400 nm) in this protocol was produced by a Hamamatsu Photonics LC5 UV spotlight source. This light source is a combination of Hamamatsu L8253 xenon lamp and an A7028–03 optical filter. The reaction is designed to be performed at ambient temperature but high pressure (40 MPa). To confine the high pressure, a microautoclave was employed. The UV irradiation was directed into the microautoclave through a window.

In 2007, this method was applied in the synthesis of [^{11}C]carbonyl esters by using alcohols as nucleophiles (Fig. 23) (Itsenko et al., 2007). The reaction

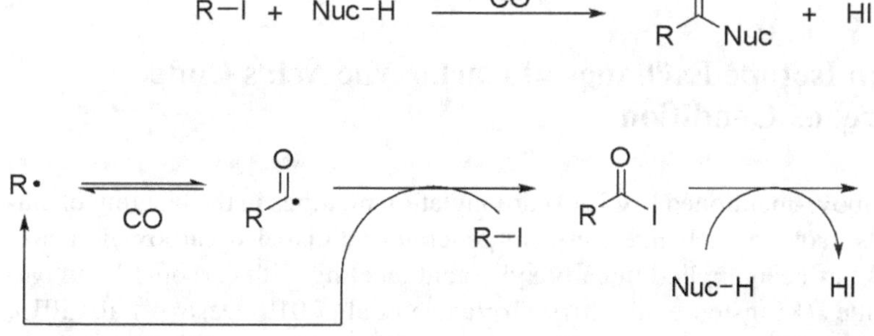

Fig. 20 Photoactivated carbonylation through radical intermediates

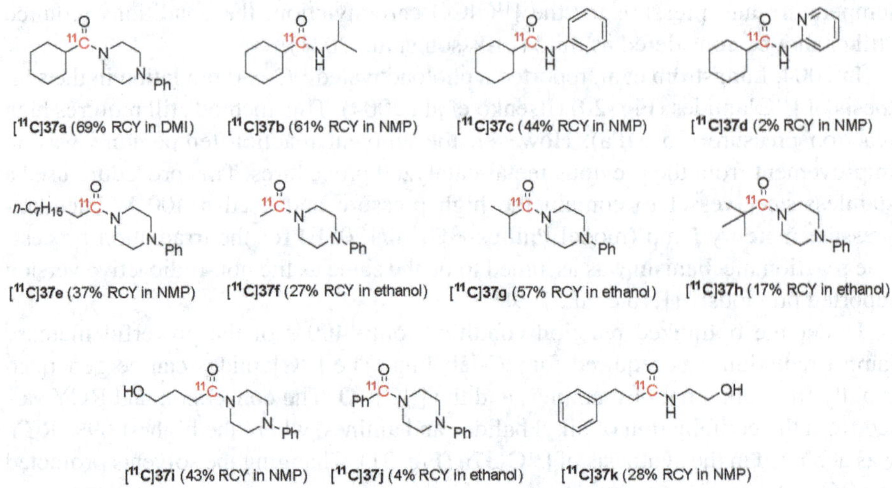

Fig. 21 Syntheses of [^{11}C]amides from amine nucleophiles via photoactivated carbonylation

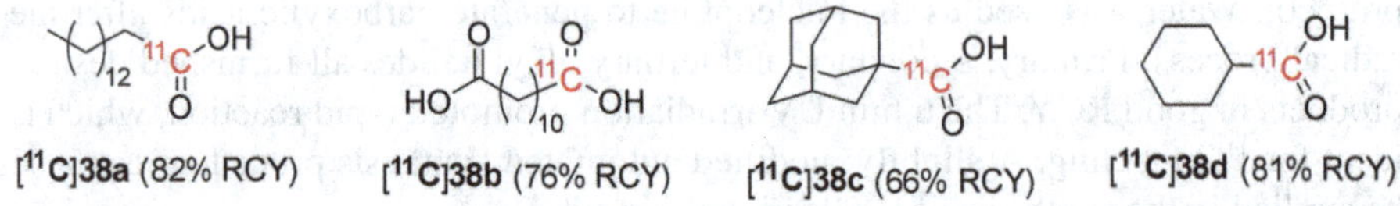

Fig. 22 Synthesis of [^{11}C]carboxylic acids from [^{11}C]CO with water as the nucleophile

efficiency was significantly improved by the presence of ketone additives and di-*tert*-butyl peroxide. Both linear and cyclic alkyl halides produced desired products in good RCY. Primary and secondary alcohols were also scanned for the substrate scope.

The photoactivation apparatus in this [^{11}C]carbonyl ester syntheses is the same as what was used in the previously mentioned [^{11}C]carboxylic acid carbonylations (Itsenko et al., 2006). The [^{11}C]carbonyl esters are formed at ambient temperature (35 °C) and high pressure (38 MPa) (Itsenko et al., 2007).

7 Carbon Isotope Exchange of Carboxylic Acids Under Photoredox Condition

Besides the above-mentioned [^{11}C]CO carbonylation methods in the labeling of carboxylic acids, isotope exchange can also generate radiolabeled carboxylic acids. This method has been applied in carboxylic acid labeling with carbon-13 and carbon-14 nuclides (Kingston et al., 2019; Tortajada et al., 2019; Destro et al., 2019; Destro et al., 2020; Kong et al., 2020; Hinsinger & Pieters, 2019). However,

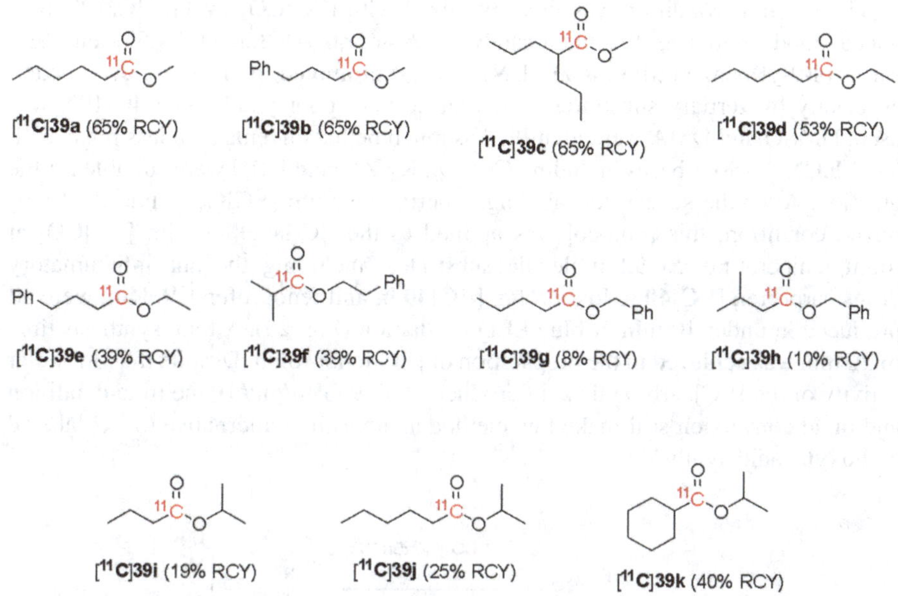

Fig. 23 Synthesis of [^{11}C]carbonyl esters with alcohols as nucleophiles

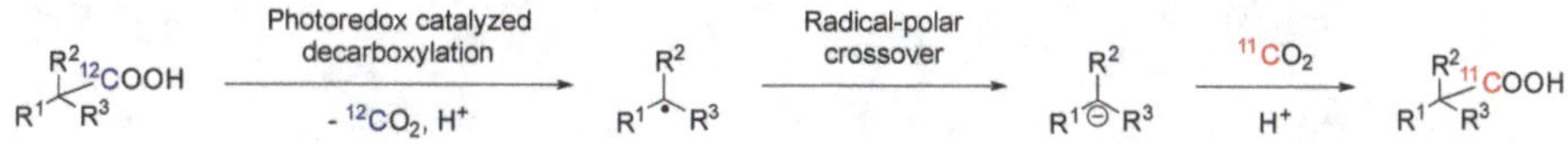

Fig. 24 Isotope exchange via the photoredox-catalyzed radical-polar crossover process

carbon-13 is a stable nuclide and carbon-14 has a half-life of 5703 years. The isotope exchange conditions designed for these two nuclides usually take hours, which is not suitable for ^{11}C-radiosynthesis. Therefore, very few ^{13}C- and ^{14}C-isotope exchange protocols have been applied to the ^{11}C-labeling. In 2020, Audisio et al. observed that some substrates require less reaction time to complete the isotope exchange. They applied their ^{13}C-/^{14}C-isotope exchange condition to ^{11}C-labeling of these specific substrates by reducing the reaction time to 10 min (Destro et al., 2020). However, harsh condition (150 °C heating) was still required, and most of the ^{11}C-labeling examples had low RCY. This method is also limited in substrate scope, because very few substrates can take advantage of such method.

By applying the organophotoredox process, Rotstein and Lundgren et al. reported an isotope exchange method with mild conditions in 2021 for both ^{11}C- and ^{13}C-labeling (Fig. 24) (Kong et al., 2021). The photoredox process can generate carbon radicals through decarboxylation. Radical-polar crossover turned the radical into reactive nucleophilic intermediate, which was used to trap the [^{11}C]CO_2 and furnish the radiolabeled carboxylic acid.

The reaction condition was first optimized with [^{13}C]CO_2, where 4CzIPN produced good results as the photocatalyst. (2r,4r,5r,6r)-3-Benzyl-2,4,5,6-tetra(9H-carbazol-9-yl)benzonitrile (4CzBnBN) catalyst promoted the reaction even better, especially for tertiary substrates which led to very poor yield when 4CzIPN was used. DMSO and DMA were identified as much better solvents than less polar THF and MeCN. Various bases including Cs_2CO_3, K_2CO_3, and DBU were suitable for the reaction. After the successful labeling experiments with [^{13}C]CO_2 under the optimized condition, this protocol was applied to the ^{11}C-labeling with [^{11}C]CO_2 at room temperature. Small molecule substrates, including the anti-inflammatory drugs carprofen **[^{11}C]40 g**, loxoprofen **[^{11}C]40 h**, and fenoprofen **[^{11}C]40i**, were all produced in under 10 min of blue LED irradiation (Fig. 25). A total synthesis time of 20 min was achieved in the preparation of [^{11}C]fenoprofen. Despite the low molar activity of the [^{11}C]carboxylic acid products (0.029 GBq/μmol), the mild condition and rapid conversion still make this method an appealing alternative to ^{11}C-labeled carboxylic acid synthesis.

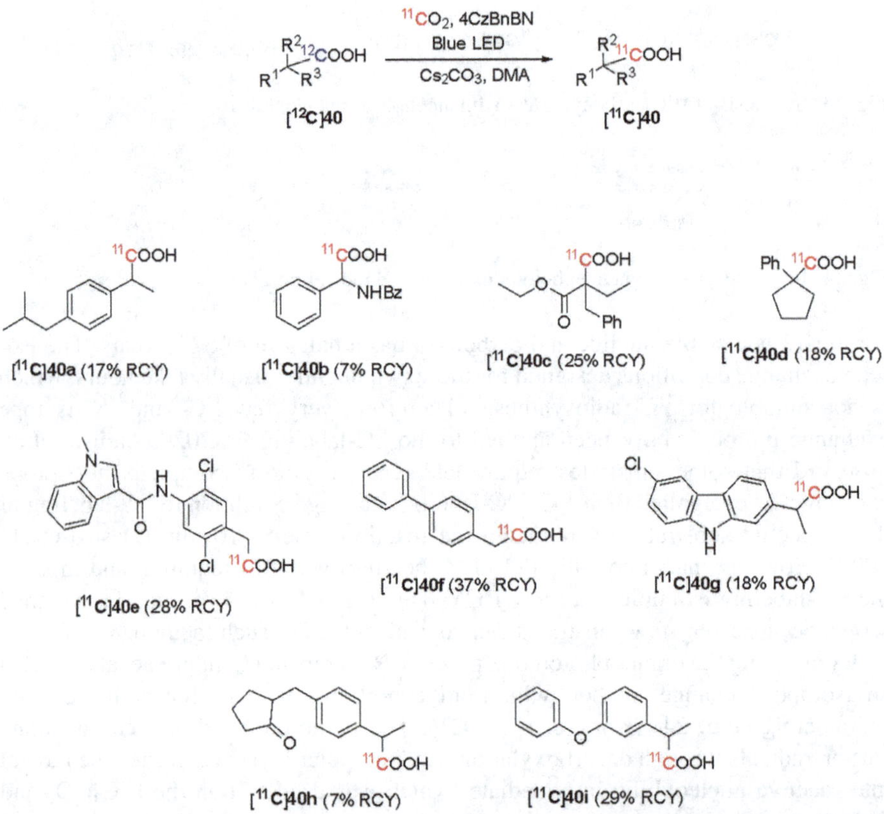

Fig. 25 ^{11}C-labeling through isotope exchange

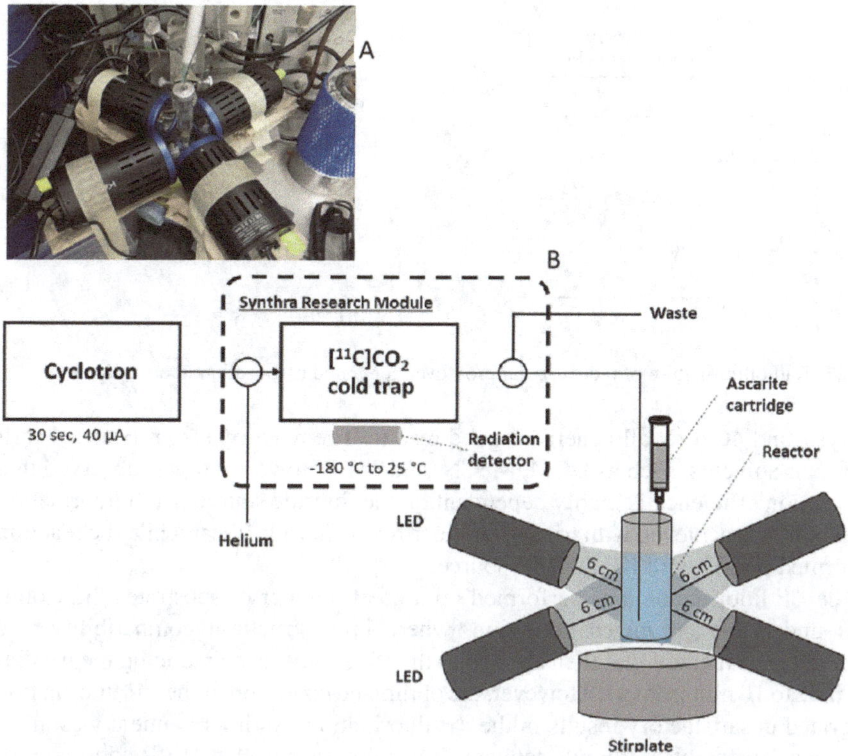

Fig. 26 Partially automated synthesis module with homemade photoreactor. A: photoreactor with four LED lights and reaction vial in the middle. B: schematic representation of the radiolabeling equipment. Reprinted from reference (Kong et al., 2021) with permission. Copyright 2021 American Chemical Society

The ^{11}C-labeling reactions were done with a homemade photoreactor which was equipped with four blue LED lamps (Kessil A160WE Tuna Blue) (Fig. 26). This photoreactor was connected to an automated Synthra research module which trapped the [^{11}C]CO_2 for the reaction.

8 Radiofluorination of Carbocations Generated from Photoredox Radical-Polar Crossover Process

Typical nucleophilic radiofluorination conditions are harsh due to the poor nucleophilicity of the [^{18}F]fluoride ion. In 2020, Doyle et al. published the radiofluorination of carbocation intermediates generated from photoredox radical-polar crossover process (Fig. 27) (Webb et al., 2020). The optimization of conditions was done with ^{19}F-based reagents. The Ir(dF-ppy)$_3$ catalyst performed the best, while other iridium

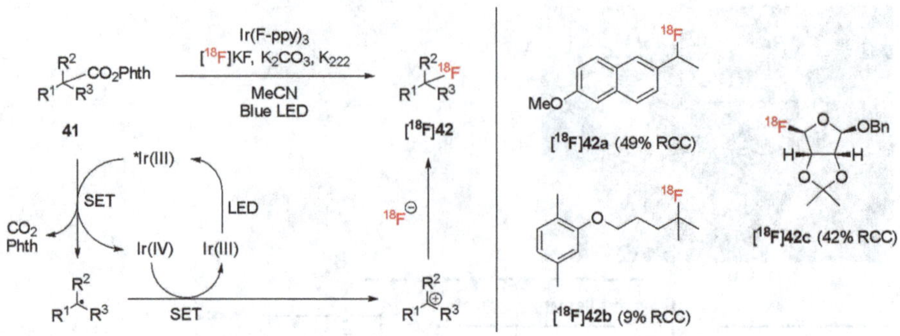

Fig. 27 Radiofluorination of radical-polar crossover generated carbocation intermediates

catalysts and 4CzIPN all generated good results. The reaction also proceeded well in various solvents, such as DCM, MeCN, and THF. However, it was observed that the reaction efficiency is highly dependent on the fluoride source. High fluorination efficiency was achieved with triethylamine trihydrofluoride, meanwhile the reaction performed poorly with KF fluoride source.

The ^{19}F-fluorination was performed on a great number of substrates (35 examples) under this optimized condition, where high structural compatibility was observed. The method was then applied to the ^{18}F-synthesis by reducing the irradiation time to 10 min from 4 h. However, the optimized condition of the ^{19}F-fluorination generated unsatisfactory results in the ^{18}F-fluorination. An improvement was made by using a different Ir(F-ppy)$_3$ catalyst. It was interesting that HFIP additive promoted the [^{19}F]KF fluorination but hindered the [^{18}F]KF radiofluorination. Under the new photoredox condition, secondary and tertiary carbocation precursors were labeled with fluorine-18. Compounds **[^{18}F]42a** and **[^{18}F]42c** were generated in good RCC. Even though **[^{18}F]42b** was produced in poor RCC, the preparations of tertiary radiofluorinated compounds are generally difficult due to the steric hindrance.

The cyclotron target water was processed in a GE Tracerlab FX2 N module in this protocol, and the dried [^{18}F]KF solution was added into the reaction vial through an automated syringe (Fig. 28). The ^{18}F-labeling reactions were then carried out in a photoreactor equipped with a blue LED lamp (Kessil H150 Blue, 34 W power).

9 Photoclick with Strain-Promoted Azide Alkyne Cyclization

Strain-promoted azide alkyne cyclization (SPAAC) employs strained alkynes to promote azide alkyne cyclization. Compared with regular azide alkyne click reactions, SPAAC requires no transition metal catalyst and can be bioorthogonal (Wang & Mooney, 2020; Dommerholt et al., 2016). Applications of SPAAC in radiochemistry and PET agent developments have demonstrated the value of this concept (Campbell-Verduyn et al., 2011; Arumugam et al., 2011; Bouvet et al., 2011;

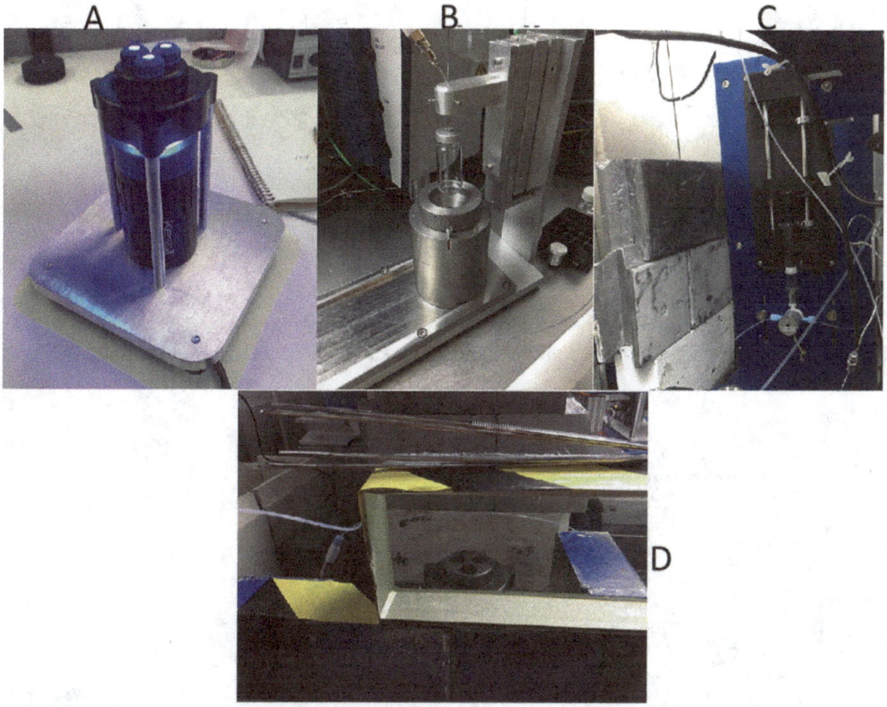

Fig. 28 Photoreactor used in the radiofluorination of radical-polar crossover intermediates. (**a**) photoreactor being irradiated with blue LED light. (**b**) pig and vial for radioactive material. (**c**) automated Hamilton syringe for radioactive material. (**d**) shielding used during manual manipulations. Reprinted from reference (Webb et al., 2020) with permission. Copyright 2020 American Chemical Society

Kettenbach & Ross, 2016; Wang et al., 2018; Lee et al., 2013; Glassner et al., 2017; Sachin et al., 2012; Kim et al., 2015; Kwon et al., 2020; Roche et al., 2019). Due to the short half-life of fluorine-18, the ideal radiochemical SPAAC needs both high reactivity and high substrate stability: the high reactivity allows for the rapid synthesis to minimize the decay loss of radioactivity, and the high substrate stability allows for the functionalization of the strained alkyne precursors in addition to a longer shelf life. Unfortunately, the highly reactive strained alkynes are usually also the less stable ones. Therefore, reported radiochemical SPAAC either suffers from short reagent shelf life and poor yields through the functionalization process or must tolerate very slow reaction rates.

It was discovered in 1981 that the cyclopropenone protected acenaphthyne can undergo decarbonylation under UV irradiation and form a triple bond (Fig. 29) (Chapman et al., 1981). In 2009, Popik et al. utilized this process in the click reaction of strained dibenzocyclooctyne (DIBO) substrate **45**, where the click reaction proceeded only after the UV irradiation (Poloukhtine et al., 2009). This process shares similar advantages as other photoclick reactions, where the precursors are

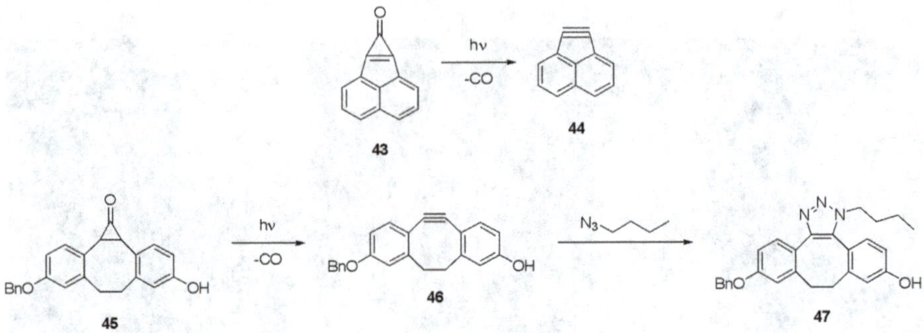

Fig. 29 Cyclopropenone protected alkynes and photoactivation of a DIBO substrate in SPAAC

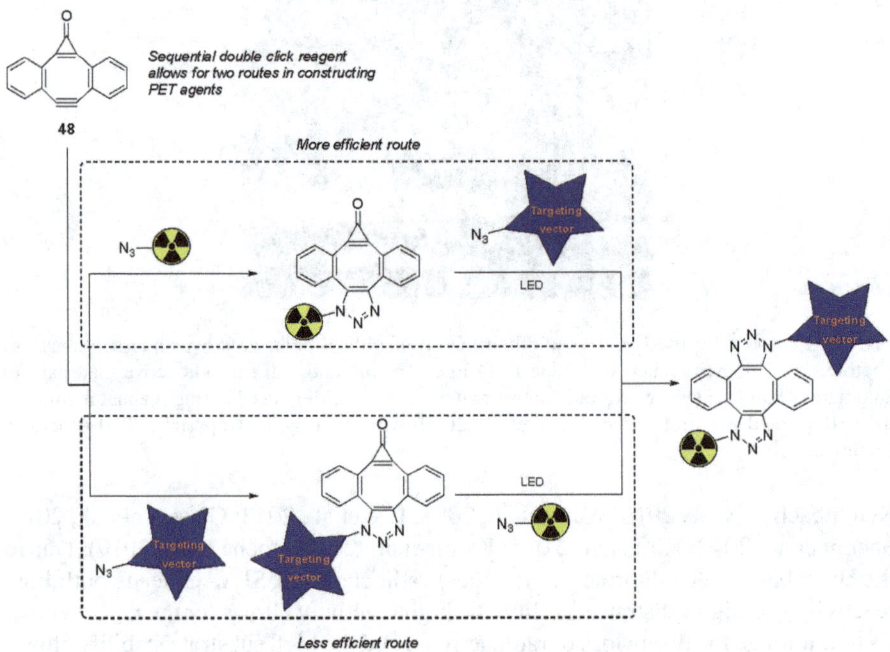

Fig. 30 Photoactivated double-click method in PET agent development

relatively inert, and the subsequent rapid click reactions can take place in a more controllable way once activated by irradiation (Kumar & Lin, 2021).

Despite the advantages of photoactivated SPAAC, it has no record of being utilized in the ^{18}F- or ^{11}C-radiosynthesis until 2022, when Li and Popik et al. reported a modular photoactivated double-click method in the preparation of new ^{18}F-labeled PET agents (Fig. 30) (Li et al., 2022). The mono-cyclopropenone-caged dibenzocyclooctadiyne (MC-DIBOD) used in this study has two-click reaction sites and was developed as a sequential double-click reagent previously (Sutton & Popik, 2016).

The two-click reaction sites differ greatly in the reaction rate, where the unprotected alkyne for the first click reaction has moderate reaction rate, and the second cyclopropenone protected alkyne demonstrated the fastest reaction rate among all DIBO compounds (Sutton & Popik, 2016). Utilizing MC-DIBOD, two routes were possible in PET agent constructions (Li et al., 2022). Based on control experiments, it was discovered that the route with radioactive monotriazole intermediate is more efficient than the other route, where the ^{18}F-labeled prosthetic group is coupled after the targeting vector.

Even though no automated synthesis module was employed, the authors did implement modular conditions in all steps (Fig. 31) (Li et al., 2022). Two ^{18}F-labeled azide synthons were prepared through the same modular direct labeling procedure. These two synthons have drastically different hydrophilicity, which later affected the biodistribution of the final PET agents. Being able to fine-tune the property of the PET agents by employing modular radiofluorinated prosthetic building blocks is one of the highlights of this strategy. The ^{18}F-labeled monotriazole intermediates **[^{18}F]51** and **[^{18}F]54** were prepared by a modular click reaction condition, which was heated to promote the conversion in compensation of the first click reaction's moderate rate.

Upon irradiation of the monotriazole **[^{18}F]51**, the second alkyne can be released and then participates in cyclization with azide-modified cancer tracers (Fig. 32). Excellent RCYs were obtained in the preparation of **[^{18}F]55a** to **[^{18}F]55d** where high concentrations of azide substrates were employed (50 mM to 55 mM). In the case where valuable tracers are limited in quantities (submillimolar concentrations), products were still generated in good RCYs with FAPI (**[^{18}F]55e** and **[^{18}F]55f**), NT (**[^{18}F]55g**), and PSMA (**[^{18}F]55h**) tracers. The mild conditions also enabled the application in the protein labeling, where azide-modified HSA protein was labeled in 78% RCY.

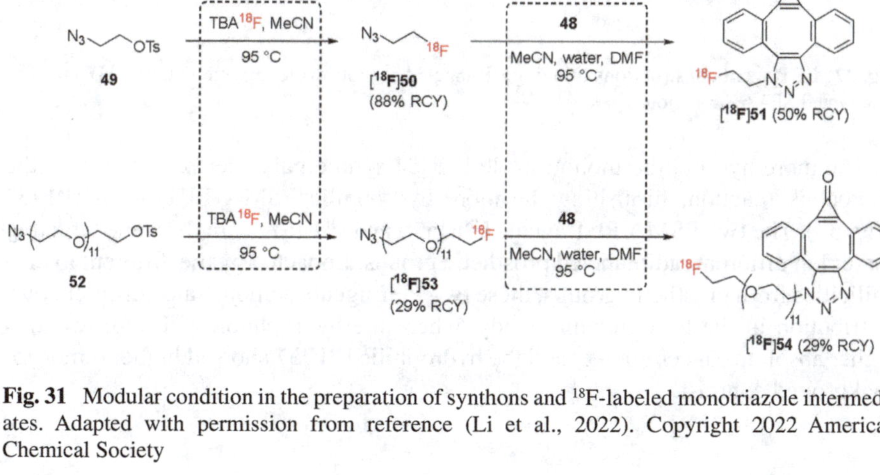

Fig. 31 Modular condition in the preparation of synthons and ^{18}F-labeled monotriazole intermediates. Adapted with permission from reference (Li et al., 2022). Copyright 2022 American Chemical Society

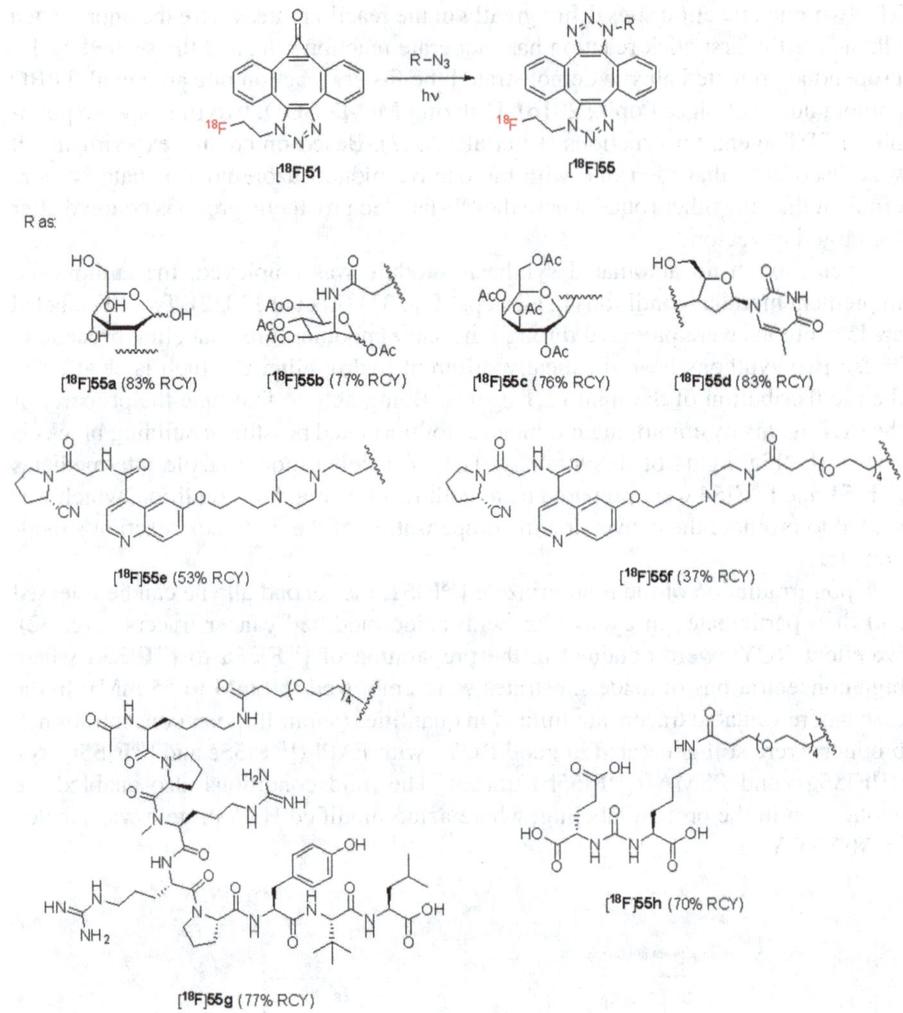

Fig. 32 PET agent constructions utilizing ^{18}F-labeled monotriazole intermediate **[^{18}F]51** via photoactivated SPAAC reaction

The more hydrophilic monotriazole **[^{18}F]54** synthon also performed well in the photoclick reaction, furnishing the more hydrophilic PSMA PET agent **[^{18}F]57** (Fig. 33). The two PSMA PET agents **[^{18}F]55h** and **[^{18}F]57** share the same targeting vector but different radiolabeled prosthetic groups. Impacted by the different hydrophilicities from prosthetic groups, these two PET agents demonstrated different biodistribution in the PET imaging study, where the hydrophobic **[^{18}F]55h** revealed higher absolute cancer uptake, and the hydrophilic **[^{18}F]57** showed higher tumor-to-background contrast.

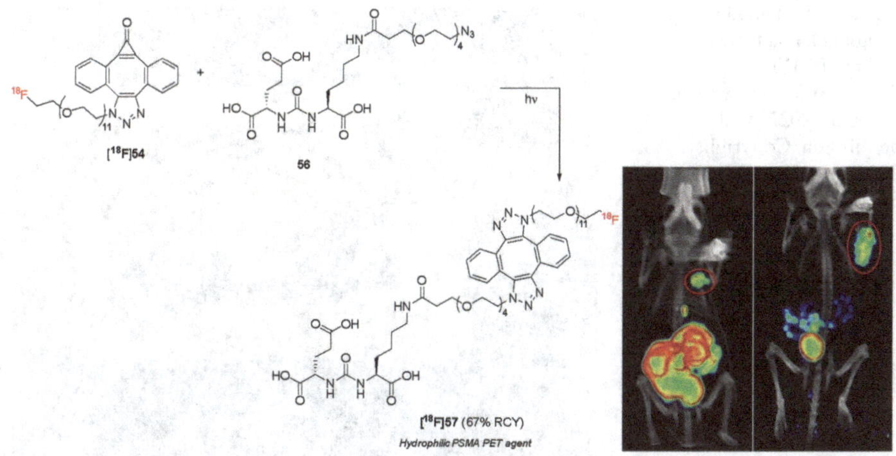

Fig. 33 Construction of hydrophilic PSMA PET agent [^{18}**F**]**57** and comparison of PET/CT images between hydrophobic [^{18}**F**]**55h** (left PET/CT image) and hydrophilic [^{18}**F**]**57** (right PET/CT image) with prostatic cancer cell line PC3-PIP bearing mice. Reprinted from reference (Li et al., 2022) with permission. Copyright 2022 American Chemical Society

The photoclick step of this protocol was done at room temperature under irradiation by a 365 nm LED light source (Fig. 34). An open-top paper box was covered in aluminum foil to increase the irradiation reflection. Reaction vial was dropped in this homemade photoreactor, and the LED light source was placed directly above it. The entire irradiation process was only 20 min, which is suitable for ^{18}F-radiosynthesis. The reaction setup does not require any fancy equipment and is very affordable.

10 The Development of ^{18}F-Labeling Methods and Applications via Organic Photocatalysis

10.1 Direct C–H Radiofluorination

Organic photochemistry is an interesting set of reactions that takes advantage of organic molecules in the excited state for single-electron redox chemistry to realize novel synthetic chemistry via simple pathways. Based on photoredox fluorination reactions, various photoredox ^{18}F-radiofluorination methods have been developed. Unlike conventional nucleophilic ^{18}F-fluorination on electron-deficient or neutral arenes, photoredox-mediated radiofluorination has emerged as a novel radiolabeling technique for a variety of electron-rich arenes with mild conditions and high functional group tolerance.

In 2019, the Nicewicz and Li groups reported a direct arene C–H ^{18}F-fluorination method for various (hetero)arene substrates with an acridinium-based organic photocatalyst **S1** (Chen et al., 2019). In this labeling reaction, a 450 nm laser (3.5 W)

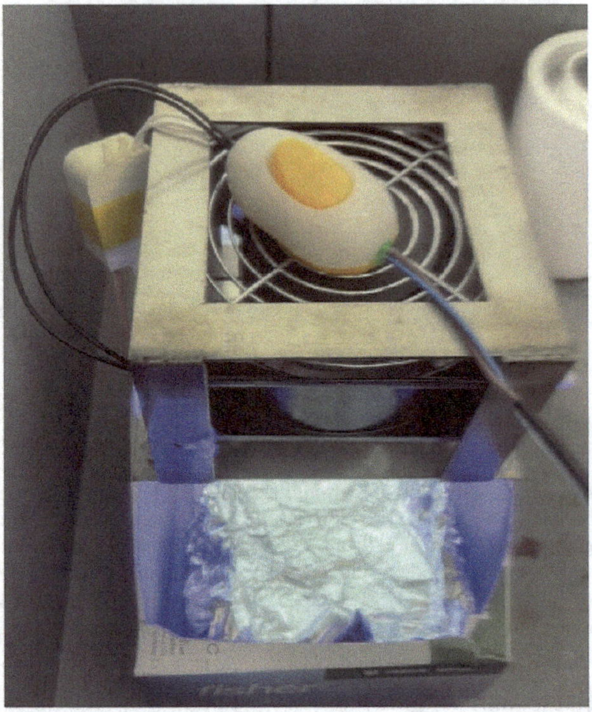

Fig. 34 Homemade photoclick reactor with 365 nm LED light. Reprinted from reference (Li et al., 2022) with permission. Copyright 2022 American Chemical Society

was applied to a mixture of the substrate, acridinium catalyst, and other components in a mixed solvent of MeCN/*t*-BuOH (4/1). Oxygen (O_2) was used as the oxidant through balloon sparging and 2,2,6,6-tetramethyl-1-piperidine 1-oxyl (TEMPO) was required as a redox co-mediator. Among the tested [^{18}F]fluoride sources, [^{18}F]TBAF (tetrabutylammonium fluoride) led to the best yield.

Through C–H ^{18}F-fluorination, a number of ^{18}F-fluorinated aryl compounds were synthesized for the first time, demonstrating the great potential of organic photocatalysis in the development of PET agents (Fig. 35). This method features a completely metal-free approach with good radiochemical yield (RCY), and starting from readily available precursors such as some natural products and active pharmaceutical ingredients (APIs). The molar activity of this labeling method was determined to be 50.8 GBq/μmol, which proved to be comparable with other no-carrier-added ^{18}F-labeling methods. In vivo PET imaging of two ^{18}F-labeled aryl amino acids ([^{18}F]-2-OMe-5-F-Phe and [^{18}F]4-(4'-F-phenyl)-Phe) was performed to study tumor accumulation in human breast cancer (MCF-7) bearing mice and to compare the results.

The prototype radiolabeling setup was simple as shown in Fig. 36, which has been adopted and constantly updated to discover other photoredox radiolabeling methods afterward.

Although the instrument setup is relatively easy to set up, it did involve the use of an expensive laser. The open-to-air reaction condition makes it easy to setup but

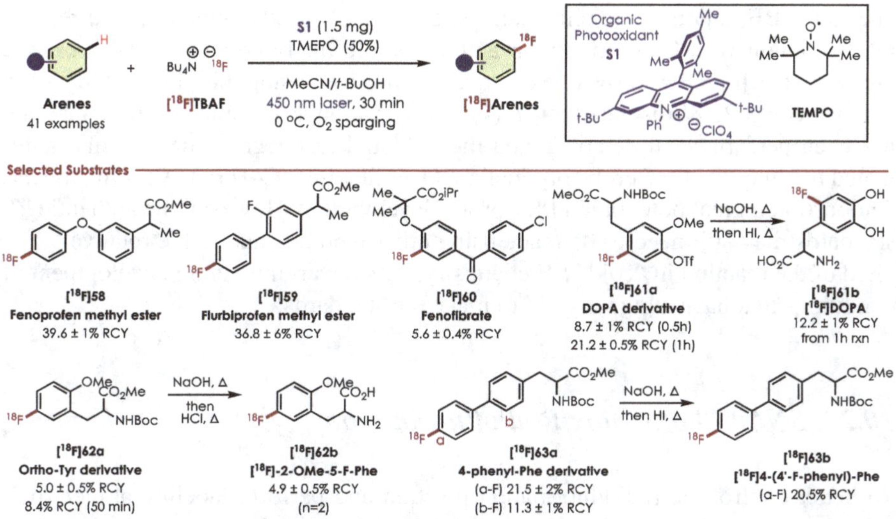

Fig. 35 Direct arene C–H ^{18}F-fluorination via organic photoredox catalysis (selected substrates, $n = 3$ when a standard deviation was given)

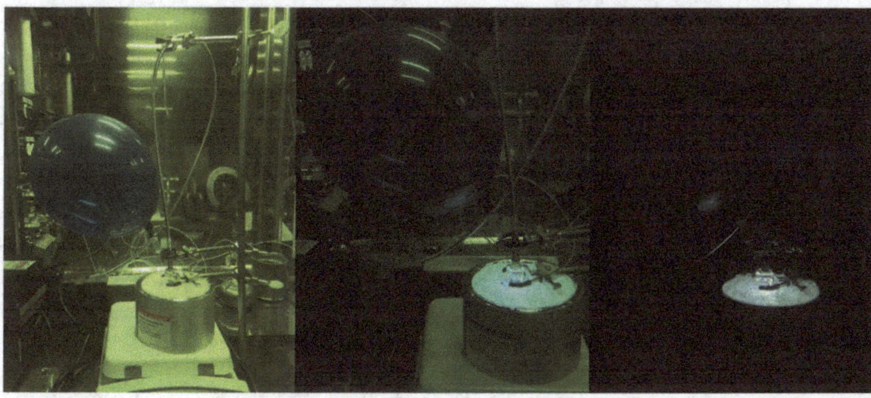

Fig. 36 Reaction setup for photoredox-catalyzed arene radiofluorination. Reproduced from reference (Chen et al., 2019) with permission from AAAS. Copyright 2019 The American Association for the Advancement of Science

also introduces the risk of losing volatile product during the reaction. To address these concerns, a modified flow-based method was introduced in 2020 after replacing oxygen with easier-to-handle peroxide oxidants (such as tert-butyl peroxyacetate) (Fig. 37) (Wang et al., 2020). The laser was also replaced by affordable LED lights. Using a sealed quartz microfluidic device (without cooling) and organic acridinium photocatalyst **S3**, over 20 substrates were tested, and the RCYs were similar to those reported previously (Fig. 35) (Chen et al., 2019). However, the molar activity of this method was determined to be reduced to 18.0 GBq/μmol, which might be due to the BF_4 anion in the photocatalyst **S3** Figs. 37. If the authors could replace the BF_4 anion with ClO_4, the molar activity would likely be enhanced. Of note, a protected [^{18}F]FDOPA derivative could be synthesized using this simple and closed system with a 22.8% isolated RCY (Fig. 38). Although no clinical grade labeling has been performed, it demonstrated the potential to integrate the unit into automated modules for PET probe production. Overall, direct C–H radiofluorination has demonstrated great potential in PET probe development. However, the authentic ^{19}F standards may still need to be made using other methods due to the relatively low yield of the reaction in "cold" ^{19}F-chemistry, which warrants future development of conditions that have a higher yield to make the standards.

10.2 SNAr Photoredox Radiofluorination

To better control the radiofluorination position and perform labeling at electronically less favored arene sites, Nicewicz, Li, and co-workers established a facile S_NAr deoxyfluorination method in 2020. Starting from phenol (4-Chlorophenoxy, in most cases) derivatives (Tay et al., 2020), both ^{19}F- and ^{18}F-fluorinated arenes can be generated in moderate to high yield using two different conditions (Figs. 39 and 40). In this method, some APIs could be easily converted into their phenol precursors in a few steps before (radio)fluorination, without complicated synthesis of the precursors and standard products. The molar activity of this method was determined to be as high as 74.7 ± 14.1 GBq/μmol, and the radiolabeling setup remained as simple as the one used in the C–H ^{18}F-fluorination reported previously (Fig. 36) (Chen et al., 2019). Different from the previously reported C–H radiofluorination method, radiofluorination can happen at relatively less electron-rich positions including the meta position, and the regioselectivity is perfectly controlled. This versatility will provide researchers with a powerful tool to screen PET agents with a subtle difference.

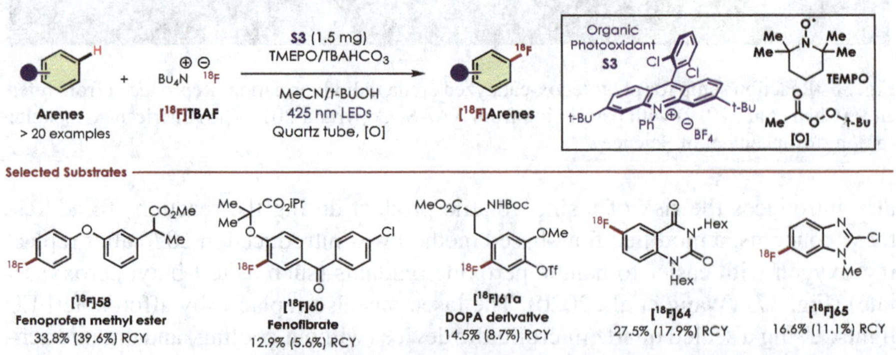

Fig. 37 Modified direct arene C–H ^{18}F-fluorination via organic photoredox catalysis (Wang et al., 2020) with some substrate examples. The RCYs of the previous system are listed in parentheses (for which O_2 was used as the oxidant, and a laser was used as the light source)

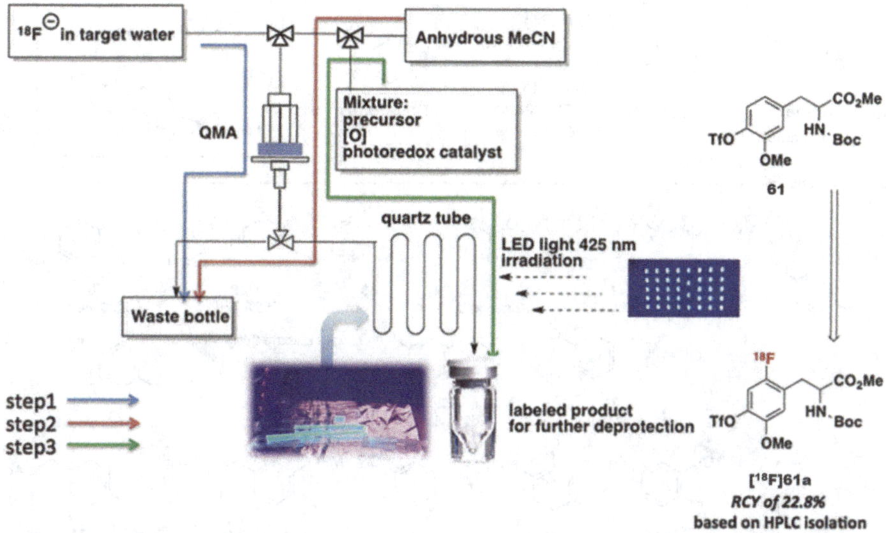

Fig. 38 Schematic workflow of preparing protected [^{18}F]FDOPA through direct C–H ^{18}F-fluorination. [O] = oxidant. Reproduced from reference (Wang et al., 2020) with permission from ACS. Copyright 2020 The American Chemical Society

Moreover, the precursor and labeled product are generally well separated, making the purification relatively easier compared to C–H radiofluorination approaches.

10.3 Halogen-^{18}F Exchange Photoredox-Mediated Radiofluorination

To further simplify the synthesis of precursors for labeling, Nicewicz, Li, and co-workers reported another novel photoredox-mediated radiofluorination method using halogen-^{18}F exchange. Initially, the authors explored the feasibility of directly converting aryl halides into ^{18}F-labeled aryl derivative in electron-rich arenes. The labeled product was formed via single-electron oxidation followed by an electron-deficient cation radical capturing the $^{18}F^-$ before reduction and expulsion of the halide nucleofuge. This radiofluorination strategy would reduce the need for lengthy, multistep precursor synthesis, greatly simplify the isolation process, and open new ways to prepare clinically significant PET agents (Fig. 41). By using the chloride precursor, high molar activity (71.25 GBq/μmol) could be achieved. Importantly, direct fluoride exchange on electron-rich arene rings was also successfully accomplished, which may allow easy conversion of fluoride-containing drugs into PET agents. A decreased molar activity (1.51 GBq/μmol) was expected because of using the inseparable ^{19}F-fluoroarene. However, the molar activities can be greatly improved by reducing the aryl fluoride precursor amount, which has been reported

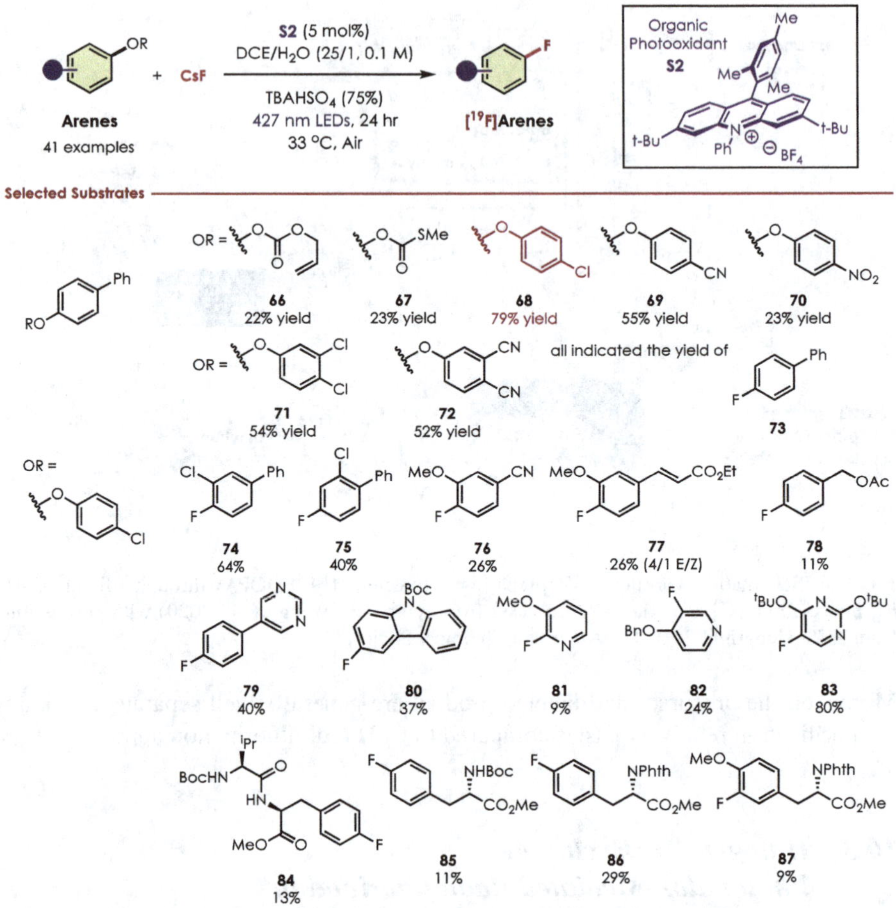

Fig. 39 ^{19}F-arene deoxyfluorination via S_NAr pathway (selected substrates)

before (Langer et al., 2003). Noticeably, a user-friendly radiofluorination condition was achieved: no pure oxygen sparging was needed, it required no special precautions on air/moisture, and LED lights (labeling setup shown in Fig. 42) could also serve as the light source to achieve high labeling yields, when photocatalyzed direct ^{18}F–^{19}F exchange was explored and conducted under nitrogen atmosphere.

In the report, a full-scale (>37.0GBq) synthesis of [^{18}F]FDOPA was developed and demonstrated (Chen et al., 2021). With the photocatalysis step followed by acidic deprotection and a single HPLC isolation, the purified [^{18}F]FDOPA product could be obtained with good HPLC-isolated RCY (28.6 ± 4.9%, n = 3). Recently, Wu, Li, and co-workers further developed the production of [^{18}F]FDOPA by employing a blue LED irradiated flow device (obtained from LED Radiofluidics), followed by a cartridge purification method with very comparable isolated RCY (23%, n = 2) (Wu et al., 2023a). It is very likely that the simplified synthesis of the [^{18}F]FDOPA

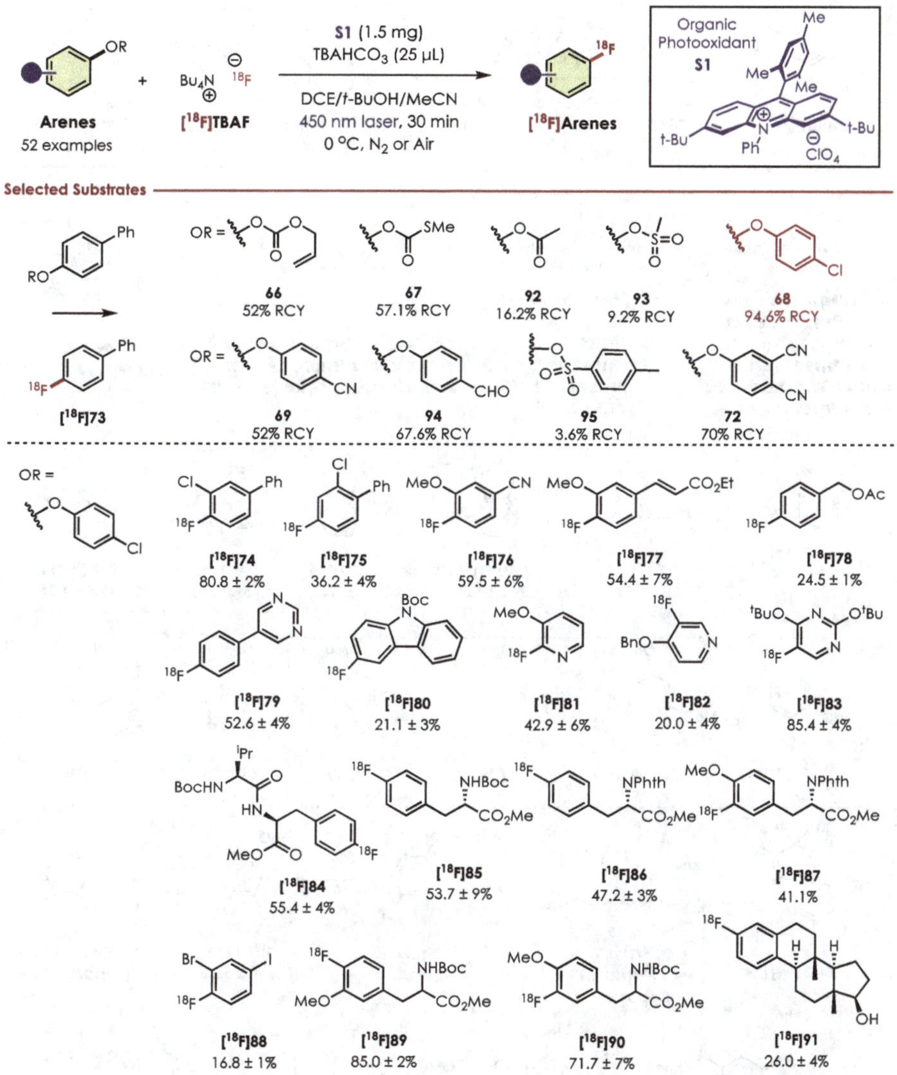

Fig. 40 ^{18}F-arene deoxyfluorination via S_NAr pathway (selected substrates, n = 3 when a standard deviation was given)

via photocatalysis could greatly increase the availability of this agent in the future. Fully automating the synthesis is still preferred in future applications.

In addition to the established agent, Wu, Li, and co-workers recently applied the technology to develop a new amino acid-based ^{18}F-based PET agent (Wu et al., 2023c). The enantio-pure tryptophan-based precursors can be easily synthesized from commercially available chemicals in only three steps. After the photoredox radiofluorination, acidic deprotection, and HPLC isolation, the desired

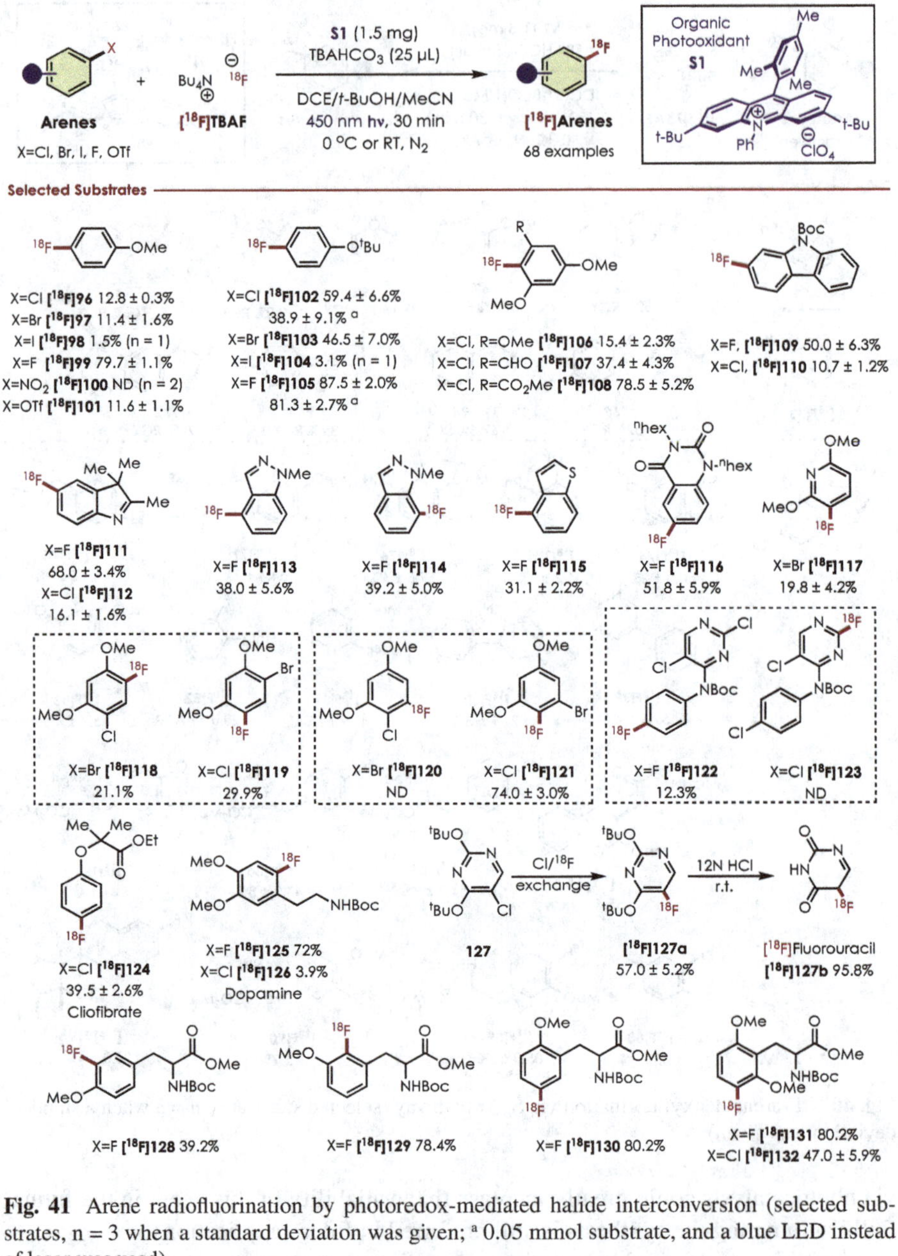

Fig. 41 Arene radiofluorination by photoredox-mediated halide interconversion (selected substrates, n = 3 when a standard deviation was given; [a] 0.05 mmol substrate, and a blue LED instead of laser was used)

tryptophan-based PET agent *L*-4[^{18}F]F-5OMe-Trp ([^{18}F]**134**) could be obtained in high RCY and RCP without racemization (as shown in Fig. 43). Utilizing a straightforward photocatalyzed radiolabeling setup (Fig. 42), the ^{19}F–^{18}F exchange occurred not only at the 4-position but also at the 6-position. However, the efficiency was

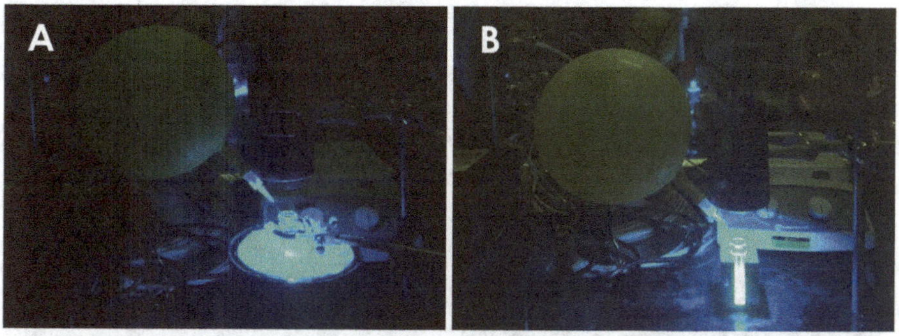

Fig. 42 Reaction setup with LED under (A) ice cooling and (B) room temperature in a hot cell. Reproduced from reference (Chen et al., 2021) with permission from Nature. Copyright 2021 The Nature Publishing Group

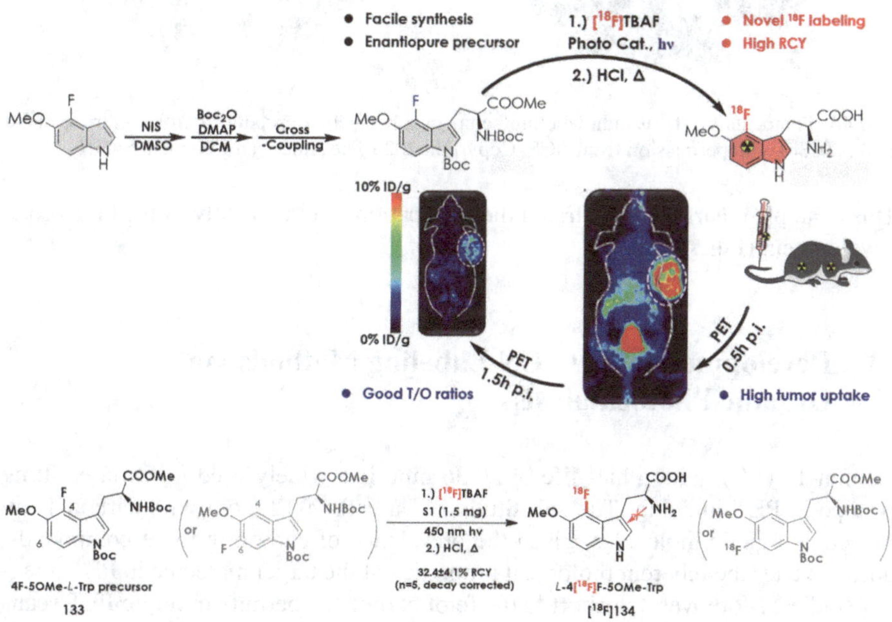

Fig. 43 The preparation of a Trp-based ^{18}F-radiotracer and its application in an imaging study of melanoma tumor-bearing mice. Reproduced from reference (Wu et al. 2023a, b, c) with permission from ACS. Copyright 2023 The American Chemical Society

significantly lower at the 6-position. Particularly, for the 6-fluorinated precursors, although the yield is low, a sufficient amount of the ^{18}F-based PET agents could still be obtained for animal imaging studies. In initial PET imaging research, prominent tumor uptake (9.58 ± 0.26%ID/g) was observed in B16F10 melanoma tumors, accompanied by a favorable tumor/muscle ratio (4.57 ± 0.20). These findings highlight the potential of using *L*-4[^{18}F]F-5OMe-Trp PET tracer for melanoma imaging.

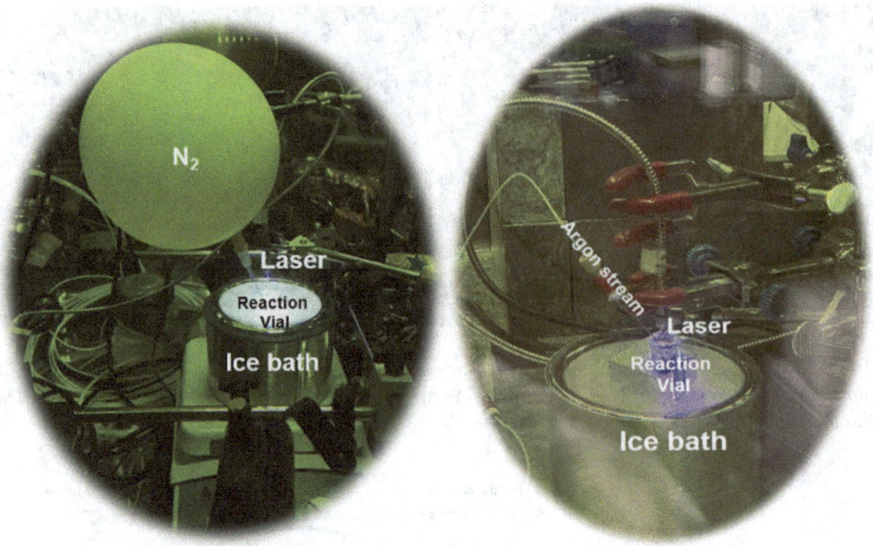

Fig. 44 Photocatalyzed ^{18}F-radiolabeling setup in a hot cell. Reproduced from reference (Wu et al., 2023c) with permission from ACS. Copyright 2023 The American Chemical Society

This example clearly demonstrated the application of photocatalysis for PET probe development (Fig. 44).

11 Development of [^{11}C]CN Labeling Methods via Organic Photocatalysis

Carbon-11 (^{11}C), with a half-life of 20.36 min, is a widely used positron-emitting isotope in PET imaging. The substitution of a carbon-12 atom with carbon-11 in bioactive organic molecules, given the prevalence of carbon in these compounds, does not alter the inherent biological properties of the target molecule in PET imaging studies. Moreover, the short half-life of carbon-11 permits multiple PET scans to be conducted on the same individual within a single day. In theory, nearly all carbon-containing organic functional groups can be labeled with carbon-11, provided that a suitable ^{11}C-synthon is available and employed. Although there are some reports on developing PET agents based on alkyl ^{11}C-nitriles, radiocyanation of arenes, particularly electron-rich arenes, has not been well explored and developed. Organic photocatalysis has garnered significant attention in recent years due to its ability to harness visible light as an energy source to drive chemical transformations. The use of organic photocatalysts offers several advantages over traditional methods, including milder reaction conditions, increased reaction rates, and enhanced selectivity. These features make organic photocatalysis an attractive strategy for ^{11}C-labeling, as it allows for the incorporation of carbon-11 into complex molecular frameworks with high efficiency. Recently, Nicewicz, Li, and coworkers reported a direct C – H radiocyanation method that introduced ^{11}C-cyanide ($^{11}CN^-$) to a variety of functional electron-rich arenes via photoredox catalysis (Fig. 45) (Chen et al., 2022). This photoredox-mediated radiocyanation could proceed in the air from readily available precursors. The reaction setup remained facile, and it achieved scalable synthesis of aryl ^{11}C-nitriles. Though this novel labeling strategy was based on the previously reported C–H cyanation method (McManus & Nicewicz, 2017), the establishment of successful ^{11}C-cyanation procedures was not straightforward due to the following obstacles. First, because of the rapid decay of ^{11}C, the ^{11}C-cyanation reaction should be finished within minutes rather than hours to obtain a useful amount of ^{11}C-labeled agent for PET imaging applications. Second, the reported cyanation used trimethylsilyl cyanide (TMSCN) as the cyanide source, which would be complicated to generate in a labeling reaction. Third, there was only a very small amount of ^{11}C-cyanide in the reaction mixture compared with an excess amount of cyanide source in normal ^{12}C-cyanation reactions. The reversed reagent ratio required additional screening of the reaction conditions, including the reagents, solvents, as well as substrate type, and loading of the additional base.

The optimal reaction conditions were identified after screening, and a series of electron-rich arenes were explored thereafter (Fig. 46). Moderate to high RCYs were achieved, and the molar activity could reach as high as 131.0 GBq/μmol when ethanol was used as the sole solvent along with small amount of TBAOH base additive. In contrast, acetonitrile would significantly lower the molar activity and has minimal influence on the RCY. Another advantage of this method is that the ^{12}C-based authentic standard products can be obtained through photoredox

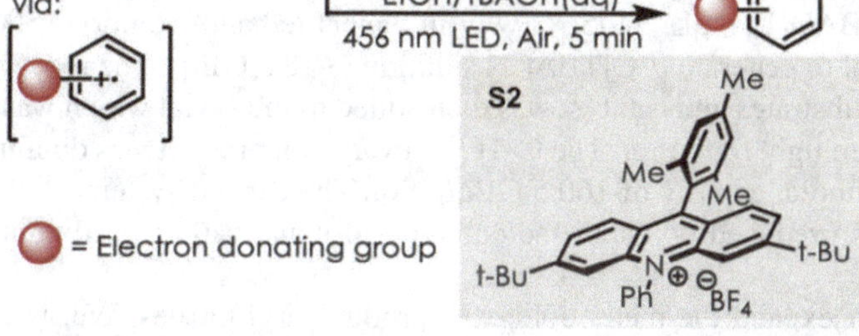

Fig. 45 Photoredox-mediated arene C–H cyanation and ^{11}C-cyanation. Reproduced from reference (Chen et al., 2022) with permission from ACS. Copyright 2022 The American Chemical Society

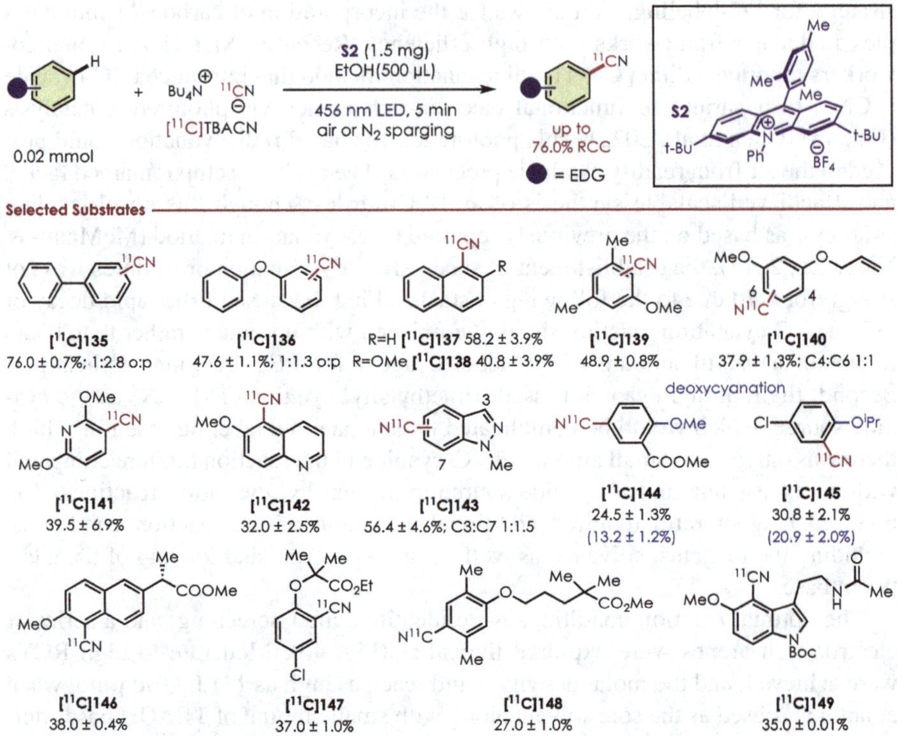

Fig. 46 Photoredox-mediated arene C–H ^{11}C-cyanation (selected substrates, n = 3 when a standard deviation was given)

chemistry as well, which may streamline the research and development process. In addition to C–H activation, competitive C–O activation was also observed when methoxy-group-bearing substances were tested, which correlated well with a previous report on ^{12}C-chemistry (Holmberg-Douglas & Nicewicz, 2019). Noticeably, an LED could be used to replace the expensive laser light source in this radiocyanation method with little influence on the RCYs.

A one-pot scale-up cyanation reaction was also performed with the model compound 1,1′-biphenyl (Fig. 47). The cyclotron produced [^{11}C]HCN was bubbled through a TBAOH (1 µL, 54–56% w/w in water) /ethanol solution (500 µL) in a capped v-vial to generate [^{11}C]TBACN solution (~18.5 GBq). The reaction mixture containing substrates and catalyst was then added to the v-vial which was irradiated with a 456 nm light for 5 min. The C–H ^{11}C-cyanation product was obtained in 58% RCC with a molar activity of 160.5 GBq/µmol. The capped system (with a ventilation line) also reduced potential radioactivity loss and can be easily automated in the future.

To further expand the methodology to produce aryl nitriles, Wu, Nicewicz, Li, and co-workers reported a riboflavin tetraacetate (RFTA, **S4**) promoted (radio)cyanation of electron-rich arenes (Wu et al., 2023b). Through direct C–O activation, the

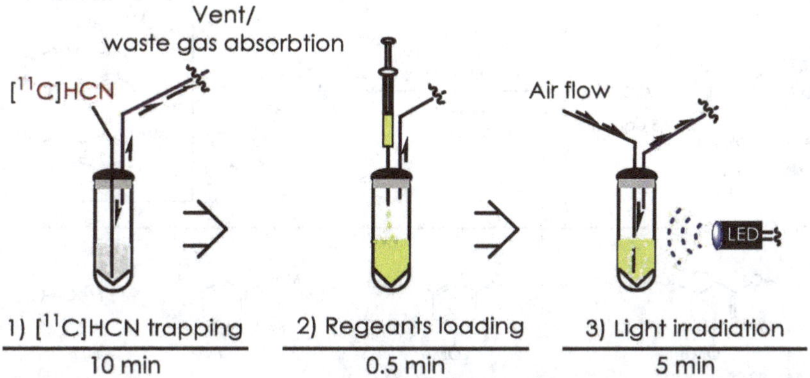

Fig. 47 One-pot scale-up C–H ^{11}C-cyanation in a hot cell. Reproduced from reference (Chen et al., 2022) with permission from ACS. Copyright 2022 The American Chemical Society

highly efficient reaction generated aryl nitriles with a variety of functionalities under mild conditions (nitrogen sparging, Fig. 48). The substrate scope includes easily available simple molecules, complex natural products, and pharmaceuticals, and the tested reactants include ^{11}C-, ^{12}C-, and ^{13}C-cyanide sources. Clearly, this simple late-stage ^{11}C-radiocyanation strategy enables the convenient transformation of off-the-shelf electron-rich aryl therapeutics into their aryl nitrile analogs as well as ^{11}C-labeled radiopharmaceuticals for PET applications (nitrogen sparging was needed, Fig. 49). In addition, in a proof-of-concept study, small animal PET imaging was performed using a ^{11}C-based PET agent generated by **S4** catalysis to evaluate its distribution profile in vivo in normal animals (Fig. 50).

The photoredox-mediated radiocyanation system based on catalyst **S2** was also explored and reported in the same manuscript (Wu et al., 2023b). For electron rich arenes, **S2** afforded lower RCYs than that of **S4**. However, **S2** was able to label compounds with less electron density when **S4** failed to provide acceptable labeling yield (Fig. 52). The reaction setup is easy as shown in Fig. 51. Noticeably, the utilization of LED in conjunction with a flow apparatus (Model: MR-LAB-VS obtained from Little Things Factory, as shown in Fig. 51b) presents a viable alternative to the costly laser light source in the radiocyanation technique, while exerting minimal impact on the RCYs. This development holds significant promise for the prospective automation of the photocatalyzed radiocyanation and the previously mentioned ^{18}F-radiofluorination methods (Fig. 52).

So far, these organic photoredox labeling methods have demonstrated their ability to generate PET agents. Additional work is needed to automate the preparation, which will facilitate the translation of this technology to cGMP. Figure 53 provides an overview of the current radiolabeling apparatus used in the Li lab. It summarizes the most efficient setups for photoredox catalysis for both ^{11}C-cyanation and ^{18}F-fluorination. The open-air systems (Fig. 53a–c) are straightforward and user-friendly during the development stage of radiochemistry methodologies. When an ice bath is added, solvent loss is minimized, especially under laser conditions.

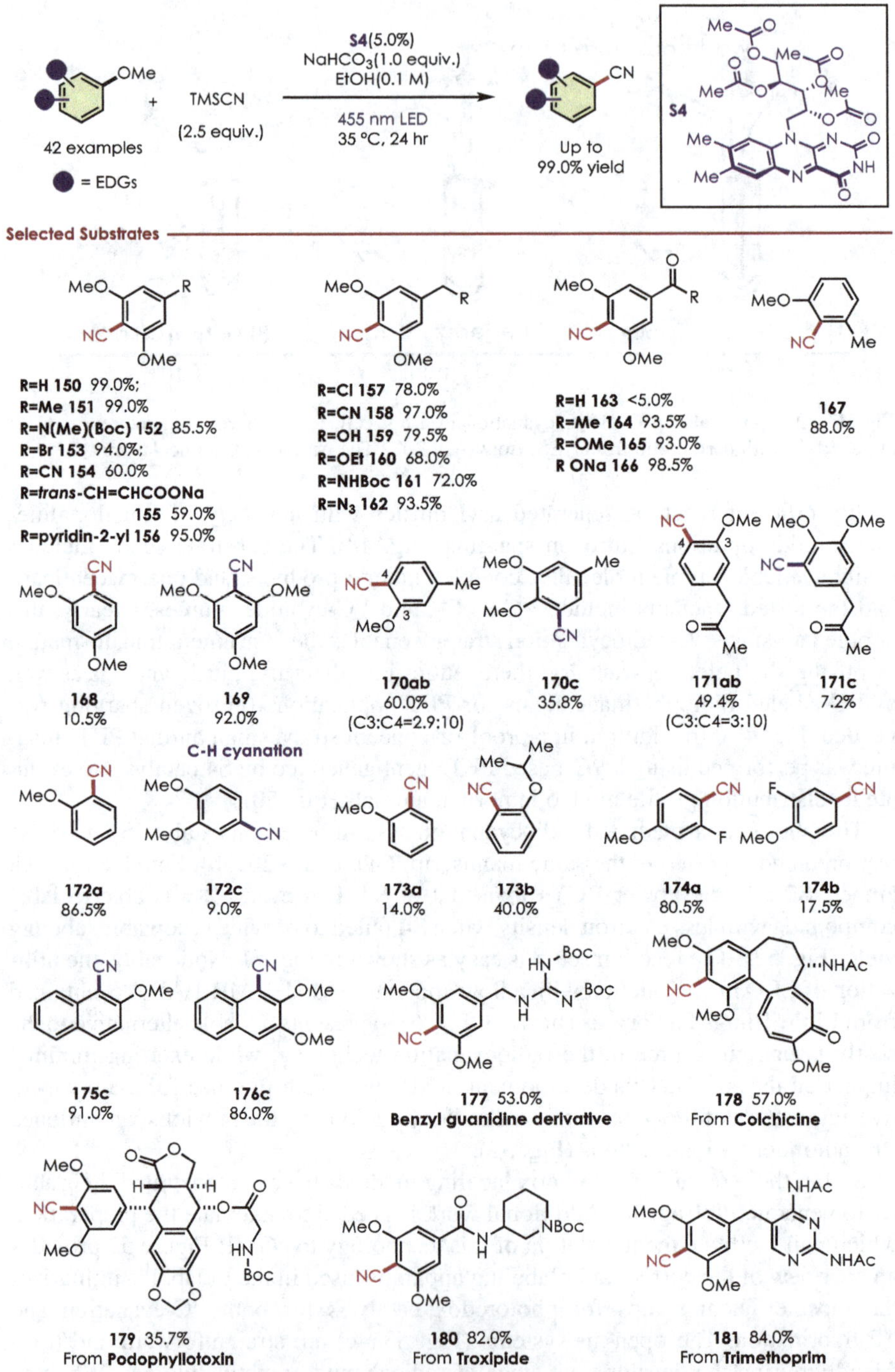

Fig. 48 Photoredox-mediated arene C–O cyanation by RFTA catalyst (selected substrates)

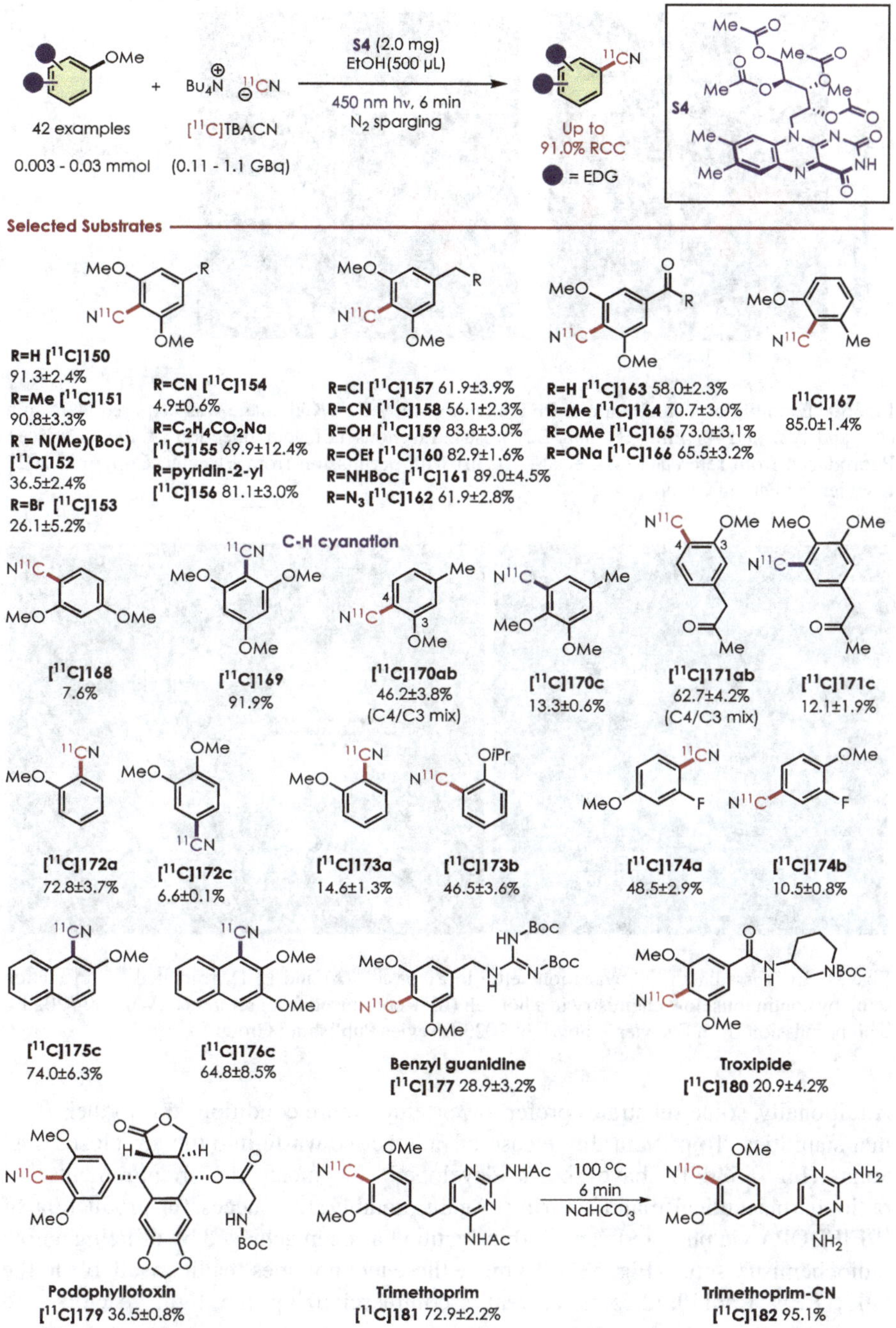

Fig. 49 Photoredox-mediated arene C–O ^{11}C-cyanation by RFTA catalyst (selected substrates, $n \geq 3$ when a standard deviation is given)

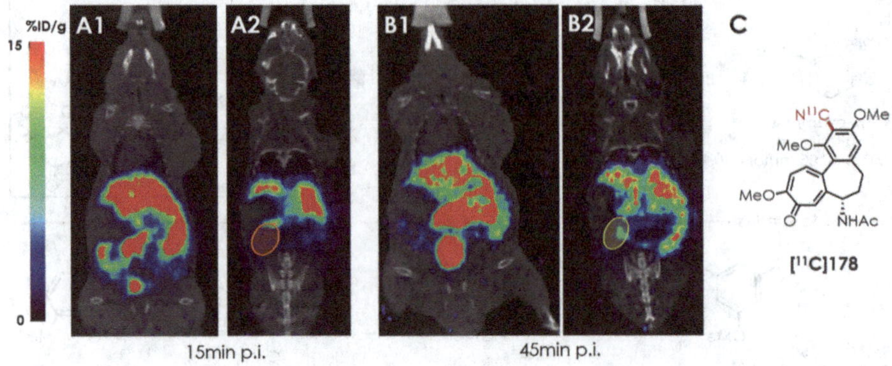

Fig. 50 Example of a small animal PET imaging study. PET/CT image was acquired at 15 min (A1 and A2) and 45 min (B1 and B2) post-intravenous tail vein injection of the ^{11}C-tracer. Reproduced from reference (Wu et al., 2023b) with permission from Elsevier. Copyright 2023 Elsevier Publishing Group

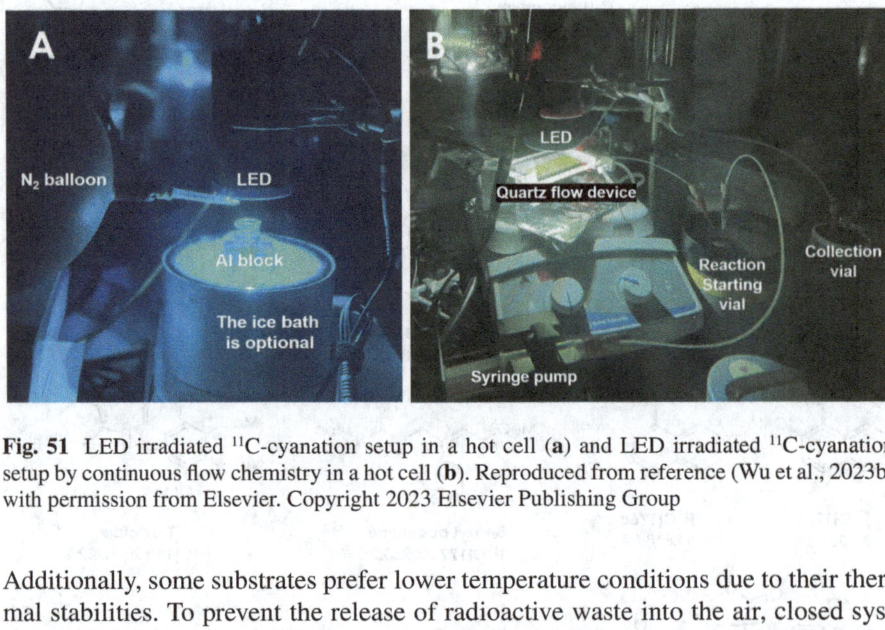

Fig. 51 LED irradiated ^{11}C-cyanation setup in a hot cell (**a**) and LED irradiated ^{11}C-cyanation setup by continuous flow chemistry in a hot cell (**b**). Reproduced from reference (Wu et al., 2023b) with permission from Elsevier. Copyright 2023 Elsevier Publishing Group

Additionally, some substrates prefer lower temperature conditions due to their thermal stabilities. To prevent the release of radioactive waste into the air, closed systems (Fig. 53d–f) have been developed to meet the requirements of radiopharmaceutical manufacturing. Again, notably, the successful production of [^{18}F]FDOPA via photoredox radiofluorination has been achieved by utilizing a flow radiochemistry setup (Fig. 53f). To make these technologies readily available to the field, LED Radiofluidics is currently commercializing the light reactors and catalysts.

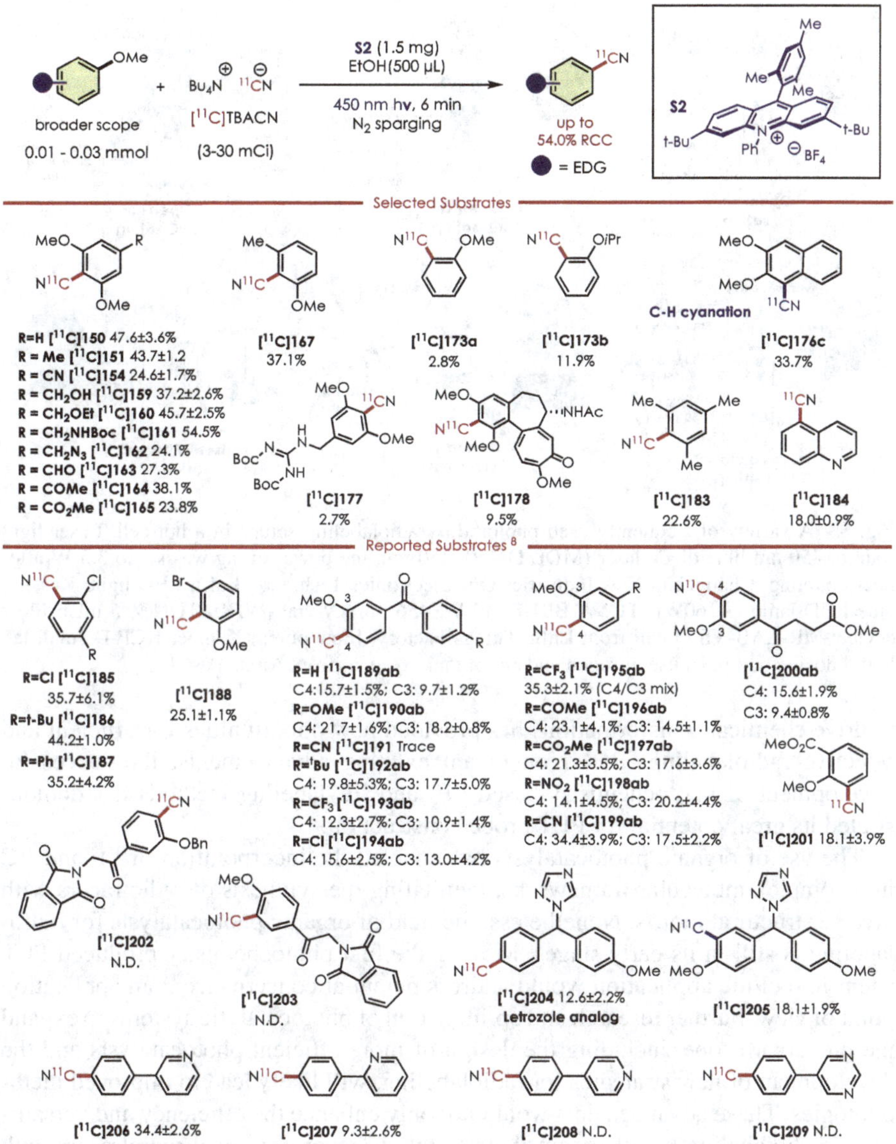

Fig. 52 Photoredox-mediated arene C–O ^{11}C-cyanation by the acridinium catalyst (selected substrates, $n = 3$ when a standard deviation is given)

12 Perspective and Conclusion

PET plays a crucial role in various fields, including medical diagnostics, drug development, and neuroscience research. Therefore, the development of efficient and versatile labeling methods is of great importance. The ability to harness visible light

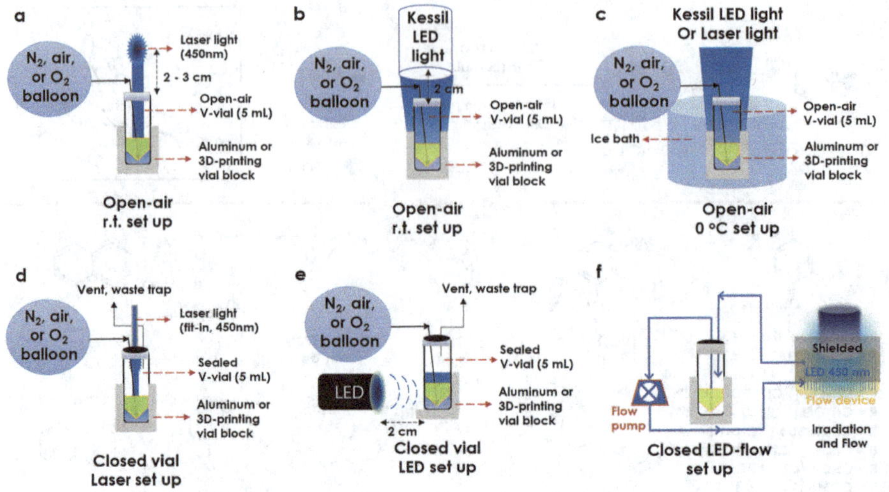

Fig. 53 A variety of frequently used photoredox radiolabeling setups in a hot cell. Laser light source: 450 nm blue diode laser (MDL-D-450, 450 nm, the power rating was set to 3.5 W after fiber coupling, Changchun New Industries Optoelectronics Tech. Co., Ltd.); LED light: Kessil® blue LED lamp (A160WE TUNA BLUE, 40 W, 456 nm). V-vial (WHEATON®, 5 mL); Flow device: MR-LAB-VS Q-cell from Little Things Factory; Flow pump: Kamoer KCP-D small lab liquid dosing pump. House nitrogen and air, or tank oxygen (from Airgas) used

to drive chemical transformations has provided new opportunities for efficient and selective radiolabeling. Based on recent exciting advancements, the successful development of photochemistry-based ^{18}F- and ^{11}C-labeling methods has demonstrated its great potential for PET probe construction.

The use of organic photocatalysts has enabled the incorporation of ^{18}F and ^{11}C into complex molecular frameworks, facilitating the synthesis of radiotracers with diverse structural motifs. Nonetheless, the field of organic photocatalysis for radiolabeling is still in its early stages. Moving the first photochemistry produced PET agent into clinic application would address potential concerns from an application point of view. Further research and optimization of photocatalytic systems to expand the precursor scope, including the design of more efficient photocatalysts and the development of new strategies for radiolabeling, will likely lead to improved methodologies. These advancements would not only enhance the efficiency and versatility of radiolabeling but also contribute to the discovery of novel radiotracers with enhanced imaging capabilities. The combination of radiolabeling via organic photocatalysis with other emerging technologies, such as machine learning and more advanced automation, has the potential to revolutionize the field of radiopharmaceutical chemistry. In general, predictive machine learning algorithms can be trained on extensive datasets of chemical structures and their corresponding reactivity data, enabling the prediction of outcomes for organic photocatalytic reactions. These models play a crucial role in identifying optimal reaction conditions, designing more efficient photocatalysts, and guiding the selection of suitable precursors for radiolabeling. Furthermore, machine learning algorithms can analyze data generated from real reaction screenings, extracting valuable insights into reaction trends. This analysis facilitates faster and more efficient optimization of photocatalytic systems, especially for radiochemistry, which often is more difficult to predict from structures. By leveraging machine learning in data mining and knowledge discovery, researchers can expedite the discovery and design of novel radiotracers, ultimately enhancing imaging capabilities, diagnostics, and therapeutic interventions in radiopharmaceutical chemistry.

Another important direction is to automate the labeling process by combining with commercial availability of apparatus or even design a dedicated photoreactor. This progress will be critical to make the technology widely available to the field, especially in a cGMP environment. The automation is well suited to perform repetitive tasks, reduce human errors, and enable easy transfer the programed conditions from one site to another. In fact, various companies have shown initial interest in this direction. Both LED Radiofluidics and Synthra have released their own prototype that can be integrated with automated modules, respectively. For example, based on the recently developed LED-flow setup (Fig. 53f), there is ongoing research and development at LED Radiofluidics to automate various photocatalyzed radiolabeling processes with flow radiochemistry. These efforts aim to create more advanced integrated and user-friendly automated synthesis platforms that enable efficient, reproducible, and high-throughput radiolabeling reaction screening. By leveraging automation, researchers can accelerate the adoption of photocatalyzed radiolabeling techniques in both research and clinical settings, facilitating its broader applications in radiopharmaceutical chemistry. By automating specific aspects of the process, researchers can potentially achieve a much safer working environment, higher efficiency, better reproducibility, and even high-throughput labeling. Here are some specific aspects that can benefit from advanced automation: 1. Reactor setup and configuration: automation can streamline the setup and configuration of reaction vessels, ensuring consistent and precise conditions for photoactivated radiolabeling. Advanced automated systems will even control parameters such as temperature, light intensity, and reaction time, optimizing the reaction conditions for different substrates. 2. Photocatalyst screening: automation can facilitate screening of photocatalysts by automating parallel small-scale radiolabeling tests of different catalysts. 3. Reaction monitoring: automation can incorporate real-time monitoring techniques, such as online radioactivity detection, radio-TLC, or UV spectroscopic analysis, to provide feedback on the progress of the radiolabeling reaction. This information can guide the optimization of reaction conditions and ensure consistent and efficient radiotracer production. 4. Radiotracer purification: automated purification systems can streamline the separation and purification of radiotracers from reaction mixtures. This can include automated HPLC systems or other purification techniques, ensuring efficient isolation of the desired radiotracer product even in lower radiochemical yield cases.

Regarding photocatalyzed radiolabeling methods in general, they offer several advantages over conventional radiosynthesis methods. Novel techniques often enable milder reaction conditions, simpler reaction pathways, and improved selectivity. By utilizing photocatalysis, radiolabeling reactions can be conducted at ambient temperatures and atmospheric pressure, eliminating the need for harsh reagents or high temperatures. This mildness is particularly beneficial for substrates with sensitive or labile functional groups that may degrade under conventional conditions. Additionally, photocatalyzed methods can lead to simplified reaction pathways, reducing the number of synthetic steps and increasing overall efficiency. While conventional radiosynthesis methods work well for a variety of tracers, photocatalyzed radiolabeling methods offer clear and significant advantages in specific categories of substrates. For example, substrates with complex or labile functional groups, as well as those requiring specific regioselectivity or stereoselectivity, can benefit from the mild reaction conditions and enhanced selectivity provided by photocatalysis. We do like to point out that one potential drawback of these method is the added chemical(s) in the reaction and their residue amounts still need to be tested. Fortunately, many organic catalyst and base additives can be readily removed through cartridges (such as aluminum and SCX). In a recent meeting abstract, cartridge-based production of [^{18}F]FDOPA was also reported (Wu et al., 2023a). Nonetheless, it would be important to move one of the photoredox-labeled agents into clinic, which would greatly promote the adoption of the technology.

It is essential to emphasize that the photoredox labeling method operates on a distinct mechanism for labeling, setting it apart from the conventional S_NAr reactions. While the latter are better suited for electron-deficient precursors, the photoredox approach excels with electron-neutral or electron-rich systems. Importantly, this method does not seek to supplant traditional labeling techniques; rather, it complements them. This complementary nature enables researchers to produce PET agents that were previously arduous to synthesize. Such a role positions the photoredox labeling method as a crucial instrument for advancing future PET agent development.

References

Alves, S., Correia, J. D. G., Gano, L., Rold, T. L., Prasanphanich, A., Haubner, R., Rupprich, M., Alberto, R., Decristoforo, C., Santos, I., & Smith, C. J. (2007). In vitro and in vivo evaluation of a novel 99mTc(CO)3-pyrazolyl conjugate of cyclo-(Arg-Gly-Asp-d-Tyr-Lys). *Bioconjugate Chemistry, 18*, 530–537.

Ametamey, S. M., Honer, M., & Schubiger, P. A. (2008). Molecular imaging with PET. *Chemical Reviews, 108*, 1501–1516.

An, P., Lewandowski, T. M., Erbay, T. G., Liu, P., & Lin, Q. (2018). Sterically shielded, stabilized nitrile imine for rapid bioorthogonal protein labeling in live cells. *Journal of the American Chemical Society, 140*, 4860–4868.

Arumugam, S., Chin, J., Schirrmacher, R., Popik, V. V., & Kostikov, A. P. (2011). [18F]Azadibenzocyclooctyne ([18F]ADIBO): A biocompatible radioactive labeling synthon for peptides using catalyst free [3+2] cycloaddition. *Bioorganic & Medicinal Chemistry Letters, 21*, 6987–6991.

Bertrand, G., & Wentrup, C. (1994). Nitrile imines: From matrix characterization to stable compounds. *Angewandte Chemie International Edition in English, 33*, 527–545.

Bohning, J. J., & Weiss, K. (1966). A kinetic study of the photochemical reaction of phenanthrenequinone with olefins. *Journal of the American Chemical Society, 88*, 2893–2898.

Bouvet, V., Wuest, M., & Wuest, F. (2011). Copper-free click chemistry with the short-lived positron emitter fluorine-18. *Organic & Biomolecular Chemistry, 9*, 7393–7399.

Campbell-Verduyn, L. S., Mirfeizi, L., Schoonen, A. K., Dierckx, R. A., Elsinga, P. H., & Feringa, B. L. (2011). Strain-promoted copper-free "Click" chemistry for 18f radiolabeling of Bombesin. *Angewandte Chemie International Edition, 50*, 11117–11120.

Cassidy, P. J., & Radda, G. K. (2005). Molecular imaging perspectives. *Journal of the Royal Society Interface, 2*, 133–144.

Chapman, O. L., Gano, J., West, P. R., Regitz, M., & Maas, G. (1981). Acenaphthyne. *Journal of the American Chemical Society, 103*, 7033–7036.

Chen, W., Huang, Z., Tay, N. E. S., Giglio, B., Wang, M., Wang, H., Wu, Z., Nicewicz, D. A., & Li, Z. (2019). Direct arene C-H fluorination with (18)F(−) via organic photoredox catalysis. *Science, 364*, 1170–1174.

Chen, W., Wang, H., Tay, N. E. S., Pistritto, V. A., Li, K. P., Zhang, T., Wu, Z., Nicewicz, D. A., & Li, Z. (2021). Arene radiofluorination enabled by photoredox-mediated halide interconversion. *Nature Chemistry*.

Chen, W., Wu, X., McManus, J. B., Bida, G. T., Li, K.-P., Wu, Z., Nicewicz, D. A., & Li, Z. (2022). Direct C–H radiocyanation of arenes via organic photoredox catalysis. *Organic Letters, 24*, 9316–9321.

Choi, J. H., Oh, D., Kim, I. S., Kim, H.-S., Kim, M., Kim, E.-M., Lim, S. T., Sohn, M.-H., Kim, D. H., & Jeong, H.-J. (2018). Light-triggered radiochemical synthesis: A novel 18F-labelling strategy using photoinducible click reaction to prepare PET imaging probes. *Contrast Media & Molecular Imaging, 2018*, 4617493.

Clovis, J. S., Eckell, A., Huisgen, R., & Sustmann, R. (1967). 1.3-Dipolare Cycloadditionen, XXV. Der Nachweis des freien Diphenylnitrilimins als Zwischenstufe bei Cycloadditionen. *Chemische Berichte, 100*, 60–70.

Deepthi, A., Acharjee, N., Sruthi, S. L., & Meenakshy, C. B. (2022). An overview of nitrile imine based [3+2] cycloadditions over half a decade. *Tetrahedron, 116*, 132812.

Deng, X., Rong, J., Wang, L., Vasdev, N., Zhang, L., Josephson, L., & Liang, S. H. (2019). Chemistry for positron emission tomography: Recent advances in 11C-, 18F-, 13N-, and 15O-labeling reactions. *Angewandte Chemie International Edition, 58*, 2580–2605.

Destro, G., Loreau, O., Marcon, E., Taran, F., Cantat, T., & Audisio, D. (2019). Dynamic carbon isotope exchange of pharmaceuticals with labeled CO_2. *Journal of the American Chemical Society, 141*, 780–784.

Destro, G., Horkka, K., Loreau, O., Buisson, D.-A., Kingston, L., Del Vecchio, A., Schou, M., Elmore, C. S., Taran, F., Cantat, T., & Audisio, D. (2020). Transition-metal-free carbon isotope exchange of phenyl acetic acids. *Angewandte Chemie International Edition, 59*, 13490–13495.

Dommerholt, J., Rutjes, F. P. J. T., & Van Delft, F. L. (2016). Strain-promoted 1,3-dipolar cycloaddition of cycloalkynes and organic azides. *Topics in Current Chemistry, 374*, 16.

Erickson, J. A., & Mcloughlin, J. I. (1995). Hydrogen bond donor properties of the difluoromethyl group. *The Journal of Organic Chemistry, 60*, 1626–1631.

Eriksson, J., Antoni, G., Långström, B., & Itsenko, O. (2021). The development of 11C-carbonylation chemistry: A systematic view. *Nuclear Medicine and Biology, 92*, 115–137.

Fan, Y., Deng, C., Cheng, R., Meng, F., & Zhong, Z. (2013). In situ forming hydrogels via catalyst-free and bioorthogonal "Tetrazole–Alkene" photo-click chemistry. *Biomacromolecules, 14*, 2814–2821.

Fu, Y., Helbert, H., Simeth, N. A., Crespi, S., Spoelstra, G. B., Van Dijl, J. M., Van Oosten, M., Nazario, L. R., Van Der Born, D., Luurtsema, G., Szymanski, W., Elsinga, P. H., & Feringa, B. L. (2021). Ultrafast photoclick reaction for selective 18F-positron emission tomography tracer synthesis in flow. *Journal of the American Chemical Society, 143*, 10041–10047.

George, M. V., & Angadiyavar, C. S. (1971). Photochemical cycloadditions of 1,3-dipolar systems. I. Additions of N,C-diphenylsydnone and 2,5-diphenyltetrazole. *The Journal of Organic Chemistry, 36*, 1589–1594.

Glassner, M., Palmieri, L., Monnery, B. D., Verbrugghen, T., Deleye, S., Stroobants, S., Staelens, S., Wyffels, L., & Hoogenboom, R. (2017). The label matters: μPET imaging of the biodistribution of low molar mass 89Zr and 18F-labeled poly(2-ethyl-2-oxazoline). *Biomacromolecules, 18*, 96–102.

Halder, R., & Ritter, T. (2021). 18F-fluorination: Challenge and opportunity for organic chemists. *The Journal of Organic Chemistry, 86*, 13873–13884.

He, M., Li, J., Tan, S., Wang, R., & Zhang, Y. (2013). Photodegradable supramolecular hydrogels with fluorescence turn-on reporter for photomodulation of cellular microenvironments. *Journal of the American Chemical Society, 135*, 18718–18721.

Hinsinger, K., & Pieters, G. (2019). The emergence of carbon isotope exchange. *Angewandte Chemie International Edition, 58*, 9678–9680.

Holmberg-Douglas, N., & Nicewicz, D. A. (2019). Arene cyanation via cation-radical accelerated-nucleophilic aromatic substitution. *Organic Letters, 21*, 7114–7118.

Holstein, J. M., Stummer, D., & Rentmeister, A. (2015). Enzymatic modification of 5′-capped RNA with a 4-vinylbenzyl group provides a platform for photoclick and inverse electron-demand Diels–Alder reaction. *Chemical Science, 6*, 1362–1369.

Huisgen, R., Seidel, M., Wallbillich, G., & Knupfer, H. (1962). Diphenyl-nitrilimin und seine 1.3-dipolaren additionen an alkene und alkine. *Tetrahedron, 17*, 3–29.

Itsenko, O., Kihlberg, T., & Långström, B. (2004). Photoinitiated carbonylation with [11C]carbon monoxide using amines and alkyl iodides. *The Journal of Organic Chemistry, 69*, 4356–4360.

Itsenko, O., Kihlberg, T., & Långström, B. (2006). Labeling of aliphatic carboxylic acids using [11C]carbon monoxide. *Nature Protocols, 1*, 798–802.

Itsenko, O., Norberg, D., Rasmussen, T., Långström, B., & Chatgilialoglu, C. (2007). Radical carbonylation with [11C]carbon monoxide promoted by oxygen-centered radicals: Experimental and DFT studies of the mechanism. *Journal of the American Chemical Society, 129*, 9020–9031.

Jacobson, O., Kiesewetter, D. O., & Chen, X. (2015). Fluorine-18 Radiochemistry, labeling strategies and synthetic routes. *Bioconjugate Chemistry, 26*, 1–18.

Jewett, J. C., & Bertozzi, C. R. (2010). Cu-free click cycloaddition reactions in chemical biology. *Chemical Society Reviews, 39*, 1272–1279.

Kazuhiro, M., Tetsuo, O., & Yoshinori, N. (1976). Photochemical reaction of 9,10-phenanthrenequinone with hydrogen donors. Behavior of radicals in solution as studied by CIDNP. *Bulletin of the Chemical Society of Japan, 49*, 791–795.

Kettenbach, K., & Ross, T. L. (2016). A 18F-labeled dibenzocyclooctyne (DBCO) derivative for copper-free click labeling of biomolecules. *MedChemComm, 7*, 654–657.

Kihlberg, T., & Långström, B. (1999). Biologically active 11C-labeled amides using palladium-mediated reactions with aryl halides and [11C]carbon monoxide. *The Journal of Organic Chemistry, 64*, 9201–9205.

Kihlberg, T., Karimi, F., & Långström, B. (2002). [11C]Carbon monoxide in selenium-mediated synthesis of 11C-carbamoyl compounds. *The Journal of Organic Chemistry, 67*, 3687–3692.

Kim, H. L., Sachin, K., Jeong, H. J., Choi, W., Lee, H. S., & Kim, D. W. (2015). F-18 labeled RGD probes based on bioorthogonal strain-promoted click reaction for PET imaging. *ACS Medicinal Chemistry Letters, 6*, 402–407.

Kingston, C., Wallace, M. A., Allentoff, A. J., Degruyter, J. N., Chen, J. S., Gong, S. X., Bonacorsi, S., & JR. & Baran, P. S. (2019). Direct carbon isotope exchange through decarboxylative carboxylation. *Journal of the American Chemical Society, 141*, 774–779.

Kong, D., Moon, P. J., Lui, E. K. J., Bsharat, O., & Lundgren, R. J. (2020). Direct reversible decarboxylation from stable organic acids in dimethylformamide solution. *Science, 369*, 557–561.

Kong, D., Munch, M., Qiqige, Q., Cooze, C. J. C., Rotstein, B. H., & Lundgren, R. J. (2021). Fast carbon isotope exchange of carboxylic acids enabled by organic photoredox catalysis. *Journal of the American Chemical Society, 143*, 2200–2206.

Koyama, H., Ikenuma, H., Toda, H., Kondo, G., Hirano, M., Kato, M., Abe, J., Yamada, T., Wakabayashi, T., Ito, K., Natsume, A., & Suzuki, M. (2017). Synthesis of PET probe O6-[(3-[11C]methyl)benzyl]guanine by Pd0-mediated rapid C-[11C]methylation toward imaging DNA repair protein O6-methylguanine-DNA methyltransferase in glioblastoma. *Bioorganic & Medicinal Chemistry Letters, 27*, 1892–1896.

Krishnan, H. S., Ma, L., Vasdev, N., & Liang, S. H. (2017). 18F-Labeling of sensitive biomolecules for positron emission tomography. *Chemistry—A European Journal, 23*, 15553–15577.

Kumar, G. S., & Lin, Q. (2021). Light-triggered click chemistry. *Chemical Reviews, 121*, 6991–7031.

Kwon, Y.-D., Lee, J. Y., La, M. T., Lee, S. J., Lee, S.-H., Park, J. H., & Kim, H.-K. (2020). Novel multifunctional 18F-labelled PET tracer with prostate-specific membrane antigen-targeting and hypoxia-sensitive moieties. *European Journal of Medicinal Chemistry, 189*, 112099.

Langer, O., Mitterhauser, M., Brunner, M., Zeitlinger, M., Wadsak, W., Mayer, B. X., Kletter, K., & Muller, M. (2003). Synthesis of fluorine-18-labeled ciprofloxacin for PET studies in humans. *Nuclear Medicine and Biology, 30*, 285–291.

Lee, S. B., Kim, H. L., Jeong, H.-J., Lim, S. T., Sohn, M.-H., & Kim, D. W. (2013). Mesoporous silica nanoparticle pretargeting for PET imaging based on a rapid bioorthogonal reaction in a living body. *Angewandte Chemie International Edition, 52*, 10549–10552.

Lemos, A. L. P., Trump, L., Lallemand, B., Pasau, P., Mercier, J., Lemaire, C., Monbaliu, J.-C., Genicot, C., & Luxen, A. (2020). Radical C–H 18F-difluoromethylation of heteroarenes with [18F]difluoromethyl heteroaryl-sulfones by visible light photoredox catalysis. *Catalysts, 10*, 275.

Li, Z., Qian, L., Li, L., Bernhammer, J. C., Huynh, H. V., Lee, J.-S., & Yao, S. Q. (2016). Tetrazole photoclick chemistry: Reinvestigating its suitability as a bioorthogonal reaction and potential applications. *Angewandte Chemie International Edition, 55*, 2002–2006.

Li, J., Kong, H., Huang, L., Cheng, B., Qin, K., Zheng, M., Yan, Z., & Zhang, Y. (2018). Visible light-initiated bioorthogonal photoclick cycloaddition. *Journal of the American Chemical Society, 140*, 14542–14546.

Li, M., Ma, X., Molnar, C. J., Wang, S., Wu, Z., Popik, V. V., & Li, Z. (2022). Modular PET agent construction strategy through strain-promoted double-click reagent with efficient photoclick step. *Bioconjugate Chemistry, 33*, 2088–2096.

Lim, R. K. V., & Lin, Q. (2011). Photoinducible bioorthogonal chemistry: A spatiotemporally controllable tool to visualize and perturb proteins in live cells. *Accounts of Chemical Research, 44*, 828–839.

Livni, E., Elmaleh, D. R., Levy, S., Brownell, G. L., & Strauss, W. H. (1982). Beta-methyl[1-11C]heptadecanoic acid: A new myocardial metabolic tracer for positron emission tomography. *Journal of Nuclear Medicine, 23*, 169–175.

McManus, J. B., & Nicewicz, D. A. (2017). Direct C-H cyanation of arenes via organic photoredox catalysis. *Journal of the American Chemical Society, 139*, 2880–2883.

Meanwell, N. A. (2018). Fluorine and fluorinated motifs in the design and application of bioisosteres for drug design. *Journal of Medicinal Chemistry, 61*, 5822–5880.

Mizuta, S., Stenhagen, I. S. R., O'duill, M., Wolstenhulme, J., Kirjavainen, A. K., Forsback, S. J., Tredwell, M., Sandford, G., Moore, P. R., Huiban, M., Luthra, S. K., Passchier, J., Solin, O., & Gouverneur, V. (2013). Catalytic decarboxylative fluorination for the synthesis of tri- and difluoromethyl arenes. *Organic Letters, 15*, 2648–2651.

Nabulsi, N. B., Mercier, J., Holden, D., Carré, S., Najafzadeh, S., Vandergeten, M.-C., Lin, S.-F., Deo, A., Price, N., Wood, M., Lara-Jaime, T., Montel, F., Laruelle, M., Carson, R. E., Hannestad, J., & Huang, Y. (2016). Synthesis and preclinical evaluation of 11C-UCB-J as a PET tracer for imaging the synaptic vesicle glycoprotein 2A in the brain. *Journal of Nuclear Medicine, 57*, 777–784.

Newton, J. J., Engüdar, G., Brooke, A. J., Nodwell, M. B., Horngren-Rhodes, H., Martin, R. E., Schaffer, P., Britton, R., & Friesen, C. M. (2022). Rapid 18F- and 19F-difluoromethylation via desulfurative fluorination of transient N-, O-, and C-linked dithioles. *Chemistry—A European Journal, 29*, e202202862.

Padwa, A., Nahm, S., & Sato, E. (1978). Photochemical transformations of small ring heterocyclic compounds. 9. Intramolecular 1,3-dipolar cycloaddition reactions of alkenyl-substituted nitrile imines. *The Journal of Organic Chemistry, 43*, 1664–1671.

Pike, V. W., Eakins, M. N., Allan, R. M., & Selwyn, A. P. (1982). Preparation of [1−11C]acetate—An agent for the study of myocardial metabolism by positron emission tomography. *The International Journal of Applied Radiation and Isotopes, 33*, 505–512.

Pipal, R. W., Stout, K. T., Musacchio, P. Z., Ren, S., Graham, T. J. A., Verhoog, S., Gantert, L., Lohith, T. G., Schmitz, A., Lee, H. S., Hesk, D., Hostetler, E. D., Davies, I. W., & Macmillan, D. W. C. (2021). Metallaphotoredox aryl and alkyl radiomethylation for PET ligand discovery. *Nature, 589*, 542–547.

Poloukhtine, A. A., Mbua, N. E., Wolfert, M. A., Boons, G.-J., & Popik, V. V. (2009). Selective labeling of living cells by a photo-triggered click reaction. *Journal of the American Chemical Society, 131*, 15769–15776.

Purser, S., Moore, P. R., Swallow, S., & Gouverneur, V. (2008). Fluorine in medicinal chemistry. *Chemical Society Reviews, 37*, 320–330.

Pysz, M. A., Gambhir, S. S., & Willmann, J. K. (2010). Molecular imaging: Current status and emerging strategies. *Clinical Radiology, 65*, 500–516.

Riss, P. J., Lu, S., Telu, S., Aigbirhio, F. I., & Pike, V. W. (2012). CuI-catalyzed 11C carboxylation of boronic acid esters: A rapid and convenient entry to 11C-labeled carboxylic acids, esters, and amides. *Angewandte Chemie International Edition, 51*, 2698–2702.

Roche, M., Specklin, S., Richard, M., Hinnen, F., Génermont, K., & Kuhnast, B. (2019). [18F]FPyZIDE: A versatile prosthetic reagent for the fluorine-18 radiolabeling of biologics via copper-catalyzed or strain-promoted alkyne-azide cycloadditions. *Journal of Labelled Compounds and Radiopharmaceuticals, 62*, 95–108.

Rubin, M. B. (1963). photochemical reactions of diketones. The 1,4-addition of ethers to 9,10-phenanthrenequinone. *The Journal of Organic Chemistry, 28*, 1949–1952.

Rubin, M. B., & Neuwirth-Weiss, Z. (1972). Photochemical reactions of diketones. V. Reaction of phenanthrenequinone and benzene. *Journal of the American Chemical Society, 94*, 6048–6053.

Rubin, M. B., & Zwitkowits, P. (1964). Photochemical reactions of diketones. II.1a,b the 1,2-addition of substituted toluenes to 9,10-phenanthrenequinone. *The Journal of Organic Chemistry, 29*, 2362–2368.

Ryu, I., Nagahara, K., Kambe, N., Sonoda, N., Kreimerman, S., & Komatsu, M. (1998). Metal catalyst-free by design. The synthesis of amides from alkyl iodides, carbon monoxide and amines by a hybrid radical/ionic reaction. *Chemical Communications*, 1953–1954.

Sachin, K., Jadhav, V. H., Kim, E.-M., Kim, H. L., Lee, S. B., Jeong, H.-J., Lim, S. T., Sohn, M.-H., & Kim, D. W. (2012). F-18 Labeling protocol of peptides based on chemically orthogonal strain-promoted cycloaddition under physiologically friendly reaction conditions. *Bioconjugate Chemistry, 23*, 1680–1686.

Sap, J. B. I., Wilson, T. C., Kee, C. W., Straathof, N. J. W., Ende, C. W. A., Mukherjee, P., Zhang, L., Genicot, C., & Gouverneur, V. (2019). Synthesis of 18F-difluoromethylarenes using aryl boronic acids, ethyl bromofluoroacetate and [18F]fluoride. *Chemical Science, 10*, 3237–3241.

Sap, J. B. I., Meyer, C. F., Ford, J., Straathof, N. J. W., Dürr, A. B., Lelos, M. J., Paisey, S. J., Mollner, T. A., Hell, S. M., Trabanco, A. A., Genicot, C., Am Ende, C. W., Paton, R. S., Tredwell, M., & Gouverneur, V. (2022). [18F]Difluorocarbene for positron emission tomography. *Nature, 606*, 102–108.

Schönberg, A., & Mustafa, A. (1944). Reactions of ethylenes with 1,2-diketones in sunlight. *Nature, 153*, 195–195.

Sessler, C. D., Rahm, M., Becker, S., Goldberg, J. M., Wang, F., & Lippard, S. J. (2017). CF2H, a hydrogen bond donor. *Journal of the American Chemical Society, 139*, 9325–9332.

Shi, H., Braun, A., Wang, L., Liang, S. H., Vasdev, N., & Ritter, T. (2016). Synthesis of 18F-difluoromethylarenes from aryl (pseudo) halides. *Angewandte Chemie International Edition, 55*, 10786–10790.

Shoup, T. M., McCauley, J. P., Lee, D. F., Chen, R., Normandin, M. D., Bonab, A. A., El Fakhri, G., & Vasdev, N. (2014). Synthesis of the dopamine D2/D3 receptor agonist (+)-PHNO via supercritical fluid chromatography: Preliminary PET imaging study with [3-11C]-(+) PHNO. *Tetrahedron Letters, 55*, 682–685.

Smith, R. T., Zhang, X., Rincón, J. A., Agejas, J., Mateos, C., Barberis, M., García-Cerrada, S., De Frutos, O., & Macmillan, D. W. C. (2018). Metallaphotoredox-catalyzed cross-electrophile Csp3–Csp3 coupling of aliphatic bromides. *Journal of the American Chemical Society, 140*, 17433–17438.

Song, W., Wang, Y., Qu, J., Madden, M. M., & Lin, Q. (2008). A Photoinducible 1,3-dipolar cycloaddition reaction for rapid, selective modification of tetrazole-containing proteins. *Angewandte Chemie International Edition, 47*, 2832–2835.

Sutton, D. A., & Popik, V. V. (2016). Sequential photochemistry of dibenzo[a,e]dicyclopropa[c,g][8]annulene-1,6-dione: Selective formation of didehydrodibenzo[a,e][8]annulenes with ultra-fast Spaac reactivity. *The Journal of Organic Chemistry, 81*, 8850–8857.

Takahashi, T., Ido, T., Hatano, K., Iwata, R., & Nakanishi, H. (1990). Synthesis of 1-11C-labeled fatty acid from [11C]HCN. *International Journal of Radiation Applications and Instrumentation Part A. Applied Radiation and Isotopes, 41*, 649–654.

Tay, N. E. S., Chen, W., Levens, A., Pistritto, V. A., Huang, Z., Wu, Z., Li, Z., & Nicewicz, D. A. (2020). 19F- and 18F-arene deoxyfluorination via organic photoredox-catalysed polarity-reversed nucleophilic aromatic substitution. *Nature Catalysis, 3*, 734–742.

Thonon, D., Goukens, E., Kaisin, G., Paris, J., Flagothier, J., & Luxen, A. (2011). Photoactivated 1,3-dipolar cycloaddition for the rapid preparation of 18F labelled radiotracers. *Tetrahedron, 67*, 5572–5576.

Tortajada, A., Duan, Y., Sahoo, B., Cong, F., Toupalas, G., Sallustrau, A., Loreau, O., Audisio, D., & Martin, R. (2019). Catalytic decarboxylation/carboxylation platform for accessing isotopically labeled carboxylic acids. *ACS Catalysis, 9*, 5897–5901.

Toubro, N. H., & Holm, A. (1980). Nitrilimines. *Journal of the American Chemical Society, 102*, 2093–2094.

Trump, L., Lemos, A., Lallemand, B., Pasau, P., Mercier, J., Lemaire, C., Luxen, A., & Genicot, C. (2019). Late-stage 18F-difluoromethyl labeling of N-heteroaromatics with high molar activity for PET imaging. *Angewandte Chemie International Edition, 58*, 13149–13154.

Trump, L., Lemos, A., Jacq, J., Pasau, P., Lallemand, B., Mercier, J., Genicot, C., Luxen, A., & Lemaire, C. (2020). Development of a general automated flow photoredox 18F-difluoromethylation of N-heteroaromatics in an AllinOne synthesizer. *Organic Process Research & Development, 24*, 734–744.

Verhoog, S., Pfeifer, L., Khotavivattana, T., Calderwood, S., Collier, T. L., Wheelhouse, K., Tredwell, M., & Gouverneur, V. (2016). Silver-mediated 18F-labeling of aryl-CF3 and aryl-CHF2 with 18F-fluoride. *Synlett, 27*, 25–28.

Wang, H., & Mooney, D. J. (2020). Metabolic glycan labelling for cancer-targeted therapy. *Nature Chemistry, 12*, 1102–1114.

Wang, Y., Rivera Vera, C. I., & Lin, Q. (2007). Convenient synthesis of highly functionalized pyrazolines via mild, photoactivated 1,3-dipolar cycloaddition. *Organic Letters, 9*, 4155–4158.

Wang, M., Mcnitt, C. D., Wang, H., Ma, X., Scarry, S. M., Wu, Z., Popik, V. V., & Li, Z. (2018). The efficiency of 18F labelling of a prostate specific membrane antigen ligand via strain-promoted azide–alkyne reaction: Reaction speed versus hydrophilicity. *Chemical Communications, 54*, 7810–7813.

Wang, L., White, A. R., Chen, W., Wu, Z., Nicewicz, D. A., & Li, Z. (2020). Direct radiofluorination of arene C-H bonds via photoredox catalysis using a peroxide as the terminal oxidant. *Organic Letters, 22*, 7971–7975.

Warnier, C., Lemaire, C., Becker, G., Zaragoza, G., Giacomelli, F., Aerts, J., Otabashi, M., Bahri, M. A., Mercier, J., Plenevaux, A., & Luxen, A. (2016). Enabling Efficient Positron Emission Tomography (PET) Imaging of Synaptic Vesicle Glycoprotein 2A (SV2A) with a robust and one-step radiosynthesis of a highly potent 18F-labeled ligand ([18F]UCB-H). *Journal of Medicinal Chemistry, 59*, 8955–8966.

Webb, E. W., Park, J. B., Cole, E. L., Donnelly, D. J., Bonacorsi, S. J., Ewing, W. R., & Doyle, A. G. (2020). Nucleophilic (Radio)Fluorination of redox-active esters via radical-polar crossover enabled by photoredox catalysis. *Journal of the American Chemical Society, 142*, 9493–9500.

Weinberg, P., Csongár, C., & Grummt, U.-W. (1989). Bis-2H-tetrazoles VII: Quantum yields of bis-2H-tetrazoles and studies of thermal consecutive reactions of bis-nitrile imines by time-resolved spectroscopy. *Journal of Photochemistry and Photobiology A: Chemistry, 50*, 11–30.

Weissleder, R., & Mahmood, U. (2001). Molecular imaging. *Radiology, 219*, 316–333.

Willmann, J. K., Van Bruggen, N., Dinkelborg, L. M., & Gambhir, S. S. (2008). Molecular imaging in drug development. *Nature Reviews Drug Discovery, 7*, 591–607.

Wu, X., Chen, W., Deng, H., Nicewicz, D. A., Li, Z., & Wu, Z. (2023a). Photoredox radiofluorination assisted by flow chemistry: A practical Manufacture Example for Cartridge based L-[18F]FDOPA production. In *SNMMI 2023*. Society of Nuclear Medicine and Molecular Imaging (SNMMI).

Wu, X., Chen, W., Holmberg-Douglas, N., Bida, G. T., Tu, X., Ma, X., Wu, Z., Nicewicz, D. A., & Li, Z. (2023b). 11C-, 12C-, and 13C-cyanation of electron-rich arenes via organic photoredox catalysis. *Chem, 9*, 343–362.

Wu, X., Ma, X., Zhong, Y., Chen, W., Xu, M., Zhang, H., Wang, L., Tu, X., Han, Z., Zhao, W., Wu, Z., Moschos, S. J., & Li, Z. (2023c). Development of [18F]F-5-OMe-tryptophans through photoredox radiofluorination: a new method to access tryptophan-based PET agents. *Journal of Medicinal Chemistry, 66*, 3262–3272.

Yuan, G., Wang, F., Stephenson, N. A., Wang, L., Rotstein, B. H., Vasdev, N., Tang, P., & Liang, S. H. (2017). Metal-free 18F-labeling of aryl-CF2H via nucleophilic radiofluorination and oxidative C–H activation. *Chemical Communications, 53*, 126–129.

Zafrani, Y., Yeffet, D., Sod-Moriah, G., Berliner, A., Amir, D., Marciano, D., Gershonov, E., & Saphier, S. (2017). Difluoromethyl bioisostere: Examining the "Lipophilic Hydrogen Bond Donor" concept. *Journal of Medicinal Chemistry, 60*, 797–804.

Zhang, P., Le, C. C., & Macmillan, D. W. C. (2016). Silyl radical activation of alkyl halides in metallaphotoredox catalysis: A unique pathway for cross-electrophile coupling. *Journal of the American Chemical Society, 138*, 8084–8087.

Zhou, Y.-P., Makaravage, K. J., & Brugarolas, P. (2021). Radiolabeling with [11C]HCN for Positron emission tomography. *Nuclear Medicine and Biology, 102-103*, 56–86.

Application of Microwaves in Radiochemistry

Hiroyuki Kimura, Yuto Kondo, Yusuke Yagi, and Takashi Kozaka

1 Characteristics of Microwave Heating

The first reports of microwave synthesis (Diels–Alder reaction, Claisen reaction, Ene reaction, etc.) were published by Gedye et al. (1986) in Canada and Giguere et al. (1986) in the USA; these syntheses involved heating reactants in glass-sealed tubes placed in a domestic microwave oven. Since then, research on the application of microwaves to organic synthesis has increased rapidly. Initially, devices converted from domestic microwave ovens were widely used. Subsequently, a compact microwave irradiation device for chemical synthesis was developed that could be used to detect and automatically control the temperature and pressure of the reaction solution. Since these microwave synthesizers have become more widely used, they have attracted attention from the viewpoint of green chemistry (Bassyouni et al., 2012).

H. Kimura (✉)
Agency for Health, Safety and Environment, Kyoto University, Kyoto, Japan

Division of Probe Chemistry for Disease Analysis, Research Center for Experimental Modeling of Human Disease, Kanazawa University, Kanazawa, Japan
e-mail: kimura.hiroyuki.4u@kyoto-u.ac.jp

Y. Kondo
Advanced Clinical Research Center, Fukushima Global Medical Science Center, Fukushima Medical University, Fukushima, Japan

Y. Yagi
Department of Radiological Technology, Faculty of Medicinal Science, Kyoto University of Medical Science, Nantan, Kyoto, Japan

T. Kozaka
Division of Probe Chemistry for Disease Analysis, Research Center for Experimental Modeling of Human Disease, Kanazawa University, Kanazawa, Japan

R. M. van Dam et al. (eds.), *Automated Technologies for the Development and Production of Radiopharmaceuticals*,
https://doi.org/10.1007/978-3-031-84632-8_14

A microwave is a type of electromagnetic wave that propagates owing to changes in electric and magnetic fields. Electromagnetic waves are classified into X-rays, ultraviolet rays, visible light, infrared rays, and radio waves depending on the frequency (wavelength). Microwaves are radio waves with frequencies ranging from 300 MHz to 300 GHz (wavelength ranging from 1 m to 1 mm, respectively). Microwaves are used in radar and satellite TV broadcasting, and the most familiar use of microwaves is in domestic ovens to heat food, liquids, or objects. Furthermore, microwaves are used in medicine and science as an energy source for clean heating, which does not require a heat-transfer medium. Since the energy of an electromagnetic wave is proportional to its frequency and inversely proportional to its wavelength ($E = h\nu = hc/\lambda$; h: Planck's constant, ν: frequency, c: speed of light, and λ: wavelength), the energy of microwaves is much lower than those of X-rays, UV–visible light, and infrared rays, which have higher frequencies. Therefore, in principle, microwave irradiation cannot activate atomic structures or directly activate or break chemical bonds.

However, microwave irradiation has been reported to promote numerous chemical reactions. In the heating method utilized in general synthetic reactions, the reaction vessel is heated by an external heat source that heats the material by convection, conduction, or radiation. In this case, the heat moves from the outside of the container near the heat source to the inside. Conversely, in microwave heating, when the object to be heated is placed in the electric field of the microwave, the molecules (permanent dipoles) constituting the object to be heated are subjected to the force of a radio wave (electric field), which displaces and polarizes the electrically neutral permanent dipoles, causing them to vibrate violently depending on the microwave frequency. In this case, frictional heat is generated between the molecules, causing the entire dielectric to heat up and become warmer, which is the general concept of "dielectric heating." Furthermore, microwave heating is characterized by its ability to heat only specific materials selectively and rapidly, with less influence from heat conduction and convection, in contrast to heating with an external heater. Therefore, microwave heating is sometimes described as "internal heating," considering the temperature distribution of the sample is different from normal heating. Conventional heating (CH) is described as "external heating" (Fig. 1).

The choice of solvent is important for reactions under microwave irradiation. Dimethyl sulfoxide, ethanol, methanol, water, *N,N*-dimethylformamide, and acetonitrile are commonly used reaction solvents in radiochemistry, and all exhibit reasonably good absorption of microwaves (Table 1). Several excellent review articles have been published (Kappe, 2004) that elaborate on the subject. The application of microwave heating to chemical reactions offers many advantages, including shorter reaction times, lower reaction temperatures, higher synthetic yields, suppression of side reactions, reaction without catalysts, reaction without solvents, heating under vacuum, reduced pressure or pressurization, and selective heating. However, there are some disadvantages, such as uneven heating due to local heating or overheating, electrical discharges when there are metal fragments in the heated object, and communication equipment malfunctions owing to leaked radio waves. Although the use of microwaves has great potential, much is still not understood regarding the

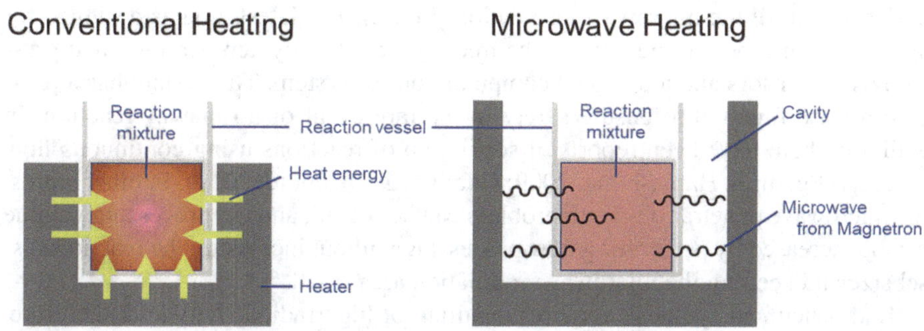

Fig. 1 Conventional heating and microwave heating

Table 1 Solvent boiling point and microwave absorbance (Kappe & Stadler, 2005)

Solvent	Boiling point (°C)	Microwave absorbance
Ethylene glycol	197	Very high
Dimethyl sulfoxide	189	High
Ethanol	78	High
Methanol	63	High
Water	100	Medium
1-Methyl-2-pyrrolidone	204	Medium
N,N-dimethylformamide	154	Medium
1,2-dichlorobenzene	180	Medium
Acetonitrile	81	Medium
Dichloromethane	40	Low
Tetrahydrofuran	66	Low
Toluene	110	Very low

mechanisms that promote chemical reactions. The elucidation of these mechanisms can enable the application of microwaves in several chemical reactions.

2 Characteristics of Microwave Reactors

The frequency of microwave heating devices in practical use ranges from 915 MHz to 2450 MHz, to avoid interference with radar and telecommunications. Microwave heating devices with a frequency of 2450 MHz, including microwave ovens, are dominant. Previously, reactions were performed using a multimode device converted from a household microwave oven. However, in these systems, complex reflections occur in the cavity, resulting in ineffective microwave irradiation of the irradiated object and poor reproducibility. Therefore, a single-mode device was developed, which allows irradiation under constant field strength (Table 2). In single-mode devices, the focus of the standing wave formed in the waveguide is aligned with the reaction vessel placed at the end of the waveguide, which makes it easier to control the microwave irradiation (Fig. 2). The latest models designed for synthesis can monitor and control the microwave intensity, temperature, and pressure using sensors and a personal computer control system. The system has a pressure release function to ensure safety in the rare event of a runaway reaction. In addition, there have been reports of scaling up of reactions using continuous-flow systems (Egami & Hamashima, 2009; Odajima & Okamoto, 2019). In other words, the microwave penetration-depth problem can be solved, and the processing volume can be increased by performing flow processing without increasing the reaction vessel size and keeping the microwave irradiation area small.

Radiochemical synthesis requires handling of high radioactivity and, therefore, uses computer-controlled automated synthesis equipment in a hot cell surrounded by lead walls to reduce worker radiation exposure. Therefore, the microwave reaction device must have a compact size to fit inside a hot cell, simple operability to enable automated synthesis, and the ability to perform microwave irradiation with a small volume of solvent, considering the amount of radioactive compounds is very small. A number of microwave radiolabeling reactions have been reported, and the reader is referred to books (Loupy, 2006, Wiley-VCH; Kappe, 2004, Wiley-VCH second, Completely Revised and Enlarged Edition) and reviews (Mats Larhed et al., 2002; Kumari et al., 2017; Alcázar et al., 2007) for more information on these reactions. Microwave synthesizers widely used in radiochemistry include the CEM Corporation (Discover®, DiscoverSP, Matthews, NC, USA), Biotage AB (Biotage

Table 2 Comparison of multimode and single-mode microwave reactors

Multimode	Single mode
Heating occurs by diffuse reflection of radio waves in a closed space (similar to a microwave oven)	Heating by standing waves (constructive interference of incident and reflected waves)
A stirring mechanism for radio waves is used to prevent uneven heating (e.g., turntable)	The object to be heated is positioned at the maximum electric or magnetic field
The electric and magnetic field distributions cannot be controlled	The electric and magnetic fields can be separately controlled

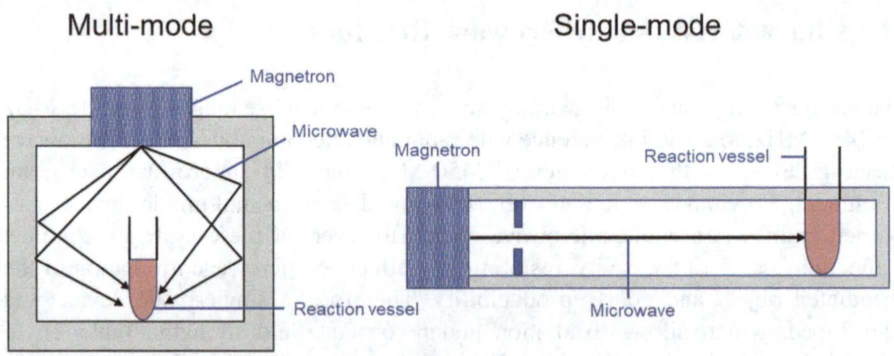

Fig. 2 Schematic of multimode and single-mode microwave reactors

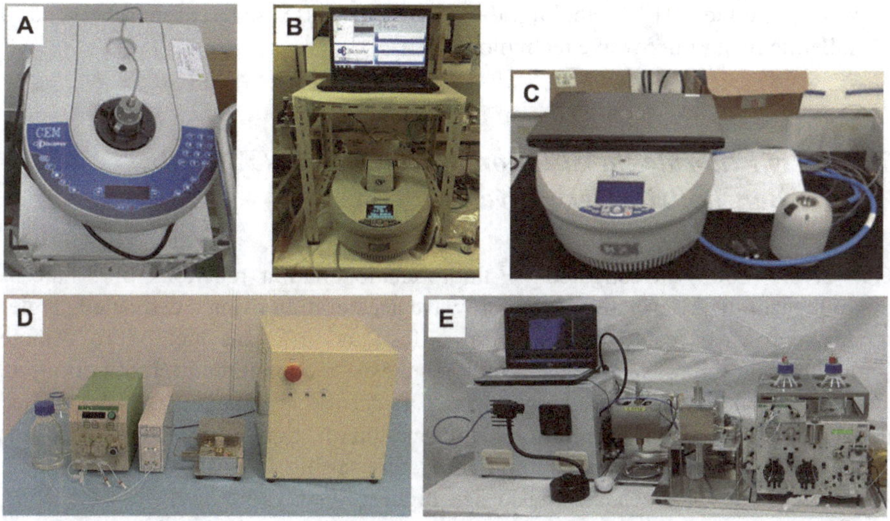

Fig. 3 Microwave synthesizers used by the author. (**a**) Discover® (CEM Corp.); (**b**) DiscoverSP (CEM Corp.); (**c**) PETWave (CEM Corp.); (**d**) resonant-type microwave reactor (SAIDA FDS Inc.); and (**e**) flow microwave reactor (SAIDA FDS Inc.)

Initiator+ Sixty or Initiator+ Eight, Uppsala, Sweden), and Resonance Instruments Inc. (Model 521-5, Stokie, IL, USA). The authors have conducted research using synthesis equipment from CEM Corporation (Discover®, DiscoverSP, PETWave), Biotage AB (Biotage Initiator+), and SAIDA FDS Inc. (resonant-type microwave reactor, flow microwave reactor) (Fig. 3). This review focuses on microwave labeling reactions using the isotopes F-18, C-11, Ga-68, and Tc-99 m, which are frequently used in radiochemistry.

3 ^{18}F-Labeling Using Microwave Heating

Fluorine-18 (F-18) has received wide attention owing to its favorable nuclear and physical characteristics, including a high positron decay ratio (97%), relatively short half-life (109.8 min), and low positron energy (maximum: 0.635 MeV). The low positron energy results in a short range (2.4 mm maximum range) that favorably increases the resolution of positron emission tomography (PET) images. (Purser et al., 2008) Because of these nuclear properties and the ability to synthesize F-18 in large quantities, many radiosynthetic methods have been developed to incorporate this radionuclide into biologically important and interesting molecules. Furthermore, owing to their short half-lives, the development of radiosynthetic schemes which introduce radionuclides in the late stage of the synthetic pathway has been emphasized. Microwaves are highly effective in decreasing the reaction time and improving the yield of ^{18}F-labeling; this section discusses several reports related to ^{18}F-labeling using microwave technology.

3.1 [^{18}F]Fluoride Substitution of Meta-Halo (F, Cl, Br, I)-Benzonitrile Derivatives

Guo et al. (2008) investigated ^{18}F-fluorodehalogenation reactions of 3-halo-5-methylbenzonitrile derivatives by microwave acceleration using a dedicated single-mode microwave reactor. To develop a fast and efficient method of remote semiautomatic microwave-induced radiosynthesis, a commercial radiofluorination chemistry module was adapted to work with a microwave apparatus with a detached resonance cavity. The nucleophilic ^{18}F-fluorination reaction is outlined in Scheme 1. Three 150 W, 20 s pulses (maximum temperature setting was 145 °C) were applied, with a 30 s pause between each pulse. The highest ^{18}F-incorporation was 64%, 13%, and 9% for the fluoro, bromo, and chloro precursors, respectively, after three pulses in DMSO. No ^{18}F-labeled product was obtained with the iodo precursor in either DMF or DMSO under any condition.

3.2 [^{18}F]Flumazenil from Nitro Precursor

Mandap et al. (2009) reported an advancement in the radiosynthesis of [^{18}F]Flumazenil to improve the radiolabeling process via microwave dielectric heating using a single-mode microwave reactor. The radiochemical yield (RCY) was 26 ± 4% [with a radiochemical purity (RCP) of >99%, a total synthesis time of 55–60 min and a molar activity of (A_m) ~295 GBq/μmol (end of synthesis)] (Scheme 2). Compared to previous studies (Ryzhikov et al., 2004; Ryzhikov et al., 2005) that used conventional heating, the microwave synthesis improved radiochemical yield and reduced the synthesis time by 20 min.

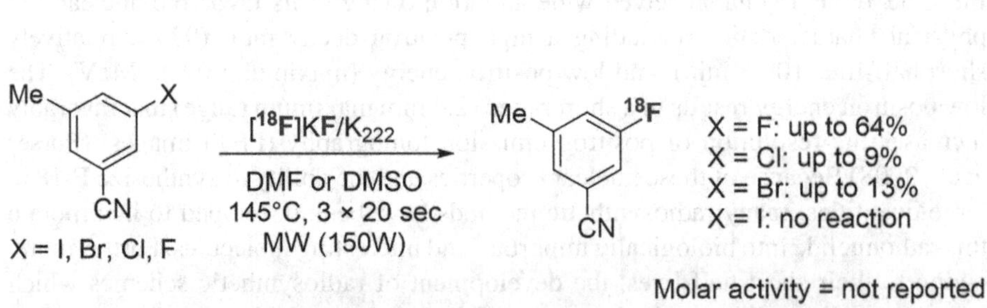

Scheme 1 Microwave-assisted ^{18}F-fluorodehalogenation reactions of 3-halo-5-methylbenzonitrile derivatives by Guo et al. (2008). Microwave equipment: Microwave (MW) irradiation was performed in a single-mode microwave reactor (Model 521, Resonance Instruments Inc.)

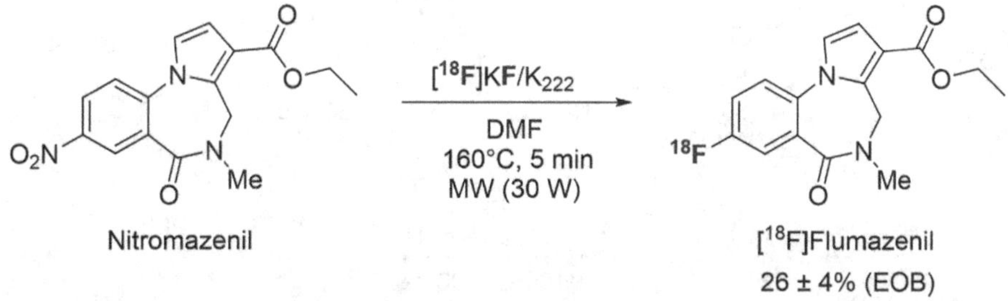

Scheme 2 Synthesis of [18F]Flumazenil from the nitro precursor using a microwave reactor by Mandap et al. (2009). Microwave equipment: Microwave irradiation was performed in a single-mode microwave reactor (Discover®, CEM Corp.) equipped with an infrared temperature control system located below the microwave cavity floor

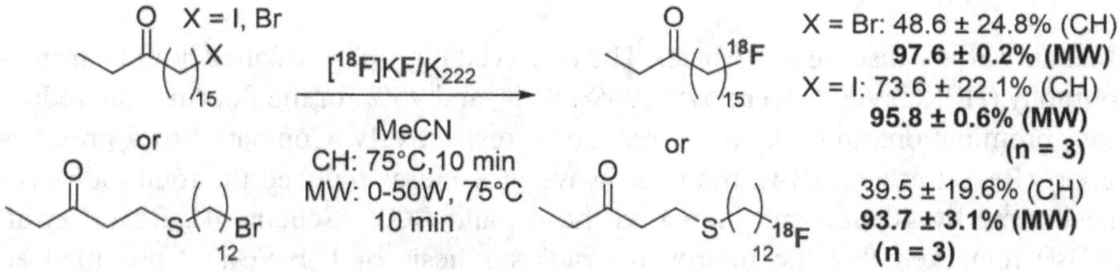

Scheme 3 [18F]Fluoride incorporation into [18F]fluoro fatty acid esters using a microwave reactor (MW) compared to conventional heating (CH) by Belanger et al. (2011). Microwave equipment: Microwave irradiation was performed in a single-mode small microwave reactor system (PETWave; CEM Corp.)

3.3 [18F]Fluoro Fatty Acid Esters

Belanger et al. (2011) reported [18F]fluoride incorporation using microwave heating and conventional heating for three different palmitate derivatives; methyl 16-bromo-hexadecanoate (BP), methyl 16-iodo-hexadecanoate (IP), and methyl 3-(12-bromododecylthio)propionate) (BTP). As shown in Scheme 3, for each derivative, microwave heating improved the RCY and repeatability (BP: 48.6 ± 4.8% → 97.6 ± 0.2%, IP: 73.6 ± 22.1% → 95.8 ± 0.6%, and BTP: 39.5 ± 19.6% → 93.7 ± 3.1%; $n = 3$).

3.4 4-[18F]Fluorobenzyltriphenylphosphonium Cation Over Multiple Steps

Ravert et al. (2014) reported rapid microwave-assisted fluorination of lipophilic 4-[18F]fluorobenzyltriphenylphosphonium cation ([18F]FBnTP), a radiotracer that is useful for studying mitochondrial dysfunction in numerous diseases, including

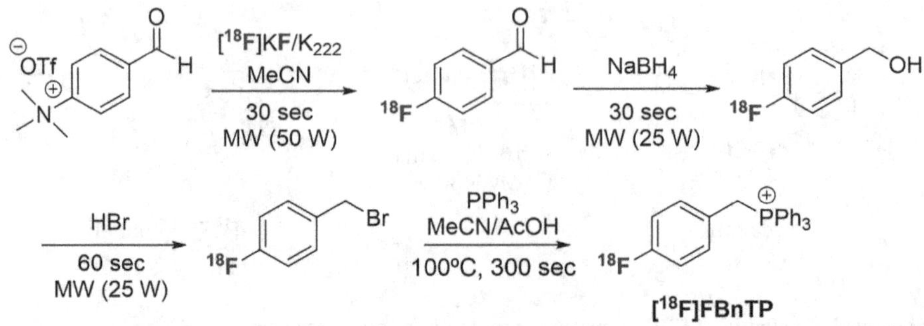

Scheme 4 4-[^{18}F]Fluorobenzyltriphenylphosphonium cation ([^{18}F]FBnTP) microwave radiosynthesis by Ravert et al. (2014). Microwave equipment: Microwave irradiation was performed in a single-mode microwave irradiation system (Model 521 fitted with a Model 521-5; Resonance Instruments) reactor cavity for a 5-mL v-vial modified for air cooling

Coronary artery disease and cancer. The observed high-performance liquid chromatography (HPLC) yields were 57%, 95%, 85%, and 95% for the fluorination, reduction, bromination, and alkylation reactions, respectively. Compared to a previous report (Ravert et al., 2014), the microwave procedures reduced the total radiosynthesis time by 40 min and increased the A_m and RCY (Scheme 4). Ravert et al. (2014) remarked that the microwave radiosynthesis of [^{18}F]FBnTP provided an automated method for the radiotracer production that simplified and eliminated steps of the original synthesis that were burdensome. Radiochemically pure products (>99%) were obtained in 8.3 ± 2.4% yield (RCY) with an average synthesis time of 52.4 ± 14 min (n = 27) and average A_m of 534.5 ± 371.4 GBq/μmol.

3.5 *Tropane Analogs [^{18}F]PR04.MZ and [^{18}F]LBT999*

Riss and Roesch (2009) reported a microwave-based high-yield direct fluorination method for the synthesis of the moderate-potency tropane analog dopamine transporter inhibitor [^{18}F]LBT999 and the recently described [^{18}F]PR04.MZ for molecular imaging of dopaminergic signaling. The RCY of [^{18}F]PR04.MZ after 45 s of heating was 60–80%, and the activity yield was 34 ± 2%. [^{18}F]LBT999 was obtained under the same conditions with a 27 ± 2% activity yield. As a result, microwave irradiation significantly improved aliphatic nucleophilic radiofluorination compared to CH (Scheme 5).

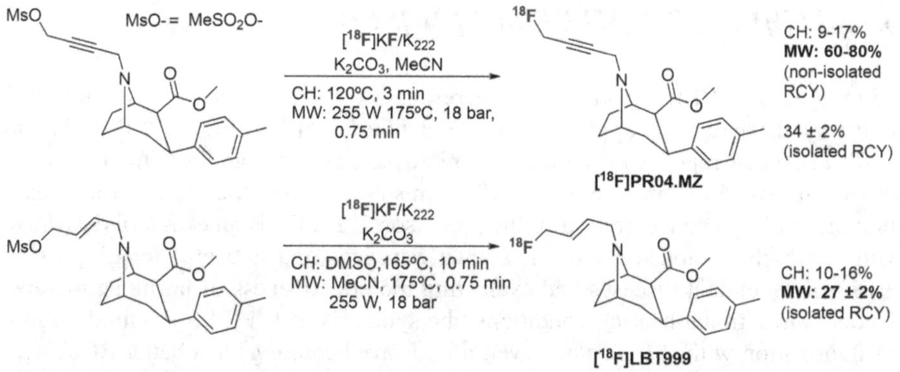

Scheme 5 Microwave radiosynthesis of [^{18}F]PR04.MZ and [^{18}F]LBT999 by Riss et al. (2009). Microwave equipment: Microwave irradiation was performed in a single-mode microwave irradiation system (Discover®, CEM Corp.)

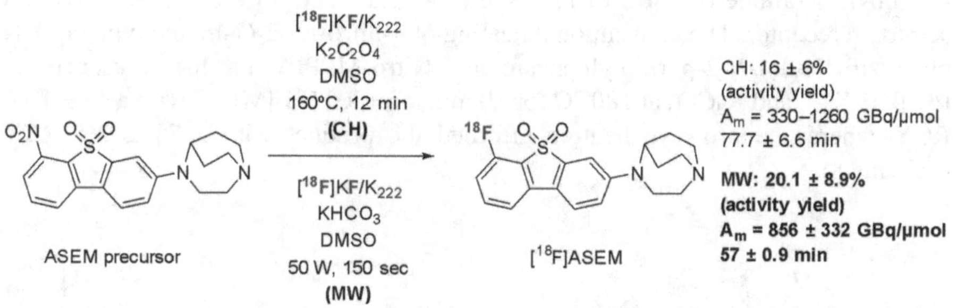

Scheme 6 [^{18}F]ASEM radiosynthesis conducted via conventional heating or microwave by Ravert et al. (2015). Microwave equipment: Microwave irradiation was performed in a single-mode microwave irradiation system (Model 521 fitted with a Model 521-5; Resonance Instruments) reactor cavity for a 5-mL v-vial modified for air cooling

3.6 3-(1,4-Diazabicyclo[3.2.2]Nonan-4-yl)-6-[^{18}F]-Fluorodibenzo[b,d] Thiophene 5,5-Dioxide ([^{18}F]ASEM)

Ravert et al. (2015) used a custom-made radiosynthesis-module-equipped microwave system (Model 5218, Resonance Instruments) for the microwave-assisted synthesis of 3-(1,4-diazabicyclo[3.2.2]nonan-4-yl)-6-[^{18}F]-fluorodibenzo[b,d] thiophene 5,5-dioxide ([^{18}F]ASEM), which is used for imaging the α7-nicotinic acetylcholine receptor. Compared to conventional heating methods, which achieved a 16 ± 6% activity yield (n = 14) with an A_m of 330–1260 GBq/µmol at EOS in 77.7 ± 6.6 min, the microwave-assisted radiosynthesis resulted in a [^{18}F]ASEM activity yield of 20.1 ± 8.9% with an average synthesis time of 57 ± 0.9 min (n = 6) and an average A_m of 856 ± 332 GBq/µmol (Scheme 6). The microwave-assisted radiosynthesis of [^{18}F]ASEM proved to be a reliable, simplified, remote, and semi-automated method that improves the yield from the original synthesis.

3.7 [18F]FES, [18F]MPPF, and [18F]SFB

Kimura et al. (2014) reported a novel resonant-type microwave reactor for PET tracer synthesis and its application in the syntheses of [18F]FES, [18F]MPPF, and [18F]SFB. Their report showed that the microwave reactor can heat small volumes of solvents (0.05–0.2 mL) used in PET chemistry, making it possible to run a reaction using a high concentration of the precursor. [18F]FES is an estradiol derivative with the highest bioactivity of all known estrogens and is useful for diagnosing estrogen-dependent diseases and evaluating the effectiveness of hormone therapy. Under conventional heating conditions, the synthesis of [18F]FES required 10 min of fluorination with [18F]KF; however, microwave heating gave a better RCY, with each reaction requiring only 1 min. Furthermore, Kimura et al. (2014) reported that the RCY of [18F]FES remained unchanged with lower amounts of precursor until the amount of the precursor was decreased by 75% (Scheme 7).

4-[18F]-Fluoro-*N*-[2-(1-methoxyphenyl)-1-piperazinyl] ethyl-*N*-2-pyridinylbenzamide ([18F]MPPF) is used as a PET tracer to evaluate the 5-HT1A serotonin receptor. The conventional heating of 4-nitro-*N*-[2-(1-methoxyphenyl)-1-piperazinyl]ethyl-*N*-2-pyridinylbenzamide (Nitro-MPPF) in the presence of [18F]KF, K_{222}, and K_2CO_3 at 180 °C for 10 min afforded [18F]MPPF with a 47 ± 10% RCY, whereas microwave heating afforded the product with a 74 ± 2% RCY (Scheme 8).

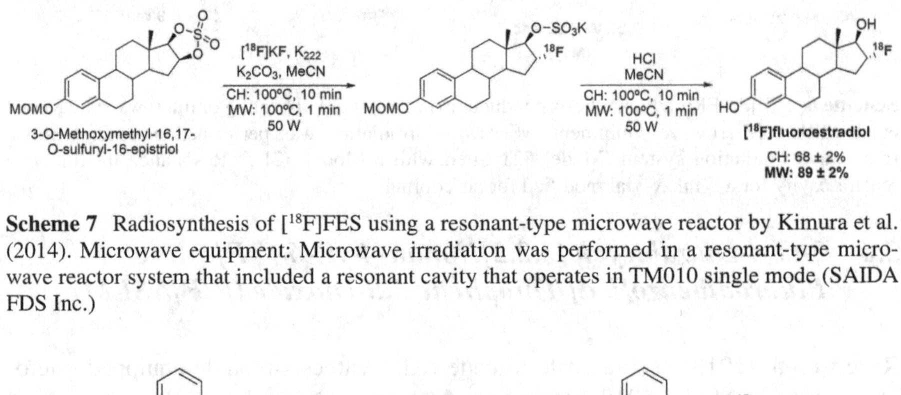

Scheme 7 Radiosynthesis of [18F]FES using a resonant-type microwave reactor by Kimura et al. (2014). Microwave equipment: Microwave irradiation was performed in a resonant-type microwave reactor system that included a resonant cavity that operates in TM010 single mode (SAIDA FDS Inc.)

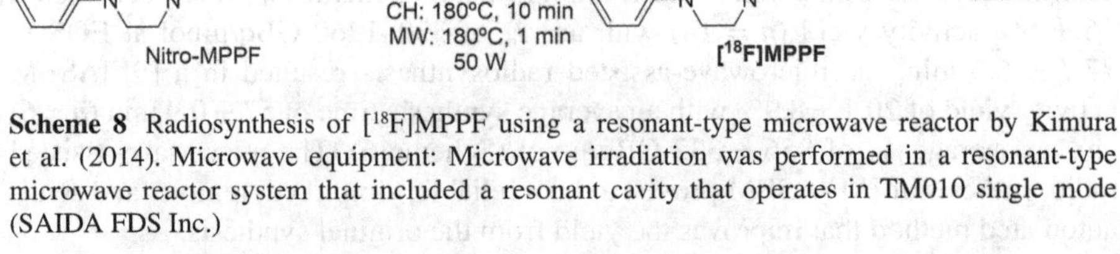

Scheme 8 Radiosynthesis of [18F]MPPF using a resonant-type microwave reactor by Kimura et al. (2014). Microwave equipment: Microwave irradiation was performed in a resonant-type microwave reactor system that included a resonant cavity that operates in TM010 single mode (SAIDA FDS Inc.)

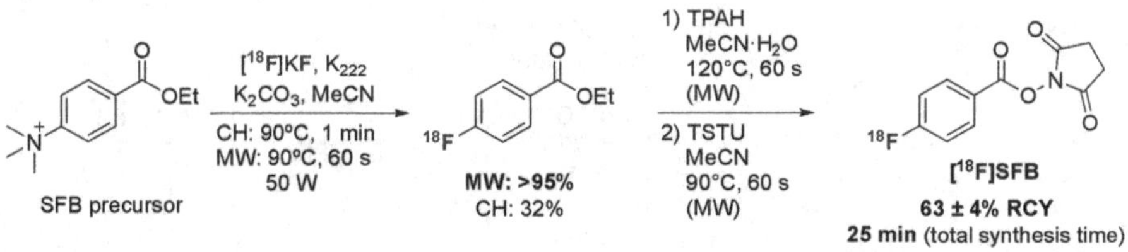

Scheme 9 Radiosynthesis of [^{18}F]SFB using a resonant-type microwave reactor by Kimura et al. (2014). Microwave equipment: Microwave irradiation was performed in a resonant-type microwave reactor system that included a resonant cavity that operates in TM010 single mode (SAIDA FDS Inc.)

N-Succinimidyl 4-[^{18}F]fluorobenzoate ([^{18}F]SFB) has been used for ^{18}F-labeling of bioactive peptides and nucleic acids. The synthesis of [^{18}F]SFB requires three steps: (1) fluorination of the SFB precursor with [^{18}F]KF in the presence of K_{222} and K_2CO_3, (2) deprotection with tetrapropylammonium hydroxide (TPAH), and (3) esterification using *N,N,N′,N′*-tetramethyl-*O*-(*N*-succinimidyl) uranium tetrafluoroborate (TSTU). For ^{18}F-fluorination, after 1 min under conventional heating conditions, the reaction proceeded to 32% conversion, whereas microwave heating gave a conversion in excess of 95% after the same reaction time. Subsequent sequential deprotection and esterification were performed under the same microwave conditions as the fluorination step (50 W, 60 s) to give the crude product. Finally, purification by solid-phase extraction (SPE) afforded [^{18}F]SFB with an RCY of 63 ± 4%, with an RCP >95% after a total synthesis time of 25 ± 5 min. Overall, this result clearly demonstrates the effectiveness of microwave heating using the resonant-type reactor while significantly improving the total synthesis of [^{18}F]SFB (Scheme 9).

3.8 [^{18}F]Pitavastatin Using the Suzuki Coupling with [^{18}F]FIB

Yagi et al. (2015) used a resonant-type microwave reactor for synthesizing the labeling reagent [^{18}F]4-fluoroiodobenzene ([^{18}F]FIB) and the hepatic organic anion-transporting polypeptide tracer [^{18}F]pitavastatin via Suzuki coupling with [^{18}F]FIB. A mixture of 4-iodophenyldiphenylsulfonium triflate, [^{18}F]KF, and K_2CO_3 in MeCN was heated under microwave irradiation at 120 °C in the presence of K_{222} to give [^{18}F]FIB with 56 ± 3% RCY. The Suzuki coupling reaction of [^{18}F]FIB with pitavastatin precursor was performed under conventional heating, which gave [^{18}F]pitavastatin with 18 ± 4% RCY. Conversely, the microwave heating condition gave 28 ± 3% RCY (Scheme 10).

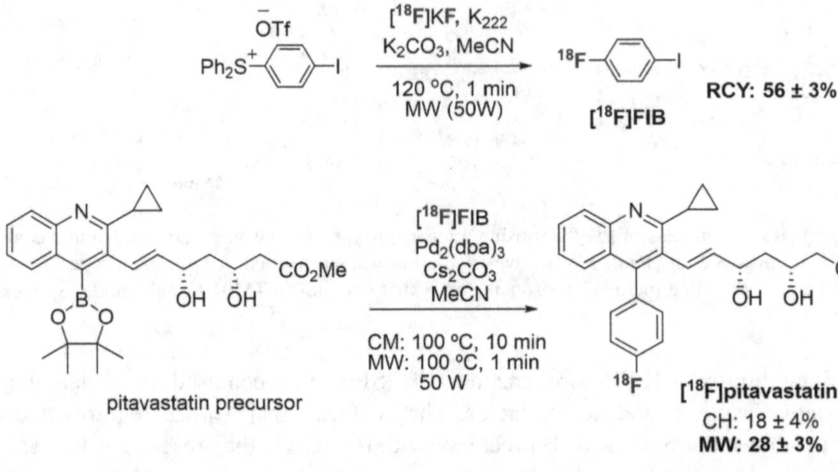

Scheme 10 Radiosynthesis of [^{18}F]4-fluoroiodobenzene ([^{18}F]FIB) and [^{18}F]pitavastatin using the resonant-type microwave reactor by Yagi et al. (2015). Microwave equipment: Microwave irradiation was performed in a resonant-type microwave reactor system comprising a resonant cavity operating in TM010 single-mode (SAIDA FDS Inc.)

3.9 [^{18}F]PTV-F1 with Multireaction Steps

Kimura et al. (2016) synthesized the hepatic organic anion-transporting polypeptide tracer [^{18}F]PTV-F1 using microwave heating in a resonant-type microwave reactor. The one-pot process had three steps, including aliphatic radiofluorination in the presence of K_{222}, deprotection under acidic conditions, and alkaline hydrolysis of the lactone compound. [^{18}F]PTV-F1 was obtained with 45 ± 3% RCY for 30 min (Scheme 11).

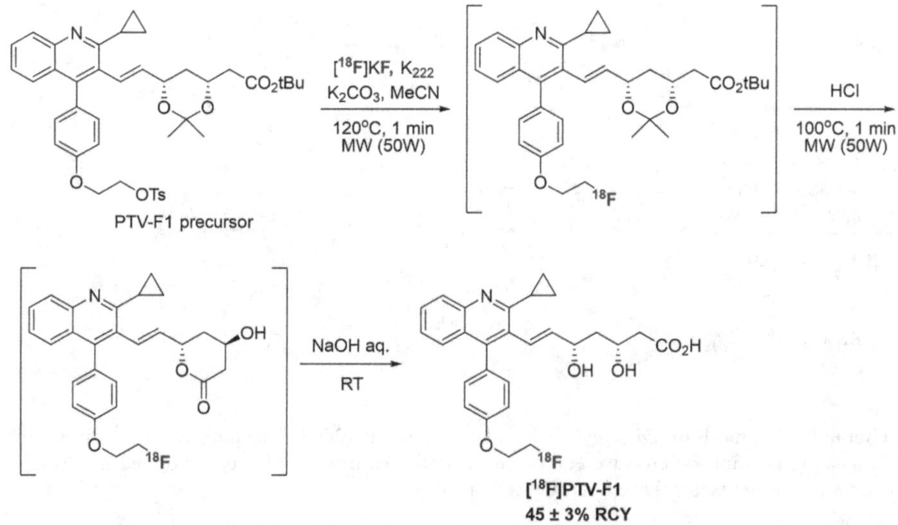

Scheme 11 Radiosynthesis of [^{18}F]PTV-F1 using a resonant-type microwave reactor by Kimura et al. (2016). Microwave equipment: Microwave irradiation was performed in a resonant-type microwave reactor system that included a resonant cavity operating in TM010 single-mode (SAIDA FDS Inc.)

4 ^{11}C-Labeling Using Microwave Heating

Owing to the ubiquity of carbon in biological molecules, radiolabeling with carbon-11 (C-11) can be achieved without changing the physicochemical or pharmacological properties of the target compound. C-11 is cyclotron-produced [^{14}N(p, α)^{11}C] and is collected either as [^{11}C]CO_2 or as [^{11}C]CH_4 depending on the trace amounts of oxygen or hydrogen gas added to the target. [^{11}C]CO_2 can be used directly. Furthermore, [^{11}C]CO_2 and [^{11}C]CH_4 can be transformed into other ^{11}C-labeled synthons, such as [^{11}C]CO, [^{11}C]CH_3OLi, [^{11}C]CH_3I, [^{11}C]CH_3OTf, and [^{11}C]HCN (McQuade et al., 2005; Scott, 2009; Taddei & Gee, 2018; Wilson et al., 2018). The half-life of C-11 is only 20.3 min, placing strict time constraints that influence the reaction parameters and labeling procedures. In several cases, compared to conventional heating methods, microwave heating yields the target compound in a shorter time. Therefore, microwave technology is used for ^{11}C-labeling. This section introduces several reports related to ^{11}C-labeling methods that use microwave technology.

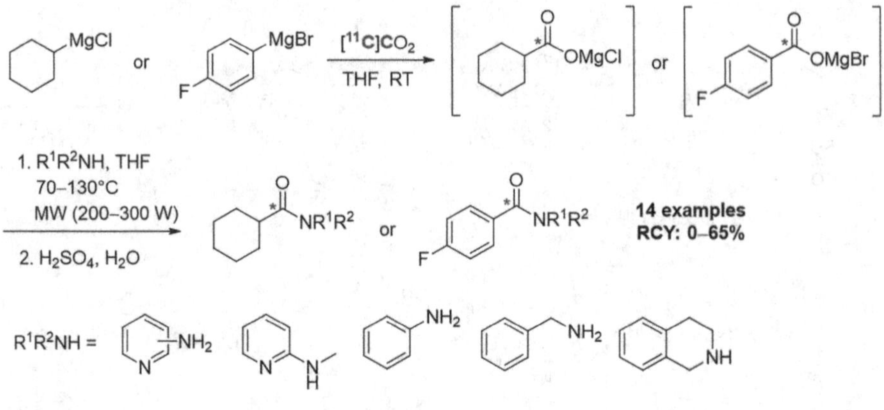

Scheme 12 Synthesis of [*carbonyl*-^{11}C]amides by Lu et al. (2003). The pink asterisk indicates the ^{11}C-labeling position. Microwave equipment: Microwave irradiation was performed in a single-mode microwave reactor (Discover®, CEM Corp.)

4.1 [Carbonyl-^{11}C]Amides

In the 1990s, Lasne et al. reported the synthesis of [^{11}C]aliphatic amides by treating magnesium halide carboxylates with amines in tetrahydrofuran (THF) in the presence of alkylmagnesium halides (Aubert et al., 1997; Perrio-Huard et al., 2000). Lu et al. (2003) reported the synthesis of [*carbonyl*-^{11}C]amides in one pot by the direct reaction of amines with [^{11}C]carboxymagnesium halides generated in situ from Grignard reagents and [^{11}C]carbon dioxide under microwave-enhanced conditions (Scheme 12). Two Grignard reagents, viz. cyclohexylmagnesium chloride and 4-fluorophenylmagnesium bromide, were used to investigate the scope of the reaction. The investigators conducted experiments wherein the RCYs of *N*-(pyridin-2-yl) cyclohexanecarboxamide and 4-fluoro-*N*-(pyridin-2-yl)benzamide were investigated as functions of time and temperature. Under identical conditions (70 °C for 10 min), the RCY of *N*-(pyridin-2-yl)cyclohexanecarboxamide with microwave heating (RCY = 55%) was higher than that with thermal heating (RCY = 37%). Additionally, under identical conditions (100 °C for 10 min), the RCY of 4-fluoro-*N*-(pyridin-2-yl)benzamide increased from 22% to 45% upon microwave heating versus thermal heating. Moreover, it was seen that cyclohexylcarboxymagnesium chloride was more reactive than 4-fluorophenylcarboxymagnesium bromide, and primary amines are significantly more reactive than secondary amines. In their study, a single-mode microwave reactor equipped with temperature and pressure sensors was used.

Lu et al. (2005) reported synthetic procedures for labeling S14506 (a 5-HT_{1A} receptor agonist) with C-11 or F-18 at three alternative positions ([*O-methyl*-^{11}C]S14506, [*carbonyl*-^{11}C]S14506, and [^{18}F]S14506) using classic radiolabeling reactions and a microwave-enhanced direct coupling of in situ generated [^{11}C]organo-carboxymagnesium bromide with a corresponding amine precursor (Scheme 13).

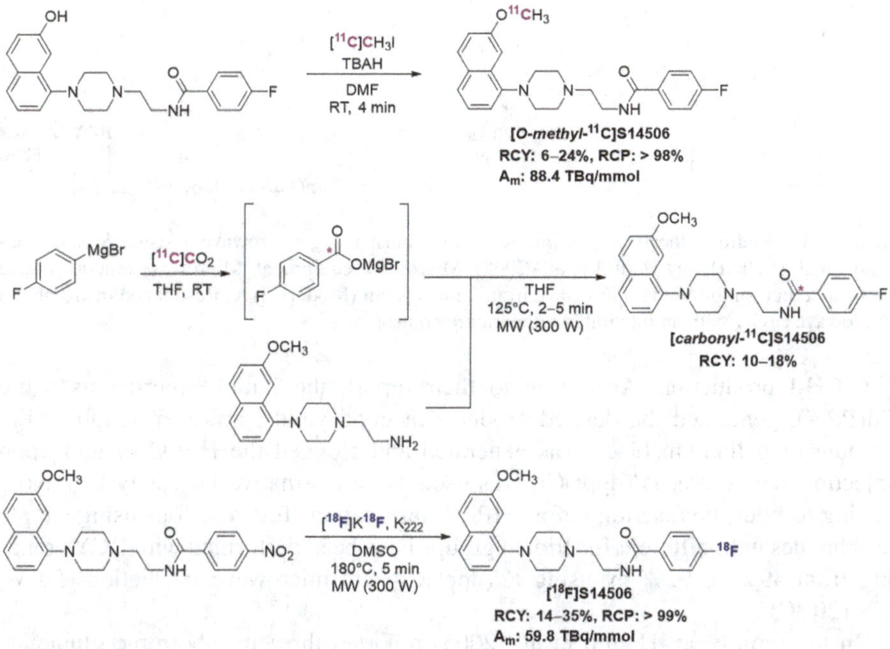

Scheme 13 Labeling of the 5-HT1A receptor agonist S14506 with C-11 or F-18 by Lu et al. (2005). Microwave equipment: Microwave irradiation was performed in a single-mode microwave reactor (Discover®, CEM Corp.)

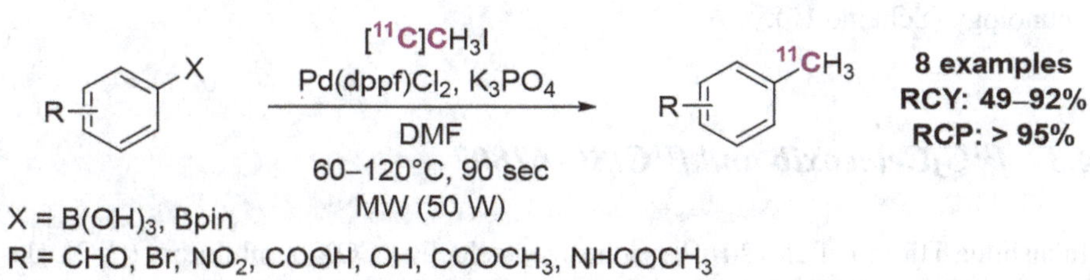

Scheme 14 Suzuki couplings with [^{11}C]methyl iodide ([^{11}C]CH_3I) using microwave technology by Hostetler et al. (2005). Microwave equipment: Microwave reactions were performed in a single-mode microwave irradiation system (Model 521A; Resonance Instruments) microwave cavity with an integrated temperature control

4.2 *Suzuki Couplings with [^{11}C]Methyl Iodide*

Hostetler et al. (2005) described the synthesis of substituted [^{11}C]toluene derivatives via Suzuki coupling with [^{11}C]methyl iodide (Scheme 14) as an alternative to Stille coupling. The avoidance of potentially toxic tin-containing byproducts is advantageous compared to Stille coupling. The investigators performed automated synthesis and the purification of the desired product in <20 min from the end of

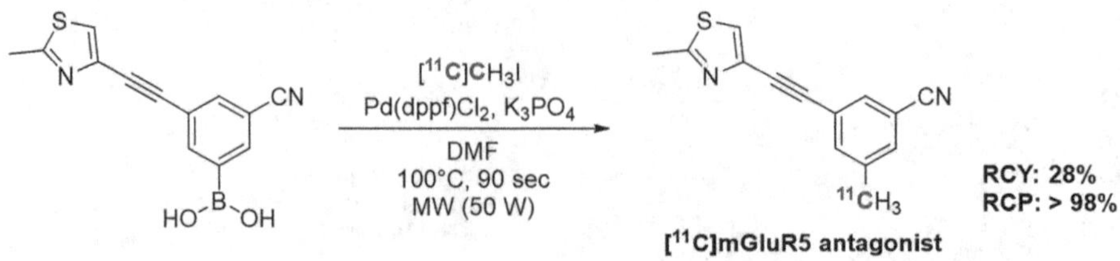

Scheme 15 Radiosynthesis of [^{11}C]mGluR5 antagonist using microwave-assisted Suzuki couplings with [^{11}C]CH_3I by Hamill et al. (2005). Microwave equipment: Microwave reactions were performed in a single-mode microwave irradiation system (Model 521A; Resonance Instruments) microwave cavity with an integrated temperature control

[^{11}C]CH_3I production. According to their report, the initial experiments using Pd(PPh_3)$_4$ generated the desired products in good yields; however, a substantial amount of palladium black was generated and clogged the HPLC system upon injection. Therefore, Pd(dppf)Cl_2 was used as an alternative Pd catalyst to avoid having to filter the reaction before HPLC purification. Test reactions using simple aryl halides with different functional groups have been performed with RCYs ranging from 49% to 92% by using Pd(dppf)Cl_2 and microwave irradiation (50 W, 60–120 °C).

In the same year, Hamill et al. (2005) reported three metabotropic glutamate receptor subtype 5 (mGluR5) PET radiotracers labeled with C-11 or F-18, and their in vitro and in vivo behaviors have been characterized in rhesus monkeys (Hamill et al., 2005). The [^{11}C]mGluR5 antagonist was synthesized via the corresponding aryl boronic acid using Suzuki coupling with [^{11}C]methyl iodide using microwave technology (Scheme 15).

4.3 [^{11}C]Celecoxib and [^{11}C]SC-62807

Takashima-Hirano et al. (2011) synthesized selective COX-2 inhibitors ([^{11}C]celecoxib) and a major metabolite of celecoxib ([^{11}C]SC-62807). The synthesis of [^{11}C]celecoxib was achieved without protecting the sulfonamide group by Pd(0)-mediated ^{11}C-methylation using a boronic precursor (Scheme 16). In their study, [^{11}C]SC-62807 was converted from [^{11}C]celecoxib via rapid oxidation in the presence of excess $KMnO_4$ using microwave technology. [^{11}C]SC-62807 was synthesized by conventional or microwave heating at 120 °C (RCY [no MW]: 32 ± 18% (n = 5), RCY [using MW]: 47 ± 11% (n = 4)), which showed that microwave irradiation increased the yield and reproducibility of the reaction. Finally, [^{11}C]SC-62807 was efficiently synthesized via oxidation of [^{11}C]celecoxib at 140 °C for 5 min under microwave irradiation with RCY 55 ± 9% (n = 3) based on [^{11}C]celecoxib (total synthesis time: 20 min).

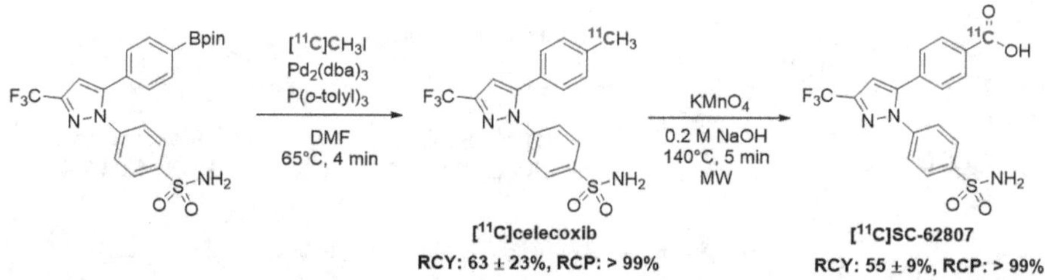

Scheme 16 Radiosynthesis of [^{11}C]celecoxib and [^{11}C]SC-62807 by Takashima-Hirano et al. (2011). Microwave equipment: Microwave irradiation was performed in a Biotage Initiator+ (Biotage AB) using a sealed vessel

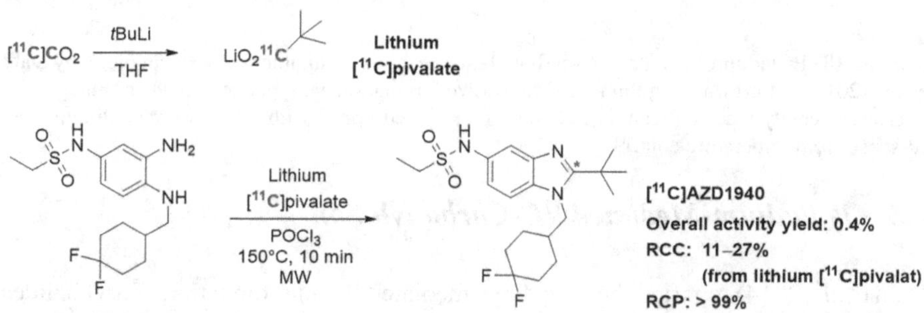

Scheme 17 Radiosynthesis of [^{11}C]AZD1940 using lithium [^{11}C]pivalate by Schou et al. (2013). The pink asterisk indicates the ^{11}C-labeling position. RCC = radiochemical conversion. Microwave equipment: Microwave irradiation was performed in a single-mode microwave irradiation system (Discover®, CEM Corp.)

4.4 [^{11}C]AZD1940

As part of the preclinical evaluation of AZD1940 (a candidate drug for the treatment of neuropathic pain), [^{11}C]AZD1940 was synthesized for a microdosing study using PET (Schou et al., 2013). Schou et al. achieved the retrosynthesis of [^{11}C]AZD1940 and tested three approaches using [^{11}C]methyl iodide, [^{11}C]ethanesulfonyl chloride, and lithium [^{11}C]pivalate. Consequently, sufficient amounts of [^{11}C]AZD1940 for microdosing studies were obtained by radiolabeling using a lithium salt of [^{11}C]pivalic acid (Scheme 17). The lithium salt of [^{11}C]pivalic acid was readily obtained via the carboxylation of *tert*-butyl lithium. Subsequent amidation and dehydration were performed in MW using an *o*-phenylene diamine analog of AZD1940 and lithium [^{11}C]pivalate in neat phosphorous oxychloride. The radiochemical conversion (RCC) (from lithium [^{11}C]pivalate) was 11–27%. The overall activity yield of [^{11}C]AZD1940 (RCP > 99%) was 0.4%, and the A_m was 13 GBq/μmol at time of administration (58 min after EOB).

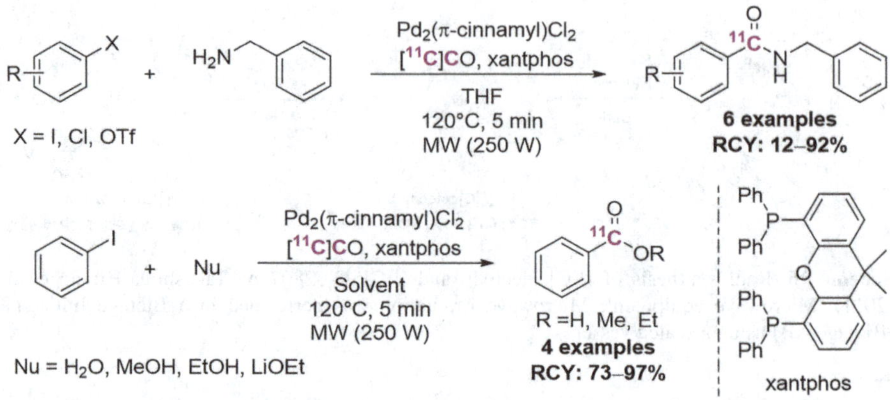

Scheme 18 Palladium-mediated ^{11}C-carbonylation reaction using microwave technology by Dahl et al. (2014). Microwave equipment: Microwave irradiation was performed in a single-mode microwave cavity reactor (Scansys Laboratorieteknik) equipped with air-pressure cooling and an IR sensor for temperature control

4.5 Palladium-Mediated ^{11}C-Carbonylation

Dahl et al. (2014) reported the palladium-mediated ^{11}C-carbonylation of aryl halides and triflates at ambient pressure using microwave heating (Scheme 18). The RCY of 3-nitrobenzyl-[*carbonyl*-^{11}C]benzamide was investigated as a function of time and temperature. Under identical conditions (100 °C for 5 min), RCY increased with microwave heating (RCY = 62%) compared to conventional heating (RCY = 46%). Additionally, microwave heating at 100 °C for 5 min using active cooling improved the RCY to 76%. Finally, the target compound was obtained with RCY of 81% under optimal conditions (120 °C for 5 min). This approach improved the RCY of the ^{11}C-aminocarbonylation of electron-deficient aryl halides compared to that obtained by conventional heating. Furthermore, the authors extended the scope of this reaction to include ^{11}C-labeled aryl acids and two ^{11}C-labeled aryl esters. The results also demonstrate the utility and potential of [^{11}C]CO as a labeling precursor for PET applications.

4.6 Carbonyl Cobalt-Mediated Aryl Acetylation Using [^{11}C] Methyl Iodide

Dahl et al. (2016) reported the direct synthesis of ^{11}C-labeled aryl methyl ketones using a microwave-assisted and carbonyl cobalt-mediated carbonylative protocol employing [^{11}C]methyl iodide (Scheme 19). Co_2CO_8 was used as both the aryl halide activator and the carbon monoxide source for the carbonylation reaction. Their experiments with iodobenzene as substrate using conventional heating

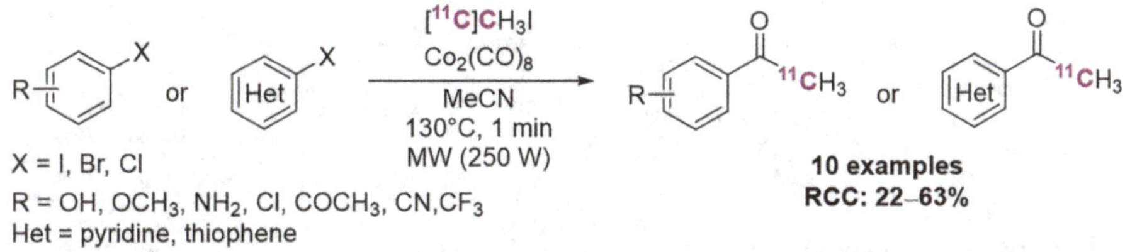

Scheme 19 Carbonyl cobalt-mediated aryl acetylation using [^{11}C]methyl iodide by Dahl et al. (2016). RCC = radiochemical conversion. Microwave equipment: Microwave irradiation was performed in a single-mode microwave cavity reactor (Scansys Laboratorieteknik) equipped with air-pressure cooling and an IR sensor for temperature control

(100 °C) produced [^{11}C]acetophenone with RCC of 40%. When using microwave heating (100 °C), an improved RCC (54%) was observed. The suitability of the methodology was demonstrated by the ^{11}C-labeling of a set of functionalized arenes and heteroarenes. Aryl chlorides with electron-withdrawing groups at the *para* position gave good RCCs (41–59%). This reaction provided a fast and simplified route to ^{11}C-labeled aryl and heteroaryl methyl ketones.

4.7 [^{11}C]AZD3241

Johnström et al. (2015) described a radiosynthesis method for [^{11}C]AZD3241 to assess the brain uptake of AZD3241 in nonhuman primates by PET microdosing. The myeloperoxidase inhibitor AZD3241 was developed as a candidate drug to delay the progression of neurodegenerative brain disorders. AZD3241 was labeled with carbon-11 at the thio-carbonyl position starting from [^{11}C]potassium cyanide in a four-step procedure using microwave-assisted heating (Scheme 20). [^{11}C]Hydrogen cyanide was produced online from [^{11}C]carbon dioxide and trapped in a solution of acetone containing potassium hydroxide and sulfur (S_8). The reaction mixture was heated under microwave irradiation, and benzoyl chloride was added to give [^{11}C]**3**. Benzoyl [^{11}C]**3** subsequently reacts with precursor (**4**). According to their report, large amounts of **4** (10–13 mg) were needed to obtain adequate conversion to [^{11}C]**5**. After the reaction mixture was purified on a solid-phase extraction (SPE) cartridge, [^{11}C]**5** was cyclized in an aqueous potassium carbonate solution containing ethanol using microwave heating to produce [^{11}C]AZD3241. The total synthesis time of [^{11}C]AZD3241 was 60 min. The A_m was 9 ± 4 GBq/μmol, and the RCP was >98%. PET measurements were performed in three cynomolgus monkeys to assess [^{11}C]AZD3241 brain uptake. Consequently, Johnström et al. confirmed adequate brain exposure to support the translation of AZD3241 to Phase 2a studies in patients.

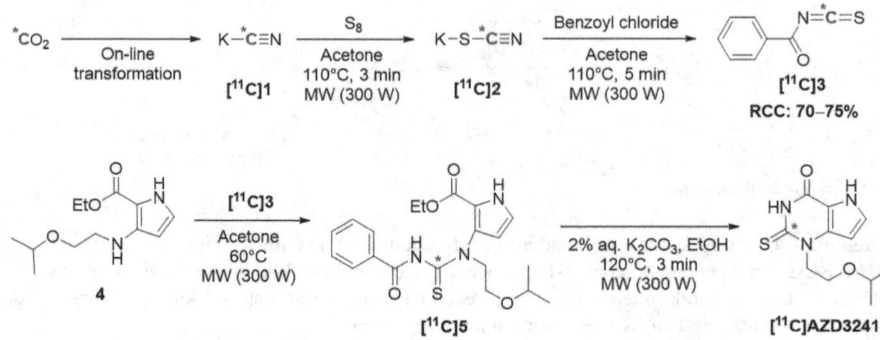

Scheme 20 Multistep radiosynthesis of [^{11}C]AZD3241 by Johnström et al. (2015). The pink asterisk indicates the ^{11}C-labeling position. Microwave equipment: Microwave irradiation was performed in a single-mode microwave irradiation system (Discover®, CEM Corp.)

5 ^{68}Ga-Labeling Using Microwave Heating

Among PET nuclides, gallium-68 (Ga-68) has a moderate half-life ($T_{1/2}$: 67.7 min) and can be produced by a ^{68}Ge/^{68}Ga generator (^{68}Ge $T_{1/2}$: 270.95 d). Therefore, PET scans can be performed even in hospitals that do not own a cyclotron. As a result, the demand for Ga-68 has been increasing in recent years, especially in Europe and the USA (Shetty et al., 2010). Considering the half-life, rapid synthesis is essential, and many studies using microwave reactors have been reported.

5.1 [^{68}Ga]Ga-DOTA-Re(Arg^{11})CCMSH

Cantorias et al. (2009) reported a ^{68}Ga-labeled DOTA-rhenium-cyclized analog of α-melanocyte-stimulating hormone (α-MSH) ([^{68}Ga]Ga-DOTA-Re(Arg^{11})CCMSH) for imaging metastatic melanoma. The labeling was performed in an ammonium acetate buffer (pH 5) for 1 min using microwave-assisted heating (power 100 W). The A_m obtained after HPLC purification was 706 GBq/μmol compared to the previously reported synthesis method (3.8 GBq/μmol, Wei et al., 2007) owing to the rapid synthesis (Scheme 21). Cantorias et al. (2009) observed significantly improved tumor uptake and pharmacokinetic properties for [^{68}Ga]Ga-DOTA-Re(Arg^{11})CCMSH compared to similar analogs, particularly higher tumor uptake and tumor-to-nontarget tissue ratios, as well as excellent clearance from the kidney and other nontarget tissues using B16/F1 tumor-bearing mouse models.

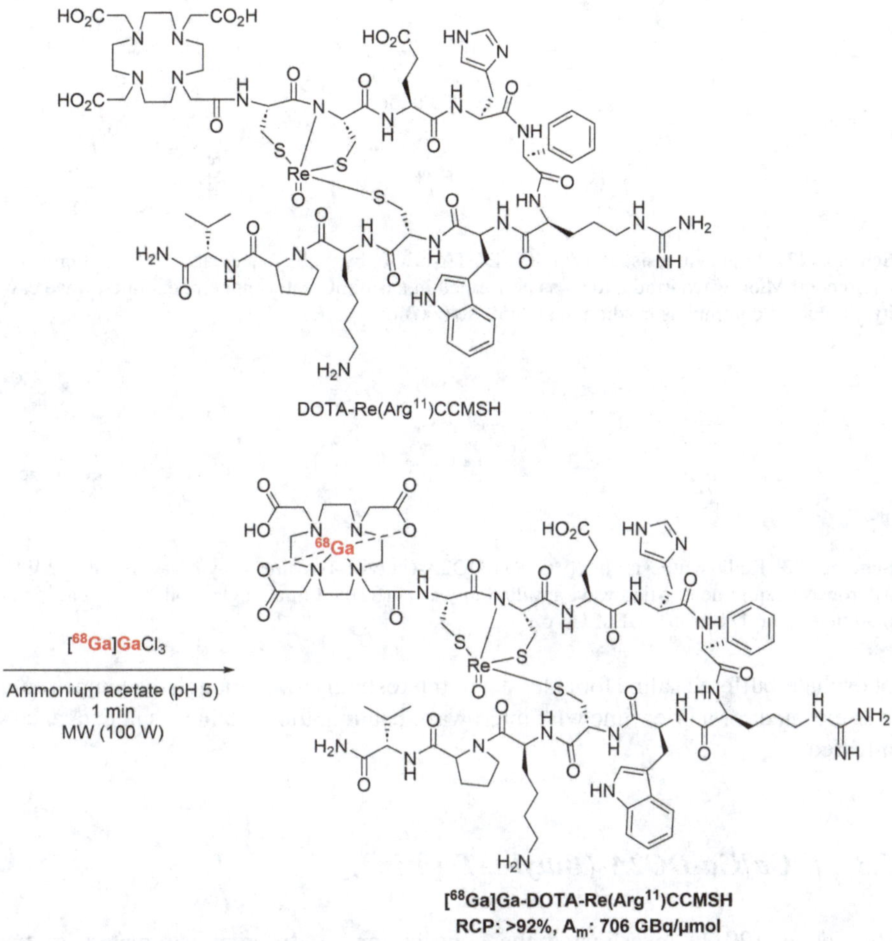

Scheme 21 Radiolabeling of DOTA-Re(Arg[11])CCMSH by Cantorias et al. (2009). Microwave equipment: Microwave irradiation was performed in a TE Mode Reaction Cavity (Model 520B-1; Resonance Instruments, Inc.)

5.2 [⁶⁸Ga]Ga-DOTA-hEGF

Velikyan et al. (2005) created a positron-emitting ^{68}Ga-labeled conjugate of human epidermal growth factor (hEGF) and evaluated the feasibility of epidermal growth factor receptor (EGFR)-expressing tumors. The microwave-assisted labeling of the conjugates was performed using either a nonconcentrated or a preconcentrated generator ^{68}Ga-eluate with respective radioactivity incorporation of 60 ± 10% ($n = 3$) and 77 ± 4% ($n = 3$) (Scheme 22). The chosen chelating moiety in hEGF was DOTA, because DOTA may provide an adequate in vivo stability of the ^{68}Ga-radiolabel. The RCP of the tracers in the study exceeded 95%. The tracer was found to be stable in

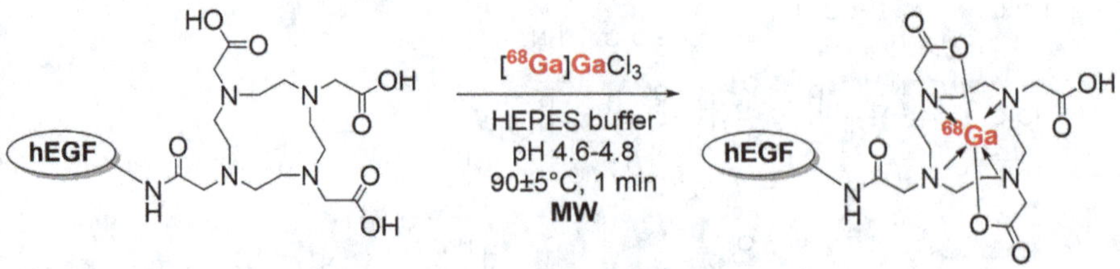

Scheme 22 Radiosynthesis of [^{68}Ga]Ga-DOTA-hEGF by Velikyan et al. (2005). Microwave equipment: Microwave irradiation was performed in a SmithCreator monomodal microwave cavity, producing continuous irradiation at 2450 MHz (Biotage AB)

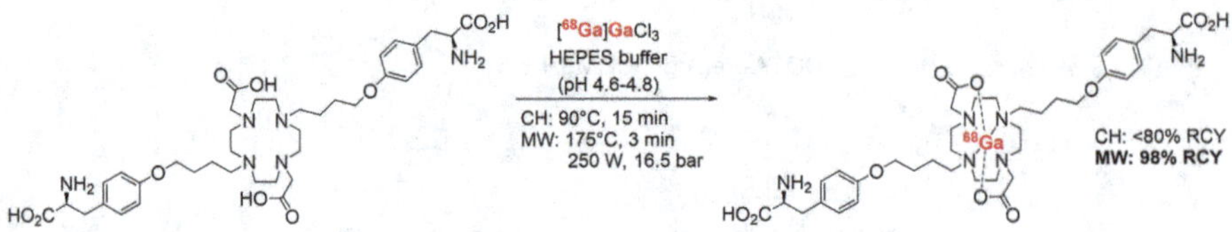

Scheme 23 Radiosynthesis of [^{68}Ga]Ga-DO2A-(butyl-L-tyrosine)$_2$ by Riss et al. (2011). Microwave Equipment: Microwave irradiation was performed in a single-mode microwave irradiation system (Discover®, CEM Corp.)

phosphate-buffered saline for at least 4 h. Interestingly, even though the temperature was elevated for a brief time with microwave heating, the protein was not adversely affected.

5.3 [^{68}Ga]Ga-DO2A-(Butyl-L-Tyrosine)$_2$

Riss et al. (2011) investigated the contribution of different parameters to the ^{68}Ga-labeling outcome of DO2A-(butyl-L-tyrosine)$_2$, wherein the tracer is used for measuring amino acid transporter overexpression. They revealed that ^{68}Ga-labeling can be improved by using a desalted precursor, purification of ^{68}Ga-eluate, optimization of the pH value, and microwave heating. In particular, microwave heating (175 °C, 3 min, 250 W, 16.5 bar) improved the RCY to >98%, compared to 80% with conventional heating (90 °C, 15 min) (Scheme 23). Though the RCY improved, they reported that the amount of labeling precursor could not be reduced using microwave labeling.

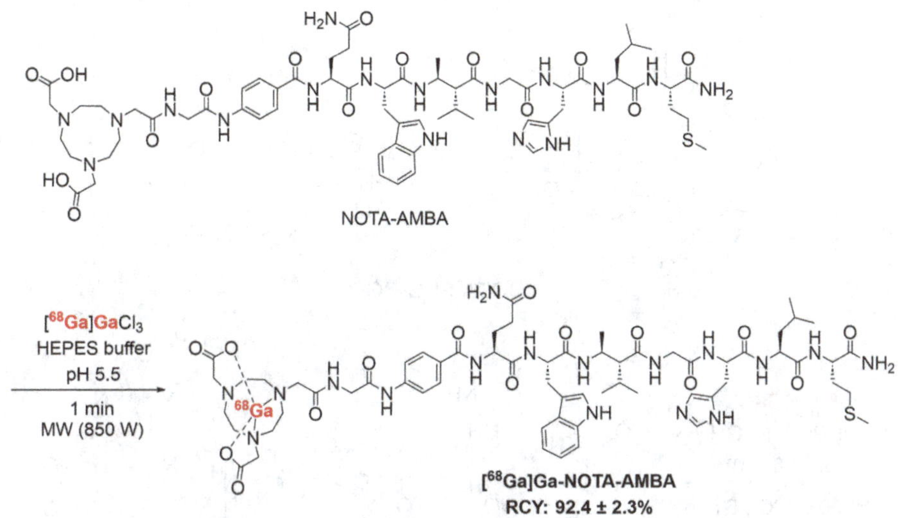

Scheme 24 Radiosynthesis of [^{68}Ga]Ga-NOTA-AMBA by Dam et al. (2016). Microwave equipment: Microwave irradiation was performed in a conventional Samsung microwave oven MW82N-B

5.4 [^{68}Ga]Ga-NOTA-AMBA

Dam et al. (2016) conducted a head-to-head evaluation of an analog of bombesin, NOTA-AMBA (NOTA-Gly-4-ABz-Gln-Trp-Ala-Val-Gly-His-Leu-Met-NH_2), labeled with the generator-produced and short-lived Ga-68 in term of labeling efficiency, biodistribution, and PET/CT imaging in prostate cancer-bearing xenograft mice. For studies with Ga-68, an RCY of 92.4 ± 2.3% and an RCP of 91.5 ± 2.4% (n = 5) were obtained (Scheme 24). SPE was applied to remove colloids, free [^{68}Ga]Ga^{3+}, and 2-[4-(2-hydroxyethyl)piperazin-1-yl]ethanesulfonic acid (HEPES) buffer, and the incorporation of radioactivity was >90% with an average RCP of 95.4 ± 1.6% (n = 5). For [^{68}Ga]Ga-NOTA-AMBA, A_m of 28 MBq/nmol at EOS was obtained. Finally, J.H. Dam et al. performed a PET study and obtained clear images using this tracer.

5.5 [^{68}Ga]Ga-DOTA-PEG-RGD and [^{68}Ga] Ga-NOTA-PEG-RGD

Blom et al. (2012) described RGD peptides with various modifications considering the amino acid sequence, and both DOTA and NOTA as chelators were labeled with Ga-68. The RGD conjugates were labeled with the [^{68}Ga]Ga(III) cation by microwave heating using a hydrochloric acid solution of the radiometal eluted from the ^{68}Ge/^{68}Ga-generator system and HEPES as a buffer. DOTA-PEG-RGD was labeled

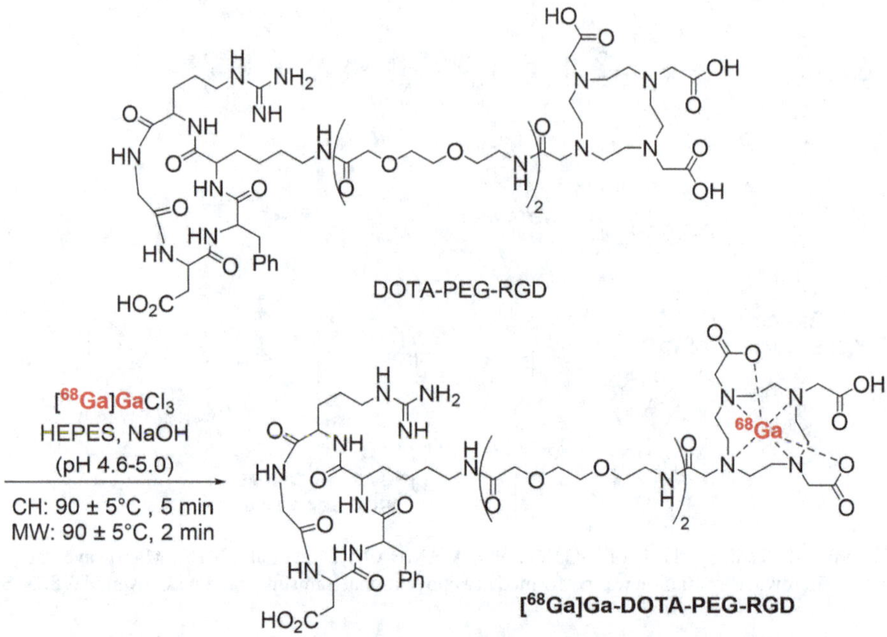

Scheme 25 Radiosynthesis of [^{68}Ga]Ga-DOTA-PEG-RGD by Blom et al. (2012). Microwave equipment: Microwave irradiation was performed in a SmithCreator™ monomodal cavity (Biotage AB) at 2450 MHz

at 90 ± 5 °C, provided by conventional heating, reaching 90% incorporation within 5 min for a 2.5 μM concentration of the precursor (Scheme 25). NOTA-PEG-RGD was labeled at room temperature and reached 95% incorporation within 5 min using the 2.5 μM precursor (Scheme 26). Similar incorporation yields were obtained when using microwave heating at 90 ± 5 °C; however, the reaction time was reduced to 2 min. The ^{68}Ga-labeling, including subsequent HPLC purification, was performed within 30 min.

5.6 *[^{68}Ga]Ga-DOTA-Biotin-Analogs*

Blom et al. (2009) reported three biotin counterparts with various alkyl, PEG, and DOTA modifications that were synthesized and labeled with Ga-68 and characterized the islets of Langerhans to monitor the survival of transplanted avidin-coated islets. The radioactivity incorporation of [^{68}Ga]Ga(III) reached 90% within 5 min with conventional heating; however, when microwave heating was applied, the reaction time was reduced to 2 min. No major differences were observed in the incorporation yield between the two heating modes. The A_m when using 10 nmol of the

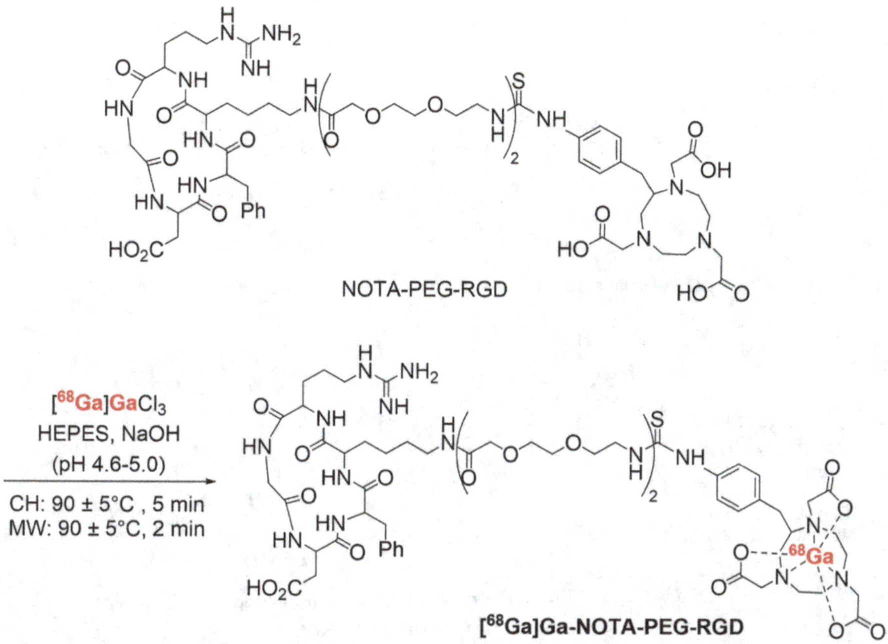

Scheme 26 Radiosynthesis of [^{68}Ga]Ga-NOTA-PEG-RGD by Blom et al. (2012). Microwave equipment: Microwave irradiation was performed in a SmithCreator™ monomodal cavity (Biotage AB) at 2450 MHz

precursor in the labeling reactions was 9 ± 3 GBq/μmol starting from 140 ± 30 MBq ($n = 4$) (Scheme 27).

5.7 *[^{68}Ga]*p*-Bn-NCS-Ga-NOTA and [^{68}Ga]Ga-DOTA-TOC*

Yagi et al. (2019) described the uses of a resonant-type microwave reactor for ^{68}Ga-labeling of model macrocyclic chelators such as *p*-Bn–NCS–DOTA and DOTA–D-Phe^1-Tyr^3-octreotide (DOTA-TOC), which are used for the diagnosis and radionuclide therapy of neuroendocrine tumors. To label the former, a 0.5–10 μM solution of the precursor (*p*-Bn-NCS-DOTA) containing 0.05 M HCl solution of [^{68}Ga]GaCl$_3$ was heated to 90 °C using a resonant-type microwave reactor or a block heater (Scheme 28). When examining labeling at a precursor concentration of 1 μM, no reaction progress was observed with conventional heating, while microwave heating enabled the reaction to proceed with over a 90% yield. For the latter, 0.7–7.0 μM solutions of the precursor (DOTA-TOC) containing 0.05 M HCl solution of [^{68}Ga]GaCl$_3$ was heated to 95 °C using microwave or conventional heating (Scheme 29). The labeling rate of the 3.5 μM solution was notably lower on using the conventional heating method than the microwave heating method (CH: 16% ± 3%, MW: >95%).

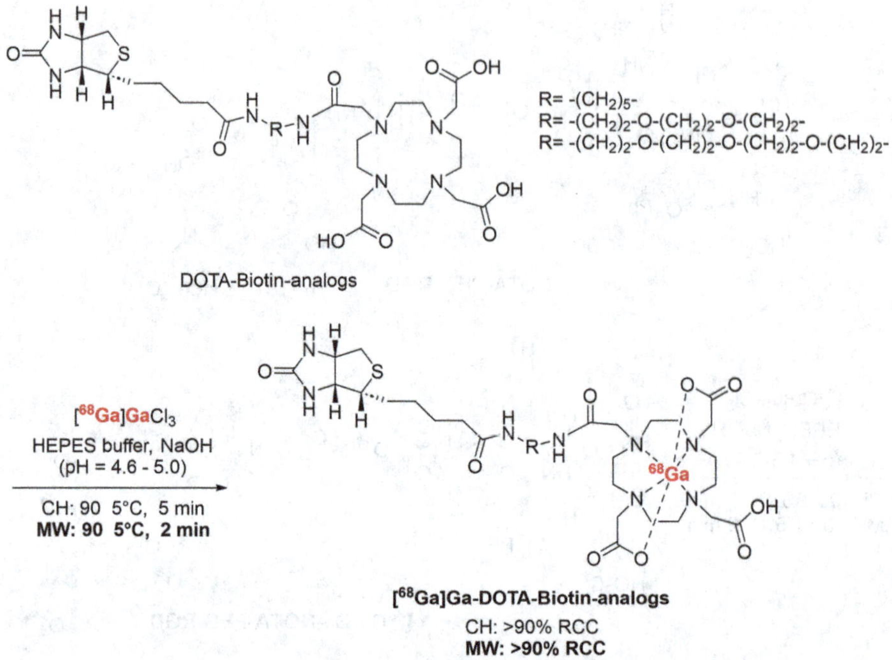

Scheme 27 ^{68}Ga-labeling of DOTA-biotin-analogs by Blom et al. (2009). Microwave equipment: Microwave heating was performed in a SmithCreator™ monomodal cavity (Biotage AB), producing continuous irradiation at 2450 MHz

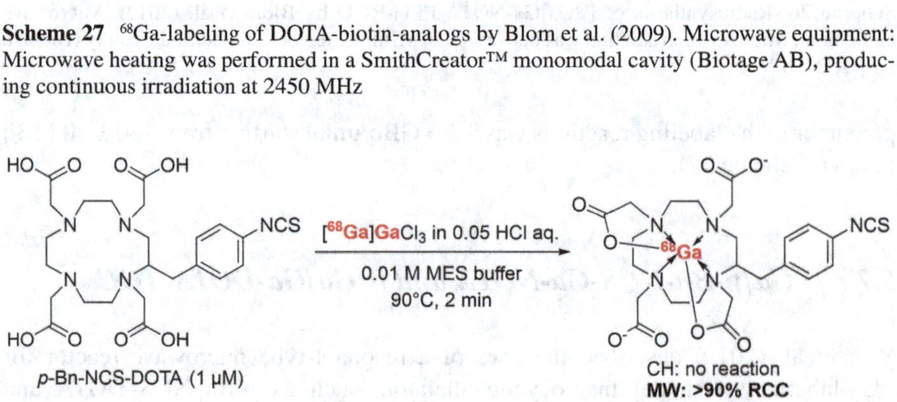

Scheme 28 Radiosynthesis of [^{68}Ga]*p*-Bn-NCS-Ga-NOTA by Yagi et al. (2019). Microwave equipment: Microwave heating was performed in a resonant-type microwave reactor (SAIDA FDS Inc.) with feedback control consisting of an electric field and a frequency-variable solid-state oscillator

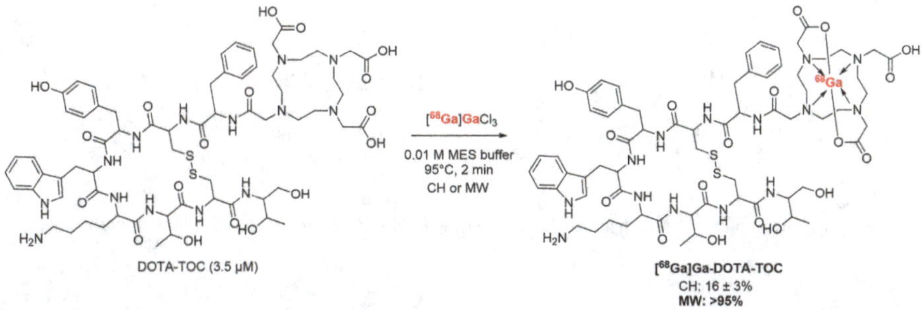

Scheme 29 Radiosynthesis of [^{68}Ga]Ga-DOTA-TOC by Yagi et al. (2019). Microwave equipment: Microwave heating was performed in a resonant-type microwave reactor (SAIDA FDS Inc.) with feedback control consisting of an electric field and a frequency-variable solid-state oscillator

6 ^{99m}Tc-Labeling Using Microwave Heating

Technetium-99m (Tc-99 m) has ideal nuclear properties for single-photon emission computed tomography (SPECT) imaging applications in terms of its half-life (6 h) and gamma-ray energy (Eγ = 140 keV). Moreover, Tc-99m (in the form of [^{99m}Tc]TcO_4^-) can be easily prepared using a ^{99}Mo/^{99m}Tc generator and rapidly converted to [^{99m}Tc][$Tc(CO)_3(H_2O)_3$]$^+$ in saline using the IsoLink kit™. Technetium (Tc; $Z = 43$) is situated in the middle of the second-row transition series of the periodic table and has no stable isotopes. Rhenium-187 (Re-187), a group 7 congener of Tc in the periodic table, is often employed as an alternative to Tc-99m considering Re generally produces complexes with physical and biodistribution properties that are similar to those of Tc-99m. Furthermore, Re-186 (Eβ^- = 1.1 MeV, $T_{1/2}$ = 90.6 h) and Re-188 (Eβ^- = 2.1 MeV, $T_{1/2}$ = 16.9 h) have attracted attention as radiotherapy nuclides (Banerjee et al., 2005; Funk et al., 2004; Kimura et al., 2012; Pillai et al., 2013). In this section, we introduce several reports related to ^{99m}Tc-labeling or $^{186/188}$Re-labeling methods using microwave technology.

6.1 *3,1,2- or 2,1,8-Metallocarboranes*

Green et al. (2006) reported microwave-assisted synthesis of 3,1,2-metallocarboranes or 2,1,8-metallocarboranes (Scheme 30). Microwave heating was used to prepare η^5-rhenium carborane complexes in aqueous reaction media and isomerization of the cage from 3,1,2 to 2,1,8 derivatives was observed concomitantly with the complexation of carboranes bearing sterically demanding substituents. The authors used *nido*-carborane ($R_1 = (CH_2)_2CO_2H$) to determine whether microwave heating could be used to perform analogous metal–carborane complexation reactions at the tracer level using Tc-99m. The *nido*-carborane combined with [^{99m}Tc][$Tc(CO)_3(H_2O)_3$]$^+$ in the presence of fluoride to yield 3,1,2-[Na]-[^{99m}Tc][$Tc(CO)_3(R_1HC_2B_9H_9)$]

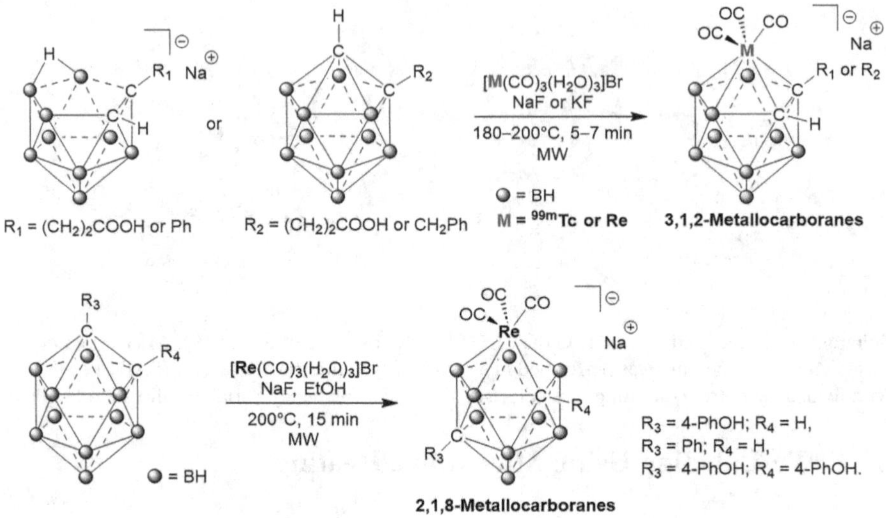

Scheme 30 Microwave-assisted synthesis of 3,1,2- or 2,1,8-metallocarboranes by Green et al. (2006). Microwave equipment: Microwave heating was performed in a Biotage® Initiator+ Sixty (Biotage AB)

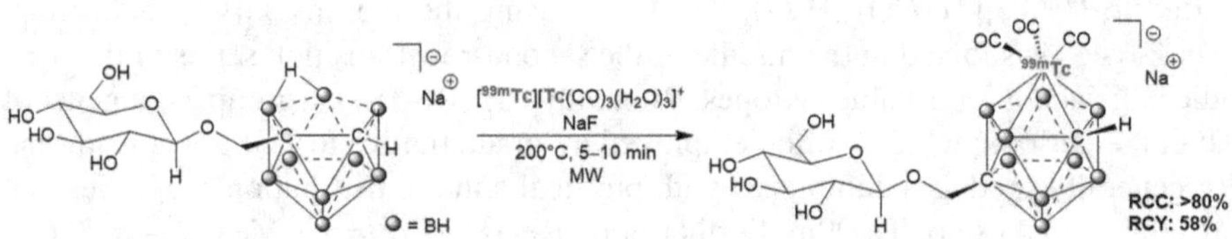

Scheme 31 Synthesis of carboranyl glycoside labeled with Tc-99m by Green et al. (2008). Microwave equipment: Microwave heating was performed in a Biotage® Initiator+ Sixty (Biotage AB)

($R_1 = (CH_2)_2CO_2H$) with 64% RCY. Green et al. (2008) achieved labeling of a *nido*-carborane linked to glucose with a ^{99m}Tc with 58% RCY via the microwave-assisted metallocarborane synthesis method (Scheme 31) (Green et al., 2008).

6.2 3,1,2- and 2,1,8-Dicarbonylnitroso-*closo*-Metallocarboranes via 3,1,2- or 2,1,8-Metallocarboranes

Armstrong et al. (2009) reported a synthesis method for 3,1,2- and 2,1,8-dicarbony lnitroso-*closo*-technetacarboranes via [^{99m}Tc][Tc(CO)$_3$]-carborane (Scheme 32). [^{99m}Tc]carborane complexes **6–9** were prepared at the tracer level using the microwave-assisted synthesis method reported in 2006 (Green et al., 2006). The authors then attempted to adapt the nitrosation methodology to the synthesis of 3,1,2- and 2,1,8-dicarbonylnitroso-*closo*-technetacarboranes at the tracer level.

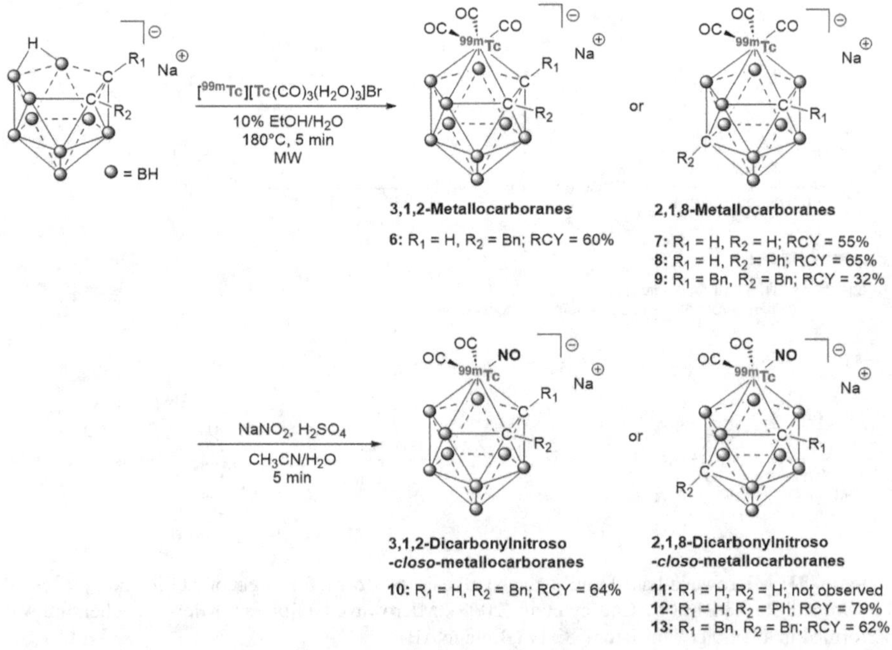

Scheme 32 Synthesis of 3,1,2- and 2,1,8-dicarbonylnitroso-*closo*-metallocarboranes via 3,1,2- or 2,1,8-metallocarboranes by Armstrong et al. (2009). Microwave equipment: Microwave heating was performed in a Biotage® Initiator+ Sixty or Initiator+ Eight (Biotage AB)

Consequently, [^{99m}Tc]carborane complexes **10**, **12**, and **13** were obtained with 62–79% RCYs; however, they did not observe complex **11** and observed slow decomposition of **10** and **12** but not **13** in aqueous solutions. Although further study of the aqueous stability of [^{99m}Tc][Tc(CO)$_2$(NO)]-carborane is necessary, these reports can contribute to the design of novel molecular imaging agents.

6.3 [^{99m}Tc]Tricarbonyl Complexes

Causey et al. (2008) reported a microwave-based multistep synthesis platform to prepare targeted organometallic ^{99m}Tc- and $^{186/188}$Re-bifunctional chelate derivatives (Scheme 33). The process, which involved microwave heating, in situ purification, and rapid solvent evaporation, produced the target compound with higher RCY (86%) than when using conventional heating and evaporation methods (RCY = 57%). The procedure was completed within 30 min. In addition, to provide a means of producing high A_m, Causey et al. developed a convenient resin capture purification method, which is based on a copper-functionalized solid support used by Ley et al. to isolate bipyridyl derivatives (Ley et al., 2001).

Harada et al. (2011) reported a synthetic protocol for [^{99m}Tc]tricarbonyl chelates using microwave technology (Scheme 34). The authors studied the effect of

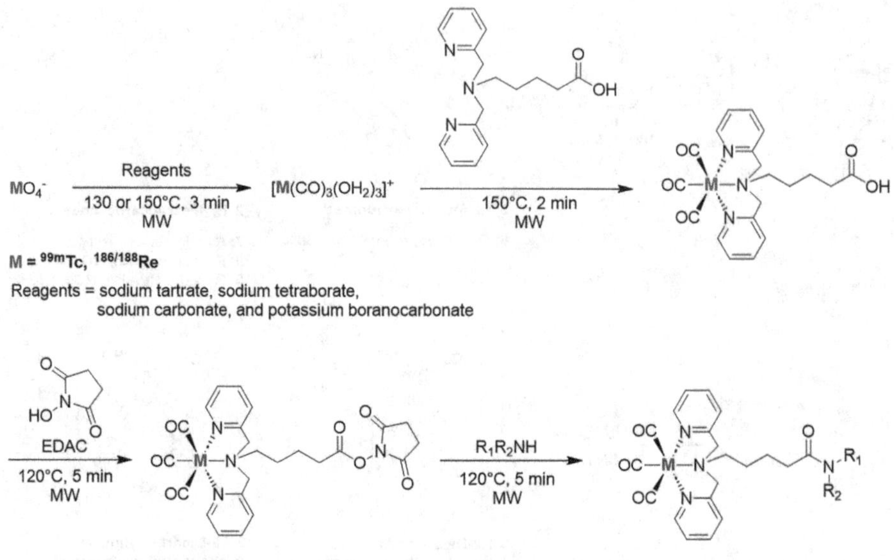

Scheme 33 Microwave-based multistep synthesis platform for organometallic complexes of Tc-99m and Re-186/188 by Causey et al. (2008). Microwave equipment: Microwave heating was performed in a Biotage® Initiator+ Sixty (Biotage AB)

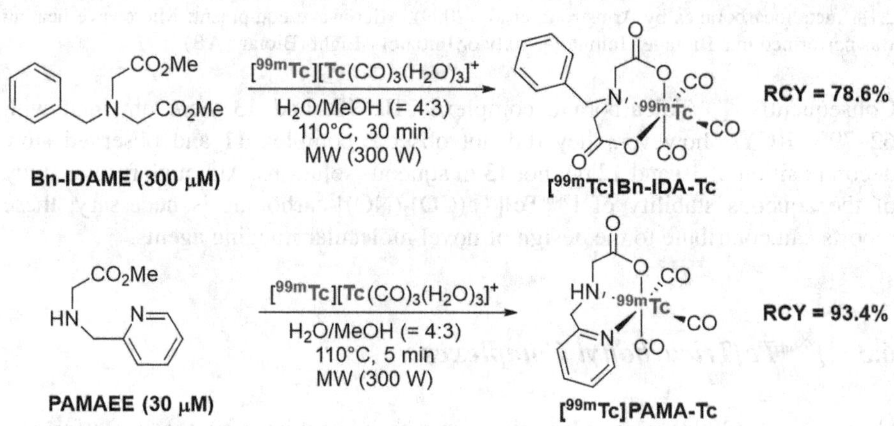

Scheme 34 Synthesis of [^{99m}Tc]tricarbonyl complexes using microwave technology by Harada et al. (2011). Microwave equipment: Microwave irradiation was performed in a single-mode microwave irradiation system (Discover®, CEM Corp.)

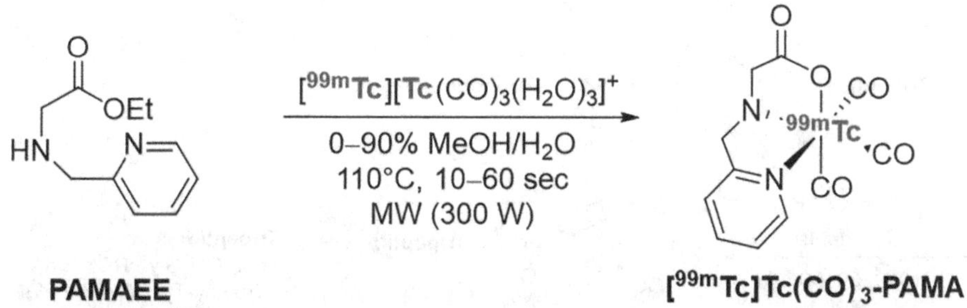

Scheme 35 Microwave-assisted synthesis of [^{99m}Tc]Tc(CO)$_3$–PAMA complexes by Kimura et al. (2012). Microwave equipment: Microwave irradiation was performed in a single-mode microwave irradiation system (Discover®, CEM Corp.)

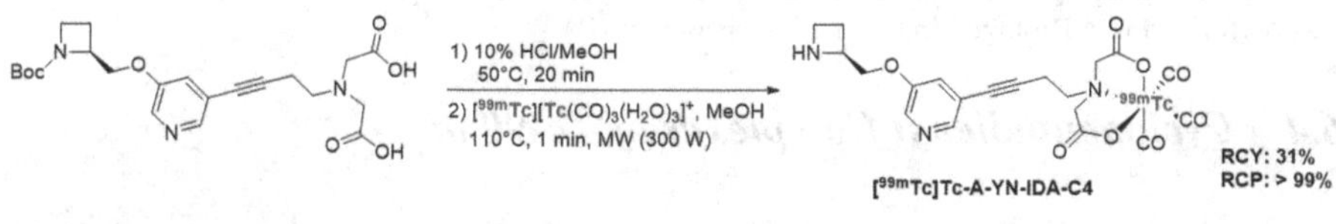

Scheme 36 Synthesis of [^{99m}Tc]Tc-A-YN-IDA-C4, an α4β2-nAChR molecular imaging probe labeled with Tc-99m by Mori et al. (2019). Microwave equipment: Microwave irradiation was performed in a single-mode microwave irradiation system (Discover®, CEM Corp.)

microwave irradiation on the synthesis of [^{99m}Tc]tricarbonyl complexes and found that the complexes could be obtained in good yields in a short time by microwave heating.

Kimura et al. (2012) reported the microwave-assisted synthesis of [^{99m}Tc]Tc(CO)$_3$–PAMA complexes (Scheme 35). The synthesis of the ^{99m}Tc-complex was performed based on the results obtained in the synthesis of the Re-complexes using PAMAEE as the ligand. The synthesis of [^{99m}Tc]Tc(CO)$_3$–PAMA with RCY (>99%) was achieved by irradiating microwaves for 10 s and using water as the solvent. Furthermore, the authors investigated the effect of MeOH-containing solvents to apply the synthetic method to a hydrophobic molecule. The reaction within 1 min of irradiation of microwaves was not significantly affected, when the MeOH content was <50%. The formation of byproducts was markedly suppressed compared to conventional methods.

Mori et al. (2019) reported [^{99m}Tc]Tc-A-YN-IDA-C4, a molecular imaging probe targeting α4β2-nAChR (nicotinic acetylcholine receptor), which was synthesized using the microwave-assisted synthesis of [^{99m}Tc]tricarbonyl complexes (Scheme 36). [^{99m}Tc]Tc-A-YN-IDA-C4 was efficiently designed using a computational scientific approach and exhibited high binding affinity for α4β2-nAChR in both the docking simulation (−19.3 kcal/mol) and binding assay ($Ki = 0.4 \pm 0.04$ nM). In their report, the microwave-assisted labeling method was used in the synthesis of other candidate compounds and provided efficient labeling.

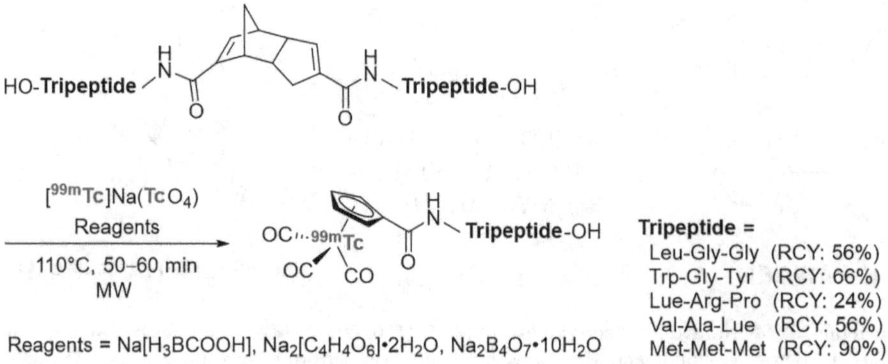

Scheme 37 Synthesis of tripeptide derivatized cyclopentadienyl complexes of Tc-99m using microwave technology by Nadeem et al. (2014). Microwave equipment: Microwave irradiation was performed in a Biotage® Initiator EXP (Biotage AB)

6.4 Cyclopentadienyl Complexes of Tc-99 m

Nadeem et al. (2014) described the synthesis of half-sandwich complexes of the type [(η^5-Cp(CONH-tripeptide))[^{99m}Tc]Tc(CO)$_3$] using microwave technology (Scheme 37). The results of their studies corroborate the feasibility of metal-mediated retro Diels–Alder reactions for the preparation of small molecules and peptides carrying a [(η^5-Cp)[^{99m}Tc]Tc(CO)$_3$] tag. The authors labeled five tripeptides that showed high binding affinities for the oligopeptide transporter PEPT2 using the described procedure. Moreover, under microwave irradiation (110 °C), five target compounds were found to produce much better yields than those by the conventional heating process.

7 Outlook

In this chapter, we explained the principle of dielectric heating by microwave irradiation and introduced examples of labeling reactions using F-18, C-11, Ga-68, and Tc-99m that make use of this approach. In a reaction system, microwave heating has the following two major differences from conventional external heating, which may lead to a significant improvement in reactivity, shortening reaction time, increasing yields, reducing precursor requirements, and decreasing byproducts. (1) Faster heating rate: Since microwave heating is dielectric heating, the temperature of the reaction solution itself rises rapidly and uniformly, and energy above the activation energy of the main reaction of interest is quickly given to the reaction system. Unlike external heating, which heats slowly and nonuniformly, the main reaction starts immediately, and it can be expected that the progress of side reactions is unlikely to occur. (2) Higher reaction temperature: Solute molecules (reactants and products) in a reaction solution are surrounded by solvent molecules and have intermolecular interactions with solvent molecules. Since the stability of the solute varies with the solvent, differences in solvents have a significant impact on reactivity and reaction rates. Microwave irradiation in a sealed vial allows the reaction temperature to be above the boiling point of the solvent, which is optimal for the reaction, so the reaction proceeds more rapidly than with an external heater.

Since the 1990s, microwave reactors have been improved in terms of performance and made less expensive, and we predict that in the future, microwaves will be used for both rapid synthesis and new reactions based on the nature of microwaves. Microwave reactors or microwave-heated flow reactors could be incorporated into radiopharmaceutical automated synthesizers, potentially enabling microwave-based reactors to substitute for the existing synthesis equipment in PET- and SPECT-probe syntheses, which require fast trace-level synthesis. We hope that this review will stimulate interest in the use of microwaves in radiochemistry.

References

Alcázar, J., Diels, G., & Schoentjes, B. (2007). Microwave assisted medicinal chemistry. *Mini Reviews in Medicinal Chemistry, 7*(4), 345–3697.

Armstrong, A. F., Lebert, J. M., Brennan, J. D., & Valliant, J. F. (2009). Functionalized carborane complexes of the $[M(CO)_2(NO)]^{2+}$ Core (M = ^{99m}Tc, Re): A new class of organometallic probes for correlated in vitro and in vivo imaging. Organometallics, 28(10), 2986–2992. https://doi.org/10.1021/om8011807

Aubert, C., Huard-Perrio, C., & Lasne, M. C. (1997). Rapid synthesis of aliphatic amides by reaction of carboxylic acids, Grignard reagents, and amines: Application to the preparation of [^{11}C]amides. *Journal of the Chemical Society, Perkin Transactions, I*, 2837–2842.

Banerjee, S. R., Maresca, K. P., Francesconi, L., Valliant, J., Babich, J. W., & Zubieta, J. (2005). New directions in the coordination chemistry of ^{99m}Tc: A reflection on technetium core structures and a strategy for new chelate design. *Nuclear Medicine and Biology, 32*(1), 1–20. https://doi.org/10.1016/j.nucmedbio.2004.09.001

Bassyouni, F. A., Abu-Bakr, S. M., & Rehim, M. A. (2012). Evolution of microwave irradiation and its application in green chemistry and biosciences. *Research on Chemical Intermediates, 38*, 283–322. https://doi.org/10.1007/s11164-011-0348-1

Belanger, A. P., Pandey, M. K., & DeGrado, T. R. (2011). Microwave-assisted radiosynthesis of [^{18}F]fluorinated fatty acid analogs. *Nuclear Medicine and Biology, 38*(3), 435–441. https://doi.org/10.1016/j.nucmedbio.2010.10.001

Blom, E., Långström, B., Velikyan, I. (2009). 68Ga-Labeling of biotin analogues and their characterization. *Bioconjugate Chemistry*, 20, 1146–1151, doi: https://doi.org/10.1021/bc800538.

Blom, E., Velikyan, I., Estrada, S., Hall, H., Muhammad, T., Ding, C., Nair, M., & Långström, B. (2012). ^{68}Ga-Labeling of RGD peptides and biodistribution. *International Journal of Clinical and Experimental Medicine*, 5(2), 165–172.

Cantorias, M. V., Figueroa, S. D., Quinn, T. P., Lever, J. R., Hoffman, T. J., Watkinson, L. D., Carmack, T. L., & Cutler, C. S. (2009). Development of high-specific-activity ^{68}Ga-labeled DOTA-rhenium-cyclized α-MSH peptide analog to target MC1 receptors overexpressed by melanoma tumors. *Nuclear Medicine and Biology, 36*, 505–513. https://doi.org/10.1016/j.nucmedbio.2009.01.017

Causey, P. W., Besanger, T. R., Schaffer, P., & Valliant, J. F. (2008). Expedient multi-step synthesis of organometallic complexes of Tc and Re in high effective molar activity. A new platform for the production of molecular imaging and therapy agents. *Inorganic Chemistry, 47*(18), 8213–8221. https://doi.org/10.1021/ic800775w

Dahl, K., Schou, M., Rahman, O., & Halldin, C. (2014). Improved yields for the palladium-mediated ^{11}C-carbonylation reaction using microwave technology. *European Journal of Organic Chemistry, 2*, 307–310. https://doi.org/10.1002/ejoc.201300989

Dahl, K., Schou, M., & Halldin, C. (2016). Direct and efficient (carbonyl)cobalt-mediated aryl acetylation using [^{11}C]methyl iodide. *European Journal of Organic Chemistry, 16*, 2775–2777. https://doi.org/10.1002/ejoc.201600085

Dam, J. H., Olsen, B. B., Baun, C., Høilund-Carlsen, P., & Thisgaard, H. (2016). In vivo evaluation of a Bombesin analogue labeled with Ga-68 and Co-55/57. *Molecular Imaging and Biology, 18*, 368–376. https://doi.org/10.1007/s11307-015-0911-z

Egami, H., & Hamashima, Y. (2009). Practical and scalable organic reactions with flow microwave apparatus. *Chemical Record, 19*(1), 157–171. https://doi.org/10.1002/tcr.201800132

Funk, T., Sun, M., & Hasegawa, B. H. (2004). Radiation dose estimate in small animal SPECT and PET. *Medical Physics, 31*(9), 2680–2686. https://doi.org/10.1118/1.1781553

Gedye, R., Smith, F., Westaway, K., Ali, H., Baldisera, L., Laberge, L., & Rousell, J. (1986). The use of microwave ovens for rapid organic synthesis. *Tetrahedron Letters, 27*(3), 279–282. https://doi.org/10.1016/s0040-4039(00)83996-9

Giguere, R. J., Bray, T. L., Duncan, S. M., & Majetich, G. (1986). Application of commercial microwave ovens to organic synthesis. *Tetrahedron Letters, 27*(41), 4945–4948. https://doi.org/10.1016/s0040-4039(00)85103-5

Green, A. E. C., Causey, P. W., Louie, A. S., Armstrong, A. F., Harrington, L. E., & Valliant, J. F. (2006). Microwave-assisted synthesis of 3,1,2- and 2,1,8-Re(I) and ^{99m}Tc(I)-metallocarborane complexes. *Inorganic Chemistry, 45*(15), 5727–5729. https://doi.org/10.1021/ic0605928

Green, A. E. C., Harrington, L. E., & Valliant, J. F. (2008). Carborane-carbohydrate derivatives—Versatile platforms for developing targeted radiopharmaceuticals. *Canadian Journal of Chemistry, 86*(11), 1063–1069. https://doi.org/10.1139/v08-157

Guo, N., Alagille, D., Tamagnan, G., Price, R. R., & Baldwin, R. M. (2008). Microwave-induced nucleophilic [^{18}F]fluorination on aromatic rings: Synthesis and effect of halogen on [^{18}F]fluoride substitution of meta-halo (F, Cl, Br, I)-benzonitrile derivatives. *Applied Radiation and Isotopes, 66*(10), 1396–1402. https://doi.org/10.1016/j.apradiso.2008.03.003

Hamill, T. G., Krause, S., Ryan, C., Bonnefous, C., Govek, S., Seiders, T. J., Cosford, N. D. P., Roppe, J., Kamenecka, T., Patel, S., Gibson, R. E., Sanabria, S., Riffel, K., Eng, W., King, C., Yang, X., Green, M. D., O'Malley, S. S., Hargreaves, R., & Burns, H. D. (2005). Synthesis, characterization, and first successful monkey imaging studies of metabotropic glutamate receptor subtype 5 (mGluR5) PET radiotracers. *Synapse, 56*, 205–216. https://doi.org/10.1002/syn.20147

Harada, N., Kimura, H., Ono, M., Mori, D., Ohmomo, Y., Kajimoto, T., Kawashima, H., & Saji, H. (2011). Effective synthesis of ^{99m}Tc tricarbonyl complexes by microwave heating. *Journal of Organometallic Chemistry, 696*(23), 3745–3749. https://doi.org/10.1002/ejoc.201600085

Hostetler, E. D., Terry, G. E., & Burns, H. D. (2005). An improved synthesis of substituted [^{11}C]toluenes via Suzuki coupling with [^{11}C]methyl iodide. *Journal of Labelled Compounds and Radiopharmaceuticals, 48*(9), 629–634. https://doi.org/10.1002/jlcr.953

Johnstroem, P., Bergman, L., Varnaes, K., Malmquist, J., Halldin, C., & Farde, L. (2015). Development of rapid multistep carbon-11 radiosynthesis of the myeloperoxidase inhibitor AZD3241 to assess brain exposure by PET microdosing. *Nuclear Medicine and Biology, 42*(6), 555–560. https://doi.org/10.1016/j.nucmedbio.2015.02.001

Kappe, C. O. (2004). Controlled microwave heating in modern organic synthesis. *Angewandte Chemie (International Ed. in English), 43*(46), 6250–6284. https://doi.org/10.1002/anie.200400655

Kappe, C. O., & Stadler, A. (2005). Microwaves in organic and medicinal chemistry. In R. Mannhold, H. Kubinyi, & G. Folkers (Eds.), *Methods and principles in medicinal chemistry* (Vol. 25, p. 93). Wiley-VCH.

Kimura, H., Mori, D., Harada, N., Ono, M., Ohmomo, Y., Kajimoto, T., Kawashima, H., & Saji, H. (2012). Microwave-assisted synthesis of organometallic complexes of $^{99m}Tc(CO)_3$ and $Re(CO)_3$: Its application to radiopharmaceuticals. *Chemical & Pharmaceutical Bulletin, 60*(1), 79–85. https://doi.org/10.1248/cpb.60.79

Kimura, H., Yagi, Y., Ohneda, N., Odajima, H., Ono, M., & Saji, H. (2014). Development of a resonant-type microwave reactor and its application to the synthesis of positron emission tomography radiopharmaceuticals. *Journal of Labelled Compounds and Radiopharmaceuticals, 57*, 680–686. https://doi.org/10.1002/jlcr.3232

Kimura, T., Yagi, Y., Arimitsu, K., Maeda, K., Ikejiri, K., Takano, J., Kusuhara, H., Kagawa, S., Ono, M., Sugiyama, Y., & Saji, H. (2016). Radiosynthesis of novel pitavastatin derivative ([^{18}F]PTV-F1) as a tracer for hepatic OATP using a one-pot synthetic procedure. *Journal of Labelled Compounds and Radiopharmaceuticals, 1–12*. https://doi.org/10.1002/jlcr.3464

Kumari, K., Vishvakarma, V. K., Singh, P., Patel, R., & Chandra, R. (2017). Microwave: An important and efficient tool for the synthesis of biological potent organic compounds. *Current Medicinal Chemistry, 24*(41), 4579–4595. https://doi.org/10.2174/0929867324666170529100929

Larhed, M., Moberg, C., & Anders, H. (2002). Microwave-accelerated homogeneous catalysis in organic chemistry. *Accounts of Chemical Research, 35*(9), 717–727. https://doi.org/10.1021/ar010074v

Ley, S. V., Massi, A., Rodriguez, F., Horwell, D. C., Lewthwaite, R. A., Pritchard, M. C., & Reid, A. M. (2001). A new phase-switch method for application in organic synthesis programs employing immobilization through metal-chelated tagging. *Angewandte Chemie, International Edition, 40*(6), 1053–1055. https://doi.org/10.1002/1521-3773(20010316)40:6<1053::aid-anie10530>3.0.co;2-d

Loupy, A. (2006). *Microwaves in organic synthesis*. Wiley-VCH Verlag GmbH & Co., KGaA.

Lu, S., Hong, J., & Pike, V. W. (2003). Synthesis of NCA [carbonyl -^{11}C]amides by direct reaction of in situ generated [^{11}C]carboxymagnesium halides with amines under microwave-enhanced conditions. *Journal of Labelled Compounds and Radiopharmaceuticals, 46*(13), 1249–1259. https://doi.org/10.1002/jlcr.786

Lu, S., Hong, J., Musachio, J. L., Chin, F. T., Vermeulen, E. S., Wikstrom, H. V., & Pike, V. W. (2005). Alternative methods for labeling the 5-HT1A receptor agonist, 1-[2-(4-fluorobenzoylamino) ethyl]-4-(7-methoxynaphthyl)piperazine (S14506), with carbon-11 or fluorine-18. *Journal of Labelled Compounds and Radiopharmaceuticals, 48*(13), 971–981. https://doi.org/10.1002/jlcr.1009

Mandap, K. S., Ido, T., Kiyono, Y., Kobayashi, M., Lohith, T. G., Mori, T., Kasamatsu, S., Kudo, T., Okazawa, H., & Fujibayashi, Y. (2009). Development of microwave-based automated nucleophilic [^{18}F]fluorination system and its application to the production of [^{18}F]flumazenil. *Nuclear Medicine and Biology, 36*(4), 403–409. https://doi.org/10.1016/j.nucmedbio.2009.01.011

McQuade, P., Rowland, D. J., Lewis, J. S., & Welch, M. J. (2005). Positron-emitting isotopes produced on biomedical cyclotrons. *Current Medicinal Chemistry, 12*(7), 807–818. https://doi.org/10.2174/0929867053507397

Mori, D., Kimura, H., Kawashima, H., Yagi, Y., Arimitsu, K., Ono, M., & Saji, H. (2019). Carborane-carbohydrate derivatives—Versatile platforms for developing targeted radiopharmaceuticals. *Bioorganic & Medicinal Chemistry, 27*(18), 4200–4210. https://doi.org/10.1016/j.bmc.2019.07.053

Nadeem, Q., Can, D., Shen, Y., Felber, M., Mahmood, Z., & Alberto, R. (2014). Synthesis of tripeptide derivatized cyclopentadienyl complexes of technetium and rhenium as radiopharmaceutical probes. *Organic & Biomolecular Chemistry, 12*(12), 1966–1974. https://doi.org/10.1039/c3ob41866a

Odajima, H., & Okamoto, T. (2019). Effective application of microwave resonant cavity system to flow chemistry. *Chemical Record, 19*(1), 204–211. https://doi.org/10.1002/tcr.201800087

Perrio-Huard, C., Aubert, C., & Lasne, M. C. (2000). Reductive amination of carboxylic acids and [^{11}C]magnesium halide carboxylates. *Journal of the Chemical Society, Perkin Transactions, 1*, 311–316.

Pillai, M. R. A., Dash, A., & Knapp, F. F., Jr. (2013). Sustained availability of ^{99m}Tc: Possible paths forward. *Journal of Nuclear Medicine, 54*(2), 313–323. https://doi.org/10.2967/jnumed.112.110338

Purser, S., Moore, P. R., Swallow, S., & Gouverneur, V. (2008). Fluorine in medicinal chemistry. *Chemical Society Reviews, 37*(2), 320–330. https://doi.org/10.1039/B610213C

Ravert, H. T., Holt, D. P., & Dannals, R. F. (2014). A microwave radiosynthesis of the 4-[^{18}F]-fluorobenzyltriphenylphosphoniumion. *Journal of Labelled Compounds and Radiopharmaceuticals*, 57, 695–698. https://doi.org/10.1002/jlcr.3241

Ravert, H. T., Holt, D. P., Gao, Y., Horti, A. G., & Dannals, R. F. (2015). Microwave-assisted radiosynthesis of [^{18}F]ASEM, a radiolabeled α7-nicotinic acetylcholine receptor antagonist. *Journal of Labelled Compounds and Radiopharmaceuticals*, 58, 180–182. https://doi.org/10.1002/jlcr.3275

Ravert, H. T., Holt, D. P., Gao, Y., Horti, A. G., & Dannals, R. F. (2015). Microwave-assisted radiosynthesis of [^{18}F] ASEM, a radiolabeled α7-nicotinic acetylcholine receptor antagonist. *Journal of Labelled Compounds and Radiopharmaceuticals, 58*, 180–182. https://doi.org/10.1002/jlcr.3275

Riss, P. J., & Roesch, F. (2009). Efficient microwave-assisted direct radiosynthesis of [^{18}F]PR04.MZ and [^{18}F]LBT999: Selective dopamine transporter ligands for quantitative molecular imaging by means of PET. *Bioorganic & Medicinal Chemistry, 17*, 7630–7634. https://doi.org/10.1016/j.bmc.2009.09.054

Riss, P. J., Burchardt, C., & Roesch, F. (2011). A methodical ^{68}Ga-labelling study of DO2A-(butyl-L-tyrosine)2 with cation-exchanger post-processed ^{68}Ga: Practical aspects of radiolabelling. *Contrast Media & Molecular Imaging, 6*, 492–498. https://doi.org/10.1002/cmmi.451

Ryzhikov, N. N., Gomzina, N. A., Fedorova, O. S., Vasil'ev, D. A., Kostikov, A. P., & Krasikova, R. N. (2004). Preparation of [^{18}F]flumazenil, a potential radioligand for PET imaging of central benzodiazepine receptors by isotope exchange. *Radiochemistry, 46*(3), 290–294. https://doi.org/10.1023/b:rach.0000031692.63830.85

Ryzhikov, N. N., Seneca, N., Krasikova, R. N., Gomzina, N. A., Shchukin, E., Fedorova, O. S., Vassiliev, D. A., Gulyás, B., Hall, H., Savic, I., & Halldin, C. (2005). Preparation of highly specific radioactivity [^{18}F]flumazenil and its evaluation in cynomolgus monkey by positron emission tomography. *Nuclear Medicine and Biology, 32*, 109–116. https://doi.org/10.1016/j.nucmedbio.2004.11.001

Schou, M., Varnäs, K., Jucaite, A., Gulyás, B., Halldin, C., & Farde, L. (2013). Radiolabeling of the cannabinoid receptor agonist AZD1940 with carbon-11 and PET microdosing in non-human primate. *Nuclear Medicine and Biology, 40*(3), 410–414. https://doi.org/10.1016/j.nucmedbio.2012.10.011

Scott, P. J. H. (2009). Methods for the incorporation of carbon-11 to generate radiopharmaceuticals for PET imaging. *Angewandte Chemie, International Edition, 48*(33), 6001–6004. https://doi.org/10.1002/anie.200901481

Shetty, D., Lee, Y.S., & Jeong, J.M. (2010). ^{68}Ga-labeled radiopharmaceuticals for positron emission tomography. Nuclear Medicine and Molecular Imaging 44(4), 233–240. https://doi.org/10.1007/s13139-010-0056-6

Taddei, C., & Gee, A. D. (2018). Recent progress in [^{11}C]carbon dioxide ([^{11}C]CO2) and [^{11}C]carbon monoxide ([^{11}C]CO) chemistry. *Journal of Labelled Compounds and Radiopharmaceuticals, 61*(3), 237–251. https://doi.org/10.1002/jlcr.3596

Takashima-Hirano, M., Takashima, T., Katayama, Y., Wada, Y., Sugiyama, Y., Watanabe, Y., Doi, H., & Suzuki, M. (2011). Efficient sequential synthesis of PET probes of the COX-2 inhibitor [^{11}C]celecoxib and its major metabolite [^{11}C]SC-62807 and in vivo PET evaluation. *Bioorganic & Medicinal Chemistry, 19*(9), 2997–3004. https://doi.org/10.1016/j.bmc.2011.03.020

Velikyan, I., Sundberg, Å. L., Lindhe, Ö., Höglund, A. U., Eriksson, O., Werner, E., Carlsson, J., Bergström, M., Långström, B., & Tolmachev, V. (2005). Preparation and evaluation of ^{68}Ga-DOTA-hEGF for visualization of EGFR expression in malignant tumors. *Journal of Nuclear Medicine, 46*(11), 1881–1888.

Wei, L., Miao, Y., Gallazzi, F., Quinn, T. P., Welch, M. J., Vāvere, A. L., & Lewis, J. S. (2007). Gallium-68-labeled DOTA-rhenium-cyclized alpha-melanocyte-stimulating hormone analog for imaging of malignant melanoma. *Nuclear Medicine and Biology, 34*(8), 945–953. https://doi.org/10.1016/j.nucmedbio.2007.07.003

Wilson, T. C., Cailly, T., & Gouverneur, V. (2018). Boron reagents for divergent radiochemistry. *Chemical Society Reviews, 47*(18), 6990–7005. https://doi.org/10.1039/c8cs00499d

Yagi, Y., Kimura, H., Arimitsu, K., Ono, M., Maeda, K., Kusuhara, H., Kajimoto, T., Sugiyama, Y., & Saji, H. (2015). The synthesis of [^{18}F]pitavastatin as a tracer for hOATP using the Suzuki coupling. *Organic & Biomolecular Chemistry, 13*, 1113. https://doi.org/10.1039/c4ob01953a

Yagi, Y., Shimizu, Y., Arimitsu, K., Nakamoto, Y., Higuchi, T., Togashi, K., & Kimura, H. (2019). Efficient gallium-68 radiolabeling reaction of DOTA derivatives using a resonant-type microwave reactor. *Journal of Labelled Compounds and Radiopharmaceuticals, 62*, 132–138. https://doi.org/10.1002/jlcr.3704

Part III
Radiopharmaceutical Analysis and Dispensing Technologies

Quality Control Automation Solutions for PET Drug Production

Arkadij Elizarov and Joshua Ashton

1 Introduction to Radiopharmaceuticals and the Need for Quality Control

The United States Pharmacopeia (USP) serves as a foundational text defining quality benchmarks for medications. As delineated in the USP General Chapter <823> titled "Positron Emission Tomography Drugs for Compounding, Investigational, and Research Uses" (Hung et al., 2011), the implementation of quality assurance (QA) and quality control (QC) are critical stages in the fabrication of Positron Emission Tomography (PET) drugs. QA represents an encompassing principle that addresses all factors affecting the identity, potency, quality, and purity of a PET drug. On the other hand, QC, a vital component of QA, is concerned with assessing materials and PET drugs to confirm their adherence to predetermined acceptance criteria.

QC roles encompass (1) inspecting every batch of incoming materials to ensure compliance with established specifications before their incorporation into PET drug preparation or testing and (2) evaluating every PET drug batch to confirm that it meets the outlined specifications prior to sanctioning its final approval or disapproval. This chapter's focus on QC automation primarily pertains to the prerequisite pre-release testing for every PET drug batch, and it excludes the examination of incoming materials, periodic non-release mandated PET drug testing, and post-release sterility testing.

A. Elizarov (✉)
Trace-Ability, Inc, Los Angeles, CA, USA
e-mail: elizarov@traceabilityinc.com

J. Ashton
LabLogic, Ltd, Sheffield, UK

R. M. van Dam et al. (eds.), *Automated Technologies for the Development and Production of Radiopharmaceuticals*,
https://doi.org/10.1007/978-3-031-84632-8_15

QC procedures vary amongst PET tracers and across different regulatory environments. Consequently, to provide a meaningful comparative platform, solutions proposed in this chapter will predominantly align with the US context, focusing particularly on 2-deoxy-2-[^{18}F]fluoro-D-glucose (FDG). Other tracers will only be addressed collectively in relation to QC aspects that FDG cannot demonstrate. It is anticipated that this narrowly focused discussion will offer a succinct narrative, enabling readers to extrapolate to their tracer of interest and regional regulatory framework.

The advent of automated systems for radiosynthesis dates back to the 1970s (Berger et al., 1979; Winicov et al., 1978) and was quickly propelled by the early achievements of PET (Comar et al., 1979; Reivich et al., 1979). However, the automation of QC has only garnered attention in recent years, despite its complexity being on par with, if not exceeding, that of manual radiosynthesis. In PET's formative years, QC procedures were not standardized among tracer manufacturers and largely hinged on scientific discernment.

One measure to assure sound judgement was through state-regulated medical practices and compounding pharmacies (Noble, 2015), imposing liability on the pharmacist. Another mechanism relied on the review of procedures by the corresponding institution's Radioactive Drug Research Committee (RDRC) (Code of Federal Regulations Title 21, Volume 5, Part 361, Section 361.1 Radioactive Drugs for Certain Research Use, 2018b). Finally, QC procedures were presented to the FDA for review via Investigational New Drug (IND) applications (Code of Federal Regulations Title 21, Volume 5, Part 312, Investigational New Drug Application, 2018a). These mechanisms resulted in a variety of QC procedures (Pike et al., 1991; Sharma, 2016; Vera-Ruiz et al., 1990) and acceptance criteria, and a lack of uniform standards. In such a diverse landscape, the prospects for QC automation development were sparse. Moreover, when most PET procedures were performed using FDG, (Kelloff et al., 2005; Lovinfosse et al., 2018; Singnurkar et al., 2017), the pressure to alter established QC procedures was negligible.

Nonetheless, more recent evolution of PET has stimulated a surge in QC automation developments worldwide, underscoring the necessity and viability of such advancements. This chapter will initially delineate the key milestones in the PET industry that have instigated QC standardization and paved the way for automated solutions.

2 The Journey of Quality Control: Evolution in PET Technology

In the unfolding story of radiopharmaceutical quality control (QC), the narrative took flight with the establishment of QC standards for the trailblazing PET tracer, sodium fluoride, [^{18}F]NaF. Its New Drug Application (NDA) received the green light from the US Food and Drug Administration (FDA) in 1972, setting a precedent for future developments (Sarikaya et al., 2017). A USP monograph, laying out the QC testing protocols, was revealed to the public in 1979 ("United States Pharmacopeia and National Formulary (USP 19-NF 14). Rockville, MD: 'Sodium Fluoride F 18 Injection' Monograph," 1979).

From this point forward, centralized QC guidance for any PET tracer was absent until the emergence of the original Fludeoxyglucose F 18 Injection (FDG) USP monograph in 1990 ("Fludeoxyglucose F 18 Injection," 2020). This landmark document marked the birth of a systematic approach to the progression of standardized QC requirements, as we recognize them today.

An essential milestone in the quality journey was achieved in 1998, with the release of the inaugural version of the USP's General Chapter <823> (US Pharmacopeial Convention. Radiopharmaceuticals for Positron Emission Tomography (PET)—Compounding, Investigational, and Research Uses. In USP 23-NF 18. Rockville, MD: USP, 1998). This detailed broad QA standards for PET drugs and specified the necessity for several tests for all PET product batches, including pH, appearance, radiochemical purity and identity, radionuclidic identity, filter integrity, endotoxin, and post-release sterility. Particularly, products with mass-dependent toxicity were subject to specific activity testing. A summary of most commonly applied QC tests is presented in Table 1 along with equipment required for their execution.

With the evolution of quality standards, the focus has also extended to toxic chemical residues and other contaminants in the final products (Tankiewicz et al., 2016). The identification of contaminants relied heavily on the synthesis and purification methods employed in the production process. As a case in point, the introduction of the phase transfer reagent Kryptofix 2.2.2 in the nucleophilic FDG production method necessitated testing for the molecule due to its toxicity profile (Hamacher et al., 1986). Similarly, the introduction of new tracers might require adjusting the QC requirements in light of any toxic substances included in their synthesis or formulations. Current FDG QC requirements, as per the 2020 updates to General Chapter <823> and the USP Monograph on Fludeoxyglucose F 18 Injection, are divided into three categories: Identity and Strength, Purity, and Safety ("Fludeoxyglucose F 18 Injection," 2020). "*Identity and Strength*" includes parameters of radionuclidic identity, radiochemical identity, and radioactivity concentration. "Purity" includes appearance, radiochemical purity, radionuclidic purity, chemical purity, concentration of residual compounds used in synthesis, residual solvents, and specific activity. "Safety" includes pH, bacterial endotoxins, filter integrity test, and (post-release) sterility.

Sterility poses a significant challenge to QC, due to the potential for severe illnesses caused by bacterially contaminated drugs, particularly in already ill patients. However, the short half-life of PET tracers necessitates their administration before the sterility test results—taking a lengthy 14 days—are available (Haka et al., 2013). The only sterility indication before release is filtration of the final product through a membrane filter and subsequent verification of the filter's integrity.

Discrepancies exist in pharmacopeial requirements across various jurisdictions such as the USP, European Pharmacopeia (EP) (Fludeoxyglucose [18F] injection.

Table 1 Typical PET QC tests

	PET drug release parameters	Typical specification	Method	Measurement device
Appearance tests	Color	Colorless	Visual inspection	Human eye
	Clarity	Clear		
	Particulates	Free from visible particulates		
Property tests	pH	4.5–7.5	pH strip	Human eye
Phase transfer catalysts	K222	<50 µg/mL	Spot test	Human eye
	TBA	<55 ppm	Spot test	Human eye
Solvents	MeCN	<0.041%	Gas chromatography	Flame ionization detector
	EtOH	<0.5%		
	DMSO	<0.5%		
	Other solvents used in radiosynthesis	*NDA-specific*		
Excipients	Sodium ascorbate	0.39–0.49 w/v%	Spot test	Human eye
	Cysteine	0.06–0.12%	Spot test	Human eye
	Other excipients used in formulation	*NDA-specific*	*NDA-specific*	*NDA-specific*
Endotoxin	Endotoxin	<175 EU/dose	Colorimetric LAL test	Endosafe PTS reader
Radiation	Radionuclidic identity	105–115 min (^{18}F products)	Radiation vs. time	Dose calibrator + clock
	Radioactivity concentration	4–300 mCi/mL	Radiation vs. volume	Dose calibrator + micropipette
Radio-chemical	Radiochemical purity	>95%	HPLC/TLC chromatogram	HPLC detector(s)/TLC scanner
	Radiochemical identity	Retention time = standard ± 10 s		
Sterile filtration confirmation	Filter integrity test	>344.8 kPa	Bubble point	Human eye

European Pharmacopoeia. 4 edition. Strasbourg, France: European Directorate for the Quality of the Medicines, 2002), British Pharmacopeia (BP) ((Monographs: Radiopharmaceutical preparation, Fludeoxyglucose [18F] injection. British Pharmacopoeia 2000. London: British Pharmacopoeia Commission, 2000), and the International Pharmacopeia managed by the World Health Organization (WHO) ("The International Pharmacopoeia: Fludeoxyglucose (18F) Injection, Document QAS/08.265/FINAL," 2009). Several reviews have undertaken the task of summarizing relevant QC procedures and drawing comparisons among various pharmacopeial monographs (Hung, 2002; Yu, 2006). The FDA has also published draft guidance detailing QC procedures for several PET tracers (Fludeoxyglucose F 18 injection. Sample Formats: Application to Manufacture Ammonia N 13 Injection,

Fluorodeoxyglucose F 18 Injection (FDG F 18) and Sodium Fluoride F 18 Injection Chemistry, Manufacturing, and Control Sections, 2011).

QC automation was propelled forward with the advent of the 21 Code of Federal Regulations (CFR) Part 212 regulations, current Good Manufacturing Practice (cGMP) for PET Drugs (Code of Federal Regulations, Title 21, Volume 4, Part 212: Current Good Manufacturing Practice for Positron Emission Tomography Drugs, 2013) in 2013. While these regulations were a call to arms for stringent control over QC procedures, they led to increasing complexities with manual processes. Manual methods were prone to subjectivity, variability, and were highly dependent on operator's expertise (Chi et al., 2014). Consequently, a large proportion of 21 CFR Part 212 violations in the first few years after implementation were traced back to lack of control over manual processes ("Inspection Observations data and information summarizing the areas of regulation cited on FDA form 483," 2018). Automation, by nature of its design, bridges this gap and reduces potential regulatory violations.

Automation also addresses the scaling up of PET procedures in the twenty-first century. With millions of PET scans being conducted annually, the potential for error is substantial, making compliance a critical concern (Schwarz et al., 2013). Especially with the FDA approval of the first proprietary PET tracer, [^{18}F]Florbetapir (Patel et al., 2018) (Amyvid), in 2012 followed by multiple others (Crişan et al., 2022), QC has grown more complex and diverse, demanding larger facilities, more equipment, and most importantly, a more skilled workforce. The shift from FDG being the primary product to a multi-tracer portfolio (Clarke, 2018; Huang, 2019) has only heightened these complexities, necessitating further developments in QC automation.

3 The Push and Pull: Driving Factors and Barriers to Automation in QC

Review of the traditional quality control (QC) methodologies reveals a pressing need for automation, along with the hurdles standing in its way. A stark contrast exists between short-lived radiopharmaceutical QC and other injectable drug types. Conventional non-radioactive drugs are produced on a large scale, with infrequent batches and abundant time for QC. Consequently, the ratio of product doses to QC analysis is remarkably higher in comparison to PET drugs. In conventional drugs, QC plays a minor role in determining the production cost of each patient dose, whereas, in PET, it is a major determinant. Thus, enhancements in QC efficiency and affordability can have an immediate, noticeable impact on the cost and availability of the product. This is of particular importance in the face of the steadily increasing number of PET procedures and the diversity of PET drugs.

As observed in Sect. 4, multiple organizations launched independent QC automation endeavors nearly simultaneously. This suggests that the previously defined milestones have stimulated a supportive environment for such innovative ventures.

While cGMP compliance and the advent of new tracers underscore the necessity for QC automation, process standardization via USP makes it feasible to meet the needs of most users with a single solution.

Additionally, the push towards QC automation has been increasingly fuelled by a number of elements:

1. Safety/Radiation Exposure: Similar to radiosynthesis automation, removing the human element from the equation significantly improves personnel safety, aligning with the ALARA (As Low as Reasonably Achievable) principle (Prasad et al., 2004). Radiosynthesis was already performed with shielding in place before automation. However, the effectiveness of shielding is limited in QC due to the myriad of manual tasks that cannot be conducted remotely (Ferguson et al., 2011). Therefore, total automation is the only certain way to eradicate the radiation exposure resulting from these procedures.
2. Human Error: Manual processes are inherently error-prone. Given the number of FDG batches produced daily, there is a high probability that at least one batch might be compromised by human error in QC. Some errors may go unnoticed, while others necessitate investigation, delaying production and ramping up costs.
3. Operational Efficiency and Throughput: As the annual number of PET scans multiplies, so does the production of FDG and new tracers. This scenario translates to more production runs per facility every day, pushing many facilities to their efficiency ceilings and hindering their capacity to start new tracer production.
4. Skillset Dependence: Radiosynthesis has advanced to a stage where any technician with basic training can undertake it. However, QC continues to be the segment of the PET tracer production process that demands the most experience and judgment from the operator.
5. Footprint: The current QC approach depends on a multitude of instruments and manual test stations, typically occupying a substantial amount of bench space. As laboratories accommodate more PET tracers, this space becomes a premium commodity.
6. Operator Variability: It is an unavoidable fact that no two operators can perform QC identically. Each operator possesses their unique precision and accuracy, introducing variability into the results.
7. Complexity: At present, QC demands an array of devices, manual test stations, sample aliquots, and a large number of data entries, complicating management even before factoring in compliance requirements. Additional complexity arises from ongoing maintenance and calibration required by the assorted QC devices.

What are the barriers to QC automation? The intricate nature of the development process poses one obstacle. The creators of automated solutions must have mastery over a wide array of analytical technologies integral to radiopharmaceutical QC. Contrary to radiosynthesis (Alexoff, 2003; Elkawad et al., 2022; Krasikova, 2013), QC automation cannot leverage prior automated devices as a foundation—because none exist. As a result, rather than minor enhancements, a bottom-up development approach has been necessary.

Cost presents another issue. Layering automation on top of the current assortment of QC instruments would naturally inflate the total solution cost. Thus, a cost-effective automated solution demands either a reduction in the number of components or a decrease in the cost per component.

Lastly, regulatory challenges loom large. Any deviation from established QC procedures mandates rigorous validation. The prevalent sentiment in the industry is that procedures not aligning with pharmacopeial monographs cannot be employed. This mindset curbs the incentive for innovation. Moreover, it is not far from the truth, as the burden of validation for novel procedures is so considerable that it practically narrows down users' options to procedures that adhere to monographs. Furthermore, some conventional QC procedures have never been properly validated. Thus, for new procedures to replace them, the proposing party would need to validate both old and new procedures in order to prove equivalence or superiority.

Even though both radiosynthesis and QC have been needed since the dawn of PET imaging, automation in the former has outpaced the latter by several decades. The disparity in radiation exposure levels is likely the reason for this preference. Yet, the complexity of the necessary automation efforts persisted as a significant deterrent for a substantial period. Traditional QC demands a variety of complex instruments that are crafted as multipurpose laboratory tools with little in common. Furthermore, it depends on a collection of manual tests (like the Kryptofix 2.2.2 spot test) that lack instrumental counterparts.

4 Reimagining Quality Control: Breakthroughs in Automation

4.1 Conventional Solutions

A laboratory equipped to carry out the procedures listed in Table 1 can be established by either sourcing each component individually or by procuring a ready-to-use laboratory package. The former approach involves purchasing each individual piece of equipment from their respective manufacturers (such as GC, TLC scanner, endotoxin measurement device, dose calibrator), setting up test stations for manual tests (like pH, Kryptofix 2.2.2), formulating analytical procedures, drafting standard operating procedure (SOP) documents, and training the staff. This also necessitates creating a system of batch records with an auditable approval path (either paper or digital). On the other hand, a turn-key laboratory solution delivers all these elements in one package. The provider supplies all the equipment, installs and qualifies it, trains the personnel, provides SOPs ready for adoption, sets up a quality management system (QMS), and connects all devices into a laboratory information management system (LIMS) that produces a batch record. Current solutions that allow the execution of tests presented in Table 1 are available from LabLogic LTD (Sheffield, UK) and Elysia-Raytest GmbH (Straubenhardt, Germany). It is worth mentioning that in addition to the analytical equipment utilized to carry out measurements on a QC sample, other equipment, including an analytical balance, a fume hood, and a refrigerator, are also necessary for a functional QC laboratory implementing traditional methods.

The innovation explored in this section is portrayed as a gradual progression with a decreasing correlation to existing procedures. Automated methods may either (1) mirror the structure of current procedures, (2) emulate the function of current procedures, (3) replicate the output of current procedures, or (4) only match the product release decision. Each subsequent option is based on a greater deviation from traditional methods than the preceding one. The examples depict the progression from (1) to (3), while the approach (4) is likely to surface in the future and is addressed towards the end of this chapter.

4.2 Shrinking the Size Through a Cabinet

One common issue with QC laboratories is their substantial size. Expansive lab benches cluttered with a variety of equipment and manual test stations do not scale well. As a result, the first and most straightforward innovative step offering a tangible improvement over traditional QC is a spatial reconfiguration of traditional equipment that diminishes the overall spatial needs while ensuring all interaction points with the equipment remain easily reachable. Typically, laboratory cabinets both under and above the equipment are utilized to store supplies or are left unused. This space could be repurposed to house more analytical equipment. Elysia-Raytest has followed this approach, culminating in a commercially available solution named the QC-Cubicle (Fig. 1). This is a custom cabinet occupying only 1 square meter of floor space, configured to hold all QC equipment and manual test stations necessary for conducting FDG QC tests according to EP. Each instrument can be used independently or contribute to a full QC process. The cabinet also incorporates local shielding to minimize user radiation exposure. On-board QC equipment includes a GC, a dose calibrator, a high-performance liquid chromatography (HPLC) device, a camera (for Kryptofix 2.2.2 and other visual tests), a multichannel analyzer (MCA), an osmometer, a TLC scanner, an endotoxin test device, and a pH strip reader. The user still needs to operate all the equipment and carry out individual tests manually just like in a conventional QC lab. Validation, calibration, and maintenance needs also remain similar. However, these processes require a considerably smaller laboratory.

Fig. 1 QC-Cubicle

4.3 Integrating Automation

The subsequent tier of innovation builds on the previous concept by incorporating automation. Such solutions offer compactly arranged conventional equipment and include an automated system to distribute the QC sample among various testing stations. The added advantage of these solutions is that they diminish reliance on operator variability and the chance of human errors. This approach has been followed by several organizations, including Cardinal Health, Siemens, and Sumitomo.

A prototype system developed by Cardinal Health (Dublin, OH, USA) (Eshima et al., 2013) was designed to enhance compliance with cGMP regulations by linking all conventional QC procedures with software. The majority of the development centered around improving the flow of communication, enabling unprecedented control and coordination of tasks. Overall, this system aimed at enhancing data integrity and eliminating the human factor in the most error-prone aspects of QC, although some manual operations were still necessary.

QC tests supported by this system included color, clarity, pH, residual solvents, residual Kryptofix 222, bacterial endotoxin, radionuclidic identity, radionuclidic purity, radiochemical identity, and radiochemical purity. System components included HPLC (with UV, radiation, and conductivity detectors), GC, Endosafe PTS device (Charles River Laboratories, Inc., Wilmington, MA, USA), pH meter, and a dose calibrator with real-time and timestamped readings.

The system produced a wide range of quality reports that included (1) analytical tests performed on the PET tracer product, (2) product yield, (3) product failure reports, (4) failure reports for systems or subsystems used in product manufacturing, and (5) operator error reports. The aim was to consider all factors influencing the quality of the PET tracer within one centralized system—a potent concept developed with a focus on cGMP compliance.

The graphical interface allowed the user to select which tests to run and establish acceptance criteria based on the PET tracers under test. Moreover, as some manual operations remained necessary, the system guided the user on what actions to perform and when, minimizing the chance of errors and omissions.

The scheduling component of the software enabled automated preparation of multiple QC instruments to be ready for analysis scheduled at a specific date and time. In this setting, it was feasible that multiple products could be tested simultaneously on different system components, with the data directed to the correct batch records for each specific product.

Another development in the "automated cabinet" category came from Siemens Molecular Imaging (Knoxville, TN, USA) (Zhang et al., 2012). This system comprised commercial instruments and novel testing devices mounted onto a frame. Configured to receive a single product sample via a sample delivery cartridge, the system employed a rotary valve downstream of the cartridge to measure the sample into multiple aliquots and distribute them to different test stations through a network of channels and valves.

Sumitomo Heavy Industries (Tokyo, Japan) has offered a system in Japan that had integrated an HPLC, pH meter, and other components interconnected by a system of channels and valves. It was integrated together with the cyclotron into a complete radiopharmaceutical production solution (Sasaki & Tanaka, 2000).

Unfortunately, the resulting complexity of the systems in this group and the need for through cleaning in-between analyses diminished the solutions' industrial potential. As a result, the Siemens system remained at a prototype stage while the Sumitomo system was only successfully applied in Japan.

4.4 Incorporating Miniaturization

The next advancement on the innovation ladder involves the miniaturization of components, paving the way for transitioning from an automated equipment cabinet to a bench-top instrument with comparable functionality. This approach has been followed by GE Healthcare (Chicago, IL, USA) and Trasis (formerly QC1) (Liege, Belgium). A diverse array of miniaturized components is necessary to make such instruments genuinely benchtop.

In an ambitious effort towards miniaturization innovation, GE Healthcare envisioned a compact system relying on many miniaturized components with a disposable cartridge, containing most of the parts of the test stations that come in contact with the sample, while the majority of each miniaturized instrument untainted by the sample remained within the stationary system (Hansteen et al., 2016). The network of fluid channels establishes a continuous pathway from the sample reservoir to every test station and ultimately to the waste container. It proposes a complete fluidic coupling of all locations, ensuring that liquids traverse through a wholly enclosed fluid path. Emulating GE's radio-synthesis automation approach with FASTLab and TRACERLab products, such systems were designed to operate with minimal cleaning or delays. Subsystems for pH, chemical purity, radiochemical purity, radionuclidic purity, and appearance were integrated. Instruments within the system included an HPLC and capillary electrophoresis, alongside the design of innovative testing devices within the cassette for endotoxin, pH, and Kryptofix 222 analyses.

A pivotal innovation was a shift from gas chromatography to head-space analysis mass spectrometer for determining organic solvent concentrations. This approach challenges conventional paradigms since for decades GC has been an undisputed gold standard for residual solvent determination not only in PET industry, but across a broad range of pharmaceutical products.

Another unique feature of the GE concept was shielding. The shield was intended to be placed within the instrument, enclosing only the cassette where all analyses took place, which would allow for a dramatic reduction in shielding weight in comparison to other approaches.

The implementation of such a solution depends on the miniaturization of HPLC columns and detectors, making them part of a disposable cassette. It also calls for the development of multiple novel miniaturized subsystems. It is plausible that one day a disposable HPLC column with detectors may become technically and commercially viable independently, which could prompt a revisit to systems like the proposed concept.

This envisioned system, though remaining a concept and not yet brought to life, illustrates an innovative approach that breaks away from traditional methodologies.

The Trasis system in development called QC1 (Fig. 2) has combined miniaturized and traditional components. It is designed to accept PET tracer samples via a sample vial. Inside the QC1 unit, there exists a mechanism that distributes the sample between the system components, each of which performs a different test. The miniaturized components encompass a GC, gamma counter, and spectrometer. The "sample hub" is engineered to conduct pH, Kryptofix 222, and radio-TLC tests. The base system announced by Trasis has a footprint of 30 by 45 cm. Trasis is also planning a future upgrade of the system to include a radio-HPLC with varied chemical detectors and a gradient pump system.

The Trasis solution utilizes two types of subsystems: (1) stationary subsystems that receive samples via tubing and hence necessitate cleaning with a reliable line-clearance procedure and (2) the sample hub that relies on disposable components.

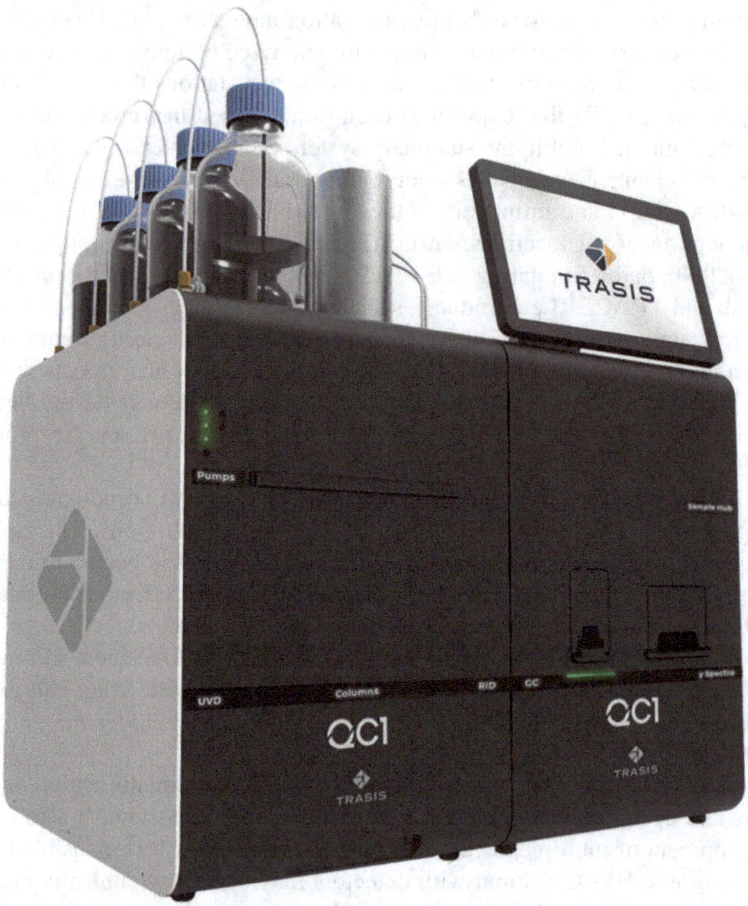

Fig. 2 QC1 base system with a future radio-HPLC concept

This approach keeps consumable usage to a minimum. System suitability tests and periodic calibration procedures are designed to be performed in an automated manner.

While the aforementioned systems primarily focus on incorporating miniaturized technologies into comprehensive QC systems, considerable academic progress is being made in other aspects (Ha et al., 2017). There is a growing focus on miniaturizing individual QC tests through microfluidics, and replacing HPLC with less complicated, compact chromatographic alternatives such as capillary electrophoresis (Ly et al., 2018). Such advances, once they mature, are predicted to lead to significant improvements in QC testing.

Indeed, it is expected that the adoption of these technologies could yield a trifecta of benefits: reducing the volume of the QC sample, shrinking the physical footprint of the testing instrument, and curtailing the amount of time required for analysis. However, it is important to note that the road from conceptualization to application can be filled with challenges, and patience and perseverance will be crucial as these technologies continue to develop and mature.

4.5 Leveraging Synergies for Component Minimization

While the innovation phases discussed previously primarily relied on the addition (e.g., cabinet, sample delivery system) or substitution of components (miniaturization), this set of advancements is aimed at simplifying systems through reduction of the number of components. The foundational idea here is to maximize the usage of each system component for multiple QC tests, thus obviating the need for other components. Solutions following this approach have been developed by Best ABT Molecular Imaging, Inc. (Louisville, TN, USA) and Trace-Ability, Inc. (Los Angeles, CA, USA).

The ultimate objective of Best ABT is the Biomarker Generator (Awasthi et al., 2014), a compact system integrating a cyclotron, radiosynthesis module, and QC module. This unified system enabled the opportunity to capitalize on synergies between the three normally discrete systems and processes.

The Best ABT system uses HPLC as the backbone of the QC system (Anzellotti et al., 2013, 2015), a direction that aligns with a similar approach previously adopted in academia (Koziorowski, 2010). Innovations in columns, detectors, temperature regulation, and the mobile phase facilitated ABT's ability to conduct the following tests on the HPLC: residual solvents (ethanol and acetonitrile), radiochemical identity and purity, and Kryptofix 2.2.2 testing. The only additional hardware required was a microelectrode measurement device for pH determination, which was paired with a syringe driver for sample delivery. Concurrently, the filter integrity test was executed on the synthesis module, a component of the same integrated Biomarker Generator system. There was also an opportunity to measure color and clarity in-line as the sample was transferred to the final product vial.

The Best ABT Biomarker Generator system is distinguished by its synthesis and QC integration. There have been no prior instances of systems where QC (even partially) depended on the synthesis hardware. However, automated sampling and HPLC injection have been previously implemented in a unique system at KFA Jülich (Crouzel et al., 1993).

Best ABT's methodology represents a fusion between disposable and reusable components. The synthesis cassette was designed for single-use. Therefore, QC functions conducted within it (appearance and filter integrity testing) also relied on the features of the disposable cassette. In contrast, chromatographic tests and pH measurement relied on enduring hardware and a system of tubing and valves for fluid transfers.

While certain tests, including radionuclidic purity/identity, sterility, and endotoxin were not incorporated, the solution was rolled out as a marketable product and deployed in the field, resulting in significant simplification of QC in comparison to conventional approaches. As per ABT's historical records, the inaugural fully integrated Biomarker Generator system, complete with automated QC, was set up and certified at Sveta Marina Hospital in Varna, Bulgaria in 2013.

Meanwhile, Trace-Ability, Inc. further pushed the boundaries in the pursuit of synergies aimed at decreasing the number of components.

Trace-Ability's guiding principle was to focus on product quality, established via the data produced by QC tests. The approach to obtain those data, as long as it was verified and trustworthy, was considered secondary. This emphasis on the objective over the means to achieve it, empowered Trace-Ability to broaden their technological palette and realize synergies absent in traditional QC approaches.

They arrived at a juncture where all necessary tests for FDG release in the United States (excluding filter integrity) could be performed on a single analytical instrument, a microplate reader (Hennessy et al., 2017; Kedar et al., 2017). Introduced in 2017, the Tracer-QC product (Fig. 3) with a footprint of the base system of 34 by 53 cm represents a conceptual shift from its predecessors.

Trace-Ability selected a plate reader as the primary analytical instrument. Plate readers have been extensively refined over decades for diagnostic (Sun et al., 2018) and other industry applications (Yang et al., 2022). Their capacity for measuring absorbed and emitted light from microplates has continuously improved due to competition amongst various manufacturers. A parallel evolution has occurred in the microplates utilized for analysis in plate readers. Trace-Ability capitalized on this development and directed their innovation towards facilitating new methods based on the capabilities of contemporary plate readers. Solutions for all QC tests are derived through light measurements (absorbance and luminescence) within a plate reader. While this makes the analytical instrument remarkably straightforward, it necessitates the invention of new tests. These novel tests and the custom plates that enable them represent the central innovation produced by Trace-Ability. FDG QC tests are categorized as follows:

- Absorbance tests, such as color and clarity, can be conducted on a pure sample in a microplate by assessing the light absorbed by the sample.
- Indicator-based absorbance tests include pH, phase transfer reagents, endotoxin, organic solvents, and excipients. These tests require the sample to be mixed with an indicator that alters color in response to the concentration of any of the mentioned components. This color change is contrasted with the color change obtained simultaneously from multiple standards, delivering a well-referenced measurement of the analyte's concentration.
- Radiation tests, encompassing radionuclidic identity, radioactivity concentration, radiochemical purity, and radiochemical identity. Most of these tests are enabled through the sample's interaction with scintillating materials within the custom microplate that emit light upon exposure to radiation. The light is gauged by the plate reader in luminescence mode and converted into the value of each of the specified properties. Radionuclidic identity is determined via half-life measurement assessing signal over time. Radiochemical identity and purity are determined via assessment of signal distribution across the surface of a TLC for

Fig. 3 Tracer-QC rHPLC

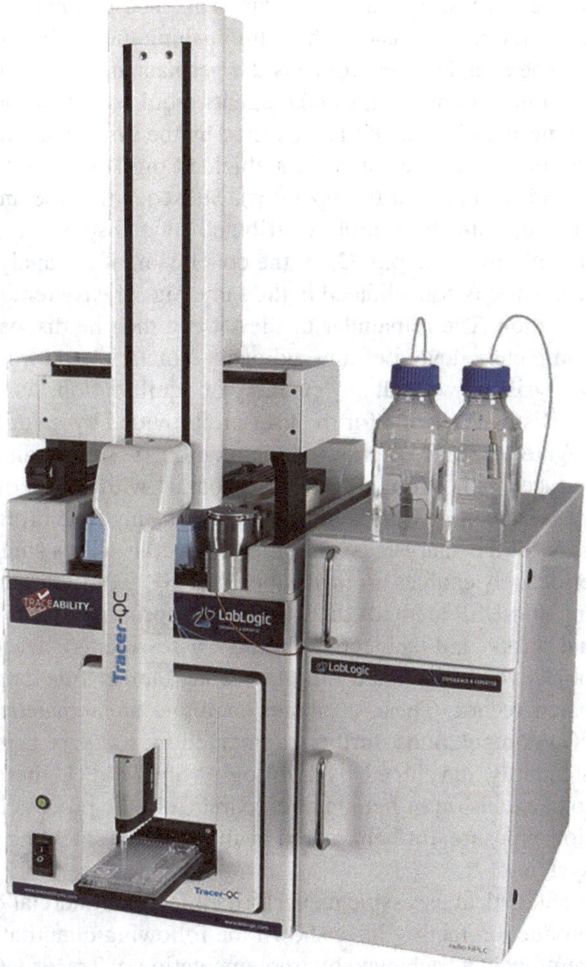

USP FDG (and via conventional HPLC for all other tracers). Radioactivity concentration is assessed by correlating radioactive signal and volume.

All these analyses are executed within a single custom microplate inside a plate reader, while sample operations (dispensing and indicator mixing) are managed by a compact automated pipettor positioned over the plate reader. This leads to a wholly automated, hands-off process from FDG sample to QC report in a small space no wider than 14 inches. Such an approach, unrestricted by conventional methods, permitted a significant simplification of the hardware platform. Without a rigid structure of channels and valves, processes can be quickly modified, appended, or discarded. The microplates provide an abundance of wells, enabling a broad spectrum of analyses to be conducted. Consequently, samples and standards are analyzed under identical conditions and simultaneously, offering in situ system suitability testing with every run. This also facilitates duplicate testing, enhancing accuracy and precision while diminishing the likelihood of invalid test results.

The sample never contacts the permanent parts of the instrument directly, since all contact components and reagents required for one analysis run are packed into a single-use kit. This kit is identified by the system and installed before analysis. The sample is then delivered in a shielded pig keyed to the instrument (which is also shielded to protect the operator). Subsequently, the analysis is automatically conducted, with the sample distributed by a disposable pipette to the test locations directly from the pig. Upon the conclusion of the analysis, the majority of radioactive waste is consolidated in the same pig for easy removal without user exposure to radiation. The remainder of the kit can then be disposed of, rendering the system completely devoid of any radioactive material and immediately ready for the next analysis run without any cleaning or equilibration.

Distinct benefits for the user are provided by an entirely disposable path which negates the risk of cross-contamination. System maintenance is also significantly reduced as no liquids come into contact with any permanent components, and the system itself possesses minimal components. The inventory of consumables necessary for the analysis is simplified to one kit with a single expiry date. This solution effectively enables users to adhere to USP quality standards for FDG, while employing non-USP methods that have been appropriately validated. By removing all manual tasks and subjective human assessments, Tracer-QC delivers completely objective results and a fully traceable, tamper-free data flow from measurement to batch record. These attributes facilitate an unmatched ease of compliance with cGMP regulations, further augmented by software compliant with 21 CFR Part 11 standards, developed by LabLogic (Sheffield, United Kingdom). Ultimately, the total exclusion of human interaction from the process leads to a reduction in radiation exposure, further assured by the compact shielding incorporated into Tracer-QC systems.

Statistical assessments performed in a commercial cGMP radio-pharmaceutical production facility have shown the following quantitative improvements over conventional QC achieved by implementation of Tracer-QC: (1) 57% reduction in staff time required for QC and QC-related activities, (2) 83% reduction in number of SOP's required for QC, (3) 69% reduction in average radiation exposure of staff. Furthermore, analysis of this manufacturer's data coupled with a site audit by an independent third party (Proceutical Ltd) has concluded that Tracer-QC implementation addresses 77% of quality risks and reduces the chance of Form 483 violations in FDA audits by 84%, dramatically improving 21 CFR Part 212 (cGMP) compliance.

All Tracer-QC methods have been validated for both "fit-for purpose" and equivalence to compendial QC methods (Marcus, 2018).

In order to accommodate the variety of PET tracers beyond FDG, the Tracer-QC system has an optional radio-HPLC subsystem. Although it introduced a shared liquid path, the level of automation ensures that HPLC cleaning and equilibration occur without user intervention. Furthermore, integrating HPLC with a precise automated liquid handler permits the quantitative use of internal standards (included within the kit) in every sample injection. This reduces the number of injections needed and delivers more reliable quantification that has been unattainable with standalone HPLCs, even those equipped with automated injectors which cannot mix sample and standards in precise ratios.

Tracer-QC rHPLC product supports a range of PET tracers by conducting the tests common to all tracers on the plate reader concurrently with running tracer-specific chromatographic tests on the rHPLC component. This platform presently enables most QC tests required for the FDA-approved and most commonly used PET tracers.

5 A Comparative Analysis: Choosing the Right Automated QC System

The preceding section outlined the evolution in PET tracer QC innovation in a sequence correlated to an increasing divergence from the traditional manual methods. The ensuing discussion concentrates on the benefits available at each stage for cGMP production of a range of PET tracers. Each stage presents some trade-offs. Each user is thus required to ascertain the net value these trade-offs present in context of their use case.

Transitioning from a standard lab to a cabinet sees a gain in floor space that is counterbalanced by the additional expense of the cabinet. If a user needs to broaden their tracer portfolio within the current facility, then the extra space is a worthwhile expenditure.

Moving to an automated cabinet solution involves an investment not only in the cabinet but also in the added automation, on top of the cost of individual instruments. However, this investment should be compensated by the savings in labor that is reduced via automation.

The shift to miniaturization may decrease the overall cost of the solution and offers further space savings compared to both prior solutions. The precise trade-off value will need to be calculated based on statistical analysis of post-deployment running costs across multiple locations.

Transitioning to a synergistic system provides overall operational simplification, user-friendliness, and compliance, concurrently offering a cost reduction.

On the whole, the instrument complexity escalates and then declines as one advances along this sequence. Supplementing existing instruments with automation increases complexity. Miniaturization may or may not decrease it. Only a synergistic approach allows for a noticeable reduction in complexity. Inherently, the synergistic approaches reduce the limits of flexibility of the QC system. However, traditional PET QC has utilized only a small fraction of the capabilities of the general-purpose instruments it relied upon thus having only potential flexibility at the cost of having everyday operational complexity. Thus, it is expected that conventional QC approaches will maintain their position in research where flexibility is a priority. Meanwhile, automated and simplified systems will be preferred in routine production where efficiency, reliability, and compliance are prioritized ahead of flexibility.

It is also crucial to acknowledge the incentives behind each of the presented developments. The GE, Sumitomo, and Best ABT solutions seem to have been motivated by integration within the total tracer-production solution. Specifications were shaped by the integration with the cyclotron and specific chemistry modules and centered around the capabilities and requirements of the latter. The Best ABT design was further refined by the dose-on-demand concept. The Siemens and Cardinal systems' specifications were driven by the needs of their respective high-volume FDG production networks (PETNET Solutions and Cardinal Health NPS). Trasis and Trace-Ability developments were influenced by broad use cases.

6 Navigating the Legal Landscape: Regulatory Compliance in Automated QC

6.1 Regulatory Advantages Derived from Automated QC

Automated QC not only provides a streamlined tool for regulatory adherence in the manufacturing of radiopharmaceuticals but also affords numerous advantages to the regulators themselves upon its widespread application. It affords regulatory authorities a concrete, verifiable record of manufacturing performance, eliminating the need to make inferences from secondary data or struggle with poorly traceable hand-drawn records. As a result, the FDA emerges as a key stakeholder and beneficiary of QC automation. Hence, QC automation serves not only as a compliance instrument for its users but also as a QA and inspection tool for regulators, offering them an unquestionable record of cGMP compliance that simplifies inspections and dispels concerns over product quality.

It is therefore vital for those developing QC automation to initiate dialogue with the FDA and their counterparts in other countries as early as possible in their product development process. They should strive to understand the needs of the regulators, manage their expectations, become familiar with their concerns, and inform them about forthcoming solutions well ahead of their launch. Ideally, by the time a solution enters the market, the regulatory agencies are well-informed, and the companies have a clear understanding of what performance evidence and regulatory submissions are expected for their products by each agency.

In the United States, Trace-Ability set an example by starting its engagement with the FDA in advance of the product launch, collaboratively identifying the agency's requirements, product performance characteristics, user specifications, and validation strategy with the Agency's Technical Committee well before these procedures were put into action.

6.2 Regulatory Approval Framework for Automated Quality Control (QC)

A review of the definitions listed in the USP General Chapter <823> could greatly assist the discussion concerning regulatory aspects of these automated QC testing methodologies. The process of "Validation" refers to the creation of documented proof that a given system, method, or procedure fulfils its purpose. "Verification" is the process of confirming that an established system, method, or process meets set standards. Performance Qualification (PQ) shows that a piece of equipment performs the necessary tasks for the production and testing of PET drugs, delivering expected outcomes. PQ should elucidate the tasks a machine must perform. If using a USP test method, it should be verified to confirm its effectiveness under actual usage conditions. Non-compendial test methods applied in PET drug testing should be trustworthy and specific, as confirmed through validation.

Introducing changes to the production process of an approved drug, such as automated QC, necessitates the submission of an NDA supplement to the FDA (Guidance for Industry Changes to an Approved NDA or ANDA, 2004; Guidance for Industry CMC Postapproval Manufacturing Changes To Be Documented in Annual Reports, 2014; Herscovitch, 2022) in the United States or analogous regulatory submissions in other countries. These submissions demand substantial proof supporting the novel method, which can be burdensome. However, if the change incorporates an already validated compendial method, then a simple verification suffices, and a CBE-30 supplement (changes taking effect within 30 days barring FDA objections) could be adequate (Guidance for Industry Changes to an Approved NDA or ANDA, 2004). Hence, many QC automation endeavors have focused on compendial methods. Yet, the internal intricacies of automated systems featuring a range of instruments connected via a network of liquid conduits and valves necessitate the validation of system robustness, absence of cross-contamination, and analytical performance of the methods under the net effect of the instrumentation changes. Therefore, the more intricate regulatory path of a prior approval supplement (PAS) is necessary for these systems, even if they utilize USP methods. The PAS mechanism is also suitable for automated systems that use non-USP methods, which must be validated before implementation.

Validation typically aims to establish the equivalence (Lebrun et al., 2008) of the new method with the method it is replacing. This approach is most suitable for automated QC systems using pharmacopeial methods. However, it's also necessary to validate that methods that are equivalent in isolation are not influenced by other system parts. The performance of a method must be demonstrated in the context of the automated equipment and process. For instance, the robustness of the sample/aliquot management system must be validated to ensure that it does not interfere with tests that might otherwise be equivalent.

The situation becomes more complicated when new methods significantly differ from traditional methods and are non-compendial. This leads to two types of challenges:

1. The inability to compare: The traditional method was historically implemented without proper validation. The spot test for Kryptofix 222, which is currently in use, does not have any record of accuracy, precision, linearity, or limit of detection (LOD). Therefore, setting criteria for a valid comparison of the new method with the old one becomes difficult without first validating the old method.
2. The new method is inferior to the old one: This may seem illogical, i.e., why would anyone opt for an inferior method? But as discussed previously, some equipment used in QC of PET tracers is general-purpose laboratory equipment, overly capable for the purpose of PET tracer QC. One example is the analysis of residual solvents. An LOD on a GC is typically in ppb (parts per billion). Meanwhile, the release test only needs to determine whether the acetonitrile concentration in the sample is above or below 400 ppm (parts per million). For this purpose, sub-ppm LOD is irrelevant. A device with a 200 ppm LOD may be adequate for this task. The purpose for which GC is used in QC of PET tracers does not necessitate ppb sensitivity. Therefore, methods should be evaluated based on how well they respond to the QC query, rather than their overall characteristics. Hence, a method with a 200 ppm LOD should be sufficient for an acetonitrile test, provided its accuracy and precision allow 95% confidence in determining a pass/fail outcome. These methods should be validated using a method other than overall equivalence or superiority.

Such a validation approach is called "fit for purpose" (Lee et al., 2006), correlating with the USP definition of validation, where the "intended requirements" for the method should be established based on the response it needs to deliver rather than overall analytical performance. Then, the validation study should be designed to assess whether the method meets these requirements. This eliminates the two challenges highlighted above.

The only automated QC solution to date that has undergone formal validation (in the United States and in Europe), with corresponding submissions to regulatory authorities, is Tracer-QC (as of the submission date of this chapter in review by the FDA and approved by the Norwegian Medicines Agency). Both validation approaches mentioned above were utilized as requested by the FDA for different analytical methods enabled by the system.

A separate part of the study focused on robustness, challenging the system with various environmental conditions, or operating near the limits of acceptable conditions. This approach relied on Tracer-QC's ability to measure multiple parameters at once and allowed for comprehensive testing of potential interferences, effectively setting very high standards for the specificity of the individual tests. The validation study, in practice, consisted of repeated analyses of multiple FDG samples spiked with a mixture of specific impurities or water. Known concentrations of the impurities were produced by meticulously preparing spiking solutions. For each run, an FDG sample was mixed with the spiking solution. FDG samples were produced according to the standard clinical production protocol. This method allowed for the analytical performance of the individual tests on the Tracer-QC platform to be measured.

The validation studies have greatly benefited from cooperation with the FDA. This allowed the development team to familiarize the Agency with the system before designing the validation studies. The studies were reviewed with the FDA before being executed. This approach allowed the Agency to scrutinize the methods and the validation approach before any labor-intensive laboratory work commenced. Consequently, when the lab work began, the developers and regulators had agreed on the validation approach. During the study, interim validation reports were also reviewed with the Agency, resulting in the identification of areas where more data was required. In the end, the FDA Technical Committee found the data sufficient to conclude that the integrated multi-parametric QC method on Tracer-QC was successfully validated. Trace-Ability now maintains a Type V DMF (#029891) with the FDA, which includes system description and validation reports. By cross-referencing this DMF, new users can implement Tracer-QC without the need for repeating complete validation. To the best of the authors' knowledge, all other QC solutions do not have DMF's and require complete method validation by every user.

The experience shows that the validation of something significantly different from compendial methods cannot be simply prescribed. It necessitates a proactive dialogue with regulators and joint development of an approval path, which is more productive than developing a validation strategy in isolation, hoping it aligns with regulators' expectations after it has been carried out. Such an approach risks producing a data set that is insufficient or irrelevant for the regulators' decision about the new technology.

6.3 Integration into Pharmacopeia

The expectation is that automated techniques will eventually become the norm and be integrated into pharmacopeial monographs. The USP standards development procedure allows any individual or group to raise an issue pertaining to the setting of standards, from the creation of a new standard to the modification of an existing one.

The USP-NF combines two compendia—the USP and the National Formulary (NF). Proposals for monographs for the USP-NF are submitted in written form to the USP, complete with reasons for the proposals and supporting data. Detailed guidance on monograph submission is provided by the USP, including advice on what data should accompany specific proposals (USP Guideline for Submitting Requests for Revision to USP–NF General Information for All Submissions, 2016). Regularly, pharmaceutical companies holding an approved NDA or an abbreviated new drug application (ANDA) for the product in question sponsor these monographs. The sponsors should include validation data for the analytical methods used for the drug product as part of their supporting data from their approved application.

Once a monograph is sponsored by an organization, it is evaluated by a USP panel composed of USP scientists and expert volunteers. This panel scrutinizes the draft to confirm the sufficiency of data supporting the proposed method, whether it is equivalent to existing methods, or fit for the purpose intended. When necessary, laboratory experiments are also conducted.

Once a new standard is crafted for the USP-NF, it is presented for a 90-day public review and commentary phase in the Pharmacopeial Forum (PF). Following this phase, all comments are reviewed by the Expert Committee, who then decide whether the standard requires further modifications. To cement the standard, independent electronic balloting is conducted by the Expert Committee members. Any standards deemed official by the USP are then established in its various compendia.

Interestingly, the USP permits the inclusion of multiple alternative analytical methods in monographs for the same drug simultaneously. As long as the sponsor maintains the monograph, it remains active, enabling users to select the most suitable method for them. This process is suitable for integrating automated QC solutions into USP standards. The monograph, or a revision of it, will be sponsored by either the manufacturer of the automated QC solution or the manufacturer of the PET drug dependent on automated QC. The review procedure, which includes comprehensive data analysis, will lead to a decision regarding the new or revised monograph.

7 QC Automation and Patient Benefits: An Undeniable Link

The move towards QC automation in PET drug production is a significant stride towards enhancing patient safety. By improving the consistency and accuracy of drug production, patients receive reliable, uniform doses, which directly influences the efficacy of disease diagnosis and confidence in treatment planning. The potential for improved therapeutic monitoring as a result of QC automation is considerable, reflecting positively on patient outcomes.

The link between QC automation and patient safety is well substantiated by empirical evidence. Numerous case studies and research findings underline how automation has notably minimized errors, improved reproducibility, and enhanced throughput. These advances collectively contribute to patient safety, underlining the imperative nature of adopting automated QC processes in PET drug production.

As we look ahead, QC automation's role is set to become even more pivotal, fueled by technological advancements. With the rise of AI-powered predictive analytics and deep learning, the promise of unprecedented robustness in QC processes is on the horizon. Such technologies can potentially push the boundaries of precision and safety in PET drug production, leading to more effective radiopharmaceuticals and, consequently, improved patient outcomes.

QC automation has evolved from being a mere logistical advantage to a necessity in PET drug production. It has brought about an era where patient safety is front and center, benefiting healthcare providers, patients, and the broader pharmaceutical industry. As technological evolution continues to accelerate, the connection between QC automation and patient safety will inevitably strengthen, ushering in a new era of radiopharmaceutical production characterized by safety, effectiveness, and precision. However, the logistical advantage should not be underestimated either. QC automation enables a PET drug manufacturer to increase their throughput and diversify the drug portfolio they offer to the clinics enabling every patient to receive the PET imaging they need in timely manner.

8 Emerging Technologies: Machine Learning and AI in Quality Control

Artificial intelligence (AI) and machine learning (ML) are currently at the forefront of the hottest topics in most business domains, though their use in QC is only just beginning.

Through their ability to comprehend incredibly vast and complex data sets, AI and ML offer unparalleled ability to not only analyze and understand the data but also predict, identify, and decipher complex patterns which grant solutions to problems given by their operators.

By creating predictive models that analyze data from isotope production and synthesis stages, they can foretell potential QC issues or failures before they happen. The ability to detect subtle, often humanly indiscernible data patterns, allows these technologies to identify problems early, based on comparisons with previous QC result datasets. Over time, AI and ML can monitor and detect the deterioration of QC tests, indicating a gradual degradation of test reliability.

While at first glance AI and ML used in this way might seem akin to existing Out-of-Spec (OOS) trending techniques, they stand apart in their ability to decipher intricate patterns. By modelling intricate, non-linear data trends, they are adept at identifying subtle anomalies, even amidst substantial noise. Furthermore, they can seamlessly adjust to emerging data trends, ensuring precise, real-time feedback and analysis, even with evolving datasets.

Such feedback can suggest the need for equipment maintenance, further investigations, or signal potential test failures, thus contributing significantly to the reliability and consistency of the radiotracer production and quality control process.

In the near future, it's conceivable that AI and ML will integrate with automated QC systems like Tracer-QC from LabLogic Systems Ltd. and Trace-Ability. These systems won't just manage routine QC tasks seamlessly but could also adeptly address unforeseen batch issues. They could either self-correct these issues or provide operators with in-depth reports, thereby reducing the need for time-consuming manual investigations. Although many current QC tools already produce detailed logs and reports, the innovator is the self-improving and adaptive nature of AI and ML. Over time, these systems enhance the precision of their predictive models with minimal human oversight.

To maximize the benefits of AI and ML in radiopharmacy, collaborations between data scientists and domain experts are vital, as is access to high-quality, curated datasets for model training and validation. Rigorous validation and verification of these models must precede their clinical deployment to ensure safety and effectiveness. Despite these challenges, the combined power of AI and ML has the potential to dramatically transform Quality Control and radiopharmacy.

9 Summary

The PET drug landscape has changed dramatically over the past decade. New tracers have entered the market and many more are in development. Production facilities are under the most severe capacity constraint that the industry has ever seen. That constraint arises primarily from shortage of skilled labor and space coupled with increasing regulatory pressure. QC automation addresses all of these issues and is therefore positioned to become the key enabler of further growth of PET with a diverse portfolio of tracers. The companies developing QC automation should focus on products that can support not only today's tracer portfolio, but also the upcoming products (in both diagnostic and therapeutic space).

It is expected that with future addition of innovative in-process controls QC will be more integrated with manufacturing, with first such attempts illustrated in this chapter.

Different solutions presented in this chapter and those in development or just emerging cannot be compared on absolute basis. Each user's PET tracer production paradigm has its unique pain points and opportunities. Thus, we encourage everyone to consider all available QC solutions available (commercially launched, validated and approved by regulators) at the time of their decision making from the perspective of resulting benefits in their specific facility. The authors hope that the arsenal of such solutions expands over time giving users more options to choose from in order to address their production bottlenecks and expand global availability of PET imaging to patients.

Disclaimer and Acknowledgements Funding for Tracer-QC development was partly provided by the US Food and Drug Administration through grant U01FD005517. The views expressed in this chapter do not necessarily reflect the official policies of the Department of Health and Human Services; neither does the mention of trade names, commercial practices, or organization imply endorsement by the United States Government.

References

Alexoff, D. L. (2003). Automation for the synthesis and application of PET radiopharmaceuticals. In *Handbook of radiopharmaceuticals* (pp. 283–305). Wiley. https://doi.org/10.1002/0470846380.ch8

Anzellotti, A. I., McFarland, A. R., Ferguson, D., & Olson, K. F. (2013). Towards the full automation of QC release tests for [18]fluoride-labeled radiotracers. *Current Organic Chemistry, 17*, 2153–2158. https://doi.org/10.2174/13852728113179990107

Anzellotti, A., Bailey, J., Ferguson, D., McFarland, A., Bochev, P., Andreev, G., Awasthi, V., & Brown-Proctor, C. (2015). Automated production and quality testing of [18F]labeled radiotracers using the BG75 system. *Journal of Radioanalytical and Nuclear Chemistry*. https://doi.org/10.1007/s10967-015-4045-1

Awasthi, V., Watson, J., Gali, H., Matlock, G., McFarland, A., Bailey, J., & Anzellotti, A. (2014). A "dose on demand" biomarker generator for automated production of [(18)F]F(−) and [(18)F]FDG. *Applied Radiation and Isotopes, 89*, 167–175. https://doi.org/10.1016/j.apradiso.2014.02.015

Berger, G., Maziere, M., Knipper, R., Prenant, C., & Comar, D. (1979). Automated synthesis of 11C-labelled radiopharmaceuticals: Imipramine, chlorpromazine, nicotine and methionine. *The International Journal of Applied Radiation and Isotopes, 30*, 393–399.

British Pharmacopoeia Commission. (2000). *Monographs: Radiopharmaceutical preparation, Fludeoxyglucose [18F] injection*. British Pharmacopoeia Commission.

Chi, Y.-T., Chu, P.-C., Chao, H.-Y., Shieh, W.-C., & Chen, C. C. (2014). Design of CGMP production of 18F- and 68Ga-radiopharmaceuticals. *BioMed Research International, 2014*, 680195. https://doi.org/10.1155/2014/680195

Clarke, B. N. (2018). PET radiopharmaceuticals: What's new, what's reimbursed, and what's next? *Journal of Nuclear Medicine Technology, 46*, 12–16. https://doi.org/10.2967/jnmt.117.205021

Code of Federal Regulations. (2013). Title 21, Volume 4, Part 212: Current Good Manufacturing Practice for Positron Emission Tomography Drugs.

Code of Federal Regulations. (2018a). Title 21, Volume 5, Part 312, Investigational New Drug Application.

Code of Federal Regulations. (2018b). Title 21, Volume 5, Part 361, Section 361.1 Radioactive Drugs for Certain Research Use. FDA.

Comar, D., Zarifian, E., Verhas, M., Soussaline, F., Maziere, M., Berger, G., Loo, H., Cuche, H., Kellershohn, C., & Deniker, P. (1979). Brain distribution and kinetics of 11C-chlorpromazine in schizophrenics: Positron emission tomography studies. *Psychiatry Research, 1*, 23–29.

Crişan, G., Moldovean-cioroianu, N. S., Timaru, D. G., Andrieş, G., Căinap, C., & Chiş, V. (2022). Radiopharmaceuticals for PET and SPECT imaging: A literature review over the last decade. *International Journal of Molecular Sciences, 23*, 5023. https://doi.org/10.3390/IJMS23095023/S1

Crouzel, C., Clark, J. C., Brihaye, C., Långström, B., Lemaire, C., Meyer, G. J., Nebeling, B., & Stone-Elander, S. (1993). Radiochemistry automation for PET. In *Radiopharmaceuticals for positron emission tomography* (pp. 45–89). Springer. https://doi.org/10.1007/978-94-015-8204-9_2

Elkawad, H., Xu, Y., Tian, M., Jin, C., Zhang, H., Yu, K., & He, Q. (2022). Recent advances in microfluidic devices for the radiosynthesis of PET-imaging probes. *Chemistry, An Asian Journal, 17*, e202200579. https://doi.org/10.1002/ASIA.202200579

Eshima, D., Husnu, M., & Stone, J. (2013). Method and system for automated quality control platform for radiopharmaceuticals. US 2013/0018618 A1 Filed July 16, 2012.

Ferguson, D., McGrath, S., O'Hara, G., & Marshall, C. (2011). Investigation of staff finger doses during quality control of FDG production. *Health Physics, 100*, 523–529. https://doi.org/10.1097/HP.0b013e3181f1e9b5

Fludeoxyglucose [18F] injection. (2002). *European pharmacopoeia* (4th ed.). European Directorate for the Quality of the Medicines.

Fludeoxyglucose F 18 injection. (2011). Sample Formats: Application to Manufacture Ammonia N 13 Injection, Fluorodeoxyglucose F 18 Injection (FDG F 18) and Sodium Fluoride F 18 Injection Chemistry, Manufacturing, and Control Sections. US Food and Drug Administration.

Fludeoxyglucose F 18 Injection. (2020). The United States pharmacopeia. The National formulary USP43-NF38, 1946.

Guidance for Industry Changes to an Approved NDA or ANDA. (2004). U.S. Department of Health and Human Services Food and Drug Administration Center for Drug Evaluation and Research (CDER) CMC.

Guidance for Industry CMC Postapproval Manufacturing Changes To Be Documented in Annual Reports. (2014). U.S. Department of Health and Human Services Food and Drug Administration Center for Drug Evaluation and Research (CDER).

Ha, N. S., Sadeghi, S., & van Dam, R. M. (2017). Recent progress toward microfluidic quality control testing of radiopharmaceuticals. *Micromachines, 8*(11), 337. https://doi.org/10.3390/MI8110337

Haka, M., Webster, E., Gutierrez, P., Young, K., & Zigler, S. (2013). Sterility testing of PET radiopharmaceuticals. *Journal of Labelled Compounds and Radiopharmaceuticals, 56*.

Hamacher, K., Coenen, H. H., & Stöcklin, G. (1986). Efficient stereospecific synthesis of no-carrier-added 2-[18F]-fluoro-2-deoxy-D-glucose using aminopolyether supported nucleophilic substitution. *Journal of Nuclear Medicine, 27*, 235–238.

Hansteen, O.H., Jakobsen, J.B., Lindgaard, S.-E., Franci, X., Roed, L., Andersen, A., Johansen, G., Knuttel, T., Aukland, T., Husbyn, M., Jakobsen, B., & Langseth, M. (2016). *Quality control devices and methods for radiopharmaceuticals*. US Patent 9,291,606 B2 filed 22 August 2011, issued 22 March 2016.

Hennessy, R. C., Stougaard, P., & Olsson, S. (2017). A microplate reader-based system for visualizing transcriptional activity during in vivo microbial interactions in space and time. *Scientific Reports, 7*, 281. https://doi.org/10.1038/s41598-017-00296-4

Herscovitch, P. (2022). Regulatory agencies and PET/CT imaging in the clinic. *Current Cardiology Reports, 24*, 1361–1371. https://doi.org/10.1007/S11886-022-01749-7/TABLES/2

Huang, Y.-Y. (2019). An overview of pet radiopharmaceuticals in clinical use: regulatory, quality and pharmacopeia monographs of the United States and Europe. *Nuclear Medicine Physics*. https://doi.org/10.5772/INTECHOPEN.79227

Hung, J. C. (2002). Comparison of various requirements of the quality assurance procedures for 18F-FDG injection*. *Journal of Nuclear Medicine, 43*, 1495–1506.

Hung, J. C., Schwarz, S. W., Zigler, S. S., & Ravichandran, R. (2011). Revision of USP general chapter radiopharmaceuticals for positron emission tomography—compounding <823>. *Pharmacopeial Forum, 37*.

Inspection Observations data and information summarizing the areas of regulation cited on FDA form 483. [WWW Document], 2018. https://www.fda.gov/iceci/inspections/ucm250720.htm.

Kedar, P., Desai, A., Warang, P., & Colah, R. (2017). A microplate reader-based method to quantify NADH-cytochrome b5 reductase activity for diagnosis of recessive congenital methaemoglobinemia. *Hematology, 22*, 252–257. https://doi.org/10.1080/10245332.2016.1252873

Kelloff, G. J., Hoffman, J. M., Johnson, B., Scher, H. I., Siegel, B. A., Cheng, E. Y., Cheson, B. D., O'shaughnessy, J., Guyton, K. Z., Mankoff, D. A., Shankar, L., Larson, S. M., Sigman, C. C., Schilsky, R. L., & Sullivan, D. C. (2005). Progress and promise of FDG-PET imaging for cancer patient management and oncologic drug development. *Clinical Cancer Research, 11*, 2785–2808. https://doi.org/10.1158/1078-0432.CCR-04-2626

Koziorowski, J. (2010). A simple method for the quality control of [18F]FDG. *Applied Radiation and Isotopes, 68*, 1740–1742. https://doi.org/10.1016/j.apradiso.2010.03.006

Krasikova, R. (2013). PET radiochemistry automation: State of the art and future trends in 18F-nucleophilic fluorination. *Current Organic Chemistry, 17*, 2097–2107. https://doi.org/10.2174/13852728113179990102

Lebrun, P., Govaerts, B., Debrus, B., Ceccato, A., Caliaro, G., Hubert, P., & Boulanger, B. (2008). Development of a new predictive modelling technique to find with confidence equivalence zone and design space of chromatographic analytical methods. *Chemometrics and Intelligent Laboratory Systems, 91*, 4–16. https://doi.org/10.1016/J.CHEMOLAB.2007.05.010

Lee, J. W., Devanarayan, V., Barrett, Y. C., Weiner, R., Allinson, J., Fountain, S., Keller, S., Weinryb, I., Green, M., Duan, L., Rogers, J. A., Millham, R., O'Brien, P. J., Sailstad, J., Khan, M., Ray, C., & Wagner, J. A. (2006). Fit-for-purpose method development and validation for successful biomarker measurement. *Pharmaceutical Research, 23*, 312–328. https://doi.org/10.1007/s11095-005-9045-3

Lovinfosse, P., Visvikis, D., Hustinx, R., & Hatt, M. (2018). FDG PET radiomics: A review of the methodological aspects. *Clinical and Translational Imaging, 6*, 379–391. https://doi.org/10.1007/s40336-018-0292-9

Ly, J., Ha, N. S., Cheung, S., & van Dam, R. M. (2018). Toward miniaturized analysis of chemical identity and purity of radiopharmaceuticals via microchip electrophoresis. *Analytical and Bioanalytical Chemistry, 410*, 2423–2436. https://doi.org/10.1007/S00216-018-0924-Y/METRICS

Marcus, C. S. (2018). How should the FDA review diagnostic radiopharmaceuticals? *Journal of Nuclear Medicine, 59*, 868–870. https://doi.org/10.2967/jnumed.117.200337

Noble, A. (2015). Regulating compounding pharmacies. *NCSL Legisbrief, 23*, 1–2.

Patel, N., Fakhry-Darian, D., Nijran, K., Khan, S., & Win, Z. (2018). Assessment and optimisation of Hermes amyloid BRASS as a quantitative diagnostic tool in reporting 18F-florbetapir (Amyvid) investigations. *Journal of Nuclear Medicine, 59*, 1011–1011.

Pike, V. W., Waters, S. L., Kensett, M. J., Bateman, D., Considine, D., Turton, D. R., Luthra, S. K., Brady, F., Shah, A., Shah, F., Silvester, D. J., et al. (1991). Radiopharmaceutical production for PET: Quality assurance practice, experiences and issues. In *New trends in radiopharmaceutical synthesis, quality assurance, and regulatory control* (pp. 433–449). Plenum. https://doi.org/10.1007/978-1-4899-0626-7_40

Prasad, K. N., Cole, W. C., & Haase, G. M. (2004). Radiation protection in humans: Extending the concept of as low as reasonably achievable (ALARA) from dose to biological damage. *The British Journal of Radiology, 77*, 97–99. https://doi.org/10.1259/bjr/88081058

Reivich, M., Kuhl, D., Wolf, A., Greenberg, J., Phelps, M., Ido, T., Casella, V., Fowler, J., Hoffman, E., Alavi, A., Som, P., & Sokoloff, L. (1979). The [18F]fluorodeoxyglucose method for the measurement of local cerebral glucose utilization in man. *Circulation Research, 44*, 127–137.

Sarikaya, I., Elgazzar, A. H., Sarikaya, A., & Alfeeli, M. (2017). Normal bone and soft tissue distribution of fluorine-18-sodium fluoride and artifacts on 18F-NaF PET/CT bone scan. *Nuclear Medicine Communications, 38*, 810–819. https://doi.org/10.1097/MNM.0000000000000720

Sasaki, M., & Tanaka, A. (2000). Quality control system and quality control method for radioactive drug. JP2000356642A Japanese patent application filed 26 December 2000.

Schwarz, S., Norenberg, J., Berridge, M., Dragotakes, S., Hung, J., Link, J., Mason, N. S., Mattmuller, S., Nickel, R. A., Packard, A., Paolino, J., Petry, N., Ponto, J., Quinton, T. M., Seifert, K. L., Swanson, D., Weiner, R. E., & Zigler, S. (2013). The future of USP monographs for PET drugs. *Journal of Nuclear Medicine, 54*, 472–475. https://doi.org/10.2967/jnumed.112.115089

Sharma, S. K. (2016). Quality control of PET radiopharmaceuticals with special reference to 18FDG. In *Progress in PET radiopharmaceuticals (quality control and theranostics)* (pp. 105–124). Nova Biomedical.

Singnurkar, A., Poon, R., & Metser, U. (2017). Comparison of 18F-FDG-PET/CT and 18F-FDG-PET/MR imaging in oncology: A systematic review. *Annals of Nuclear Medicine, 31*, 366–378. https://doi.org/10.1007/s12149-017-1164-5

Sun, Y.-P., Zhang, W., Zhao, Q.-J., Cao, J.-L., Zhang, L.-J., Xiang, J.-Y., Si, J.-Y., Lin, X., Chen, L., Zheng, Z.-Z., & Xia, N.-S. (2018). An optimized high-throughput fluorescence plate reader-based RSV neutralization assay. *Journal of Virological Methods, 260*, 34–40. https://doi.org/10.1016/J.JVIROMET.2018.07.004

Tankiewicz, M., Namieśnik, J., & Sawicki, W. (2016). Analytical procedures for quality control of pharmaceuticals in terms of residual solvents content: Challenges and recent developments. *TrAC Trends in Analytical Chemistry, 80*, 328–344. https://doi.org/10.1016/j.trac.2015.09.008

The International Pharmacopoeia. (2009). *Fludeoxyglucose (18F) Injection, Document QAS/08.265/FINAL* (pp. 1–6).

United States Pharmacopeia and National Formulary (USP 19-NF 14). (1979). "Sodium Fluoride F 18 Injection" Monograph.

US Pharmacopeial Convention. (1998). Radiopharmaceuticals for Positron Emission Tomography (PET)—Compounding, Investigational, and Research Uses <823>. In USP 23-NF 18. USP.

US Pharmacopeial Convention. (2016). USP Guideline for Submitting Requests for Revision to USP–NF General Information for All Submissions. US Pharmacopeial Convention.

Vera-Ruiz, H., Marcus, C. S., Pike, V. W., Coenen, H. H., Fowler, J. S., Meyer, G. J., Cox, P. H., Vaalburg, W., Cantineau, R., & Helus, F. (1990). Report of an International Atomic Energy Agency's advisory group meeting on "Quality control of cyclotron-produced radiopharmaceuticals". *International Journal of Radiation Applications and Instrumentation. Part B, 17*, 445–456.

Winicov, H., Schainbaum, J., Buckley, J., Longino, G., Hill, J., & Berkoff, C. E. (1978). Chemical process optimization by computer—A self-directed chemical synthesis system. *Analytica Chimica Acta, 103*, 469–476. https://doi.org/10.1016/S0003-2670(01)83110-X

Yang, J., Wu, Y., Wang, H., Yang, W., Xu, Z., Liu, D., Chen, H. J., & Zhang, D. (2022). An improved automated high-throughput efficient microplate reader for rapid colorimetric biosensing. *Biosensors, 12*, 284. https://doi.org/10.3390/BIOS12050284

Yu, S. (2006). Review of F-FDG synthesis and quality control. *Biomedical Imaging and Intervention Journal,* 2(4), e57. https://doi.org/10.2349/biij.2.4.e57

Zhang J, Elizarov, A. M., Miraghaie, R., Ball, C. E., Kolb, H. C., Ball, C. E., Elizarov, A. M., Kolb, H. C., Miraghaie, R., & Zhang, J. (2012). *Apparatus and method for automated quality control.* 8,214,159 B2, filed 3 December 2009, issued 3 July 2012.

Current and Emerging Applications of Electrophoresis in the Radiopharmaceutical Field

Jason Jones, Raisa Krasikova, and R. Michael van Dam

1 Introduction

High-resolution analytical techniques for determination of chemical and radiochemical composition are critical in the radiopharmaceutical field (Coomber, 1975; Gillings et al., 2020; Wieland et al., 1986). They are employed to assess synthesis performance (e.g., conversion of isotope to intermediate or product) during radiosynthesis development, to perform quality control (QC) tests during the production of radiopharmaceuticals intended for human use (Decristoforo et al., 2007; Molavipordanjani & Hosseinimehr, 2018; International Atomic Energy Agency, 2018; Elizarov, 2020; Ha et al., 2017a, 2017b), and to assess radiometabolites (Ghosh et al., 2020).

J. Jones
Department of Molecular and Medical Pharmacology, David Geffen School of Medicine, University of California Los Angeles, Los Angeles, CA, USA

Crump Institute for Molecular Imaging, University of California Los Angeles, Los Angeles, CA, USA

DropletPharma Corp, Los Angeles, CA, USA

R. Krasikova
Laboratory of Radiochemistry, N.P. Bechtereva Institute of Human Brain, Russian Academy of Science, St. Petersburg, Russia

R. M. van Dam (✉)
Department of Molecular and Medical Pharmacology, David Geffen School of Medicine, University of California Los Angeles, Los Angeles, CA, USA

Crump Institute for Molecular Imaging, University of California Los Angeles, Los Angeles, CA, USA
e-mail: mvandam@mednet.ucla.edu

R. M. van Dam et al. (eds.), *Automated Technologies for the Development and Production of Radiopharmaceuticals*,
https://doi.org/10.1007/978-3-031-84632-8_16

In addition to liquid chromatography methods such as high-performance liquid chromatography (HPLC) and ultra-performance liquid chromatography (UPLC) (Franck et al., 2009), and planar chromatography methods such as thin-layer chromatography (TLC) (Sherma & DeGrandchamp, 2015; Carpenter, 1986), several techniques based on electrophoresis have also been used in the analysis of radiopharmaceuticals.

First described about a hundred years ago, the techniques of electrophoresis have evolved dramatically over the years to improve many aspects of performance (Righetti, 2005; Vesterberg, 1993). At its core, electrophoresis is the separation of analytes by applying an electric field applied across a separation medium (Bočekxs, 1993; Li, 1992). Analyte molecules migrate based on their charges and electrophoretic mobilities, which are related to their size and shape, as well as the nature of the buffer solution and separation medium. Electrophoretic methods offer high flexibility in separation conditions and analyte compatibility (small molecules to large biomolecules; charged and neutral species), and can be configured to operate with very high performance (i.e., high resolution, fast separation, high sensitivity, low sample consumption). Named after the type of separation medium, the most common implementations of electrophoretic separations are gel electrophoresis (GE) (Magdeldin, 2012), paper electrophoresis (PE) (Wolstenholme & Millar, 2009), capillary electrophoresis (CE) (Li, 1992), and microchip electrophoresis (MCE) (Henry, 2006). In all of these methods, the separation medium is saturated or filled with an ionic buffer to allow for current flow. The buffer can also be used to control the pH and charge state of analytes and additives can be used to enhance separation in certain applications. In many respects, GE and PE bear similarity to TLC, and CE and MCE are similar to HPLC.

In this chapter, we describe each of these different electrophoretic methods and highlight selected examples of applications involving radiopharmaceuticals.

2 Gel Electrophoresis

2.1 Introduction

Gel electrophoresis using agar gels was first introduced by Gordon et al. (1950) for the separation of proteins. Subsequent developments in different types of gels (i.e., agarose, starches, and then polyacrylamide) resulted in significantly improved separation resolution. Separation is performed within the gel immersed in an ionic buffer. The most common technique today is sodium dodecyl sulfate polyacrylamide gel electrophoresis (SDS-PAGE), the name reflecting the type of buffer and type of gel. Samples are deposited into "well" features created within the gel (Fig. 1a), and the separation field is created by applying a voltage along the length of the gel (Fig. 1b) (Magdeldin, 2012). Multiple adjacent wells can be loaded, enabling multiple lanes of separation in parallel. Gel electrophoresis is commonly used to

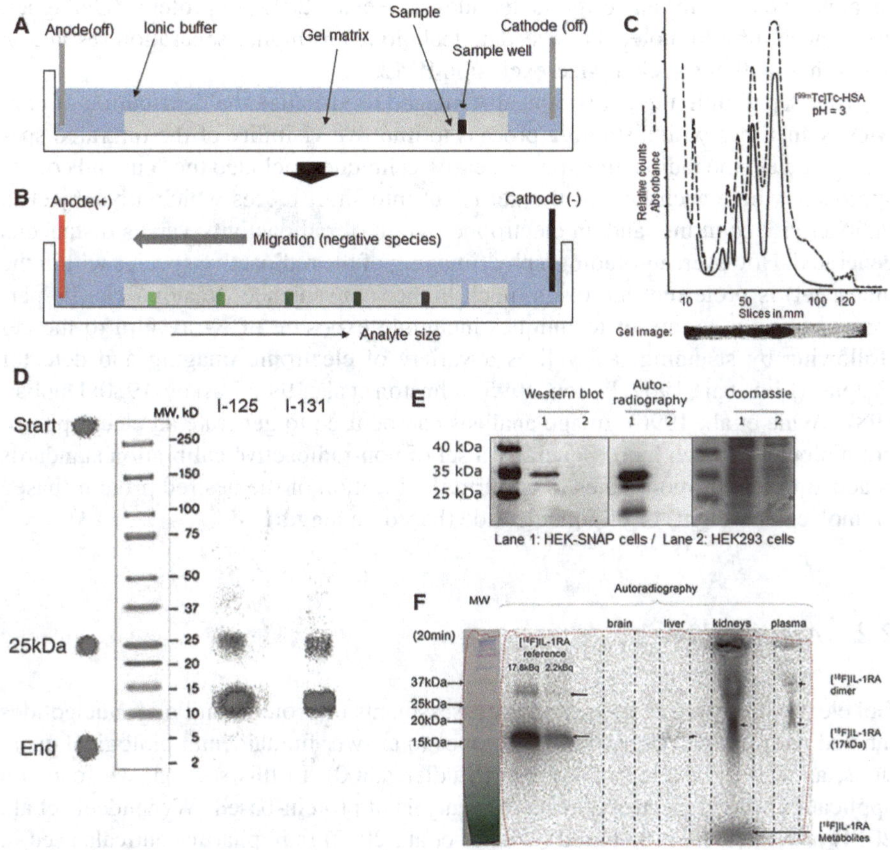

Fig. 1 (**a**) Schematic of apparatus and mechanism of gel electrophoresis. The gel is immersed in an ionic buffer and the sample is loaded into the sample well. (**b**) Application of the electric field leads to migration through the gel and separation by electrophoretic mobility. Proteins or nucleic acids are typically separated under conditions that confer a negative charge, and the mobility is inversely related to molecule size. (**c**) Example of radiochemical purity analysis via SDS-PAGE (bottom) and result of cut-and-count readout (top) of human serum albumin (HSA) labeled with Tc-99m using a commercially-available labeling kit. Only 37% of the recovered activity is in the form of the monomer. (Adapted from Jánoki et al. 1986 with permission Springer ©1986). (**d**) Example of radiochemical purity analysis via SDS-PAGE and phosphorimaging of an anti-HER2 nanobody labeled with I-125 or I-131, showing the presence of monomer and dulsulfide-bridged dimer. (Adapted from Pruszynski et al. 2013 with permission Elsevier ©2013). (**e**) Example of confirmation of binding of novel reporter probe [^{18}F]pFBG to SNAPTag reporter protein target by analysis of cell lysates via SDS-PAGE and phosphorimaging readout. (Adapted from Stotz et al. 2021 with permission MDPI ©2021). (**f**) Example of radiometabolite analysis of [^{18}F]IL-1RA in different tissues 20 min post injection showing the presence of significant metabolites in the kidneys. Analysis was performed via SDS-PAGE and readout via phosphorimaging. (Adapted from Cawthorne et al. 2011 with permission Wiley ©2011)

separate nucleic acids and oligonucleotides (Lee et al., 2012) or proteins (Gallagher, 2012) according to molecular size. The technique has higher separation resolution than other methods such as size-exclusion HPLC.

After separation, the gel is typically imaged to visualize the distribution of analytes, sometimes with a staining process to improve visibility of the migrated species. To detect the radioactive species, early techniques included the "cut and count" approach, where each lane of the gel is cut into short pieces which are subject to radioactivity counting, and an electropherogram of radioactivity versus distance is generated. However, autoradiography (imaging of the radioactive species within the intact gel) is preferred due to its much higher convenience. Imaging can be performed using a variety of techniques including exposure of X-ray film to the gel (following by scanning) as well as a variety of electronic imaging and detector systems (Lim et al., 1969; Patton, 1995; Johnston et al., 1990; Laskey, 1980; Dunbar, 1987; Aynié et al., 1996). Image analysis can be used to generate an electropherogram plot along each lane. Generally a set of non-radioactive calibration standards is added to one or more lanes to confirm the location of the desired protein (based on molecular weight) or oligonucleotide (based on length).

2.2 *Applications*

Gel electrophoresis has been used for assessments of proteins and oligonucleotides labeled with a wide range of radioisotopes to answer fundamental biological questions, as well to develop new drugs (Studier, 2000). In this section, we focus on applications of gel electrophoresis for analysis of protein-based (Wynendaele et al., 2014) or oligonucleotide-based (Younes et al., 2002) radiopharmaceuticals used in nuclear medicine. As gel electrophoresis is used extensively in the radiopharmaceutical field, primarily in the development of novel protein-based radiopharmaceuticals, we describe selected examples to illustrate the types of applications and isotopes that have been reported. We focus here mainly on works that performed radiometric analysis of the gel.

2.2.1 Assessment of Radiochemical Purity

Due to the high resolution when separating large biomolecules, gel electrophoresis is used extensively to analyze the chemical or radiochemical purity of labeled biomolecules. For example, it is possible to distinguish the monomeric and dimeric forms of a protein and detect the presence of aggregates. Such studies are valuable not only for quality control but also for labeling or purification method development. Comparison with a molecular weight ladder can be used to identify the product species.

For example, Jánoki et al. (1986) used SDS-PAGE as a QC method to analyze human serum albumin or fibrinogen labeled using commercially available kits with

Tc-99m and I-125. In contrast to TLC analysis which showed only a single apparent product peak, the SDS-PAGE analysis had 2–5 bands (using the cut and count readout method) revealing that the labeled samples in fact comprised multiple species (monomer, dimer, trimer, polymer, and colloids) suggesting that TLC was therefore insufficient for assessing the radiochemical purity of these compounds (Fig. 1c).

In another example, Decristoforo et al. (1997) showed that electronic autoradiography via a microchannel plate analyzer (Instant Imager, Canberra Packard) could be used for rapid and convenient readout of radioactivity in gels for QC testing or research purposes. They assessed the detection sensitivity and dynamic range for different radionuclides and assessed reproducibility and resolution. Multi-lane SDS-PAGE under both reducing and non-reducing conditions was used to analyze several samples including a ^{99m}Tc-labeled anti-CEA monoclonal antibody (MAb BW431/26). Though bands corresponding to fragments and degradation products were apparent in the images, the authors remarked that quantification was difficult, limited by the gel readout resolution.

Potamianos et al. (2006) compared different chromatographic techniques as potential QC methods for assessment of radiochemical purity when performing enzymatic radiolabeling of polyclonal immunoglobulin G (IgG) with I-125 using lactoperoxidase (LPO). Paper electrophoresis (described in Sect. 3 of this Chapter) and size exclusion chromatography both suggested a 90% purity of the labeled protein, but SDS-PAGE analysis with Coomassie stain readout and also autoradiography (via exposure of the gel to photographic film) revealed the existence of self-iodination of LPO (in addition to aggregates, and light or heavy chain fragments of IgG), and enabled estimation of the true radiochemical purity (79%). Quantitation was performed via cutting and counting the bands of interest.

Numerous groups have reported the use of SDS-PAGE analysis to assess the success and performance of novel protein labeling methods or the purity of novel radiopharmaceuticals.

In an example illustrating the value of gel electrophoresis in optimizing the purification process after radiolabeling, Chianelli et al. (1997) used SDS-PAGE in conjunction with readout via exposure to X-ray film for analyzing the radiochemical purity of ^{99m}Tc-labeled IL-2. They compared the radioactivity recovery and final radiochemical purity of different purification methods based on reverse-phase purification (tC2 cartridges) or gel-filtration purification (PD10 column). While gel filtration afforded higher product recovery, a significant amount of micro-aggregates remained. The authors moreover noted the superior resolution and more accurate determination of radiochemical purity via SDS-PAGE compared to HPLC. While HPLC analysis suggested a radiochemical purity >99%, the SDS-PAGE analysis revealed the presence of a small amount (~6%) of dimers.

In an illustrative example using gel electrophoresis to analyze a novel radiopharmaceutical, Tavaré et al. (2009) developed a novel radiopharmaceutical for imaging cell death based on C2A, the phosphatidylserine-binding domain of rat synaptotagmin I. To avoid interference with binding they engineered a new protein, C2AcH, containing a cysteine for site-specific covalent modification and hexahistidine tag for site-specific radiolabeling with $[^{99m}Tc][Tc(CO)_3(OH_2)_3]^+$. In this work,

SDS-PAGE in conjunction with phosphorimaging (Cyclone, Perkin-Elmer) was used to confirm the purity (100%) after radiolabeling and purification on a PD 10 size exclusion column.

Wuest et al. (2008) developed a novel prosthetic group, [^{18}F]FDG-maleimide hexyloxime ([^{18}F]FDG-MHO) for mild thiol-specific labeling of peptides and proteins. As an example, the single cysteine containing protein annexin A5 was labeled, with radiochemical purity analyzed via SDS-PAGE with phosphorimaging readout (BAS 5000, Fujix) before (44–58%) and after (>95%) purification via size-exclusion chromatography. The complex mixture evident in the crude product highlights the need for a high-resolution separation method.

Pruszynski et al. (2013) used SDS-PAGE with phosphorimaging readout (Cyclone Plus, PerkinElmer) to determine the purity of a radioiodinated anti-HER2 nanobody (5F7GGC), labeled with I-125 or I-131. For the ^{125}I-labeled compound, the product comprised 63% monomer and 37% dimer, while for the ^{131}I-labeled version, the product comprised 69% monomer and 31% dimer (Fig. 1d). An equilibrium mixture is expected due to the presence of a carboxy-terminal cysteine in the nanobody.

As an example of a radiotherapeutic from the same group, Pruszynski et al. (2018) performed ^{225}Ac-labeling of a DOTA-conjugated anti-HER2 nanobody. SDS-PAGE analysis with phosphorimaging readout (Cyclone Plus, PerkinElmer) confirmed successful labeling without the formation of aggregates with only a single band visible at the desired mass. Using the same technique, the stability of the compound in PBS at room temperature and in human serum at 37 °C was assessed and found to be stable (>90% intact) up to 10 days.

In addition to protein separation, gel electrophoresis excels at separation of oligonucleotides.

In one example, Hilger et al. (1998) introduced a N_3S chelator into an L-Selectin-binding RNA, and subsequently performed radiolabeling with Tc-99m. After measuring the radiochemical yield to be 95% with HPLC and TLC, PAGE analysis confirmed a single major band corresponding to the desired product.

2.2.2 Biological Evaluation of Radiopharmaceuticals

Gel electrophoresis is valuable not only in the development and assessment of the radiopharmaceutical labeling process, but in evaluation of its biological properties such as stability, metabolism, and pharmacokinetics.

In an early study looking at pharmacokinetics, Beutler et al. (1985) explored the in vivo fate of cachectin/tumor necrosis factor (TNF) in mice by radiolabeling with I-125, monitoring the radioactivity concentration in blood and tissues via gamma counting, and assessing the radiochemical species present via SDS-polyacrylamide/acrylaide gradient gel electrophoresis (read out using an intensifying screen). The labeled hormone was cleared from the plasma with a half-life of 6–7 min, with most of the hormone taken up by liver, kidneys, skin, and gastrointestinal tract.

Electrophoretic analysis suggests that the hormone is very rapidly degraded to low molecular weight species after binding.

In a later study, Song et al. (2002) remarked that despite the significantly higher resolution of gel electrophoresis compared to size-exclusion HPLC, the technique had largely been used only as a qualitative tool in protein pharmacokinetic studies. They performed a feasibility study on the use of SDS-PAGE with phosphorimaging readout (Storm 860, Molecular Dynamics) as a general quantitative approach for pharmacokinetic studies of protein drugs. Using multiple examples of proteins labeled with I-125, the authors studied the impact of protein loading, autoradiography exposure times, and image analysis methods on the assay results. The results were compared against the common approach of trichloroacetic acid (TCA) precipitable counts (measured via gamma counter) and found to be more accurate estimate of bioavailability, because TCA precipitation cannot distinguish whether the radioactivity is intact drug, radiolabeled degradation product, liberated radiolabel, or endogenous proteins that have incorporated liberated radiolabels.

Gel electrophoresis analysis can also be valuable for assessing the serum stability of novel radiotracers.

For example, Corsetti et al. (2004) developed a new tracer for diagnosis of non-iodine taking up differentiated thyroid cancer metastases. To prepare the tracer, rhTSH (Thyrogen) was radioiodinated with either I-125 or I-123 using an enzymatic method with lactoperoxidase/glucose oxidase. SDS-PAGE analysis with autoradiography readout (by exposure to photographic film) was used to evaluate the stability of the tracers in saline and serum. The images showed high purity of the labeled protein though a small amount of the dimer was present. In vitro stability studies showed a radioiodine release of 4% after 4 h in both saline and serum.

Zarschler et al. (2014) assessed stability of Cu-64 in various chelators (TETA, DOTA, NOTA, Cyclam, DiamSar, EDTA) under physiological conditions by performing human superoxide dismutase and serum stability assays. SDS-PAGE with phosphorimaging readout (BAS-1800II, Raytest) was chosen due to the ability to analyze multiple samples simultaneously (in contrast to HPLC) by running several lanes in parallel. The intensity of bands was measured to determine the amount of transchelation of the radiometal to EDTA or to serum proteins, respectively, in the assay mixtures, and [^{64}Cu]Cu-diamSar was found to be the most kinetically stable complex. They suggest their approach could be used to measure the properties of other chelators and radiometals.

Using gel electrophoresis, it is also possible to accurately assess interaction of the radiopharmaceutical with the target.

For example, Fan et al. (2018) describe the development of novel targeted radiotherapeutics using a NTSR1 receptor-targeting moiety, a DOTA chelator for Lu-177, and a cysteine cathepsin trapping agent to prolong the time in tumor. SDS-PAGE with phosphorimaging readout (Typhoon FLA 9500, GE Lifesciences) was used to conduct assays at multiple stages of the development, including to quantify the impact of the targeting moiety on cysteine cathepsin B binding, to perform competition binding assays, and to determine the timescale of retention of cysteine cathepsin B-conjugate adducts.

Stotz et al. (2021) report a novel ^{18}F-based reporter gene imaging system based on the SNAPTag genetic construct, which leads to expression of a protein on the cell surface that can be detected by binding of benzylguanine ligands. The authors compared [^{18}F]pFBG, [^{18}F]mFBG, and [^{18}F]FBBG as potential ligands, finding [^{18}F]pFBG to have the highest cell uptake in transfected cells. SDS-PAGE with phosphorimaging readout (Storm 840, Molecular Dynamics) was used to analyze cell lysates and confirm that the radioactivity corresponded to the size of the ^{18}F-ligand bound to the SNAPTag (Fig. 1e).

Electrophoretic separations are also valuable for assessment of radiometabolites.

As an example, Cawthorne et al. (2011) labeled an IL-1 antagonist, IL-1RA, with F-18 to determine biodistribution and metabolism, which had not been previously reported, despite the significant clinical interest for treatment of chronic diseases and stroke. After size-exclusion HPLC purification, SDS-PAGE with phosphorimaging readout (BAS-1800II, FujiFilm) was used to assess radiochemical purity and the authors reported a minor dimer fraction, though radiochemical purity was >95%. The same technique was also used to determine metabolites in rat tissue extracts from various organs at different time points (Fig. 1f). Radiometabolites were found within plasma as early as 20 min post-injection (with 35% low molecular weight metabolites). Interestingly, there was a discrepancy between radiometabolite results using SDS-PAGE and size-exclusion HPLC; the authors speculate that it may be due to the low amount of sample activity used in the gel versus HPLC.

2.3 Summary

These examples give a sense of the diverse ways in which gel electrophoresis aids in labeling and verification of novel protein- and oligonucleotide-based compounds, using a number of different SPECT, PET, and therapeutic isotopes. It is also a valuable tool in studying the in vitro and in vivo properties of these labeled compounds.

To further improve resolution, gel electrophoresis can also be performed in capillaries (see Sect. 4 in this chapter) using a technique known as capillary gel electrophoresis (CGE), but to our knowledge the technique has not yet been reported using radiometric analysis.

3 Paper Electrophoresis

3.1 Introduction

Paper electrophoresis (PE), often referred to as zone paper electrophoresis, is one of the simplest and cheapest electrophoretic techniques that was introduced in the 1950s for the quantitative determination of serum proteins (Kunkel & Tiselius, 1951; Jencks et al., 1955). The method is based on the differences in the electrophoretic mobility of the charged compounds and is particularly suitable for the separation of small charged molecules (for example, amino acids and proteins of small size). The PE apparatus consists of a high-voltage supply for two electrodes (anode and cathode) that are immersed into two electrolyte buffer reservoirs (Fig. 2a). The separation is usually performed on filter paper as a supporting medium, typically on commercially-available chromatography paper (Whatman No. 1 or Whatman-3MM). The paper strip, typically 30 cm × 2 cm, pre-saturated with a buffer, is placed with a special holder with one end soaked to each reservoir while the analytical sample (5–10 μL) is spotted in the center of the strip. Under applied voltage (150–600 V), a current is generated and the charged species formed migrate toward the electrodes, i.e., cations toward the anode and anions toward the cathode (Fig. 2b).

After completing electrophoresis (typically 15–60 min), the strip is dried and the separated species are then detected by different post-processing procedures. Some common methods include staining followed by photographic imaging (Minden et al., 2009) or fluorescence (Van den Bergh & Arckens, 2004). In the case of radiolabeled compounds, the spots of the separated components are easily evaluated by measuring radioactivity distribution over the length of the strip. It can be achieved by simple "cut and count" technique, in which the strip is cut into equal segments

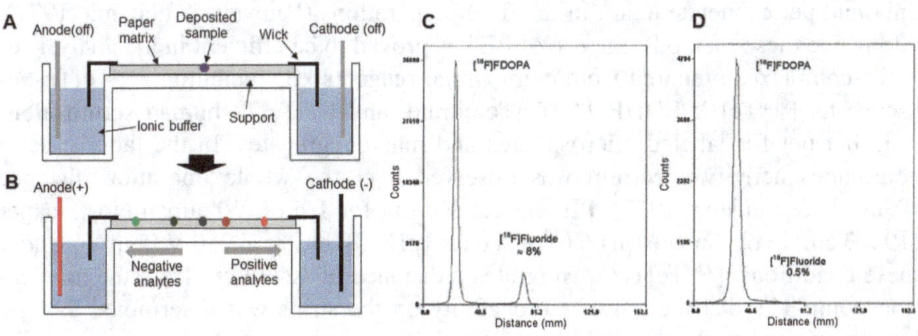

Fig. 2 (**a**) Apparatus for paper electrophoresis. A paper matrix is initially saturated with buffer and fluidically connected to anode and cathode buffer reservoirs. (**b**) After depositing the sample, voltage is applied to separate the analytes, and the resulting spatial distribution can be analyzed. (**c**, **d**) Examples of paper electropherogram showing radiochemical purity of 6-L-[^{18}F]FDOPA after purification on different semi-preparative HPLC columns: conventional silica-based RP C18 column (**c**) and polymer-based RP column (**d**). (**c** and **d** adapted from Füchtner et al. 2002 with permission Elsevier © 2002)

and the radioactivity of each piece is measured in a gamma-counter with NaI scintillation crystal (Najafi & Hutchison, 1985). More sophisticated equipment has also been employed, including radio-TLC scanners or radiographic imaging (Voytas & Ke, 2002). Imaging has been done via autoradiography apparatus (Microchannel plate analyzer, Instant Imaging, Canberra Packard, USA) (Füchtner et al., 2002; Kuntner et al., 2011), a Storage Phosphor System Cyclone Plus (Perkin-Elmer Life and Analytical Sciences) (Koźmiński et al., 2021), a 2π-scanner (Berthold, Germany) (Muhammad U. Akbar et al., 2019), and others.

3.2 Applications

Already in the 1950s, PE in conjunction with radiodetection of labeled species started to be applied in radiopharmaceutical chemistry research (Gabrieli et al., 1954). As for routinely produced radiopharmaceuticals, the method was recommended for determination of negatively charged radiochemical impurities, such as $[^{99m}Tc]TcO_4^-$, $[^{18}F]$fluoride or $[^{123/124}I]$iodide (International Atomic Energy Agency, 2018). Under electrophoresis conditions, these ionic species move rapidly towards the anode while uncharged radiotracer molecules remain at the application spot.

Most applications of PE have been directed to radiopharmaceuticals labeled with Tc-99m ($T_{1/2} = 6.0$ h) available from a $^{99}Mo/^{99m}Tc$ generator in the form of pertechnetate anion, $[^{99m}Tc]TcO_4^-$. ^{99m}Tc-based radiopharmaceuticals can be easily prepared via one or two step synthesis using commercially available reagents kits, an approach that greatly facilitates both routine application and implementation of new radiotracers into nuclear medicine practice. For analysis of radiochemical purity, several chromatographic approaches have been applied (TLC, paper chromatography, ion exchange chromatography) allowing for a quantitative determination of unbound pertechnetate anion in the final preparations (Pauwels & Feitsma, 1977). Alongside these methods, the use of PE has proved to be efficient in the analysis of ^{99m}Tc-complexes prepared from commercial reagents kits including $[^{99m}Tc]$Tc-Fe-ascorbate, $[^{99m}Tc]$Tc-EDHP, $[^{99m}Tc]$Tc-citrate, and $[^{99m}Tc]$Tc-human serum albumin, but not for labeled microspheres and macroaggregates. In the latter case, a continuous activity spectrum was observed over the whole migration distance (Pauwels & Feitsma, 1977). PE was carried out for 1 h on Whatman No. 1 paper (19 × 3 cm) in phosphate (pH 6.0) or acetate (pH 4.8) buffer at 250 V (5 mA). Under these conditions, $[^{99m}Tc]TcO_4^-$ migrates a distance of 12–13 cm from the application point. The distribution of radioactivity on the strips was determined using a home-designed scanner based on a NaI scintillation detector. Despite free pertechnetate and other components of the radiopharmaceuticals being well-separated by PE, the method was time consuming and technically challenging and it was not considered by the authors to be suitable for daily routine application. Planar chromatography methods such as radioTLC and currently available iTLC are more convenient than PE for evaluation of radiochemical purity of radiopharmaceuticals.

Despite this, the PE method continues to be an indispensable research tool in radiopharmaceutical development. The reason for this is that electrophoresis provides information on the charge state of complexes (anionic, cationic, or neutral) that cannot be easily determined by other methods. For example, neutral charge is needed to facilitate penetration of the blood-brain barrier, which is important for brain perfusion agents; on the other hand, positively charged molecules are often used as radiotracers for myocardial perfusion. Furthermore, in combination with other analytical data, PE may provide an additional information on the oxidation state of Tc-99m (Blower et al., 1991; Shukla et al., 2004) as well as structural information (Marzilli et al., 1979; Donald Burns et al., 1981) for carrier-free ^{99m}Tc-labeled radiotracers. This is especially important in the lack of availability of well-defined analogues labeled with long-lived ("pseudo-stable") technetium-99 ($T_{1/2}$ = 211,000 years).

Extensive interest in the PE technique is evidenced by the publication of numerous papers. Most of them are devoted to diagnostic radiometals but the technique has also been used for radiohalogens (I-123/124/131, F-18) and radiotherapeutic agents (Lu-177, At-211). Some selected examples of the application of PE in research and routine quality control of radiopharmaceuticals are given below.

3.2.1 ^{99m}Tc-Labeled Radiopharmaceuticals

^{99m}Tc-based radiopharmaceuticals are the most widely used diagnostic agents for clinical nuclear medicine studies using gamma scintigraphy and single photon emission computed tomography (SPECT). The radionuclide may form complexes with a great variety of ligands enabling the design of new radiotracers for clinical application. The use of PE was reported by several groups aimed to determine the charge of new carrier-free ^{99m}Tc-labeled complexes with potential clinical application, as summarized in Table 1.

In addition to ^{99m}Tc-labeled diagnostic radiotracers listed in Table 1, there has also been research focused on PE characterization of different labeled tumor drugs, including [^{99m}Tc]Tc-mitomycin-C (Bokhari et al., 2015), [^{99m}Tc]Tc-epirubicin (Hina et al., 2018), [^{99m}Tc]Tc-cyclophosphamide (Bokhari et al., 2018), as well as ^{99m}Tc-labeled insulin (Muhammad Usman Akbar et al., 2016). The neutral nature of these labeled species was confirmed by PE that was carried out by using a commercial electrophoresis chamber (Deluxe, Gelman) in phosphate buffer of pH6.0 (300 V, 1 h).

3.2.2 Radiohalogen-Labeled Radiopharmaceuticals

One of the first applications of the PE in radiopharmaceutical research has been electrophoretic investigation of commercially available ^{131}I-labeled human serum albumin (IHSA), an intracranial tumor-seeking agent (Gabrieli et al., 1954). This therapeutic radionuclide ($T_{1/2}$ = 8.0 d, 100% β^-, γ 0.364 MeV) was also used at that

Table 1 Charge determination of ^{99m}Tc-labeled complexes by PE

Radiotracer	Diagnostics area	Buffer	Voltage/ field	Time (min)	Charge of the complex	References
[^{99m}Tc]Tc–nitrido dithiocarbamate complexes	Brain perfusion	PB pH 7.4	10 V/ cm	60	Neutral	Mallia et al. (2006)
[^{99m}Tc]TcN(DMCHDTC)$_2$	Myocardial perfusion	PB pH 7.4	150 V	90	Neutral	Zhang et al. (2009)
[^{99m}Tc]TcN-tertiary butyl xanthate	Myocardial perfusion	PB pH 7.5	10 V/ cm	60	Neutral	Mathur et al. (2006)
[^{99m}Tc][TcN(PNP5)(HPDTF)]+	Myocardial perfusion	PB pH 7.0	150 V	15	Cationic	Lu et al. (2007)
[^{99m}Tc] Tc-N,N′-dimethyldiaminedithiol	Myocardial perfusion	PB pH 7.4	150 V	240	Cationic	Jeong et al. (2002)
[^{99m}Tc] Tc-hydroxyacetyltriglycine ([^{99m}Tc]Tc-HAG3)	Renal function	PB, various pH	300 V	30	Anionic	Vanbilloen et al. (1997)
[^{99m}Tc] TcN-methylpyridoxal	Renal function	0.1M Tris buffer pH 7.4	600 V	30	Cationic (slightly-positive charge)	Karube et al. (1994)
[^{99m}Tc]Tc-clindamycin	Infection	PB pH 6.8	300 V	30–60	Neutral	Hina et al. (2014)
[^{99m}Tc]Tc-Dactinomycin	Infection	PB pH 6.8	300 V	45	Neutral	Rizvi et al. (2018)
[^{99m}Tc]Tc-ciprofloxacin ([^{99m}Tc]Tc-CIP)	Infection (diabetic foot syndrome)	PB pH 7.4	200 V	45	Neutral	Koźmiński et al. (2021)
^{99m}Tc-labeled benzoic acid complex	Spleen and liver imaging	PB pH 6.0	300 V	60	Neutral	Akbar et al. (2018)

PB phosphate buffer, *DMCHDTC* 2,3-dimethylcyclohexyl dithiocarbamato-, *PNP5* bis-(dimethoxypropylphosphinoethyl)ethoxyethylamine, *HPDTF* potassium 4-(cyclohexylpiperazin-1-yl)-dithioformate

time for diagnostic purposes. The PE study revealed two different electrophoretic patterns that were assigned to two forms of IHSA with different metabolic rates. These findings helped to interpret the results of biological studies on the disappearance of the iodinated HSA in the blood stream of rats, which did not follow the expected clearance rate. Motaleb et al. (2017) reported on the use of PE for determination of the radiochemical yield (RCY) of ^{131}I-labeled trazodone, a promising brain imaging agent. On the electrophoretic diagram, the labeled product and free [^{131}I]iodide were well separated moving to distances of 3 and 12 cm, respectively, from the spotting point. Along with other methods (radioTLC and HPLC), PE data were used to establish optimal labelling conditions providing RCY >90%.

One of the few reported applications of PE in routine quality control of radiotracers has been the determination of radiochemical purity of ^{123}I-labeled vasoactive intestinal peptide (VIP), a SPECT radiotracer for the visualization of primary and recurrent colorectal adenocarcinomas and metastatic sites. In addition to HPLC, radiochemical purity was accessed by zone paper electrophoresis (Whatman 3 MM paper, 0.1 barbital buffer, pH 8.6, 300 V for 10 min). The percentage of unbound [^{123}I]iodide (<3%) remained stable over at least 24 h which secured the use of this SPECT radionuclide with half-life of 13.31 h. PE was also applied for evaluation of radiochemical purity of ^{124}I-labeled chitosan, prepared for small-animal preclinical PET studies of chitosan-based ophthalmic formulations (Kuntner et al., 2011).

The PE technique was found to be useful in the quality control of 6-L-[^{18}F]FDOPA (Füchtner et al., 2002), a widely-used PET radiotracer for diagnosis of Parkinson disease, neuroendocrine tumors and gliomas. The radiotracer was produced by electrophilic radiofluorination of a stannylated precursor using gaseous [^{18}F]F_2 and recovered from the reaction mixture by semi-preparative HPLC using a conventional silica-based reversed phase (RP) C18 column. Radio HPLC analysis using a RP C18 silica-based column confirmed over 99% radiochemical purity of 6-L-[^{18}F]FDOPA. However, PE analysis (using 25 mM acetate buffer, pH 4.7, 300 V, 15 min) revealed about 8% of free [^{18}F]fluoride that migrated to the anode clearly separated from 6-L-[^{18}F]FDOPA (Fig. 2c). The presence of this anionic impurity (7%) was confirmed by analytical radio HPLC on a polymer based RP column. Based on these findings, an alternate purification procedure using a semi-preparative polymer-based RP HPLC column was developed, which provided 6-L-[^{18}F]FDOPA with radiochemical purity of 99.5%, according to PE (Fig. 2d) (Füchtner et al., 2002).

3.2.3 Radiotherapeutic Agents

The PE technique has been employed to a limited extent for characterization of new compounds labeled with lutetium-177 ($T_{1/2} = 6.6$ d; 100% β^-, E_β (max) 0.497 MeV; maximum tissue range ~ 2.5 mm), one of the most important theranostic radioisotope used for the treatment of neuroendocrine tumors, prostate cancer, and other oncological disorders (George & Samuel, 2023). Due to the co-emitted γ-rays (208 keV), Lu-177 is well-suited for gamma scintigraphy and SPECT imaging for tumor localization and calculation of the radiotherapeutic dose. Akbar et al. (2019) suggested ^{177}Lu-labeled diethylenetriamine penta-acetic acid (DTPA) as a new radiopharmaceutical for renal function imaging and treatment of renal tumors. The neutral nature of [^{177}Lu]Lu-DTPA was confirmed by PE (carried out using a commercial electrophoresis chamber (Deluxe, Gelman) in phosphate buffer of pH 6.0 (300 V, 1 h)). The same authors employed similar PE conditions for the confirmation of the neutral charge of the ^{177}Lu-labeled antiobiotic kanamycin (Muhammad Usman Akbar et al., 2017). The formation of negatively charged complexes was demonstrated by PE for three new bone-seeking and palliative agents, [^{177}Lu]Lu-ethylenediaminetetramethylene phosphonic acid (EDTMP), [^{177}Lu]Lu-methylene diphosphonate (MDP), and [^{177}Lu]Lu-pyrophosphate (PYP) (Abbasi, 2015). Particular mention should be made that the presence of a negative charge facilitates the affinity of the complex to the positively charged Ca^{2+} in mineral bone component hydroxyapatite.

Another example in the field of targeted radiotherapy is the application of PE to developing a new approach for labelling biologically active compounds with α-particle emitter At-211 ($T_{1/2}$ = 7.2 h). This was achieved via attachment of At-211 (and I-131/125) anions to the soft metal cations Rh(III) and Ir(III), which are complexed by the ligand 1,5,9,13- tetrathiacyclohexadecane-3,11-diol (16aneS4-diol). Four complexes Rh[16aneS$_4$-diol][$^{131/125}$I]I/[^{211}At]At and Ir[16aneS$_4$-diol][$^{131/125}$I]I/[^{211}At]At were synthesized and analyzed using PE (10 V/cm for 25–60 min, 0.02 M phosphate buffer at pH 4.0 and 7.4). The cationic nature of the newly prepared radiopharmaceuticals was confirmed by their adsorption on a cation exchange resin (Pruszyński et al., 2008; Pruszyński et al., 2015). In the following study (Lyczko et al., 2017), the obtained ^{211}At-labeled or ^{131}I-labeled complexes were conjugated with substance P (5–11), to prepare a potential radiopharmaceutical for glioma treatment and distant metastases. Paper electrophoresis analysis confirmed the positive charge of the complexes, in which radioactivity migrated toward the cathode, similar to previously obtained rhodium complexes (Pruszyński et al., 2015). Free I-131 and At-211 migrated toward the anode as anions. Positive charge of the complexes may be linked to the observed high level of kidney uptake as the glomerular capillaries have a negatively charged surface layer.

4 Capillary Electrophoresis

4.1 Introduction

Since Jorgenson and Lukacs first introduced modern capillary electrophoresis (CE) in 1981 (Jorgenson & Lukacs, 1981), it has been widely recognized as a powerful, efficient, and high resolution separation technique with diverse applications. CE is based on the migration of ions or solutes in a fused silica capillary (typically 50 or 75 μm internal diameter) filled with ionic aqueous buffer (background electrolyte; BGE) with the use of applied voltage along the capillary axis. Capillaries are typically kept in a relatively straight conformation and typically have a length of ~25–60 cm. The sample is loaded (injected) into one end of the capillary and a detector is positioned near the other end (Fig. 3a).

In capillary zone electrophoresis (CZE), this simplest form of CE, the presence of charges on the inner surface of the capillary (e.g., negatively charged silanol groups when using silica capillaries) leads to accumulation near the capillary surface of oppositely charged ions in the buffer solution. When the separation field is

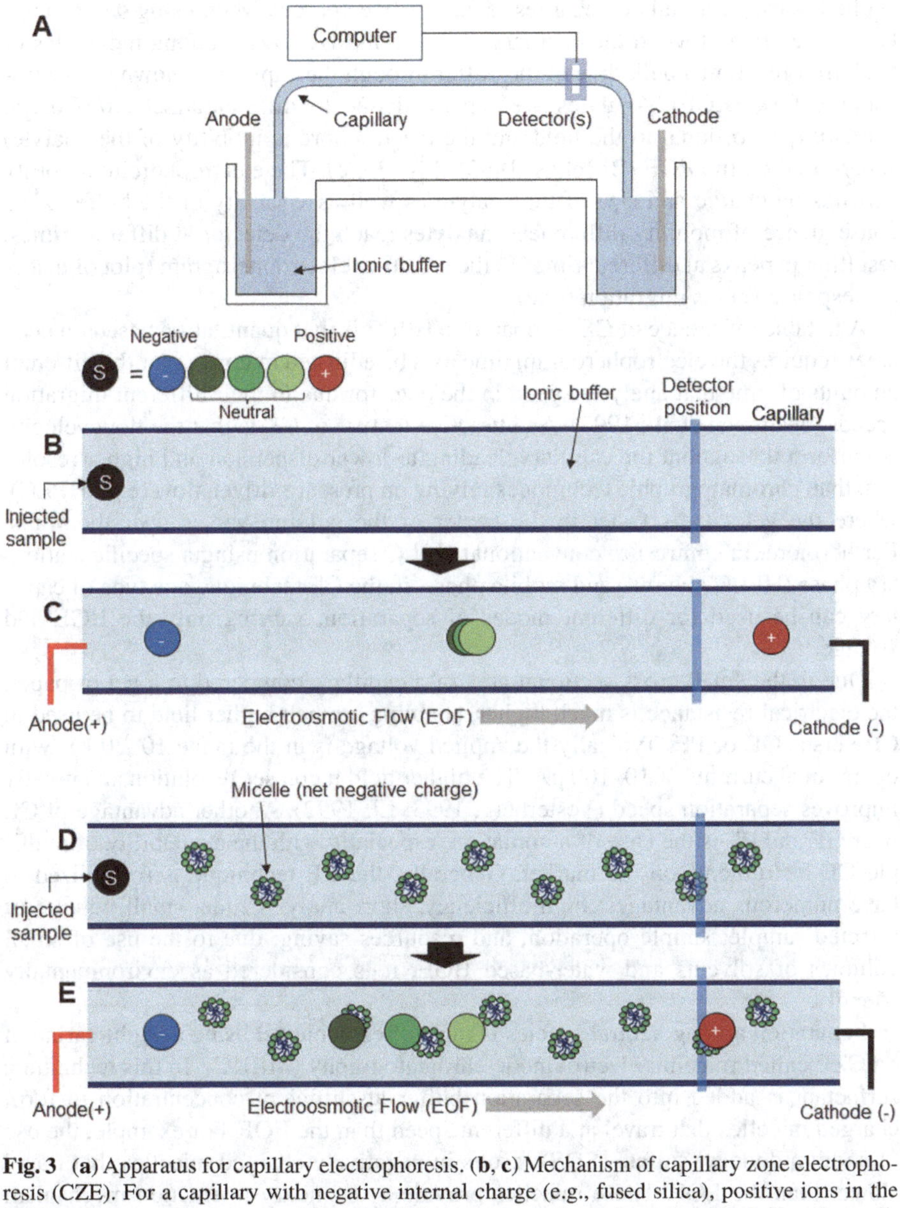

Fig. 3 (**a**) Apparatus for capillary electrophoresis. (**b**, **c**) Mechanism of capillary zone electrophoresis (CZE). For a capillary with negative internal charge (e.g., fused silica), positive ions in the buffer solution accumulate near the surface. When the separation voltage is applied along the capillary, an electroosmotic flow (EOF) is generated towards the detector, and analytes are separated based on electrophoretic mobility. The charge dependence of the electrophoretic mobility causes cations to travel faster than the EOF and anions to travel slower. (**d**, **e**) The mechanism of micellar electrokinetic chromatography (MEKC) is similar to CZE, except that addition of surfactant to the buffer results in formation of charged micelles (in this case negative charge). When the separation potential is applied, the micelles move slower than the EOF and retard the migration of analytes (relative to EOF) based on the strength of interaction between the analytes and micelles. This enables separation of neutral species

applied, these accumulated charges (e.g., positive charges when using silica capillaries) are driven toward the capillary outlet. Cohesive forces among molecules of the buffer result in a bulk flow of the buffer through the capillary, known as electroosmotic flow (EOF). Analytes are separated due to differential electrophoretic velocity (proportional to the field and the electrophoretic mobility of the analyte) superposed on the EOF (Bočekxs, 1993) (Fig. 3b, c). The electrophoretic mobility depends on charge and size of the analyte as well as viscosity of the buffer. As a consequence of mobility differences, analytes reach the detector at different times, resulting in peaks at different times in the resulting electropherogram (plot of detector response versus migration time).

A notable difference of CE compared to HPLC is that quantitation based on peak areas requires the electropherogram time axis be adjusted to correct for the different amounts of time that analytes spend in the detector due to their different migration speeds (Mammen et al., 1997). An attractive feature of CE is that the flow velocity is uniform throughout the capillary, leading to lower dispersion and higher resolution than chromatographic techniques relying on pressure-driven flow (e.g., HPLC), where the velocity is faster in the center of the column versus near the walls. Furthermore, in contrast to conventional HPLC separation using a specific stationary phase (HPLC column) and mobile phase, in the CE technique one type of capillary can be used for different modes of separation, varying only the BGE and voltage.

Due to the small cross-sectional area of a capillary compared to a gel or paper, the electrical resistance is much higher, enabling a much higher field to be used in CE versus GE or PE. Typically the applied voltage is in the range 10–30 kV with operational currents of 10–100 μA. The higher field increases resolution and greatly improves separation speed (Vesterberg, 1993; Li, 1992). Another advantage of CE over GE and PE is the ease of automation, especially with the availability of multiple CE instruments on the market. Generally, the CE technique is recognized to have numerous advantages: high efficiency, short analysis time, small volume of injected sample, simple operation, and resources saving; due to the use of small volumes of solvents and water-based BGE, it is considered as environmentally friendly.

Separation among neutral species is most often achieved using a slight variation on CZE called micellar electrokinetic chromatography (MEKC). In this technique, surfactant is added into the separation buffer at sufficient concentration to form charged micelles that travel at a different speed than the EOF. For example, the use of sodium dodecyl sulfate (SDS) results in negatively charged micelles that travel slower than the EOF. These micelles provide a secondary interaction mechanism (e.g., degree of hydrophobic interaction with micelles) that differentially impacts the migration of different analytes, resulting in separation among neutral analytes (Li, 1992) (Fig. 3d, e).

There are additional ways to control the separation in CE. For example, the EOF can be controlled via buffer additives, or eliminated altogether (Rodriguez & Li, 1999; Li et al., 2010; Corradini, 2023). Sometimes pressure is applied in conjunction with voltage to shorten the overall separation time. Additional control over the separation can be achieved by filling the capillary with a stationary phase such as a gel, enabling separations similar to gel electrophoresis via capillary gel electrophoresis (CGE) (Dolník, 1994; Gahoual et al., 2016), or using monolithic or packed column stationary phases, and performing separations similar to HPLC via capillary electrochromatography (CEC) (Altria, 1999; Eeltink et al., 2003). Chiral selectors can be used to enable CE separation based on chirality (Eeckhaut & Michotte, 2009; Yu & Quirino, 2019). The various modes of CE are currently applied in the analysis of small inorganic ions, low molecular weight molecules, chiral analysis of large biomolecules and play an essential role in pharmaceutical development and the drug industry.

Detection of analytes in CE is usually performed on-line using by monitoring the signal from a single-point detector near the outlet of the capillary over time, though methods based on fraction collection can also be used. Example modes of on-line detection include UV absorbance, laser induced fluorescence (LIF) (Szökő & Tábi, 2010), capacitively coupled contactless conductivity detectors (C4D) (da Silva et al., 2013), electrochemical (Yu et al., 2012; Fedorowski & LaCourse, 2015), and refractive index (Deng & Li, 1998; Felhofer et al., 2010). For analysis of radiopharmaceuticals, UV absorbance detection is typically employed in combination with radiation detection. UV detection is often implemented with a cross junction, in which two optical fibers are aligned perpendicularly on opposite sides of the capillary, one connected to an illumination source, and one connected to a spectrophotometer. Absorption at one or more wavelengths is monitored. In such a setup, the optical path length is very short (equal to capillary inner diameter, often 50 or 75 μm), but high sensitivity comparable to HPLC can still be achieved (Cheung et al., 2014).

General approaches for radiation detection in CE were reviewed by Poirier (1995). Most commonly, the detector comprises a scintillator, optically coupled to a photomultiplier tube (PMT) (Pentoney et al., 1989b) or a semiconductor detector (Pentoney et al., 1989a) placed in close proximity to the capillary. Detection of SPECT isotopes (e.g., Tc-99m) relies on detection of gamma rays and detection of PET isotopes (e.g., C-11, Ga-68, F-18) relies on detection of annihilation gamma rays or positrons. A wide variety of radioisotopes have also been monitored via radio-CE for non-radiopharmaceutical applications. Such isotopes (e.g., C-14, P-32, Eu-152, Cs-137) are detected via gamma rays or high-energy beta particles (Kaniansky et al., 1989; Pentoney et al., 1989a, 1989b; Klunder et al., 1997). An alternative detection approach based on fraction collection was used for radionuclides with low-energy beta emissions (e.g., S-35, H-3), in which the capillary output was moved in a rasterized pattern on a filter membrane during separation. The membrane was impregnated with a solid scintillator, allowing the radioactivity distribution to be imaged with a CCD camera, and enabling an electropherogram to be generated (Tracht et al., 1994).

4.2 Applications

The first application of CE for the analysis of radiopharmaceuticals was reported in the late 1980s, initially using custom designed radioactivity detection systems. With an initial focus on the analysis of ^{99m}Tc-labeled SPECT tracers, Altria et al. (1990) developed a gamma detector based on a thallium-doped sodium iodide (NaI(Tl)) scintillator coupled to a PMT, which was placed adjacent to the capillary. The detection volume was restricted to a short segment of the capillary by inserting lead shielding containing a slit between the capillary and detector. Detection efficiency was <50% due to the 2π geometry (i.e., detector covered only about half of the possible directions that radiation could be emitted from the capillary). The authors calibrated the detector response using [^{99m}Tc]NaTcO$_4$, and in the analysis of [^{99m}Tc]Tc-dimercaptosuccinic acid ([^{99m}Tc]Tc-DMSA), were able to distinguish species containing different Tc oxidation states (III or V) due to the different charge of the resulting molecule.

Jankowsky et al. (1999a) customized a commercial NaI scintillator/PMT detector to achieve a more sensitive 4π geometry, by threading the capillary between two NaI scintillators. Lead collimators with 1.5 mm diameter openings were used to confine the detection volume. With this design, a detection efficiency up to 85.4% was achieved. The setup was used with pressure-driven CE to determine pK$_a$ values of well-known ^{99m}Tc-labeled radiotracers: [^{99m}Tc]Tc-mercaptoacetylglycylglycylglycine ([^{99m}Tc]Tc-MAG3) and [^{99m}Tc]Tc-ethylenecysteine dimer ([^{99m}Tc]Tc-EC). In this study, the electrophoretic velocities of the complexes were measured under different pH values of the BGE; they were referred to the electrophoretic velocities of tetraphenyle arsonium chloride as a positively charged marker (Fig. 4a). However, the determination of pK$_a$ values from the measured data required very sophisticated calculations limiting practical application of this methodology. The setup was also used to analyze a wide range of products prepared using Tc-99m labeling kits (Jankowsky et al., 1999a, 1999b) for the presence of the major radiochemical impurity, namely, unbound [^{99m}Tc]TcO$_4^-$, the chemical form of the radionuclide originating from the isotope generator. Cationic ^{99m}Tc-radiopharmaceuticals ([^{99m}Tc]Myoview, [^{99m}Tc]Tc-Q12 and [^{99m}Tc]Tc-MIBI), and neutral complexes ([^{99m}Tc]Tc-HMPAO and [^{99m}Tc]Tc-EC) were analyzed by conventional CZE. Some anionic species ([^{99m}Tc]Tc-MAG3, [^{99m}Tc]Tc-DMSA, and [^{99m}Tc]Tc-MDP) were not detectable in conventional CZE, and therefore pressure-driven CZE (50 mbar) was applied to study these analytes in the same BGE. Pressure-driven CE offers the possibility to detect both cations as well as highly mobile metal anions within one run. The authors observed the expected differences in migration times depending on the charge and size of each species. The majority of samples were radiochemically pure, but in a few cases, impurities were observed, suggesting the potential for CE to be used for QC testing of ^{99m}Tc-labeled radiopharmaceuticals.

Poirier (1995) used BGO as a scintillator in a similar geometry and demonstrated the separation of [^{99m}Tc]Tc-sestamibi from colloid and [^{99m}Tc]TcO$_4^-$. An

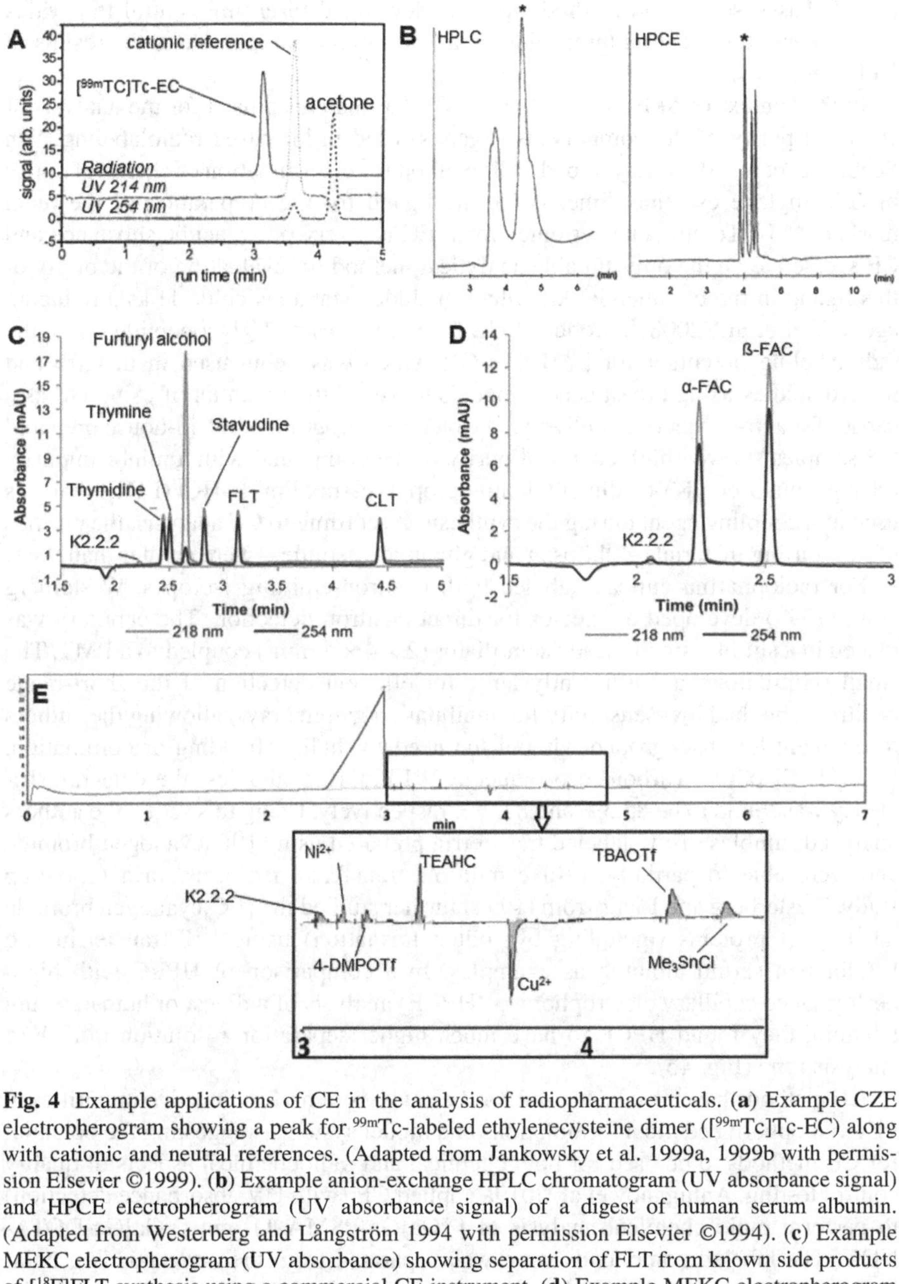

Fig. 4 Example applications of CE in the analysis of radiopharmaceuticals. (**a**) Example CZE electropherogram showing a peak for ^{99m}Tc-labeled ethylenecysteine dimer ([^{99m}Tc]Tc-EC) along with cationic and neutral references. (Adapted from Jankowsky et al. 1999a, 1999b with permission Elsevier ©1999). (**b**) Example anion-exchange HPLC chromatogram (UV absorbance signal) and HPCE electropherogram (UV absorbance signal) of a digest of human serum albumin. (Adapted from Westerberg and Långström 1994 with permission Elsevier ©1994). (**c**) Example MEKC electropherogram (UV absorbance) showing separation of FLT from known side products of [^{18}F]FLT synthesis using a commercial CE instrument. (**d**) Example MEKC electropherogram showing separation of ß-FAC from the impurity α-FAC formed during [^{18}F]FAC synthesis. (**c** and **d** adapted from Cheung et al. 2014 with permission from Elsevier © 2014). (**e**) Example CZE electropherogram showing resolution of seven non-radioactive impurities at the ppm level in a standard solution. Peaks: (I) Kryptofix 222–5 ppm; (II) Ni^{2+}–2 ppm; (III) 4-DMPOTf—200 ppm; (IV) TEAHC—62.5 ppm; (V) Cu^{2+}–25 ppm (VI) TBAOTf—90 ppm; (VII) Me_3SnCl—25 ppm. (Adapted from Antuganov et al. 2019b with permission Elsevier © 2019)

MEKC-based separation method was also developed to ensure neutral impurities were separated (revealing three additional impurities), comparable to the results of HPLC analysis.

In the context of SPECT radiotracers, CE has also been used for the analysis of chemical purity of the components (ligands) used in kit-based radiolabeling. Van Schepdael et al. (1997) developed a CE method for determination of chemical purity of *L,L*-ethylene cysteine dimer (EC), the ligand for the preparation of the renal marker [^{99m}Tc]Tc-ethylene cysteine dimer. EC is a very polar, acidic substance and CE seemed to be the only suitable analytical method for evaluation of the purity of this ligand in the commercial kit, which included stannous chloride as a reducing agent. Wei et al. (2008) introduced CE for the analysis of glycopeptide (GP), the radiolabeling precursor for [^{99m}Tc]Tc-GP, which was being used in in vitro and in vivo studies using breast cancer models to assess the potential of using GP as a carrier for anti-cancer drug delivery. The electropherogram of the in-house prepared GP samples showed high chemical purity of the compound with a minor impurity arising from 3-ethyl-3-(3-dimethylaminopropyl) carbodiimide-HCl (EDC) that was used as a coupling agent during the synthesis. According to CE analysis, the purities of the starting materials—chitosan and glutamate peptide—were greater than 95%.

For radiopharmaceuticals labeled with positron-emitting isotopes, Westerberg et al. (1993) developed a detector for direct positron detection. The capillary was placed in a slit in a small plastic scintillator ($2 \times 4 \times 4$ mm^3) coupled to a PMT. The small scintillator was sufficiently large for efficient detection of the short-range positrons, but had low sensitivity for annihilation gamma rays, allowing the authors to maintain low background without the need for bulky shielding or collimation. Using [^{11}C]sodium carbonate or aqueous [^{18}F]fluoride samples, the detector efficiency was found to be 85.3% and 87.2%, respectively. Using this setup, the authors analyzed samples of ^{11}C-labeled transferrin prepared using [^{11}C]cyanogen bromide and were able to partially resolve multiple transferrin isoforms. In a follow-up study, Westerberg and Långström (1994) further studied the [^{11}C]cyanogen bromide labeling of proteins (including byproduct formation) using [^{11}C]transferrin and [^{11}C]human serum albumin as examples. In a comparison of HPLC with high-performance capillary electrophoresis (HPCE) analysis of a digest of human serum albumin, they found HPCE to have much higher separation resolution in a short analysis time (Fig. 4b).

CE methods have also been used for the analysis of non-radioactive impurities in PET radiopharmaceutical formulations and model systems, suggesting the potential for CE methods to be used for both chemical and radiochemical aspects of quality control testing. Antuganov et al. 2019a applied CE (with UV absorbance detection) to perform rapid chemical analysis of [^{68}Ga]Ga-PSMA-11 and [^{68}Ga]Ga-DOTA-NOC. The labeled compounds were well separated from the corresponding unlabeled precursors within 11–12 min, and analysis of HEPES buffer could be performed in 4.5 min. With the ability to detect analytes at ppm levels, the authors proposed the technique exhibited the potential for miniaturized QC testing. In another example related to ^{68}Ga-labeled compounds, a compact device based on CE (with C4D detection) was developed to measure the presence of Ge-68 in the eluate from a Ga-68 generator, i.e., breakthrough of the parent isotope (Weinberg, 2014). The separation method was able to separate Ge^{4+} from Ga^{3+} and detect a higher amount of breakthrough when a commercial generator was eluted with 1M HCl instead of the recommended 0.1M HCl.

In the context of ^{18}F-radiopharmaceuticals, Cheung et al. (2014) investigated the ability to separate multiple neutral impurities using MEKC. One model system explored was a set of 5 structurally-related neutral impurities known to occur during the synthesis of [^{18}F]FLT. After optimization of the separation conditions, baseline separation among all impurities and FLT was obtained (Fig. 4c). A second model system was prepared from impurities present in the synthesis of [^{18}F]FAC, and the MEKC separation was able to readily resolve the desired β-FAC from the α-FAC isomer and from Kryptofix 2.2.2 (Fig. 4d). Furthermore, analyte detection sensitivity of the CE setup was compared to that of HPLC and found to be similar. Studies with radioactive samples were not reported, but additional work was performed to miniaturize the separation using microchip electrophoresis (described in the next section). In another example, Antuganov et al. 2019b developed a validated CE based method (using UV absorbance detection) for the rapid analysis of a set of multiple potentially toxic impurities (including Kryptofix 222, Ni^{2+}, 4-DMPOTf, TEAHC, Cu^{2+}, TBAOTf, Me_3SnCl) encountered in ^{18}F-radiosyntheses (Fig. 4e). The method was used for QC testing of tracers for clinical use ([^{18}F]FDG, [^{18}F]FMISO and 6-[^{18}F]fluoro-L-DOPA) as well as research tracers produced via copper-mediated radiofluorination. For example, in 2021, Antuganova et al. (2021) developed and validated a CE method for determination of 2-chloro-2-deoxy-D-glucose (ClDG), 2-fluoro-2-deoxy-D-glucose (FDG) and glucose in clinical [^{18}F]FDG samples. The species were well-resolved (minimum chromatographic resolution 2.1), with migration times of 6.04, 6.10, and 6.69 min, respectively.

In addition to the applications highlighted above, CE methods (without radiometric detection) are also used in screening and preparation of precursors for radiolabeling with PET isotopes, especially for larger molecules based on peptides (Pastorino et al., 2021), proteins (He et al. 2020), or nucleic acids (Gijs et al., 2016).

In other work related to PET radiopharmaceuticals, Huang et al. (2003) developed an electrolytic method for removal of impurities from recycled O-18 water, and used CE for quantifying the remaining anions in the purified solution and gas chromatography (GC) for quantifying residual organic solvents. With limit of quantitation (LOQ) levels of 0.5 ppm or lower for multiple impurities, they developed a QC method for the purified O-18 water, and showed that [^{18}F]fluoride produced from recycled O-18 water that passed their QC tests produced similar results for the synthesis of [^{18}F]FDG as and when using newly purchased O-18 water.

CE is also valuable in relation to therapeutic and theranostic isotopes. In particular, due to the ability to probe molecular charge and size, CE has been valuable for detailed studies of radiometal complexes where there are multiple oxidation states and diverse coordination chemistries. Such studies generally make use of non-radioactive isotopes and rely on UV or other non-radiometric detection. As an example, 1-hydroxyethylidenediphosphonic acid (HEDP) labeled with the beta-emitting isotopes Re-186 or Re-188 shows promise for palliative treatment of bone metastases, but the complexation is quite complicated. Rohde (1996) used multiple methods (CZE, MEKC, and CGE) to study the influence of various parameters on the reaction process. Kohlíčková-Koudelková et al. (2002) and Koudelková and Jedináková-Křížová (2003), using CZE with UV and proton-induced X-ray emission (PIXE) detection, studied the complexation reaction and the kinetics of decomposition. The results of the study elucidated the optimal pH range for the Re-HEDP complex preparation (pH 2–7), composition of the reaction mixture (Re:HEDP at least 51:500) and time stability (2–5 days). The results were comparable with a radio-TLC study using [^{186}Re]Re-HEDP. Additionally, their analysis revealed the presence of at least three different Re-HEDP complex compounds (mixture of Re(III), Re(IV), and Re(V) states) and possible incorporation of Sn (from one of the reagents) into the structure. Focusing on a different complex, Kaliba et al. (2017) developed a CZE method as a research tool to separate the various species in rhenium complexes with 1,2,3-trihydroxybenzene. CE is also beneficial in the study of the kinetics and stability of radiometals with chelators that can be used to label peptides and proteins. As just one example, in a study relating to scandium, which has multiple isotopes that can be used in nuclear medicine (Sc-43, Sc-44, Sc-44m, Sc-47), Huclier-Markai et al. (2015) used CZE to with ICP-MS detection to measure the stability of Sc(III) complexes with the chelators DOTA and DTPA under a variety of conditions.

5 Microchip Electrophoresis

5.1 Introduction

Significant advantages have been realized by implementing the separation within a microchannel rather than a capillary (Castro & Manz, 2015; Dawod et al., 2017). The basic mechanics of microchip electrophoresis (MCE) are similar to CE, but the overall setups can be much smaller, and the microfabrication techniques used to fabricate the microfluidic chips give exquisite control over channel geometry of the injector and detection point (Breadmore, 2012; Ly et al., 2018), as well as the possibility to integrate functional elements such as detectors or electrodes directly in the device. In addition, microchips can be designed to inject smaller sample volumes than CE, leading to improved separation performance, often in a shorter distance and time (Caruso et al., 2020). MCE has been used in a wide range of applications (Ragab & El-Kimary, 2021) and is of high interest for analysis and quality control of pharmaceuticals (Dispas et al., 2021).

MCE chips are primarily made of quartz or glass and sometimes polymers (Castro & Manz, 2015; Dispas et al., 2021). The simplest structure of an MCE chip contains a straight microchannel for separation intersected by a short perpendicular channel for sample loading (Karlinsey, 2012). If a longer separation channel length is needed, the separation channel can be designed with a serpentine pattern (Molho et al., 2001; Zhuang et al., 2011). The chip can be much smaller than the typical many 10s of cm length of the capillary in a CE setup. Introduction of the sample is usually achieved via electrokinetic (voltage-driven) injection, wherein an electric field is used to drive the sample through the loading channel and into the channel junction, after which the portion of sample within the junction can be separated by applying the separation field along the separation channel. Additional sample loading strategies have been developed to increase repeatability and quality of the injections (Wenclawiak & Püschl, 2006; Karlinsey, 2012; Saito et al., 2012), with microvalve-based injection methods providing particularly high accuracy. When integrating microvalves for sample injection, the elastomer polydimethylsiloxane (PDMS) is generally used for fabrication of the MCE chip. Unfortunately, due to poor and unstable charge accumulation on the interior surfaces of PDMS, which can hinder the strength and stability of the EOF, separation resolution can be compromised (Lacher et al., 2003; Coltro et al., 2008). To address this issue, an architecture known as "hybrid MCE" has been used, in which a silica capillary is connected between microchips for injection and detection (Dimov et al., 2010). This combines the advantages of sophisticated microchip-based injectors and detectors with the stable surface charge of a fused silica capillary, though such devices are more complex to fabricate due to the need for in-plane capillary to chip connections (requiring tapering of the capillary ends) to minimize dead volume (Bings et al., 1999; Chiou & Lee, 2004).

5.2 Applications

In relation to the CE-based analysis of Ge-68 breakthrough in the eluate of a Ga-68 generator described earlier, Weinberg et al. described in a patent application some potential microchip-based embodiments of their system, using isotachophoresis in combination with conductivity detection and optionally using valves to collect the Ge-68 fraction for off-chip radiometric detection (Weinberg et al., 2013). (Isotachophoresis is a CE modality in which the sample is injected between leading and trailing electrolytes, and analytes are separated based on ionic mobility.)

The first demonstration of MCE for application to radiopharmaceutical analysis was by Ly et al. (2018), who designed a proof-of-concept MCE device for the separation of FLT and known byproducts of [^{18}F]FLT synthesis using MEKC. A hybrid MCE design was used, comprising a PDMS injection chip for hydrodynamic injection using pneumatically-actuated microvalves, a PDMS detection chip incorporating a Z-shaped channel with 0.5 mm optical path for UV absorbance detection, and a 60 cm long capillary (75 μm inner diameter) between them (Fig. 5a). Though the UV detection sensitivity was comparable to HPLC, the preliminary hybrid MCE chip did not achieve the desired separation resolution due to suboptimal geometry of the capillary to chip connections. The setup was improved in work reported by Ha et al. (2017a, 2017b), in which a collinear connection was made between the

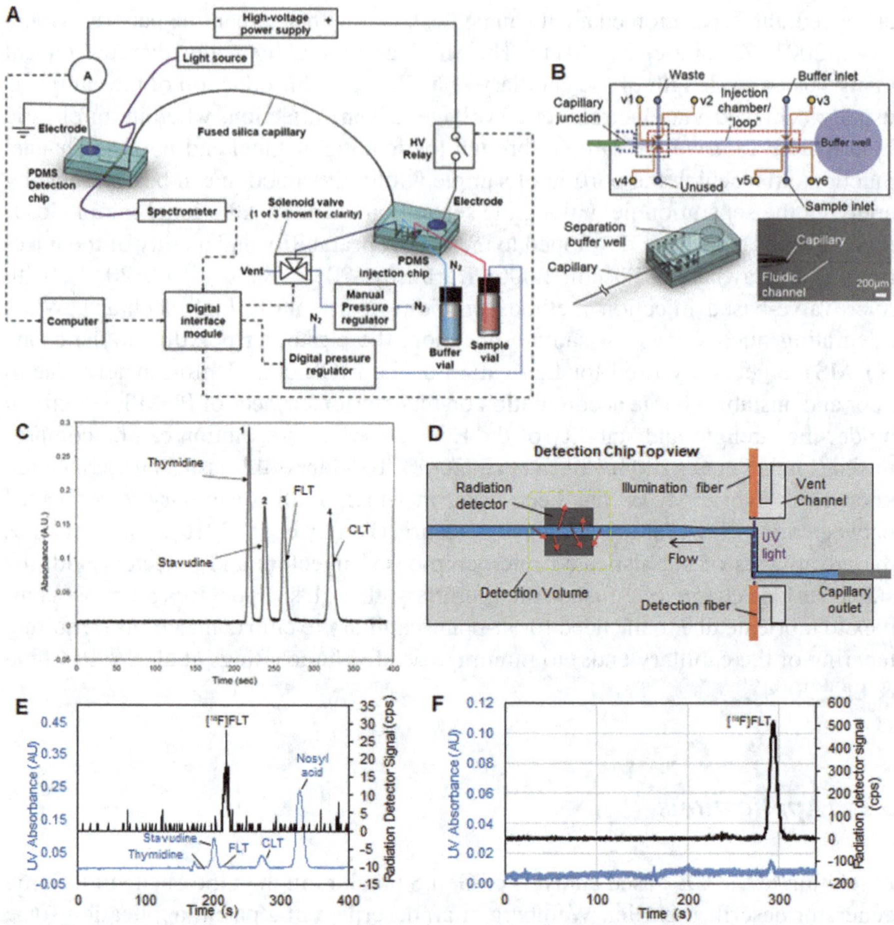

Fig. 5 (**a**) Schematic of hybrid MCE setup. (Reproduced from Ly et al. 2018 with permission of Springer ©2018). (**b**) Detailed design of volumetric injection chip for improved injection repeatability. (**c**) Example electropherogram from the hybrid-MCE chip showing separation of FLT, thymidine, stavudine, and CLT. (**b**, **c** reproduced from Ha et al. 2017a, 2017b with permission of Elsevier ©2017). (**d**) Detailed design of modified MCE detection chip containing an improved UV detection cell (with reduced volume) and an added radiation detector based on an APD for direct positron detection. (**e**) Example analysis from the modified hybrid-MCE setup of an unpurified [^{18}F]FLT sample. (**f**) Example analysis of a purified [^{18}F]FLT sample. (**d**, **e**, **f** adapted from Jones et al. 2020 with permission of Elsevier ©2020)

channel within the PDMS injection chip and the capillary to reduce dead volume. This design dramatically improved peak symmetry and separation efficiency despite using only a 20 cm long capillary instead of 60 cm. Ha et al. furthermore developed a modified injection chip based on the concept of an HPLC injection valve, in which the sample was loaded into a fixed volume channel segment (with ~4 nL volume) that was then connected into the separation path by opening microvalves, resulting in highly uniform injected volume (Fig. 5b). As measured by the area of peaks on the electropherogram for repeated injections, the relative standard deviation of the injected volume was 1.04% (n = 11), addressing one of the major criticisms regarding use of MCE in quantitative quality control of radiopharmaceuticals (Dispas et al., 2021). An example electropherogram showing MEKC separation of a model sample (FLT and known byproducts of [^{18}F]FLT synthesis) using this device is shown in Fig. 5c.

More recently, Jones et al. (2020) combined this injection concept with a new detection chip (Fig. 5d). This chip contained a Z-shaped cell with 2 mm long optical path length for UV absorbance detection, in which the channel width within the detection region was narrowed to maintain a low detection volume to avoid compromising separation resolution. A 2 × 2 mm^2 solid state radiation detector (avalanche photodiode; APD) was also included in this chip. The small detector provided very high sensitivity to positrons and was nearly immune to annihilation gamma rays, enabling localized detection of positron-emitting isotopes without the need for shielding or collimation. The efficiency of the radiation detector was not reported, but would be <50% due to its 2π design. Jones et al. demonstrated that the system could perform simultaneous chemical and radiochemical analysis, and that baseline resolution between FLT and side products could be achieved (Fig. 5e, f). The LOQ for the UV absorbance detector was excellent (~μM range), and the LOQ of the radiation detector was 114 MBq/mL, which is suitable for testing large, concentrated batches of radiopharmaceuticals. Quantitative analysis of radiochemical purity in smaller batches may require further improvement in sensitivity, perhaps by using a 4π detector design, using sample preconcentration (Quirino & Haddad, 2008; Cong et al., 2016; John et al., 2021), or using adaptive flow methods that slow the migration speed when analytes are within the detector region Pentoney et al., 1989a, 1989b).

6 Perspectives and Outlook

Though not as widely used as liquid chromatography and planar chromatography methods for the analysis of radiopharmaceutical compounds, electrophoresis methods can provide excellent performance across a wide range of types of analytes. In addition to leveraging the strong dependence on molecular charge and size of the separation mechanism, additional interactions can be introduced to enhance the separation. Below we briefly summarize the main applications, advantages, and limitations of each type of electrophoresis.

6.1 Gel Electrophoresis

The apparatus needed for gel electrophoresis is simple and inexpensive, and due to the extensive use of gel electrophoresis in various fields of biochemistry and biological studies, the apparatus is widely available, making this technique very accessible. In the radiopharmaceutical field, gel electrophoresis has found wide use when analyzing samples comprising radiolabeled proteins and nucleic acids. For such biomolecules, gel electrophoresis provides superior separation resolution compared to other techniques such as size exclusion liquid chromatography methods. The technique is primarily used upstream of the radiochemistry, for example in preparation and characterization of the labeling precursor. However, analysis of radiolabeled species in the gel after separation has been shown using cut-and-count and autoradiographic methods. The availability of apparatus with built-in radiodetection capabilities would further increase accessibility of this technique. Though the technique is relatively slow, multiple samples can be run in parallel in adjacent lanes, increasing throughput. Lack of automation can be cumbersome in some situations, yet gel electrophoresis serves as an indispensable tool for characterization of high molecular weight radiopharmaceuticals.

6.2 Paper Electrophoresis

Paper electrophoresis has been used for diverse applications, including radiopharmaceutical and reaction development as well as for quality control of prepared radiopharmaceuticals. Though planar chromatography methods such as TLC and iITLC available today are more convenient than PE for routine quality control of radiochemical purity of radiopharmaceuticals, paper electrophoresis serves as an indispensable research tool in radiopharmaceutical studies providing information on the charge of the labeled molecules and complexes that cannot be easily determined by other methods. From the view of evaluating whether a specific radiolabeled complex is anionic, cationic, or neutral, electrophoresis is the best technique currently available. The information on the charge state can be used for predicting in vivo properties of newly developed radiotracers and for interpreting data on the biodistribution, specific uptake and clearance, blood brain barrier (BBB) permeability, and others. An advantage of PE in radiopharmaceutical research is the possibility of running multiple samples in parallel lanes to increase throughput. The availability of commercial PE apparatus equipped with radio detection systems will greatly contribute to wider implementation of this technique into routine practice and research.

6.3 Capillary Electrophoresis

Capillary electrophoresis has been reported in all aspects of the radiopharmaceutical field, from analyzing labeling precursors, studying reaction mechanisms, optimizing reaction conditions, and performing quality control of the final radiopharmaceutical products.

CE offers a number of benefits compared to liquid chromatography and planar chromatography methods. (1) First, separations are often very fast, which is advantageous when working with short-lived isotopes. Examples of short migration times observed include 3.0 ± 0.2 min for FLT (Cheung et al., 2014), 2.48 ± 0.05 min for ß-FAC (Cheung et al., 2014), ~3.6 min for [^{99m}Tc]Tc-ethylenecysteine dimer (Jankowsky et al., 1999a, 1999b), and between 3.0–4.5 min for a variety of commonly-encountered impurities in F-18 radiochemistry (Antuganov et al. 2019a, 2019b). (2) The low sample volume of CE is also attractive as it reduces the proportion of the formulated radiopharmaceutical batch (sometimes only 1–2 mL total volume) that must be dedicated to QC testing, and also reduces radiation exposure during the QC tests and minimizes the need for shielding. Injected sample volume for CE and MCE typically ranges from a few to tens of nL (Ha et al., 2017a, 2017b; Jankowsky et al., 1999b; Kuczewski et al., 2003), compared to the tens to hundreds of μL used for radio-HPLC. (3) CE-based techniques are exceptionally versatile. Methods are available that exhibit excellent separation resolution for analytes composed of small ions, small molecules, peptides, or large molecules such as proteins and nucleic acids. There is a growing recognition of the value of CE for analysis of conventional therapeutics (Rastogi et al., 2019) and the same arguments would also apply to analysis of radiopharmaceutical compounds.

In terms of sensitivity, UV absorbance detection in CE setups exhibits similar performance to HPLC systems despite the small capillary diameter and thus optical path length, and therefore can be used to perform the same QC testing (Cheung et al., 2014). Radiation detection is fundamentally challenging in CE due to the low sample volume (i.e., low amount of radioactivity) and short separation times (i.e., short amount of sample-detector interaction time); however, a number of different types of efficient radiation detectors have successfully been implemented. Characterization of radiation detection limits would need to be more thoroughly investigated to confirm suitability for radiochemical purity testing.

Despite the success and potential of CE methods, their use is currently rather limited due to the lack of installed CE instruments in radiochemistry labs. The development of smaller and lower cost CE systems (e.g., sacrificing some flexibility to increase simplicity) could encourage wide adoption by lowering the barrier for laboratories to begin using CE. Furthermore, it seems that most research groups working in this area have built or customized their radiation detectors, requiring unique knowledge and skills. The development of commercial CE-optimized radiation detectors would ease adoption by non-experts. To address potential regulatory concerns of introducing new or alternative testing methods, further development of CE-based methods for analyzing a wide variety of radiopharmaceuticals, particularly validated methods for use in clinical production settings would go a long way in demonstrating versatility and value of electrophoretic separations. Acceptance may be swiftest in areas where CE can provide more detailed information than typical HPLC or TLC methods, for example, enabling confirmation of oxidation state, or providing improved separation to confirm purity of complex samples (e.g., peptides or proteins).

6.4 Microchip Electrophoresis

The introduction of microchip electrophoresis into the field of radiopharmaceuticals is very new, but due to several advantages over CE, one would expect its use to grow to those areas where CE has been used. MCE systems are far more compact than CE systems and could lower the barrier for labs to acquire instrumentation. The advent of improved injection methods (with higher repeatability), and the ability to easily design more complex detection cells (e.g., Z-shaped channel for optical detectors), enables MCE to improve on the performance of CE.

An additional potential advantage is the possibility of using MCE chips in a disposable manner, which could save significant time in terms of instrument maintenance, and provide the flexibility for back-to-back use without the need to develop and validate cleaning protocols. Ultimately, integration of MCE with other types of miniaturized analysis (Ha et al., 2017b) could greatly simplify analysis workflows such as QC testing.

As described in a recent review, a number of commercial MCE systems are now on the market (Dispas et al., 2021); however, the systems are not designed for analysis of radiopharmaceuticals (i.e., they lack radiation detection). Unlike CE systems the small size of MCE would make it difficult for investigators to add their own detectors and thus use of MCE within the radiopharmaceutical field may be quite limited until commercial MCE systems with options for integrated radiation detection become available.

Disclosures R.M. van Dam is a founder and shareholder of SOFIE, Inc. J. Jones and R.M. van Dam are founders and shareholders of DropletPharma Corp.

References

Abbasi, I. (2015). Studies on the labeling of ethylenediaminetetramethylene phosphonic acid, methylene diphosphonate, sodium pyrophosphate and hydroxyapatite with lutetium-177 for use in nuclear medicine. *World Journal of Nuclear Medicine, 14*(02), 95–100. https://doi.org/10.4103/1450-1147.155752

Akbar, M. U., Bokhari, T. H., Roohi, S., Zia, K. M., Zuber, M., Parveen, N., & Ali, N. (2016). Development, optimization, and biovalidation of 99mTc–insulin complex. *Russian Journal of Bioorganic Chemistry, 42*(5), 491–496. https://doi.org/10.1134/S1068162016050125

Akbar, M. U., Bokhari, T. H., Khalid, M., Ahmad, M. R., Roohi, S., Hina, S., Mehmood, S., Sohaib, M., & Jabbar, T. (2017). Radiolabeling, quality control, and biological characterization of 177Lu-labeled kanamycin. *Chemical Biology & Drug Design, 90*(3), 425–431. https://doi.org/10.1111/cbdd.12960

Akbar, M. U., Bokhari, T. H., Roohi, S., Zia, K. M., Ahmad, M. R., Hina, S., & Sohaib, M. (2018). Preparation, optimization and pharmacological evaluation of 99mTc-4-Hydroxy-3-(2-Hydroxy-3-methoxybenzylideneamino) benzoic acid complex: As a novel potential radiopharmaceutical agent with hepatobiliary excretion. *Molecular Biology Reports, 45*(6), 2717–2723. https://doi.org/10.1007/s11033-018-4402-6

Akbar, M. U., Bokhari, T. H., Ahmad, M. R., Zia, K. M., Roohi, S., Ayub, N., & Sohaib, M. (2019). Radiosynthesis, radiochemical, and biological characterization of 177Lu-labeled diethylenetriamine penta-acetic acid. *Journal of Cellular Biochemistry, 120*(9), 14510–14517. https://doi.org/10.1002/jcb.28712

Altria, K. D. (1999). Overview of capillary electrophoresis and capillary electrochromatography. *Journal of Chromatography A, 856*(1–2), 443–463. https://doi.org/10.1016/S0021-9673(99)00830-4

Altria, K. D., Simpson, C. F., Bharij, A. K., & Theobald, A. E. (1990). A gamma-ray detector for capillary zone electrophoresis and its use in the analysis of some radiopharmaceuticals. *Electrophoresis, 11*(9), 732–734. https://doi.org/10.1002/elps.1150110910

Antuganov, D., Antuganova, Y., & Zykova, T. (2019a). Rapid analysis of chemical purity in 68Ga-PSMA-11 and 68Ga-DOTA-NOC by capillary electrophoresis. In *Proceedings of the 23rd International Symposium on Radiopharmaceutical Sciences (ISRS 2019)* (Vol. 62, p. S582). Journal of Labelled Compounds and Radiopharmaceuticals. https://doi.org/10.1002/jlcr.3725

Antuganov, D., Antuganova, Y., Zykova, T., & Krasikova, R. (2019b). Use of capillary electrophoresis for the determination of impurities in preparations of fluorine-18 labelled PET radiopharmaceuticals. *Journal of Pharmaceutical and Biomedical Analysis, 173*(September), 68–74. https://doi.org/10.1016/j.jpba.2019.05.016

Antuganova, Y., Nadporojskii, M., & Antuganov, D. (2021). Determination of 2-Chloro-2-deoxy-D-glucose, 2-Fluoro-2-deoxy-D-glucose, and glucose in [18F]Fluorodeoxyglucose by capillary electrophoresis. *Nuclear Medicine and Biology, 96–97*(S1), S80.

Aynié, I., Vauthier, C., Foulquier, M., Malvy, C., Fattal, E., & Couvreur, P. (1996). Development of a quantitative polyacrylamide gel electrophoresis analysis using a multichannel radioactivity counter for the evaluation of oligonucleotide-bound drug carrier. *Analytical Biochemistry, 240*(2), 202–209. https://doi.org/10.1006/abio.1996.0350

Beutler, B. A., Milsark, I. W., & Cerami, A. (1985). Cachectin/tumor necrosis factor: Production, distribution, and metabolic fate in vivo. *The Journal of Immunology, 135*(6), 3972–3977. https://doi.org/10.4049/jimmunol.135.6.3972

Bings, N. H., Wang, C., Skinner, C. D., Colyer, C. L., Thibault, P., & Jed Harrison, D. (1999). Microfluidic devices connected to fused-silica capillaries with minimal dead volume. *Analytical Chemistry, 71*(15), 3292–3296.

Blower, P. J., Singh, J., & Clarke, S. E. M. (1991). The chemical identity of pentavalent technetium-99m-dimercaptosuccinic acid. *Journal of Nuclear Medicine, 32*(5), 845–849.

Bočekxs, P. (1993). Capillary electrophoresis: Theory and practice. Paul D. Grossman and Joel C. Colburn (Eds.) Academic Press, Inc., San Diego, California, USA, 1992, 352 pp., Price US $ 69.95, ISBN 0–12–304250-X. *Electrophoresis, 14*(1), 559–559. https://doi.org/10.1002/elps.1150140185

Bokhari, T. H., Akbar, M. U., Roohi, S., Hina, S., Sohaib, M., & Rizvi, F. A. (2015). Synthesis, characterization and biological evaluation of 99mTc-labeled mitomycin C. *Journal of Radioanalytical and Nuclear Chemistry, 303*(3), 1779–1784. https://doi.org/10.1007/s10967-014-3743-4

Bokhari, T. H., Roohi, S., Hina, S., Akbar, M. U., Sohaib, M., & Iqbal, M. (2018). Synthesis, quality control, and bio-evaluation of 99mTc-cyclophosphamide. *Chemical Biology & Drug Design, 91*(2), 456–462. https://doi.org/10.1111/cbdd.13107

Breadmore, M. C. (2012). Capillary and microchip electrophoresis: Challenging the common conceptions. *Journal of Chromatography A, 1221*(January), 42–55. https://doi.org/10.1016/j.chroma.2011.09.062

Carpenter, A. P. (1986). Radioanalytical techniques: ITLC, TLC, mini-columns, and electrophoresis. In D. M. Wieland, M. C. Tobes, & T. J. Manger (Eds.), *Analytical and chromatographic techniques in radiopharmaceutical chemistry* (pp. 39–70). Springer. https://doi.org/10.1007/978-1-4612-4854-5_2

Caruso, G., Musso, N., Grasso, M., Costantino, A., Lazzarino, G., Tascedda, F., Gulisano, M., Lunte, S. M., & Caraci, F. (2020). Microfluidics as a novel tool for biological and toxicological assays in drug discovery processes: Focus on microchip electrophoresis. *Micromachines, 11*(6), 593. https://doi.org/10.3390/mi11060593

Castro, E. R., & Manz, A. (2015). Present state of microchip electrophoresis: State of the art and routine applications. *Journal of Chromatography A, 1382*(February), 66–85. https://doi.org/10.1016/j.chroma.2014.11.034

Cawthorne, C., Prenant, C., Smigova, A., Julyan, P., Maroy, R., Herholz, K., Rothwell, N., & Boutin, H. (2011). Biodistribution, pharmacokinetics and metabolism of Interleukin-1 receptor antagonist (IL-1RA) using [18F]-IL1RA and PET imaging in rats. *British Journal of Pharmacology, 162*(3), 659–672. https://doi.org/10.1111/j.1476-5381.2010.01068.x

Cheung, S., Ly, J., Lazari, M., Sadeghi, S., Keng, P. Y., & Michael van Dam, R. (2014). The separation and detection of PET tracers via capillary electrophoresis for chemical identity and purity analysis. *Journal of Pharmaceutical and Biomedical Analysis, 94*(June), 12–18. https://doi.org/10.1016/j.jpba.2014.01.023

Chianelli, M., Signore, A., Fritzberg, A. R., & Mather, S. J. (1997). The development of technetium-99m-labelled interleukin-2: A new radiopharmaceutical for the in vivo detection of mononuclear cell infiltrates in immune-mediated diseases. *Nuclear Medicine and Biology, 24*(6), 579–586. https://doi.org/10.1016/S0969-8051(97)00021-8

Chiou, C.-H., & Lee, G.-B. (2004). Minimal dead-volume connectors for microfluidics using PDMS casting techniques. *Journal of Micromechanics and Microengineering, 14*(11), 1484.

Coltro, W. K. T., Lunte, S. M., & Carrilho, E. (2008). Comparison of the analytical performance of electrophoresis microchannels fabricated in PDMS, glass, and polyester-toner. *Electrophoresis, 29*(24), 4928–4937. https://doi.org/10.1002/elps.200700897

Cong, Y., Katipamula, S., Geng, T., Prost, S. A., Tang, K., & Kelly, R. T. (2016). Electrokinetic sample preconcentration and hydrodynamic sample injection for microchip electrophoresis using a pneumatic microvalve. *Electrophoresis, 37*(3), 455–462. https://doi.org/10.1002/elps.201500286

Coomber, D. I. (Ed.). (1975). *Radiochemical methods in analysis*. Springer US. https://doi.org/10.1007/978-1-4613-4401-8

Corradini, D. (2023). Buffering agents and additives for the background electrolyte solutions used for peptide and protein capillary zone electrophoresis. *TrAC Trends in Analytical Chemistry, 164*(July), 117080. https://doi.org/10.1016/j.trac.2023.117080

Corsetti, F., Chianelli, M., Cornelissen, B., Van de Wiele, C., D'Alessandria, C., Slegers, G., Mather, S. J., et al. (2004). Radioiodinated recombinant human TSH: A novel radiopharmaceutical for thyroid cancer metastases detection. *Cancer Biotherapy and Radiopharmaceuticals, 19*(1), 57–63. https://doi.org/10.1089/108497804773391685

da Silva, J. A. F., do Lago, C. L., de Jesus, D. P., & Coltro, W. K. T. (2013). Capacitively coupled contactless conductivity detection (C4D) applied to capillary electrophoresis (CE) and microchip electrophoresis (MCE). In C. D. García, K. Y. Chumbimuni-Torres, & E. Carrilho (Eds.), *Capillary electrophoresis and microchip capillary electrophoresis* (pp. 145–160). Wiley. http://onlinelibrary.wiley.com/doi/10.1002/9781118530009.ch8/summary

Dawod, M., Arvin, N. E., & Kennedy, R. T. (2017). Recent advances in protein analysis by capillary and microchip electrophoresis. *Analyst, 142*(11), 1847–1866. https://doi.org/10.1039/C7AN00198C

Decristoforo, C., Zaknun, J., Kohler, B., Oberladstaetter, M., & Riccabona, G. (1997). The use of electronic autoradiography in radiopharmacy. *Nuclear Medicine and Biology, 24*(4), 361–365. https://doi.org/10.1016/S0969-8051(97)00055-3

Decristoforo, C., Zolle, I., Rakiás, F., Imre, J., Jánoki, G., & Hesslewood, S. R. (2007). Quality control methods of 99mTc pharmaceuticals. In I. Zolle (Ed.), *Technetium-99m pharmaceuticals: Preparation and quality control in nuclear medicine* (pp. 123–150). Springer. https://doi.org/10.1007/978-3-540-33990-8_9

Deng, Y., & Li, B. (1998). On-column refractive-index detection based on retroreflected beam interference for capillary electrophoresis. *Applied Optics, 37*(6), 998–1005. https://doi.org/10.1364/AO.37.000998

Dimov, I. K., Riaz, A., Ducrée, J., & Lee, L. P. (2010). Hybrid integrated PDMS microfluidics with a silica capillary. *Lab on a Chip, 10*(11), 1468–1471. https://doi.org/10.1039/B925132D

Dispas, A., Emonts, P., & Fillet, M. (2021). Microchip electrophoresis: A suitable analytical technique for pharmaceuticals quality control? A critical review. *TrAC Trends in Analytical Chemistry, 139*(June), 116266. https://doi.org/10.1016/j.trac.2021.116266

Dolník, V. (1994). Capillary gel electrophoresis. *Journal of Microcolumn Separations, 6*(4), 315–330. https://doi.org/10.1002/mcs.1220060402

Donald Burns, H., Sowa, D. T., Worley, P., Vaum, R., & Marzilli, L. G. (1981). Electrophoretic determination of charge on carrier-free 99mTc-Labeled complexes. *Journal of Pharmaceutical Sciences, 70*(4), 436–439. https://doi.org/10.1002/jps.2600700424

Dunbar, B. S. (1987). Use of autoradiography in polyacrylamide gel electrophoresis. In B. S. Dunbar (Ed.), *Two-dimensional electrophoresis and immunological techniques* (pp. 117–122). Springer US. https://doi.org/10.1007/978-1-4613-1957-3_7

Eeckhaut, A. V., & Michotte, Y. (2009). *Chiral separations by capillary electrophoresis*. CRC Press.

Eeltink, S., Rozing, G. P., & Wim, T. K. (2003). Recent applications in capillary electrochromatography. *Electrophoresis, 24*(22–23), 3935–3961. https://doi.org/10.1002/elps.200305638

Elizarov, A. M. (2020). Automation of PET radiopharmaceutical quality control. In *Handbook of radiopharmaceuticals* (pp. 457–489). John Wiley & Sons, Ltd. https://doi.org/10.1002/9781119500575.ch14

Fan, W., Zhang, W., Alshehri, S., & Garrison, J. C. (2018). Increasing time on target: Utilization of inhibitors of cysteine cathepsins to enhance the tumor retention of receptor-targeted agents. *Chemical Communications, 54*(80), 11268–11271. https://doi.org/10.1039/C8CC05982A

Fedorowski, J., & LaCourse, W. R. (2015). A review of pulsed electrochemical detection following liquid chromatography and capillary electrophoresis. *Analytica Chimica Acta, 861*, 1–11.

Felhofer, J. L., Blanes, L., & Garcia, C. D. (2010). Recent developments in instrumentation for capillary electrophoresis and microchip-capillary electrophoresis. *Electrophoresis, 31*(15), 2469–2486. https://doi.org/10.1002/elps.201000203

Franck, D., Nann, H., Davi, P., Schubiger, P. A., & Ametamey, S. M. (2009). Faster analysis of radiopharmaceuticals using ultra performance liquid chromatography (UPLC®) in combination with low volume radio flow cell. *Applied Radiation and Isotopes, 67*(6), 1068–1070. https://doi.org/10.1016/j.apradiso.2009.02.077

Füchtner, F., Angelberger, P., Kvaternik, H., Hammerschmidt, F., Peric Simovc, B., & Steinbach, J. (2002). Aspects of 6-[18F]Fluoro-L-DOPA preparation: Precursor synthesis, preparative HPLC purification and determination of radiochemical purity. *Nuclear Medicine and Biology, 29*(4), 477–481. https://doi.org/10.1016/S0969-8051(02)00298-6

Gabrieli, E. R., Goulian, D., Kinersly, T., & Collet, R. (1954). Zone paper electrophoresis studies on radio-iodinated human serum albumin. *The Journal of Clinical Investigation, 33*(2), 136–141. https://doi.org/10.1172/JCI102880

Gahoual, R., Beck, A., Leize-Wagner, E., & François, Y.-N. (2016). Cutting-edge capillary electrophoresis characterization of monoclonal antibodies and related products. *Journal of Chromatography B, Analytical Characterization of Biopharmaceuticals, 1032*(October), 61–78. https://doi.org/10.1016/j.jchromb.2016.05.028

Gallagher, S. R. (2012). One-dimensional SDS gel electrophoresis of proteins. *Current Protocols in Molecular Biology, 97*(1), 10.2A.1-10.2A.44. https://doi.org/10.1002/0471142727.mb1002as97

George, S. C., & Samuel, E. J. J. (2023). Developments in 177Lu-based radiopharmaceutical therapy and dosimetry. *Frontiers in Chemistry, 11*, 1218670. https://doi.org/10.3389/fchem.2023.1218670

Ghosh, K. K., Padmanabhan, P., Yang, C.-T., Mishra, S., Halldin, C., & Gulyás, B. (2020). Dealing with PET radiometabolites. *EJNMMI Research, 10*(1), 109. https://doi.org/10.1186/s13550-020-00692-4

Gijs, M., Aerts, A., Impens, N., Baatout, S., & Luxen, A. (2016). Aptamers as radiopharmaceuticals for nuclear imaging and therapy. *Nuclear Medicine and Biology, 43*(4), 253–271. https://doi.org/10.1016/j.nucmedbio.2015.09.005

Gillings, N., Todde, S., Behe, M., Decristoforo, C., Elsinga, P., Ferrari, V., Hjelstuen, O., et al. (2020). EANM guideline on the validation of analytical methods for radiopharmaceuticals. *EJNMMI Radiopharmacy and Chemistry, 5*(1), 7. https://doi.org/10.1186/s41181-019-0086-z

Gordon, A. H., Keil, B., Šebesta, K., Knessl, O., & Šorm, F. (1950). Electrophoresis of proteins in agar jelly. *Collection of Czechoslovak Chemical Communications, 15*, 1–16. https://doi.org/10.1135/cccc19500001

Ha, N. S., Ly, J., Jones, J., Cheung, S., & Michael van Dam, R. (2017a). Novel volumetric method for highly repeatable injection in microchip electrophoresis. *Analytica Chimica Acta, 985*(September), 129–140. https://doi.org/10.1016/j.aca.2017.05.037

Ha, N. S., Sadeghi, S., & Michael van Dam, R. (2017b). Recent progress toward microfluidic quality control testing of radiopharmaceuticals. *Micromachines, 8*(11), 337. https://doi.org/10.3390/mi8110337

He, K., Zeng, S., & Qian, L. (2020). Recent progress in the molecular imaging of therapeutic monoclonal antibodies. *Journal of Pharmaceutical Analysis, 10*(5), 397–413. https://doi.org/10.1016/j.jpha.2020.07.006

Henry, C. S. (2006). *Microchip capillary electrophoresis: Methods and protocols*. Springer Science & Business Media.

Hilger, C. S., Willis, M. C., Wolters, M., & Pieken, W. A. (1998). Synthesis of Tc-99m-labeled, modified RNA. *Tetrahedron Letters, 39*(51), 9403–9406. https://doi.org/10.1016/S0040-4039(98)02193-5

Hina, S., Rajoka, M. I., Roohi, S., Haque, A., & Qasim, M. (2014). Preparation, biodistribution, and scintigraphic evaluation of 99mTc-clindamycin: An infection imaging agent. *Applied Biochemistry and Biotechnology, 174*(4), 1420–1433. https://doi.org/10.1007/s12010-014-1075-z

Hina, S., Bokhari, T. H., Roohi, S., Rajoka, M. I., & Sohaib, M. (2018). Labeling of epirubicin with technetium-99m: Optimization, biodistribution and scintigraphic imaging in tumor bearing mice. *Pakistan Journal of Pharmaceutical Sciences, 31*(2), 517–524.

Huang, B. X., Channing, M. A., Plascjak, P. S., Kiesewetter, D. O., Der, M., Ma, Y., & Eckelman, W. C. (2003). Routine quality control of recycled target [18O]water by capillary electrophoresis and gas chromatography. *Nuclear Medicine and Biology, 30*(7), 785–790. https://doi.org/10.1016/S0969-8051(03)00083-0

Huclier-Markai, S., Alliot, C., Sebti, J., Brunel, B., & Aupiais, J. (2015). A comparative thermodynamic study of the formation of scandium complexes with DTPA and DOTA. *RSC Advances, 5*(121), 99606–99617. https://doi.org/10.1039/C5RA16736A

International Atomic Energy Agency. (2018). *Quality control in the production of radiopharmaceuticals*. International Atomic Energy Agency. https://www.iaea.org/publications/13422/quality-control-in-the-production-of-radiopharmaceuticals.

Jankowsky, R., Friebe, M., Noll, B., & Johannsen, B. (1999a). Determination of dissociation constants of 99mTechnetium radiopharmaceuticals by capillary electrophoresis. *Journal of Chromatography. A, 833*(1), 83–96.

Jankowsky, R., Noll, B., & Johannsen, B. (1999b). Capillary electrophoresis of 99mtechnetium radiopharmaceuticals. *Journal of Chromatography B: Biomedical Sciences and Applications, 724*(2), 365–371. https://doi.org/10.1016/S0378-4347(98)00577-5

Jánoki, G. A., Kórösi, L., Spett, B., & Kocsár, L. T. (1986). Protein based radiopharmaceuticals: Applicability of polyacryl-amide gel electrophoresis in the quality control of labelled protein preparation. In P. H. Cox, S. J. Mather, C. B. Sampson, & C. R. Lazarus (Eds.), *Progress in radiopharmacy* (pp. 570–576). Springer. https://doi.org/10.1007/978-94-009-4297-4_32

Jencks, W. P., Jetton, M. R., & Durrum, E. L. (1955). Paper electrophoresis as a quantitative method. Serum proteins. *Biochemical Journal, 60*(2), 205–215. https://doi.org/10.1042/bj0600205

Jeong, J. M., Lee, Y.-S., Kim, Y. J., Lee, D. S., Chung, J.-K., Suh, Y.-G., & Lee, M. C. (2002). Cationic complex: 99mTc-N,N′-dimethyldiaminedithiol. *Nuclear Medicine and Biology, 29*(6), 711–717. https://doi.org/10.1016/S0969-8051(02)00312-8

John, A. S., Sidek, M. M., Thang, L. Y., Sami, S., Tey, H. Y., & See, H. H. (2021). Online sample preconcentration techniques in nonaqueous capillary and microchip electrophoresis. *Journal of Chromatography A, 1638*(February), 461868. https://doi.org/10.1016/j.chroma.2020.461868

Johnston, R. F., Pickett, S. C., & Barker, D. L. (1990). Autoradiography using storage phosphor technology. *Electrophoresis, 11*(5), 355–360. https://doi.org/10.1002/elps.1150110503

Jones, J., Ha, N. S., Barajas, A. G., Chatziioannou, A. F., & Michael van Dam, R. (2020). Integration of high-resolution radiation detector for hybrid microchip electrophoresis. *Analytical Chemistry, 92*(4), 3483–3491. https://doi.org/10.1021/acs.analchem.9b04827

Jorgenson, J. W., & Lukacs, K. D. A. (1981). Zone electrophoresis in open-tubular glass capillaries. *Analytical Chemistry, 53*(8), 1298–1302. https://doi.org/10.1021/ac00231a037

Kaliba, D., Štícha, M., Jelínek, I., Poláková, J., & Radová, J. (2017). Application of capillary electrophoresis to the separation of rhenium complex of 1,2,3-Trihydroxybenzene. *Monatshefte Für Chemie—Chemical Monthly, 148*(9), 1619–1624. https://doi.org/10.1007/s00706-017-1982-0

Kaniansky, D., Rajec, P., Švec, A., Marák, J., Koval', M., Lúčka, M., Franko, Š., & Sabanoš, G. (1989). On-column radiometric detector for capillary isotachophoresis. *Journal of Radioanalytical and Nuclear Chemistry, 129*(2), 305–325. https://doi.org/10.1007/BF02039828

Karlinsey, J. M. (2012). Sample introduction techniques for microchip electrophoresis: A review. *Analytica Chimica Acta, 725*(May), 1–13. https://doi.org/10.1016/j.aca.2012.02.052

Karube, Y., Iwamoto, K., & Takata, J. (1994). Cationic technetium-99m complexes of N-substituted pyridoxal derivatives as renal function agents. *Journal of Nuclear Medicine, 35*(10), 1691–1697.

Klunder, G. L., Andrews, J. E., Grant, P. M., Andresen, B. D., & Russo, R. E. (1997). Analysis of fission products using capillary electrophoresis with on-line radioactivity detection. *Analytical Chemistry, 69*(15), 2988–2993. https://doi.org/10.1021/ac970042e

Kohlíčková-Koudelková, M., Jedináková-Křížová, V., & Deyl, Z. (2002). Study of perrhenate reduction by capillary electrophoresis. *Electrophoresis, 23*(2), 245–248. https://doi.org/10.1002/1522-2683(200202)23:2<245::AID-ELPS245>3.0.CO;2-D

Koudelková, M., & Jedináková-Křížová, V. (2003). Capillary electrophoretic and thin-layer chromatographic characterization of rhenium complexation with 1-hydroxyethylidenediphosphonic acid. *Journal of Chromatography A, 990*(1), 317–323. https://doi.org/10.1016/S0021-9673(02)01798-3

Koźmiński, P., Gawęda, W., Rzewuska, M., Kopatys, A., Kujda, S., Dudek, M. K., Halik, P. K., Królicki, L., & Gniazdowska, E. (2021). Physicochemical and biological study of 99mTc and 68Ga radiolabelled ciprofloxacin and evaluation of [99mTc]Tc-CIP as potential diagnostic radiopharmaceutical for diabetic foot syndrome imaging. *Tomography, 7*(4), 829–842. https://doi.org/10.3390/tomography7040070

Kuczewski, B., Marquardt, C. M., Seibert, A., Geckeis, H., Kratz, J. V., & Trautmann, N. (2003). Separation of plutonium and neptunium species by capillary electrophoresis–inductively coupled plasma-mass spectrometry and application to natural groundwater samples. *Analytical Chemistry, 75*(24), 6769–6774. https://doi.org/10.1021/ac0347213

Kunkel, H. G., & Tiselius, A. (1951). Electrophoresis of proteins on filter paper. *Journal of General Physiology, 35*(1), 89–118. https://doi.org/10.1085/jgp.35.1.89

Kuntner, C., Wanek, T., Hoffer, M., Dangl, D., Hornof, M., Kvaternik, H., & Langer, O. (2011). Radiosynthesis and assessment of ocular pharmacokinetics of 124I-labeled chitosan in rabbits using small-animal PET. *Molecular Imaging and Biology, 13*(2), 222–226. https://doi.org/10.1007/s11307-010-0352-7

Lacher, N. A., de Rooij, N. F., Verpoorte, E., & Lunte, S. M. (2003). Comparison of the performance characteristics of poly(dimethylsiloxane) and pyrex microchip electrophoresis devices for peptide separations. *Journal of Chromatography A, 1004*(1–2), 225–235. https://doi.org/10.1016/S0021-9673(03)00722-2

Laskey, R. A. (1980). The use of intensifying screens or organic scintillators for visualizing radioactive molecules resolved by gel electrophoresis. In *Methods in enzymology, nucleic acids part I* (Vol. 65, pp. 363–371). Academic Press. https://doi.org/10.1016/S0076-6879(80)65047-2

Lee, P. Y., Costumbrado, J., Hsu, C.-Y., & Kim, Y. H. (2012). Agarose gel electrophoresis for the separation of DNA fragments. *Journal of Visualized Experiments, 62*(April), e3923. https://doi.org/10.3791/3923

Li, S. F. Y. (1992). *Capillary electrophoresis: Principles, practice and applications*. Elsevier.

Li, J., Han, H., Wang, Q., Liu, X., & Jiang, S. (2010). Polymeric ionic liquid as a dynamic coating additive for separation of basic proteins by capillary electrophoresis. *Analytica Chimica Acta, 674*(2), 243–248. https://doi.org/10.1016/j.aca.2010.06.044

Lim, R., Huang, J. J., & Davis, G. A. (1969). Autoradiography with acrylamide gel slab electrophoresis. *Analytical Biochemistry, 29*(1), 48–57. https://doi.org/10.1016/0003-2697(69)90006-2

Lu, J., Kong, D., Yang, Z., Yang, S., Fan, W., Jia, H., & Wang, X. (2007). Preparation of 99mTc-Nitrido asymmetrical heterocomplex with 4-(Cyclohexylpiperazin-1-Yl)-Dithioformate and its biological evaluation as a potential myocardial imaging agent. *Applied Radiation and Isotopes, 65*(10), 1134–1139. https://doi.org/10.1016/j.apradiso.2007.04.021

Ly, J., Ha, N. S., Cheung, S., & Michael van Dam, R. (2018). Toward miniaturized analysis of chemical identity and purity of radiopharmaceuticals via microchip electrophoresis. *Analytical and Bioanalytical Chemistry, 410*(9), 2423–2436. https://doi.org/10.1007/s00216-018-0924-y

Lyczko, M., Pruszynski, M., Majkowska-Pilip, A., Lyczko, K., Was, B., Meczynska-Wielgosz, S., Kruszewski, M., et al. (2017). 211At Labeled substance P (5–11) as potential radiopharmaceutical for glioma treatment. *Nuclear Medicine and Biology, 53*(October), 1–8. https://doi.org/10.1016/j.nucmedbio.2017.05.008

Magdeldin, S. (2012). *Gel electrophoresis—Principles and basics*. https://doi.org/10.5772/2205

Mallia, M. B., Mathur, A., Subramanian, S., Banerjee, S., Kothari, K., Koiry, S. P., Sarma, H. D., & Venkatesh, M. (2006). Synthesis and evaluation of ether containing 99mTc–Nitrido dithiocarbamate complexes as brain perfusion imaging agent. *Applied Radiation and Isotopes, 64*(3), 361–367. https://doi.org/10.1016/j.apradiso.2005.09.003

Mammen, M., Colton, I. J., Carbeck, J. D., Bradley, R., & Whitesides, G. M. (1997). Representing primary electrophoretic data in the 1/time domain: Comparison to representations in the time domain. *Analytical Chemistry, 69*(11), 2165–2170. https://doi.org/10.1021/ac961123y

Marzilli, L. G., Worley, P., & Donald Burns, H. (1979). A new electrophoretic method for determining ligand: Technetium stoichiometry in carrier free 99mTc-radiopharmaceuticals. *Journal of Nuclear Medicine, 20*(8), 871–876.

Mathur, A., Mallia, M. B., Subramanian, S., Sharmila Banerjee, H. D., Sarma, K. K., & Venkatesh, M. (2006). Preparation and in vivo evaluation of 99mTcN-tertiary butyl xanthate as a potential myocardial agent. *Applied Radiation and Isotopes, 64*(6), 663–667. https://doi.org/10.1016/j.apradiso.2005.12.008

Minden, J. S., Dowd, S. R., Meyer, H. E., & Stühler, K. (2009). Difference gel electrophoresis. *Electrophoresis, 30*(S1), S156–S161. https://doi.org/10.1002/elps.200900098

Molavipordanjani, S., & Hosseinimehr, S. J. (2018). Fundamental concepts of radiopharmaceuticals quality controls. *Pharmaceutical and Biomedical Research, 4*(3), 1–8. https://doi.org/10.18502/pbr.v4i3.538

Molho, J. I., Herr, A. E., Mosier, B. P., Santiago, J. G., Kenny, T. W., Brennen, R. A., Gordon, G. B., & Mohammadi, B. (2001). Optimization of turn geometries for microchip electrophoresis. *Analytical Chemistry, 73*(6), 1350–1360. https://doi.org/10.1021/ac001127+

Motaleb, M. A., Ibrahim, I. T., Sayyed, M. E., & Awad, G. A. S. (2017). 131I-trazodone: Preparation, quality control and in vivo biodistribution study by intranasal and intravenous routes as a hopeful brain imaging radiopharmaceutical. *Revista Española de Medicina Nuclear e Imagen Molecular, 36*(6), 371–376. https://doi.org/10.1016/j.remn.2017.03.001

Najafi, A., & Hutchison, N. (1985). Electrophoretic analysis of different technetium-99m (SnCl2) methylene diphosphonate complexes. *Journal of Nuclear Medicine, 26*(5), 524–530.

Pastorino, S., Baldassari, S., Ailuno, G., Zuccari, G., Drava, G., Petretto, A., Cossu, V., et al. (2021). Two novel PET radiopharmaceuticals for endothelial vascular cell adhesion molecule-1 (VCAM-1) targeting. *Pharmaceutics, 13*(7), 1025. https://doi.org/10.3390/pharmaceutics13071025

Patton, W. F. (1995). Biologist's perspective on analytical imaging systems as applied to protein gel electrophoresis. *Journal of Chromatography A, 698*(1), 55–87. https://doi.org/10.1016/0021-9673(94)00987-K

Pauwels, E. K. J., & Feitsma, R. I. J. (1977). Radiochemical quality control of 99mTc-labeled radiopharmaceuticals. *European Journal of Nuclear Medicine, 2*(2), 97–103. https://doi.org/10.1007/BF00253680

Pentoney, S. L., Zare, R. N., & Quint, J. F. (1989a). Semiconductor radioisotope detector for capillary electrophoresis. *Journal of Chromatography A, 480*(January), 259–270. https://doi.org/10.1016/S0021-9673(01)84293-X

Pentoney, S. L., Zare, R. N., & Quint, J. F. (1989b). On-line radioisotope detection for capillary electrophoresis. *Analytical Chemistry, 61*(15), 1642–1647. https://doi.org/10.1021/ac00190a011

Poirier, M. J. (1995). Capillary electrophoresis using radiometric detection for the evaluation of radiopharmaceuticals. M.S., University of Massachusetts Lowell. https://www.proquest.com/docview/304253593/abstract/A5EEBE7B54934FC7PQ/1

Potamianos, S., Varvarigou, A., Costopoulos, B., Mitsokapas, N., & Archimandritis, S. (2006). Establishing a quality control protocol for the radiochemical evaluation of radioiodinated biomolecules with application in nuclear medicine. *Journal of Radioanalytical and Nuclear Chemistry, 242*(2), 437–444. https://doi.org/10.1007/bf02345575

Pruszyński, M., Bilewicz, A., & Zalutsky, M. R. (2008). Preparation of Rh[16aneS4-Diol]211At and Ir[16aneS4-Diol]211At complexes as potential precursors for astatine radiopharmaceuticals. Part I: Synthesis. *Bioconjugate Chemistry, 19*(4), 958–965. https://doi.org/10.1021/bc700413r

Pruszynski, M., Koumarianou, E., Vaidyanathan, G., Revets, H., Devoogdt, N., Lahoutte, T., & Zalutsky, M. R. (2013). Targeting breast carcinoma with radioiodinated anti-HER2 nanobody. *Nuclear Medicine and Biology, 40*(1), 52–59. https://doi.org/10.1016/j.nucmedbio.2012.08.008

Pruszyński, M., Łyczko, M., Bilewicz, A., & Zalutsky, M. R. (2015). Stability and in vivo behavior of Rh[16aneS4-diol]211At complex: A potential precursor for astatine radiopharmaceuticals. *Nuclear Medicine and Biology, 42*(5), 439–445. https://doi.org/10.1016/j.nucmedbio.2014.12.011

Pruszynski, M., D'Huyvetter, M., Bruchertseifer, F., Morgenstern, A., & Lahoutte, T. (2018). Evaluation of an anti-HER2 nanobody labeled with 225Ac for targeted α-particle therapy of cancer. *Molecular Pharmaceutics, 15*(4), 1457–1466. https://doi.org/10.1021/acs.molpharmaceut.7b00985

Quirino, J. P., & Haddad, P. R. (2008). Online sample preconcentration in capillary electrophoresis using analyte focusing by micelle collapse. *Analytical Chemistry, 80*(17), 6824–6829. https://doi.org/10.1021/ac801258r

Ragab, M. A. A., & El-Kimary, E. I. (2021). Recent advances and applications of microfluidic capillary electrophoresis: A comprehensive review (2017–mid 2019). *Critical Reviews in Analytical Chemistry, 51*(8), 709–741. https://doi.org/10.1080/10408347.2020.1765729

Rastogi, S., Shukla, S., Kalaivani, M., & Singh, G. N. (2019). Peptide-based therapeutics: Quality specifications, regulatory considerations, and prospects. *Drug Discovery Today, 24*(1), 148–162. https://doi.org/10.1016/j.drudis.2018.10.002

Righetti, P. G. (2005). Electrophoresis: The march of pennies, the march of dimes. *Journal of Chromatography A, Separation Science: Past, Present and Future, 1079*(1), 24–40. https://doi.org/10.1016/j.chroma.2005.01.018

Rizvi, S. F. A., Tariq, S., & Mehdi, M. (2018). Synthesis and preclinical evaluation of bactericidal agent isolated from soil bacterium (Streptomyces): A novel diagnostic technique to enhance anticancer efficacy through radiosynthesis. *Nuclear Medicine Communications, 39*(12), 1081. https://doi.org/10.1097/MNM.0000000000000916

Rodriguez, I., & Li, S. F. Y. (1999). Surface deactivation in protein and peptide analysis by capillary electrophoresis. *Analytica Chimica Acta, 383*(1), 1–26. https://doi.org/10.1016/S0003-2670(98)00485-1

Rohde, E. (1996). Application of capillary electrophoresis to analytical problems in nuclear medicine and forensic science. Ph.D., University of Cincinnati. https://www.proquest.com/docview/304238642/abstract/E6CA80AE46E94844PQ/1

Saito, R. M., Coltro, W. K. T., Jesus, D., & Dosil, P. (2012). Instrumentation design for hydrodynamic sample injection in microchip electrophoresis: A review. *Electrophoresis, 33*(17), 2614–2623. https://doi.org/10.1002/elps.201200089

Sherma, J., & DeGrandchamp, D. (2015). Review of advances in planar radiochromatography. *Journal of Liquid Chromatography & Related Technologies, 38*(3), 381–389. https://doi.org/10.1080/10826076.2014.941265

Shukla, S. K., Atzei, G., Sedda, A. F., Rossi, G., & Cipriani, C. (2004). Paper chromatographic and paper electrophoretic study of the solution chemistry of Tc—99m—methylendiphosphonate and of Tc—99m—dimercaptosuccinate for improving the tumour affinity of Tc-99-m during scintigraphic imaging of cancer. *Annali di Chimica, 94*(9–10), 761–767. https://doi.org/10.1002/adic.200490094

Song, D., Ma, S., & Khor, S.-P. (2002). Gel electrophoresis–autoradiographic image analysis of Radiolabeled protein drug concentration in serum for pharmacokinetic studies. *Journal of Pharmacological and Toxicological Methods, 47*(1), 59–66. https://doi.org/10.1016/S1056-8719(02)00203-4

Stotz, S., Bowden, G. D., Cotton, J. M., Pichler, B. J., & Maurer, A. (2021). Covalent 18F-radiotracers for SNAPTag: A new toolbox for reporter gene imaging. *Pharmaceuticals, 14*(9), 897. https://doi.org/10.3390/ph14090897

Studier, F. W. (2000). Slab-Gel Electrophoresis. *Trends in Biochemical Sciences, 25*(12), 588–590. https://doi.org/10.1016/S0968-0004(00)01679-0

Szökő, É., & Tábi, T. (2010). Analysis of biological samples by capillary electrophoresis with laser induced fluorescence detection. *Journal of Pharmaceutical and Biomedical Analysis, 53*(5), 1180–1192. https://doi.org/10.1016/j.jpba.2010.07.045

Tavaré, R., De Rosales, R. T. M., Blower, P. J., & Mullen, G. E. D. (2009). Efficient site-specific Radiolabeling of a modified C2A domain of Synaptotagmin I with [99mTc(CO)3]+: A new radiopharmaceutical for imaging cell death. *Bioconjugate Chemistry, 20*(11), 2071–2081. https://doi.org/10.1021/bc900160j

Tracht, S., Toma, V., & Sweedler, J. V. (1994). Postcolumn radionuclide detection of low-energy .Beta. Emitters in capillary electrophoresis. *Analytical Chemistry, 66*(14), 2382–2389. https://doi.org/10.1021/ac00086a026

Van den Bergh, G., & Arckens, L. (2004). Fluorescent two-dimensional difference gel electrophoresis unveils the potential of gel-based proteomics. *Current Opinion in Biotechnology, 15*(1), 38–43. https://doi.org/10.1016/j.copbio.2003.12.001

Van Schepdael, A., Verbeke, K., Van Nerom, C., Hoogmartens, J., & Verbruggen, A. (1997). Capillary electrophoretic analysis of ethylene Dicysteine, a precursor of the radiopharmaceutical 99mTc ethylene Dicysteine. *Journal of Chromatography B: Biomedical Sciences and Applications, 697*(1), 251–254. https://doi.org/10.1016/S0378-4347(97)00027-3

Vanbilloen, H. P., Dezutter, N. A., Cleynhens, B. J., & Verbruggen, A. M. (1997). Characteristics and biological behaviour of 99mTc-labelled Hydroxyacetyltriglycine, a potential alternative to 99mTc-MAG3. *European Journal of Nuclear Medicine, 24*(11), 1374–1379. https://doi.org/10.1007/s002590050163

Vesterberg, O. (1993). A short history of electrophoretic methods. *Electrophoresis, 14*(1), 1243–1249. https://doi.org/10.1002/elps.11501401188

Voytas, D., & Ke, N. (2002). Detection and quantitation of Radiolabeled proteins and DNA in gels and blots. *Current Protocols in Immunology Appendix, 3*(November), A.3J.1–A.3J.10. https://doi.org/10.1002/0471142735.ima03js50

Wei, I. C., Tsao, N., Huang, Y. H., Ho, Y. S., Wu, C. C., Yu, D. F., & Yang, D. J. (2008). 99mTc-Glycopeptide: Synthesis, biodistribution and imaging in breast tumor-bearing rodents. *Applied Radiation and Isotopes, 66*(3), 320–331. https://doi.org/10.1016/j.apradiso.2007.10.002

Weinberg, I. (2014). *Semi-automated Lab-on-a-Chip for dispensing GA-68 radiotracers*. DOE-WMP-0007568. Weinberg Medical Physics LLC. https://doi.org/10.2172/1122933.

Weinberg, I. N., Stepanov, P., Mario, G. U., Probst, R., & Zavarzin, V. (2013). *Rapid compact assay for parent radionuclides in generators*. United States US20130146458A1, filed January 10, 2013, and issued June 13, 2013. https://patents.google.com/patent/US20130146458A1/en

Wenclawiak, B. W., & Püschl, R. J. (2006). Sample injection for capillary electrophoresis on a micro fabricated device/on Chip CE injection. *Analytical Letters, 39*(1), 3–16. https://doi.org/10.1080/00032710500460932

Westerberg, G., & Långström, B. (1994). Labelling of proteins with 11C in high specific radioactivity: [11C]albumin and [11C]transferrin. *Applied Radiation and Isotopes, 45*(7), 773–782. https://doi.org/10.1016/0969-8043(94)90128-7

Westerberg, G., Lundqvist, H., Kilár, F., & Bengt L»ngström. (1993). B+-selective Radiodetector for capillary electrophoresis. *Journal of Chromatography A, 645*(2), 319–325. https://doi.org/10.1016/0021-9673(93)83392-6

Wieland, D. M., Tobes, M. C., & Manger, T. J. (Eds.). (1986). *Analytical and chromatographic techniques in radiopharmaceutical chemistry*. Springer. https://doi.org/10.1007/978-1-4612-4854-5

Wolstenholme, G. E. W., & Millar, E. C. P. (2009). *Paper electrophoresis*. John Wiley & Sons.

Wuest, F., Berndt, M., Bergmann, R., van den Hoff, J., & Pietzsch, J. (2008). Synthesis and application of [18F]FDG-Maleimidehexyloxime ([18F]FDG-MHO): A [18F]FDG-based prosthetic Group for the Chemoselective 18F-Labeling of peptides and proteins. *Bioconjugate Chemistry, 19*(6), 1202–1210. https://doi.org/10.1021/bc8000112

Wynendaele, E., Bracke, N., Stalmans, S., & De Spiegeleer, B. (2014). Development of peptide and protein based radiopharmaceuticals. *Current Pharmaceutical Design, 20*(14), 2250–2267.

Younes, C. K., Boisgard, R., & Tavitian, B. (2002). Labelled oligonucleotides as radiopharmaceuticals: Pitfalls, problems and perspectives. *Current Pharmaceutical Design, 8*(16), 1451–1466.

Yu, R. B., & Quirino, J. P. (2019). Chiral selectors in capillary electrophoresis: Trends during 2017–2018. *Molecules, 24*(6), 1135. https://doi.org/10.3390/molecules24061135

Yu, H., Xiaoyu, X., Sun, J., & You, T. (2012). Recent Progress for capillary electrophoresis with electrochemical detection. *Central European Journal of Chemistry, 10*(3), 639–651. https://doi.org/10.2478/s11532-012-0005-6

Zarschler, K., Kubeil, M., & Stephan, H. (2014). Establishment of two complementary in vitro assays for Radiocopper complexes achieving reliable and comparable evaluation of in vivo stability. *RSC Advances, 4*(20), 10157–10164. https://doi.org/10.1039/C3RA47302C

Zhang, J. B., Song, Z. X., & Wang, X. B. (2009). Synthesis of a Bis(2,3-Dimethylcyclohexyl-Dithiocarbamato)-Nitrido 99mTc complex: A potential tracer for myocardial imaging. *Applied Radiation and Isotopes, 67*(4), 577–580. https://doi.org/10.1016/j.apradiso.2008.11.006

Zhuang, Z., Mitra, I., Hussein, A., Novotny, M. V., Mechref, Y., & Jacobson, S. C. (2011). Microchip electrophoresis of N-Glycans on serpentine separation channels with asymmetrically tapered turns. *Electrophoresis, 32*(2), 246–253. https://doi.org/10.1002/elps.201000461

Vanhoenacker, A., Verbeke, K., Van Hoorebeke, [illegible], Bormans, G., & Verbruggen, A. (1993). Capillary electrophoretic analysis of ... Dosimetric ... of the ... [illegible] ... Journal of ... [illegible]

[illegible] ... of ... [illegible] ... https://doi.org/10.[illegible]

[illegible] (1998). [illegible] ... https://doi.org/10.[illegible]

[illegible] (2012). Determination and quantification of radiolabeled ... and DNA [illegible] ... https://doi.org/10.[illegible]

Wei, L., [illegible] ... Wang, D. [illegible] ... imaging ... https://doi.org/10.[illegible]

[illegible] (2014). [illegible] ... Chemistry ... [illegible]

[illegible] (2017). [illegible] ... https://doi.org/10.[illegible]

[illegible] ... (2008). Sample ... capillary electrophoresis of ... [illegible] ... https://doi.org/10.[illegible]

[illegible] (1994). [illegible] ... https://doi.org/10.[illegible]

[illegible] ... Radiolabeling ... [illegible] ... https://doi.org/10.[illegible]

[illegible] ... (1986). [illegible] ... Springer ... https://doi.org/10.1007/978-1-[illegible]

Weisel, [illegible] ... (2009). [illegible] ... Wiley & Sons.

Wust, F., [illegible] ... Bergmann, R., [illegible] (2003). Synthesis and ... [illegible] ... https://doi.org/10.[illegible]

[illegible] (2018). Development of peptide ... [illegible]

[illegible] (2022). [illegible] ... https://doi.org/10.[illegible]

[illegible] ... https://doi.org/10.[illegible]

[illegible] ... (2021). [illegible] ... https://doi.org/10.[illegible]

Zanzonico, [illegible] ... (2019). [illegible] ... https://doi.org/10.[illegible]

Zhang, [illegible] ... (2011). [illegible] ... https://doi.org/10.[illegible]

Zhang, [illegible] ... Jakobsen, S. (2019). [illegible] ... https://doi.org/10.[illegible]

Radiopharmaceutical Dispensing Systems

Mario Malinconico, Aureliana Esposito, William Tieu, and Michela Aurilio

1 Introduction

Nuclear medicine is a medical specialty that includes applications of radioactive drug substances in disease diagnosis and treatment processes, i.e., radiopharmaceuticals. Due to their radioactive nature, several constraints become self-evident when one considers the manufacturer of radiopharmaceuticals in a GMP environment. As the radionuclide typically utilized in the drug substance possesses a short half-life (i.e., 110 min for F-18, 68 min for Ga-68, and so on), there are numerous constraints on the production workflow in both commercial and non-commercial (hospital, institution, and academic) settings. In a commercial setting, typically only one or a small number of products are made as high radioactivity bulk products in multi-dose vials or finish fill vial/syringes distributed to the clinical network. Conversely, in non-commercial settings, a larger number of in-house preparations of radiopharmaceuticals are produced. In both instances, multi-batch production is typical on a working day, with the dispenser used for all products. As such, the understanding of

M. Malinconico (✉)
COMECER, Castel Bolognese, RA, Italy
e-mail: mmalinconico@comecer.com

A. Esposito
Nuclear Medicine Division, Istituto Nazionale Tumori-IRCCS-Fondazione G. Pascale, Naples, Italy

W. Tieu
The University of Adelaide, Adelaide, SA, Australia

M. Aurilio
Hospital Pharmacy Division, Istituto Nazionale Tumori-IRCCS-Fondazione G. Pascale, Naples, Italy

R. M. van Dam et al. (eds.), *Automated Technologies for the Development and Production of Radiopharmaceuticals*,
https://doi.org/10.1007/978-3-031-84632-8_17

the user requirements and their alignment with the different options for the dispenser module is critical and is the focus of this chapter.

The frequent handling and access to the "bulk" radiopharmaceutical solution for the splitting up of the doses, increases radiation exposure to the body and hands of personnel and, increases the risk of contamination and infection for the patient. Dispensing individual patient doses and manual injection involves high doses to extremities (Dhanse et al., 2000) and eye lenses (Wrzesień & Albiniak, 2018), regardless of the technique for withdrawal of activity from a vial (Whitby & Martin, 2005), even when heavy syringe shields are used. In some cases, the external radiation causes an exposure to body and fingertips exceeding yearly limits (Dhanse et al., 2000). The dose limit for the exposed workers according to all European and International regulations is 20 mSv body dose with 500 mSv extremity dose. Moreover, the typical eye dose, where lenses are radiosensitive and less protected, are limited to 20 mSv (International Atomic Energy Agency, 2008).

There are two key points that have driven the use of dispensers and infusion devices for radiopharmaceuticals. The first is the need to protect operators from radiation burden to the whole body and hands. As multiple doses of radiopharmaceuticals are dispensed from a single batch or multiple batch productions over 8 h, an automated dispenser is needed to reduce the operator's radiation exposure (Plascjak et al., 1997).

Numerous automated or semi-automated dispensers are commercially available, in this chapter we will describe deeply the COMECER ones. These are used to distribute the whole activity in multiple vials or to prepare syringes containing patient doses. Recently, automated systems capable of directly administrating the radiopharmaceutical to patients have been developed (Boschi & Todde, 2022).

Regarding the second key point, it was demonstrated that the implementation of an automated radiopharmaceutical dispenser and infusion device, in association with an infection prevention practices protocol (Murray et al., 2005), facilitates maintaining sterility and mitigates the infection risk (Karanfilovska et al., 2020).

Furthermore, the use of dispensers for radiopharmaceuticals eliminates the possibility that misadministrations will happen. Even though nuclear medicine misadministrations are uncommon (Yenson et al., 2005; Zhai et al., 2015), automation of the fractioning and dispensing processes decreases the risk of operator mistakes, especially on days when clinical demand is heavy.

2 Dispensing Systems

Dispensing systems are usually made of: a main unit ("dispenser"), responsible for high precision liquid handling and interface to the container to be filled; one or more dose calibrators, to measure the radiopharmaceutical bulk radioactivity and/or the radioactivity in the final container; depending on the dispensing system model, a formulation unit might be part of the instrument, to check the incoming synthesis radioactivity concentration (RAC) and dilute if necessary; a single disposable sterile fluid pathway; a printer, to print the container and lead pot label; and a control platform (hardware/software) to manage, monitor and track the system and process in a GMP way (Alexoff, 2003). The dispensing system must be housed inside a properly shielded and GMP-qualified environment.

2.1 Dispenser

The dispenser is the "heart" of the whole process. It is a complex system able to manage the liquid handling and the formulation process. Additionally, depending on the primary container (i.e., vial, syringe, or cartridge), the dispenser is responsible for (de)crimping or (un)capping or moving or other mechanical manipulations of the primary container. As a whole, a qualified and validated dispenser system is responsible for the accuracy, repeatability, and reliability of the dispensing process (Lecchi et al., 2012).

There are many dispensers on the market (Fig. 1) with many variables within each manufacturer and model (Skovorodko et al., 2020). As with any key equipment acquisition, when selecting a dispenser, the end user must consider their specific operational needs. In understanding the specific user requirements, the user may, from among the many different dispensers available on the market, narrow and identify critical parameters and functions that match the specific operational needs. The following list summarizes the most common considerations:

- Is terminal sterilization or are aseptic process required? (the Dispensing Process)
- What is the type of container that will be filled? (the Primary Container)
- What is the batch size, radioactivity level, and dispensing frequency? The amounts handled will dictate the level of automation required for the dispenser (manual handling, semi-automated with primary container automatic handling, fully automated)
- Type of dispensing process to be performed: fixed concentration dispensing, final product dilution in primary container, other?
- Performance and accuracy: average filling volume and requested accuracy, throughput (how many vials/syringes per minute)

The following sections describe in further detail these key considerations.

2.2 Dispensing Process

There are two main methods for ensuring sterility of the drug dispensing process that are applicable to the radiopharmaceutical field. The first approach is aseptic dispensing by means of terminal filtration, and the second approach is terminal sterilization. The justification for the choice between the two approaches should include an extensive and science-based benefit risk evaluation and it should be

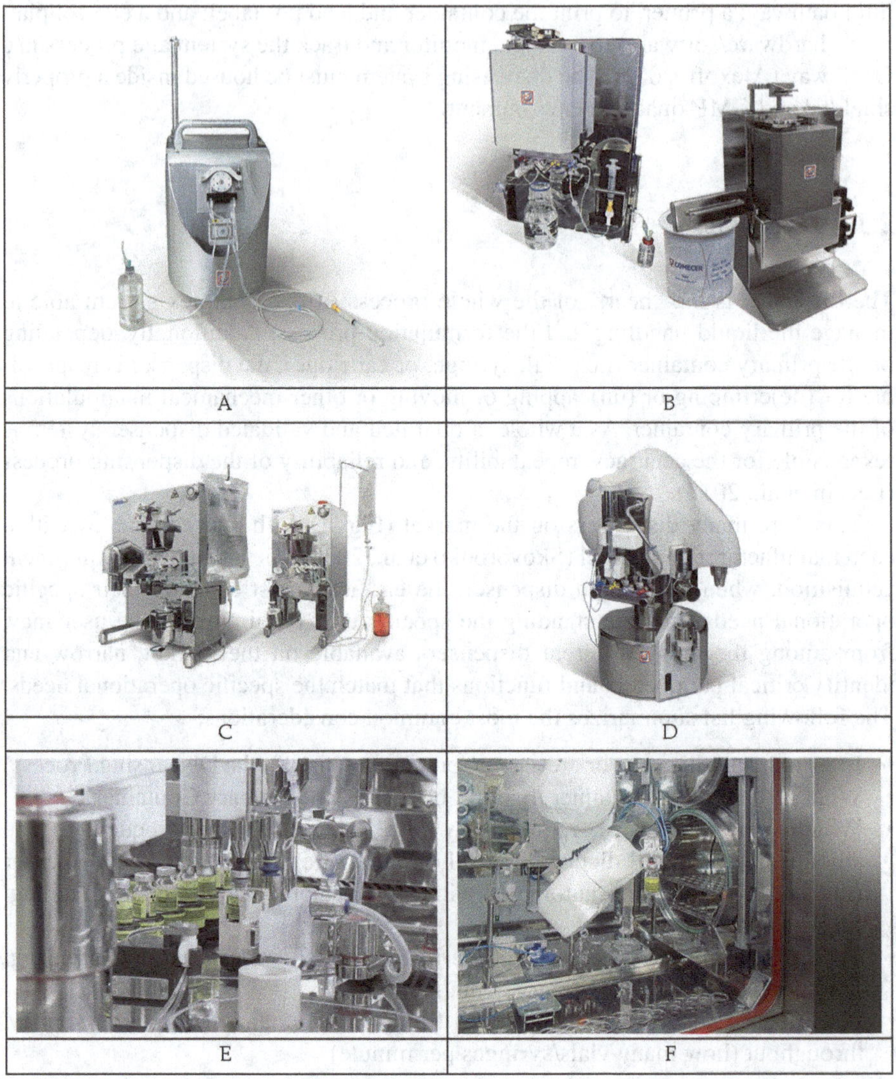

Fig. 1 Examples of dispensing systems available on the market: (**a**) Comecer TIMO-2 semi-automatic syringes dispensing system; (**b**) Comecer FEBO SERIES semi-automatic syringe/cartridge dispensing system with an integrated dose calibrator; (**c**) Comecer ARGO SERIES semi-automatic vial dispensing system; (**d**) Comecer CLIO semi-automatic vial and syringe dispensing system; (**e**) Comecer HELIOS fully automatic vial and syringe dispensing system; (**f**) Comecer THEODORICO 2 robotic vial dispensing system. Other companies are available on the market with similar products, to cite two more, ELSE with its JOKER-5-V3, a fully automatic system for dispensing liquid radiotracers from multi-dose Vials to 5 mL or 10 mL Patient Syringes; or VON GAHLEN with their automatic (VG-ADD, OVDD-SA) and semi-automatic (CVDD-SA) vial dose dispensing systems

demonstrated that suitable development efforts have been made to enable the better sterilization method.

Aseptic dispensing by terminal filtration is based on the principle that the product is dispensed in an aseptic environment being filtered by a membrane with a pore size no larger than 0.22 μm (Fig. 2a, b). This filter is normally included in a single-use aseptic fluid pathway ("cassette" or "kit"), usually supplied by the manufacturer of the dispenser. By aseptic environment, it is meant a shielded isolator with laminar airflow (LAF) ventilation, qualified according to cGMP class "A," connected with a pharmaceutical pre-chamber qualified as grade "B," equipped with a qualified bio-decontamination system (e.g., Vapor Phase Hydrogen Peroxide—VPHP Generator) (World Health Organization, 2011) (Fig. 2c).

The terminal sterilization method is based on the possibility of dispensing the product in a non-aseptic environment, and sterilizing it at the end of the dispensing process. There are different possible sterilization methods available. The most

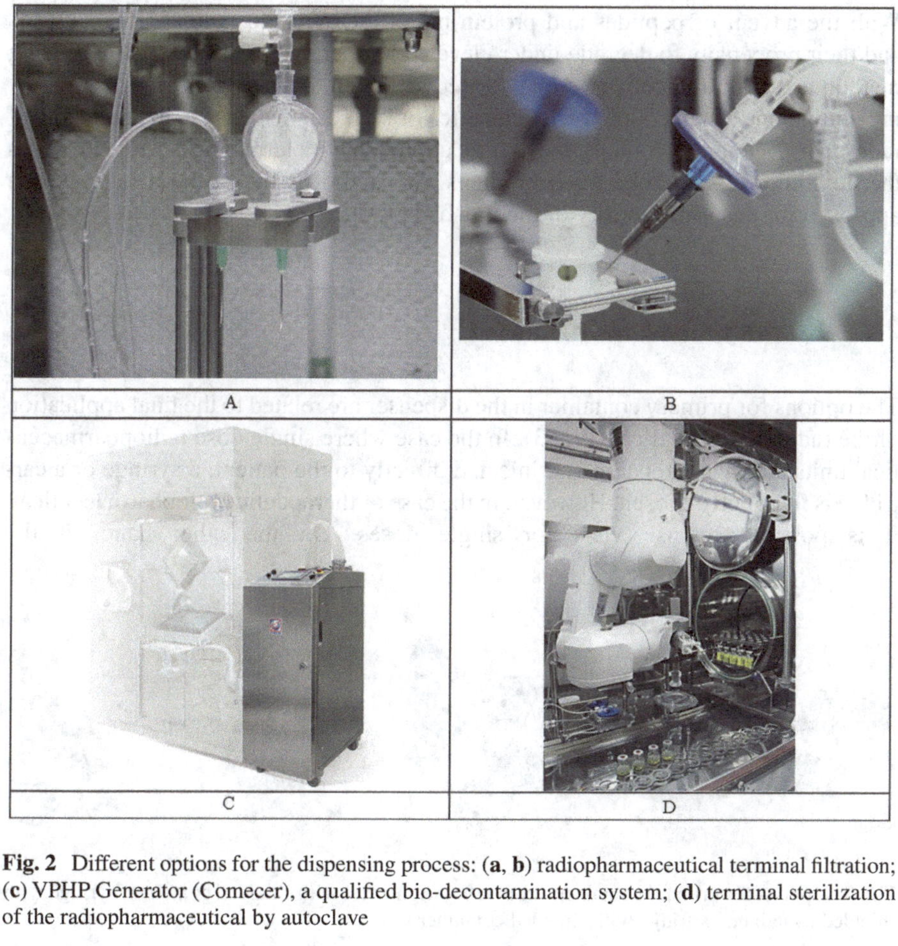

Fig. 2 Different options for the dispensing process: (**a**, **b**) radiopharmaceutical terminal filtration; (**c**) VPHP Generator (Comecer), a qualified bio-decontamination system; (**d**) terminal sterilization of the radiopharmaceutical by autoclave

commonly used for radiopharmaceuticals is thermal sterilization by means of a saturated steam autoclave cycle (Karesh, 1989). The preference for the autoclave method is due to easy access to this technology, reduced space, compatibility with some radiopharmaceuticals (such as [^{18}F]FDG) and proven efficiency (Fig. 2d). On the other hand, some radiopharmaceuticals are susceptible to thermal degradation and the autoclave method described in pharmacopeia, which specify 130 °C for on average 20 min, is not compatible (Karwath et al., 2005). Moreover, the products that contain short-lived radioisotopes (given the total 1-hour sterilization process) are not compatible with the autoclave method time. For these products the sterilization process must be performed using the final filtration through a 0.22 μm filter and, integrity control of the filter used for dispensing should be performed before the release for use of the finished product (Aerts et al., 2014).

However, it is worth noting that the terminal sterilization with autoclave method is the first choice for sterile radioisotope Active Pharmaceutical Ingredients (APIs) preparation, where radioisotopes typically have long half-lives (>1 h), are produced at high multi-dose levels, and are not typically susceptible to thermal degradation. With the advent of peptides and protein-based theragnostic radiopharmaceuticals and their propensity to degrade under elevated temperatures, the aseptic dispensing method is the most widely used one. Moreover, given that a single piece of dispensing equipment, when coupled with cassettes or kits, can dispense a diverse range of radiopharmaceutical products, the aseptic approach with terminal filtration provides the flexibility to switch between products. As such, it is by far the most common approach for ensuring sterility of the finished product.

2.3 Primary Container

The options for primary container in the dispenser are related to the final application of the radiopharmaceutical (Fig. 3). In the case where single dose radiopharmaceutical units are to be prepared and injected directly to the patient, a syringe or a cartridge is frequently chosen. However, in the case of therapeutic radiopharmaceutical, it is possible to use vials for single doses. On the other hand, if the

Fig. 3 Options for primary container. From left to right: vial with shielded container, syringe with shielded container, cartridge with shielded container

radiopharmaceuticals are to be dispensed as multidose units, then vials are invariably the only option available.

The choice between syringe, cartridge, open/closed vial, can also be dictated by the final use of the production center. In the hospital setting where positron emission tomography (PET) scans are performed, nuclear medicine departments commonly require dispensing and calibration of radiopharmaceuticals from multidose vials behind lead shielding. Manual drawing up from a multidose vial for immediate use in the clinical environment does not require special facilities, but is generally discouraged due to infection risks (Karanfilovska et al., 2020). In particular, a multidose vial requires further dilution prior to injection into the patient i.e., the process involves a more complex calculation and several aseptic manipulations.

If the production center wants to deliver multi-dose radiopharmaceutical, usually an automatic or semi-automatic dispenser for vials will be chosen. If the production center has its own hospital in the same building or very close to it, the choice of the dispenser can be oriented toward a multipurpose dispenser, e.g., able to dispense in every kind of container (from vials to syringes/cartridges).

When using vials, the production center can choose to work with an open vial or a closed vial. Open vial dispensing is based on the introduction of pre-sterilized open vials and is apyrogenic. For small batches (from 1 to 200 vials), the radiopharmacy standard practice is introducing these vials pre-stoppered (i.e., with the closure silicone septa in place over the vial), with the aluminum seal pre-positioned over the stopper and not crimped: It will be the dispenser that will take care of opening and closure during the process. For batches over ~200 vials, the market standard is considering using either vials in trays or ready-to-use (RTU) in tubs, while stoppers and caps are fed separately into the filling environment and automatically positioned only after vial filling.

2.4 Dose Calibrator

The function of a dose calibrator is to measure the activity of a radiopharmaceutical dose that will be administrated. The activity delivered to patients must be accurately known both to fulfil radioprotection requirements and to assure good quality imaging and the successful therapy.

Briefly, the central component of the dose calibrator is a gas-filled ionization chamber, which converts gamma emissions into electrical signals and consequently provides an indirect means to measure the radioactivity (potency) of a sample (Ceccatelli et al., 2007) (Fig. 4). Typically, a dose calibrator contains dedicated hardware that provides both power and communication to the ionization chamber. The hardware interfaces with the dose calibrator software. Here the software may be supplied as a part of the whole dispenser software package or maybe a distinct software package supplied on a personal computer or notebook. In addition to the ionization chamber, built-in dose calibrators will typically have a removable internal liner and dipper. The dipper is tasked with lowering the radioactive sample into the

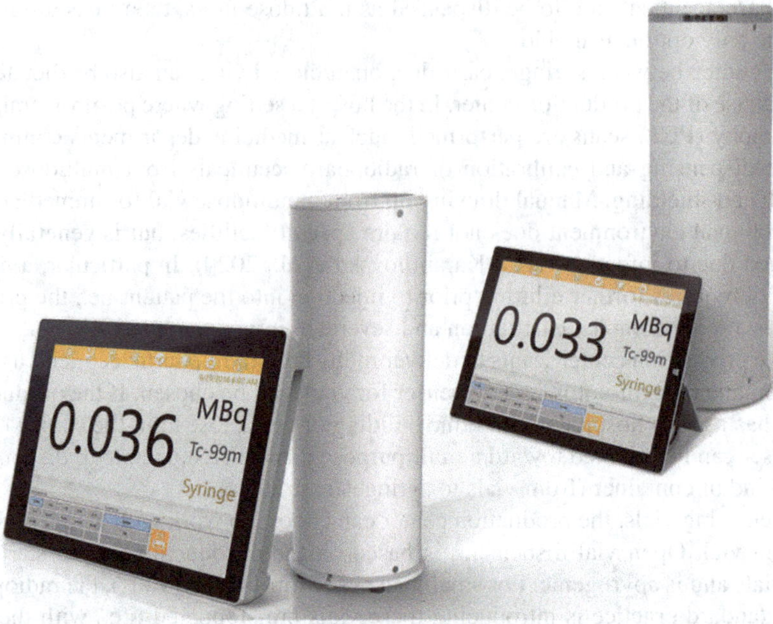

Fig. 4 Example of a commercial dose calibrator (Comecer VDC-606). There are two main ion chambers available: the RDC-VIK-202 (left; to measure up to 74 GBq of F-18) and RDC-VIK-203 (right; to measure up to 740 GBq of F-18)

dose calibrator well. Whilst this task looks simple, the positioning of a sample is critical to ensuring accuracy of measurement. Finally, for dispensers where high activities are expected, the dose calibrator is usually built into a hot cell floor with extensive shielding. Moreover, some dispensers are designed to further minimize external radiation interference through additional shielding within the dipper.

The accuracy of a radionuclide activity calibrator should be assessed at installation and at appropriate intervals thereafter. The instrument should be calibrated in accordance with nationally recognized standards or the manufacturer's instructions. A system suitability test should include the measurement of a reference radionuclide standard source of suitable emission energy (Elsinga et al., 2010).

Moreover, the accuracy of a measurement is dependent of the location, size, and shape of the container of the measured source (Zimmerman & Cessna, 2000) and such factors must be considered when measuring a source. The user always has to be aware of the fact that a radionuclide can contain impurities of other nuclides. These impurities can influence the measurement of the primary nuclide if the contribution of radioactive impurities is sufficiently large. The supplier of the radionuclide should provide information on its purity.

2.5 Formulation Unit

The formulation unit is composed of an ionization chamber with integrated scale. It can be stand-alone or mounted inside the shielded compartment for the dose calibrator of the cell containing the dispenser. The weight of the (synthesized) bulk radiopharmaceutical is determined and the bulk activity is measured, enabling straightforward calculation of the bulk concentration. The dispensing process software will make a calculation of the volume that needs to be prepared given the number of orders (where each order is one vial or syringe) that need to be dispensed out of the bulk activity.

2.6 Fluid Pathway

All of the dispensing platforms on the market are based on the implementation of a single use, pre-sterilized fluid path, which is the main component that the product comes in contact with, in order to minimize risks of microbial, particulate and endotoxin/pyrogen contamination. Part of the tubing set is also the terminal sterilizing filter with a membrane pore size of 0.22 μm, and typically a connection to perform the filter integrity testing routine at the end of the batch (and/or also at the beginning if Pre-Use Post Sterilization Integrity Test, "PUPSIT") is required.

The tubing sets are normally pre-sterilized for aseptic filling applications, normally using sterilization methods such as ethylene oxide or gamma irradiation. The sterile tubing sets are packed in a double bag, that enables material introduction according to cGMP procedures, i.e., to preserve sterility.

The selection and validation of the fluid pathway components often becomes the most critical one, as it is required to undergo to a strong validation process, which may include (but is not limited to): leachable, extractables, bioburden, sterility, cytotoxicity, and other validation studies.

The system can be composed of a single simple tubing set to be installed on a peristaltic pumping system or a single simple syringe acting as a syringe pump, or can be a more complex system including stopcock valves (3-way), unidirectional valves, connectors, etc. The design of the tubing set is also sometimes strictly linked to the type of container to be filled. There might be a considerable difference between a tubing path designed to fill open or closed vials, versus one designed for syringes, versus a multi-purpose one.

What is important to evaluate on these systems is the ease of installing and removing them for line clearance, and minimizing the dead volume of the whole set to avoid that residual drops of active product cause excessive radiation dose to operators.

These fluid paths are generally disposed of as a radioactive waste after use and replaced with new ones for every batch. Sometimes there is the possibility, under certain circumstances, to re-use the fluid path for a second batch after a clean-in place procedure to be developed and validated by the production center, for example, using a cleaning and drying cycle with an ethanol solution, and replacing the sterilizing filter where required.

2.7 Printer and Labeling

At the end of the radiopharmaceutical dispensing process, one of the most important steps is the final label (containing information of the dispensed activity, volume dispensed, timing and date of dispensing and shipment) to be attached on the vial/syringe as well as to the final shielding container where the vial/syringe will be placed for transport.

The label on the primary container should include a radioactivity symbol, name of the manufacturer, the name of the radiopharmaceutical preparation, the batch number, radioactivity content at a stated date and time, radioactive concentration (in MBq or mCi per mL of the solution), solution identifier (i.e., saline, etc.) and total volume, intended purpose (diagnostic or for therapeutic), and administration route (Knapp & Dash, 2016).

The labeling on the primary container can be handwritten or computer-generated. In the first case, because of radiation exposure concerns, it is a common practice to prepare much of the label in advance (Elsinga et al., 2010), but this method requires a lot of time spent to follow all the GMP procedures: introducing pre-written labels through the pre-chamber, sterilizing them, and finally labeling them manually to the primary containers. Alternatively, labeling of the primary container can be performed automatically. The labeling is automatically performed just before delivering the vial into the shielded container. The labeling is started when the operator performs the delivery of a filled vial; after the label is printed the vial is rotated to apply the label via a peeling plate. All the support ribbon tape is automatically recovered inside the cassette, so no additional waste material is produced during the labeling process. The vial compartment is mounted on a motorized sliding guide in order to allow an easy way to receive the vial and, after the application of the label, sending the vial to the delivery position (over the shielding container).

Regarding the lead pot label, it is printed using a standalone external printing machine, placed outside the shielded isolator and connected through the dispensing system software, to receive the data to print on label. The labels are generated automatically and applied on the transport shielded container manually by the operator.

3 Future Outlook

In the last several years, the demand of therapeutic/theragnostic radiopharmaceuticals, most notably radiolabeled somatostatin analogs for neuroendocrine tumors and PSMA-ligand for prostate cancer, has been increasing very quickly (Vogel et al., 2021). Many production centers need to dispense high concentrated activities (mostly free radioisotopes such as Ac-225 or Lu-177) in very small volumes. For applications like this, a very high accuracy micro dispensing system (from 5 μL to 500 μL) is needed. Few of them are already available on the market (Fig. 5).

On the other hand, speaking about diagnostic radiopharmaceuticals, the number of patients with cancer per year is increasing, reaching almost 20 million of people per year (Ferlay et al., 2021). This means that the requests for PET radiopharmaceutical doses (for example [^{18}F]FDG) are increasing as well, but considering the short half-life of the isotopes used, it is becoming more common that a daily multi-batch production is required in nuclear medicine centers. Also in this case, many dispensing systems are available: from semi-automatic to fully automatic ones, designed to satisfy the request of operators and the market. Typically, the major providers of a radiopharmaceutical need fully automated dispensing systems, and usually they dispense into vials, while their end users need to dispense in syringes/cartridges.

The most recent revision of the European Union GMP Guidelines, Eudralex Volume 4, Annex 1 (European Commission, 2023) has introduced specific

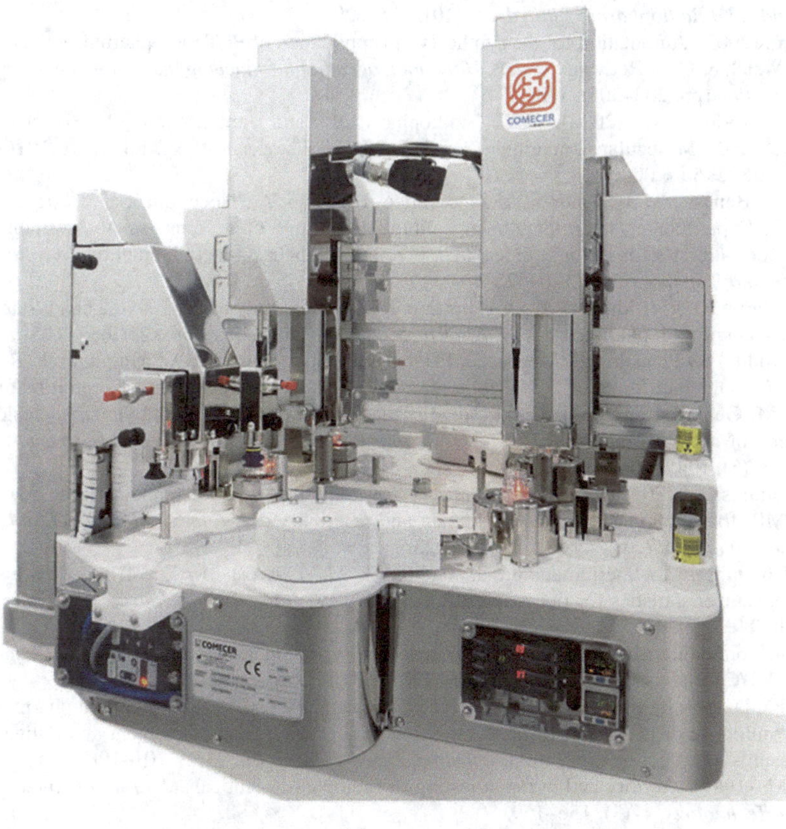

Fig. 5 Mikros (Comecer), a micro-dispensing system for carrier-free radioisotopes (e.g., Lu-177, Ac-225, Tb-161)

requirements for single use systems (e.g., dispensing sets or cassettes) with a strong focus on risk management and contamination control strategy (Kiss et al., 2023).

Future developments for dispensing systems include finding easy ways to allow multi-batch productions while avoiding radiation exposure to operators, and thus increasing use of kits, which will allow more dispensing processes coming from more production batches, all in accordance with GMP guidelines. It is important to find dispenser and kit to be used which are both validated or able to be validated.

Acknowledgments We thank Federico Vissani and Davide Scilimati for bibliographic research.

References

Aerts, J., Ballinger, J. R., Behe, M., Decristoforo, C., Elsinga, P. H., Faivre-Chauvet, A., Mindt, T. L., Kolenc, P. P., Todde, S. C., & Koziorowski, J. (2014). European Association of Nuclear Medicine. Guidance on current good radiopharmacy practice for the small-scale preparation of radiopharmaceuticals using automated modules: A European perspective. *Journal of Labelled Compounds and Radiopharmaceuticals, 57*(10), 615–620.

Alexoff, D. L. (2003). Automation for the synthesis and application of PET radiopharmaceuticals. In M. J. Welch & C. S. Redvanly (Eds.), *Handbook of radiopharmaceuticals. Radiochemistry and application* (pp. 283–305). Wiley.

Boschi, S., Todde, S. (2022). PET radiopharmacy and automation. In Nuclear Medicine and Molecular Imaging -1, 227–234). Elsevier https://doi.org/10.1016/B978-0-12-822960-6.00004-1.

Ceccatelli, A., Benassi, M., D'Andrea, M., De Felice, P., Fazio, A., Nocentini, S., & Strigari, L. (2007). Experimental determination of calibration settings of a commercially available radionuclide calibrator for various clinical measurement geometries and radionuclides. *Applied Radiation and Isotopes, 65*(1), 120–125.

Dhanse, S., Martin, C. J., Hilditch, T. E., & Elliott, A. T. (2000). A study of doses to the hands during dispensing of radiopharmaceuticals. *Nuclear Medicine Communications, 21*(6), 511–519.

Elsinga, P., Todde, S., Penuelas, I., Meyer, G., Farstad, B., Faivre-Chauvet, A., Mikolajczak, R., Westera, G., Gmeiner-Stopar, T., & Decristoforo, C. (2010). Radiopharmacy Committee of the EANM. Guidance on current good radiopharmacy practice (cGRPP) for the small-scale preparation of radiopharmaceuticals. *European Journal of Nuclear Medicine and Molecular Imaging, 37*(5), 1049–1062.

European Commission. (2023). EudraLex volume 4—EU guidelines for good manufacturing practices (GMP) for medicinal products for human and veterinary use. *Annex I: Manufacture of sterile medicinal products*. Revised August 2023.

Ferlay, J., Colombet, M., Soerjomataram, I., Parkin, D. M., Piñeros, M., Znaor, A., & Bray, F. (2021). Cancer statistics for the year 2020: An overview. *International Journal of Cancer*, 149: 778–789.

International Atomic Energy Agency. (2008). Radiation protection in newer medical imaging techniques: PET/CT, Safety reports series, ISSN 1020-6450; no. 58.

Karanfilovska, D., Martin, R., Barton, H., Yap, K. S., & Cheng, A. (2020). Use of a radiopharmaceutical multidose dispenser for positron emission tomography: Risk assessment and mitigation measures for infection prevention. *Infection, Disease & Health, 25*(2), 101–106.

Karesh, S. M. (1989). Sterility and pyrogen testing of radiopharmaceuticals. *Journal of Nuclear Medicine Technology, 17*(3), 156–159.

Karwath, P., Sartor, J., Gries, W., Wodarski, C., Dittmar, C., Biersack, H. J., & Guhlke, S. (2005). Steam sterilization and automatic dispensing of [18F] fludeoxyglucose (FDG) for injection. *Applied Radiation and Isotopes, 62*(4), 577–586.

Kiss, O. C., Scott, P. J. H., Behe, M., et al. (2023). Highlight selection of radiochemistry and radiopharmacy developments by editorial board. *EJNMMI Radiopharmacy and Chemistry, 8*, 6.

Knapp, F. F., & Dash, A. (2016). Translation of radiopharmaceuticals from bench to bedside: Regulatory and manufacturing issues. In *Radiopharmaceuticals for therapy* (pp. 323–343). Springer. https://doi.org/10.1007/978-81-322-2607-9_17

Lecchi, M., Lucignani, G., Maioli, C., Ignelzi, G., & Del Sole, A. (2012). Validation of a new protocol for ^{18}F-FDG infusion using an automatic combined dispenser and injector system. *European Journal of Nuclear Medicine and Molecular Imaging, 39*, 1720–1729.

Murray, T., Elliott, A. T., & Hilditch, T. E. (2005). Long-lived radiopharmaceuticals: Dispensing from multidose vials. *Nuclear Medicine Communications, 26*(6), 555–556.

Plascjak, P. S., Kim, K., Meyer, W., Divel, J., Der, M., & Eckelman, W. C. (1997). Automated radiopharmaceutical dispenser. *Applied Radiation and Isotopes, 48*, 345–348.

Skovorodko, K., Bareikė, M., Gudelis, A., & Gricienė, B. (2020). Occupational exposure in a PET/CT facility using two different automatic infusion systems. *Physica Medica, 77*, 169–175.

Vogel, W. V., van der Marck, S. C., & Versleijen, M. W. J. (2021). Challenges and future options for the production of lutetium-177. *European Journal of Nuclear Medicine and Molecular Imaging, 48*(8), 2329–2335.

Whitby, M., & Martin, C. J. (2005). A multi-centre study of dispensing methods and hand doses in UK Hospital radiopharmacies. *Nuclear Medicine Communications, 26*(1), 49–60.

World Health Organization. (2011). Annex 6, WHO Technical Report Series, No. 961.

Wrzesień, M., & Albiniak, Ł. (2018). ^{68}Ga-DOTA-TATE—a source of eye lens exposure for nuclear medicine department workers. *Journal of Radiological Protection 38*(4), 1512–1523.

Yenson, T., Larcos, G., & Collins, L. T. (2005). Radiopharmaceutical maladministrations in New South Wales. *Nuclear Medicine Communications, 26*(11), 1037–1041.

Zhai, W., Yu, W., Xi, J., & He, W. (2015). The utility of automatic dispenser in nuclear medicine. *Journal of Nuclear Medicine, 56*(3), 1169.

Zimmerman, B. E., & Cessna, J. T. (2000). Experimental determinations of commercial 'dose calibrator' settings for nuclides used in nuclear medicine. *Applied Radiation and Isotopes, 52*(3), 615–619.

Part IV
Technologies for Radiopharmaceutical Development

High-Throughput Techniques in Radiochemistry

Gregory D. Bowden, Yingqing Lu, Jason Jones, R. Michael van Dam, Giancarlo Pascali, and Peter J. H. Scott

1 Introduction

The rise of nuclear medicine as a driving force of precision medicine using targeted imaging and radioligand therapy has placed an unprecedented and growing demand on radiopharmacies and radiochemists to deliver established and novel

G. D. Bowden
Department of Radiology, University of Michigan, Ann Arbor, MI, USA

Werner Siemens Imaging Center, Department of Preclinical Imaging and Radiopharmacy, Eberhard Karls University Tuebingen, Tuebingen, Germany

Cluster of Excellence iFIT (EXC 2180) "Image-Guided and Functionally Instructed Tumor Therapies", Eberhard Karls University, Tuebingen, Germany

Y. Lu · R. M. van Dam
Crump Institute for Molecular Imaging, University of California, Los Angeles, Los Angeles, CA, USA

Department of Molecular & Medical Pharmacology, David Geffen School of Medicine, University of California, Los Angeles, Los Angeles, CA, USA

Physics and Biology in Medicine Interdepartmental Graduate Program, University of California, Los Angeles, Los Angeles, CA, USA

J. Jones
DropletPharma Corp, Los Angeles, CA, USA

G. Pascali
Australian Nuclear Science and Technology Organisation, Lucas Heights, NSW, Australia

School of Chemistry, University of New South Wales, Kensington, NSW, Australia

Brain and Mind Centre, University of Sydney, Sydney, NSW, Australia

P. J. H. Scott (✉)
Department of Radiology, University of Michigan, Ann Arbor, MI, USA

Department of Medicinal Chemistry, University of Michigan, Ann Arbor, MI, USA

Department of Pharmacology, University of Michigan, Ann Arbor, MI, USA
e-mail: pjhscott@med.umich.edu

R. M. van Dam et al. (eds.), *Automated Technologies for the Development and Production of Radiopharmaceuticals*,
https://doi.org/10.1007/978-3-031-84632-8_18

radiopharmaceuticals (Dolgin, 2018; Lapi et al., 2024). Additionally, preclinical PET and SPECT imaging techniques are increasingly valued as powerful tools for medical research, drug discovery, and lead optimization (Ghosh et al., 2022). This increased demand for radiopharmaceutical production will strain existing radiopharmaceutical infrastructure and the currently limited pool of skilled scientists and pharmacists (Scott et al., 2024). Radiochemists must, therefore, adapt and invent new and innovative experimental techniques and production strategies to meet the growing demand for new radiopharmaceuticals and ensure a sustainable future for the field of nuclear medicine (Bowden et al., 2023).

Over the last several decades, high-throughput experimentation (HTE) technologies have revolutionized chemical biology and medicinal chemistry by accelerating drug discovery and development cycles while reducing overall costs (Macarron et al., 2011). HTE techniques are also being adopted and adapted for process and discovery chemistry programs, enabling chemists to optimize, explore, and understand new chemical and reaction space (Taylor et al., 2023; Cernak et al., 2016; Mahjour et al., 2021). Within the contexts of radiochemistry and radiopharmacy, HTE techniques promise to accelerate radiopharmaceutical discovery and development programs significantly (Webb et al., 2024; Rios et al., 2022; Jones et al., 2023). Accessible, user-friendly, high-throughput techniques and equipment designed specifically for radiochemical applications would allow radiochemists to address pertinent radiochemical problems more rapidly, efficiently, and cost-effectively than currently possible.

This chapter will discuss the traditional radiopharmaceutical development pipeline and how implementing radiochemistry-specific HTE techniques at various stages of the pipeline could complement or replace existing workflows, accelerating all aspects of radiochemical discovery, synthesis, and production. However, developing and adapting HTE techniques to the unique constraints of working with short-lived radionuclides poses several significant chemical and technical challenges. This chapter will also discuss these challenges and how recent innovations in high-throughput radiochemistry and high-throughput radiochemical engineering have been implemented to address them.

1.1 *The Case for High-Throughput Chemistry in Radiopharmaceutical Development*

Radiosynthesis, the process of attaching a radionuclide to a pharmaceutically active entity, is often considered the primary bottleneck in the radiopharmaceutical development pipeline. This is due to the significant chemical and technical challenges and constraints associated with producing and handling short-lived radionuclides. Their limited half-lives and emission of harmful ionizing radiation mean that almost all radiochemical reactions and assays cannot be handled with standard laboratory workflows or equipment. This has made radiopharmaceutical development, using traditional radiochemical workflows and equipment, a slow and expensive process that does not align with a rapid, iterative "design, make, test" approach to radiopharmaceutical development (Amor-Coarasa et al., 2019). As such, the cumbersome and costly radiopharmaceutical development cycle is one of the rate-limiting factors slowing much-needed advancements in clinical nuclear medicine. With the increasing clinical and preclinical demand for both established and novel radiotracers (Cardinal Health, 2023), radiochemists must fundamentally reevaluate the current radiopharmaceutical research and production workflows (Bowden et al., 2023). Implementing high-throughput techniques for both radiopharmaceutical research and development, as well as clinical radiopharmaceutical production, will help address and improve upon many of the inefficiencies that result in the radiopharmaceutical development bottleneck. In radiochemical research, miniaturized high-throughput methods will drastically increase the rate at which complex optimization problems can be solved, while also reducing resource usage, labor, and the radiation exposure for experimenters. The quick identification of reliable and optimal radiosynthesis conditions for novel radiopharmaceuticals will help researchers assess project feasibility earlier on in the development pipeline and thus reduce the risk of project failure, accelerating the progression of new radiopharmaceutical concepts through preclinical evaluation and into clinical trials. Additionally, high-throughput screening techniques will provide unprecedented access to diverse radiochemical compound libraries, aiding and accelerating de novo radiopharmaceutical discovery.

Predictive data modeling techniques like artificial intelligence (AI) and machine learning (ML) are already beginning to be applied in nuclear medicine and drug discovery (Saboury et al., 2023; Jayatunga et al., 2022; Struble et al., 2020). However, leveraging these tools for radiopharmaceutical development will require large, reliable, and accurate radiochemical datasets that currently do not exist (Webb & Scott, 2021). High-throughput techniques are by nature highly standardized, and their adoption could, over time, generate datasets large and reliable enough for generalized predictive AI/ML process modeling. Finally, the lessons learned from the implementation of HTE in research will also inform new and optimized techniques, chemistries, and radiosynthesizer architectures that will help increase the production capacities of GMP radiopharmacies, thus future-proofing nuclear medicine as the demands for an increasingly diverse selection of clinical theranostic radiopharmaceuticals continue to grow.

1.2 Process Strategies in Radiosynthesis

Most chemical processes can be abstracted into a series of core tasks (Steiner et al., 2019). First, the experiment must be planned and set up, i.e., the stoichiometries calculated, the precursors and reagents weighed out, stocks solutions prepared, and the necessary reaction equipment (vials, hotplates, synthesis modules) set up appropriately for the research question at hand. Once the experiment has been run over a set time, the required data must be acquired, recorded, processed, and analyzed before resetting the experimental apparatus. Finally, the data is interpreted and applied to inform the next set of experiments, and this cycle repeats until a conclusion about the research question can be made (Fig. 1). Radiochemical experiments complicate this paradigm by introducing short-lived radionuclides, like ^{18}F ($t_{1/2}$ = 110 min) and ^{11}C ($t_{1/2}$ = 20 min). The production and chemical processing of the radionuclide into a reactive form (e.g., the azeotropic drying of [^{18}F]fluoride processing, the conversion of [^{11}C]CO_2 into [^{11}C]MeI or [^{11}C]CN, or purification of a radiometal) adds a significant time constraint that limits how much data can be obtained before the radioactivity decays below detectable levels, and the experiment must be reset.

1.2.1 Current Workflows in Radiochemistry

Time, defined by the half-life of the radionuclide being used, is the principal technical limitation that governs and constrains almost all aspects of research with short- to medium-lived radionuclides. The half-life dictates the number of chemical or biological experiments that can be conducted before the radionuclide becomes undetectable. Higher initial radioactivity allows for longer synthesis procedures, larger chemical or biological experimental regimes, or multiple doses for animal or patient imaging studies, but handling large quantities of radioactive material is challenging (e.g., from a radiation safety perspective) and costly. To address these issues, most radiopharmaceuticals, particularly those labeled with ^{18}F and ^{11}C, are synthesized in large batches using specialized equipment (i.e., automated radiosynthesizers and lead-walled "hot cells") (Amor-Coarasa et al., 2019; Bruton & Scott, 2020). However, these require manual resetting, cleaning, and recharging with refreshed reagents between batches. As the radioactivity must decay to safe levels before operators can access the hot cell (typically a delay of 10 half-lives), it causes expensive infrastructure to sit idle for much of the day. For large batch ^{18}F-radiosyntheses, with a half-life of 110 min, this limits each synthesis module

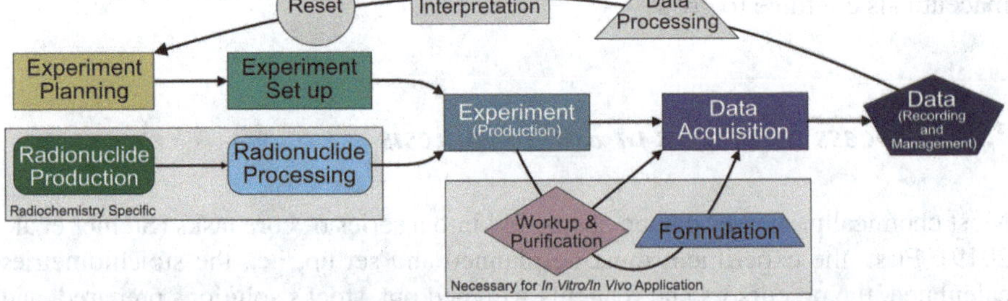

Fig. 1 A generalized abstraction of the radiochemical workflow. When using conventional radiochemistry instruments, one reaction is performed per "reset," and the reset time is dictated by the need to wait for radioactive decay between experiments (typically 12–24 h when using ^{18}F)

and its associated infrastructure (lead hot cell) to one production per day. Shorter-lived radionuclides like ^{11}C (20 min) allow for quicker turnover, achieving 2–3 production runs daily, but this remains highly inefficient.

Because most radiopharmacies tend to prioritize their increasingly busy clinical production schedules, there is often reduced equipment and hot cell capacity remaining for radiochemical research. As such, research radiochemists have had to develop strategies to manually handle radioactivity and acquire radiochemical data for the study and optimization of new radiochemical methodologies and radiosynthetic processes, as well as the development of novel preclinically relevant radiotracers for preliminary biological evaluation. The constraints of using short-lived radionuclides have meant that throughput has always been a significant concern of most research radiochemists, i.e., how much empirical data one can acquire from a single batch of processed radionuclide before the activity decays below practically detectable limits.

Despite their limitations, manual reactions using standard synthetic and analytical tools (radio-HPLC or radio-TLC) are typically highly flexible and easy to manipulate and analyze. Therefore, most radiochemical investigators have used manual and semi-manual approaches, sometimes performed in rapid overlapping sequences to increase throughput to a few experiments per batch (e.g., 4–20 runs). However, the initial amount of activity must be kept relatively low to limit radiation exposure. Unfortunately, results obtained with significantly reduced amounts of radioactive material may not always represent a reaction's performance at a larger scale. Therefore, the results obtained from such experiments may be unreliable and difficult to replicate when transferred to automated large-scale production equipment (Mossine et al., 2018).

These traditional "low throughput" radiochemical workflows have been employed for all aspects of radiochemistry research, including radiochemical methodology research and novel radiopharmaceutical development. They are widely used for reaction solvent and additive screenings, substrate scope studies, and complex molecule synthesis optimization studies. These studies often require large amounts of empirical data to formulate a good hypothesis and understanding of the process involved, making them labor-intensive and prone to systematic errors. In practice, due to their high time and labor costs, these studies often do not include enough replicates and are not performed with sufficient statistical rigor to guarantee reproducibility. More extensive scientific investigations (10+ runs) are almost always carried out over multiple experimental sessions, which risks introducing other sources of variability into the data that complicates the comparison of results acquired on different days with different reagent batches and, sometimes, even different equipment setups.

1.2.2 Non-radioactive Approaches

One possible way to increase radiochemical throughput is to remove the time constraint entirely by performing experiments without radioactive material. This not only provides more time for setting up, performing, and analyzing multiple experiments but also avoids the need for lengthy decontamination and cleaning protocols between experiments, decoupling the timing of experiments with the availability of radioisotopes and equipment. However, this approach has some significant limitations. In radiochemical experiments, the radioisotope is almost always present in substoichiometric (often pM–nM) quantities relative to the precursor, reagents, and even minor chemical impurities. However, the detection sensitivities of most chemical assays are much lower than those of radiochemical assays. Therefore, significantly higher loadings of the non-radioactive isotope are required for the products of a reaction to be detectable. This means that the effects of stoichiometry and concentration on a reaction's outcome are often not comparable between standard chemical reactions and their radiochemical counterparts (e.g., chemical yield is not necessarily a predictor of radiochemical yield).

To avoid the above issues, Zhang et al. (2018) showed that with high sensitivity detection using LC-MS/MS, it is possible to perform reactions with nanomole amounts of the stable isotope [^{19}F]KF that mimic the normal trace-level concentrations of [^{18}F]KF used in radiosyntheses. Using a conventional synthesis module, up to 10 automated experiments could be conducted in 24 h (with LC-MS/MS analysis completed overnight). The authors first validated their method by synthesizing [^{18}F]fallypride under different conditions and comparing the resulting RCY (measured by radio-HPLC) to the results of non-radioactive syntheses under analogous conditions and measured via LC-MS/MS. The method was then applied to the Cu-mediated synthesis of [^{18}F]MDL100907, a 5-HT$_{2a}$ receptor ligand. First, 14 experiments were conducted with [^{19}F]KF under different conditions. The optimal condition (with predicted 18% conversion) was verified in a radiosynthesis with [^{18}F]fluoride and found to have 19–22% ($n = 3$) isolated RCY, providing a significant improvement upon prior reported synthesis yield of 3%.

This approach provides the possibility to conduct more experiments in a shorter time frame. It avoids the cost incurred by using a radioisotope for the majority of the experiments, with the exception of the final verification of [^{19}F]KF-optimized conditions using [^{18}F]fluoride. Such an approach could also enable investigators without access to radiochemistry facilities to contribute to the radiochemistry field. A limitation of this approach is that LC-MS/MS is absent from virtually all radiochemistry laboratories, though academic laboratories may have access via core facilities on campus. The method also requires regular access to a production radiosynthesis module and significant amounts of precursor and reagents, which can be costly.

1.2.3 Sequential and Parallelized HTE Strategies

To address the issues discussed above, new radiochemistry-specific technologies and techniques are needed to increase experimental throughput. Once established, new experimental design approaches and data modeling algorithms will need to be developed to leverage these technologies in scientifically and clinically meaningful ways. As time is the main technical and experimental limitation in radiochemistry, the challenge is to develop equipment, methods, and workflow strategies to efficiently carry out the core radiochemistry tasks (Fig. 1) and maximize the amount of meaningful data that can be collected. Two core HTE strategies have emerged to achieve this aim: sequential and parallelized HTE. The choice of approach is highly dependent on the variables under consideration, the system being studied, and the overall scientific question.

A sequential strategy involves performing reactions one after the other (Fig. 2). To be feasible, the experimental protocol should be optimized for speed, maximizing the number of data points acquired in a given time (Coley et al., 2020; Taylor et al., 2023). As these sequential experimental regimes are run continuously, all aspects of the workflow, including setting up, running, and analyzing the experiment and resetting (cleaning) the equipment between each experiment, should ideally be automated to reduce experimenter time and labor and minimize radiation exposure. Sequential approaches are, therefore, often technically challenging to implement, requiring the skilled engineering of bespoke integrated hardware solutions. However, a sequential experimental workflow has the significant advantage of being easily adapted and changed extemporaneously (Hart et al., 2020). The chemist (or automated software) can adjust the input parameters of new experiments as new information from subsequent experiments comes to light, making the sequential approach ideal for scientific questions that require iterative investigation, such as some of the iterative self-optimizing algorithms (Nambiar et al., 2022).

Alternatively, a parallelized strategy increases throughput by performing multiple experiments simultaneously (Fig. 3). The benefit of this approach is that it can

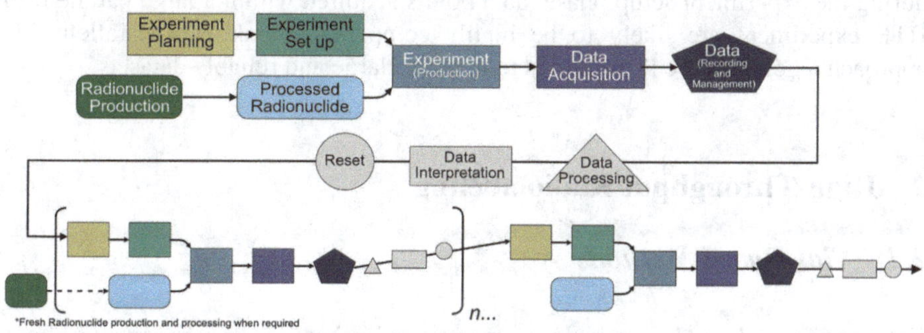

Fig. 2 Sequential workflows allow for experiments to be designed and modified between runs, providing a high level of empirical flexibility. In some cases, the "Radionuclide Processing" step is performed only at the beginning, followed by multiple experimental runs

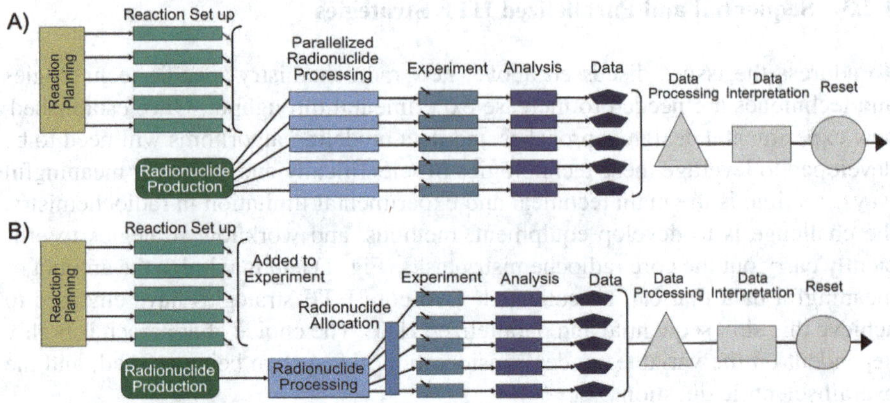

Fig. 3 Parallelized experiments provide higher throughput at the expense of greater protocol complexity, higher upfront cost and time commitments, and increased risk. Aliquots of the radionuclide may be distributed to experiments (**a**) before, or (**b**) after, processing the radionuclide

be expanded to study more extensive experimental regimes with varied numbers of categorical or continuous input factors, allowing flexibility in the research questions it aims to address. Parallelized experimental workflows are, however, always carried out in large experimental "blocks" that are set up and carried out all at once and, therefore, require significant upfront time and cost commitments. Moreover, the challenge of accurately setting up tens to hundreds of experiments simultaneously comes with its own significant technical challenges. Apart from extensive categorical variable screening studies, where failed experiments are expected and accepted, the success of this approach requires carefully preconceived experimental plans based on well-defined scientific goals and solid preliminary data. Iterations to the workflow can be made from block to block, but overall, the parallelized experimental strategy does not have the same workflow flexibility as the sequential approach, increasing the risk of acquiring redundant "junk" data (data that does not provide any new information). On the other hand, if systematic errors are not introduced during the experiment setup phase, data points acquired within a large parallelized HTE experiment are likely to be highly comparable, making a parallelizable approach a good choice if the aim is to generate large and reliable datasets.

2 High-Throughput Radiolabeling

2.1 *Vial-Based Methods*

The use of small single-use vials for manual radiochemical experiments has been standard practice for decades. Typically, once a reaction has been prepared in a vial or test tube (with magnetic stir bar), a prepared aliquot of processed radionuclide is added, the reaction vessel is fitted with a septa cap and set to react at the desired temperature in an aluminum heating block. After a predefined reaction time, the reaction is quenched, sampled, and analyzed via radio-TLC or HPLC. This workflow is widely used because it is intuitively similar to standard organic chemistry experiment setups, is operationally simple, and uses cheap, commercially available equipment already present in most labs. However, setting up and analyzing multiple reactions using this approach is labor-intensive and time-consuming, with ten reactions taking approximately 2–8 h to set up, run, and analyze with standard liquid handling approaches (single-channel pipettes) and radioanalytical techniques (radio-HLPC and radio-TLC). Thus alternative approaches are also being considered to streamline workflows (vide infra).

Since this approach is analogous to the batch reaction approach used by most automated synthesizers, it is, in theory, relatively easy to scale up and automate using such systems. However, there are some nuances to be aware of. Chief amongst these is that such parallelized approaches depend heavily on effective reaction miniaturization to keep reagent costs low and to enable a large number of experiments to be conducted from each batch of the radioisotope. Consequently, care must be taken to ensure that the results of these studies are scalable and meaningful at full (production) scale. As an example, the problem of scaling miniaturized experiments to fully automated batch production is particularly well-known for copper-mediated radiofluorination (CMRF) chemistry (Zlatopolskiy et al., 2015; Preshlock et al., 2016). Many of the initial studies were conducted using production scale quantities of reagents and precursors with small aliquots of traditionally prepared [^{18}F]KF and Kryptofix 2.2.2. The test scale manual experiments worked remarkably well, producing the desired products in good radiochemical yields; however, the chemistry did not perform as well when translated to automated radiosynthesizers using a single whole batch of processed [^{18}F]fluoride (Makaravage et al., 2016; Mossine et al., 2015; Tredwell et al., 2014). The authors of these publications made the assumption that [^{18}F]F$^-$ is present in such low stoichiometric amounts and that adding less (or more) radionuclide would not significantly change the reaction stoichiometry (or reaction performance). However, the higher concentrations of carbonate and Kryptofix salts in full scale [^{18}F]fluoride preparations were found to poison the copper-mediator and reduced reaction performance on scale up, meaning yields obtained from manual (higher throughput) test/screening experiments were not directly predictive of full scale performance.

Full scale conditions could be approximated by adding additional salts to the [^{18}F]F$^-$ aliquots used during screening but this adds complexity to the workflow. Another solution to the problem is to perform a separate ^{18}F-processing step for every reaction in the experimental session (Fig. 3a); however, this also adds significant time, labor, and radiation exposure to the experimental protocol (Bowden et al., 2021). Since the original work, several papers have described alternative low or no-base ^{18}F-processing methods that decouple the effects of ^{18}F-processing from the CMRF reaction's performance (Bowden et al., 2021; Zischler et al., 2017; Zlatopolskiy et al., 2015; Bratteby et al., 2021). This allows a single batch of processed [^{18}F]Fluoride to be used in multiple reactions, making CMRF chemistry truly miniaturizable and more amenable to parallelized experimentation (Fig. 3b). Being mindful of and solving these kinds of scalability/translatability problems from the outset will be vital when designing and developing HTE methods for reaction screening and optimization, particularly for studying new radiochemical methodologies. Any data obtained during the research phase of a radiochemistry project must be meaningful when applied to solve real-world scientific or production problems.

2.2 *Multi-well Plates*

As noted above, conducting manual radiochemical experiments in single-use vials has been standard practice for decades. However, this has traditionally limited manual experimental sessions from a single batch of processed radionuclide (e.g., ^{18}F) to between 4 and 20 runs. To address this issue, Webb and colleagues set out to increase the throughput of this approach by adapting commercially available miniature 24- and 96-well high-throughput chemistry equipment (purchased from Analytical Sales and Services, Inc., Ledgewood, New Jersey) for radiochemical use (Fig. 4a) (Webb et al., 2024). They identified the need for a parallelized workflow to (1) rapidly set up and run multiple radiochemical reactions simultaneously, (2) separate the organic ^{18}F-product fractions (i.e., [^{18}F]fluoride-incorporated) from the unincorporated [^{18}F]fluoride fractions, and (3) analyze and compare the activity concentrations of each fraction pair to determine the radiochemical conversion (RCC).

The reaction plate consisted of commercially available 1 mL glass reaction vials arranged into the standard 96-well format, with each vial containing a magnetic stir bar (Fig. 4a). These vials could then be transferred to a 96-well format aluminum heating block fitted with a thermocouple and placed on a standard laboratory magnetic stirring hot plate. However, the authors found that the thermal mass of the aluminum heating block and the reaction vials took over 30 min to reach reaction temperature (120 °C) if heated from room temperature. To overcome this issue, the authors developed a method to rapidly transfer the 96-reaction vials between preheated and room-temperature aluminum blocks using a machined aluminum "transfer plate" with the vial pushed through a thin sheet of PTFE film, which produced enough friction to hold them in place. Once the reagents and activity had been transferred into the reaction plate, the vials were sealed with a silicon rubber capping mat and transferred to a preheated aluminum heating block, after which a final aluminum "top plate" was fitted on top of the capping mat and bolted down using wingnuts and threaded screws. After the reaction, the wingnuts, top plate, and capping mat were carefully removed, and the 96 vials were picked up by the transfer plate and placed into a room-temperature aluminum block to cool and stop the reaction. The setup and execution of HTE experiments in this manner are fundamentally similar to the traditional vial methods, making this approach operationally intuitive and accessible.

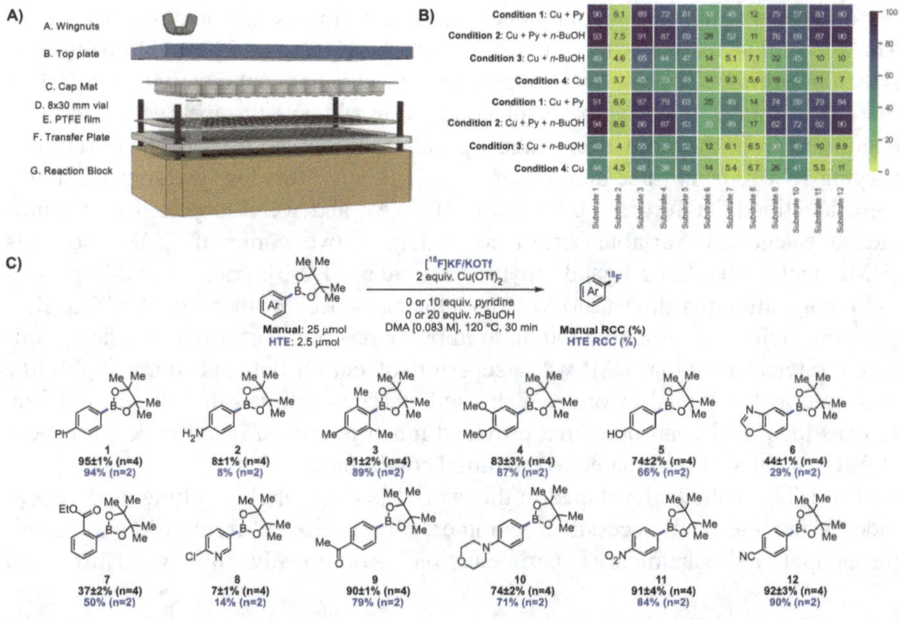

Fig. 4 (**a**) Commercially available 96- and 24-well format high-throughput chemistry apparatus was adapted for radiochemistry. (**b**) In an initial validation experiment, a 96-well plate was divided into two identical replicates (first four rows and second four rows), (**c**) 12 substrates were subjected to four sets of commonly used copper-mediated radiofluorination conditions at 120 °C. The reaction RCC values were measured manually via radio-TLC. (Figure adapted from Webb et al. (2024) and used with permission. © American Chemical Society)

To develop the workflow and validate the strategy, the authors used their approach to study the effects of four sets of copper-mediated radiofluorination conditions on 12 boronate pinacol ester substrates (Fig. 4b, c). CMRF reactions are complex multicomponent systems that depend on several highly substrate-dependent categorical variables and are notoriously difficult to optimize using traditional low-throughput experimentation methods. To set up the 96 reactions (12 precursors × 4 conditions × 2 replicates), the authors prepared separate reaction stock solutions of the copper mediator and its additives ($Cu(OTf)_2$, ligand, and co-solvent), the boronated precursor, and the processed [^{18}F]fluoride in the reaction solvent (DMA). The stock solutions were then dispensed into 1 mL glass vials arranged in rows and columns in separate "staging plates." With the stock solutions prepared this way, 25 μL aliquots of each reagent solution could be quickly dispensed into the 96-well "reaction plate" within ~20 min using a multichannel pipette. Notably, in terms of reagent concentration and usage, each reaction was performed at one-tenth the scale of a standard CMRF radiosynthesis.

The authors demonstrated the utility of the new HTE protocol by studying the CMRF reactions of an informer library of 19 drug-like boronic acid ester substrates. Informer libraries are selections of readily available compounds that represent complex drug-like structures and are used to study and evaluate chemical reaction performance in high-throughput process chemistry studies (Dreher & Krska, 2021). The aim of this study was to demonstrate that the HTE technique could be used to screen a multitude of drug-like compounds for radiochemical reactivity and further fine optimization. Initially, these substrates were subjected to the best-performing conditions from the validation studies (pyridine, DMA, no *n*-BuOH) and returned varying RCCs, from trace amounts to >80%. Finally, two low-yielding substrates were identified for further optimization (Fig. 5), and were subjected to a multiplexed categorical variable screen consisting of two common CMRF solvents (DMA and DMI), three ligands (pyridine, imidazo[1,2-b]pyridazine (IMPY), and 4-dimethylaminopyridine (DMAP)) and the presence or absence of *n*-BuOH, a reaction additive commonly thought to improve reaction performance. The optimization screen found that DMI was a superior solvent for both substrates if pyridine was used as the ligand. Moreover, the authors demonstrated that these follow-up screens identified conditions that provided much-improved reaction performances at full manual scale compared to the initial conditions.

The HTE methods developed in this work are particularly exciting as they provide a flexible and accessible, commercially available framework for high-throughput radiochemical experimentation. Additionally, the validation and

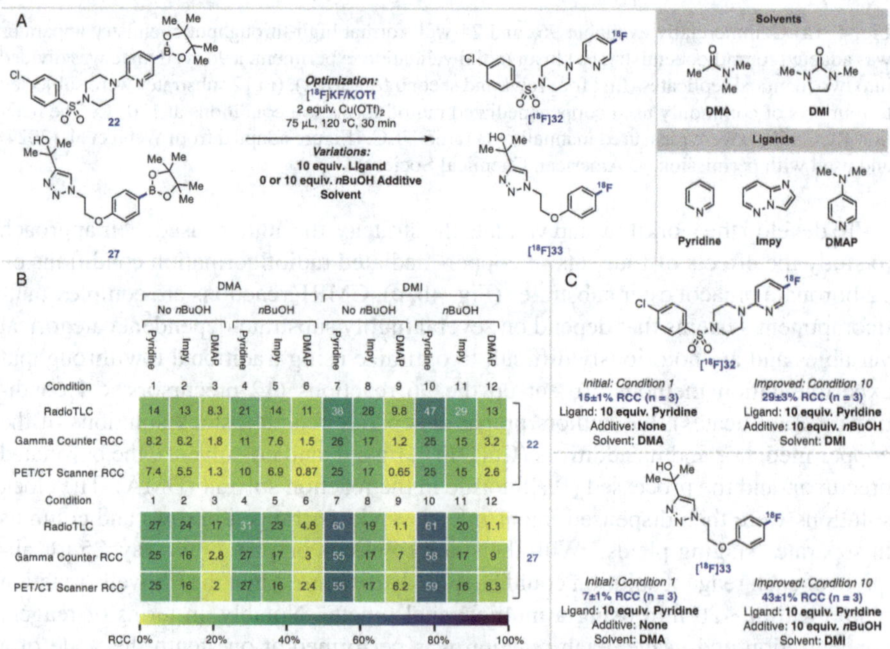

Substrate	Measurement	1	2	3	4	5	6	7	8	9	10	11	12
	Solvent	DMA	DMA	DMA	DMA	DMA	DMA	DMI	DMI	DMI	DMI	DMI	DMI
	Additive	No nBuOH	No nBuOH	No nBuOH	nBuOH	nBuOH	nBuOH	No nBuOH	No nBuOH	No nBuOH	nBuOH	nBuOH	nBuOH
	Ligand	Pyridine	Impy	DMAP	Pyridine	Impy	DMAP	Pyridine	Impy	DMAP	Pyridine	Impy	DMAP
22	RadioTLC	14	13	8.3	21	14	11	38	28	9.8	47	29	13
22	Gamma Counter RCC	8.2	6.2	1.8	11	7.6	1.5	26	17	1.2	25	15	3.2
22	PET/CT Scanner RCC	7.4	5.5	1.3	10	6.9	0.87	25	17	0.65	25	15	2.6
27	RadioTLC	27	24	17	31	23	4.8	60	19	1.1	61	20	1.1
27	Gamma Counter RCC	25	16	2.8	27	17	3	55	17	7	58	17	9
27	PET/CT Scanner RCC	25	16	2	27	16	2.4	55	17	6.2	59	16	8.3

Fig. 5 (**a**) Two substrates from the informer library were identified for optimization. (**b**) A multiplexed HTE study was used to study the effects of two different solvents, ligands, and additives on reaction performance. (**c**) Improved reactions were identified and demonstrated to improve reaction performance at full manual scale conditions. (Figure reproduced from Webb et al. (2024) and used with permission. © American Chemical Society)

informer library optimization studies included in this work demonstrate the ability of this work to answer multiple pertinent radiochemical questions in a standardized and scalable manner.

2.3 Flow Chemistry Methods

The optimization of radiofluorination reactions was the driving motive behind one of the first examples for which a programmatic high-throughput approach was reported (Pascali et al., 2014). In this case, the task was facilitated by the use of a microfluidic flow system, NanoTek, at that time commercially available from Advion. In detail, this system is made up of at least three modules with the following functions: (1) to prepare the dried radiofluoride complex (batch approach), (2) to deliver precursor and secondary reagent solutions, (3) to deliver and microfluidically react the radiolabeling solution with precursor and secondary reagent. Modules 2 and 3 comprise syringe pumps able to reliably deliver fluids at rates down to 5 μL/min and against pressures up to ~400 psi. These syringe pumps (P1, P2, and P3) are connected to 8-port valve heads with a rotor featuring a "bridging" internal grooving; such solution allows the system to fill a storage loop with the desired solution and to deliver small aliquots of it towards the reaction side of the system (Fig. 6). One of the pumps (i.e., P3) has a slightly different set-up, allowing it to either deliver stored reagent from the loop or deliver "sweep" solvent towards the pump output. The reactor is made up of a few meters (1–4 m) of fused silica tubing of small inner diameter (e.g., 100 μm), tightly coiled into a conductive epoxy; this element is placed into localized heaters that can rapidly heat up to 220 °C or cool down via passive cooling. The microfluidic flow environment guarantees uniformity of temperature across all the fused silica coil. Up to four such microreactors can be installed in the system to perform one-step (P1 + P3) or two-step ([P1 + P3] + P2) processes.

To achieve high-throughput operation, such a system is unique in the fact that all the relevant tubing, from precursor/reagent/radiofluoride vial to collection vial, is accurately measured, and their volume is recorded in the controlling software. This allows for the system to "load" solutions up to the entrance of the microreactor before actual reactions occur, thus reducing the time required for the next steps. Afterward, the system can precisely meter the amount of inert solvent needed to: (a) deliver the desired aliquots of solutions from storage loops, contemporarily (one-step synthesis processes) or with the appropriate time gap (two-step synthesis processes) to ensure precise merging; (b) deliver just enough additional solvent to push the reaction bolus moving through the fluidic system (i.e., "sweep" step) until the collection point (where the sample is retrieved for offline analysis), conserving the overall desired flow rate (Fig. 6).

This approach is nicely implemented in the controlling software and called "Automatic Discovery." Once the necessary preparation steps are done, the operator simply needs to position a collection vial; declare the amounts of reagents to be

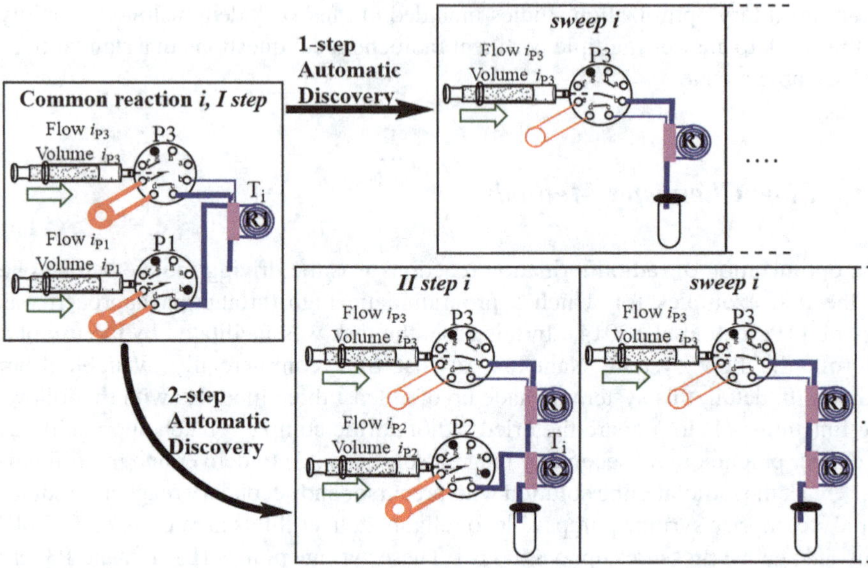

Fig. 6 System schematic for the "Automatic Discovery" strategy adopted by the Advion NanoTek equipment. In typical usage, pump P3 handles the radiolabeling solution and the "sweep" solvent, P1 handles the radiolabeling precursor and P2 handles a reagent for a secondary reaction, if applicable. The contents of the storage loops (in orange) are pushed by dispensing reaction solvent in them. The combined reaction bolus is pushed through the reactor(s) (R1 and R2) via sweep solvent by P3. (Adapted from Pascali et al. (2014))

used in the reaction and their relative flow ratio, the reactor temperature, and the flow rate; and press "start." Multiple reports have demonstrated that reactions performed in flow microfluidics can already provide high yields with residence times (i.e., the time spent in the heated zone) of only tens of seconds, compared to the 5–30 min typically required in batch environment. This fact, combined with the possibility to use as little as 5 μL of reagents per reaction, an overall system volume as low as ~100 μL (of which 15–30 μL is typically the microreactor volume in a one-step process) and few minor time improvements implemented by the software, means that different reaction mixtures (i.e., reactions performed under different conditions) can be collected every 2–3 min, with the time gap expanding to 5–7 min if substantial temperature differences are set between successive runs. However, larger boluses of reagents can also be used, as large as the whole content of the storage loops, thus increasing the process time accordingly. For a typical experimental campaign, a total of 500 μL of initial radiofluoride and precursor solutions are available, and it is possible to perform ~40 reactions in less than 4 h. These reactions are then fed into an analytical system able to automatically analyze multiple samples, most typically a radio-HPLC (sequential analysis) with autosampler or a multi-lane radio-TLC (parallel analysis).

The wealth of data obtainable using this strategy, and the reliability and repeatability of the results, allowed the authors to devise a two-tier optimization approach

(Fig. 7). In the first tier, a set of fixed reaction parameters (i.e., that cannot be varied in the same microfluidic campaign) are chosen and temperature (*T*), flow rate (FR), and reagent ratios (RR) are modified in a predetermined manner. In particular, the

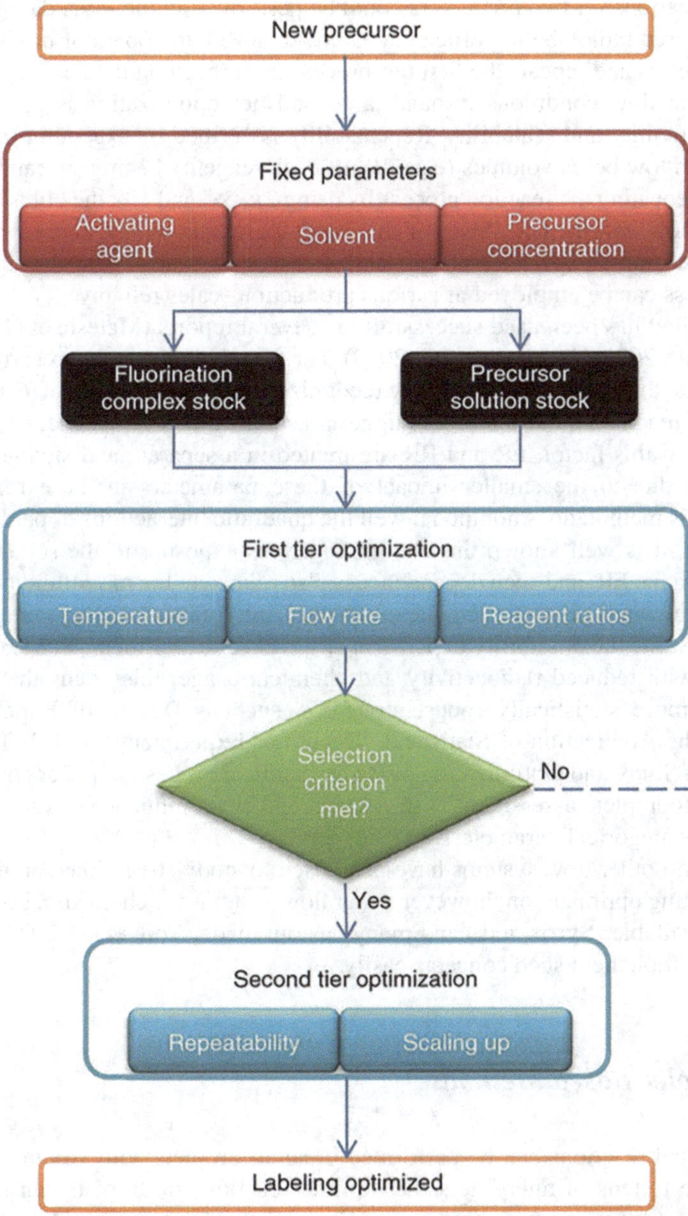

Fig. 7 Two-tiered optimization strategy for radiofluorinations in flow microfluidic reactors. (Reproduced from Pascali et al. (2014))

first 18 runs gradually increase *T* (e.g., 90–190 °C) keeping FR and RR constant; the next five experiments are executed using the best *T* and modifying only FR; the last five runs are run modifying the RR but maintaining *T* and FR constant. With a total of 28 runs, it is forecasted that such optimization tier can be executed in 1 day, and even optimizations for two precursors could be performed in the same day.

If the desired radiolabeling efficiency is not achieved, the operator modifies the fixed parameters and repeats the first tier process until reaching the desired criteria. With the best flow conditions in hand, a second tier optimization is planned, to verify repeatability and scalability. Repeatability is verified by executing the same reaction with low bolus volumes (e.g., 10 μL each reagent) 12 times; scalability is verified by repeating the reaction process by using 2×, 5×, and 10× the bolus amount used in the low-volume test. Also for this case, the whole second tier can be performed in the same experimental day, and it serves to demonstrate that the optimized process can be employed at various production scales reliably.

This method has been used successfully in several reports (Matesic et al., 2017; Klenner et al., 2020a, b; Zhang et al., 2019). Far from being exactly prescriptive in all the details, this optimization strategy recognizes the crucial impact of *T* in radiofluorination reactions, and therefore suggests acquiring a complete set of experiments varying this factor. FR and RR are treated in a separate and smaller set of experiments due to the smaller impact of these parameters in these reactions. However, this method does not model well the quadratic interactions of parameters; for example, it is well known that the homogenous exposure to the set *T* can be dependent on the FR, and such dependence is also related to the microfluidic reactor and heating element geometry. On the other hand, microfluidic flow systems of this kind demonstrate the possibility to perform large sets of experiments in a short time frame and with reduced radioactivity and chemical usage; this opens the way to employing more statistically robust strategies, such as Design of Experiments (chapter "The Application of Statistical "Design of Experiments" (DoE) Towards the Efficient Study and Optimization of Radiochemical Processes"), that could provide a more complete assessment of the impact and cross-influence of various continuous and categorical parameters.

To date, no other flow systems have been used to study streamlined approaches to radiolabeling optimization; however, other flow systems for chemistry, both commercially available (Syrris, n.d.) and many custom made (Volk et al., 2023), could in principle implement such concepts easily.

2.4 Droplet-Based Methods

Radiochemical reactions can be performed in small droplets with volumes on the order of one to tens of microliters. Performing reactions in droplets can increase reaction rates (due to increased reagent and/or radioisotope concentration), shorten process times (due to high heat transfer and low thermal mass), reduce reagent consumption, etc. (Wang & van Dam, 2020). Such approaches are described in more detail in chapters "Progress in Droplet-Based Methods for Simple, Low-Cost Radiopharmaceutical Production" and "Digital Microfluidics and Radiochemistry: Current State and Future Trends". Open droplet formats pioneered by van Dam et al. are particularly suitable for high-throughput manipulations due to compatibility with robotic pipetting systems (Jones et al., 2023). The microchips used for these droplet reactions are relatively simple in their design: silicon wafer substrates are first coated with hydrophobic Teflon AF and then circular regions are etched away, forming hydrophilic surface tension traps for the reaction sites. The silicon substrate provides high thermal conductivity and high chemical inertness. Each chip is mounted on top of a high-power ceramic heater connected to a closed loop temperature controller, affixed with thermally-conductive compound to provide an efficient thermal interface.

Creating multiple reaction sites on a single substrate (Fig. 8a) enables multiple evaporation or reaction processes to proceed on all sites in parallel, providing a straightforward way to increase throughput. For example, to optimize conditions for a one-step ^{18}F-radiosynthesis, all reaction sites can be loaded with radioisotope (premixed with phase-transfer catalyst, or with phase-transfer catalyst added in a second step), and all sites can be dried in parallel. Next all sites are loaded with precursor, potentially with varied concentration or reaction solvent, and reactions are carried out in parallel at the desired temperature. Subsequently, crude reactions are collected from the chip for analysis. Rios et al. first demonstrated proof-of-concept of using such chips to perform an accelerated optimization of the synthesis of [^{18}F]Fallypride (Rios et al., 2019). Using a chip design with a 2 × 2 array of reaction sites, they explored 20 distinct conditions ($n = 2$ replicates each) within a few days, including eight values for the amount of $TBAHCO_3$, four values for precursor volume, and eight values for precursor concentration. At the optimal conditions (240 nmol $TBAHCO_3$, 6 mL reaction volume, and 39 mM precursor concentration), the ^{18}F-fluorination efficiency was 96.0 ± 0.5% ($n = 2$) and crude RCY was 87 ± 3% ($n = 2$). The results were higher and more consistent than prior droplet syntheses of [^{18}F]Fallypride. Using Cerenkov luminescence imaging (CLI) after different stages of the reaction process, cross-contamination on both 2 × 2 and 4 × 4 array chips were remarkably low (<0.4%). A study of reaction performance repeatability was conducted by synthesizing [^{18}F]Fallypride in parallel at multiple sites on the same chip, under identical radiosynthesis conditions (i.e., with optimal base condition). The results showed minimal standard deviations (1–2%) in fluorination conversion, collection efficiency, and crude RCY, highlighting the uniformity of reactions across the chip.

The method was then extended by Rios et al. (2022) to enable operation of up to four chips in parallel, as a means to further increase throughput (up to 64 reactions in parallel; Fig. 8b), and to enable studies of different reaction temperatures or different reaction times. The platform was used to optimize the synthesis for four different tracers: [^{18}F]fallypride (a probe targeting dopamine D2/D3 receptors), [^{18}F]PBR06 (a probe targeting translocator protein (TSPO)), [^{18}F]FEPPA (TSPO ligand), and [^{18}F]flumazenil (for imaging $GABA_A$ receptor). The first three involve S_N2 reactions at a tosylate leaving group and the latter involves an S_NAr reaction at

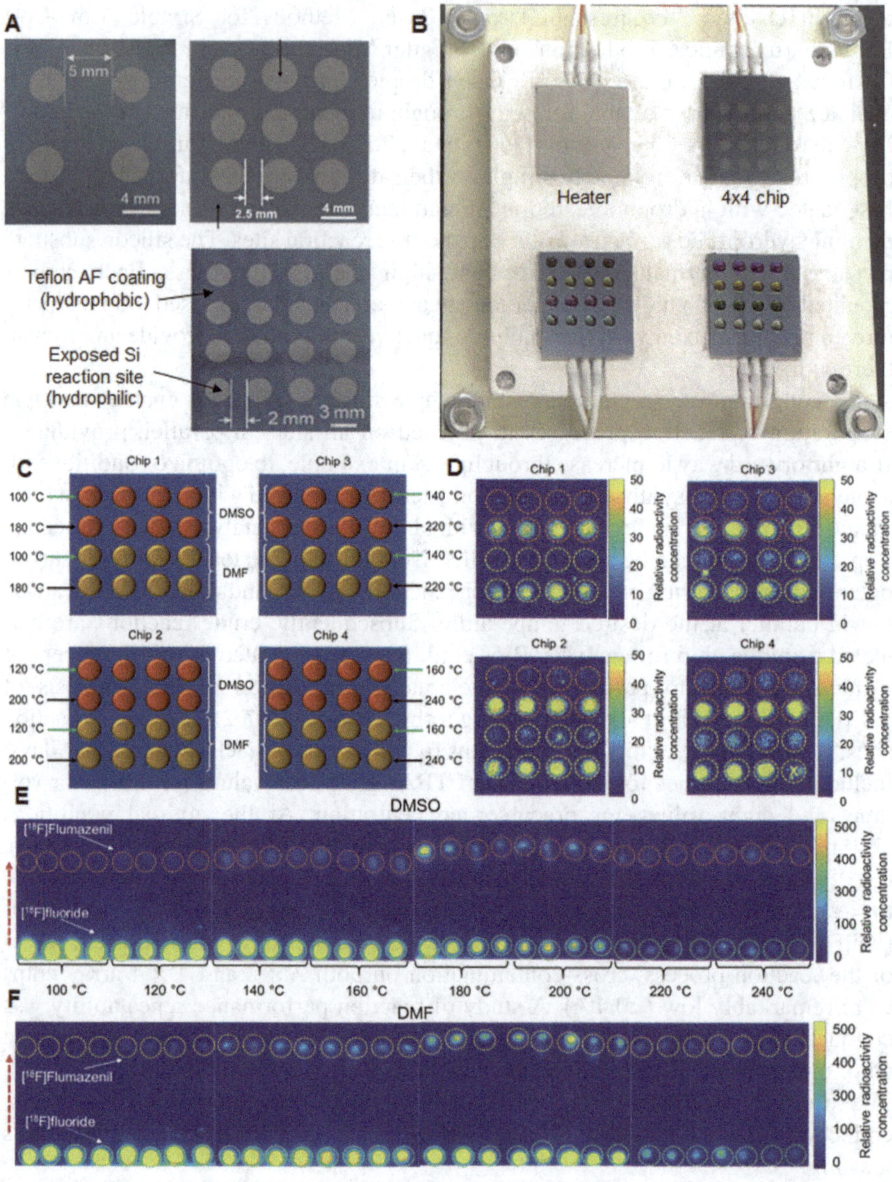

Fig. 8 High-throughput optimization via arrays of droplet reactions. (**a**) Photographs of droplet array chips (2 × 2, 3 × 3, 4 × 4). (**b**) Photograph of the platform for independent heating of four chips. (**c**) Experimental setup for one batch of experiments that explored the influence of reaction temperature (eight values) and solvent (two types) for the synthesis of [^{18}F]Flumazenil. Half of the reaction sites were used first to explore four different temperatures in the first set of 32 simultaneous reactions, and then the other half of the sites were used for the remaining four temperatures. (**d**) CLI images show the distribution of the residual activity on each chip after collecting the crude products. (**e, f**) CLI images of developed TLC plates (each containing eight samples) for reactions that used DMSO and DMF as the reaction solvent. Dashed circles indicate the ROIs used for analysis. The dashed red arrow indicates the direction of solvent movement during development. White dotted lines represent the boundary of each multi-sample plate. (Adapted from Rios et al. (2022) with permission from Springer Nature and from Lu et al. (2023a) with permission from the Royal Society of Chemistry)

a nitro group. Over 800 reactions were conducted within only 15 experiment days, to investigate the influence of a wide range of reaction parameters: type of phase transfer catalyst/base, amount of phase transfer catalyst/base, precursor amount, solvent, reaction temperature, and reaction time. An example experimental set up for one batch of 64 experiments that explored the influence of reaction temperature (eight values) and solvent (two types) for the synthesis of [^{18}F]Flumazenil, including CLI assessment of residual activity losses on reaction chips and multi-lane radio-TLC readout of fluorination efficiency, is summarized in Fig. 8c–f. Using high-throughput droplet array chips, 85 different radiosynthesis conditions were explored to optimize the labeling of [^{18}F]Flumazenil. In contrast, prior publications, using conventional technologies for optimization explored far few conditions (Ryzhikov et al., 2005; Massaweh et al., 2009; Mandap et al., 2009; Nasirzadeh et al., 2016). Additionally, diverse droplet radiolabeling conditions were investigated to optimize the production of [^{18}F]PBR06 (74 conditions), [^{18}F]Fallypride (64 conditions), and [^{18}F]FEPPA (8 conditions). In general, optimization studies yielded conditions that provided comparable or even higher activity yields than conventional methods, while requiring 2–3× shorter synthesis time, and ~10–100× less precursor. The optimized droplet-based conditions offered remarkably higher activity yield for [^{18}F]PBR06, [^{18}F]Fallypride and [^{18}F]FEPPA, and short synthesis times (all within 30–35 min) (Rios et al., 2022). Though improved yield was not achieved for [^{18}F]Flumazenil, the total preparation time was significantly shortened, i.e., 35 min compared to the 75–80 min required for conventional preparation (Ryzhikov et al., 2005). The combination of reduced reaction time and reduced precursor consumption per reaction provides considerable efficiencies in time and cost during the optimization process and could be especially valuable at the early stages of new tracer development.

High-throughput optimization via droplet arrays was also used to study Cu-mediated radiofluorination in droplet-based reactions, in particular for [^{18}F]FBnTP, a PET probe for myocardial perfusion imaging (Werner et al., 2019) and lung cancer investigations (Han et al., 2023; Momcilovic et al., 2019), as well as [^{18}F]YH149, an innovative tracer targeting monoacylglycerol lipase (MAGL) (He et al., 2022). In the study of [^{18}F]YH149 using 3 × 3 or 4 × 4 array chips, a total of 117 experiments were conducted over a span of 5 days to explore 36 distinct reaction conditions, while consuming a total precursor amount of <15 mg (Lu et al., 2023a). The outcome of this optimization was a substantial improvement in reaction performance, achieving a significant improvement in RCY (52 ± 8%, $n = 4$) compared to the original macroscale report (4.4 ± 0.5%, $n = 5$).

While the above results demonstrate that droplet reactions can be performed with a relatively simple setup using manual pipetting, the needed large number of pipetting operations to add reagents, collect crude products, and transfer crude products to TLC plates for analysis are highly tedious, susceptible to human error, and involve some radiation exposure. To address this, Jones et al. (2023) developed a fully-automated robotic system for droplet-based radiochemistry optimization (Fig. 9a–c). The platform performs liquid transfers using piezoelectric dispensers or pipetting of isotope/reagents to specific reaction sites, heating/evaporation steps,

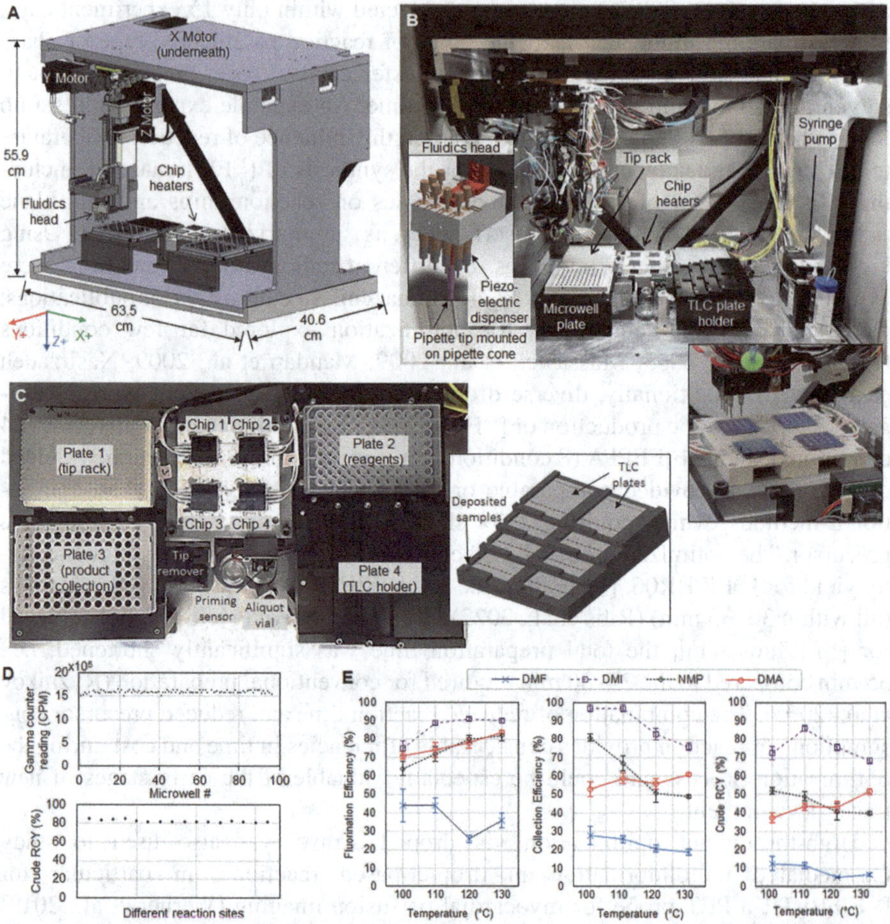

Fig. 9 Robotic system for high-throughput optimization via droplet arrays. (**a**) CAD model of the optimization platform showing the geometry and major components. (**b**) Photograph of the system inside a minicell showing detail of the multiple-heater platform with four 4 × 4 chips installed for 64 parallel reactions (inset). (**c**) Photograph overhead of the work area of the robotic system including the stacked TLC plate holder (inset) that allows up to 64 samples to be spotted on multi-lane TLC plates. (**d**) Top: Uniformity of reagent dispensing (measured via gamma counting of [^{18}F]fluoride solution into 96 wells of a microwell plate. Bottom: Uniformity of [^{18}F]Fallypride synthesis across 16 reaction sites of a multi-reaction chip. (**e**) Results of a single synthesis optimization experiment exploring impact of temperature and reaction solvent ($n = 4$ replicates per condition) on the synthesis of [^{18}F]FBnTP. (Adapted from Jones et al. (2023) with permission from Elsevier)

collecting crude products from individual reaction sites into microwell plates, and spotting samples of crude products onto TLC plates. The workspace contains a four-heater platform and four microplate nests used for (1) a rack of pipette tips, (2) a microwell plate for reagents, (3) a microwell plate for crude products, and (4) a custom stacked holder for multiple multi-lane TLC plates. The platform was designed to fit within a compact minicell (68 × 50 × 61 cm^3). It is operated via a user-friendly LabVIEW interface, reading user-written script files that define the optimization study to be performed.

Cross-contamination and uniformity tests were performed using the robotic system, and for a set of 16 identical reactions, synthesis of [^{18}F]Fallypride was highly uniform with crude RCY of 83 ± 1% (n = 16) (Fig. 9d). To further validate the automated system, an optimization of [^{18}F]Fallypride synthesis was performed (varying $TBAHCO_3$ and precursor amounts) and results were highly consistent with those of a prior manual study. The robotic system was then used to perform a novel optimization of the Cu-mediated synthesis of [^{18}F]FBnTP. A set of 64 reactions including varied reaction solvents and temperatures was performed (Fig. 9e). Notably, the choice of solvent exhibited a critical impact on the synthesis performance. The optimal condition emerged as 110 °C in 1,3-dimethyl-2-imidazolidinone (DMI), yielding a high fluorination conversion of 89 ± 1% (n = 4) and crude RCY of 86 ± 2% (n = 4) (Jones et al., 2023), while in comparison, the original report showed a fluorination conversion of ~60% (Zhang et al., 2018). The isolated RCY after purification and formulation was high (66 ± 6%, n = 3) with high radiochemical purity (100%), and the overall preparation time was only 42 ± 1 min (n = 3), showing significant improvements over a reported implementation on a conventional module which had RCY of 9.1 ± 2% (n = 7) and synthesis time of 60 ± 5 min (Lin et al., 2023). Performing the set of experiments, including the time needed for manual measurements (i.e., activity of collected crude products and chips, and CLI of chips and TLC plates) and analysis, required only 4 h. Overall, the robotic system allows large numbers of reactions to be rapidly studied (up to 64 data points per batch) with the convenience of automated operation while consuming only low amounts of precursor.

Though proof-of-concept of scaling up droplet conditions by increasing volumes to typical automated vial-based modules has been demonstrated (Lu et al., 2023a), the preferred method to scale to higher activity levels is to perform the production using droplet reactions. Scale-up to multiple human doses has been demonstrated by using concentration methods to increase the amount of [^{18}F]fluoride (without increasing the amount of salts/base) in a single droplet reaction (Wang et al., 2020a; Rios et al., 2022), or performing multiple droplet reactions in parallel (Lu et al., 2024b).

2.5 Comparison of High-Throughput Labeling Approaches

The development of technologies specifically designed to perform significant numbers of reactions sequentially (flow chemistry) or simultaneously (arrays of vials or droplets) has opened up unprecedented throughput that offers many practical advantages:

1. A single batch of radioisotope can be split among a large number of reactions. In addition to saving money, this ability to split a single isotope batch across many experiments can reduce the impacts of batch-to-batch variations (e.g., for isotopes such as [^{18}F]fluoride). Further, for isotopes with supply limitations, e.g., Ga-68 or Ac-225, high-throughput techniques enable more data to be generated from each batch of the radioisotope.
2. Performing more experiments with the same "lot" of radioisotope, or aliquot of prepared reagents, can reduce the impacts of batch-to-batch isotope variation, or batch-to-batch variations in reagents (e.g., that might occur as reagents degrade over time, or for which there are variations in concentration due to measurement errors).
3. Reducing the cost and effort of experiments can be used to lower the cost and timescale of an overall study, or to increase the rigor of results (e.g., more replicates), and can increase the granularity of parameter values, or the number of parameters, being investigated in optimization studies.

Despite these common advantages, the different platforms have advantages and limitations (Table 1).

3 Accelerating Radiochemical Analysis

3.1 Overview

Reaction analysis is a significant challenge facing the development of high-throughput methods for radiochemistry. Most radiochemical experimental workflows make use of either radio-TLC or radio-HPLC to analyze radiochemical reactions or process outcomes. Radio-TLC is very effective for separating and quantifying the organic and inorganic radioactive components of a reaction mixture, making it the go-to method for quantifying the radiochemical conversion of reactions involving ionic radionuclide sources such as [^{18}F]fluoride or [^{11}C]cyanide. Radio-TLC is also relatively quick; once spotted, plates can be developed and analyzed within minutes, and reactions of interest can be quickly earmarked for further analysis. However, as every reaction needs to be spotted by hand before measurement, standard single-lane radio-TLC is labor-intensive and thus not well suited to the analysis of large run numbers. On the other hand, radio-HPLC can resolve the organic components of a reaction mixture, making it useful for studying reactions involving organic prosthetic groups or radionuclide formulations (e.g., [^{11}C]MeI). Peak retention time is highly reproducible and HPLC can, therefore, be used to confirm product identity and molar activity and determine byproduct formation (chemical/radiochemical purity). Many HPLC systems are also equipped with autosamplers, allowing a sequence of runs to be carried out without the experimenter being present. However, radio-HPLC is a comparatively slow method, with most analysis methods taking 15–30 min or more (when factoring in time for

Table 1 Comparison of features and capabilities of different platforms for high-throughput reactions

Feature	Flow-based approaches	Vial-based approaches	Droplet-based approaches
Parallel/sequential reactions	Sequential	Parallel	Parallel
Open/closed reaction	Closed (in capillary)	Closed (if vial capped)	Open
Typical reaction volume	10s of μL	50–250 μL	10 μL (typ.)
Typical reagent consumption per reaction	0.01–0.5 μmol	0.1–5 μmol	0.005–0.5 μmol
Ability to vary reaction time	Yes	Not in same campaign	Yes (if use multiple chips)
Ability to vary reaction temperature	Yes	Not in same campaign	Yes (if use multiple chips)
Ability to vary reagent concentrations	Yes	Yes	Yes
Ability to vary preparation of isotope (e.g., type of phase transfer catalyst (PTC))	Not in same campaign	Yes, if multiple aliquots of radionuclide are prepared	Yes
Ability to vary reaction solvent	Not in same campaign	Yes	Yes
Ability to study multi-step reactions	Yes	Yes	Yes
Ability to work with gas-phase reagents or intermediates/products	Possible	Possible	Not possible
Ability to support substrate-screening studies	Not in same campaign	Possible	Possible
Ability to scale-up	Increase reagent and isotope boluses	Load more isotope per reaction	Load more isotope per droplet reaction or increase number of droplet reactions in parallel
Ability to transfer to conventional vial-based system	Difficult	Easy	Currently being studied

column cleaning and re-equilibration). These methods are appropriate for traditional low-throughput radiochemistry techniques; however, as the amount of detectable analyte exponentially drops due to radioactive decay, higher run numbers quickly become limited by the sensitivity of the detector (Fig. 10). Implementing HTE thus requires new approaches to analyze large numbers of samples in a short time frame before the amounts of radioactive analytes drop below detectable limits (Webb et al., 2024).

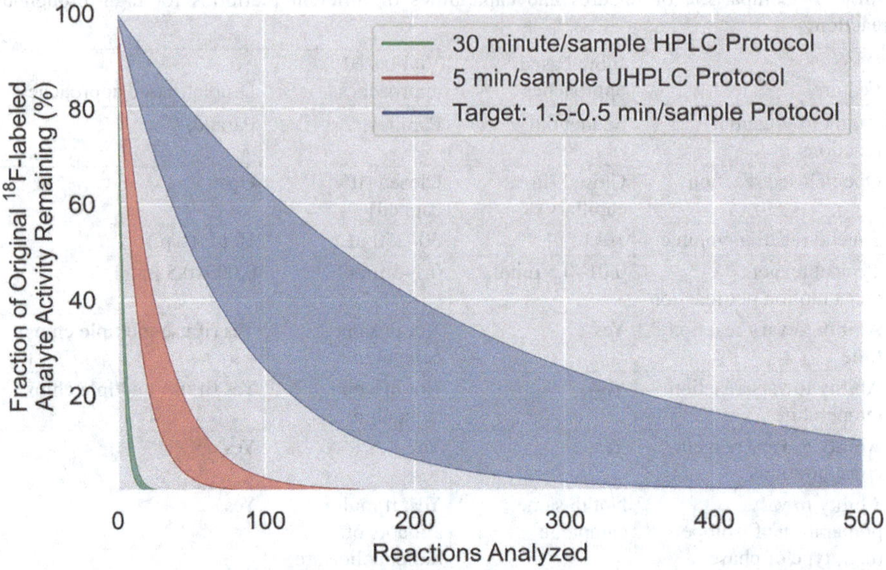

Fig. 10 Currently, most radiochemical analyses are done in sequence, which cumulatively takes a lot of time, preventing the analysis of large numbers of experiments before samples decay to undetectable levels

3.2 UPLC

Liquid chromatography analyses can be accelerated by the use of ultra-high-performance liquid chromatography (UHPLC) or ultra-performance liquid chromatography (UPLC). These methods rely on very high pressure pumps (15,000–20,000 psi versus <6000 psi for HPLC), which enables the use of columns with very small pore sizes (e.g., 1.7 μm compared to 5 μm of typical HPLC systems). With small pore size, high separation resolution can be achieved in a short column length, enabling fast methods with short retention times compared to HPLC.

Franck et al. (2009) demonstrated that UPLC in conjunction with a low-volume flow cell for radiation detection could be used to speed up LC analysis of radiopharmaceuticals. A method that took 15 min for analysis of [^{18}F]FMISO on HPLC (using 2.1 mm × 50 mm column with 1.7 μm particle size) took only 2 min on the UPLC setup (using 4.6 mm × 300 mm column with 10 μm particle size). More recently, UPLC has been used for analysis and QC testing of additional ^{18}F-labeled compounds, including [^{18}F]MC225 (Garcia-Varela et al., 2021) and [^{18}F]AlF-NOTA-octreotide (Tshibangu et al., 2020).

Similarly, utilizing a 15 μL cell volume, Kryza and Janier (2013) developed the first radio-UPLC method for determination of the radiochemical purity of ^{99m}Tc-radiopharmaceuticals. This method enables rapid analysis in <3 min, a substantial improvement compared to the 25–40 min required by package insert-recommended techniques such as silica gel strips or radio-HPLC. Three representative tracers, including [^{99m}Tc]Tc-mercaptoacetyltriglycine ([^{99m}Tc]Tc-MAG3), widely used for renal function determination, as well as [^{99m}Tc]Tc-tetrofosmin and [^{99m}Tc]Tc-sestamibi for myocardial perfusion imaging, were tested. Samples at 15 min and 6 h post-preparation were subjected to analysis using package-insert methods (radio-HPLC or TLC) and radio-UPLC, and no significant differences were observed. Radio-UPLC proved to be up to 10 times faster than the standard TLC method, accounting for the time invested in drying the sample spot, developing the TLC plate, and monitoring the dryness of the TLC plate. Schmitt et al. (2021) developed and validated UHPLC methods for QC testing of additional ^{99m}Tc-radiopharmaceuticals, attempting to develop a more sensitive, yet rapid, alternative to radio-TLC.

Rezes et al. explored the application of UPLC for quality control of [^{11}C]methionine, used for imaging malignant tumor cells. The examination of the radiochemical purity of [^{11}C]methionine using HPLC takes ~16 min (Giglio et al., 2018), but UPLC has shown the potential to significantly reduce the analysis time to <2 min (Rezes et al., 2020). This swift analysis facilitates a rapid release of the final product, preventing substantial radioactivity loss within the short half-life of carbon-11 (20 min). The optimized UPLC protocol enables simultaneous analysis of both the ^{11}C-labeled and non-radioactive ingredients.

Though we are not aware of any published studies leveraging UPLC in the context of high-throughput experimentation, and though the instrumentation is not widely available in radiochemistry laboratories, there is a clear potential for UPLC to be used for high-throughput sample analysis.

3.3 Multi-lane Thin-Layer Chromatography (TLC)

As inorganic [^{18}F]fluoride is always the limiting reagent of any radiofluorination, radio-TLC on silica plates using a scanning detector for readout is widely considered the gold standard for quickly estimating ^{18}F-radiosynthesis radiochemical conversion. Thin-layer chromatography is used to separate the unincorporated inorganic [^{18}F]fluoride and incorporated organic ^{18}F-labelled product (or intermediate), and the scanner is used to obtain a chromatogram of the radioactivity as a function of distance along the plate. Analysis of the chromatogram enables calculation of the proportion of the total plate radioactivity in each chemical form. Unfortunately, radio-TLC using standard single-lane TLC equipment is a laborious and time-consuming technique where each reaction needs to be manually spotted onto a precut TLC plate, developed in an appropriate mobile phase, and then analyzed, one at a time, using a TLC scanner (Fig. 11a). Typically, the scanning process alone takes 3–10 min per reaction/plate, thus making this conventional technique impractical for the analysis of large numbers of reactions.

However, these limitations can be overcome by extending the approach to multi-lane methods (Wang et al., 2020b), where multiple samples are spotted side by side,

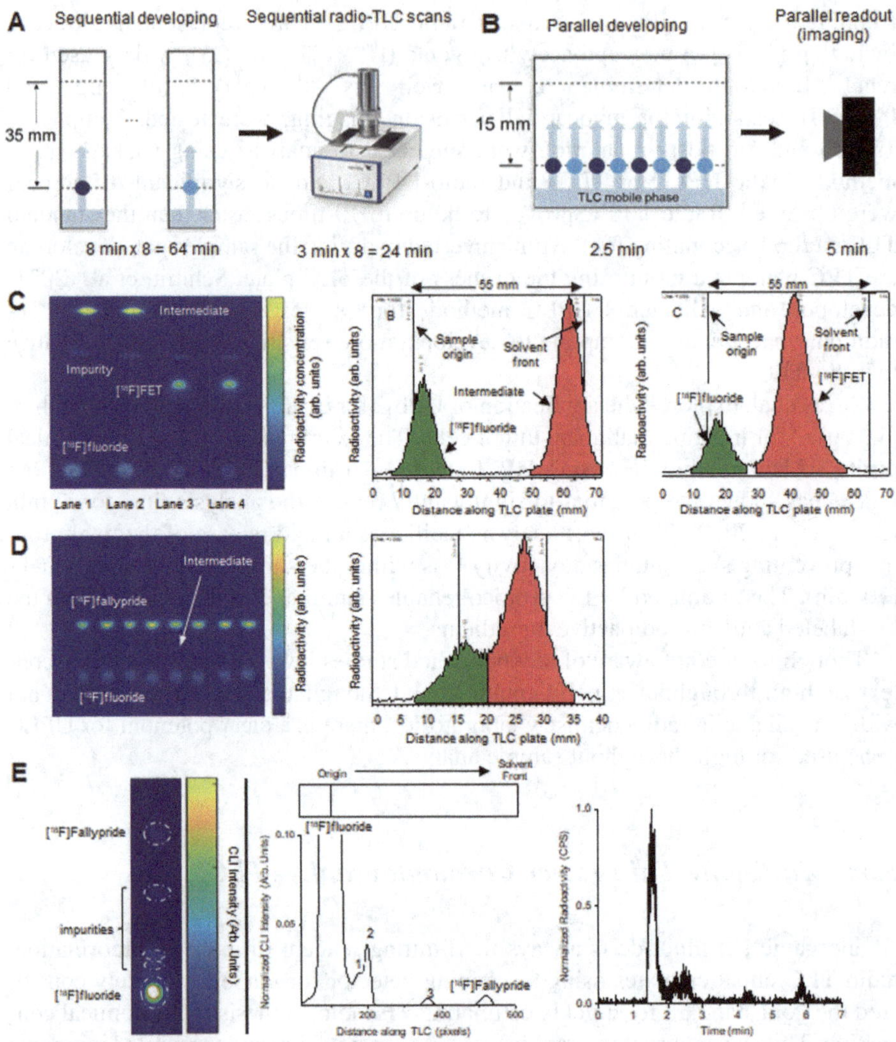

Fig. 11 (**a**) With conventional TLC approaches, eight samples are spotted onto eight separate TLC plates, developed, and individually read out via a radio-TLC scanner. Processed sequentially analysis of the eight plates could take 88 min. (**b**) Using multi-lane techniques, the separation of all samples is performed in parallel (and can be accelerated by using shorter separation distance, leveraging the high readout resolution) and readout is performed in parallel. Analysis of the same eight samples would take 7.5 min. (**c**) Left: Cerenkov luminescence image of developed along 35 mm distance of TLC plate. There are two samples of [^{18}F]FET intermediate after fluorination (Lanes 1,2), and two samples of [^{18}F]FET crude product after deprotection (Lanes 3,4). Middle: Radio-TLC scan of a sample of [^{18}F]FET intermediate (same sample) separated along a 55 mm distance. Right: Radio-TLC scan of a sample of [^{18}F]FET crude product. (**d**) Left: Cerenkov luminescence image of eight samples from the same crude [^{18}F]Fallypride product with a separation distance of just 15 mm. Right: Radio-TLC scan of the same sample developed over a 15 mm distance (different horizontal scale than in part **c**) (Adapted from Wang et al. (2020b) with permission of Elsevier). (**e**) Left: Cerenkov luminescence image of TLC plate spotted with a sample of crude [^{18}F]Fallypride containing numerous radioimpurities. Center: Line profile of intensity (radioactivity) along the plate. Right: Analytical HPLC chromatogram of the same sample, showing the same bands (Adapted from Laferriere-Holloway et al. (2023) with permission of Elsevier)

and are separated in parallel lanes during the development process (Fig. 11b). Using image-based (2D) detection methods enables readout of the whole set of samples simultaneously, followed by image analysis to quantify the proportion of radioactive species in each band of each sample (lane). Wang et al. demonstrated that samples of 0.5 μL could be spotted at 5 mm intervals, allowing eight samples to be spotted along a 50 mm wide TLC plate. Separation was performed by placing the plate in a conventional developing chamber, and readout was performed via optical imaging. Specifically, the distribution of radioactivity was measured by placing the TLC plate inside a light-tight box and taking a long exposure (e.g., 5 min) with a cooled CCD or CMOS camera. Conversion of radioactive emissions (beta particles) to light is achieved by placing a glass sheet over the TLC plate (which generates light due to Cerenkov emissions) or a scintillator sheet over the TLC plate (which generates light due to excitation of the scintillator by the beta particles). The latter provides significantly higher sensitivity and is able to detect faint bands (low-abundance species) (Fig. 11c). Wang et al. demonstrated the analysis of crude samples of [^{18}F]Fallypride, [^{18}F]FET, and [^{177}Lu]Lu-PSMA-617. Comparing the accuracy and precision of the imaging-based readout to a conventional radio-TLC scanner (using cut-and-count gamma counter measurements as a gold standard), they found that the imaging-based readout had superior accuracy and precision, largely attributed to the higher readout resolution. Several of the analyzed samples showed clearly defined additional bands not corresponding to either [^{18}F]fluoride or the desired ^{18}F-incorporated product or intermediate, i.e., corresponding to other organic radioimpurities. The high readout resolution enabled separation distances to be kept to a minimum (typically 35 mm, but as low as 15 mm) (Fig. 11d), which minimizes the time needed for sample separation, and the size of the TLC plates (allowing more plates to be read out at a time within a fixed field of view), both speeding the process and enabling more samples to be analyzed in a given time.

Achieving high chromatographic separation resolution (i.e., well-separated bands) requires optimization of the developing conditions. Laferriere-Holloway et al. (2023) showed that a well-known TLC mobile-phase optimization technique (PRISMA) could be adapted for use in analysis of radioactive samples. The technique is performed by preparing a mixture of many of the desired species to be separated, including the crude radiofluorination product, precursor, and reference standard. A series of separations are performed, each one analyzed to determine band positions of radioactive species through Cerenkov luminescence imaging, and also imaged under UV illumination to determine positions of non-radioactive species. Sequential sets of experiments are performed to determine the best set of solvents to mix, solvent strength, additives, and solvent ratios, where "best" is the condition with the best separation resolution among all species. Following the prescribed protocol, the optimal mobile phase can be found in about a half day. In the same work, optimal mobile phases were developed for multiple ^{18}F-compounds. To highlight the excellent separation resolution, crude samples of [^{18}F]Fallypride were prepared by reacting under highly basic conditions to generate large numbers of side-products, and were analyzed via radio-TLC and also radio-HPLC. Under isocratic radio-HPLC analysis, six radioactive peaks were observed; the same number of distinct bands were observed on the TLC plate (with only 35 mm separation distance) using the optimized TLC mobile phase (Fig. 11e).

This high-throughput multi-lane radio-TLC method has been used extensively in the context of high-throughput droplet radiochemistry methods (Rios et al., 2019, 2022; Lu et al., 2023a, b, 2024a, b). Crude samples can be manually spotted to lanes on the TLC plate using manual pipetting, and multi-channel pipetting could likely be used. In the automated robotic system for HTE developed by Jones et al. (2023), a TLC plate "rack" is included in the work envelope of the pipetting system. The rack can be loaded with eight TLC plates, each with a capacity of eight lanes. The robotic system can be programmed to automatically transfer the 64 crude products generated in the droplet reactions to the TLC plates. Though developing the plates and readout are still performed manually outside the system, this technique makes it practical to rapidly analyze large numbers of samples. For the separation distances and mobile phases used to analyze a variety of droplet reactions of ^{18}F-labeled compounds, developing and readout of a set of 64 compounds can be performed within 15–30 min by using multiple TLC developing chambers and by imaging multiple TLC plates at a time in the readout system. Though the above studies used a custom-built imaging system and purpose-built readout systems are not commercially available, the spatial distribution of Cerenkov or scintillation light can be imaged and quantified via other commercially available devices that may be found in or near a radiochemistry laboratory such as small-animal optical imaging systems and 2D radiation detection systems (Wang et al., 2020b).

3.4 *Parallel Solid-Phase Extraction (SPE)*

For large numbers of reactions (24+), reaction analysis via traditional methods of radio-HPLC and radio-TLC quickly becomes unfeasible. In conjunction with their new HTE radiochemistry method described above (Sect. 2.2), Webb et al. (2024) also developed a new parallelized analysis method to quickly and reliably determine the reaction performance of each of 96 well-based reactions before the radioactive analytes decayed below detectable limits based on solid-phase extraction (Fig. 12a). They reasoned that 96-well cartridge-based solid phase extraction (SPE) could be used to separate the organic and inorganic ^{18}F-containing fractions of each reaction in parallel, in a similar fashion to the common usage of TLC plates for assessing radiochemical conversion. To achieve this, each reaction well was diluted with water, and the contents were transferred to a 96-well SPE cartridge stacked on top of a 96-well collection plate. A vacuum apparatus was then used to pull the aqueous solution through the cartridge array, trapping the organic products while allowing the inorganic fractions to pass through. The cartridge array was then again washed with water, and the combined inorganic fractions were collected. The SPE cartridges were transferred to a new collection plate, and the process was repeated, this time eluting the organic fractions from the cartridge array with organic solvents. The reliability of this procedure was confirmed by manually checking the contents

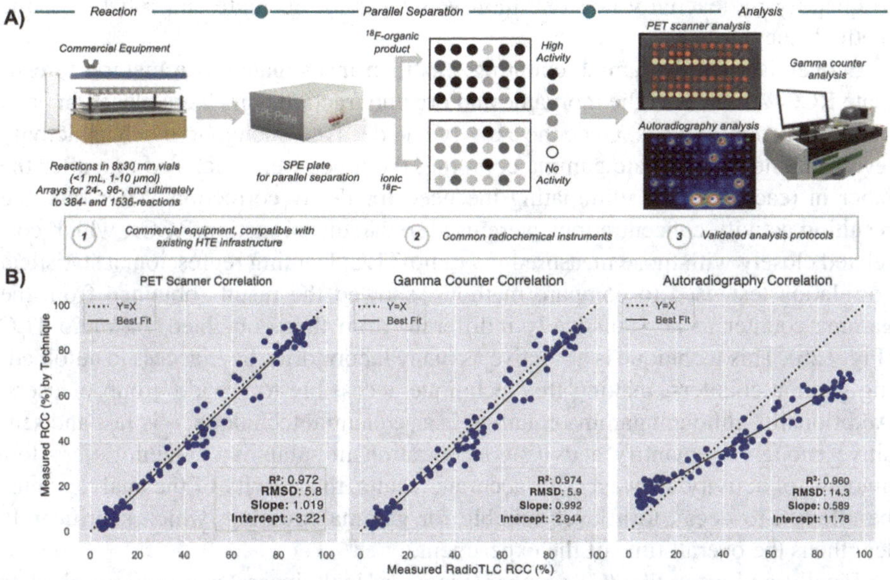

Fig. 12 (**a**) The high-throughput experimentation workflow for radiochemistry. (**b**) After SPE separation, RCC was calculated using the PET scanner, gamma counter, and autoradiography methods, and these results were correlated with the manually acquired radio-TLC %RCC data. (Figure reproduced from Webb et al. (2024) and used with permission. © American Chemical Society)

of each inorganic and organic fraction by TLC for the presence of organic or inorganic contaminations that might skew the RCC calculation.

With the SPE separation protocol in hand, the authors set out to identify an efficient and accurate method to quantify the activity concentrations of the resulting inorganic and organic fractions, comparing autoradiography, gamma counting, and volumetric activity quantification using a PET scanner as methods to quantify the activity concentrations of 100 μL aliquots of the organic and inorganic fraction solution pairs (Fig. 12). These data were used to calculate the observed RCCs for each method, which were subsequently compared to those obtained using a standard radio-TLC protocol. To assess the accuracy of all methods, the authors manually measured the reaction RCCs of each of the 96 reactions via the standard (and laborious) radio-TLC protocol before quenching and diluting the reaction mixtures for SPE separation.

Autoradiography was performed by placing a rack containing pairs of glass vials with equal volumes of the inorganic and organic SPE fractions for each reaction directly onto an autoradiography plate. Once the plate was developed, the activity concentrations of the inorganic and organic fraction pairs were quantified and used to calculate the RCCs. Unfortunately, using this approach, autoradiography was found to overestimate the reaction RCCs when compared to radio-TLC, especially for higher-performing reactions (Fig. 12b). The authors concluded that this protocol to quantify radioactivity concentration was the least accurate and reliable of the methods they tested.

Sequential sample gamma counting was then investigated as a method to estimate RCC. Aliquots of the inorganic and organic fraction pairs were placed in pairs of gamma counter tubes. Once the activity had decayed enough to reach an activity level suitable for accurate gamma counting, the tubes were analyzed one after the other in reaction pairs, eliminating the need for decay correction. After this, the resulting activity concentrations were used to calculate reaction RCCs, which correlated closely with those measured via radio-TLC. Deming regression, a statistical correlation test used to compare methods, showed the results obtained from the gamma counter to be statistically indifferent from those obtained via radio-TLC (Fig. 12b). This technique is attractive as many laboratories have access to autoloading gamma counters, making this technique accessible to a wide group of users. Additionally, although gamma counting is a sequential technique, it is fast and sensitive enough to quantify activity concentration in samples containing very low amounts of activity. However, for accurate results, the activity of the analytes must be allowed to decay to a level suitable for gamma counting, which significantly lengthens the overall time of the experiment.

Finally, an Inveon PET/CT scanner (Siemens Healthineers) was used to calculate the activity concentrations of 24 reaction pairs (48 vials) at a time. After CT attenuation correction, a 10-min PET frame was acquired for each rack (2 racks per 96 experiments), and regions of interest (ROI) of equal volume were drawn for each organic and inorganic reaction sample. From the acquired data pairs, the RCCs for each reaction were calculated and compared, using linear regression, to the ground truth radio-TLC data, showing that the two methods were again highly comparable (Fig. 12b). The PET scanner analysis method had, by far, the greatest experimental throughput of all the methods tested, making it capable of quickly acquiring large radiochemical experimental datasets. A limited number of "hit" reactions could then be quickly identified for further analysis (e.g., by-product analysis via radio-HPLC) before the activity decayed below detectable limits. However, depending on the PET reconstruction software and computer hardware available, the manual processing of the PET data can take a significant amount of time and effort. Automated approaches to overcome this bottleneck are currently being developed. Additionally, the use of this technique is limited to the study of positron-emitting radionuclides and given that not all radiochemistry labs have ready access to a PET scanner, this would further restrict the accessibility and usability of this method.

3.5 Comparison of High-Throughput Analysis Methods

The development of analytical methods capable of analyzing many samples in a short period of time and specifically tailored for high throughput workflows has been essential for the introduction of HTE for radiochemistry. The key aspect of such analysis is the need for speed—short-lived radioactive products have to be analyzed quickly meaning traditional high throughput analytical methods, such as dozens of HPLC analyses that can be automated to run overnight, are not compatible. Several methods have been introduced around the same principles:

1. Rapid analysis is key for radiochemistry HTE. This can be accomplished by analyzing samples in series, but short analysis times are essential (i.e., even at 5 min analysis time, 96 samples would take 8 h to analyze!). Given this, there have been efforts to develop methods to analyze reactions in parallel.
2. Parallel analyses (e.g., using multi-lane TLC, or scanning 96-array SPE cartridges in a PET scanner) offer the convenience of rapid data acquisition compatible with short-lived products, but can create a data analysis bottle neck (e.g., defining 96 × ROIs to analyze one plate). Thus, as these techniques become more refined, automated data analysis for ease of processing will be imperative if such techniques are to be widely used.
3. Ideally, as HTE analysis methods continue to be refined, they will utilize standard "off-the-shelf" consumables and commonplace equipment so that they are compatible with most radiochemistry labs that might want to utilize HTE radiochemistry in the future.

The various analytical methods described to date offer different pros and cons (Table 2).

4 Future Considerations

4.1 *Algorithms and Computational Methods*

One of the more challenging activities in radiopharmaceutical synthesis research and development is navigating and optimizing the vast number of complex and interacting variables that influence a radiochemical reaction's performance. The varied HTE techniques discussed in this chapter for efficiently and rapidly performing and analyzing large numbers of reactions will be crucial in facilitating such studies. However, even with these advanced tools, exploring complex radiochemical processes will require radiochemists to adopt and adapt intelligent experimental design strategies and optimization algorithms to build informative, relevant, and useful models of radiochemical space (Taylor et al., 2023).

The traditional "one-variable-at-a-time" (OVAT) optimization approach employed by most radiopharmaceutical development programs holds all variables constant while adjusting one variable of interest through a range of settings until a maximum is observed (Bowden et al., 2019). These settings then inform the next set of experiments to study the next variable of interest, and the process is repeated until all variables of interest have been studied in this way, at which point the process is considered optimized. The method is notoriously inefficient, slow, and unreliable, requiring great experimental effort while being prone to only finding local optima.

Table 2 Comparison of features and capabilities of different analytical methods for high-throughput reactions

Feature	UPLC approaches	TLC approaches	SPE approaches
Parallel/ sequential analysis	Sequential	Sequential/parallel	Parallel
Acquisition time for 96 samples	≥180 min	~65 min (using 8-lane TLCs)	20 min
Approx. data analysis time	1 h	1 h	2–4 h
Specialized equipment needed for separation	UPLC with appropriate column	No	No
Specialized equipment needed for readout	UPLC with UV/ radiation detectors	Radio-TLC scanner, phosphorescence imager, optical imager, etc.	Gamma counter, PET scanner etc.
Resolution of radioactive species	Individual species	Aqueous $^{18}F^-$ from ^{18}F-labeled organic species Individual species possible with mobile phase optimization (PRISMA)	Aqueous $^{18}F^-$ from ^{18}F-labeled organic species
Concurrent analysis of chemical purity	Yes	No	No
Issues determining RCC	$^{18}F^-$ may not be eluted from column, leading to overestimation of RCC	^{18}F-labeled organic species may not separate, leading to overestimation of RCC	^{18}F-labeled organic species may not separate, leading to overestimation of RCC

Because the OVAT approach only studies one variable at a time, it can only examine small unidimensional slices of reaction space, which also means that it is incapable of resolving variable interactions (where the setting of the variable affects the behavior of others) and thus often provides incomplete information about the system being optimized. However, the OVAT approach can be systematized to rapidly screen process variable effects for magnitude and significance. The work by Ahmadi et al. used an efficient, self-contained, and automated digital microfluidics (DMF) system to perform an OVAT screening of several variables related to the performance of a DMF [^{18}F]FDG synthesis experiment (Ahmadi et al., 2023). Using this limited data set, they were able to use and compare several linear and non-linear regression techniques to construct a machine learning derived model capable of predicting moderately improved synthesis conditions (by optimizing main effects only). However, the authors noted that the size of their dataset likely prevented the accurate modeling of higher-order interaction and quadratic terms.

The OVAT approach is intuitively simple and easy to implement, and as such, it is widely used to solve more simple optimization problems. However, its limitations are well understood. Several optimization approaches and algorithms exist to help build more thorough yet efficient experimental regimes and facilitate predictive modeling. One example of such an approach is the "Design of Experiments" (DoE) (chapter "The Application of Statistical "Design of Experiments" (DoE) Towards the Efficient Study and Optimization of Radiochemical Processes"), which is used to build efficient experimental plans that study multiple variables simultaneously, providing optimal datasets for response modeling, reducing the overall experimental effort while providing more information from fewer data points (Bowden et al., 2019). However, while experimentally efficient, these studies require a high degree of precision in their execution with reproducible, standardized protocols, making them labor-intensive and sensitive to experimenter mistakes and systematic errors. The HTE methods discussed in this chapter are highly standardized (from run to run) and are, thus, by nature, particularly well-suited to addressing some of the technical difficulties that are often encountered when trying to implement advanced experimental design and process modeling tactics like DoE.

While DoE has been a pillar of the process industry for decades, interest has also grown in the use and development of other advanced and highly efficient multivariate optimization algorithms built off recent developments in artificial intelligence and machine learning (Taylor et al., 2023). Several local (e.g., simplex and gradient of steepest ascent/descent algorithms) and global (e.g., SNOBFIT and Bayesian (TSEMO) algorithms) self-optimizing techniques have been well investigated as tools for the study and optimization of diverse chemical systems (Soritz et al., 2022). While there are significant differences in how these algorithms operate, they are almost all iterative in nature and require constant interaction between experiment, analysis, and data interpretation and modeling, with the results of previous experiments being used to inform the design and settings of the next. These algorithms, therefore, are not well suited to the parallel HTE systems described in this chapter. They require rapid and robust sequential experimental setups and many of the initial studies were conducted using fully automated flow chemistry systems. The application of these protocols and algorithms for radiochemistry-specific problems will necessitate the further development of the fully automated sequential high-throughput radiochemistry techniques discussed in this chapter.

Significant progress has been made in applying iterative machine learning-based algorithms to specific chemical optimization problems (Struble et al., 2020). However, using large predictive AI models to predict and inform chemical discovery and synthesis in a more generalized manner remains a significant challenge (Saebi et al., 2023). This is largely due to the lack of large, reliable, and reproducible chemical reaction datasets that cover pharmaceutically relevant regions of chemical space. The high-throughput experimentation (HTE) techniques discussed in this chapter will be crucial in creating large standardized radiochemical datasets needed to build reliable and predictive AI models specific to radiochemistry. Since radiopharmacists and radiochemists work with much smaller subsections of broader chemical and reaction space, the lessons learned from implementing HTE and AI models for enhanced radiopharmaceutical research and development may have a significant impact on the broader pharmaceutical and process industries (Bowden et al., 2023).

4.2 Standardization and Data Recording with High-Throughput Experimentation

The HTE techniques described in this chapter hold enormous potential to revolutionize all aspects of radiopharmaceutical research, development, and production. However, leveraging these techniques to their full extent will require the establishment of open, collaborative, and standardized systems for designing and conducting experiments, as well as for recording, managing, and curating the resulting datasets. Systematizing how, when, and why a data point is acquired is vital to the quality and future usability of the resulting datasets. Reaction performance data can be measured and recorded in several ways and at several time points. Choosing which data points are acquired is dictated by the scientific question at hand and is a critical part of any experimental design. However, many laboratories use slightly different equipment and protocols for carrying out experiments and measuring and recording empirical data. This often results in the presence of hidden variables in published datasets, making data acquired in different labs difficult to compare (Saebi et al., 2023). In chemistry, this has led to a crisis of irreproducibility in the published literature and made the combined use of published datasets for the training of predictive models extremely difficult (Salley et al., 2023).

In radiochemistry, this problem is particularly pertinent as there are a multitude of metrics used to assess process performance, with each lab favoring different measures and techniques relevant to their specific scientific aims. If, for example, the scientific goal of the experiment is to quantify the performance of the chemistry, the reaction mixture can be sampled and analyzed directly; this is often recorded as radiochemical conversion (RCC). As a further example, the reaction mixture could also be subjected to a purification and/or formulation procedure before analysis. In this case, the decay-corrected product radioactivity output of radiosynthesis vs. the starting radioactivity as a percentage is defined as the radiochemical yield (RCY) and is intended to measure the performance and efficiency of an entire radiochemical process. However, precise descriptions of "product output" (this could be crude, purified, or formulated product, depending on the context) and "starting activity" (this could be radioactivity at EOB, radioactivity delivered to the hot cell, processed radionuclide, etc.) are not included in the formal standardized definition. Many labs thus have their own subtly different protocols for measuring and recording these data, making combining multicenter datasets difficult. There are also similar systematic inconsistencies in how different labs measure molar activity (MA), as well as radiochemical purity (RCP) and chemical purity; all measures of process performance and product quality are important to GMP production.

In addition to variations in reported measurements, different labs have different equipment to perform experiments, which can affect the actual reaction conditions experienced by the reaction mixture. For example, whether in manual or automated systems, heating characteristics of different reactors (due to geometry, material, heating method, and whether the approach is based on flow or batch techniques) can impact the temperature profile. As another example, loss characteristics of the reagent delivery mechanism and method can impact the amount and concentration of reagents present during reaction steps. Detailed reporting of the equipment and techniques used will increase the ability to compare data among different labs.

The standardized nature of high-throughput radiochemical experimentation will allow us to rapidly and efficiently generate large quantities of high quality reliable radiochemical data. The relatively small and well-connected radiopharmaceutical sciences community is well-situated to leverage these data to revolutionize the discovery and development of clinically impactful radiopharmaceuticals. However, the community will also need to carefully consider how the collection, recording, and management of these data across multiple sites should be standardized and controlled to ensure that the resulting datasets are comparable, robust, and ultimately useful for data mining and AI/ML modeling. Designing a standardized collaborative framework for the curation, quality control, management, and sharing of HTE radiochemical-generated datasets will be crucial to bringing nuclear medicine and radiopharmacy into the age of big data.

5 Summary

Nuclear medicine is experiencing unprecedented growth, with some suggesting the market size could be as large as $30–50 billion by 2030. This growth is happening faster than new infrastructure can be added or the workforce can be expanded, putting demand on both diagnostic and therapeutic radiopharmaceutical manufacturing sites in industry and academia. In particular, both screening of new radiotracers and optimizing the synthesis of leads for translation are being forced to take a back seat to growing routine clinical production demands, which is not good for the future of the field. To overcome these challenges, new workforce paradigms are required both for initial production of new radiopharmaceuticals for preclinical evaluation, and to optimize the synthesis conditions for translation of promising candidates into early phase clinical studies.

The high-throughput technologies introduced in this chapter offer great potential in this regard. As the methods discussed in this chapter illustrate, HTE methods enable the parallel synthesis of numerous new tracers for rapid screening, as well as high-throughput evaluation of synthesis conditions to both rapidly select a production method and optimize reactions conditions within said method. Such techniques are still in their infancy, and promising chemistry solutions as well as high-throughput radiochemical reaction analysis have been reported recently. Future efforts need to better define HTE approaches for purification and formulation of radiopharmaceuticals, and also work out interfacing radiochemistry HTE with established approaches for high-throughput evaluation of drugs (e.g., biochemical/binding assays, cell-based assays, proteomics, high content screening etc.) (Blay et al., 2020; Aldewachi et al., 2021). The combination of radioactive drugs with HTE methods may also offer scope for development of new HTE approaches, given the possibility of rapidly screening plates containing radiopharmaceuticals using PET scanners and autoradiography approaches.

Artificial intelligence techniques such as machine learning also offer enormous potential for rapid optimization of radiosynthetic methods. However, one of the challenges with such techniques is that they require large amounts of data to train the models but, as noted above, generating such data in the traditional radiochemistry lab setting is becoming increasingly challenging. HTE methods are thus also likely to prove useful for rapid generation of large amounts of synthesis data with which to populate machine learning algorithms. Key to realizing this will be introduction of guidelines and best practices for collecting synthesis data (e.g., what to collect, how to report), as well as community initiatives (e.g., databases hosted by professional societies) to facilitate sharing of data amongst key stakeholders and/or those developing machine learning approaches for radiochemistry. Since we are at the beginning of this AI revolution today, the time to define these approaches is now as they will inform how quickly and successfully we can bring this technology to bear on building the future of radiochemistry.

Disclosures Dr. van Dam is a shareholder of SOFIE Inc. and of DropletPharma Corp. Dr. Jones is a shareholder and employee of DropletPharma Corp. Dr. Scott acknowledges the NIH (R01EB021155) for research support.

References

Ahmadi, F., Simchi, M., Perry, J. M., Frenette, S., Benali, H., Soucy, J.-P., Massarweh, G., & Shih, S. C. C. (2023). Integrating machine learning and digital microfluidics for screening experimental conditions. *Lab on a Chip, 23*, 81–91. https://doi.org/10.1039/D2LC00764A

Aldewachi, H., Al-Zidan, R. N., Conner, M. T., & Salman, M. M. (2021). High-throughput screening platforms in the discovery of novel drugs for neurodegenerative diseases. *Bioengineering, 8*(2), 30. https://doi.org/10.3390/bioengineering8020030

Amor-Coarasa, A., Kelly, J. M., & Babich, J. W. (2019). 3D-printed automation for optimized PET radiochemistry. *Science Advances, 5*(9), eaax4762. https://doi.org/10.1126/sciadv.aax4762

Blay, V., Tolani, B., Ho, S. P., & Arkin, M. R. (2020). High-throughput screening: Today's biochemical and cell-based approaches. *Drug Discovery Today, 25*(10), 1807–1821. https://doi.org/10.1016/j.drudis.2020.07.024

Bowden, G. D., Pichler, B. J., & Maurer, A. (2019). A Design of Experiments (DoE) approach accelerates the optimization of copper-mediated ^{18}F-fluorination reactions of arylstannanes. *Scientific Reports, 9*(1), 11370. https://doi.org/10.1038/s41598-019-47846-6

Bowden, G. D., Chailanggar, N., Pichler, B. J., & Maurer, A. (2021). Scalable ^{18}F processing conditions for copper-mediated radiofluorination chemistry facilitate DoE optimization studies and afford an improved synthesis of [^{18}F]olaparib. *Organic & Biomolecular Chemistry, 19*(32), 6995–7000. https://doi.org/10.1039/D1OB00903F

Bowden, G. D., Scott, P. J. H., & Boros, E. (2023). Radiochemistry: A hot field with opportunities for cool chemistry. *ACS Central Science, 9*(12), 2183–2195. https://doi.org/10.1021/acscentsci.3c01050

Bratteby, K., Shalgunov, V., Battisti, U. M., Petersen, I. N., Lopes, S., van den Broek, T., Ohlsson, N. G., Erlandsson, M., & Herth, M. M. (2021). Insights into elution of anion exchange cartridges: Opening the path toward aliphatic ^{18}F-radiolabeling of base-sensitive tracers. *ACS Pharmacology & Translational Science, 4*(5), 1556–1566. https://doi.org/10.1021/acsptsci.1c00133

Bruton, L., & Scott, P. J. H. (2020). Automated synthesis modules for PET radiochemistry. In *Handbook of radiopharmaceuticals* (pp. 437–456). John Wiley & Sons, Ltd. https://doi.org/10.1002/9781119500575.ch13

Cardinal Health. (2023). *FDA-approved radiopharmaceuticals*. Cardinal Health. https://www.cardinalhealth.com/content/dam/corp/web/documents/fact-sheet/cardinal-health-fda-approved-radiopharmaceuticals.pdf.

Cernak, T., Dykstra, K. D., Tyagarajan, S., Vachal, P., & Krska, S. W. (2016). The medicinal chemist's toolbox for late stage functionalization of drug-like molecules. *Chemical Society Reviews, 45*(3), 546–576. https://doi.org/10.1039/C5CS00628G

Coley, C. W., Eyke, N. S., & Jensen, K. F. (2020). Autonomous discovery in the chemical sciences Part I: Progress. *Angewandte Chemie International Edition, 59*(51), 22858–22893. https://doi.org/10.1002/anie.201909987

Dolgin, E. (2018). Radioactive drugs emerge from the shadows to storm the market. *Nature Biotechnology, 36*(12), 1125–1127. https://doi.org/10.1038/nbt1218-1125

Dreher, S. D., & Krska, S. W. (2021). Chemistry informer libraries: Conception, early experience, and role in the future of cheminformatics. *Accounts of Chemical Research, 54*(7), 1586–1596. https://doi.org/10.1021/acs.accounts.0c00760

Franck, D., Nann, H., Davi, P., Schubiger, P. A., & Ametamey, S. M. (2009). Faster analysis of radiopharmaceuticals using Ultra Performance Liquid Chromatography (UPLC®) in combination with low volume radio flow cell. *Applied Radiation and Isotopes, 67*(6), 1068–1070. https://doi.org/10.1016/j.apradiso.2009.02.077

Garcia-Varela, L., Attia, K., Sembrano, J. C., Jacquet, O., Antunes, I. F., Kwizera, C., Visser, T. J., Dierckx, R. A. J. O., Elsinga, P. H., & Luurtsema, G. (2021). A new approach to produce [^{18}F]MC225 via one-step synthesis, a PET radiotracer for measuring P-Gp function. *EJNMMI Radiopharmacy and Chemistry, 6*(1), 24. https://doi.org/10.1186/s41181-021-00139-8

Ghosh, K. K., Padmanabhan, P., Yang, C.-T., Ng, D. C. E., Palanivel, M., Mishra, S., Halldin, C., & Gulyás, B. (2022). Positron emission tomographic imaging in drug discovery. *Drug Discovery Today, 27*(1), 280–291. https://doi.org/10.1016/j.drudis.2021.07.025

Giglio, J., Rosas, G., Basso, M., Boné, A., Savio, E., & Engler, H. (2018). An alternative methodology for the determination of the radiochemical purity of ^{11}C-methionine. *EJNMMI Radiopharmacy and Chemistry, 3*(1), 15. https://doi.org/10.1186/s41181-018-0053-0

Han, M., Bushong, E. A., Segawa, M., Tiard, A., Wong, A., Brady, M. R., Momcilovic, M., et al. (2023). Spatial mapping of mitochondrial networks and bioenergetics in lung cancer. *Nature, 615*(7953), 712–719. https://doi.org/10.1038/s41586-023-05793-3

Hart, T., Schultz, V. L., Thomas, D., III, Kulesza, T., & Jensen, K. F. (2020). Development of a versatile modular flow chemistry benchtop system. *Organic Process Research & Development, 24*(10), 2105–2112. https://doi.org/10.1021/acs.oprd.0c00160

He, Y., Schild, M., Grether, U., Benz, J., Leibrock, L., Heer, D., Topp, A., et al. (2022). Development of high brain-penetrant and reversible monoacylglycerol lipase PET tracers for neuroimaging. *Journal of Medicinal Chemistry, 65*(3), 2191–2207. https://doi.org/10.1021/acs.jmedchem.1c01706

Jayatunga, M. K. P., Xie, W., Ruder, L., Schulze, U., & Meier, C. (2022). AI in small-molecule drug discovery: A coming wave? *Nature Reviews Drug Discovery, 21*(3), 175–176. https://doi.org/10.1038/d41573-022-00025-1

Jones, J., Do, V., Lu, Y., & van Dam, R. M. (2023). Accelerating radiochemistry development: Automated robotic platform for performing up to 64 droplet radiochemical reactions in a morning. *Chemical Engineering Journal, 468*, 143524. https://doi.org/10.1016/j.cej.2023.143524

Klenner, M. A., Zhang, B., Ciancaleoni, G., Howard, J. K., Maynard-Casely, H. E., Clegg, J. K., Massi, M., Fraser, B. H., & Pascali, G. (2020a). Rhenium(i) complexation–dissociation strategy for synthesising fluorine-18 labelled pyridine bidentate radiotracers. *RSC Advances, 10*(15), 8853–8865. https://doi.org/10.1039/D0RA00318B

Klenner, M. A., Fraser, B. H., Moon, V., Evans, B. J., Massi, M., & Pascali, G. (2020b). Telescoping the synthesis of the [^{18}F]CABS13 Alzheimer's disease radiopharmaceutical via flow microfluidic rhenium(I) complexations. *European Journal of Inorganic Chemistry, 2020*(37), 3554–3564. https://doi.org/10.1002/ejic.202000433

Kryza, D., & Janier, M. (2013). Radio-UHPLC: A tool for rapidly determining the radiochemical purity of technetium-99m radiopharmaceuticals? *Applied Radiation and Isotopes, 78*, 72–76. https://doi.org/10.1016/j.apradiso.2013.04.005

Laferriere-Holloway, T. S., Rios, A., Lu, Y., Okoro, C. C., & van Dam, R. M. (2023). A rapid and systematic approach for the optimization of radio thin-layer chromatography resolution. *Journal of Chromatography A, 1687*, 463656. https://doi.org/10.1016/j.chroma.2022.463656

Lapi, S. E., Scott, P. J. H., Scott, A. M., Windhorst, A. D., Zeglis, B. M., Abdel-Wahab, M., Baum, R. P., et al. (2024). Recent advances and impending challenges for the radiopharmaceutical sciences in oncology. *The Lancet Oncology, 25*(6), e236–e249. https://doi.org/10.1016/S1470-2045(24)00030-5

Lin, M., Pham, C., Mukhopadhyay, U., & Manning, H. C. (2023). CGMP compliant one-step, one-pot automated [^{18}F]FBnTP production for clinical imaging of mitochondrial activity. *Journal of Nuclear Medicine, 64*(Suppl 1), P22.

Lu, Y., He, Y., Schibli, R., Linjing, M., & van Dam, R. M. (2023a). Proof-of-concept optimization of a copper-mediated ^{18}F-radiosynthesis of a novel MAGL PET tracer on a high-throughput microdroplet platform and its macroscale translation. *Lab on a Chip, 23*(21), 4652–4663. https://doi.org/10.1039/D3LC00735A

Lu, Y., Sharma, G., Da Pieve, C., Kramer-Marek, G., Miller, P., & van Dam, R. M. (2023b). Proof of principle: [^{18}F]AlF radiolabelling using a droplet microreactor. *Nuclear Medicine and Biology, 126–127*(November), 108758. https://doi.org/10.1016/j.nucmedbio.2023.108758

Lu, Y., Chen, K., & van Dam, R. M. (2024a). Rapid, high-yield radiosynthesis of 2′-Deoxy-2′-[^{18}F]Fluoro-5-Methyl-1-β-D-Arabinofuranosyluracil ([^{18}F]FMAU) in a microdroplet reactor. *Journal of Nuclear Medicine, 65*(Suppl 2), 241087–241087.

Lu, Y., Collins, J., Lin, K.-S., & van Dam, R. M. (2024b). Scalable droplet-based radiosynthesis of [^{18}F]fluorobenzyltriphenylphosphonium cation ([^{18}F]FBnTP) via a 'numbering up' approach. *Lab on a Chip, 24*(4), 728–737. https://doi.org/10.1039/D3LC01068F

Macarron, R., Banks, M. N., Bojanic, D., Burns, D. J., Cirovic, D. A., Garyantes, T., Green, D. V. S., et al. (2011). Impact of high-throughput screening in biomedical research. *Nature Reviews Drug Discovery, 10*(3), 188–195. https://doi.org/10.1038/nrd3368

Mahjour, B., Shen, Y., & Cernak, T. (2021). Ultrahigh-throughput experimentation for information-rich chemical synthesis. *Accounts of Chemical Research, 54*(10), 2337–2346. https://doi.org/10.1021/acs.accounts.1c00119

Makaravage, K. J., Brooks, A. F., Mossine, A. V., Sanford, M. S., & Scott, P. J. H. (2016). Copper-mediated radiofluorination of arylstannanes with [^{18}F]KF. *Organic Letters, 18*(20), 5440–5443. https://doi.org/10.1021/acs.orglett.6b02911

Mandap, K. S., Ido, T., Kiyono, Y., Kobayashi, M., Lohith, T. G., Mori, T., Kasamatsu, S., Kudo, T., Okazawa, H., & Fujibayashi, Y. (2009). Development of microwave-based automated nucleophilic [^{18}F]fluorination system and its application to the production of [^{18}F]flumazenil. *Nuclear Medicine and Biology, 36*(4), 403–409. https://doi.org/10.1016/j.nucmedbio.2009.01.011

Massaweh, G., Schirrmacher, E., la Fougere, C., Kovacevic, M., Wängler, C., Jolly, D., Gravel, P., Reader, A. J., Thiel, A., & Schirrmacher, R. (2009). Improved work-up procedure for the production of [^{18}F]flumazenil and first results of its use with a high-resolution research tomograph in human stroke. *Nuclear Medicine and Biology, 36*(7), 721–727. https://doi.org/10.1016/j.nucmedbio.2009.05.008

Matesic, L., Kallinen, A., Greguric, I., & Pascali, G. (2017). Dose-on-demand production of diverse ^{18}F-radiotracers for preclinical applications using a continuous flow microfluidic system. *Nuclear Medicine and Biology, 52*, 24–31. https://doi.org/10.1016/j.nucmedbio.2017.05.004

Momcilovic, M., Jones, A., Bailey, S. T., Waldmann, C. M., Li, R., Lee, J. T., Abdelhady, G., et al. (2019). In vivo imaging of mitochondrial membrane potential in non-small-cell lung cancer. *Nature, 575*(7782), 380–384. https://doi.org/10.1038/s41586-019-1715-0

Mossine, A. V., Brooks, A. F., Makaravage, K. J., Miller, J. M., Ichiishi, N., Sanford, M. S., & Scott, P. J. H. (2015). Synthesis of [^{18}F]arenes via the copper-mediated [^{18}F]fluorination of boronic acids. *Organic Letters, 17*(23), 5780–5783. https://doi.org/10.1021/acs.orglett.5b02875

Mossine, A. V., Brooks, A. F., Bernard-Gauthier, V., Bailey, J. J., Ichiishi, N., Schirrmacher, R., Sanford, M. S., & Scott, P. J. H. (2018). Automated synthesis of PET radiotracers by copper-mediated ^{18}F-fluorination of organoborons: Importance of the order of addition and competing protodeborylation. *Journal of Labelled Compounds and Radiopharmaceuticals, 61*(3), 228–236. https://doi.org/10.1002/jlcr.3583

Nambiar, A. M. K., Breen, C. P., Hart, T., Kulesza, T., Jamison, T. F., & Jensen, K. F. (2022). Bayesian optimization of computer-proposed multistep synthetic routes on an automated robotic flow platform. *ACS Central Science, 8*(6), 825–836. https://doi.org/10.1021/acscentsci.2c00207

Nasirzadeh, M., Vaulina, D. D., Kuznetsova, O. F., & Gomzina, N. A. (2016). A novel approach to the synthesis of [^{18}F]flumazenil, a radioligand for PET imaging of central benzodiazepine receptors. *Russian Chemical Bulletin, 65*(3), 794–800. https://doi.org/10.1007/s11172-016-1376-1

Pascali, G., Matesic, L., Collier, T. L., Wyatt, N., Fraser, B. H., Pham, T. Q., Salvadori, P. A., & Greguric, I. (2014). Optimization of nucleophilic ^{18}F radiofluorinations using a microfluidic reaction approach. *Nature Protocols, 9*(9), 2017–2029. https://doi.org/10.1038/nprot.2014.137

Preshlock, S., Calderwood, S., Verhoog, S., Tredwell, M., Huiban, M., Hienzsch, A., Gruber, S., et al. (2016). Enhanced copper-mediated ^{18}F-fluorination of aryl boronic esters provides eight radiotracers for PET applications. *Chemical Communications, 52*(54), 8361–8364. https://doi.org/10.1039/C6CC03295H

Rezes, R., Csoma, S. L., Németh, E., & Jószai, I. (2020). Fast liquid chromatographic determination of radiochemical and chemical purity of [^{11}C]methionine by UPLC technique. *Journal of Radioanalytical and Nuclear Chemistry, 324*(3), 1237–1244. https://doi.org/10.1007/s10967-020-07179-5

Rios, A., Wang, J., Chao, P. H., & van Dam, R. M. (2019). A novel multi-reaction microdroplet platform for rapid radiochemistry optimization. *RSC Advances, 9*(35), 20370–20374. https://doi.org/10.1039/C9RA03639C

Rios, A., Holloway, T. S., Chao, P. H., De Caro, C., Okoro, C. C., & van Dam, R. M. (2022). Microliter-scale reaction arrays for economical high-throughput experimentation in radiochemistry. *Scientific Reports, 12*(1), 10263. https://doi.org/10.1038/s41598-022-14022-2

Ryzhikov, N. N., Seneca, N., Krasikova, R. N., Gomzina, N. A., Shchukin, E., Fedorova, O. S., Vassiliev, D. A., et al. (2005). Preparation of highly specific radioactivity [^{18}F]flumazenil and its evaluation in cynomolgus monkey by positron emission tomography. *Nuclear Medicine and Biology, 32*(2), 109–116. https://doi.org/10.1016/j.nucmedbio.2004.11.001

Saboury, B., Bradshaw, T., Boellaard, R., Buvat, I., Dutta, J., Hatt, M., Jha, A. K., et al. (2023). Artificial intelligence in nuclear medicine: Opportunities, challenges, and responsibilities toward a trustworthy ecosystem. *Journal of Nuclear Medicine, 64*(2), 188–196. https://doi.org/10.2967/jnumed.121.263703

Saebi, M., Nan, B., Herr, J. E., Wahlers, J., Guo, Z., Zurański, A. M., Kogej, T., et al. (2023). On the use of real-world datasets for reaction yield prediction. *Chemical Science, 14*(19), 4997–5005. https://doi.org/10.1039/D2SC06041H

Salley, D., Sebastián Manzano, J., Kitson, P. J., & Cronin, L. (2023). Robotic modules for the programmable chemputation of molecules and materials. *ACS Central Science, 9*(8), 1525–1537. https://doi.org/10.1021/acscentsci.3c00304

Schmitt, C., Fouque, J., Huguet, S., Da Costa, E., Branquinho, S. B., Rezai, K., & Madar, O. (2021). Single radio UHPLC analysis for the quality control of technetium-99m radiolabelled radiopharmaceuticals. *Applied Radiation and Isotopes, 176*(October), 109874. https://doi.org/10.1016/j.apradiso.2021.109874

Scott, A. M., Zeglis, B. M., Lapi, S. E., Scott, P. J. H., Windhorst, A. D., Abdel-Wahab, M., Giammarile, F., et al. (2024). Trends in nuclear medicine and the radiopharmaceutical sciences in oncology: Workforce challenges and training in the age of theranostics. *The Lancet Oncology, 25*(6), e250–e259. https://doi.org/10.1016/S1470-2045(24)00037-8

Soritz, S., Moser, D., & Gruber-Wölfler, H. (2022). Comparison of derivative-free algorithms for their applicability in self-optimization of chemical processes. *Chemistry–Methods, 2*(5), e202100091. https://doi.org/10.1002/cmtd.202100091

Steiner, S., Wolf, J., Glatzel, S., Andreou, A., Granda, J. M., Keenan, G., Hinkley, T., et al. (2019). Organic synthesis in a modular robotic system driven by a chemical programming language. *Science, 363*(6423), eaav2211. https://doi.org/10.1126/science.aav2211

Struble, T. J., Alvarez, J. C., Brown, S. P., Chytil, M., Cisar, J., DesJarlais, R. L., Engkvist, O., et al. (2020). Current and future roles of artificial intelligence in medicinal chemistry synthesis. *Journal of Medicinal Chemistry, 63*(16), 8667–8682. https://doi.org/10.1021/acs.jmedchem.9b02120

Syrris. (n.d.). *Asia flow chemistry system*. Syrris. Retrieved January 17, 2024, from https://www.syrris.com/product/asia-flow-chemistry/.

Taylor, C. J., Pomberger, A., Felton, K. C., Grainger, R., Barecka, M., Chamberlain, T. W., Bourne, R. A., Johnson, C. N., & Lapkin, A. A. (2023). A brief introduction to chemical reaction optimization. *Chemical Reviews, 123*(6), 3089–3126. https://doi.org/10.1021/acs.chemrev.2c00798

Tredwell, M., Preshlock, S. M., Taylor, N. J., Gruber, S., Huiban, M., Passchier, J., Mercier, J., Génicot, C., & Gouverneur, V. (2014). A general copper-mediated nucleophilic ^{18}F fluorination of arenes. *Angewandte Chemie International Edition, 53*(30), 7751–7755. https://doi.org/10.1002/anie.201404436

Tshibangu, T., Cawthorne, C., Serdons, K., Pauwels, E., Gsell, W., Bormans, G., Deroose, C. M., & Cleeren, F. (2020). Automated GMP compliant production of [^{18}F]AlF-NOTA-octreotide. *EJNMMI Radiopharmacy and Chemistry, 5*(1), 4. https://doi.org/10.1186/s41181-019-0084-1

Volk, A. A., Epps, R. W., Yonemoto, D. T., Masters, B. S., Castellano, F. N., Reyes, K. G., & Abolhasani, M. (2023). AlphaFlow: Autonomous discovery and optimization of multistep chemistry using a self-driven fluidic lab guided by reinforcement learning. *Nature Communications, 14*(1), 1403. https://doi.org/10.1038/s41467-023-37139-y

Wang, J., & van Dam, R. M. (2020). High-efficiency production of radiopharmaceuticals via droplet radiochemistry: A review of recent Progress. *Molecular Imaging, 19*, 1–21. https://doi.org/10.1177/1536012120973099

Wang, J., Chao, P. H., Slavik, R., & van Dam, R. M. (2020a). Multi-GBq production of the radiotracer [^{18}F]fallypride in a droplet microreactor. *RSC Advances, 10*(13), 7828–7838. https://doi.org/10.1039/D0RA01212B

Wang, J., Rios, A., Lisova, K., Slavik, R., Chatziioannou, A. F., & van Dam, R. M. (2020b). High-throughput radio-TLC analysis. *Nuclear Medicine and Biology, 82–83*, 41–48. https://doi.org/10.1016/j.nucmedbio.2019.12.003

Webb, E. W., & Scott, P. J. H. (2021). Potential applications of artificial intelligence and machine learning in radiochemistry and radiochemical engineering. *PET Clinics, 16*(4), 525–532. https://doi.org/10.1016/j.cpet.2021.06.012

Webb, E. W., Cheng, K., Winton, W. P., Klein, B. J. C., Bowden, G. D., Mami Horikawa, S., Liu, W., et al. (2024). Development of high-throughput experimentation approaches for rapid radiochemical exploration. *Journal of the American Chemical Society, 146*(15), 10581–10590. https://doi.org/10.1021/jacs.3c14822

Werner, R. A., Chen, X., Rowe, S. P., Lapa, C., Javadi, M. S., & Higuchi, T. (2019). Moving into the next era of PET myocardial perfusion imaging: Introduction of novel ^{18}F-labeled tracers. *The International Journal of Cardiovascular Imaging, 35*(3), 569–577. https://doi.org/10.1007/s10554-018-1469-z

Zhang, X., Dunlow, R., Blackman, B. N., & Swenson, R. E. (2018). Optimization of ^{18}F-syntheses using ^{19}F-reagents at tracer-level concentrations and liquid chromatography/tandem mass spectrometry analysis: Improved synthesis of [^{18}F]MDL100907. *Journal of Labelled Compounds and Radiopharmaceuticals, 61*(5), 427–437. https://doi.org/10.1002/jlcr.3606

Zhang, B., Fraser, B. H., Klenner, M. A., Chen, Z., Liang, S. H., Massi, M., Robinson, A. J., & Pascali, G. (2019). [^{18}F]Ethenesulfonyl fluoride as a practical radiofluoride relay reagent. *Chemistry, 25*(32), 7613–7617. https://doi.org/10.1002/chem.201900930

Zischler, J., Kolks, N., Modemann, D., Neumaier, B., & Zlatopolskiy, B. D. (2017). Alcohol-enhanced Cu-mediated radiofluorination. *Chemistry, 23*(14), 3251–3256. https://doi.org/10.1002/chem.201604633

Zlatopolskiy, B. D., Zischler, J., Krapf, P., Zarrad, F., Urusova, E. A., Kordys, E., Endepols, H., & Neumaier, B. (2015). Copper-mediated aromatic radiofluorination revisited: Efficient production of PET tracers on a preparative scale. *Chemistry, 21*(15), 5972–5979. https://doi.org/10.1002/chem.201405586

The Application of Statistical "Design of Experiments" (DoE) Towards the Efficient Study and Optimization of Radiochemical Processes

Gregory D. Bowden and Andreas Maurer

1 Optimization and Experimental Design

1.1 The Importance of Optimization

Nuclear medicine is a discipline of ever-increasing clinical relevance, and thus, the demand for imaging and therapeutic agents continues to expand rapidly. However, the production of radiopharmaceuticals under "Good Manufacturing Practice" (GMP) conditions is expensive and requires a relatively high investment in terms of infrastructure and personnel. With the growing clinical demand for new and established radiopharmaceuticals, the economic and scientific efficiency at which radiopharmaceuticals can be developed, evaluated, and manufactured will be critical to the future of nuclear medicine. Furthermore, with the limited capacity of GMP radiopharmacies to produce prospective radiotracers for clinical translation, the establishment of high-yielding and "developable" radiosyntheses from early in the tracer development pipeline is crucial for bringing new radiotracers from bench to bedside as quickly as possible.

An important historical example of how "developability and synthesizability" can affect the adoption and success of a clinical tracer is the synthesis of 2-[^{18}F]fluoro-2-deoxy-D-glucose ([^{18}F]FDG.) When it was first developed, [^{18}F]FDG was produced from a technically challenging electrophilic reaction with highly toxic elemental [^{18}F]F_2 that afforded the radiotracer in quantities sufficient for

G. D. Bowden (✉) · A. Maurer
Werner Siemens Imaging Center, Department of Preclinical Imaging and Radiopharmacy, Eberhard Karls University, Tuebingen, Germany
Cluster of Excellence iFIT (EXC 2180) "Image-Guided and Functionally Instructed Tumor Therapies", Eberhard Karls University, Tuebingen, Germany
e-mail: g.bowden@hzdr.de

R. M. van Dam et al. (eds.), *Automated Technologies for the Development and Production of Radiopharmaceuticals*,
https://doi.org/10.1007/978-3-031-84632-8_19

first-in-human studies but not large-scale production (Fowler, 2020). However, as the clinical importance of [^{18}F]FDG became apparent and demand for the tracer grew, a high-yielding, high specific activity, and automatable synthesis of [^{18}F]FDG became desperately needed. This breakthrough came in 1986 with the development of a no-carrier-added stereospecific synthesis of [^{18}F]FDG from nucleophilic [^{18}F]fluoride that produced the radiotracer in the excellent yields required for mass production and widespread clinical application (Hamacher et al., 1986). [^{18}F]FDG has since become the workhorse of nuclear medicine and has potentially saved millions of lives. The modern synthesis of [^{18}F]FDG serves as a strong argument as to why the development of optimized and efficient radiosyntheses for novel radiotracers is important to the patient. The sooner a promising tracer can be brought into routine GMP production for clinical application, the more significant its potential impact.

Chemistry and process optimization are thus essential activities within the fields of radiochemistry and radiopharmacy. Radiochemists use optimization studies to characterize novel reaction methodologies, fine-tune purification protocols, and optimize radiosyntheses for performance, reliability, and efficiency. In radiochemical methodology development, optimization studies are performed to identify the critical experimental variables (also termed factors), study the scope and limitations of a reaction, and learn how to best manipulate the reaction conditions. As these new radiochemical methodologies are adapted into routine use, they must be optimized to suit specific radiochemical problems. Useful radiochemical reactions are ideally widely applicable and tolerant to a broad set of substrates; however, in reality, a general set of optimal conditions that are appropriate for every new radiopharmaceutical precursor seldom exist. The synthesis of every new radiopharmaceutical thus represents a unique set of radiochemical, synthetic, purification, and formulation optimization challenges that need to be overcome before the process can be effectively implemented for production at scale.

Unfortunately, traditional process and chemistry optimization methods are notoriously time and resource-intensive tasks that are further complicated by the practical implications imposed by using short-lived radionuclides such as ^{11}C (20 min) and ^{18}F (110 min). The short half-lives of many clinically relevant radionuclides mean that the concepts of radiosynthesis time and yield are related. Higher activity yields (%AY) are a function of both the radiolabeling efficiency (the decay corrected radiochemical yield, RCY%) and the duration of the synthesis from the delivery of the radioactivity. This necessitates that high-yielding radiotracer productions be both efficient and fast; thus, the process needs to be well optimized with respect to both the radiolabeling and deprotection chemistry and the synthesis time. For more complex, multistep radiosynthetic processes, time is often a factor that cannot be easily controlled. For example, it may take several minutes to transfer a solution over a solid phase extraction (SPE) cartridge or to properly separate certain chemical or radiochemical impurities from the product via semi-preparative HPLC. The chemical performance of a given radiochemical reaction is thus often the focus of most optimization efforts, as a well-optimized radiolabeling reaction provides an increased margin of error in later steps, which improves the overall robustness and reliability of the whole radiosynthetic process.

In addition to radiosynthesis yield and robustness, proper optimization can affect tracer quality and quality control. Tracer quality is a crucial consideration for new radiotracers for preclinical experiments (from a perspective of good scientific practice), clinical trials, and, later, for routine clinical use (from the standpoint of patient benefit and safety). Additionally, many radiotracers are "made-to-order," often for a single dose, and thus quality control, which must take place after every synthesis, represents a significant source of cost, time, and practical effort. Changes in the batch-to-batch production quality of a product stem from controllable and uncontrollable variances in the manufacturing process. The "Quality by design" (QbD) approach to product development aims to ensure quality by identifying, understanding, and managing all the sources of variance critical to meeting a predefined quality standard (Yu et al., 2014). In the context of radiopharmaceutical development, applying QbD principles through rigorous optimization allows one to establish a solid scientific basis from which to build quality into the manufacturing process from the beginning of the development pipeline, saving significant time and costs later on.

Process optimization is a vital yet often superficially considered activity in radiopharmaceutical research and development. However, good optimization data allow us to study and better exploit new radiochemical methodologies and reliably produce impactful radiopharmaceuticals efficiently and with good quality. However, the traditional methods of process optimizations are inherently flawed, time-consuming, and resource-intensive. The remainder of this chapter will discuss the benefits and drawbacks of both conventional and alternative statistical approaches to process optimization within the context of novel radiopharmaceutical and radiosynthesis development. It will also discuss the basic theory of statistical experimental design and how it can be effectively implemented into routine use. Finally, the chapter will conclude with a review of how experimental design has been implemented in radiopharmaceutical development efforts and what potential future impact these approaches may have on the field.

1.2 Optimization Studies

The outputs (responses, the dependent variables, Y_i) of any process are a function of any number (k) of input factors. These input factors (the independent variables) feature both controllable (X_k) and uncontrollable factors (U_k), and the effects of these factors combine and interact to influence the response in various ways. The relationship between the input factors and the response can be represented as a "response surface" that can be used to describe the behavior of the process in multidimensional experimental space (Fig. 1).

Traditionally, most radiochemical optimization studies are carried out using a "one-variable-at-a-time" (OVAT) approach, whereby a single experimental factor

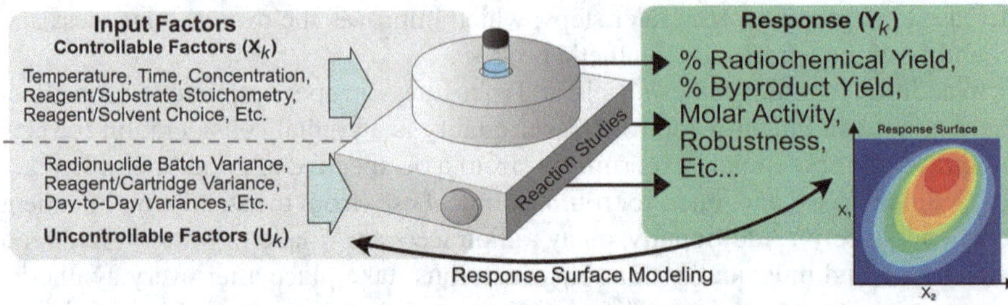

Fig. 1 The response of a given process (*Y*) can be affected by *k* input factors, which are comprised of both controllable (*X*) and uncontrollable variables (*U*). Response surface modeling is a method for analyzing the relationship between multiple input variables and the responses

(X_i) is adjusted over a range of settings while all other controllable variables are held constant (Murray et al., 2016). Once an optimum has been identified, the first variable is held constant at its optimal setting while the next variable is investigated. This process is repeated one variable at a time until all variables of interest have been evaluated. This approach is intuitive, and experiments can be easily adjusted "on-the-fly" to suit new data or the experimenter's intuition; thus, it is widely used in academic radiopharmaceutical laboratories for the study and optimization of both batch and microfluidic reaction setups (Pascali et al., 2014; Rios et al., 2022). However, the OVAT approach is inefficient, laborious, and resource-intensive. Additionally, it provides incomplete information about the system under investigation, while the results can be difficult to interpret correctly (Lendrem et al., 2015). As only one variable is investigated at a time, OVAT is unable to identify or quantify factor interactions (when the setting of one factor affects the behavior of another), and, as a result, OVAT studies are heavily dependent on arbitrarily chosen starting conditions and the order in which the factors are studied (Tye, 2004). This approach is thus prone to finding local optima, often missing the "true" set of optimal conditions entirely (Fig. 2a). Moreover, good experimental practice demands that replicate experiments be performed to measure experimental error, and thus, over many variables, the number of individual experimental runs can quickly become impractical. Ultimately, the OVAT approach results in a low-resolution unidimensional snapshot of each variable's behavior at a relatively high experimental cost, which, in the context of radiochemistry, can translate to the excessive use of expensive reagents, kits, machine time, and an unnecessarily high radiation dose to the radiochemist.

"Design of Experiments" (DoE, also known as factorial experimental design) is an alternative systematic and statistics-based approach to the optimization study that is seen as standard practice by process chemists and engineers across a diverse spectrum of industries. However, its use in academic research laboratories has been limited, likely due to its use of statistics and perceived "steep learning curve" (Tye, 2004). (It is one of the goals of this chapter to dispel this myth. While the statistical and mathematical methods used by DoE have posed a significant barrier to entry for many classically trained chemists, several convenient and well-designed software

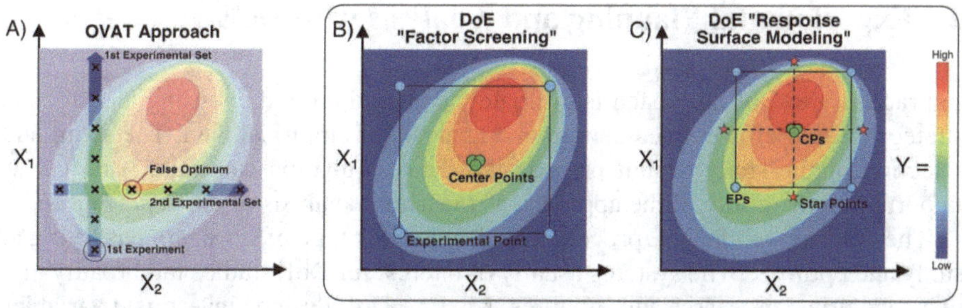

Fig. 2 (**a**) The OVAT approach studies one variable at a time. In this example, X_1 is the first variable to be investigated, while X_2 is held at an arbitrarily chosen value. The arbitrarily chosen order in which these experiments are conducted leads to the identification of a false optimum. The DoE Approach aims to explore a larger region of the response surface by conducting experiments throughout the defined experimental space. DoE studies typically take place in phases. (**b**) Factor screening: "Low resolution" study designs are first used to screen as many variables as possible to determine which have the greatest effects on the response. (**c**) Response surface modeling (RSM): Once the most important factor effects have been identified, more focused study designs are used to build an accurate and predictive high-resolution model of the response surface so that optimal conditions can be identified. (Figure modified from Bowden et al. (2019) and Murray (2019))

packages are available that feature user-friendly wizards and guides to assist one through the DoE design and analysis process.) DoE aims to extract as much information from a system as possible by statistically modeling the response surface using a limited number of individual experimental runs (Fig. 2b, c). This is typically achieved by using predefined experimental matrices or "designs" to select an optimal set of experiments that aim to investigate the effects of multiple variables simultaneously. After the set of empirical data has been acquired, multivariate analysis methods, e.g., multiple linear regression (MLR), are used to quantitatively determine which variables have significant effects on the measured responses, determine if any factor interactions may be present, and subsequently generate a predictive model of the response surface. Additionally, DoE studies can be constructed in "blocks" to better quantify and understand uncontrollable and hidden variables, such as day-to-day variances in reagent or activity batches, operators, laboratory conditions, different equipment, etc. Moreover, a global estimation of the error in the model can be easily obtained without replicating experiments at every experimental point. Once a reasonable model has been fitted, model statistics can help identify outliers in the data that can be repeated or excluded if need be. DoE holds significant advantages over the OVAT approach; however, DoE studies need to be designed and set up appropriately to ensure that meaningful and useful data is acquired to address the scientific question at hand adequately. Arguably, the most important part of the DoE process occurs before a single experiment is conducted.

2 Experimental Planning and Analysis with DoE

In practice, the DoE approach is ultimately centered around choosing an effective strategy for the efficient acquisition and analysis of empirical data. The setup and selection of the study design precede the data acquisition step, and it is just as important to the utility of the approach as the actual analysis of the data (Fig. 3).

The first step of the DoE process is to decide what type of system one is trying to study and optimize. The systems usually of interest for DoE studies fall broadly into two categories: processes and mixtures. Mixture DoE designs investigate variables whose settings are interdependent, specifically, the ratios of ingredients in a mixture or formulation (i.e., if one uses more of one component, one is forced to use less of another). The nature of most radiopharmaceutical formulations means that there are not many use cases for mixture designs; thus, they will not be discussed further here. On the other hand, process designs study variables whose settings are independent of each other (i.e., temperature and reagent quantity can be varied independently). Process DoE designs are typically of greater importance to most radiochemical workflows and thus will be the focus of the rest of this chapter; however, the specific choice of design is highly dependent on the question being asked and the resources at hand.

2.1 The Design Objective

The setup and choice of a DoE design are usually made with a specific design objective in mind. The need for a well-defined design objective comes from the fact that DoE designs are usually built to balance the amount and detail of the obtained information with the number of experimental runs. Therefore, choosing an experimental design and design space appropriate to the scientific question is vitally

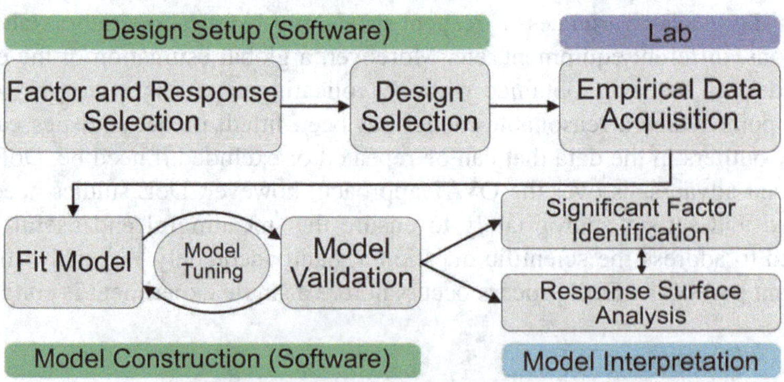

Fig. 3 The DoE process, from experimental design to model construction and interpretation, will be covered in this chapter

important. DoE studies are often carried out in phases to address different design objectives. First, a "factor screening" study is performed to cover as many variables (controllable and uncontrollable) as efficiently as possible with the aim of identifying significant factors of interest (those that have a statistically significant influence on the response over the given factor range) while eliminating non-significant factors from future studies (Fig. 2b). These studies are designed to be as broad as possible while keeping the number of experiments low; thus, they often do not provide sufficient data to build a detailed understanding of the system. (In many cases, however, they can be "good enough" for some experimental aims.) Once the most significant variables have been identified, the second phase of the DoE process, "response surface modeling" (RSM), aims to model the response surface using a more focused and detailed experimental design (Fig. 2c). Once empirically validated, the response surface model can be used to estimate a set of optimal conditions, guide further optimization studies, or aid in practical decision-making.

2.2 *Defining the Design Space (Choosing Variables)*

The DoE process starts with the selection of response variables, i.e., the process outcomes being measured (e.g., RCY, molar activity, byproduct yield, etc.). Importantly, one needs to consider not only "What do I want to measure?" but also "How do I want to measure it?" Any analytical method used for DoE should be definitive, quick, and as operationally simple as possible. The analysis itself can add significantly to the experimental effort, and laborious protocols for measuring the response can be problematic, especially if multiple responses are of interest. This is particularly important in the case of experiments involving short-lived radioisotopes, where radioactive decay can reduce the signal-to-noise ratios of certain measurements.

The choice of response will also influence the experimental variables one may want to manipulate and, thus, the experimental setup itself. This includes the definition of the design space, which is the combination of all the factor ranges across all the factors (both quantitative and qualitative) under investigation. DoE studies are constructed to include experimental points (Eps) that are spread out through multidimensional design space in such a way as to maximize the coverage of the region of interest while providing an optimal dataset for the construction of a statistical response surface model. The error in the system is calculated from a set of replicate center points (CPs) located at the center of the design space (Fig. 4a). Together, this predefined set of experiments is termed the experimental design (also termed experimental matrix). More experiments within the design generally translate to a more detailed model at a greater experimental cost.

The choice of the factors and their ranges (the minimum and maximum settings of each variable) is heavily dependent on the objective of the study (Sect. 2.1) and the nature of the chemical system and experimental setup; however, in all cases, they are critically important as they define the design space under investigation. Any

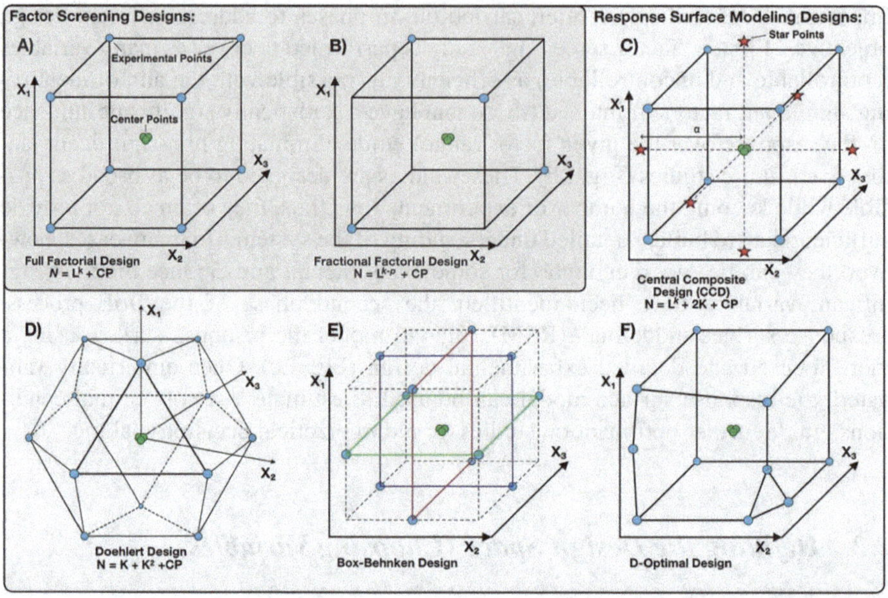

Fig. 4 Graphical representations of several commonly used experimental designs ($k = 3$, where k is the number of parameters under investigation): (**a**) Full factorial design with each variable measured at two settings, (**b**) 2_{III}^{3-1} ($k = 3$, and the design generator $p = 1$) fractional factorial design, (**c**) central composite design, (**d**) Doehlert matrix, (**e**) Box-Behnken design, (**f**) D-optimal design. Each design is composed of a combination of experimental points (EP, blue circles), center points (CP, green circles), or star points (red stars)

statistical conclusions can only be drawn within the bounds of the defined factor ranges. Ideally, the process should still work at all points in the experiment set, as any “failed” experiments effectively translate to lost information. The selection of the factor ranges can be assisted by pilot experiments, literature research, the chemist’s intuition, or any combination of these. Additionally, extreme experimental factor ranges should be avoided in lower-resolution designs where useful data points could be missed. On the other hand, excessively narrow factor ranges may make it difficult to model important trends in the response surface, such as response surface curvature.

In most cases, variables can be classified as either quantitative (continuous) or qualitative (categorical, discrete). Continuous variables have both a scale and a direction. Continuous factors thus have an infinite number of possibly optimal settings that need to be modeled using linear regression. Qualitative variables, however, will typically have multiple settings that do not stand in relation to each other, e.g., argon or air atmosphere, ethanol or acetonitrile as the solvent. These variables are, therefore, not modelled linearly but as a comparison between the settings. It is thus often easier to make decisions about categorical factors as, usually, only one setting is optimal or the factor is non-significant. While most classical response surface optimization DoE designs are very good at handling continuous process variables such as temperature or reagent amount, the introduction of many categorical variables with more than two settings may be problematic, requiring many experiments to achieve a reasonable model. Care should thus be taken during the factor screening phase of the DoE process to screen out as many non-significant categorical variables as possible. Dimensional reduction techniques like principal component analysis (PCA) have been used to convert categorical variables with very large numbers of settings (e.g., the many possible solvents or Lewis bases) into continuous variables that can be modeled using DoE. The use of techniques is excellently described in the following papers (Murray et al., 2013, 2016); however, an in-depth discussion of these methods is beyond the scope of this chapter.

In some instances, it is difficult to identify all variables that may influence the system. This is particularly true in radiochemistry, where unavoidable systematic error is often introduced by factors beyond the experimenter's control, such as the use of different cyclotron targets (i.e., if an external production facility is used for isotope production), different batches of reagents or cartridges, and other day-to-day variations. Uncontrollable factors can often be reduced or eliminated through careful planning, e.g., by performing all experiments with a single batch of precursor or by using only one specific piece of equipment. However, this is often not practically feasible when many experiments need to be conducted either in rapid sequence or in parallel. "Design blocking" helps detect and quantify uncontrollable or hidden sources of variance by arranging experiments into "like" blocks. For example, radioactivity from cyclotron target A is suspected of giving different results relative to cyclotron target B, but the experimenter cannot choose which target is used on a given day. To find out, the DoE experiments can be randomized into separate blocks for each day of the experiment, with each block containing the same set of center points so that the blocks can be compared. If any significant differences between the blocks are detected, blocking factors, mathematical terms that correct for these deviations, can be introduced into the regression model.

2.3 Choosing a Design Structure

DoE methods make use of a multitude of "classical" experimental design structures as well as non-standard "optimal" designs. It would be impractical to cover all possible designs in detail here, and there are several excellent papers (Murray et al., 2016; Tye & Whittaker, 2004), books and e-books (Box et al., 2005; Dunn, 2018; NIST/SEMATECH, 2022), reviews (Dejaegher & vander Heyden, 2011; Leardi, 2009; Lendrem et al., 2015; Tye, 2004), and other resources available that discuss these topics in-depth (see section "Useful Resources"). However, it is important to have an overview of some of the designs that may be applicable to radiochemistry, their structures, and how and when to use them.

2.3.1 The Full Factorial Design

The full factorial design is a classical experimental matrix that serves as the mathematical basis for most of the classical DoE designs discussed in this chapter. A typical 2-setting full factorial design features experiments conducted at the highest (+1) and lowest (−1) values of each variable's predefined range in combination with all the other variable settings (excluding the center point setting (0)). The design space can thus be represented graphically as a hypercube of k dimensions (each dimension is a variable of interest); thus, a study with three variables is represented by a 3D cube (Fig. 4a) with an experiment point (EP) on each of its vertices. In addition to the experimental points, the design also features 3–5 replicate experiments termed center points (CP) positioned at the center of the design space. These are used to calculate the pure experimental error for the whole model as well as test for and model curvature in the response surface. Each variable is therefore studied at three settings $(-1, 0, +1)$, which provides enough information to model the behavior of all the main effects, all factor interactions, and estimate any curvature in the response surface. The number of single experiments in a full factorial design is given by: Runs $= L^k + \mathrm{CP}$, where L is the number of settings (for most fractional and full factorial designs, $L = 2$). Unfortunately, this means that as the number of variables increases, the number of required runs explodes exponentially, and thus, full factorial designs are impractical for screening or modelling higher numbers of factors ($k > 5$).

2.3.2 Factor Screening Designs

When setting out to optimize a new radiochemical reaction, the number of variables for consideration can quickly become overwhelming. Factor screening DoE studies are designed to rapidly and efficiently screen a broad selection of variables to gain a broad overview of how the process of interest behaves. Their purpose is to deliver a crude linear model of the system to determine which variables have significant effects on the process and where the design space should be focused in future more detailed experiments. Screening designs are typically constructed to be as economical as possible, often only modeling the main effects of the system. Many screening designs are available, such as the "Placket-Burnham designs" or "definitive screening designs," and they all have subtly different use cases. Fractional factorial designs are some of the most widely applicable screening designs available. Understanding how they are structured and applied is vital to understanding some of the decisions one needs to make when setting up any DoE study.

2.3.2.1 Fractional Factorial Designs

For the purposes of factor screening, a full factorial design is often excessive, and, in many cases, useful conclusions (such as an estimation of the main effects and some factor interactions) can still be drawn from a carefully selected fraction of the experiments prescribed by the full factorial design. Fractional factorial designs are, therefore, modified full factorial designs that aim to reduce the total number of experiments by excluding a certain fraction of the full factorial experimental points (Fig. 4b). Well-designed fractional factorial DoE studies are well suited to factor screening and, in the case of higher resolution designs, can provide an adequate model for limited response surface optimization.

The elimination of a particular fraction of experiments is done at the cost of design "resolution," whereby certain factor effects become "confounded" with each other. "Confounding" occurs when a factor effect or factor interaction effect is estimated by the same combination of experimental observations as another effect, producing a spurious relationship between the two (one cannot distinguish from which factor the effect comes from). These two effects can no longer be reliably estimated and are thus said to be confounded. However, if the confounded effects are non-significant or are unimportant to the question at hand, they can be considered an acceptable loss of information. The amount of confounding and how one introduces confounding into a design's structure are essential to the topic of fractional factorial experimental design.

Fractional factorial designs are generated by choosing a predefined aliasing (confounding) structure, which determines which factor effects will be confounded and what information (relative to a full factorial design) will be lost. Ultimately, the choice of aliasing structure is based on the number of factors under investigation, the desired number of experiments, and the design resolution (RES) required to meet the study's goals (Table 1). The number of non-center point experiments (N) in fraction factorial design is given by $N = L^{k-p}$ where L is the number of settings, k is the number of factors, and p is the number of generators used by the aliasing structure (Fig. 4b). Generators are mathematical expressions that are used to eliminate a certain fraction of runs from the full factorial design (see (Box et al., 2005; Dunn, 2018; NIST/SEMATECH, 2022) for how generators are derived). The generators also describe which factors will be confounded with each other. They are predefined by the user to ensure orthogonality and balance in the design. In practice, modern DoE software packages can help select an appropriate aliasing structure. Thus, a design investigating the effects of five factors over two settings with two generators will require $2^{5-2} = 8$ experiments, plus an additional 3 center points for a total of only 11 runs.

Design resolution (RES) is defined by the severity of confounding for a given fractional factorial design (Table 1). Higher resolution designs provide a "better" picture of a process's behavior but at a higher experimental cost. For a fractional factorial design to be useful, it is important that, at a minimum, the main effects are not confounded with each other and can, therefore, be reliably estimated (defined as a RES III design, see Table 1 for resolution definitions). Fractional factorial designs

Table 1 Fractional factorial designs balance the quality of information with increased resolution at the cost of more experimental effort

Number of Runs N \ Number of Factors k (Increasing Aliasing and Greater Loss of Information; More Factors Investigated →)	3	4	5	6	7	8	9	10	11
4 (↓ Increasing Experimental Cost; More Information (wasteful))	2^{3-1}_{III} (1/2)								
8	2^3 (Full)	2^{4-1}_{IV} (1/2)	2^{5-2}_{III} (1/4)	2^{6-3}_{III} (1/8)	2^{7-4}_{III} (1/16)				
16	2^3 (Full*2)	2^4 (Full)	2^{5-1}_{V} (1/2)	2^{6-2}_{IV} (1/4)	2^{7-3}_{IV} (1/8)	2^{8-4}_{IV} (1/16)	2^{9-5}_{III} (1/32)	2^{10-6}_{III} (1/64)	2^{11-7}_{III} (1/128)
32	2^3 (Full*4)	2^4 (Full*2)	2^5 (Full)	2^{6-1}_{VI} (1/2)	2^{7-2}_{IV} (1/4)	2^{8-3}_{IV} (1/8)	2^{9-4}_{IV} (1/16)	2^{10-5}_{IV} (1/32)	2^{11-6}_{IV} (1/64)
64	2^3 (Full*8)	2^4 (Full*4)	2^5 (Full*2)	2^6 (Full)	2^{7-1}_{VII} (1/2)	2^{8-2}_{V} (1/4)	2^{9-3}_{IV} (1/8)	2^{10-4}_{IV} (1/16)	2^{11-5}_{IV} (1/32)
128	2^3 (Full*16)	2^4 (Full*8)	2^5 (Full*4)	2^6 (Full*2)	2^7 (Full)	2^{8-1}_{VIII} (1/2)	2^{9-2}_{VI} (1/4)	2^{10-3}_{V} (1/8)	2^{11-4}_{V} (1/16)

RES <III:	Main effects are confoundedwith each other. Resolution too low to be useful.	RES III:	Main effects are not confounded.Main effects confounded with 2-factorinteractions. Useful for Initial factor screening
RES IV:	Main effects are not confoundedwith each other or with 2-factor interactions. 2-factorinteraction confoundedwith each other. Useful for detailed factor screening and low level optimization.	RES V (+)	Main effects are not counfoundedwith each other or 2-factor interactions. 2-factorinteractions not confoundedwith each other. Can perform optimization studies with large sets of factors. resolves complexeffects for RSO studies
Full	Full factorial design	Full	Full factorial with repetitions

Table adapted from "Statistics for Experimenters (Box et al., 2005) (2nd edition, p 272) and "Process Improvement Using Data" (Dunn, 2018) (Creative Commons Attribution-ShareAlike 4.0 Unported (CC BY-SA 4.0))

The number in parenthesis is the full factorial fraction. The legend defines the properties of RES III–V(+) fractional factorial designs

are denoted as L^{k-p}_{RES},and thus our 2-setting 5-factor 2-generator fractional factorial design is thus denoted as a 2^{5-2}_{III} design that can resolve all main effect terms and is thus a design that is economical and suitable for factor screening. Ultimately, it is up to the experimenter to decide which design resolution will achieve the required information at an acceptable cost.

2.3.3 Response Surface Modeling (RSM) and Optimization

Once factor screening has identified significant experiment factors, a "response surface modeling" (RSM) study can then be constructed to generate a predictive model of the response surface that incorporates main effects, factor interactions, and quadratic effects (curvature). RSM designs (also called second-order designs) are typically more focused on finding and modeling detail in the response surface and are thus less focused on experimental economy. In most cases, RSM designs require more experiments per factor; therefore, it is essential that these higher-resolution studies are performed after non-significant factors have been eliminated by appropriate factor screening studies. Two and three-setting full factorial designs, while inefficient at higher factor numbers, can be used to resolve all the model terms required to generate a response surface model. Well-chosen RES V+ fractional factorial designs are also capable of providing sufficient information for the construction of a reasonable response surface model. However, several design types are specifically suited to the task of response surface modeling and optimization once the number of factors has been reduced.

The Box-Behnken design (BBD, Fig. 4e) features experimental points arranged on the edges of the design space cube to inscribe a sphere around the design center. The BBD can also be considered a series of 2-factor full factorial designs that run perpendicular to each other in the design space. Thus, at low factor numbers, the BBD can build detailed quadratic models of a system while remaining experimentally efficient. The BBD is a particularly useful design when combinations of factor extremes should be avoided (to ensure that the process still works), as is often the case in many chemical systems. Doehlert designs are another commonly used RSM design (Fig. 4d). They are expandable designs that use a polyhedron design space built from a set of connectable hyper-triangles (called simplexes). At its simplest (2D) level, the design space is represented by a hexagon composed of six experimental points and two center points.

2.3.3.1 Central Composite Designs

One of the more widely used classical designs for RSM is the central composite design (CCD, Fig. 4c)). The CCD is constructed around a full factorial design with extra axial experimental points (also termed "star points") positioned on each of the cube's faces at a distance $\alpha \geq 1$ from the cube center (α is dependent on the number of factors and their ranges.) The value α is chosen to maintain the property of rotatability, where the variance at any point in the model is dependent only on its distance from the model center (where the center points estimate the true error.) Each variable is thus studied at five different settings ($-\alpha$, -1, 0, +1, $+\alpha$), making the CCD an excellent design choice for the detailed exploration of curvature in the response surface. There are three special cases of CCD, each with a different α value. A circumscribed CCD (CCCD) has $|\alpha| = (2^k)^{1/4}$ so that the combination of the factorial and star points circumscribe a hypersphere around the enclosed design space. The CCCD has the advantage that it maximizes the useful volume of the design space while remaining rotatable; however, this often means that the star points can be impractically far from the design center. The face-centered CCD (FCCD) addresses this problem by situating the star points on the faces of the factorial cube ($\alpha = \pm 1$), but this then means that each factor is only investigated at three settings (-1, 0, +1). The orthogonal CCD (OCCD), which uses an α value that is adjusted to the number of factors and factor ranges, is thus commonly recommended as a reasonable compromise between the CCCD and FCCD.

2.3.3.2 Optimal Designs (D-Optimal)

The classical DoE designs mentioned above are usually the go-to choice for most DoE studies as they are reliable, intuitive, and easy to implement with commonly available DoE software. However, they are not suitable for all design spaces. For example, all classical designs require that the design spaces are balanced regular (hyper)polyhedrons such as cubes, spheres, or simplexes, but in reality, factor ranges can often be constrained in such a way as to make the design space irregular. This is a common problem when formulation factors (that have an upper or lower limit, i.e., 0–100%) are investigated. For example, a central composite design may require that at setting $-\alpha$, a negative amount of a reagent or a negative volume of a solvent is used; this, of course, makes no practical sense. To get around this problem, one may adjust the factor ranges or shift the design space to eliminate the constraints; however, this is often not feasible or may unfavorably shift the region of investigation away from the region of interest.

Optimal designs were specifically developed to address some of these issues. Optimal designs are constructed algorithmically using computer software that tries to select a set of experiments that are optimal with respect to a certain statistical criterion (Fig. 4f). For example, in the case of the "D-optimal" design, the software will try to maximize the "Determinate" matrix of a candidate set of experiments chosen from all possible experiments within the defined design space. In simple terms, the algorithm will try to select experiments that maximize the volume of the design space, and the resulting model can be fitted regardless of the presence of design space irregularities. The one drawback of optimal designs is that they require one to make some assumptions about what terms the model will contain before the experiments are selected. These assumptions should be carefully considered when constructing the design space. However, in practice, this issue only becomes problematic when many factors must be investigated with a very limited number of experiments (factor screening is essential!).

D-Optimal designs are thus extremely flexible and lend themselves well to situations where classical designs break down. They can model combinations of process and mixture variables simultaneously. They are also better at handling multilevel ($L > 2$) discrete variables (such as reagent choice or reactor vessel material) in combination with continuous factors, something that can become complicated (or expensive) with classical designs. Additionally, as one can define the experimental cost (number of runs) relative to the assumed model, they are useful for studies with special practical limitations or requirements, such as a fixed number of runs.

2.4 DoE in the Radiochemistry Lab

The question "How do I perform the experiment?" is one that should be considered throughout the DoE process. The rigorously systematic DoE approach can drastically simplify the planning of reaction studies; however, one still needs to go into the lab and collect high-quality data cost-effectively. In the context of radiochemistry, short-lived radioactive nuclides complicate many protocols and procedures typically considered trivial in a standard chemistry lab. Therefore, some thought does need to be given to how radiochemical DoE studies can be performed reliably, safely, and in a reasonable amount of time. Every experimental workflow will, of course, be unique to the reaction or process being optimized, but there are some general practical considerations specific to radiochemistry that, in our experience, one should be aware of when planning radiochemical DoE studies.

As cyclotron-produced radionuclides are often expensive to produce, one typically tries to maximize the number of reactions performed from a single batch or production. This carries three major implications: (1) A method or protocol in place that allows experiments to be carried out in parallel (or at least in quick succession) is highly beneficial. (2) The radionuclide should be prepared and processed so that it can be easily and accurately aliquoted and dispensed for parallel experiments, with minimal uncontrollable variation in the composition of each reaction solution. And (3) care should be taken to ensure that the reactions under investigation are scalable from test reactions using small aliquots of activity to full batch productions where aliquots are not diluted.

One possible way around potential scalability issues are to entirely avoid a parallel reaction setup by performing DoE experiments at full production scale using an appropriate automated synthesizer. This can potentially introduce several confounding variables or uncontrollable sources of error, as well as being costly in terms of time and resources. However, this is a viable possibility when chemistry for parallel reactions may be exceedingly difficult or time-consuming to develop or implement (as may be the case with some ^{11}C chemistry). Indeed, one of the purposes of DoE is to reduce cost and effort in exactly these circumstances.

Another critical point of consideration is the management of systematic error and the standardization of the experimental protocols. Experimental variance (used to calculate the pure error of the model) is measured by the center points, and it is usually assumed that the variance will vary uniformly throughout the model (the property of design rotatability.) However, this requires that the experimental error in each measurement is consistent throughout the data set. It is thus of utmost importance that the protocol used to perform each experiment be as consistent as possible. Moreover, the run order of the DoE study must be randomized to further reduce the influence of systematic error.

2.5 Model Construction, Data Interpretation, and Validation

With a set of empirically acquired data in hand, the next step is the construction of a useful model. In most classical DoE studies, the relationship between the input factors and one or more response variables, the response surface (RS), is described as a polynomial (Fig. 5, Eq. 1), comprised of a series of model terms. Each factor effect term is modeled as the product of the effect setting X_k and the effect coefficient β_k, which is a measure of the magnitude of the effect. The effect coefficients are used to determine whether that effect is statistically significant. Main effects terms (represented by the product of effect coefficient β_k and X_k) are the direct influence of a factor on the outcome being measured, and they typically have the greatest overall effect. However, factors can also interact with each other so that the setting of one factor can influence the behavior of another; these effects are termed factor interactions ($\beta_{ij}X_iX_j$). Finally, factors can also exhibit quadratic behaviors ($\beta_{kk}X_k^2$), which introduce curvature into the response surface. Higher order effects such as three-factor interactions ($X_hX_iX_j$) and cubic effects (X_k^3) are seldom significant, and most DoE models thus ignore them; however, modeling these terms is indeed possible with a sufficiently large dataset.

However, before the model can be fitted, one should check the acquired dataset for outliers, mistakes, or other obvious problems (NIST/SEMATECH, 2022). A good way to do this is to look at the data set as a scatter plot of the response vs. the run order; this will help one identify obvious anomalies and any obvious systematic errors. Additionally, one should check that the response data is normally distributed, and if it is not (fails a skewness test), it can be transformed if necessary. If the data is suitable, the next step is to fit the data to a theoretical model using MLR or partial least squares (PLS) techniques. The theoretical model often includes all the main effect terms, factor interaction effects, and quadratic effect terms that can be studied

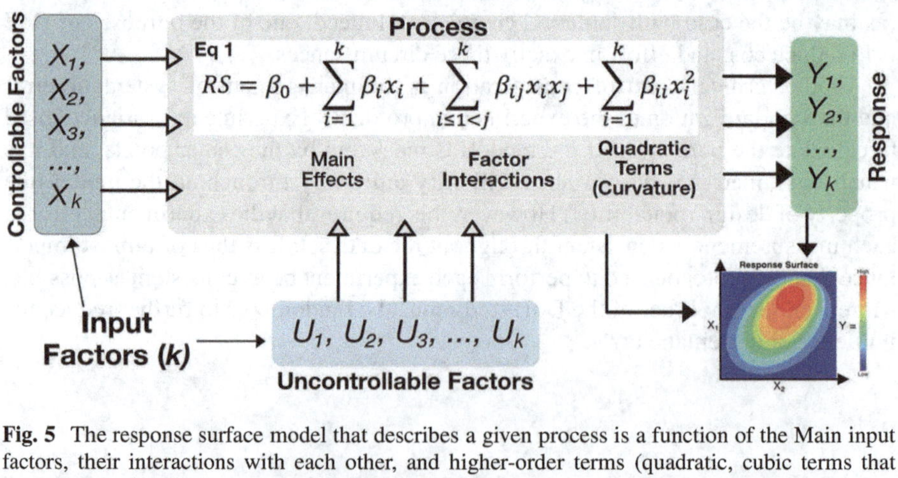

Fig. 5 The response surface model that describes a given process is a function of the Main input factors, their interactions with each other, and higher-order terms (quadratic, cubic terms that describe curvature in the response surface)

by the chosen experimental design (Fig. 6b). With a theoretical model fitted, the next step is to refine and simplify the model (model tuning) to improve the goodness of fit (R^2) and predictive accuracy of the model (Q^2) with respect to the acquired data. This is done by checking each effect coefficient for statistical significance, eliminating non-significant terms (in order of increasing magnitude) from the model while adding terms that improve the model's statistical characteristics. In practice, most modern DoE software will graphically represent the scaled regression coefficients as a bar chart, allowing one to visually gauge which factors are the most influential by their relative magnitude (Fig. 6c). Error bars are used to define the uncertainty range of the factor coefficient (at the 95% confidence interval), and if they cross the $y = 0$ line, that term can be considered non-significant and therefore removed from the model.

Finally, the assumptions of the new model are tested by looking at the goodness of fit (R^2), the residuals (the difference between the observed and predicted values), and the predictive accuracy of the model (Q^2, i.e., how well the predicted values match the actual values). If the model can no longer be simplified and refined without degradation of the fit and predictiveness statistics, then the model data can be used to address the question posed by the study. For example, this may be a decision about which factor effects should be investigated further, or the model could then be

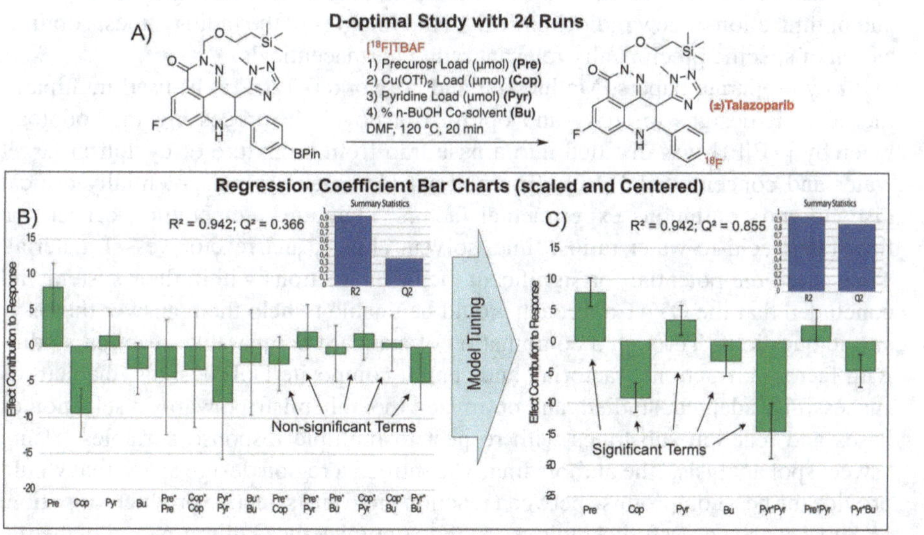

Fig. 6 An example of model tuning from the D-Optimal DoE RSM optimization of [18F]talazoparib (Bowden et al., 2021b). (**a**) The four-factor D-optimal design optimized the precursor load, copper triflate load, pyridine load, and percent of co-solvent. The bar chart represents the relative magnitudes of the scaled and centered regression coefficients (β), allowing one to easily compare them within the defined confidence intervals. (**b**) The theoretical model contains multiple non-significant terms that need to be sequentially eliminated (from least significant to more significant) to refine the model. (**c**) Eliminating non-significant terms generally results in a model with better model fit and model predictiveness

used to plot the response surface and to make conclusions about the radiochemical process under scrutiny or predictions about which set of optimal conditions could be used for a radiochemical production. If more than one response is measured, "sweet spot analysis" could be used to balance gains against costs e.g., a high product yield against byproduct formation or poor specific activity; or, find conditions that produce the product in sufficient (though perhaps not optimal) yield while using a minimal amount of expensive precursor or reagent.

It is important to remember that any model, no matter how precise or predicative it claims to be, is just an approximation of reality. George E. P. Box, one of the pioneers of DoE statistics, is famously quoted as having written: "All models are wrong; some models are useful" (Box et al., 2005, p. 440). Ultimately, it falls to the chemist to use their judgment about how much to trust and best utilize the information provided by their model.

3 Examples of DoE in Radiochemistry and Radiopharmacy

Despite the power of DoE, examples of its use in the radiopharmaceutical sciences are sparse. However, several groups, including our own, have applied DoE-based optimization methods to various radiochemical problems, ranging from the study and optimization of new radiochemical methodologies to the radiosynthesis optimization of specific preclinically relevant radiopharmaceuticals.

In two separate papers, Mathiessen and co-workers (2011a, b) used multiparametric DoE designs to study and optimize a novel ^{18}F processing methodology whereby [^{18}F]FH was distilled into a base trap from a mixture of cyclotron target water and concentrated H_2SO_4. The authors identified several potentially critical discrete and continuous experimental factors, including temperature, carrier gas flow rates, acid-to-water ratios, time, solvent choice, and reactor vessel material. They noted the potential for significant factor interactions within their system and concluded that the OVAT approach would be unable to help them answer their scientific questions. Through a combination of standard comparative reaction studies with factorial, fractional factorial, and central composite DoE designs, the authors successfully adapted, studied, and optimized their chemistry towards a selection of bases and reaction substrates with respect to multiple response variables. Using "sweet spot analysis," the authors finally identified a region design space that would provide the best compromise between reaction yield and selective product formation.

Salvé et al. successfully employed DoE to optimize the Gallium-68 radiolabeling of the pegylated RGD dimers NOTA-PRGD$_2$ and NODAG-PRGD$_2$ via both batch and microfluidic reactor strategies (Salvé et al., 2018). The authors specifically used the DoE approach to target robust, high-yielding, and automatable reaction conditions that could be used to produce the integrin radiotracers at production scale under cGMP conditions. A combination of both DoE (I-optimal design) and Bayesian design space characterization methods was used to optimize the quantity of peptide (two levels, given as peptide concentration (μg/mL)), reaction pH (three levels), the labeling temperature (five levels), and reaction time (five levels) required to maximize the process robustness (this is the probability that a set of reaction conditions will provide a yield above a certain threshold).

In a more recent example, Wagner et al. used DoE to assist in the optimization of an automated radiosynthesis of [^{18}F]F-Al-PSMA-11 (Wagner et al., 2022). The authors used prior knowledge obtained from previous OVAT experiments to guide factor selection for the construction of a Doehlert DoE study. The DoE study was then ultimately used to find the optimal conditions for the formation of the [^{18}F]AlF-chelator complex attached to the PSMA-11 ligand.

Farrag and coworkers also used a fractional factorial optimization approach to study the radioiodination of [^{125}I]omeprazole, a candidate SPECT tracer for imaging proton pump expression in the stomach (Farrag et al., 2019). The authors aimed to demonstrate the utility of statistics-based optimization approaches for cost-effectively developing high-yielding radiochemical processes and ultimately compared their results against the previously described optimized synthesis (97% RCY) of [^{125}I]omeprazole. Their use of a five-factor RES V fractional factorial design allowed them to home in on the optimal region of design space, after which they used a particle swarm optimization algorithm to refine their results further.

Our work using DoE began as an attempt to understand and optimize radiosyntheses using the then recently published family of copper-mediated radiofluorination (CMRF) reactions (Makaravage et al., 2016; Mossine et al., 2015; Tredwell et al., 2014). Our initial attempts to optimize these reactions via the OVAT approach proved unsuccessful, given the reaction's complex multi-component and multifactor nature. Furthermore, at the time, there was not a lot of literature available on the subject, and our optimization attempts were scattershot and arbitrary at best. We thus decided to apply DoE to better understand the complex reactions and how best to apply them to make new tracers. Our first paper explored the DoE method by optimizing the copper-mediated radiofluorination of several aryl stannane precursors (Bowden et al., 2019). With this work, we could establish both the experimental design and practical workflows required for implementing DoE more widely in our tracer development projects.

One of the compounds we studied contained several aromatic amines and heteroaromatic nitrogen atoms, which had previously been found to inhibit CMRF reactions significantly. One of the central conclusions from this work was that, in some cases, these deleterious effects could be "optimized out" to provide the target radiotracer with sufficient radiochemical yields without the need for complicated protecting group strategies or multistep syntheses. This could potentially save significant tracer development time by simplifying both precursor syntheses and the translation of the radiosynthesis to automated radiosynthesizers. However, each precursor requires a unique set of optimal conditions, and thus a fast, efficient, reliable, and scalable DoE-based workflow to rapidly and systematically screen for optimal reaction conditions would be invaluable.

However, this preliminary foray into the use of DoE revealed the limitations of our initial methodology. The QMA cartridge-based processing method we used to prepare the [^{18}F]fluoride for the CMRF reaction introduced carbonate salts into the reaction mixture, poisoning the copper mediator. This effect was concentration-dependent; when we conducted our DoE experiments using small aliquots of the prepared [^{18}F]KF, the reactions provided good data, which we used to construct the DoE models. However, when we attempted to scale up our reactions to run on automated synthesizers using a full batch of the processed [^{18}F]KF, the conclusions drawn from our models broke down, and we could not produce the radiolabeled product in the expected yields.

Thus, we developed a convenient carbonate-free ^{18}F processing method that not only skipped the lengthy azeotropic drying step but was also fully scalable between small-scale DoE experiments and full-scale productions (Bowden et al., 2021a). To demonstrate the utility of this method, we carried out a manual DoE optimization study on the CMRF synthesis of the PARP radiotracer [^{18}F]olaparib using small aliquots of processed activity. The optimal conditions we discovered were then successfully automated to yield comparable radiochemical yields (up to 80% RCY) at full automated production scale using a full batch of processed [^{18}F]TBAF, improving upon the previously published synthesis results by up to sevenfold.

We then set our sights on the radiosynthesis and enantiomeric resolution of the novel PARP tracer [^{18}F]talazoparib (Bowden et al., 2021b). Talazoparib is a chiral molecule, and its enantiomer is inactive against the PARP enzyme-DNA complex. The challenge was thus to produce enantiomerically pure [^{18}F]talazoparib for preliminary preclinical investigation. Developing a synthesis to produce a chiral precursor, either through a stereospecific synthesis or late-stage enantiomeric resolution, was deemed too time-intensive and high risk as we were uncertain if the chiral centers would survive the radiosynthesis without racemization. To get the tracer into preclinical studies as quickly as possible, we decided to simply modify the published synthesis of talazoparib to produce the racemic precursor. We then planned to separate the two enantiomers in a second chiral-HPLC step after the radiosynthesis; however, this meant that our theoretical maximum RCY would be only 50%. It was, therefore, critical that the radiolabeling step (the first step of the synthesis) be as high-yielding as possible. Using the knowledge obtained from our previous work, we used a D-optimal DoE study to optimize the CMRF of [^{18}F]talazoparib with respect to precursor load, copper load, pyridine load, and the %n-BuOH as a co-solvent (Fig. 7a). This more than doubled the yield of the initial trials (15–18% RCY) to incorporate $57 \pm 7.5\%$ ($n = 6$) of the radioactivity into the target product. The radiosynthesis and the subsequent deprotection and purification steps were then fully automated to afford enantiomerically pure [^{18}F]talazoparib ($13 \pm 3.4\%$ RCY ($n = 4$)) in moderate to high molar radioactivities (Fig. 7b).

In our experience, DoE has proved to be an invaluable tool for establishing and studying copper-mediated radiofluorination chemistry in our labs. We have used the information obtained from our DoE models to aid decision-making and guide the development of several novel radiotracers, in addition to those described here, around many of the pitfalls of radiosyntheses development, thus expediting the tracer development pipeline. Moreover, we are learning how to better leverage DoE and other data-based approaches to efficiently build quality and reliability into our novel radiosyntheses from early in the tracer development cycle.

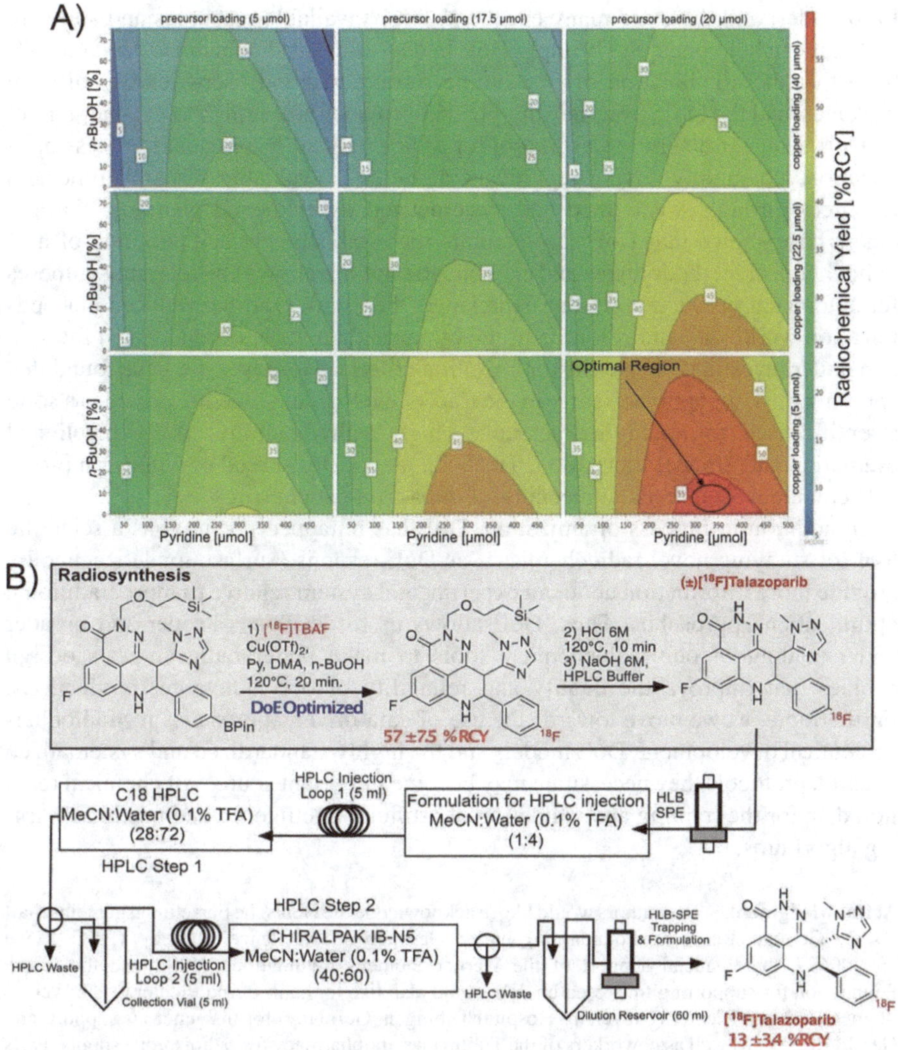

Fig. 7 (**a**) A 4D D-optimal response surface model of the CMRF of [^{18}F]talazoparib. Plotted using Modde Go 12. (**b**) The results of the RSM optimization study were automated and used to produce enantiomerically pure [^{18}F]talazoparib. (Adapted with permission from Bowden et al. (2021b). Copyright (2021) American Chemical Society)

4 Outlook

The potential benefits of applying the DoE approach within the field of radiochemical research are far-reaching. Unfortunately, DoE's learning curve and heavy use of statistics seem to be one of the main factors holding it from more widespread adoption; however, the approach has been well-established in multiple fields of inquiry for decades, and there are many excellent, freely available resources and software tool that can help one quickly apply DoE in day-to-day radiochemical research. We hope that the combination of free online resources, good user-friendly software packages, and the basic overview this chapter provides are enough to get one started.

DoE is an extremely powerful tool for radiochemical reaction and process optimization, potentially saving significant amounts of valuable research time and resources through streamlined and systematized experimental planning. We and others have shown that DoE can facilitate the establishment and adoption of new radiochemical methodologies and expedite the optimization of novel radiosyntheses for automated production. In our experience, the use of DoE has reduced our optimization workloads and resource usage by more than 50%, depending on the system under investigation and the optimization criteria. Moreover, we have found that applying the DoE approach early in the tracer development pipeline provides a solid scientific basis to quickly deliver novel, high-quality radiotracers for preclinical evaluation and clinical translation. Thus, the use of DoE-based optimization protocols could have a direct and meaningful impact on patient care.

In addition to process optimization, DoE has untapped potential as a scientific tool for studying novel radiochemistry, as DoE response surface models generally provide more information about an experimental system relative to more traditional optimization approaches. Thus, DoE allows us to develop a deeper and broader understanding of our radiochemical tools to make better radiosynthesis design choices that improve the quality and reliability of our radiotracer productions. Furthermore, as we move towards the use of data-driven approaches for radiopharmaceutical development, DoE models and the highly standardized and systematized reaction protocols they necessitate may become important sources of chemical reaction data for the training and validation of artificial intelligence and machine learning algorithms.

Acknowledgments The authors would like to acknowledge the Deutsche Forschungsgemeinschaft (DFG, German Research Foundation) under Germany's Excellence Strategy (EXC 2180) 390900677, the financial support of the Werner Siemens Foundation (WSS), and the Leuze Foundation for supporting this research. We would also like to thank Bernd Pichler of the Werner Siemens Imaging Center (University Hospital Tübingen, Germany) for his generous support, and Gerald Reischl, as well as coworkers of the Tübingen radiopharmacy for radioisotope supply. Parts of this review were featured in the doctoral thesis of Dr. Gregory Bowden (Bowden, 2020).

Appendix: DoE Software and Useful Resources

Software

There is a wide array of computational options available for anyone wishing to implement DoE principles and workflows into their research routine. Widely available statistics packages such as JMP or Minitab feature DoE-specific software tools that are well-supported and documented with plenty of reference material. Moreover, many academic institutes have site-wide software license agreements that make these software packages a good entry point to experiment with and learn DoE.

There are also several excellent user-friendly and well-designed DoE-specific software packages, such as Modde (Sartorius), Fusion Pro DoE (S-Matrix), and Design Expert (Stat-Ease) that are designed to make DoE as accessible as possible. We highly recommend Modde as it features several "wizards" that walk one through design choice and construction, model construction and validation, and data analysis. Modde is also well-documented, and the company has many webinars and tutorials available on its YouTube channel. Modde is also available in two versions: Modde Pro and a more affordable and beginner-friendly "lite" version titled Modde GO. Free trials of both Modde and Design Expert are available, and these can be invaluable for learning the DoE process before making a financial commitment. While proprietary software packages can be costly, they represent the lowest barrier to entry to DoE and should thus definitely be considered as a quick and easy option to learn and implement DoE in one's radiochemical research.

DoE-specific statistics software is not a necessity to implement DoE. Many of the classical DoE designs can be set up, analyzed, and run in spreadsheet software like Microsoft Excel. For more powerful and flexible alternatives, there are also free and open-source DoE plugins and libraries for many commonly used scientific computing languages such as Python, MATLAB, Scilab, and R.

Useful Resources

YouTube:

Design of Experiments, Sartorius Stedim Data Analytics; https://www.youtube.com/watch?v=UTYRpStJDE8

There are hundreds of other excellent YouTube tutorials and lectures on DoE and how to implement it.

Udemy Courses:

Design and Analysis of Experiments | DoE; Design of Experiments: from ANOVA to factorial Designs using Excel and R; Rosane Rech: https://www.udemy.com/course/design-of-experiments-i/

Design of Experiments for Optimization; DoE using R: Response Surface Methodology, Lack-of-fit, Central Composite Designs, Box-Behnken Designs; Rosane Rech: https://www.udemy.com/course/design-of-experiments-for-optimisation/

References

Bowden, G. D. (2020). *The design, synthesis, and radiosynthesis optimization of novel approaches towards the in vivo PET imaging of beta cell mass*. Doctoral Thesis, Eberhard Karls Universität Tübingen.

Bowden, G. D., Pichler, B. J., & Maurer, A. (2019). A Design of Experiments (DoE) approach accelerates the optimization of copper-mediated ^{18}F-fluorination reactions of arylstannanes. *Scientific Reports, 9*(1), 11370. https://doi.org/10.1038/s41598-019-47846-6

Bowden, G. D., Chailanggar, N., Pichler, B. J., & Maurer, A. (2021a). Scalable ^{18}F processing conditions for copper-mediated radiofluorination chemistry facilitate DoE optimization studies and afford an improved synthesis of [^{18}F]olaparib. *Organic & Biomolecular Chemistry, 19*(32), 6995–7000. https://doi.org/10.1039/D1OB00903F

Bowden, G. D., Stotz, S., Kinzler, J., Geibel, C., Lämmerhofer, M., Pichler, B. J., & Maurer, A. (2021b). DoE optimization empowers the automated preparation of enantiomerically pure [^{18}F]talazoparib and its in vivo evaluation as a PARP radiotracer. *Journal of Medicinal Chemistry, 64*(21), 15690–15701. https://doi.org/10.1021/acs.jmedchem.1c00903

Box, G. E. P., Hunter, J. S., & Hunter, W. G. (2005). *Statistics for experimenters* (2nd ed.). Wiley Interscience.

Dejaegher, B., & vander Heyden, Y. (2011). Experimental designs and their recent advances in set-up, data interpretation, and analytical applications. *Journal of Pharmaceutical and Biomedical Analysis, 56*(2), 141–158. https://doi.org/10.1016/j.jpba.2011.04.023

Dunn, K. (2018). *Process improvement using data*. Kevin Dunn.

Farrag, N. S., Abdel-Halim, H. A., & Abdel Moamen, O. A. (2019). Facile radiolabeling optimization process via design of experiments and an intelligent optimization algorithm: Application for omeprazole radioiodination. *Journal of Labelled Compounds and Radiopharmaceuticals, 62*(6), 280–287. https://doi.org/10.1002/jlcr.3734

Fowler, J. S. (2020). ^{18}F-FDG radiosynthesis: A landmark in the history of PET. *Journal of Nuclear Medicine, 61*(Suppl 2), 105S–109S. https://doi.org/10.2967/jnumed.120.250191

Hamacher, K., Coenen, H. H., & Stöcklin, G. (1986). Efficient stereospecific synthesis of no-carrier-added 2-[^{18}F]-fluoro-2-deoxy-D-glucose using aminopolyether supported nucleophilic substitution. *Journal of Nuclear Medicine, 27*(2), 235–238.

Leardi, R. (2009). Experimental design in chemistry: A tutorial. *Analytica Chimica Acta, 652*(1–2), 161–172. https://doi.org/10.1016/j.aca.2009.06.015

Lendrem, D. W., Lendrem, B. C., Woods, D., Rowland-Jones, R., Burke, M., Chatfield, M., Isaacs, J. D., & Owen, M. R. (2015). Lost in space: Design of experiments and scientific exploration in a Hogarth Universe. *Drug Discovery Today, 20*(11), 1365–1371. https://doi.org/10.1016/j.drudis.2015.09.015

Makaravage, K. J., Brooks, A. F., Mossine, A. V., Sanford, M. S., & Scott, P. J. H. (2016). Copper-mediated radiofluorination of arylstannanes with [^{18}F]KF. *Organic Letters, 18*(20), 5440–5443. https://doi.org/10.1021/acs.orglett.6b02911

Mathiessen, B., Jensen, A. T. I., & Zhuravlev, F. (2011a). Homogeneous nucleophilic radiofluorination and fluorination with phosphazene hydrofluorides. *Chemistry, 17*(28), 7796–7805. https://doi.org/10.1002/chem.201100458

Mathiessen, B., Jensen, M., & Zhuravlev, F. (2011b). [^{18}F]Fluoride recovery via gaseous [^{18}F]HF. *Journal of Labelled Compounds and Radiopharmaceuticals, 54*(13), 816–818. https://doi.org/10.1002/jlcr.1931

Mossine, A. V., Brooks, A. F., Makaravage, K. J., Miller, J. M., Ichiishi, N., Sanford, M. S., & Scott, P. J. H. (2015). Synthesis of [^{18}F]arenes via the copper-mediated [^{18}F]fluorination of boronic acids. *Organic Letters, 17*(23), 5780–5783. https://doi.org/10.1021/acs.orglett.5b02875

Murray, P. M. (2019). *What is experimental design?* Catalysis Consulting Co. http://www.catalysisconsulting.co.uk/what-is-experimental-design.html

Murray, P. M., Tyler, S. N. G., & Moseley, J. D. (2013). Beyond the numbers: Charting chemical reaction space. *Organic Process Research & Development, 17*(1), 40–46. https://doi.org/10.1021/op300275p

Murray, P. M., Bellany, F., Benhamou, L., Bučar, D.-K., Tabor, A. B., & Sheppard, T. D. (2016). The application of design of experiments (DoE) reaction optimisation and solvent selection in the development of new synthetic chemistry. *Organic & Biomolecular Chemistry, 14*(8), 2373–2384. https://doi.org/10.1039/C5OB01892G

NIST/SEMATECH. (2022, September 22). E-handbook of statistical methods. https://www.Itl.Nist.Gov/Div898/Handbook/Pri/Section4/Pri41.Htm.10.18434/M32189.

Pascali, G., Matesic, L., Collier, T. L., Wyatt, N., Fraser, B. H., Pham, T. Q., Salvadori, P. A., & Greguric, I. (2014). Optimization of nucleophilic ^{18}F radiofluorinations using a microfluidic reaction approach. *Nature Protocols, 9*(9), 2017–2029. https://doi.org/10.1038/nprot.2014.137

Rios, A., Holloway, T. S., Chao, P. H., de Caro, C., Okoro, C. C., & van Dam, R. M. (2022). Microliter-scale reaction arrays for economical high-throughput experimentation in radiochemistry. *Scientific Reports, 12*(1), 10263. https://doi.org/10.1038/s41598-022-14022-2

Salvé, M., Avohou, H. T., Monbaliu, J. C. M., Lebrun, P., Lemaire, C., Damblon, C., de Tullio, P., Hubert, P., Hustinx, R., & Luxen, A. (2018). NOTA-PRGD2 and NODAGA-PRGD2: Bioconjugation, characterization, radiolabelling, and design space. *Journal of Labelled Compounds and Radiopharmaceuticals, 61*(6), 487–500. https://doi.org/10.1002/jlcr.3613

Tredwell, M., Preshlock, S. M., Taylor, N. J., Gruber, S., Huiban, M., Passchier, J., Mercier, J., Génicot, C., & Gouverneur, V. (2014). A general copper-mediated nucleophilic ^{18}F fluorination of arenes. *Angewandte Chemie International Edition, 53*(30), 7751–7755. https://doi.org/10.1002/anie.201404436

Tye, H. (2004). Application of statistical 'design of experiments' methods in drug discovery. *Drug Discovery Today, 9*(11), 485–491.

Tye, H., & Whittaker, M. (2004). Use of a Design of Experiments approach for the optimisation of a microwave assisted Ugi reaction. *Organic and Biomolecular Chemistry, 2*(6), 813–815. https://doi.org/10.1039/b400298a

Wagner, L., Bastogne, T., Karcher, G., Acherar, S., & Collet, C. (2022). Automated ^{18}F-Al-radiolabelling optimization of PSMA-11 for PET imaging using a design of experiments approach. *Nuclear Medicine and Biology, 108–109*, S138–S139. https://doi.org/10.1016/S0969-8051(22)00300-6

Yu, L. X., Amidon, G., Khan, M. A., Hoag, S. W., Polli, J., Raju, G. K., & Woodcock, J. (2014). Understanding pharmaceutical quality by design. *The AAPS Journal, 16*(4), 771–783. https://doi.org/10.1208/s12248-014-9598-3

Next-Generation Techniques for In Vitro Characterization of Cancer Radiopharmaceuticals

Syamantak Khan and Guillem Pratx

1 Introduction

The use of radioisotopes in medicine has a long history dating back to the 1940s (Sawin & Becker, 2009). In the last decades, this field has advanced with unprecedented acceleration and has transformed the standard of care for diagnosis, staging, therapy, and treatment monitoring in oncology. Radiopharmaceuticals have emerged as a promising tool to diagnose and treat various advanced cancers, which previously lacked effective treatment options.

Radioiodine is one of the oldest radiopharmaceuticals used in the clinic and is primarily used to investigate and treat thyroid diseases (Sawin & Becker, 2009). Presently, iodine-123 is commonly used as a diagnostic radiopharmaceutical for imaging thyroid cancer and its metastatic spread to other organs (Yaakob et al., 1999), whereas iodine-131 is used for its ability to ablate disseminated thyroid cancer cells throughout the body (McCready, 2017). Technetium-99m (^{99m}Tc) is the most common radioactive isotope tracer used for SPECT (single-photon emission computerized tomography) imaging of the brain, bones, lungs, kidneys, thyroid, heart, gallbladder, liver, spleen, bone marrow, salivary and lachrymal glands, blood pool, and sentinel nodes (Banerjee et al., 2001). Fluorine-18 (^{18}F), a positron emitter with a half-life of 110 min, has revolutionized the field of medical imaging with PET. Its application ranges from oncology and neurology to cardiology. In particular, [^{18}F]fluorodeoxyglucose (FDG) has emerged as one of the most sensitive tools to image malignancy in human subjects (Almuhaideb et al., 2011). The discovery of the Warburg effect (Bhattacharya et al., 2016) has led to its extensive use as the main PET tracer to map glucose metabolism in primary and metastatic tumors. In the last

S. Khan · G. Pratx (✉)
Department of Radiation Oncology, Stanford University, Stanford, CA, USA
e-mail: pratx@stanford.edu

R. M. van Dam et al. (eds.), *Automated Technologies for the Development and Production of Radiopharmaceuticals*,
https://doi.org/10.1007/978-3-031-84632-8_20

decade, FDG-PET drastically improved cancer diagnosis, staging, and subsequent treatment and has emerged as a reliable tool for assessing treatment response after chemotherapy and radiotherapy.

Gallium-68 (^{68}Ga), with a shorter half-life of 68 min, is another radioisotope that has proved extremely useful thanks to the advancement of radiometal conjugation chemistry, which has allowed the development of several cancer-specific radiopharmaceuticals. [^{68}Ga]gallium-PSMA for imaging prostate cancer (Han et al., 2018) and [^{68}Ga]gallium-DOTATOC or DOTATATE for imaging neuroendocrine tumors (Poeppel et al., 2011) are two promising ^{68}Ga-labeled radiotracers recently approved by the FDA. A new generation of radiopharmaceuticals, capable of imaging cancer-specific targets such as bombesin receptors (Mansi et al., 2016), somatostatin receptors (Geijer & Breimer, 2013), αvβ6 integrin (Feng et al., 2020), and fibroblast activation protein (Kratochwil et al., 2019) are now in clinical trials. Tracers have been developed to assess tumor-infiltrating lymphocytes' activity, which measures T-cell activation during immunotherapy (Levi et al., 2019).

The ability to specifically target various cancer biomarkers has opened the door to "theranostics," combining therapeutic radioisotopes with a diagnostic tool to enable therapy and visualization. More than half of cancer patients receive some form of ionizing radiation during cancer treatment. Unlike external beam radiation therapy, radiopharmaceuticals are designed to deliver ionizing radiation directly to the internal organs of interest in a molecularly targeted manner. The delivery of these radiopharmaceuticals can be imaged using a gamma camera, positron emission tomography (PET), or SPECT in the patients. Some of the examples include iodine-131 for thyroid cancer (McCready, 2017), [^{177}Lu]lutetium-DOTATATE (Lutathera) for neuroendocrine tumors (Jia et al., 2022), [^{177}Lu]lutetium-PSMA-617 (Pluvicto) for metastatic castration-resistant prostate cancer (Sartor et al., 2021), [^{90}Y]yttrium-ibritumomab tiuxetan (Zevalin) for lymphoma (Witzig, 2004), and [^{89}Sr]strontium chloride (Metastron) and [^{153}Sm]samarium lexidronam (Quadramet) for palliative care of bone metastasis (Bączyk et al., 2007). One of the most exciting developments in theranostics is using alpha emitters with high linear energy transfer (LET) to localize energy deposition in tumor cells while sparing the surrounding healthy tissue. Radium-223 is the first α-emitter approved by the FDA to treat castration-resistant prostate cancer with metastatic bone lesions (Parker et al., 2013). Several other promising radionuclides, such as actinium-225, bismuth-213, lead-212, and astatine-211, are being assessed in clinical trials (Eychenne et al., 2021).

The emerging potential of radiopharmaceuticals in oncology calls for more translational research to identify new radiotracer candidates and further characterize existing molecules. The traditional approach of clinical translation relies largely on preclinical testing of new radiotracers in animal models. These models are instrumental in studying the in vivo biodistribution and pharmacokinetics of novel radiotracers. Nevertheless, they have a higher turnaround time and complexity, limiting their use for rapid tracer discovery. In this regard, advanced in vitro cancer models are now showing great promise, allowing high-resolution and high-throughput characterization of radiopharmaceuticals. In particular, various in vitro techniques, including 3D cultures, stem cell- and patient-derived organoids, microfluidic cell culture, organ-chip technologies, and tissue engineering, can now model various aspects of human disease with high reliability and reproducibility. Consequently, these advancements present unprecedented opportunities to study radiopharmaceuticals in vitro using an array of specialized imaging and analytical techniques. This chapter discusses several strategies that can transform next-generation radiopharmaceutical research by leveraging the power of in vitro models.

2 Principles of Radiation Detection

2.1 History

Thermoluminescence, the gold leaf electroscope, and photographic film are some of the oldest tools capable of detecting radiation. However, the discovery of X-ray by Wilhelm C. Rontgen and his acquisition of the first medical photograph on 22 December 1895 laid the foundation for the scientific field of radiation detection. The spinthariscope was the first instrument to detect individual nuclear disintegration, invented by Crookes in 1903. In 1911, C.T.R. Wilson built a cloud chamber, another instrument that made single radiation events visible. These early discoveries opened the door for modern radiation detection and quantification techniques employing ionization counters, scintillation detectors, or chemical reactions (Fig. 1).

2.2 Principles of Radiation Detection

2.2.1 Ionization Detectors

One of the most direct ways of detecting ionizing radiation is directly quantifying the ionization of a medium resulting from the application of radiation. The fundamental principle is to collect the ionization changes generated inside the chamber by applying an electrical voltage to the medium and measuring the resulting electric current pulses. The first generation of ionization detectors was based on gas-filled chambers. For instance, Victor Hess provided the first evidence for cosmic rays in 1910 by using an ionization chamber during a balloon flight. Later, Geiger and Muller discovered a gas-filled counter that responded to individual radiation-induced events by giving a high-level output signal. Their discovery made way for the hand-held radiation survey GM counter, one of the most used radiation detection instruments worldwide. In addition, the ionization medium can also be a semiconductor solid, which is much denser than gases and can be used to fabricate compact detectors. Silicon diodes and germanium ion-drift detectors are widely used for charged-particle spectrometry and gamma-ray measurements.

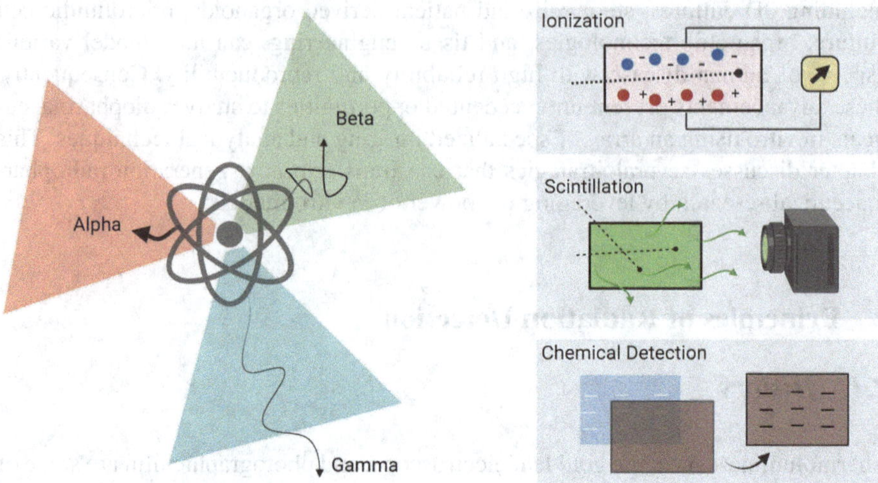

Fig. 1 Schematic showing the three main types of radiation and three main detection techniques

2.2.2 Scintillation Detectors

Scintillation counters are widely used in radiation protection, radiation spectrometry, and radionuclide imaging. The technique involves detecting and measuring ionizing radiation using a "scintillator" material, which generates photons in response to incident radiation. The scintillator consists of a transparent medium, such as an inorganic crystal, an organic composite (plastic), or an organic solution. Scintillation photons are detected by a sensitive photodetector, such as a photomultiplier tube (PMT), a charge-coupled device (CCD) camera, or a photodiode. A good scintillator has typically high density and atomic number, can convert a significant fraction of the incident radiation energy into light, and creates short light pulses ($\lesssim$100 ns), which enables faster data acquisition with low statistical uncertainty in position. Cesium iodide (protons and alpha), sodium iodide (gamma), zinc sulfide (alpha), and lithium iodide (neutron) are examples of widely used scintillator materials. Cherenkov detectors are another class of detectors that emit light when ionized particles pass through a transparent medium and are widely used in high-energy physics experiments.

2.2.3 Photographic Emulsions and Other Chemical Methods

Photographic emulsions were previously widely used in medicine, industry, and research to detect and quantify radiation exposure. Most commonly, the emulsion consists of silver halide crystals dispersed in gelatin. The silver film emulsion is sensitive to radiation, and once developed, exposed areas increase in optical density in response to incident radiation. Other chemical substrates that undergo a fundamental change in their physical properties as a function of the absorbed radiation dose are also used to quantify radiation. For instance, gel dosimetry exploits radiation-induced free radical formation and recombination, resulting in the chemical crosslinking of polymers. Similarly, Fricke dosimetry measures radiation-induced redox change, converting ferrous (Fe^{2+}) ions to ferric (Fe^{3+}) ions.

3 Emerging In Vitro Radiometry Techniques

3.1 Single-Cell Characterization

Current in vivo imaging techniques cannot determine the cellular origin of a radioactive signal. This limitation creates a severe bottleneck in understanding the biological underpinnings of a disease or therapeutic response concerning the binding of a radiotracer. For instance, most tumors have a complex tissue microenvironment consisting of many heterotypic cells. Therefore, a higher radiotracer uptake in tumors observed using PET scans provides limited biological information. Several research groups have invented and used novel strategies to fill this unmet need.

3.1.1 Flow Cytometry Sorting of Radiolabeled Cells

One of the common strategies for dissecting the role of individual tissue components is cell sorting using flow cytometry. This approach can be applied with high throughput to analyze radiotracer uptake and binding mechanism in dissociated tissue, thus separating the contributions to the overall signal of individual cellular compartments. In a recent landmark study, Reinfeld et al. (2021) used PET radiotracers to measure glucose and glutamine metabolism by various cells in the tumor microenvironment. They collected spleens and tumors from mice 40 min after radiotracer administration and dissociated the tissues into a single-cell suspension. Cells were sorted and separated using flow cytometry, and the radioactivity was quantified with a gamma counter (Fig. 2a–d). Strikingly, their data indicate that myeloid cells had the most avidity for intratumoral glucose, followed by T cells and cancer cells, across a range of cancer models. Cancer cells were found to be addicted to glutamine instead. They showed that glutamine metabolism suppresses glucose uptake without glucose being a limiting factor in the tumor microenvironment. This result is consistent with studies by multiple other research groups, which found that activation of inflammatory pathways and tumor infiltration of lymphocytes or myeloid-derived immune cells, particularly macrophage, is associated with higher tumor SUV (Hesketh et al., 2019; Floberg et al., 2021; Park et al., 2021).

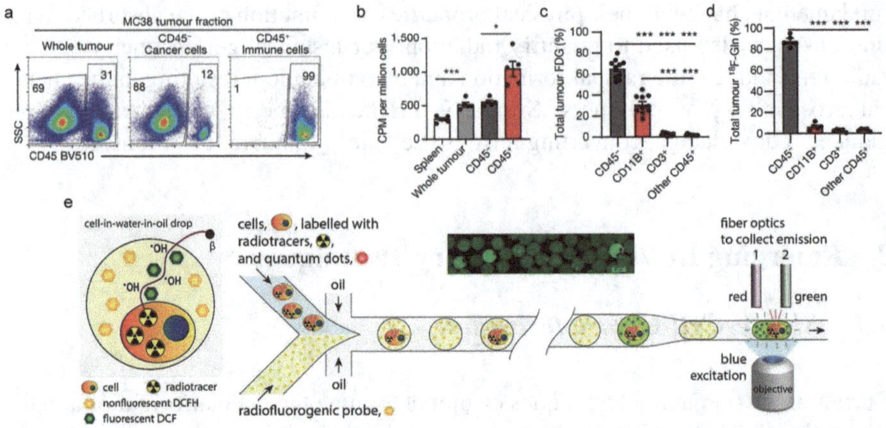

Fig. 2 Flow cytometry and radiocytometry for single-cell studies. (**a**) Representative flow cytometry analysis of the whole MC38 mice tumor, followed by sorting of immune ($CD45^+$) and cancer ($CD45^-$) cell fractions. (**b**) Cellular FDG uptake analysis shows that the $CD45^+$ immune cell fraction has higher avidity for glucose than cancer cells ($CD45^-$). (**c, d**) Contribution of different cell populations to total MC38 tumor uptake for FDG ($n = 10$ mice) and (2*S*,4*R*)4-[^{18}F]fluoroglutamine ($n = 5$ mice). Unlike glucose, glutamine is more specifically partitioned into the cancer cells. (**e**) Illustration showing the principle of radiofluorogenic conversion, droplet generation, incubation, and reading. A single radioactive cell is encapsulated in a water-in-oil drop. As the radiotracer decays, the emitted beta radiation creates ROS, such as hydroxyl radicals, mediating the radiofluorogenic conversion of non-fluorescent DCFH into fluorescent DCF. Cells are prelabeled with red-emitting quantum dots to identify droplets containing cells. A single blue excitation was used for dual-color detection of DCF (green) and quantum dots (red). (Figures panels adapted with permission from Reinfeld et al. (2021) and Ha et al. (2021))

3.1.2 Single-Cell Droplet Microfluidics

Microfluidic technologies enable single-cell manipulation and analysis and are promising solutions for radiopharmaceutical assays. While conventional flow cytometry studies typically analyze tens of thousands of cells, they cannot directly probe radiopharmaceuticals. Ha et al. (2021) have developed a droplet microfluidics method called flow radiocytometry (FRCM) that broadens the scope of current cytometry methods to include beta-emitting radiotracers as probes for single-cell studies. This technology uses radiofluorogenesis, converting non-fluorescent dichloro-dihydro-fluorescein (DCFH) into fluorescent dichlorofluorescein (DCF), to translate the radioactivity of single cells into a fluorescent signal that is then read out using a high-throughput optofluidic device (Fig. 2e). Their recent work quantitated FDG uptake in single human breast cancer cells. They successfully assessed the metabolic flux of glucose and its heterogeneity at the cellular level. By converting stochastic radioactive decay into a continuous fluorescence signal, FRCM significantly reduced the time required to measure the radioactivity of each cell down to 100 ms/cell, enabling higher throughput than other single-cell radiometric assays. This speed is >100 times higher than the microscopy-based single-cell detection methods discussed later in this chapter and permits analysis of thousands of cells over a short time by increasing cellular radiolabeling efficiency; however, the authors showed that radioactivity of 10–20 Bq/cell was sufficient to achieve very high throughput in their study.

3.1.3 Microscopy Techniques

Precise spatial mapping of the radioactive sources is critical for various biological studies. Autoradiography can achieve very high spatial resolution, but, in general, the method requires long exposure times and fixed samples, restricting its suitability for rapid survey work or dynamic systems studies. George T. Reynolds' "scintillo-microscope" is one of the earliest pioneering technologies in radioisotope microscopy. He developed an image intensifier and applied it to the photographic recording of weak radioactive sources, e.g., Phosphorus-32 and Carbon-14. The beta particles of the radioactive decay resulted in scintillation, which was focused on the cathode of the intensifier using a pair of lenses (Reynolds, 1968). However, a fundamental limitation of this system was the thickness of the scintillator and the scattering of the emitted particles that created a bottleneck for achieving high resolution in imaging biological samples.

3.1.3.1 Radioluminescence Microscopy

Pratx et al. (2012) developed radioluminescence microscopy (RLM), an in vitro radionuclide imaging technique, to resolve radiopharmaceutical distribution at the level of individual cells (Fig. 3a–c). The method allows dynamic imaging of live cells, cell aggregates, tissue, and other biological specimens (Kim et al., 2017). RLM consists of a scintillator plate close to the sample, a high numerical aperture optical train that maximizes light collection, and a highly sensitive EMCCD camera detecting scintillation photons. It is capable of imaging a range of alpha and beta-emitting radioisotopes. Live cells are cultured sparsely on a thin scintillator plate to achieve single-cell resolution and incubated with a radiotracer. The light flashes produced following beta decay are measured using a single-photon sensitive EMCCD camera. The microscope is set up to focus on the topmost region of the scintillator, which is closest to the imaged sample and thus maximizes the spatial resolution achievable in localizing the impinging positrons. Background events occurring deeper within the scintillator (e.g., gamma-ray interactions) are out of focus and thus do not contribute to the acquired images.

Moreover, imaging scintillation tracks of ionizing particles and applying a reconstruction algorithm (ORBIT) developed by their lab allows high-resolution and quantitative radioactivity measurements in biological samples (Pratx et al., 2013; Almasi & Pratx, 2019). By automatically processing and counting the scintillation flashes from a series of image frames, ORBIT can quantify fewer than 1000 radioactive molecules per cell, albeit with variable acquisition time, depending on the

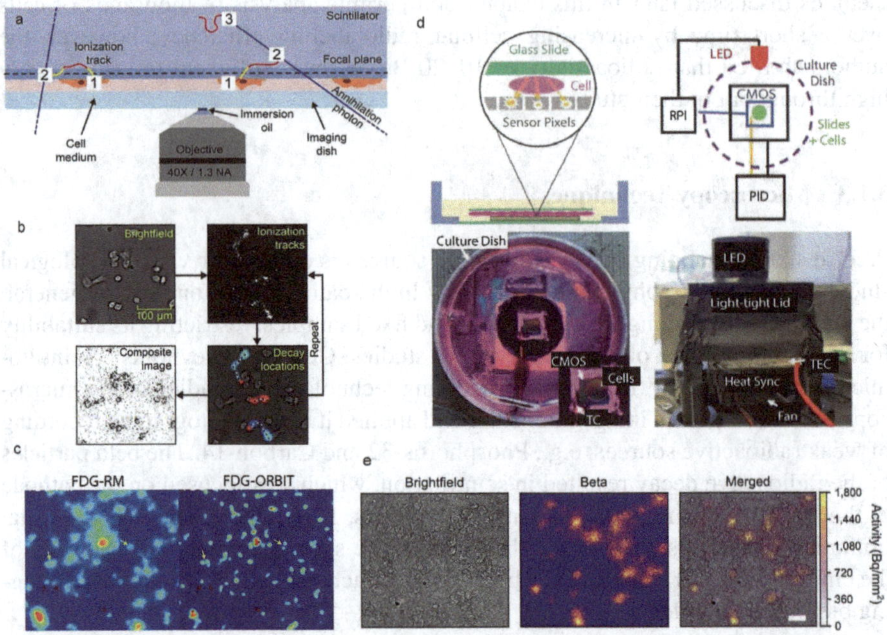

Fig. 3 Microscopy-based imaging of radioisotopes with (**a–c**) RLM and (**d, e**) LRM. (**a**) The detection principle of RLM is based on a thin scintillator plate placed on the top of an objective lens. (**b**) In one approach called ORBIT, images are reconstructed by acquiring, processing, and counting individual events. (**c**) Representative RLM images of cellular FDG uptake, comparing analog signal integration (left) and digital event-counting reconstruction (right). (**d**) High-resolution beta imaging of FDG in breast cancer cells using the LRM. (**e**) MDA-MB-231 cells were imaged using bright-field and beta modes. Images are cropped to 600 μm × 600 μm from the entire 3.7 mm × 2.8 mm field of view. The total imaging time for beta imaging was 65 min. The scale bar is 50 μm. (Figure panels adapted with permission from Pratx et al. (2012, 2013) and Klein et al. (2022))

half-life of a radiotracer. The unique capabilities of RLM have revealed previously unattainable information on single-cell response to clinical radiotracers such as FDG, [^{18}F]Fluorothymidine (FLT), and hexadecyl-4-[^{18}F]fluorobenzoate (HFB). For instance, FDG uptake was highly heterogeneous at the cellular level, even in standard cell cultures. FLT presents a bimodal distribution, where the two subpopulations correspond to distinct phases in the cell cycle. RLM also found that HFB, a lipophilic tracer developed initially for in vivo cell tracking applications, had a strong tendency to accumulate in dead or compromised cells, which hindered its practical use for tracking live cells. The method can also be used to measure the uptake of radiolabeled drugs at the single-cell level (Natarajan et al., 2015).

One can further improve detection sensitivity and speed by using a double scintillator configuration instead of a single scintillator, thus capturing ionizing particles from both directions (Kim et al., 2020). RLM can also be combined with bright-field and fluorescence imaging modalities for better colocalization of radiotracer distribution inside biological specimens and validation with established fluorescent markers. This unique capability enables multiparametric studies to correlate radiotracer uptake with other biological characteristics of cells. Further applications of RLM in emerging in vitro models are discussed later.

3.1.3.2 Lensless Radiomicroscopy (LRM)

To further advance the use of single-cell radiometry, Klein et al. (2022) have recently reported the development of a low-cost lensless radiomicroscope for cellular-resolution radionuclide imaging of beta- and alpha-emitting radionuclides (Fig. 3d, e). The instrument images radiopharmaceuticals by directly detecting the ionizing charged particles via a consumer-grade complementary metal-oxide semiconductor (CMOS) array. They reported a spatial resolution of 5 μm for bright-field imaging and 30 μm for ^{18}F positron imaging. Their palm-sized instrument has a large field of view (~1 cm^2) and can be constructed from off-the-shelf parts for less than $100. This price tag is approximately 500× less than RLM, yet it can acquire radionuclide and bright-field images with comparable performance. Moreover, with its large field of view, the LRM can measure radiotracer uptake of >5000 cells, a 100× increase over RLM's field of view. However, the limitations of the lensless method are that it is unable to image activities higher than 10.8 Bq/px and it lacks fluorescence imaging capability. The authors foresee this new technology to make in vitro single-cell radionuclide imaging more accessible in biological research due to the low cost and easy availability of its components.

3.2 Radiopharmaceutical Transport Kinetics

In the last two decades, novel microfluidic-based radiobioassays have been developed that improve performance for measuring dynamic radiotracer behavior with in vitro models (Liu & Lan, 2019). Radiotracer characterization, single-cell imaging, drug and metabolic assays, cellular pharmacokinetics, and dynamic PET imaging are some areas that significantly advanced with the development of microfluidic integration. The minimized and integrated detection scheme of these radiobioassays allows real-time tracking of dynamic uptake, high-sensitivity radiotracer imaging, and quantitative interpretation of imaging results in an unprecedented manner. Some of the key innovations are discussed here.

3.2.1 Betabox

Vu et al. (2011) developed a beta camera for imaging charged particles emitted from radiotracers in vitro using a solid-state position-sensitive avalanche photodiode (PSAPD) detector. They have integrated this system with a microfluidic chip that enables static and dynamic imaging of live cell cultures. The Betabox, as this device is called, was used to study the kinetics of responses of model lymphoma and glioblastoma cancer cell lines to FDG uptake after drug exposure (Dooraghi et al., 2013). Those responses correlated well with kinetic changes in the cell cycle or receptor tyrosine kinase signaling. The kinetic analysis enabled by the platform provided a rapid (1 h) drug screening assay. Betabox's spatial resolution ranges from 0.4 mm (FWHM at the center of the field of view) to 1 mm at a distance of 5 mm away from the center, yielding a useful field of view of approximately $10 \times 10\ \text{mm}^2$. Compared to RLM, the resolution is more limited due to the need for a layer of polydimethylsiloxane (PDMS) between the imaged cells and the sensitive detector volume. The high spatial resolution allows radioassays of small-cell populations down to the single-cell level. This powerful tool enables digital control of live cell cultures and simultaneously allows in vitro quantitative radioassays for multiple samples in parallel (Fig. 4a–c).

3.2.2 Continuously Infused Microfluidic Radioassay (CIMR)

Liu et al. (2016) developed a CIMR system to measure FDG uptake kinetics. CIMR automates the tracer medium's loading, unloading, and cleaning in the culture chamber. The system includes two parallel measurement chambers to subtract the signal from the unbound tracer. Thus, unlike the Betabox, the measurements can deliver direct and continuous cellular-uptake information without additional intervention during the medium exchange. The radioactivity from the culture chambers is measured by a positron camera built from a Timepix detector. Timepix consists of a single-particle-counting 300 mm thick silicon pixel detector bonded to a complementary metal oxide-semiconductor readout chip (Wang et al., 2014). The CIMR system can be extended to other positron-emitting radiotracers, providing a valuable platform for studying cellular pharmacokinetics (Esposito et al., 2011).

3.2.3 Microfluidics-Coupled Radioluminescence Microscopy

Kim et al. (2021) demonstrated microfluidics-coupled radioluminescence microscopy (μF-RLM) based on the RLM system as previously described. μF-RLM can detect radiotracer uptake in microfluidic cell culture chambers with spatial resolution better than 30 μm (Fig. 4d–f). Using this tool, the authors studied the effect of flow rate and lactic acidosis on the FDG uptake of cells. The continuous flow of the radiotracer led to fourfold higher uptake than static incubation in their study, while lactic acidosis inhibited FDG uptake by cancer cells. The critical step of this development is integrating an inorganic scintillator plate ($CdWO_4$) into a PDMS-based microfluidic chip, which was achieved by bonding the microfluidics device directly to the scintillator plate. The construction of μF-RLM uses standard microfluidics technology and therefore opens avenues for investigating the efficacy of radiopharmaceuticals using emerging in vitro cancer models by this straightforward

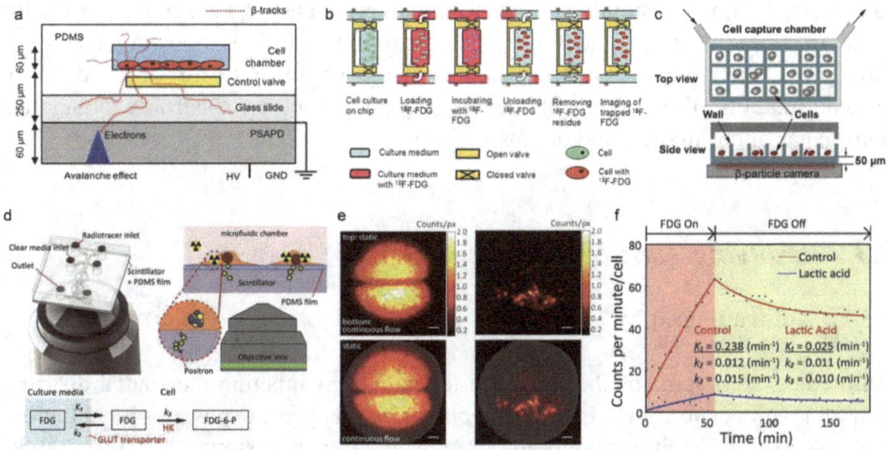

Fig. 4 Microfluidics-based assay for measuring uptake kinetics. (**a–c**) Real-time radioassay using Betabox. (**a**) Schematic cross-section of the β-camera integrated with a microfluidic chip. (**b**) Chip operation for FDG uptake radioassay. From left to right: Valves are opened to replace the culture medium with a culture medium containing FDG. Cells are incubated in FDG to start uptake. Valves are opened again to wash away extracellular FDG using a fresh culture medium. The channel network is washed with culture medium to remove FDG residing outside the cell chamber. Finally, the cells containing FDG are imaged with the β-camera. (**c**) Design of the cell-capture chamber. (**d–f**) Real-time radioassay with μF-RLM. (**d**) Schematic diagrams of μF-RLM setup, the procedure for radiolabeling cells, and the two-tissue pharmacokinetic model. (**e**) Radioluminescence images showing FDG bolus (left column) and radiolabeled cells (right column) during the washing phase with clear media. (**f**) A plot of average values of decay counts vs. time combined with a compartmental two-tissue model. The solid lines represent the tracer kinetic model estimated using nonlinear least-square optimization. (Figure panels adapted with permission from Vu et al. (2011), Wang et al. (2013) and Kim et al. (2021))

integration of microfluidics. The technology could be applied to image more advanced tumor chips cultured in 3D inside microfluidic perfusion chambers.

Among the other emerging techniques, silicon photomultipliers (SiPMs) based detection and characterization of radiotracers in miniaturized microfluidic devices (Taggart et al., 2016; Tarn et al., 2018) are promising candidates for studying cellular uptake kinetics with radioisotopes.

3.2.4 PET-Based Assays

To increase the throughput of radiobioassays, some research groups have tested the feasibility of using modern high-sensitivity PET scanners to measure cell tracer uptake. In one of the notable works, Liu et al. (2020) developed a continuously infused microfluidic radioassay system within a custom-built miniaturized dual-panel PET scanner. Their parallel infusion setup can image the cellular pharmacokinetics of three samples in one measurement. This system thus allows for different samples to be imaged simultaneously to obtain replicates of the same experiments or to measure response under different conditions. Besides boosting the assay throughput, their imaging method creates a new opportunity for standard tomographic imaging of emerging 3D cancer models, including spheroids, organoids, non-planar microfluidics, and organs-on-chips.

3.3 Resolving Tissue Heterogeneity

3.3.1 Autoradiography

Autoradiography remains the reference technique for imaging the spatial distribution of a radioactive source inside a specimen by the direct exposure of a radiation-sensitive media. Traditional autoradiography uses film emulsions, e.g., silver halide, as discussed before in Sect. 2.2.3. The resolution of this imaging system depends on the thickness of the emulsion, the size of the silver halide grains dispersed in the emulsion, the optical densitometer performance, and the strength of the development process. Autoradiography is used for a range of applications, including quantifying proteins or DNA separated by Polyacrylamide Gel Electrophoresis (PAGE), imaging the heterogeneous distribution of radionuclides in tissue slices, and measuring radiolabeled protein drug concentration in serum for pharmacokinetic studies. The development of microautoradiography (MAR), which includes high-resolution microscopic imaging of the detection plates, significantly advanced this tool to achieve radionuclide imaging with a very high spatial resolution (down to a few μm), at the level of resolving individual strands of a DNA (Huberman & Riggs, 1966). For imaging tissue samples, thin tissue sections (5–10 μm) are exposed directly to a radiochromic emulsion. However, to separate the tissue section from the emulsion before imaging, a micron-thick polyphenylene sulfide film is often inserted in between (Hirose et al., 2014).

In 1990, Yamada et al. (1990) first used MAR to localize FDG in mouse brain neurons. In another study, they measured the intratumoral distribution of FDG and [^{3}H]deoxyglucose, which showed a similar distribution pattern (Kubota et al., 1992). They observed that newly formed granulation tissue around the tumor and macrophages were more FDG avid than the viable tumor cells, an early observation consistent with the recent finding by Reinfeld et al. (2021) and Hesketh et al. (2019). Their study also provided insights into the differences and similarities between glucose utilization and DNA synthesis using a dual tracer FDG and [^{3}H]thymidine distribution analysis. With further developments of several fast digital autoradiography techniques (e.g., storage phosphor, electronic direct detection, thin phosphor layer, scintillator, and gaseous chamber), many research groups have used autoradiography in biological studies with a wide range of radiotracers.

3.3.2 Multimodal Imaging

Combining MAR with in vitro fluorescence imaging is a powerful technique for high-content imaging. Initially, Lee et al. (1999) first reported the integration of fluorescent in situ hybridization within microautoradiography. The combined technique has been described with several acronyms, such as MAR-FISH, STARFISH, and MicroFISH, but the basic principle is identical. The innovation in multimodal imaging has emerged rapidly with several digital detection techniques, which allow rapid data acquisition of radioscintigraphic and fluorescence signals at a comparable timescale. For instance, Khan et al. (2021a) have used RLM to image the heterogeneous FDG uptake in an engineered tumor-stroma model and in patient-derived tumor organoids (Fig. 5a–c). Their simultaneous bright-field, fluorescence, and

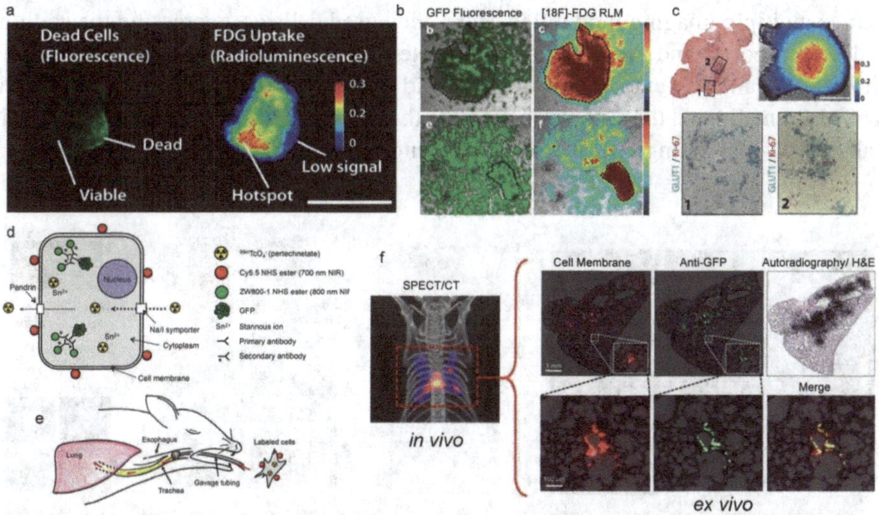

Fig. 5 Multimodal imaging studies. (**a**) RLM can resolve the heterogeneity of radiotracer uptake in complex tissue and co-localize with other standard imaging techniques. FDG uptake in tumor organoids is consistent with live/dead fluorescence imaging using Sytox Green. (**b**) FDG uptake of GFP green expressing HUVEC endothelial network and A549 tumor cell spheroid (nonfluorescent). (**c**) RLM images can be co-registered with standard histopathological analysis (H&E staining) and fluorescence in-situ hybridization for gene-expression analysis (RNAscope). (**d**) Multimodality labeling: Summary of various NIR fluorescent and radioactive molecules used to label lymphangioleiomyomatosis cells. Cy5.5 NHS ester (700-nm NIR) was conjugated to the cell membrane. ZW800-1 (800-nm NIR)–conjugated secondary antibody was used for immunofluorescence staining of GFP. ^{99m}Tc pertechnetate was accumulated via sodium iodide symporter, and stannous ion (Sn^{2+}) was used to improve the retention of the radioisotope in the cell. (**e, f**) In vivo and ex vivo multifunctional imaging. (**e**) Schematic representation of in vivo mouse model. (**f**) In vivo maximal-intensity-projection SPECT/CT image (left) and ex vivo color, 700-nm fluorescence, color-NIR merge (700-nm signal is pseudocolored in red), and maximal-intensity-projection SPECT/CT images (right) of the excised lung. The white rectangle delineates excised tissue samples studied in Same-slide fluorescence, autoradiography, and H&E. (Figure panels adapted with permission from Khan et al. (2021a, b) and Inoue et al. (2014))

radioluminescence imaging technique provided high-resolution information on the distribution of FDG in the engineered tissue microstructure. The microvascular stromal compartment as a whole took up a significant fraction of the FDG, comparable to the uptake of the tumor component. Even using a 200-μm-thick tissue construct, they achieved an average spatial resolution of 64 ± 4 μm for imaging FDG. These RLM images can be co-registered with histology and other high-content imaging.

Multiple other research groups have come up with innovations in multimodal imaging for radiotracer imaging. In the pursuit of determining the molecular and cellular basis for in vivo radioscintigraphic images, Inoue et al. (2014) developed a fiducial marker simultaneously compatible with 2-color near-infrared (NIR) fluorescence (700 and 800 nm), autoradiography, and conventional hematoxylin-eosin (H&E) histology (Fig. 6d–f). By definition, a fiducial marker acts as a reference marker in radiographic imaging to identify and orient a common region of interest across multiple imaging modalities. Here, they used cells co-expressing the sodium iodide symporter and green fluorescent protein, which were also labeled with NIR fluorophore and ^{99m}Tc, to obtain an in vivo fiducial following in vivo administration. This design allowed the authors to validate the in vivo SPECT imaging microscopically by co-registering NIR fluorescence, autoradiography, and H&E staining.

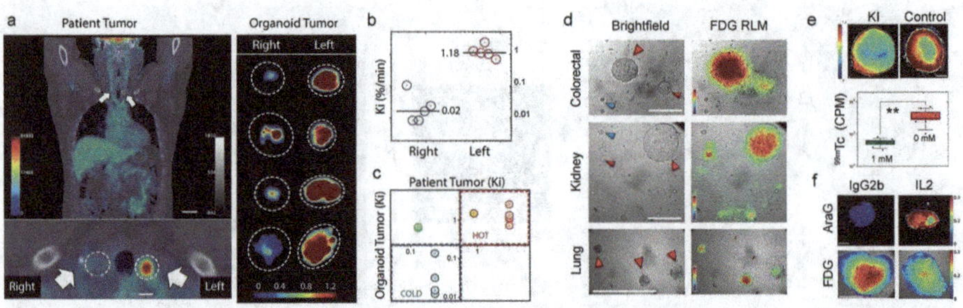

Fig. 6 Positron emission microscopy of patient-derived tumor organoids. (**a**) Two papillary thyroid carcinoma nodules (white arrows) show contrasting levels of FDG uptake. Tumor organoids derived from these nodules retained the two original nodules' metabolic identities. The dotted lines show the spatial extent of the individual organoids. The color bar shows radioactivity (Bq/pixel). Scale bar: 1 mm. (**b**) Scatter plot of FDG influx constant (Ki) for organoids ($n = 5$) derived from left and right nodules. (**c**) Scatter plot showing the correlation between the Ki of organoids ($n = 13$) and tumors of origin ($n = 4$, shown in four different colors; Pearson's $r = 0.756$, $P = 0.0032$). (**d**) FDG imaging of organoids derived from lung, kidney, and colorectal cancers. (**e**) ^{99m}Tc imaging and quantification of organoids derived from carcinoma thyroid carcinoma with (left) or without (right) competitive inhibition by 1 mM potassium iodide. (**f**) F-AraG imaging of HNSCC with (right) or without IL2 treatment. Corresponding FDG imaging is shown in the lower panel. (Figure panels are adapted with permission from Khan et al. (2021b))

3.4 Translational Cancer Models

While in vitro cancer models have many advantages, they also suffer from a few limitations. First, they comprise established, highly differentiated cell lines that do not reflect the heterogeneity and genetic complexity of patient tumors. They lack tumor immune microenvironment, heterotypic interactions between various cell types, complex metabolic signaling, tumor vasculature, and gradients of glucose oxygen and various other nutrients, which often determine the drug sensitivity and behavior of a tumor. Advances in stem cell biology, microfluidics, and cell culture have led to the development of sophisticated bioengineered microscale organotypic models that promise to fill this gap.

A significant recent development by Khan et al. (2021b) is imaging patient-derived organoids with clinical radiotracers. They used the RLM platform to image organoids derived from various patient cancers, including head and neck, kidney, lung, and colon (Fig. 6). These organoids recreate the complex structures and functions of the original patients' tumors, offering a superior translational model of solid tumors. In combination with bright-field and fluorescence imaging, the authors demonstrated imaging of organoids with ~100-fold higher resolution than clinical PET, using clinical PET radiotracers. A computational method called Fourier ring correlation estimated an average limit of resolution of 60 ± 4 μm over the entire thick specimen. Due to the high count rate from the large specimen, there was sufficient scintillation signal for fast and direct analog measurements using a single EMCCD exposure of 10–300 s, as opposed to the previous approach that quantified individual decays. Using this technique, they quantified FDG influx into head and neck squamous cell carcinoma (HNSCC) organoids and discovered that these in vitro models faithfully recapitulate the glycolytic activity of the original tumor. Using a mathematical model, they computed tumors' FDG influx rate (Ki) from FDG-PET images and organoids from RLM and gamma counting measurements. Even two different organoid lines from two tumor nodules within the same patient retained their different metabolic signatures, as measured by FDG avidity (Fig. 6a–c).

Furthermore, using organoids derived from thyroid carcinomas, they investigated sodium-iodine symporter activity using ^{99m}Tc-pertechnetate and found that the iodine uptake pathway is also functionally conserved in these models. For instance, competitive inhibition of NIS using potassium iodide caused a ~10-fold decrease in ^{99m}Tc uptake (Fig. 6e). The radiotracer imaging results were further quantified and validated by traditional gamma counting. They have also tested 2′-deoxy-2′-[^{18}F]fluoro-9-β-D-arabinofuranosylguanine (F-AraG), a radiotracer specific to activated immune cells, demonstrating that this tracer could be imaged in organoids which retain a tumor microenvironment including immune cells (Fig. 6f). These studies open new possibilities for drug discovery, new radiotracers discovery, personalized treatment regimens, and incorporation of clinical imaging biomarkers using organoids. Furthermore, combined with PET in clinical trials, this method can integrate unified image-derived biomarkers in concurrent preclinical arms, thus opening new doors for future co-clinical trials of imaging agents using organoid models.

4 Outlook

In vitro cell culture models have been one of the foundations of drug discovery and cancer research for decades. The exciting development of more advanced in vitro cancer models has now opened new doors for screening drugs and radiopharmaceuticals with stronger predictive power. Nevertheless, the challenges of using in vitro cancer models are worth mentioning. First, it is hard to model whole-body pharmacokinetics and systemic responses of multi-organ systems in vitro. Another long-standing challenge is to model the human immune system. Toward overcoming these critical barriers, organ-chip and organoid technologies have shown great promise. For instance, in one development, novel multi-organ "patient-on-a-chip" models allowed high-throughput pharmacokinetics analysis (Herland et al., 2020). Organoids from lymphoid organs successfully modeled human adaptive immune response in vitro (Wagar et al., 2021). Additionally, cancer organoids cultured at the air-liquid interface have been shown to better preserve the tumor immune microenvironment (Neal et al., 2018). In a different study, lymphoid follicles were developed in a microfluidic organ-chip, replicating human immune functions and vaccine responses in vitro (Goyal et al., 2022). With these and many other exciting discoveries in the last decade, increasing sophistication of in vitro techniques is highly anticipated in the emerging landscape of precise cancer modeling.

One exciting new in vitro modeling capability that could be harnessed for radiopharmaceutical applications is engineering microvascular networks (Zhao et al., 2021). The primary benefit is that the tissue culture can be vascularized, allowing more realistic flow-enabled tissue architecture and radiopharmaceutical perfusion to the tissue. Second, relevant biomarkers related to angiogenesis or vasculature growth could be rapidly assayed in these models following radionuclide therapies. Third, the vascular permeability of certain radiotracers could be screened. Furthermore, as tumor neovasculature is an emerging radiotracer target in solid tumors, the ability to study them in vitro could be critical. For instance, overexpression of PSMA on newly formed tumor vessels (Van de Wiele et al., 2020) can be targeted using PSMA-based theranostic agents. In this context, recapitulation of PSMA overexpression in vitro when vasculatures are co-cultured with tumor tissue can be promising and is worth further investigation.

In their recent introduction, patient-derived organoid models have opened tremendous new opportunities for radiopharmaceutical research. Diagnostic tracers can be used in parallel in patients and matched organoids to measure responses to a specific treatment, creating a framework similar to that of co-clinical trials. For radionuclide therapies, tracer distribution and partitioning in the tumor microenvironment can be measured with higher resolution using organoids, which would also allow for the estimation of microdosimetry of radionuclide therapies. Heterogeneity is another critical challenge in oncology that can be modeled effectively with organoids derived from different patients (inter-patient) and regions of a particular patient (intra-patient) resected tumor. Patient-derived tissue cultures can test if certain prescribed drugs are likely to alter the target expressions (e.g., NIS for thyroid cancer, PSMA for prostate cancer), and radionuclide retention in patients. For example, many thyroid cancer patients have dedifferentiated tumors that no longer trap iodine and are refractory to the therapy. These models can also be used for high-throughput identification of approved drugs, which might trigger redifferentiation of such tumors, enhance the target expression, restore radioiodine uptake, and consequently sensitize tumors for a better therapeutic outcome.

Finally, the rapidly evolving field of theranostics, as mentioned in the introduction, can benefit immensely from the advancement of in vitro techniques. Many of the detection, imaging, and assay techniques discussed in this chapter can also be employed for imaging therapeutic radiopharmaceuticals. For instance, patient-derived models can be used to assess the specific therapeutic potential of a drug, especially considering the heterogeneity of uptake and response among patient cohorts. High-resolution imaging techniques such as microautoradiography or RLM can provide high-content spatial information on drug distribution using patient biopsy or using patient-derived models, which can be validated with pathological investigations. In summary, these techniques can provide an excellent toolbox for translational and clinical research, complementing clinical tools and improving clinical decision-making.

Acknowledgments GP is supported by NIH grants R61CA278459, R01CA268514, and R01EB030367. SK is supported by NIH grant K99DE032094. Figure 1 was created with BioRender.com.

References

Almasi, S., & Pratx, G. (2019). High-resolution radioluminescence microscopy image reconstruction via ionization track analysis. *IEEE Transactions on Radiation and Plasma Medical Sciences, 3*(6), 660–667.

Almuhaideb, A., Papathanasiou, N., & Bomanji, J. (2011). 18F-FDG PET/CT imaging in oncology. *Annals of Saudi Medicine, 31*(1), 3–13.

Bączyk, M., et al. (2007). Efficacy of samarium 153 and strontium 89 treatment for bone metastases in prostate cancer patients: Monotherapy vs. treatment combined with external beam radiotherapy. Preliminary report. *Reports of Practical Oncology and Radiotherapy, 12*(4), 211–216.

Banerjee, S., Pillai, M. R. A., & Ramamoorthy, N. (2001). Evolution of Tc-99m in diagnostic radiopharmaceuticals. *Seminars in Nuclear Medicine, 31*, 260–277.

Bhattacharya, B., Mohd Omar, M. F., & Soong, R. (2016). The Warburg effect and drug resistance. *British Journal of Pharmacology, 173*(6), 970–979. https://doi.org/10.1111/bph.13422

Dooraghi, A. A., et al. (2013). Betabox: A beta particle imaging system based on a position sensitive avalanche photodiode. *Physics in Medicine & Biology, 58*(11), 3739.

Esposito, M., et al. (2011). Energy sensitive Timepix silicon detector for electron imaging. *Nuclear Instruments and Methods in Physics Research Section A: Accelerators, Spectrometers, Detectors and Associated Equipment, 652*(1), 458–461. https://doi.org/10.1016/j.nima.2011.01.148

Eychenne, R., et al. (2021). Overview of the most promising radionuclides for targeted alpha therapy: The "hopeful eight". *Pharmaceutics, 13*(6), 906. https://doi.org/10.3390/pharmaceutics13060906

Feng, X., et al. (2020). Clinical translation of a ^{68}Ga-labeled integrin αvβ6–targeting cyclic radiotracer for PET imaging of pancreatic cancer. *Journal of Nuclear Medicine, 61*(10), 1461–1467. https://doi.org/10.2967/jnumed.119.237347

Floberg, J. M., et al. (2021). Standardized uptake value for ^{18}F-fluorodeoxyglucose is a marker of inflammatory state and immune infiltrate in cervical cancer. *Clinical Cancer Research, 27*(15), 4245–4255. https://doi.org/10.1158/1078-0432.CCR-20-4450

Geijer, H., & Breimer, L. H. (2013). Somatostatin receptor PET/CT in neuroendocrine tumours: Update on systematic review and meta-analysis. *European Journal of Nuclear Medicine and Molecular Imaging, 40*, 1770–1780.

Goyal, G., et al. (2022). Ectopic lymphoid follicle formation and human seasonal influenza vaccination responses recapitulated in an organ-on-a-chip. *Advanced Science, 9*(14), 2103241. https://doi.org/10.1002/advs.202103241

Ha, B., et al. (2021). Flow radiocytometry using droplet optofluidics. *Biosensors and Bioelectronics, 194*, 113565.

Han, S., et al. (2018). Impact of ^{68}Ga-PSMA PET on the management of patients with prostate cancer: A systematic review and meta-analysis. *European Urology, 74*(2), 179–190. https://doi.org/10.1016/j.eururo.2018.03.030

Herland, A., et al. (2020). Quantitative prediction of human pharmacokinetic responses to drugs via fluidically coupled vascularized organ chips. *Nature Biomedical Engineering, 4*(4), 421–436. https://doi.org/10.1038/s41551-019-0498-9

Hesketh, R. L., et al. (2019). Magnetic resonance imaging is more sensitive than PET for detecting treatment-induced cell death-dependent changes in glycolysis. *Cancer Research, 79*(14), 3557–3569. https://doi.org/10.1158/0008-5472.CAN-19-0182

Hirose, A., et al. (2014). A microautoradiographic method for fresh-frozen sections to reveal the distribution of radionuclides at the cellular level in plants. *Plant and Cell Physiology, 55*(6), 1194–1202. https://doi.org/10.1093/pcp/pcu056

Huberman, J., & Riggs, A. (1966). Autoradiography of chromosomal DNA fibers from Chinese hamster cells. *Proceedings of the National Academy of Sciences of the United States of America, 55*, 599–606. https://doi.org/10.1073/pnas.55.3.599

Inoue, K., et al. (2014). Microscopic validation of macroscopic in vivo images enabled by same-slide optical and nuclear fusion. *Journal of Nuclear Medicine, 55*(11), 1899–1904.

Jia, A. Y., et al. (2022). Lutetium-177 DOTATATE: A practical review. *Practical Radiation Oncology, 12*(4), 305–311. https://doi.org/10.1016/j.prro.2022.02.002

Khan, S., Kim, S., et al. (2021a). High-resolution radioluminescence microscopy of FDG uptake in an engineered 3D tumor-stoma model. *European Journal of Nuclear Medicine and Molecular Imaging, 48*(11), 3400–3407. https://doi.org/10.1007/s00259-021-05364-6

Khan, S., Shin, J. H., et al. (2021b). High-resolution positron emission microscopy of patient-derived tumor organoids. *Nature Communications, 12*(1), 5883. https://doi.org/10.1038/s41467-021-26081-6

Kim, T. J., Türkcan, S., & Pratx, G. (2017). Modular low-light microscope for imaging cellular bioluminescence and radioluminescence. *Nature Protocols, 12*(5), 1055–1076. https://doi.org/10.1038/nprot.2017.008

Kim, T. J., et al. (2020). Single-cell radioluminescence microscopy with two-fold higher sensitivity using dual scintillator configuration. *PLoS One, 15*(7), e0221241. https://doi.org/10.1371/journal.pone.0221241

Kim, T. J., et al. (2021). Microfluidics-coupled radioluminescence microscopy for in vitro radiotracer kinetic studies. *Analytical Chemistry, 93*(10), 4425–4433.

Klein, J. S., Kim, T. J., & Pratx, G. (2022). Development of a lensless radiomicroscope for cellular-resolution radionuclide imaging. *Journal of Nuclear Medicine, 64*, 479.

Kratochwil, C., et al. (2019). ^{68}Ga-FAPI PET/CT: Tracer uptake in 28 different kinds of cancer. *Journal of Nuclear Medicine, 60*(6), 801–805.

Kubota, R., et al. (1992). Intratumoral distribution of fluorine-18-fluorodeoxyglucose in vivo: High accumulation in macrophages and granulation tissues studied by microautoradiography. *Journal of Nuclear Medicine, 33*(11), 1972–1980.

Lee, N., et al. (1999). Combination of fluorescent in situ hybridization and microautoradiography—A new tool for structure-function analyses in microbial ecology. *Applied and Environmental Microbiology, 65*(3), 1289–1297. https://doi.org/10.1128/AEM.65.3.1289-1297.1999

Levi, J., et al. (2019). Imaging of activated T cells as an early predictor of immune response to anti-PD-1 therapy. *Cancer Research, 79*(13), 3455–3465. https://doi.org/10.1158/0008-5472.CAN-19-0267

Liu, Z., & Lan, X. (2019). Microfluidic radiobioassays: A radiometric detection tool for understanding cellular physiology and pharmacokinetics. *Lab on a Chip, 19*(14), 2315–2339. https://doi.org/10.1039/C9LC00159J

Liu, Z., et al. (2016). A continuously infused microfluidic radioassay system for the characterization of cellular pharmacokinetics. *Journal of Nuclear Medicine, 57*(10), 1548–1555.

Liu, Z., et al. (2020). A mini-panel PET scanner-based microfluidic radiobioassay system allowing high-throughput imaging of real-time cellular pharmacokinetics. *Lab on a Chip, 20*(6), 1110–1123. https://doi.org/10.1039/C9LC01066A

Mansi, R., et al. (2016). Bombesin-targeted PET of prostate cancer. *Journal of Nuclear Medicine, 57*(Suppl 3), 67S–72S.

McCready, V. R. (2017). Radioiodine—The success story of nuclear medicine. *European Journal of Nuclear Medicine and Molecular Imaging, 44*(2), 179–182. https://doi.org/10.1007/s00259-016-3548-5

Natarajan, A., et al. (2015). Multiscale framework for imaging radiolabeled therapeutics. *Molecular Pharmaceutics, 12*(12), 4554–4560. https://doi.org/10.1021/acs.molpharmaceut.5b00392

Neal, J. T., et al. (2018). Organoid modeling of the tumor immune microenvironment. *Cell, 175*(7), 1972–1988.e16. https://doi.org/10.1016/j.cell.2018.11.021

Park, S., et al. (2021). Relationship of the standard uptake value of ^{18}F-FDG-PET-CT with tumor-infiltrating lymphocytes in breast tumors measuring $\geq$1 cm. *Scientific Reports, 11*(1), 12046. https://doi.org/10.1038/s41598-021-91404-y

Parker, C., et al. (2013). Alpha emitter radium-223 and survival in metastatic prostate cancer. *New England Journal of Medicine, 369*(3), 213–223.

Poeppel, T. D., et al. (2011). ^{68}Ga-DOTATOC versus ^{68}Ga-DOTATATE PET/CT in functional imaging of neuroendocrine tumors. *Journal of Nuclear Medicine, 52*(12), 1864–1870.

Pratx, G., et al. (2012). Radioluminescence microscopy: Measuring the heterogeneous uptake of radiotracers in single living cells. *PLoS One*. https://doi.org/10.1371/journal.pone.0046285

Pratx, G., et al. (2013). High-resolution radioluminescence microscopy of ^{18}F-FDG uptake by reconstructing the β-ionization track. *Journal of Nuclear Medicine, 54*(10), 1841–1846.

Reinfeld, B. I., et al. (2021). Cell-programmed nutrient partitioning in the tumour microenvironment. *Nature, 593*(7858), 282–288.

Reynolds, G. T. (1968). Scintillomicroscope for radioactive tracer detection. *Review of Scientific Instruments, 39*(3), 298–302.

Sartor, O., et al. (2021). Lutetium-177–PSMA-617 for metastatic castration-resistant prostate cancer. *New England Journal of Medicine, 385*(12), 1091–1103.

Sawin, C. T., & Becker, D. V. (2009). Radioiodine and the treatment of hyperthyroidism: The early history. *Thyroid, 7*(2), 163–176. https://doi.org/10.1089/thy.1997.7.163

Taggart, M. P., et al. (2016). Development of radiodetection systems towards miniaturised quality control of PET and SPECT radiopharmaceuticals. *Lab on a Chip, 16*(9), 1605–1616. https://doi.org/10.1039/C6LC00099A

Tarn, M. D., et al. (2018). Plastic scintillator-based microfluidic devices for miniaturized detection of positron emission tomography radiopharmaceuticals. *Chemistry, 24*(52), 13749–13753. https://doi.org/10.1002/chem.201802395
Van de Wiele, C., et al. (2020). PSMA expression on neovasculature of solid tumors. *Histology and Histopathology, 35*(9), 919–927. https://doi.org/10.14670/HH-18-215
Vu, N. T., et al. (2011). A β-camera integrated with a microfluidic chip for radioassays based on real-time imaging of glycolysis in small cell populations. *Journal of Nuclear Medicine, 52*(5), 815–821. https://doi.org/10.2967/jnumed.110.078725
Wagar, L. E., et al. (2021). Modeling human adaptive immune responses with tonsil organoids. *Nature Medicine, 27*(1), 125–135. https://doi.org/10.1038/s41591-020-01145-0
Wang, J., et al. (2013). Fast Metabolic Response to Drug Intervention Through Analysis on a Miniaturized, Highly Integrated Molecular Imaging System. *Journal of Nuclear Medicine, 54*(10), 1820–1824. https://doi.org/10.2967/jnumed.112.118497
Wang, Q., et al. (2014). Evaluation of Timepix silicon detector for the detection of ^{18}F positrons. *Journal of Instrumentation, 9*(05), C05067. https://doi.org/10.1088/1748-0221/9/05/C05067
Witzig, T. E. (2004). Yttrium-90-ibritumomab tiuxetan radioimmunotherapy: A new treatment approach for B-cell non-Hodgkin's lymphoma. *Drugs of Today (Barcelona, Spain: 1998), 40*(2), 111–119. https://doi.org/10.1358/dot.2004.40.2.799423
Yaakob, W., et al. (1999). The usefulness of iodine-123 whole-body scans in evaluating thyroid carcinoma and metastases. *Journal of Nuclear Medicine Technology, 27*(4), 279–281.
Yamada, S., et al. (1990). Localization of [^{18}F]fluorodeoxyglucose in mouse brain neurons with micro-autoradiography. *Neuroscience Letters, 120*(2), 191–193. https://doi.org/10.1016/0304-3940(90)90035-8
Zhao, X., et al. (2021). Review on the vascularization of organoids and organoids-on-a-chip. *Frontiers in Bioengineering and Biotechnology, 9*, 637048. https://doi.org/10.3389/fbioe.2021.637048

Epilogue: Overcoming the Challenges of Translating Innovative Technologies into Standard Practice

Volker Hessel, R. Michael van Dam, Yuji Kuge, and Giancarlo Pascali

Introduction

This book has highlighted the outstanding works of many research groups and companies focused on improving the methods and processes that currently represent standard practice in radiopharmaceutical sciences, with the intention to make them more efficient, affordable, sustainable, and robust.

Technological innovations, with broadly similar objectives, are also being made in many other chemistry fields, such as chemical and pharmaceutical manufacturing. However, despite achieving measurable improvements, such innovations face numerous obstacles to adoption within well-established chemical manufacturing plants and processes. By considering the progression of technology in these other fields, we can learn how to streamline the translation of radiochemistry technologies into routine use.

V. Hessel
School of Chemical Engineering, University of Adelaide, Adelaide, Australia

R. M. van Dam
Crump Institute for Molecular Imaging, University of California Los Angeles (UCLA), Los Angeles, CA, USA
Department of Molecular and Medical Pharmacology, David Geffen School of Medicine, University of California Los Angeles (UCLA), Los Angeles, CA, USA

Y. Kuge
Global Center for Biomedical Science and Engineering, Faculty for Medicine, and Central Institute of Isotope Science, Hokkaido University, Sapporo, Japan

G. Pascali
Heath RTG, Australian Nuclear Science and Technology Organisation, Sydney, Australia

School of Chemistry, University of New South Wales, Sydney, Australia

R. M. van Dam et al. (eds.), *Automated Technologies for the Development and Production of Radiopharmaceuticals*,
https://doi.org/10.1007/978-3-031-84632-8

In this epilogue, we have interviewed Prof. Volker Hessel, one of the most prominent experts in flow chemistry and an active supporter of this and other transformative technologies in chemical manufacturing. Prof. Hessel has extensive experience spanning from the ideation of new approaches all the way to their application into real industrial practice. In our discussion below about translating flow chemistry and radiochemistry technologies from academic research into industrial practice, we can glean useful insights and inspiration on the avenues, obstacles, and opportunities that could help drive the successful adoption of novel technologies for radiopharmaceutical manufacturing and development.

Discussion

In the following discussion, questions by the editors (R. Michael van Dam, Yuji Kuge, Giancarlo Pascali) are prefixed with "**Editors:**", responses by Prof. Hessel are prefixed by "**VH:**" and displayed in italics, and comments by the editors are prefixed with "**Editors' Comments:**".

Editors: These are exciting times to investigate novel chemical reaction technologies, as the interface with various other fields, such as physics, engineering, and biology, is becoming increasingly frequent and productive. In a way, scientific research is moving away from a "many silos" concept toward a more integrated approach. These interactions are bringing about several interesting and promising concepts with substantial repercussions in the fields of chemistry and chemical engineering. Could you point out some of these concepts that you find particularly unexpected and promising?

VH: *Continuous-flow processing on a small scale—engineers term it "microreactors" and chemists name it "flow chemistry"—is impressive and has made its way from the research lab to industrial use and large-scale production, at least for pharmaceutical processing (medicinal chemistry). Another development is plasma technology, which has potential for CO_2 conversion; H_2 generation; NH_3, NOx, and CH_4 upgrading; biomass; and more. Designer solvents, ionic liquids, and (natural) deep eutectic solvents have all demonstrated improved separation performances. The list is certainly not complete as there are also many efficient microwave-, radiofrequency-, and ultrasound-assisted reaction techniques, and a revival of photosynthesis and electrochemistry (the "electrification of chemistry"), that are all becoming increasingly promising in terms of widening the chemical space* (Hessel et al., 2013).

Editors' Comments: Indeed, many of these same developments in the broader chemistry field are also making an impact in radiochemistry. Microwave activation (chapter "Application of Microwaves in Radiochemistry"), photoactivation (chapter "Photoactivated Radiochemistry for the Construction of ^{18}F- and ^{11}C-Labeled Compounds"), and electrochemistry (chapter "Recent Developments in Electrochemical Radiofluorination: Methods and Applications"), as well as novel catalytic approaches, are all expanding our ability to efficiently synthesize an increasing variety of radiolabeled molecules. Microreactor technology is also being pursued by a number of groups (chapters "Democratizing Radiopharmaceutical Production: The RAPID Chip," "Progress in Droplet-Based Methods for Simple, Low-Cost Radiopharmaceutical Production," and "Digital Microfluidics and Radiochemistry: Current State and Future Trends") as a way to reduce costs, increase labeling efficiency, and increase modularity and versatility.

Editors: It can be noted that the radiopharmaceutical science field has been somewhat resistant to changing the *status quo* with regard to product manufacturing, and this may be related to a number of reasons that may have a parallel with traditional chemical manufacturing. We would like to analyze some of these potential reasons with you. First and foremost: What elements should we evaluate when we compare the advantages of new processes with established ones?

VH: *With regard to certain radionuclides for imaging and therapy, I understand that much of the established large-scale production is made in reactors installed in the 1950s and 1960s, which need increasing maintenance and are at the end of their lifetime. Using microfluidic approaches to separate such radioisotopes could modernize and simplify target processing and radionuclide purification for such facilities. However, the need of a high number of extraction stages to achieve adequate separation undermines the benefits and has still to be fully explored and optimized in this field. This need could potentially be reduced, removed, or modified with the implementation of complementary process innovation* (e.g., *using complexants, designer solvents), which certainly would then have sustainability benefits over the older technologies and even resin-based approaches. Operations must be made as safe or safer, and, in this respect, flow chemistry separations provide a sizeable advantage, as well as being proven to be cost-efficient and versatile.*

Editors' Comments: The point described here is of paramount importance, especially given the rapidly growing impact and increasing need for diverse radionuclides for diagnostic and therapeutic use. To address supply chain issues, alternative production methods are being developed, each of which will require new methods to perform automated target processing and radionuclide separation. In this book, there is a background perspective on such radionuclide productions (chapters "Cyclotron Radioisotope Production: Developments and Challenges in the New Era" and "Nuclear Reactor-Based Production of Medical Radionuclides") and a few examples of the impact of novel separation technologies in the field, such as for improving radionuclide separation by using flow equipment (chapters "New Trends in the Automation of Solvent Extraction for the Production of Radionuclides: From Macro- to Microscale Systems" and "Recent Advances in Liquid–Liquid Extraction in Flow of Radiometals for Positron Emission Tomography") or electrochemistry (chapter "Rapid and Effective Purification for Radioactive Cu and Ga by Means of Analytical Chemistry Approaches"), and new modular approaches to automatically process the solid targets often used to produce such radionuclides (chapter "New Trends in the Automation of Solid Target Dissolution for Radionuclide Production: A Universal Target Dissolution System for Cassette-Based Radiosynthesis Modules").

Editors: The short half-life of radiopharmaceuticals necessitates tight scheduling of their production and use, and the need for repeated production (e.g., daily) requires high *reliability* of production (and associated equipment). However, any new technology carries with it some risk until proven over time, creating a barrier to change. What kind of obstacles hindered the adoption of flow chemistry systems in chemical manufacturing? How were they overcome? How long did flow chemistry take to gain wide acceptance?

VH: *Flow chemistry is starting to take the stage at industrial scale after >20 years of development. Following this, many companies now exist that can install flow production processes and production micro/milli-structured reactors to fit any manufacturing need. A main obstacle was initially the repeated clogging of flow reactor systems, but that was solved in many ways, like somewhat widening the inner diameter or tailoring the components of the chemical mixture. Scale-up needed an extension of the initial "numbering up" concepts toward a more holistic "scale-out" approach. Finally, it was evident that it was not sufficient to simply copy the chemistry approach from batches, but rather to develop an altogether new chemistry concept—namely, "novel process windows"—to boost chemical reactivity. Only in this way have microreactors become economically productive and competitive with standard batch approaches* (Porta et al., 2016).

Editors' Comments: At the time of writing, it has been nearly 20 years since the initial application of microreactors to the radiopharmaceutical field, yet the technology has not yet found widespread use, despite demonstrating a number of cost and performance advantages. Similar challenges will also face the adoption of other novel reaction modalities (such as electrochemistry in chapter "Recent Developments in Electrochemical Radiofluorination: Methods and Applications" and photochemistry in chapter "Photoactivated Radiochemistry for the Construction of ^{18}F- and ^{11}C-Labeled Compounds"), which are earlier in their development. To understand the lack of wide adoption in the field, we need to consider the full suitability for the task at hand (e.g., Is it sufficiently reliable for routine use? Is it possible to produce sufficient doses at a time?) as well as the availability of the technology. Although scalability has been demonstrated for many microfluidic approaches, to date, there are only very few options for commercial microreaction technology, thus creating a barrier to adoption. The most convincing demonstration of reliability will be to show the advantages of microfluidics in the context of clinical production in academic or industry radiopharmaceutical production. In addition to having a reliable technology, reliable processes are needed. Here, newly developed statistical and unbiased optimization methods (described in chapters "High-Throughput Techniques in Radiochemistry" and "The Application of Statistical "Design of Experiments" (DoE) Towards the Efficient Study and Optimization of Radiochemical Processes") can assist by identifying which crucial parameters must be finely controlled in order to maximize performance and reliability.

Editors: The *market size* of each radiochemical production site (radiopharmacy) is small if compared to traditional chemical plants, due to the lower global demand for doses, reduced variety of products, and limited distribution distance of each site (i.e., limited by half-life). These facts make it challenging to bring in new technologies to substitute or upgrade synthetic and analytical equipment. Furthermore, to distribute a new radiopharmaceutical requires multiple satellite production sites, enhancing the financial barriers to change. In which ways does this situation hamper the adoption of new technologies, and how can change be incentivized?

VH: *Flow chemistry is a small-scale, modular, and mobile production technology; to my belief, this feature fully supports overcoming what has been mentioned here as challenges* (Hessel et al., 2005). *However, an obstacle might be represented by the limits in using radiopharmaceuticals in standard laboratories and the radiation licenses required to operate them; this fact, and the constraints associated, reduces the ability for many engineering researchers to be involved and hinders the establishment of innovation hubs where extremely innovative technologies can be developed into mature products.*

Editors' Comments: Dr. Hessel's response further confirms that the microfluidic approach is a good fit for the small-scale batches characteristic of radiopharmaceuticals (compared to pharmaceuticals). In addition to the performance and flexibility advantages offered by microfluidics, it is anticipated that by innovating the radiosynthetic armamentarium, commercial entities will be able to leverage other aspects (e.g., small size and low cost) to expand their manufacturing capacity at existing sites to a wider portfolio of radiopharmaceutical products, providing a nontrivial advantage that could help drive adoption. Another point is illustrated in chapters "Basics and Progress of PET Radiopharmaceutical Synthesizers from a Company's Viewpoint" and "Radiopharmaceutical Dispensing Systems," highlighting that new technologies and commercial products do not remain static, but innovation continues as the needs of customers in the radiopharmaceutical manufacturing industry evolve. With regard to the barrier of radiation safety, the very nature of microfluidic radiochemistry systems, which are smaller and easier to shield than their conventional counterparts, could help; this compact and inexpensive technology could reduce the radiation hazard and facilitate wider adoption of microfluidic radiochemistry systems at a wider range of research centers than is possible with conventional technologies. Another technology that may aid adoption is 3D printing of radiochemistry systems (chapter "3D Printed Microreactors for Radiochemical Reactions"), potentially empowering more research groups to build reaction systems (with the advantage of copying a "standard" design used by others) at a fraction of the cost of commercial radiosynthesizers.

Editors: Radiopharmaceutical products are to be ultimately used in clinical procedures, and for this reason, they are heavily *regulated*, both on the pharmaceutical and radiation protection sides. This requirement makes the implementation of changes cumbersome, not only in terms of paperwork, but also in terms of regulators being able to evaluate risk and benefits in a way that is unbiased by ongoing practice. The fact that such products are produced in small batches, but with high repetition, makes the regulatory controller even more cautious. Big pharma production was able to introduce drastic changes to their technologies, such as the use of flow instead of batch chemistry in many reactive steps, being evidently able to convince regulators that benefits outweighed risks. What was the winning strategy in those cases?

VH: *The winning strategy was that the pharmaceutical industry was strongly supportive of flow chemistry in the overarching belief in green chemistry. At that* *point, one of the most influential regulatory bodies, the FDA, noticed this drive and worked hand in hand with their industry to make these approaches an accepted reality* (Poechlauer et al., 2013). *Such massive and coordinated push is needed to change technologies. The arguments that were noticed and convinced the FDA to support the move also hold true for radiopharmaceutical production—high production quality, high reproducibility, and ease of cleaning after production campaigns of flow chemistry.*

Editors' Comments: Prof. Hessel argues that demonstrating the unique advantages of a novel technological innovation remains the key element to convince regulators of the benefits and opportunity for a change. This requires the continued effort from researchers to publish the achievements that demonstrate the improved efficacy of the innovation reported, and several examples are reported in this book. Since a primary concern of the FDA is patient safety, a particular challenge in the radiopharmaceutical industry is the wide variety of different manufacturing equipment and the large number of sites/operators making radiolabeling compounds. One would expect that movement toward technologies with improved standardization, with reduced usage of toxic solvents and reagents (e.g., microreactors), and more sensitive QC approaches would be of high interest. The fact that a completely new approach to QC of radiopharmaceuticals has been approved by the FDA (chapter "Quality Control Automation Solutions for PET Drug Production") demonstrates that such changes are possible and that regulators are open to them.

Editors: In order to devise new ways to perform chemical processes, we need to think "outside the box," and then, once the innovation is created, come back "in the box" to understand its crucial details, so it can become the new norm. What is the role of *education and training* in this process? Radiopharmaceutical science courses are seldom taught in academia, which exacerbates concerns about having a workforce trained to not only use, but also further improve and develop novel technologies. How have these concerns been addressed in the flow chemistry field?

VH: *Teaching of flow chemistry is still rare as few universities have courses featuring these approaches: in a way, teachers are more conservative than researchers are. There is a common belief that microreactor engineering is just (a small) part of reactor engineering and not a subject of its own. We still teach chemical engineering with the principles gained up to the 1950s and the reactor concepts of the 1960s–1980s. The "education" success in flow chemistry is not due to teaching, but due to the formation of a community which is part of larger movements, such as green chemistry. Therefore, current training and education happen in the workshops within this community and, although not well coordinated or comprehensive, are still driving the field forward* (Plutschack et al., 2017). *However, as in the case of radiopharmaceutical sciences, an academic recognition of these fields and topics deserving a dedicated education would be extremely important.*

Editors' Comments: We believe that this book tries to partly fill this gap, providing educators the possibility to see different technologies or the application of novel reaction concepts under the unified umbrella of radiopharmaceutical production. In addition, we and others have established and advocated for the development of specialized programs that strive to bring together sufficient expertise in chemical engineering and microreactor engineering that are most suited to dealing with the unique challenges of radiochemistry (small volumes, small space, short process times), radiopharmaceutical analysis techniques, physics of radioisotopes and radiation, and biological and biochemical aspects of nuclear medicine imaging and therapy.

Editors: Producing and using radiopharmaceuticals are an expensive business due to the (cGMP) infrastructure and logistics needed to deal with such short half-life products, with their high clinical value further driving the price upwards. However, it looks like that many innovations with great potential to reduce costs and provide other advantages may find a difficult path to adoption in the absence of sizable financial investments to support the development. How can one know the best way to demonstrate the value of a new technology, up to the point of reaching a critical mass of commercial interest?

VH: *I believe that it will be possible to finance the implementation of flow chemistry or similar approaches, if the business is seen as "expensive" (as is the case for radiopharmaceutical production), also because microreactor systems are affordable (compared to the overall expense) and their benefits are reported and accessible. Flow chemistry has just not been tested to the extent needed to demonstrate a performance ready for industrial use, including scale-up to commercially relevant productivity* (Ott et al., 2016).

Editors' Comments: Indeed, in the radiopharmaceutical field, short-term adoption of novel technologies will certainly require demonstrations that are matched to the needs of industry, including the capability to manufacture specific radiopharmaceuticals at the desired scales and then to provide improvements (e.g., cost) that incentivize a change from existing approaches to the new technology. Microfluidic approaches also have the potential to more significantly transform the manufacturing landscape by balancing in different ways the capital costs to build/expand facilities, operating costs, costs of transportation/decay, etc.

Editors: A particular aspect of radiopharmaceutical production is the extremely small quantity of product that is produced in each batch (pmol to μmol per batch), and reduction of volumes via microfluidics has been a key focus in the development of novel technologies for synthesis, purification, and formulation. Are there other areas of chemistry where small batch sizes and small volumes are particularly relevant that the radiochemistry field could learn from or perhaps that could benefit from developments in the radiopharmaceutical field?

VH: *Flow chemistry gives superb results for nanomaterial synthesis; this is also a field where small volumes are commonly used. In these cases, quality and functional control dominates over sheer production of large volumes, and such features are delivered in a much more reliable manner by microreactor systems than conventional systems. Lab-on-a-chip technologies, and in particular droplet microfluidics, are able to handle very small volumes, which is a major advantage for biological, biochemical, and analytical applications.*

Editors' Comments: In addition to the precision control offered by microreactors, which can improve reaction yields and product consistency, the radiopharmaceutical field can also benefit from the ability of some types of microfluidic systems to handle small total volumes, enabling minimization of precursor and reagents, shortening process times, simplifying purification, and improving specific activity. Microfluidic techniques that excel at handling small total reagent volumes include microscale batch reactors (chapter "Democratizing Radiopharmaceutical Production: The RAPID Chip") and droplet-based reactors (chapters "Progress in Droplet-Based Methods for Simple, Low-Cost Radiopharmaceutical Production" and "Digital Microfluidics and Radiochemistry: Current State and Future Trends"). Microscale methods are also advantageous for the analysis of radiopharmaceuticals by minimizing the quantity consumed for testing (chapter "Current and Emerging Applications of Electrophoresis in the Radiopharmaceutical Field") and can play unique roles in the discovery of radiopharmaceuticals by enabling precision control and study of small biological systems such as organoids and single cells (chapter "Next-Generation Techniques for In Vitro Characterization of Cancer Radiopharmaceuticals").

Editors: A number of technologies are starting to emerge, such as artificial intelligence/machine learning and 3D printing, with a potential impact across a large variety of disciplines. Also, we are slowly realizing the need to consider issues such as energetic/environmental sustainability in chemical manufacturing, especially as demand for a particular chemical (or radiopharmaceutical) increases, and to solve problems that did not previously exist (e.g., drug manufacturing in extremely remote regions, such as space). Where do you think new chemical technologies can make a difference, or can get inspiration to progress further, and potentially provide value in the radiochemistry field?

VH: *Specialty and niche applications such as remote and space manufacturing can sharpen our view on what "frontiers" can be opened by new technologies such as flow chemistry. They are a great opportunity for training and creativity, and they will exist as fields in their own right, yet they will stay small. These new chemical technologies need to enter mainstream business, but this will only be effective and accepted when they are able to transform industrial manufacturing on a broader scale. A good example is to have multiple distributed chemical plants instead of a world-scale plant, with the advantage of drastically reducing transport costs and environmental impacts and risks. Producing close to the customer or the resource (biomass) will be game changing and likely cannot be accomplished with traditional technology.*

Editors' Comments: Prof. Hessel's last point about distributed production is fundamental to the practice of radiochemistry, where the short shelf life (limited by short half-life of radionuclides) demands manufacturing at multiple locations close to the customer/patient. This necessity for multiple manufacturing plants to supply radiopharmaceuticals to the full patient population has led to numerous innovations and miniaturization and automation as covered in this book. In these applications, perhaps what has been learned in the context of radiopharmaceuticals can provide guidance/strategies to distributed manufacturing in the broader pharmaceutical and chemical industries.

Outlook

Development and adoption of new chemical technologies need the closing of ranks of academic research, industry, as well as government (e.g., regulation). This is finally a matter of belief and trust. It is a matter of willingness for change. External drivers help to make game changes. Currently, we have a number of global drivers, including the global warming reduction, resilience of resources and services, epidemics, and more. The magnitude of these events can be used to create momentum for the establishment of new technologies. Different from game changes in the past, which focused on one or a very few innovations, e.g., the steam engine, today's innovation pool is huge including techniques such as 3D printing, artificial intelligence, quantum materials, and many more. A concerted action might be a way to maximize impact and to be able to provide the focus the former "industrial revolutions" had.

References

Hessel, V., Lob, P., & Lowe, H. (2005). Development of microstructured reactors to enable organic synthesis rather than subduing chemistry. *Current Organic Chemistry, 9*(8), 765–787. https://doi.org/10.2174/1385272053764953

Hessel, V., et al. (2013). Novel process windows for enabling, accelerating, and uplifting flow chemistry. *ChemSusChem, 6*(5), 746–789. https://doi.org/10.1002/CSSC.201200766

Ott, D., Borukhova, S., & Hessel, V. (2016). Life cycle assessment of multi-step rufinamide synthesis—From isolated reactions in batch to continuous microreactor networks. *Green Chemistry, 18*(4), 1096–1116. https://doi.org/10.1039/C5GC01932J

Plutschack, M. B., et al. (2017). The Hitchhiker's guide to flow chemistry'. *Chemical Reviews, 117*, 11796–11893. https://doi.org/10.1021/acs.chemrev.7b00183

Poechlauer, P., et al. (2013). Pharmaceutical roundtable study demonstrates the value of continuous manufacturing in the design of greener processes. *Organic Process Research and Development, 17*(12), 1472–1478. https://doi.org/10.1021/op400245s

Porta, R., Benaglia, M., & Puglisi, A. (2016). Flow chemistry: Recent developments in the synthesis of pharmaceutical products. *Organic Process Research & Development, 20*(1), 2–25. https://doi.org/10.1021/acs.oprd.5b00325

R. M. van Dam et al. (eds.), *Automated Technologies for the Development and Production of Radiopharmaceuticals*,
https://doi.org/10.1007/978-3-031-84632-8

Index

R. M. van Dam et al. (eds.), *Automated Technologies for the Development and Production of Radiopharmaceuticals*,
https://doi.org/10.1007/978-3-031-84632-8